Materials Processing and Manufacturing Science

Materials Processing and Manufacturing Science

Rajiv Asthana

Ashok Kumar

Narendra B. Dahotre

AMSTERDAM • BOSTON • HEIDELBERG • LONDON
NEW YORK • OXFORD • PARIS • SAN DIEGO
SAN FRANCISCO • SINGAPORE • SYDNEY • TOKYO

Academic Press is an imprint of Elsevier

ELSEVIER

Academic Press is an imprint of Elsevier

30 Corporate Drive, Suite 400, Burlington, MA 01803, USA
525 B Street, Suite 1900, San Diego, CA 92101-4495, USA
84 Theobald's Road, London WC1X 8RR, UK

This book is printed on acid-free paper. ∞

Library of Congress Cataloging-in-Publication Data: on file

British Library Cataloguing in Publication Data
A catalogue record for this book is available from the British Library

ISBN 13: 978-0-7506-7716-5
ISBN 10: 0-7506-7716-3

For all information on all Academic Press publications
visit our Web site at www.books.elsevier.com

Printed in the United States of America
05 06 07 08 09 10   9 8 7 6 5 4 3 2 1

To our parents, spouses and children,

and

To students of materials processing everywhere.

Contents

Index   **615**

Contributors

Rajiv Asthana is a professor of engineering and technology at the University of Wisconsin–Stout, a Baldridge National Quality Award–winning institution, where he has been teaching since 1995. He received his B.S. and M.S. degrees from the Indian Institute of Technology (Kharagpur) and his Ph.D. from the University of Wisconsin–Milwaukee. He has been a NASA Faculty Fellow, an NRC post-doctoral research associate, and a NASA Project Scientist at NASA Glenn Research Center (Cleveland); a visiting scientist at Foundry Research Institute (Krakow) under a National Academy of Sciences Award; a visiting associate professor at the University of Wisconsin–Milwaukee; and a scientist with the Council of Scientific & Industrial Research (India). He is the author of the book *Solidification Processing of Reinforced Metals* (Trans Tech, Switzerland, 1998) and an author or coauthor of 110 refereed publications. Dr. Asthana is the associate editor of *J. Materials Engineering & Performance*, coeditor of a forthcoming issue of *Current Opinion in Solid State and Materials Science* on high-temperature capillarity, and on the editorial advisory board of *Bulletin of Polish Academy of Sciences*. He has been an invited speaker and session-highlight speaker at international conferences and a reviewer for sixteen international journals and numerous edited volumes. Dr. Asthana is a member of the American Society for Materials, the American Ceramic Society, the American Foundry Society, The Minerals, Metals & Materials Society, and the American Society for Engineering Education. He also received the Outstanding Scholar/Researcher Award of UW–Stout and a NASA Certificate of Recognition and Award for research contributions.

Ashok Kumar is a professor in the Department of Mechanical Engineering and the Center for Nanomaterials and Microelectronics Research at the University of South Florida. Dr. Kumar received his B.S. and M.S. degrees from the Indian Institute of Technology (Kanpur) and his Ph.D from North Carolina State University, Raleigh. He has been an associate professor at the University of South Florida, an assistant professor of Electrical and Computer Engineering at the University of South Alabama, and a post-doctoral fellow at North Carolina State University. He was a cluster director of the Advanced Materials Program of the State of Alabama for the NASA EPSCoR program and in that capacity supervised materials research initiatives of a cluster of five research universities, six colleges, and two HBCUs. He has received research grants from NSF, NASA, DOE, and private industries. He has authored or coauthored over 115 refereed publications; given 80 presentations and invited lectures; edited 7 books, supervised Ph.D., M.S., and post-doctoral students; and organized symposia at international conferences. Dr. Kumar is a member of the Materials Research Society, the American Physical Society, ASM International, TMS, the American Vacuum Society, the American Ceramic Society, and IEEE. He also received an NSF Career award, which is the highest honor given by the National Science Foundation to young university faculty.

Narendra B. Dahotre is a professor with joint appointment at Oak Ridge National Laboratory and the Department of Materials Science and Engineering of the University of Tennessee–Knoxville. Dr. Dahotre is also a senior faculty member of the Center for Laser Applications at the University of Tennessee Space Institute–Tullahoma. He received his B.S. degree from University of Poona (India), and his M.S. and Ph.D. degrees from Michigan State University. He has been chair and vice-chair of the Center of Excellence for Laser Applications at the University of Tennessee; a visiting research fellow in the Optoelectronic Division of the Electrotechnical Laboratory, Tsukuba, Japan; and a Post Doctoral Fellow/Instructor at the University of Wisconsin–Milwaukee. He has been working in the field of laser materials processing for the past 25 years with funding from organizations such as NSF, DOE, the U.S. Air Force, Ford, Honda, ALCOA, and Johnson Controls. He has received 15 U.S. patents on laser processing; edited 12 books; published over 100 refereed papers and book chapters; supervised several Ph.D., M.S., and post-doctoral students; and received numerous honors and awards, including the ALCOA Foundation Research Award and UT Vice President's Award for Research Excellence. He is on the editorial boards of several journals, is an Honorary Technical Consultant to Asean Tribology Center (Philippines), and is on the Board of Technical Advisors, Center for Laser Processing of Materials, NFTDC (India). In 2004, he was elected a Fellow of the American Society for Materials.

Preface

Competitive manufacturing relies on judicious selection of materials and processes to convert these materials into useful products, structures, and devices. Transforming materials into value-added products requires knowledge of manufacturing technology, processing science, and the material's response to external stimuli as it is coaxed to adopt the desired shape, structure, and other attributes.

This book focuses on the interrelationship among the "*structure and behavior of materials*" (materials science), the techniques of "*how to make things*" (manufacturing technology), and the "*theory of how things are made*" (processing science). It emphasizes a fundamental understanding of a range of processes used in manufacturing. This is important because diverse manufacturing techniques often exhibit an underlying commonalty of process mechanics, the study of which aids premeditated design (as opposed to serendipitous development) of new techniques. Such understanding also aids in adapting current manufacturing practices to technological constraints imposed by the discovery of new materials.

In our assessment, most books that deal with the topics covered here tend to be either predominantly "vocational" or unabashedly "scientific." They are written either for students majoring in the physics and chemistry of materials or for those training to become skilled craftspeople. Books that pursue a cross-disciplinary focus to processing usually become elementary surveys that sacrifice technical depth for greater breadth of coverage. Vocational books on manufacturing pay cursory attention to the process science knowledge base and at best view it as information that must be presented for the sake of completeness rather than as building blocks that are integral to the manufacturing enterprise. In contrast, most 'science'-oriented books chiefly focus on the science of materials behavior and usually exclude any coverage of processing technology. The essential connectivity between materials science, processing science, and manufacturing technology is seldom emphasized.

The above is not to criticize the many eminently valuable books written to satisfy different objectives; such books have served and continue to serve students with varied academic backgrounds and career aspirations. This book is intended to fill a niche at the interface of materials science, processing science, and manufacturing technology. Fundamental materials phenomena are pervasive and manifest themselves in manufacturing processes in ways that are usually difficult to capture and assimilate using the knowledge base and paradigms specific to a single discipline. We believe that cultivating a mental orientation and habits of thought that permit integration of concepts, theories, techniques, and visions from a multiplicity of disciplines can be learned, taught, and profitably used in solving materials problems. This book had its genesis in this premise.

In the past, concerns have been raised about a lack of integration in technical curricula. A National Academies Report* states, "...*The area of synthesis and processing has suffered neglect in our universities and industry. A particularly compelling need is to provide undergraduates with a thorough grounding in the science and engineering of processing and its relation to manufacturing ... New courses and textbooks are needed at both the undergraduate and graduate levels ... These textbooks should also explicitly address the complementary approaches of physics, chemistry and engineering.*"

The book provides a contextual background in the elements of processing science and manufacturing technology. It customizes the content for diverse material classes and manufacturing processes. It is intended to cater to the needs of students who possess a basic, college-level background in physics, chemistry, and math through elementary calculus. It should also serve as a resource for those pursuing advanced graduate studies and research but possessing limited background in materials processing. The book does not follow an evolutionary approach usually needed for establishing the foundation of an undergraduate course, but it should be useful for in-depth treatment of selected topics. Above all, we hope that the book shall kindle students' interest to pursue advanced independent study in materials processing.

The book covers an expanded range of materials and processes in greater depth than has been customary in materials processing books. Chapter 1 reviews the foundational topics in materials science and engineering. The discussion is elementary and is intended to be a brief refreshment. An excessively discursive treatment has been avoided; in fact, the discussion is occasionally rather dense because prior knowledge of the content is assumed. Chapter 2 covers the industrial casting techniques and fundamental concepts of solidification science. Concepts in advanced solidification processing such as single crystal growth and semi-solid forming are also covered. Chapter 3 introduces the ceramic forming and powder metallurgy techniques. Wherever feasible, the underlying process physics is quantitatively described. Chapter 4 presents the basic concepts and elementary theory of selected surface, subsurface, and interface phenomena important in materials processing. Chapter 5 introduces the theory and practice of coating and surface modification technologies with emphasis on laser surface engineering. Chapter 6 focuses on the role of processing in the structure and properties of engineered composites, chiefly metal-matrix composites, particularly the high-temperature Ni-base composites. Chapter 7 introduces the theory and practice of semiconductor processing, including integrated circuits, silicon wafer manufacture, crystal growth, thermal oxidation, ion implantation, and lithography. The final chapter, written in a somewhat different key, covers emerging nanomaterials and their processing; because of the highly dynamic nature of the field, this chapter is written as a research report summarizing the latest findings. Overall, the book's topical coverage is not intended to be comprehensive, and the reader will note glaring omissions (metal working, polymer processing, etc.). The book, however, includes emerging materials and processes that are scantily covered in most similar books.

The book evolved out of lectures given over a decade or more by each author on one or more topics covered in the book, although some material derives from the scholarly writings for the professional community and from the authors' own research activities. Overall, the book is an

---

*Materials Science & Engineering for the 1990s: *Maintaining Competitiveness in the Age of Materials*, The National Academies Press, 1989 (http://books.nap.edu/books/0309039282/html).

outcome of the authors' combined teaching and scholarly efforts of nearly thirty-five years at different institutions. We are thankful to students from various disciplines whose educational needs in materials processing served as the driving force for this book. We hope that the book shall facilitate learning by the current and future generations of students.

We owe special gratitude to Elsevier's Senior Editor Joel Stein, Associate Editor Shoshanna Grossman, Production Editor Matt Heidenry, and Project Manager Brandy Lilly for their valuable help and guidance in completing this book, their patience with our pace of writing, and their encouragement at every activation barrier. We wish to thank the University of Wisconsin–Stout, the University of South Florida, and the University of Tennessee–Knoxville for institutional support. We are most indebted to our families—spouses, children, and parents—for their support and for the personal sacrifices that were mandated by this scholarly undertaking.

Rajiv Asthana
Ashok Kumar
Narendra B. Dahotre

Acknowledgments

The authors gratefully acknowledge the permission of the following organizations and publishers to use copyrighted materials from the cited sources.

American Ceramic Society

American Ceramic Society Bulletin

American Chemical Society

J. American Chemical Soc.
Macromolecules
Nano Letters
Analytical Chemistry

American Foundry Society

Modern Castings
Transactions of the American Foundry Society
Aluminum Casting Technology, D. L. Zalensas, ed., 1997
Basic Principles of Gating and Risering, Cast Metals Institute, 1985

American Institute of Physics

J. Applied Physics
Applied Physics Letters

American Society for Materials, International

Advanced Materials & Processes
Metals Handbook, vols. 4 and 9
Liquid Metals and Solidification, 1958
Binary Alloy Phase Diagrams, T. B. Massalski, ed., 1990
Atlas of Isothermal Transformation & Cooling Transformation Diagrams, H. Boyer, 1977
Functions of the Alloying Elements in Steels, E. C. Bain, 1939

Chapman and Hall

Ceramic-Matrix Composites, R. Warren, ed., 1992

CRC Press

Tribology: Friction and Wear of Engineering Materials, I. M. Hutchings, 1999

Elsevier

Concise Encyclopedia of Composite Materials, A. Kelly, ed., 1994
Mechanical Testing of Engineering Ceramics at High Temperatures, B. F. Dyson, R. D. Lohr, and R. Morrel, eds, 1989
Metal Matrix Composites: Thermomechanical Behavior, M. Taya and R. J. Arsenault, 1989
Wettability at High Temperatures, N. Eustathopoulos, M.G. Nicholas and B. Drevet, 1999
The Coming of Materials Science, R. W. Cahn, 2001
Advances in Particulate Materials, A. Bose, 1995
Castings, J. Campbell, 1999
Physical Metallurgy Principles, R. W. Cahn and P. Haasen, eds., 1983
Journal of the European Ceramic Society
Intermetallics
Composites Science and Technology
Tribology International
Materials Science & Engineering
Chemical Engineering Science
Journal of Colloid and Interface Science

Foundry Research Institute (Krakow)

Proceedings of International Conference on HTC, 29 June-2 July, 1997, Krakow, Poland, eds. N. Eustathopoulos and N. Sobczak, 1998

Institute of Materials (London)

An Introduction to the Solidification of Metals, W. C. Winegard, 1964

Materials Research Society

MRS Symp. Proc. Vol. 120

McGraw-Hill

Electroplating: Fundamentals of Surface Finishing, F. A. Lowenheim, 1978
Transformations in Metals, P. G. Shewmon, 1969
Essentials of Materials Science, A. G. Guy, 1976

Metal Powder Industries Federation (Princeton, NJ)

Powder Metallurgy Design Manual, 1998
Powder Metallurgy Science, R. M. German, 1984

Mir Publishers (Moscow)

Solid State Physics, G. I. Epifanov, 1979 (English translation)

National Physical Laboratory (London)

The Relation Between the Structure and Mechanical Properties of Metals, vol. II, Symposium No. 15, 1963

Prentice Hall

Introduction to Materials Science for Engineers, J. F. Shackelford, 1985

Springer

Composite Materials: Science and Engineering, K. K. Chawla, 1988
Journal of Materials Science

Taylor Knowlton

Materials in Art and Technology, R. Trivedi, 1998

The European Powder Metallurgy Association (Shrewsbury, England)

EPMA Educational Aid

The Minerals, Metals & Materials Society (TMS)

Journal of Materials (JOM)
Solidification of Metal-Matrix Composites, P. Rohatgi, ed., 1990
Diffusion in Solids, P. G. Shewmon, 1989

Wiley

Materials Science & Engineering: An Introduction, W. D. Callister, Jr., 2000
An Introduction to Materials Engineering and Science for Chemical and Materials Engineers, B. S. Mitchell, 2004
The Science & Engineering of Thermal Spray Coating, L. Pawlowski, 1995
Transport Phenomena in Materials Processing, S. Kou, 1996
Principles of Ceramic Processing, J. S. Reed, 1995
Foundry Engineering, H. F. Taylor, M. C. Flemings, and J. Wulff, 1959
Materials and Processes in Manufacturing, E. P. DeGarmo, J. T. Black, R. A. Kohser, and B. E. Klanecki, 2003
Solidification and Casting, G. J. Davis, 1979
Properties of Materials, vol. 1, Structure, W. G. Moffat, G. W. Pearsall, and J. Wulff, 1964
The Structure and Properties of Materials, vol. 3, Mechanical Behavior, H. W. Hayden, W. G. Moffat, and J. Wulff, 1965
The Structure and Properties of Materials, vol. 4, Electronic Properties, R. M. Rose, L. A. Shepard, and J. Wulff, 1966
Introduction to Materials Science & Engineering, K. M. Ralls, T. H. Courtney and J. Wulff, 1976

Machine Design, Penton Media Inc., 1300 East Ninth St., Cleveland OH 44114

Industrial Heating: Journal of Thermal Technology, Industrial Heating, Manor Oak One, Suite 450, 1910 Cochran Road, Pittsburgh, PA 15220

Phillips Plastics Corporation, Metal Injection Molding, 422 Technology Drive East, Menomonie, WI 54751

Investment Casting Institute, 136 Summit Avenue, Montvale, NJ 07645-1720

Amsted Industries, Two Prudential Plaza, 180 North Stetson Street, Suite 1800, Chicago, IL 60601

North American Die Casting Association, North American Die Casting Association, 241 Holbrook Dr, Wheeling, Illinois 60090-5809 USA

Pratt & Whitney, Corporate Headquarters, 400 Main Street, East Hartford, CT 06108

United States Steel Corporation, 600 Grant St, Pittsburgh, PA 15219-2702

Technology Research News
The Nature Group of Publications, Co.
The American Scientist
Science
The Chemical Society of Japan

The authors gratefully acknowledge the permission from the following organizations and publishers to use copyrighted materials.

American Ceramic Society, Westerville, OH

American Chemical Society, Washington, DC

American Foundry Society, Schaumburg, IL

American Institute of Physics, Melville, NY

American Society for Materials (ASM International), Materials Park, OH

Amsted Industries, Chicago, IL

Applied Science Publishers, Ltd., Barking Essex, U.K.

Cast Metals Institute, Schaumburg, IL

Chemical Society of Japan, Tokyo, Japan

Chapman and Hall, London, U.K.

CRC Press, Boca Raton, FL

Elsevier, Boston, MA

European Powder Metallurgy Association (Shrewsbury, England)

Foundry Research Institute, Krakow, Poland

Howard Taylor Trust, Boston, MA

Institute of Materials, London, U.K.

Investment Casting Institute, Montvale, NJ

John-Wiley, New York, NY

Materials Research Society, Warrendale, PA

McGraw-Hill, New York, NY

Metal Powder Industries Federation, Princeton, NJ

Modern Casting, Schaumburg, IL

Mir Publishers, Moscow, Russia

National Physical Laboratory, London, U.K.

Nature Publishing Group, CCC, Danvers, MA

North American Die Casting Association, Wheeling, IL

Pearson Education, Inc., Upper Saddle River, NJ

Penton Media Inc., Cleveland, OH

Phillips Plastics Corporation, Menomonie, WI

Pratt & Whitney, East Hartford, CT

Royal Society of Chemistry, Cambridge, U.K.

Springer, New York, NY

Taylor Knowlton, Ames, IA

The Minerals, Metals & Materials Society, Warrendale, PA

United States Steel Corporation, Pittsburgh, PA

1   Materials Behavior

Introduction

Innovative materials and processes to produce them are enabling technologies. Materials that are multifunctional, smart, and possess physical and engineering properties superior to the existing materials are constantly needed for continued technical advances in a variety of fields. In modern times, the development, processing, and characterization of new materials have been greatly aided by novel approaches to materials design and synthesis that are based on a fundamental and unified understanding of the processing-structure-properties-performance relationships for a wide range of materials.

The subject matter of materials and manufacturing processes is very broad, and integrates the understanding derived from the study of materials science and engineering, process engineering, physical sciences, and the applied knowledge about practical manufacturing technologies. There are several complementary ways to approach this subject matter, and one that this book follows is from the viewpoint of materials science and engineering, which is the study of structure, processing, and properties, and their interrelationship. Perhaps more than any other technical discipline (with the exception perhaps of computer science and engineering), the discipline of materials science and engineering (MSE) builds a bridge between scientific theory and engineering practice. This is clearly reflected in its widely accepted title; we generally talk of mechanical engineering and electrical engineering rather than mechanical science and engineering, or electrical science and engineering! MSE has intellectual roots in physical sciences, but ultimately it represents the marriage of the "pure" and the "applied" and of the "fundamental" and the "practical."

In this chapter, we shall briefly review some of the foundational topics in materials science and engineering in order to develop a better understanding of the topics related to manufacturing processes that are covered in later chapters. We shall, however, first present some examples of innovations in materials and processes—taken from a National Academies Report—that touch upon our everyday lives. These examples also highlight how premeditated design based on the scientific method has led to technical innovations (with the exception perhaps of the tungsten filament for which the technological advance preceded a scientific understanding of the materials behavior).

Process Innovation as Driver of Technological Growth

Single-Crystal Turbine Blades

Turbine blades for gas turbine engines are made out of Ni-base high-temperature superalloys that retain their strength even at 90% of their melting temperature. This has permitted an increase in the fuel inlet temperatures and increased engine efficiency (the efficiency increases about 1% for every 12°F increment in the fuel inlet temperature). However, even super-alloys become susceptible to creep and failure at high fuel-combustion temperatures under the centrifugal force generated by a rotational speed of 25,000 revolutions per minute. Early approaches succeeded in strengthening the superalloys by adding C, B, and Zr to the superalloys. These elements segregate at and strengthen the grain boundaries, thus providing resistance to creep and fracture. Unfortunately, these additives also lower the melting temperature of the superalloy.

In the 1960s, the problem was addressed from a different angle. It was demonstrated that eliminating grain boundaries that were oriented perpendicular to the centrifugal stress could increase the blade's service life. This is because such grain boundaries experience greater stress for deformation and fracture than boundaries oriented parallel to the blade axis. During casting of the blades, directional solidification was initiated with the help of a chill, which led to large, columnar grains oriented parallel to the blade axis (i.e., direction of the centrifugal stress). The method increased the high-temperature strength of superalloy turbine blades by several hundred percent.

A breakthrough in further enhancing the high-temperature strength was subsequently achieved by eliminating all grain boundaries, resulting in the growth of the entire blade as a single crystal. Single crystals of semiconducting materials (Si and Ge) had already been grown using special techniques (crystal pulling, floating-zone directional solidification, etc., see Chapter 2). The key innovation in the growth of single-crystal turbine blades centered around extremely slow directional cooling and design of a "crystal selector," a pigtail-shaped tortuous opening at the base of the casting mold that would annihilate all but one grain. This single grain would then grow into the liquid alloy when the mold was slowly withdrawn out of the hot zone of the furnace. Since 1982, single-crystal turbine blades have become a standard element in the hot zone of gas turbine engines.

Copper Interconnects for Microelectronic Packages

Faster and more efficient microcircuits require an increasing number of transistors to be inter-connected on chips. At first, Al metal proved convenient as the interconnect material for microelectronic packages, although in terms of electrical resistivity Cu was known to be far superior (with about 40% less resistance than Al). Tiny Cu microwires could also withstand higher current densities so they could be packed closer together for increased chip efficiency and miniaturization. However, Cu had a major drawback over Al; Cu readily diffuses into silicon wafer. In addition, depositing and patterning Cu microcircuitry proved more difficult than Al. With continued push toward miniaturization, the limitations of Al relative to its resistivity and current density became more pronounced. Research on depositing Cu interconnects continued, and around 1997, a viable technology for Cu interconnects was unveiled that relied on the development of a reliable diffusion barrier for Cu.

Tungsten Filament for Light Bulbs

The development of tungsten filament wire for use in incandescent lamps is a well-known example of process innovation that drove major technological advance. More than a century ago, carbon filaments were used in light bulbs. However, carbon filaments were fragile, brittle, short-lived, and reacted with the residual gas in the bulb, leading to soot deposition and diminished lumens. Tungsten was known to provide better light output than carbon, and its extremely high melting point and tendency to retain strength at high temperatures suggested longer filament life. However, making tungsten into a filament (to increase its light-emitting surface area) was nearly impossible because of its extremely poor ductility. William Coolidge at General Electric was able to make long W filaments by heating the metal ingot and pulling the hot metal piece through a series of wire-drawing dies. The W metal that was used for wire drawing was first obtained by reducing tungsten oxide to tungsten metal in a clay crucible. Interestingly, W metal obtained via reduction in other (non-clay) crucibles was brittle and not amenable to wire drawing. Only around the 1960s did researchers find the scientific reason for this anomaly. Potassium from the clay crucibles had dissolved into the metal during the reduction process and made the metal ductile. Potassium then turned into tiny bubbles during high-temperature processing. These bubbles elongated into tubes during wire drawing. After annealing, the tubes pinched off into a series of tiny bubbles that anchored the tungsten grain boundaries (whose movement would otherwise cause filament failure).

Tailor-Welded Blanks

In the past, structural automobile body parts were made by cutting steel sheets into starting shapes or blanks. These steel sheets had a specific thickness, protective coating, and metallurgical structure required for the application. The blanks were then stamped into the three-dimensional forms of the finished body parts. Areas such as side panels and wheel housings required selective reinforcement with heavier steel for safety or to withstand stresses. These composite assemblies were made by first making individual parts and then welding these parts together into finished assemblies.

A manufacturing innovation of the 1980s, called tailor-welded blanks (TWBs), considerably simplified auto body assembly. The key was the incorporation of the heterogeneous material properties needed for auto parts into a single blank that could be formed into the finished shape with a single set of forming dies. The blanks were tailor-made by laser-welding flat steel sheets with different thicknesses, strengths, and coatings. As laser welding was already a mature technology, TWBs saw rapid and wide industrial acceptance.

As the preceding example of tailor-welded blanks shows, most applications of engineering materials require welding (or fastening and adhesive bonding) of materials into parts, devices, or structural elements, and these into assemblies, packages, or structural systems. In some applications, however, material synthesis and fabrication of the device or component may occur concurrently and seamlessly so that the boundaries between materials and devices based on them can no longer be distinguished as separate entities. An example is the junction between negative-type and positive-type extrinsic semiconductors; junctions between these semiconductors for use in transistors are synthesized at the same time as the semiconducting materials themselves (see Chapter 7 for a discussion of semiconducting materials and devices).

Thus, innovations in materials and processes also bring about evolutionary changes in prevailing manufacturing paradigms.

Atomic Bonding in Materials

The origin of the physical and mechanical behaviors of materials can be traced to the interatomic forces in solids. The two fundamental forces between atoms are the attractive forces due to the specific type of chemical bonding in a particular solid, and the repulsive forces due to overlapping of the outer electron shells of neighboring atoms. The magnitude of these forces decreases as the separation between atoms increases. The net force between atoms is the sum of the attractive and repulsive forces, and varies with the distance between atoms as shown in Figure 1-1. The net force approaches zero at a distance (typically, a few angstroms) where these two forces exactly balance each other, and a mechanical equilibrium is reached. Because the net

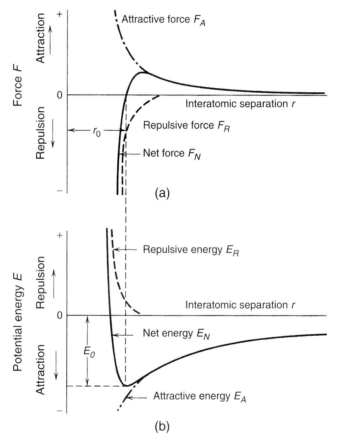

FIGURE 1-1 *(a) Variation of the repulsive force, attractive force, and net force between two isolated atoms as a function of the interatomic separation, and (b) variation of the repulsive, attractive, and net potential energies as a function of the interatomic separation. (W. D. Callister, Jr., Materials Science & Engineering: An Introduction, 5th ed., Wiley, New York, 2000, p. 19).*

force, F_N, is related to the total energy, E, by

$$E = \int_\infty^r F_N . dr,$$

where r is the distance, the net energy is a minimum at the equilibrium separation, r_0, between atoms. In other words, $(dE/dr) = 0$ at $r = r_0$. The binding energy between atoms is the net energy corresponding to this equilibrium separation.

In a real solid composed of a large number of mutually interacting atoms, the estimation of binding energies from interatomic forces becomes complex, although in principle a binding energy can be specified for the atoms of any real solid. Properties such as thermal expansion, stiffness, and melting temperatures are derived from the shape of the interaction energy curve (Figure 1-1), the magnitude of minimum (binding) energy, and the nature of chemical bond between atoms of the solid.

Atoms in solids exhibit three types of chemical bonds: ionic, covalent, and metallic. Ionic bonds are universally found in solids that are made of a metallic and a nonmetallic element (e.g., NaCl, Al_2O_3). In ionic bonding, the outer (valence) electrons from the metallic element are transferred to the nonmetallic element, causing a positive charge on the atoms of the former and a negative charge on the atoms of the latter. These ionized atoms develop electrostatic (Coulombic) forces of attraction and repulsion between them, which decrease with increasing separation between atoms. Ceramic materials exhibit predominantly ionic bonding.

In covalent solids, the electrons between neighboring atoms are shared, and the shared electron belongs to both atoms. Molecules such as Cl_2, F_2, HF, and polymeric materials are covalently bonded. Covalent bonds are directional in the sense that the bond forms only between atoms that share an electron. In contrast, ionic bonds are nondirectional; that is, the bond strength is same in all directions around an ion. Many solids are partially ionic and partially covalent; the larger the separation of two elements in the periodic table (i.e., the greater the difference in the electronegativity of the two elements), the greater will be the degree of ionic bonding in compounds of the two elements.

The third type of primary bond, metallic bond, occurs in metals and alloys, and is due to the freely drifting valence electrons that are shared by all the positively charged ions in the metal. Thus, an "electron cloud" or "electron sea" permeates the entire metal, and provides shielding against mutual repulsion between the positive ions. The electron cloud also acts as a "binder" to hold the ions together in the solid via electrostatic attractive forces. Table 1-1 shows the relationship between the binding energy and melting temperatures of some solids; solids of high bond energy exhibit high melting points.

In addition to the primary or chemical bonds, secondary bonds exist between atoms and influence properties such as surface energies. Secondary bonds are weaker than primary bonds and have energies of a few tens of kJ/mol as opposed to a few hundred kJ/mol or higher for the primary bonds. Nevertheless, secondary bonds are ubiquitous; they are present between all atoms and molecules, but their presence can be masked by the stronger chemical bonds. The genesis of secondary bonds lies in the transient and permanent dipole moments of atoms or molecules. The constant thermal vibration of an atom causes transient (short-lived) distortion of an otherwise spatially symmetric electron distribution around the nucleus, leading to a separation of the centers of positive and negative charges (Figure 1-2). This results in a transient induced dipole that induces a dipole in a neighboring atom by disturbing its charge symmetry, and

TABLE 1-1 Melting Temperatures and Bond Energies

Material	T_m, $°K$	Bond Energy, $kJ \cdot mol^{-1}$
NaCl	1074	640
MgO	3073	1000
Si	1683	450
Al	933	324
Fe	1811	406
W	3683	849
Ar	84	7.7
Cl_2	172	31
NH_3	195	35
H_2O	273	51

Source: Adapted from W. D. Callister, Jr., *Materials Science & Engineering: An Introduction*, 5th ed., Wiley, New York, 2000.

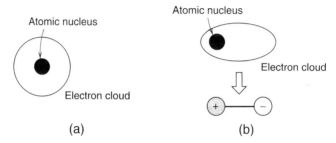

(a)   (b)

FIGURE 1-2 *Schematic representation of (a) an electrically symmetric atom, and (b) an induced atomic dipole due to a net shift in the centers of the positive and negative charges.* (W. D. Callister, Jr., Materials Science & Engineering: An Introduction, 5th ed., Wiley, New York, 2000, p. 25).

so forth. The resulting electrostatic forces fluctuate with time. Weak secondary forces that have their origin in such induced atomic dipoles are called van der Waals bonds. Certain molecules, called polar molecules, possess a permanent dipole moment because of an asymmetric charge distribution in their atoms and molecules. Such molecules can induce dipoles in neighboring nonpolar molecules, causing an attractive force or bond to develop between the molecules.

Crystal Structure

Upon slow cooling, the disordered structure of a liquid transforms into an ordered structure characteristic of crystalline solids. In crystalline solids, the atoms are arranged in three-dimensional periodic arrays over large distances; these arrays could be relatively simple as in common metals or extremely complex as in polymeric materials. Certain materials, however, do not exhibit the long-range atomic order characteristic of crystalline solids. Such materials form either a completely amorphous or partially crystalline structure when they are cooled from the liquid state. Many complex polymers comprised of long-chain molecules show only partial crystallinity under normal cooling because of the entanglement of chain segments that create pockets of atomic disorder. Similarly, most metals and alloys that would normally crystallize under slow cooling may exhibit an amorphous or glassy structure under ultra-fast cooling conditions, which restrict atomic diffusion needed to form a periodic atomic array, thus causing the random structure of the liquid to be "frozen" in the solid state. Our current knowledge of the atomic arrangements in solids is largely derived from the use of x-rays as a tool to probe the crystal structure. The periodically arranged atoms scatter the x-rays of a wavelength comparable to the spacing between atoms, and give rise to the phenomena of diffraction or specific phase relationships between scattered waves.

Many physical attributes of crystalline solids are determined by the type of geometric arrangement of their constituent atoms. A useful model to visualize the atomic packing in crystalline solids is to first liken each atom as a "hard sphere," and then identify the smallest repeating cluster of atoms (unit cell) that could be stacked in three dimensions to generate the long-range atomic order. For each metallic element, the hard spheres (with a characteristic atomic radius) represent a positive ion in a sea of electrons. Many metallic elements crystallize in one of three basic geometric forms or crystal structure: face-centered cubic (FCC) (Figure 1-3), body-centered cubic (BCC) (Figure 1-4), and hexagonal close-packed (HCP) (Figure 1-5).

Many features of atomic packing depend on the crystal directions and crystal planes. For example, the packing density of atoms (or atomic density) and interatomic voids depend upon crystal directions and crystal planes. The packing density and the void content in the crystal structure influence the alloying behavior, diffusion processes, plastic deformation, and various other material properties. Thus, knowledge about the crystal structure is important in understanding the materials behavior.

Crystallographic directions are specified in terms of a line between two points in the unit cell, and denoted as [uvw] where u, v, and w are the projections along the x, y, and z axes, respectively (with reference to the origin of the coordinate system, conveniently located at a corner in the unit cell). Crystal planes or atomic planes are specified relative to the unit cell as (hkl) and for HCP (Figure 1-5) as (hklm). The intercepts made by a crystal plane along the x,

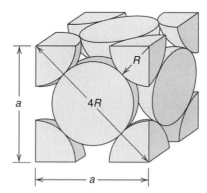

FIGURE 1-3 *A hard-sphere unit cell representation of the face centered cubic (FCC) crystal structure. R is the atomic radius, and a is the side of the unit cell.* (W. D. Callister, Jr., Materials Science & Engineering: An Introduction, 5th ed., Wiley, New York, 2000, p. 32).

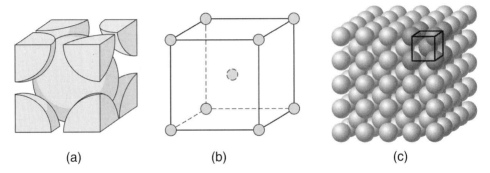

(a)   (b)   (c)

FIGURE 1-4 *(a) A hard-sphere unit cell representation of the body-centered cubic (BCC) crystal structure, (b) a reduced-sphere unit cell, and (c) an aggregate of many atoms with the BCC arrangement.* (W. D. Callister, Jr., Materials Science & Engineering: An Introduction, 5th ed., Wiley, New York, 2000, p. 34).

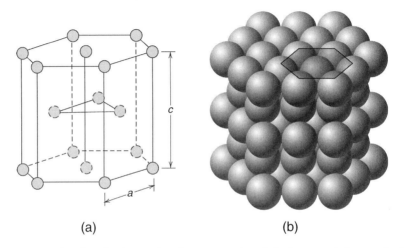

(a)   (b)

FIGURE 1-5 *(a) A reduced-sphere unit cell representation of the hexagonal close-packed (HCP) crystal structure, and (b) an aggregate of many atoms with the HCP arrangement.* (W. D. Callister, Jr., Materials Science & Engineering: An Introduction, 5th ed., Wiley, New York, 2000, p. 35).

y, and z axes are written in terms of the intercepts within the unit cell (i.e., normalized with respect to the length of the sides of the cube to obtain integral values for h, k, and l), and the reciprocal of these intercepts is then written in a reduced form (i.e., in terms of the smallest integers). Figure 1-6 illustrate some examples of crystal planes and crystal directions.

A large number of physical and mechanical properties of materials depend on the crystallographic orientation along which the property is measured, and significant differences can occur along different directions. For example, the modulus of elasticity of metals is orientation-dependent. The modulus of Fe along [100], [110], and [111] directions is 125 GPa, 211 GPa, and 273 GPa, respectively. Similarly, the modulus of Cu along [100], [110], and [111] directions has been measured to be 67 GPa, 130 GPa, and 191 GPa, respectively.

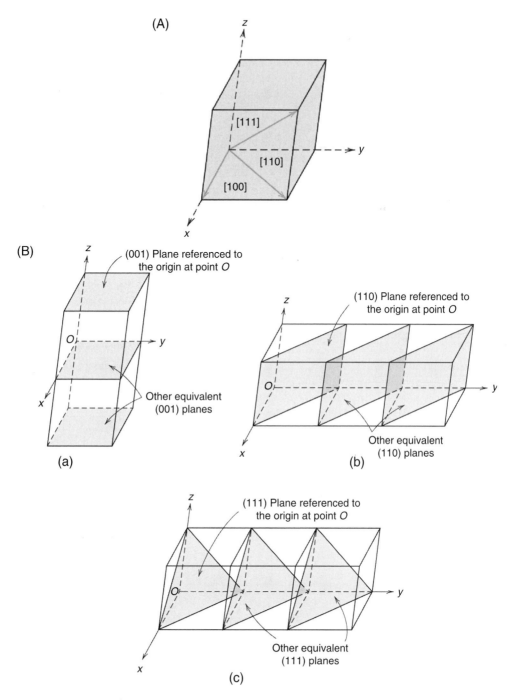

FIGURE 1-6 *(A) The [100], [110], and [111] crystallographic directions in a unit cell.* (W. D. Callister, Jr., Materials Science & Engineering: An Introduction, 5th ed., Wiley, New York, 2000, p. 41) *(B) Representation of a series each of (a) (001), (b) (110), and (c) (111) crystallographic planes.* (W. D. Callister, Jr., Materials Science & Engineering: An Introduction, 5th ed., Wiley, New York, 2000, p. 44).

Defects in Crystalline Solids

Perfect long-range atomic order does not exist in crystalline solids even in most carefully prepared materials. Various types of imperfections or irregularities exist in the atomic arrangement in all solid materials. The most common imperfections in crystalline solids include vacancies, interstitials, solute (or impurity) atoms, dislocations, grain boundaries, and surfaces and interfaces. Many physical and mechanical properties of crystalline solids are determined by the nature of the defects, and their distribution and concentration in the material.

Vacancies or vacant atomic sites (Figure 1-7) exist in solids at all temperatures; their concentration increases exponentially with temperature according to the Boltzmann distribution function, $N = N_0 \exp(-Q/RT)$, where N is the equilibrium number of vacancies per unit volume of the material at an absolute temperature, T, N_0 is the total number of atomic sites per unit volume (which depends upon the crystal structure), Q is the activation energy to form the vacancy (related to the energy barrier that must be surmounted to dislodge an atom from its normal position and create a vacancy), and k is the Boltzmann's constant (13.81×10^{-24} J/K). Vacancies increase the disorder (entropy) in the crystal, thus making their presence a thermodynamic necessity. Unlike a vacancy, an interstitial defect forms when either a host atom or an impurity atom resides in a preexisting void in the crystal lattice. Figure 1-7 shows the formation of a self-interstitial when a host atom is dislodged from its normal site and forced into the (smaller) void between atoms.

Most practical engineering materials are alloys rather than pure elemental solids. Alloys are solid solutions and form when an impurity atom either substitutes a host atom (substitutional solid solution) or enters the interstices of the parent lattice (interstitial solid solution). Figure 1-8 shows substitutional and interstitial impurity atoms in a crystal lattice. Solute atoms in a host crystal lattice can also form a compound (e.g., intermetallic compounds in which the different types of atoms are combined in a fixed or nearly fixed proportion). High solubility of impurity

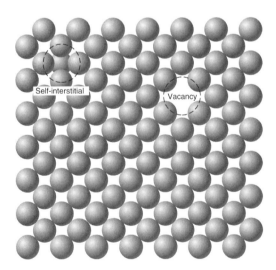

FIGURE 1-7 *Schematic illustration of two types of point defects in crystalline solids: a vacancy and a self-interstitial.* (W. G. Moffat, G. W. Pearsall, and J. Wulff, Properties of Materials, vol. 1, Structure, Wiley, 1964, p. 77). Reprinted with permission from Janet M. Moffat, 7300 Don Diego NE, Albuquerque, NM.

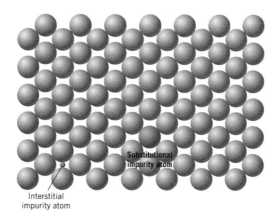

Substitutional
impurity atom

Interstitial
impurity atom

FIGURE 1-8 *Schematic illustration of an interstitial and a substitutional impurity atom in crystalline solids.* (W. G. Moffat, G. W. Pearsall, and J. Wulff, Properties of Materials, vol. 1, Structure, Wiley, 1964, p. 77). Reprinted with permission from Janet M. Moffat, 7300 Don Diego NE, Albuquerque, NM.

atoms in substitutional solid solutions is favored when a set of criteria, called the Hume-Rothery rules, are satisfied. These rules stipulate that solid solutions form when one or more of the following criteria are met: (1) the difference in the atomic radii of the two atom types should be less than 15%, (2) crystal structure of the two metals should be the same, (3) the electronegativity difference between the two atom types should be small (metals widely separated in the periodic table, i.e., those exhibiting large electronegativity difference, are more likely to form an intermetallic compound rather than a solid solution), and (4) a metal of higher valence will dissolve more readily in the host metal than a metal of valence lower than the host metal. A classic substitutional solid solution is Cu-Ni, which exhibits complete solubility. In contrast to substitutional solid solutions, in an interstitial solid solution, the need for impurity atoms to fit in the interstices of the host lattice limits the solubility (usually <10%). In iron-carbon alloys, the much smaller carbon atoms occupy the interstitial positions in the iron lattice and form an interstitial solid solution.

Dislocations are linear defects in crystals that form in a region where a plane of atoms terminates abruptly in the lattice, as shown in Figure 1-9. This figure shows an edge dislocation where an atomic plane is shown missing in the bottom half of the crystal. Due to the disturbance in the periodicity of the lattice near a dislocation line, there is some distortion (stress) around the atomic planes, which in turn influences the physical and mechanical behaviors of the material. A screw dislocation forms when a shear stress causes the atomic planes across a region within the crystal to be shifted one atomic spacing relative to the other planes (Figure 1-10). In reality, mixed dislocations (comprised of edge and screw components) are more common than either pure edge or screw dislocations. Transmission electron microscopy (TEM) techniques permit visual observation of the dislocations. Figure 1-11 shows TEM photomicrographs of dislocation lines in a deformed intermetallic compound, NiAl, alloyed with a small amount of chromium; the dislocation lines are tangled and pinned by secondary phases. All crystalline materials contain dislocations, and it is virtually impossible to produce a dislocation-free crystal even under the most stringent processing conditions. Plastic deformation, phase transformations (e.g., solidification), thermal stresses, and irradiation increase the concentration of dislocations in the solid.

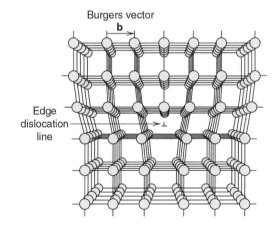

FIGURE 1-9 *The atom positions around an edge dislocation; extra half-plane of atoms shown in perspective.* (A. G. Guy, Essentials of Materials Science, McGraw-Hill, New York, 1976, p. 153).

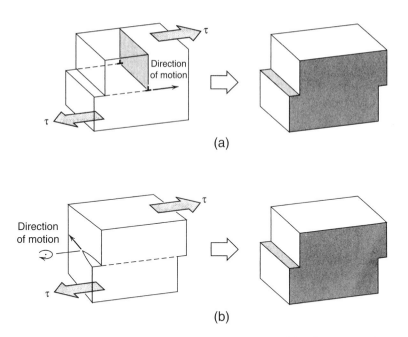

FIGURE 1-10 *The formation of a step on the surface of a crystal by the motion of (a) an edge dislocation, and (b) a screw dislocation. For the edge dislocation, the dislocation line moves in the direction of the applied shear stress, and for the screw dislocation, the dislocation line moves perpendicular to the direction of the shear stress.* (H. W. Hayden, W. G. Moffat, and J. Wulff, The Structure and Properties of Materials, vol. 3, Mechanical Behavior, Wiley, New York, 1965, p. 70). Reprinted with permission from Janet M. Moffat, 7300 Don Diego NE, Albuquerque, NM.

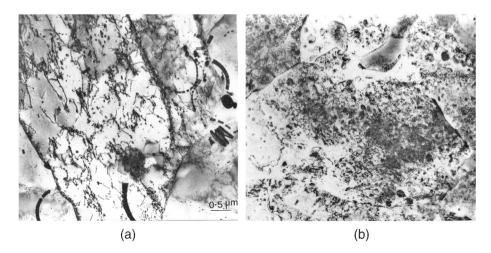

(a)   (b)

FIGURE 1-11 *(a) Transmission electron micrograph of an extruded Ni-48.3Al-1W alloy showing dislocation networks, and (b) transmission electron micrograph of an extruded Ni-43Al-9.7Cr alloy showing dislocation networks.* (R. Tiwari, S. N. Tewari, R. Asthana and A. Garg, J. Materials Science, 30, 1995, 4861–70).

The dislocation density (i.e., number of dislocations per unit area) in real crystalline solids is very large, on the order of 10^{12} dislocations per square meter in annealed metals, and 10^{15} to 10^{16} in cold-worked metals. Dislocations can be mobile under an applied stress, and hindrances to dislocation motion such as grain boundaries, secondary precipitates, inclusions, and other dislocations that form tangles and impede one another's motion lead to strengthening. Dislocations move on slip planes along certain preferred (close-packed) directions under stress, and emerge at the crystal surface in the form of a step. Such a process of dislocation exhaustion should in fact promote the solid's progression toward crystallographic perfection through migration of dislocations toward the surface and their elimination from the crystal lattice by formation of a step on an external surface of the crystal. However, this process of dislocation exhaustion is more than compensated by the generation of new dislocations during deformation. An important mechanism of dislocation generation, called the Frank-Reed source, mimics the evolution of a soap bubble at the end of a capillary under air pressure. A dislocation line D-D′, with ends pinned by solute atoms (or immovable points of intersection with other dislocations), steadily grows under an applied stress, τ, as shown in Figure 1-12, until it bends into a semicircle. Beyond this stage, the dislocation continues to grow at a decreasing stress, a closed dislocation loop forms by joining at points 6-6′ and 7-7′, and the dislocation loop grows until it reaches the solid's surface where it forms a step. In the process, an internal dislocation, D-D′, is generated at which the preceding mechanism repeats itself, thus generating new dislocation loops and increasing the overall dislocation density in the material.

Surfaces and interfaces represent discontinuity in the ordered arrangement of atoms and are also crystal defects. The free surface of a solid is associated with an excess energy due to the unsaturated atomic bonds at the surface. This excess energy is the surface energy (or surface tension) of the solid and is a measure of the driving force needed to minimize the free surface of the solid (in reality, a solid's free surface is actually an interface between the solid and the surrounding atmosphere or vapor). Grain boundaries in crystalline solids separate grains or regions having different crystallographic orientations. Both grain boundaries and

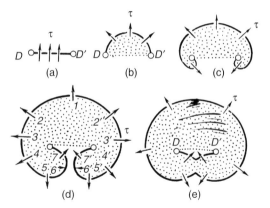

FIGURE 1-12 *Schematic illustration of the operation of a Frank-Reed dislocation source. (a) Initial position of the dislocation line D-D', (b) bending of the dislocation line under an applied stress, (c) and (d) continuing development of a symmetric dislocation loop, (e) formation of external dislocation loop spreading across the crystal and of internal dislocation D-D' returning to the original position.* (G. I. Epifanov, Solid State Physics, Mir Publishers, Moscow, 1979, p. 68, English translation).

interfaces have some atomic mismatch (lattice strain) because of the different crystallographic orientations of the neighboring regions and are, therefore, associated with an excess energy. Small- or low-angle grain boundaries form when the mismatch is small and can be accommodated by an array of dislocations. A special type of grain boundary, called a twin boundary, forms when the atoms across the boundary are located to form a mirror image of the other side. Twin boundaries form across definite crystallographic planes when either shear forces are applied or the material is annealed following plastic deformation. Figure 1-13 shows a schematic illustration of low-angle boundaries and twin boundaries.

Annealing

Annealing is a heat treatment that is applied to cold-worked metals to allow the structure and properties of the pre–cold-worked state to be regained. This occurs through the temperature-sensitive processes of recovery, recrystallization, and grain growth. During recovery, some of the physical properties (thermal and electrical conductivities) are recovered, although mechanical properties do not revert to the pre–cold-worked state. No observable microstructural changes occur during recovery. During the next stage of recrystallization, strain-free grains nucleate within the cold-worked material and slowly consume the entire cold-worked structure. Complete restoration of mechanical properties to their pre–cold-worked state occurs during recrystallization. Figure 1-14 shows the metallurgical structure of a cold-worked Sn-Pb alloy that was recrystallized at room temperature.

Highly cold-worked metals recrystallize faster, and the recrystallization temperature decreases with increasing degree of cold-working. Pure metals recrystallize faster than alloys, and alloying raises the recrystallization temperature. The recrystallization temperature, T_{cr}, is defined as the temperature at which the cold-worked structure fully recrystallizes in 1 h. For pure metals, $T_{cr} \sim 0.3 T_m$ (T_m is the absolute melting temperature), but for alloys, $T_{cr} \sim 0.7 T_m$. High-melting-point metals have a high T_{cr}. Many deformation processes use hot-working to shape parts; hot-working is done above T_{cr}. Note that room temperature deformation of

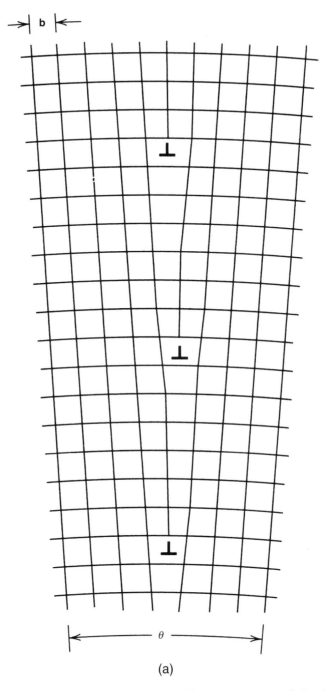

b

θ

(a)

FIGURE 1-13 *(a) A low-angle grain boundary formed by the arrangement of edge dislocations. This type of low-angle boundary is called a tilt boundary. (b) A twin boundary and the adjacent atom positions. The atoms on one side of the boundary are mirror images of the atoms on the other side.* (W. D. Callister, Jr., Materials Science & Engineering: An Introduction, 5th ed., Wiley, New York, 2000, p. 80).

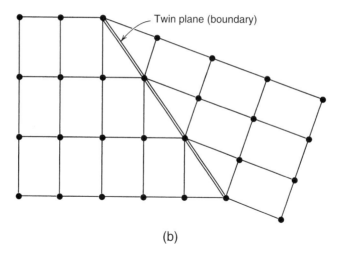

(b)

FIGURE 1-13 *Continued.*

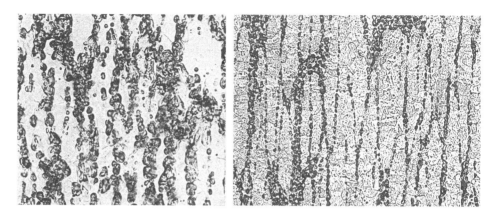

FIGURE 1-14 *Photomicrographs showing recrystallized grains in Sn-Pb alloys with two different volume fractions of the eutectic, which were mechanically deformed (swaged) and then recrystallized at room temperature.* (R. Asthana, unpublished research, NASA Glenn Research Center, Cleveland, OH, 1995).

Pb, Sn, and Sn-Pb alloys ($T_{cr} \sim -4°C$) is actually hot-working, whereas the deformation of $W(T_{cr} \sim 1200°C)$ at 1000°C is cold-working.

The new stress-free recrystallized grains continue to grow if a high temperature is maintained for a long period. This is because there is a distribution of grain sizes in the recrystallized material, and the need to decrease the grain boundary area provides the driving force for grain growth (or grain coarsening). Grain coarsening occurs by competitive dissolution of small grains and growth of larger grains in the distribution through mass transport via atomic diffusion. Grain coarsening in many polycrystalline metallic and ceramic materials follows the relationship: $d^n - d_0^n = Kt$, where d is the initial grain diameter (at $t = 0$, i.e., at the onset of grain growth), and K and n are time-independent constants, with n being greater or equal to 2. The coefficient K is estimated from the curve-fitting of experimental d versus t data.

Diffusion in Crystalline Solids

Diffusion in crystalline solids involves movement of atoms in steps within a crystal lattice. It could involve either one type of atoms (e.g., self-diffusion of like atoms in a pure metal) or different types of atoms (e.g., interdiffusion or impurity diffusion). The process of diffusion requires breaking of existing bonds by an atom, its migration to a vacant site in the lattice via an atomic jump process, and formation of chemical bonds with its new neighbors. Small atoms (such as C, N, etc.) diffuse in a crystal via interstitial positions (interstitial diffusion), whereas larger substitutional impurity atoms diffuse by jumping into vacancies whose concentration exponentially increases with temperature. This makes diffusion easier at high temperatures. In addition, at high temperatures atoms have high thermal (vibrational) energy, which also facilitates their migration.

Atomic diffusion at a constant temperature causes concentration variations with time and position. The concentration of diffusing atoms is specified by the diffusion flux, J, which is defined as the mass, M (or concentration, C), diffusing per unit time across a plane of area, A, normal to the diffusion direction. The diffusion flux, $J = (1/A)(dM/dt)$, where t is the time. If a steady-state is reached, then J becomes independent of time, and a linear concentration gradient of diffusing atoms is attained as shown for the case of gaseous diffusion across a thin metal foil in Figure 1-15. The steady-state diffusion process is described by Fick's first equation, which in one dimension reads, $J = -D(dC/dx)$, where C is the concentration and D is the diffusion coefficient. D represents the mobility of the diffusing atoms and has the dimensions of $(length)^2/time$. If the diffusion flux and the concentration gradient at a point change with time, then the diffusion is non-steady (Figure 1-16), and the concentration, C, is related to the

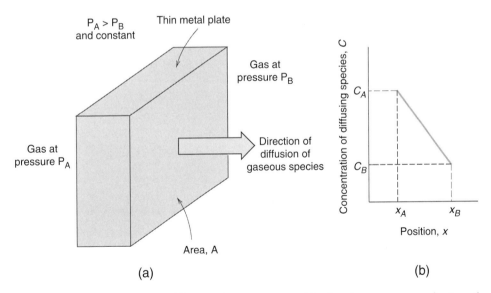

FIGURE 1-15 *(a) Steady-state diffusion of a gas across a thin plate in a pressure gradient, and (b) a linear concentration profile for the diffusion situation in (a). (W. D. Callister, Jr., Materials Science & Engineering: An Introduction, 5th ed., Wiley, New York, 2000, p. 96).*

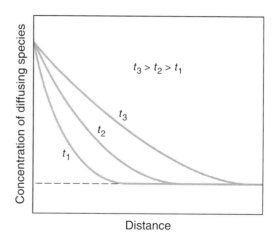

Distance

FIGURE 1-16 *Time modulation of the concentration distribution of a diffusing species.* (W. D. Callister, Jr., Materials Science & Engineering: An Introduction, 5th ed., Wiley, New York, 2000, p. 98).

position, x, and time, t, by Fick's second equation, whose one-dimensional form at a constant temperature is

$$\frac{\partial C}{\partial t} = \frac{\partial}{\partial x}(D\frac{\partial C}{\partial x}). \tag{1-1}$$

If D is constant, then it can be taken out of the partial derivative in Equation 1-1, thus yielding the simpler equation $\frac{\partial C}{\partial t} = D\frac{\partial^2 C}{\partial x^2}$. In many situations, however, D is a strong function of concentration, C, and Equation 1-1 must be solved. Fick's equations are analogous to the Fourier equations for heat conduction through solids (with the diffusion coefficient replaced with the thermal diffusivity, and C replaced with temperature, T). Many mathematical solutions to the above equation have been derived for boundary conditions that are encountered in a variety of physical processes. An important practical situation involving doping of semiconducting materials to control their electronic properties involves diffusion of impurity atoms, and the applicable solution to the Fick's equation is discussed in Chapter 7.

The diffusion coefficient, D, is very sensitive to temperature, and for atomic diffusion in solids, $D = D_0\exp(-Q/RT)$, where Q is the activation energy for diffusion, R is the gas constant, and D_0 is a pre-exponential term, called the frequency factor, which depends on the atomic vibration frequency and the crystal structure. The activation energy, Q, represents the energy consumed in distorting the local lattice to permit an atomic jump. Table 1-2 summarizes the values of Q and D_0 for several materials, and Figures 1-17 and 1-18 show the dependence of the diffusion coefficient on temperature for some common elements. By taking the natural logarithm of the preceding expression for D, a linear relationship between $\ln D$ and inverse absolute temperature, T, is obtained, which has the form, $\ln D = \ln D_0 - (Q/R)(1/T)$. Thus, by plotting experimentally measured D values at different temperatures, it is possible to obtain the activation energy for diffusion and the frequency factor.

TABLE 1-2   Diffusion Data for Metals and Semimetals

Diffusing Atom	Host Material	D_0, $m^2 \cdot s^{-1}$	Q, $kJ \cdot mol^{-1}$
Fe	α-Fe	2.8×10^{-4}	251
Fe	γ-Fe	5.0×10^{-5}	284
C	γ-Fe	2.3×10^{-5}	148
Cu	Cu	7.8×10^{-5}	211
Zn	Cu	2.4×10^{-5}	189
Cu	Al	6.5×10^{-5}	136
Cu	Ni	2.7×10^{-5}	256
W	W	1.88×10^{-4}	586
Al	Al	4.7×10^{-6}	123
Si	Si	20×10^{-4}	424
Ge	Ge	25×10^{-4}	318
Cr	Cr	9.7×10^{-2}	435

Note: D_0, frequency factor; Q, activation energy for diffusion.

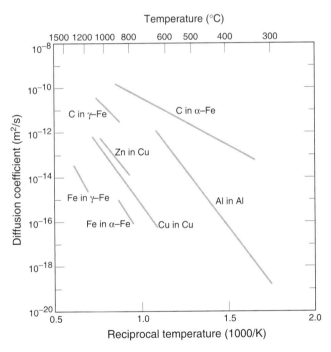

FIGURE 1-17   *Plot of the logarithm of the diffusion coefficient versus the reciprocal of absolute temperature for C and Fe in α- and γ-Fe, Zn in Cu, Al in Al, and Cu in Cu.* (W. D. Callister, Jr., Materials Science & Engineering: An Introduction, 5th ed., Wiley, New York, 2000, p. 103).

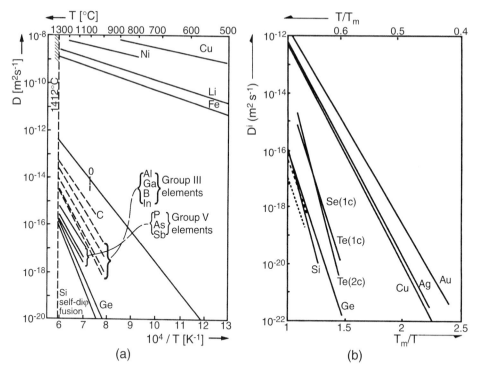

FIGURE 1-18 *(a) Plot of the logarithm of the diffusion coefficient versus the reciprocal of absolute temperature for various solutes in silicon.* (P. G. Shewmon, Diffusion in Solids, 2nd ed., The Minerals, Metals & Materials Society, Warrendale, PA, 1989, p. 175). *(b) Plot of the logarithm of the diffusion coefficient versus the reciprocal of the absolute homologous temperature (T_m/T) for Si, Ge, and noble metals.* (P. G. Shewmon, Diffusion in Solids, 2nd ed., 1989, p. 175). Reprinted with permission from The Minerals, Metals & Materials Society, Warrendale, PA (www.tms.org).

Figure 1-18a shows the diffusion coefficient versus $1/T$ plots for various solutes in common metals and in the semiconducting element Si. This figure shows that at a given temperature, Group III elements (Al, Ga, B, In) and Group V elements (P, As, Sb) diffuse faster than the Si atoms in Si crystals. As the slopes of lines for Group III and Group V solutes in Si in Figure 1-18a are roughly the same, the activation energies for diffusion are approximately identical. Figure 1-18a also shows that solutes such as Ni, Cu, Li, and Fe diffuse much faster in Si (and have lower activation energies) than other common solutes.

There are several basic differences in atomic diffusion in semiconductors such as Si and Ge, and in common metals such as Ni and Cu. Silicon and germanium form a diamond cubic crystal lattice characteristic of diamond (see Chapter 3), which is more open than the crystal lattice of close-packed metals. Second, the energy to form vacancies in Si and Ge is higher relative to the thermal energy at the melting point (i.e., Q/kT_m is large, where T_m is the melting point), and the energy to form vacancies and self-interstitials are more nearly equal in Si and Ge than in metals. Third, atoms occupy interstitial positions much more often in Si and Ge than in metals; therefore, interstitials play a more important role in self-diffusion in semiconductors. In addition, the presence or absence of bonding between solute atoms and Si or Ge determines the mobility of solute atoms. For example, oxygen atom is small and occupies interstitial sites in Si but bonds with Si and diffuses with a relatively high activation energy (2 eV), whereas the larger Ni and Cu atoms that form no bonds with Si move with a lower activation energy (0.5 eV). Another difference between metals and semiconductors is related to the difference in the dislocation density. In a carefully grown Ge or Si crystal, the dislocation density may average 10^4 per m^2 or less, whereas in metals, the dislocation density is high, 10^9 per m^2 even in well-annealed condition. Thus, in a metal, the distance a vacancy must diffuse to find a dislocation is much shorter than in a semiconductor (dislocations pin vacancies). As a result, whereas in a metal the equilibrium concentration of vacancies is maintained throughout the crystal, in Si or Ge, the vacancy concentration may deviate from the equilibrium value over a large fraction of the crystal.

The self-diffusion in Si and Ge at their respective melting points is orders of magnitude slower than that in metals at their melting points, as seen from Figure 1-18b. This difference between metals and semiconductors increases yet more at lower temperatures due to the relatively larger activation energies for Si and Ge. Experiments show that self-diffusion in Ge and Si is dominated by vacancy motion at low temperatures, but at high temperatures it is dominated by the motion of interstitials.

In the manufacture of Si-based devices, it is customary to oxidize the surface at intermediate temperatures to form an insulating silica layer. The growth of this layer speeds up the diffusion of Group III elements (B, Al, Ga, and P), and it slows down the diffusion of Group V elements (Sb and As). The diffusion of Group V solutes in Si depends on their atomic radius. Phosphorus (the smallest in the group) diffuses primarily by an interstitial mechanism, whereas the largest (Sb) diffuses by a vacancy mechanism. The role of diffusion in the manufacture of silicon-based devices is discussed in greater depth in Chapter 7.

Mechanical Behavior

For many crystalline solids, mechanical deformation is elastic at low applied normal stresses and follows Hooke's law, according to which $\sigma = E\varepsilon$, where σ is the applied stress, ε is the elastic strain, and E is the modulus of elasticity or Young's modulus, which represents the stiffness of the material. The elastic strain is completely recovered on removal of the stress, and there is no permanent deformation. For certain materials (many polymers and concrete), the stress–strain relationship is not linear, and it is a standard practice to characterize the stiffness or modulus of such materials either at a particular value of the strain or as an average over a range of strain values.

The origin of elastic modulus or stiffness lies in the strength of the interatomic bonds. As per the potential energy curve between neighboring atoms in a solid shown in Figure 1-1, the equilibrium separation is r_0 and the energy is $E(r_0)$. If an external stress increases the distance between the two atoms by a small amount, "x," then the potential energy increases to $E(r)$ where $r = r_0 + x$. The work done for displacement through "x" is therefore, $E(x) = E(r) - E(r_0)$. Expanding $E(r)$ into a Taylor series in terms of "x" yields

$$E(r) = \left(\frac{\partial E}{\partial r}\right)_0 \cdot x + \frac{1}{2} \cdot \left(\frac{\partial^2 E}{\partial r^2}\right)_0 \cdot x^2 + \frac{1}{6} \cdot \left(\frac{\partial^3 E}{\partial r^3}\right)_0 \cdot x^3 + \cdots \qquad (1\text{-}2)$$

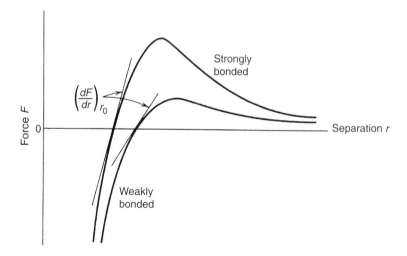

FIGURE 1-19 *Interatomic force versus separation plot for strongly bonded and weakly bonded atom pairs. The slope of the force–separation curve is proportional to the elastic modulus.* (W. D. Callister, Jr., Materials Science & Engineering: An Introduction, 5th ed., Wiley, New York, 2000, p. 120).

If all the terms higher than the quadratic are neglected due to the small value of the slope (dE/dr) near r_0, then we obtain

$$E(x) = \frac{1}{2}\left(\frac{\partial^2 E}{\partial r^2}\right)_0 \cdot x^2 = 0.5\beta x^2, \tag{1-3}$$

where β specifies the strength of the interatomic bond. The force needed to displace the atoms through "x" is therefore $F = -dE(x)/dx = -\beta x$, i.e., the force to increase the separation between neighboring atoms is directly proportional to the displacement. This relationship may be considered as a microscopic analogue of Hooke's law; summing up the force between all atom pairs and adding all the increments in atomic displacements, one would arrive at an equation substantially similar to Hooke's law. Figure 1-19 shows the interatomic force versus distance curves for strongly bonded and weakly bonded solids. The larger slope of the strongly bonded solid indicates the greater stiffness (higher modulus) of this material.

In a manner similar to the case of normal stress considered above, the deformation of solids under low shear and torsional stresses is also elastic; for example, shear strain, γ, is proportional to shear stress, τ, so that $\tau = G\gamma$, where G is the shear modulus.

The initial portion of a typical tensile stress–strain curve for a metal is shown in Figure 1-20. The initial linear regime is the elastic behavior described by Hooke's law just discussed. At a certain stress value (elastic limit), a transition occurs to a nonlinear behavior and the onset of plastic (permanent) deformation. The transition stress is the yield strength of the material, often not precise or distinct on the curve, as in the example of Figure 1-20. Often the stress rises to a peak value (upper yield point), and then drops to a lower value (lower yield point) about which the stress fluctuates before rising again, this time in a nonlinear fashion. The yield

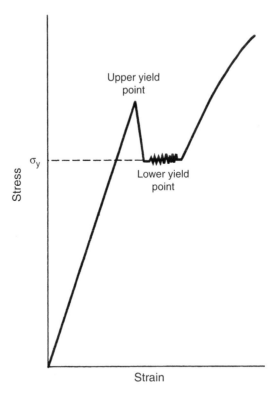

FIGURE 1-20 *The yield point phenomenon in steels that results in an upper yield point and a lower yield point.* (W. D. Callister, Jr., Materials Science & Engineering: An Introduction, 5th ed., Wiley, New York, 2000, p. 124).

strength in such a case is taken as the stress corresponding to 0.2% offset strain. A straight line is drawn parallel to the linear portion of the stress–strain curve starting at a strain of 0.002, and the intersection of the line with the stress–strain curve gives the yield strength. After the onset of plastic deformation, the stress needed to further deform the material continues to increase, reaches a maximum (tensile strength, point M in Figure 1-21), and then continuously decreases until the material fractures at the fracture stress or breaking stress. The deformation beyond the tensile stress is confined to a small region of the sample or "neck," which is a region of highly localized deformation. Fracture occurs at the necked region at the breaking stress (Figure 1-21).

Two important measures of the energy absorbed by a material during deformation are resilience and toughness. The resilience of a material is the energy absorbed during elastic deformation. It is estimated from the area under the linear portion of the stress–strain diagram under uniaxial tension, i.e.,

$$U = \int_0^{\varepsilon_y} \sigma \, d\varepsilon, \tag{1-4}$$

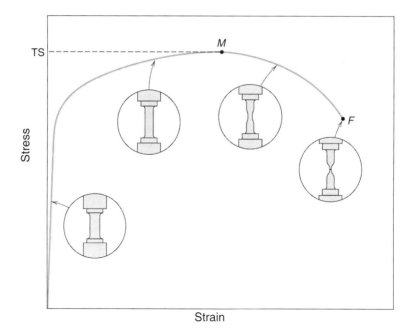

FIGURE 1-21 *Typical engineering stress–strain behavior to fracture, point F. The tensile strength, TS, is indicated by point M. The circular inserts represent the geometry of the deformed specimen at various points along the curve.* (W. D. Callister, Jr., Materials Science & Engineering: An Introduction, 5th ed., Wiley, New York, 2000, p. 126).

where U is the resilience and ε_y is the yield strain. If the material obeys Hooke's law, the expression for resilience becomes

$$U = \frac{1}{2}\sigma_y\varepsilon_y = \frac{\sigma_y^2}{2E}, \tag{1-5}$$

which shows that materials with high yield strength and low elastic modulus will be highly resilient. Toughness of a material is the total energy absorbed by a material until it fractures, and is estimated from the area under the entire stress versus strain curve of a material.

A distinction is usually made between engineering stress and strain, and true stress and strain. Engineering stress is obtained by dividing the applied load by the original cross-sectional area of the test specimen. As the load-bearing area continuously decreases and specimen length continuously increases during tensile deformation, a more logical approach would be to define the true stress and true strain in terms of the instantaneous area and length. True stress is defined as $\sigma_t = F/A_i$, and true strain as $\varepsilon_t = \ln(l_i/l_0)$, where A_i, l_i, and l_0 are the instantaneous area, instantaneous length, and initial length, respectively. Figure 1-22 shows a schematic plot of true stress versus true strain superimposed on the corresponding plot of engineering stress versus engineering strain. Because the total volume of the material is conserved during loading, $A_0 l_0 = A_i l_i$, where A_0 is the original cross-section of the sample. This relationship allows one to relate the true stress and true strain to engineering stress and engineering strain, and it can

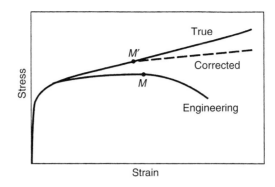

FIGURE 1-22 *A comparison of the engineering stress-strain and true stress-strain behaviors. Necking begins at point M on the engineering curve, which corresponds to point M' on the true curve. The corrected true stress–strain curve takes into account the complex stress state within the neck region.* (W. D. Callister, Jr., Materials Science & Engineering: An Introduction, 5th ed., Wiley, New York, 2000, p. 132).

TABLE 1-3   Values of Exponents n and K in the Flow Stress Equation, $\sigma_T = K\varepsilon_T^n$, for Different Alloys

Alloy	n	K, MPa
Low-C steel (annealed)	0.26	530
4340 alloy steel (annealed)	0.15	640
304 stainless steel (annealed)	0.45	1275
2024 Al alloy (annealed)	0.16	690
Brass (70Cu-30Zn, annealed)	0.49	895

be readily shown that $\sigma_t = \sigma(1 + \varepsilon)$ and $\varepsilon_t = \ln(1 + \varepsilon)$, where the engineering stress, σ, and engineering strain, ε, are given from $\sigma = F/A_0$, and $\varepsilon = (l_i - l_0)/l_0$. The preceding equations for true stress and true strain are, however, valid only up to the onset of necking.

For a large number of metals and alloys, the region between the inception of plastic deformation (yield point) and the onset of necking (tensile strength) on a true stress–true strain diagram can be described by a power law relationship of the form $\sigma_t = K\varepsilon_t^n$, where K and n are material-specific constants, and n is called the strain-hardening exponent. The strain-hardening exponent, n, is a measure of the ability of a metal to work harden (i.e., strengthen through plastic deformation). A large value of n indicates that strain hardening will be large for a given amount of plastic deformation. Thus, annealed copper ($n = 0.54$) will strain-harden more than annealed Al ($n = 0.20$) as would 70/30 brass ($n = 0.49$) compared to 4340 alloy steel ($n = 0.15$). Table 1-3 gives the values of n and K for common metals and alloys.

The rate of deformation, or strain rate, is a parameter of importance to the mechanical behavior of metals. High strain rate increases the flow stress of the metal and the temperature of the deformed metal because adiabatic conditions could exist due to rapid loading. A general relationship between flow stress, σ, and strain rate ($\dot{\varepsilon}$) at constant temperature and strain is $\sigma = C(\dot{\varepsilon})^m$, where m is the strain-rate sensitivity of the metal. Metals with a high strain-rate sensitivity ($0.3 < m < 1.0$) do not exhibit localized necking. Superplastic alloys (capable of 100–1000% elongation) have a high value of strain-rate sensitivity. This behavior occurs in materials with a very fine grain size ($\sim 1\,\mu m$) and at temperatures of about 0.4 T_m, where T_m is the absolute melting point of the metal. The major advantage of superplastic materials is the ease of shaping a part because of the very low flow stress required for deformation at low strain rates. Superplasticity is lost at strain rates above a critical value ($0.01\ s^{-1}$).

Strengthening of Metals

The fundamental approach to strengthening metals is to devise methods that increase the resistance to the motion of dislocations responsible for the plastic deformation. Reducing the grain size by rapid cooling of a casting, or by adding inoculants or "seed" crystals to a solidifying alloy to promote nucleation and grain refinement (see Chapter 2) is one method to strengthen metals. This is because fine-grained materials have a large-grain boundary area, and the motion of dislocation is hindered by grain boundaries because of crystallographic misorientation between grains, and because of the discontinuity of slip planes between neighboring grains. The strengthening by grain size reduction leads to an increase in the yield strength of the material. The grain size dependence of the yield strength, σ_0, of monolithic crystalline solids is described by the Hall-Petch equation according to which, $\sigma_0 = \sigma_i + k \cdot D^{-1/2}$, where σ_i is a friction stress that opposes dislocation motion, k is a material constant, and D is the average grain diameter. Thus, a plot of the yield strength as a function of the inverse square root of the average grain diameter will yield a straight line. Figure 1-23 shows an example of the effect of grain size on the yield strength of a brass alloy, which is consistent with the Hall-Petch equation. The Hall-Petch equation correctly describes the grain size dependence of yield stress for micrometer size grains but fails at grain diameters smaller than about 10 nm. This is because the equation was derived for relatively large dislocation pile-ups at grain boundaries, with about 50 dislocations in the pile-up. Clearly, at nanometric grain sizes, the pile-ups would contain fewer dislocations, thus violating the key assumption of the Hall-Petch equation.

The properties and behavior of materials change, often dramatically and in surprising ways, when the grain size approaches nanoscale, with the lower limit approaching the size of several crystal units. At nanometric grain sizes the fraction of the disordered interfacial grain boundary area becomes very large, and the grain diameter approaches some characteristic physical length such as the size of the Frank-Reed loop for dislocation slip. Furthermore, interfacial defects such as grain boundaries, triple points, and solute atoms segregated at an interface begin to increasingly influence the physical and mechanical behaviors. In addition, interfacial defects from processing such as micropores, and minute quantities of oxide contaminants and residual binders mask the grain size effects in nanomaterials, making it very difficult to isolate the true effect of grain-size reduction in such materials.

The other methods to strengthen metals include alloying, work hardening, and precipitation hardening. Pure metals are almost always soft, and alloying strengthens a metal. This is because the solute atoms (interstitial or substitutional) locally distort the crystal lattice, and generate stresses that interact with the stress field around dislocations, which hinder the dislocation motion. Impurity atoms segregate around the dislocation line in configurations that

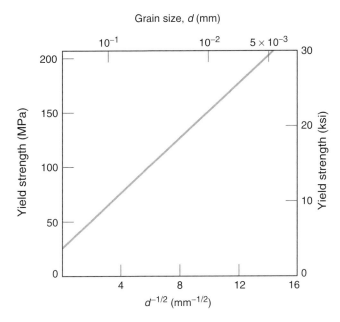

FIGURE 1-23 *The influence of grain size on the yield strength of a 70Cu-30Zn brass alloy. The theoretical relationship expressing the influence of grain size on yield strength is the Hall-Petch equation.* (Adapted from H. Suzuki, The Relation Between the Structure and Mechanical Properties of Metals, vol. II, National Physical Laboratory Symposium No. 15, 1963, p. 524). Reprinted with permission from National Physical Laboratory, Tedclington, Middlesex, U.K.

lower the total energy. This causes the dislocation to experience a drag due to the impurity "atmosphere" surrounding it, which must be carried along if the dislocation must move to cause deformation. Besides alloying, plastic deformation can also strengthen metals, a process called work-hardening or strain-hardening. Plastic deformation rapidly increases the density of dislocations, whose stress fields inhibit the motion of other dislocations. Frequently, dislocation tangles form, which provide considerable resistance to continued plastic deformation. The strain-hardening exponent, n, in equation $\sigma_t = K\varepsilon_t^n$ introduced in the preceding section is a measure of the ability of a metal to work harden; metals with a large value of n strain harden more than metals with a low value of n for a given amount of plastic deformation or degree of cold work. Certain alloys can be strengthened via a special heat treatment that creates finely dispersed hard second-phase particles, which resist the motion of dislocations. This is discussed in the section on heat treatment. Finally, hard second-phase particles and fibers may be added to metals from outside for strengthening by creating a composite material; this is discussed in Chapter 6.

Fracture Mechanics

The theoretical fracture strength of brittle materials, calculated on the basis of atomic bonding considerations, is very high, typically on the order of $E/10$, where E is the modulus of elasticity. For example, the theoretical strengths of aluminum oxide (Al_2O_3) and zirconia (ZrO_2) are approximately 39.3 GPa and 20.5 GPa, respectively ($E_{Al2O3} = 393$ GPa, and $E_{ZrO2} = 205$ GPa). The actual strengths are in the range 0.80–1.50 GPa for zirconia and 0.28–0.70 GPa

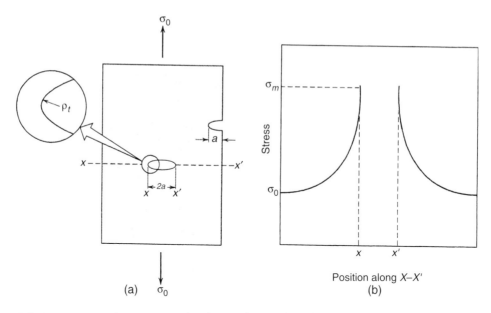

FIGURE 1-24 *(a) The geometry of surface and internal cracks, and (b) schematic stress profile along the line X–X' in (a), demonstrating stress amplification at the crack tip.* (W. D. Callister, Jr., *Materials Science & Engineering: An Introduction*, 5th ed., Wiley, New York, 2000, p. 191).

for alumina. The actual fracture strength of most brittle materials is 10 to 1000 times lower than their theoretical strength. The discrepancy arises because of minute flaws (cracks) that are universally present at the surface and in the interior of all solids. These cracks locally amplify or concentrate the stress, especially at the tip of the crack. The actual stress experienced by the solid near the crack tip could exceed the fracture strength of the material even when the applied stress is only a fraction of the fracture strength. For a long, penny-shaped crack of length $2a$, with a tip of radius ρ_t, the distribution of stress in the material is schematically shown in Figure 1-24. The maximum stress, σ_m, occurs at the crack tip, and is given from

$$\sigma_m = 2\sigma_0 \left(\frac{a}{\rho_t} \right)^{1/2} \tag{1-6}$$

where σ_0 is the applied stress, and the stress ratio (σ_m/σ_0) is called the stress concentration factor, K. Thus, long cracks with sharp tips will raise the stress more than small cracks with large tip radii. For fracture to occur, a crack must extend through the solid, a process that creates additional free surface within the material. The elastic strain energy released on crack propagation is consumed in creating the new surface that has an excess energy (surface energy) associated with it. The critical stress, σ_c (i.e., the fracture strength), required to propagate a crack of length $2a$ through a solid of elastic modulus, E, and surface energy, γ_{sv}, is given from the Griffith equation, applicable to brittle solids, which is

$$\sigma_c = \left(\frac{2E\gamma_{sv}}{\pi a} \right)^{1/2}. \tag{1-7}$$

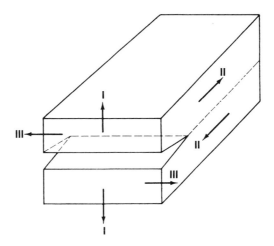

FIGURE 1-25 *The three modes of crack displacement: Mode I (crack opening or tensile mode), Mode II (sliding mode), and Mode III (tearing mode).*

In the Griffith equation, the parameter a is the depth of a surface crack, or half-length of an internal crack. For metals (and some polymers), fracture is accompanied by plastic deformation, so that a portion of the stored mechanical energy is used up in the plastic deformation accompanying crack propagation. For such materials, the Griffith equation has been modified to account for the material plasticity in terms of a plastic energy, γ_p, so that

$$\sigma_c = \left[\frac{2E(\gamma_{sv} + \gamma_p)}{\pi a}\right]^{1/2}. \tag{1-8}$$

A useful parameter to characterize a material's resistance to fracture is the fracture toughness of the material. Fracture toughness is most commonly specified for the tensile fracture mode, or Mode I, in which a preexisting crack extends under a normal tensile load. Two other modes of fracture are encountered, albeit to a lesser degree, in solids. These are shear failure and failure by tearing (Figure 1-25). Mode I fracture toughness defined for the tensile fracture mode is also called the plain-strain fracture toughness, K_{IC}, and is given from

$$K_{IC} = Y\sigma\sqrt{\pi a}, \tag{1-9}$$

where Y is a geometric factor. Table 1-4 gives the K_{IC} values of several materials. Brittle materials have a low K_{IC}, whereas ductile materials have a high K_{IC}. The mode I toughness, K_{IC}, depends also on the microstructure, temperature, and strain rate. Generally, fine-grained materials exhibit a high K_{IC}, and high temperatures and low strain rates increase the K_{IC}.

The engineering value of K_{IC} is in calculating the critical crack size that would cause catastrophic failure in a component under a given applied stress, σ. This critical size is given from

$$a_c = \frac{1}{\pi}\left(\frac{K_{IC}}{Y\sigma}\right)^2. \tag{1-10}$$

TABLE 1-4 Mode I Fracture Toughness of Selected Metallic, Polymeric and Ceramic Materials

Material	$K_{Ic}(MPa \cdot m^{1/2})$
Cu alloys	30–120
Ni alloys	100–150
Ti alloys	50–100
Steels	80–170
Al alloys	5–70
Polyethylene	1–5
Polypropylene	3–4
Polycarbonate	1–2.5
Nylons	3–5
GFRP	20–60
alumina	3–5
Si_3N_4	4–5
MgO	3
SiC	3
ZrO_2 (PSZ)	8–13
Soda glass (Na_2O-SiO_2)	0.7–0.8
Concrete	0.2–1.4

Alternatively, a maximum stress, σ, can be specified for a given material containing cracks of known size such that the material will not fail. This stress is

$$\sigma = \frac{K_{IC}}{Y\sqrt{(\pi a)}}.$$ (1-11)

Sophisticated tests and analytical methods have been developed to measure the K_{IC} of materials in various geometrical configurations. A qualitative assessment of the relative fracture toughness of different materials, especially at high strain rates, is readily obtained from a simple test, called the Charpy V-notch test. The test utilizes a standard specimen with a V-notch machined on one of its faces. A pendulum weight is allowed to fall from a fixed height and strike the sample securely positioned on a fixture to expose its back-face (i.e., the face opposite the V-notch) to the striking pendulum. The energy needed to fracture the specimen is obtained from the difference in the height of the swinging pendulum before and after the impact. A semiquantitative assessment of toughness can also be obtained from the measurement of total crack length in a material when a microscopic indentation is made on its surface under a fixed load. The plastic yielding and stress accommodation in a ductile material will prevent cracking, whereas fine, hairline cracks will emanate from the indentation in a brittle material (Figure 1-26). Examination of the fracture surface of a material can also reveal whether the failure is predominantly brittle or ductile. The scanning electron micrographs in Fig. 1-27a show that the intermetallic compound, NiAl, fractures in a predominantly intergranular manner, whereas the same material with tungsten alloying exhibits transgranular fracture (Fig. 1-27b), which is indicative of a predominantly ductile mode of fracture in this material.

FIGURE 1-26 *Qualitative assessment of fracture toughness from the material's response to indentation. Cracks emanate from the indentation area in a brittle material, whereas no cracks are seen in a tough material.* (R. Asthana, R. Tiwari and S. N. Tewari, Materials Science & Engineering, A336, 2002, 99–109).

Fatigue

Failure caused by a fluctuating stress (periodic or random) acting on a material that is smaller in magnitude than the material's tensile or yield strength under static load, is called the fatigue failure. Parameters used to characterize fatigue include: stress amplitude, σ_a, mean stress, σ_m, and stress range, σ_r. These parameters are defined as follows: $\sigma_a = \frac{\sigma_{max} - \sigma_{min}}{2}$, $\sigma_m = \frac{\sigma_{max} + \sigma_{min}}{2}$, and $\sigma_r = \sigma_{max} - \sigma_{min}$. Fatigue behavior is most conveniently characterized by subjecting a specimen to cyclic tensile and compressive stresses until the specimen fails. The results (Figure 1-28) of a series of such tests depict the number of cycles to failure, N, as a function of the applied stress, S (usually expressed as stress amplitude, σ_a). The general response of most materials is represented by one of the two S–N curves shown in Figure 1-28. Some iron and titanium alloys attain a limiting stress, called the fatigue limit or endurance limit (Fig. 1-28a), below which they survive an infinite number of stress cycles. In contrast, many nonferrous alloys do not exhibit a fatigue limit, and the number of cycles to failure, N, continuously decreases with increasing stress, S (Fig. 1-28b). For these materials, fatigue strength is specified as the value of stress at which the material will fail after a specified number of cycles, N (usually, 10^7). Alternatively, fatigue life of a material can be defined as the number of cycles to failure at a specified stress level. Fatigue life is influenced by a large number of variables. For example, minute surface or bulk imperfections (e.g., cracks, microporosity) in the material, design features that could raise the stress locally, surface treatments, and second-phase particles all influence the fatigue life.

In reality, the S–N curves of the type shown in Figure 1-28 are characterized by a marked scatter in the data, usually caused due to the sensitivity of fatigue to metallurgical structure, sample preparation, processing history, and the test conditions. As a result, for design purpose,

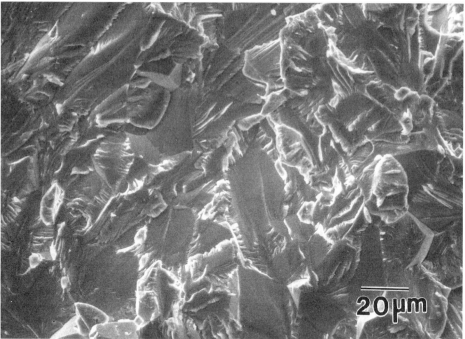

FIGURE 1-27 *(a) Scanning electron micrograph (SEM) view of the fracture surface of an extruded Ni-46Al alloy, compression tested at 300 K, showing intergranular fracture.* (R. Tiwari, S. N. Tewari, R. Asthana and A. Garg, Materials Science & Engineering, A 192/193, 1995, 356–363). *(b) SEM view of the fracture surface of a directionally solidified and compression deformed Ni-48.3Al-1W alloy showing evidence of transgranular fracture.* (R. Asthana, R. Tiwari and S. N. Tewari, Materials Science & Engineering, A336, 2002, 99–109).

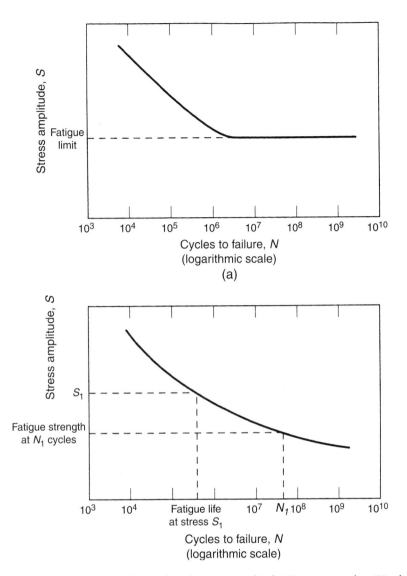

FIGURE 1-28 *Fatigue response depicted as the stress amplitude (S) versus number (N) of cycles to failure curves. In (a) the fatigue limit or endurance limit is reached at a specific N value, and in (b) the material does not reach a fatigue limit.* (W. D. Callister, Jr., Materials Science & Engineering: An Introduction, 5th ed., Wiley, New York, 2000, p. 212).

fatigue data are statistically analyzed and presented as probability of failure curves. Fatigue failures occur in three distinct steps: crack initiation, crack propagation, and a rather catastrophic final failure. Introducing compressive stresses in the surface of the part (through surface hardening such as carburizing or shot peening) improves the fatigue life because the compressive stresses counter the tensile stresses during service.

Creep

Creep is the time-dependent deformation at elevated temperatures under constant load or constant stress. Creep deformation occurs in all type of materials, including metals, alloys, ceramics, plastics, rubbers, and composites. The deformation strain versus time behavior for high-temperature creep under constant load (Figure 1-29) is characterized by three distinct regions, following an initial instantaneous (elastic) deformation region. These three creep regimes are (1) primary creep with a continuously decreasing creep rate; (2) secondary or steady-state creep, during which the creep rate is constant; and (3) tertiary creep, during which creep rate is accelerated, leading to eventual material failure. In primary creep, the decrease in the creep rate is caused by the material's strain hardening. In the steady-state regime, usually the longest in duration, equilibrium is attained between the competing processes of strain hardening and recovery (softening). The creep failure toward the end of the third stage is caused by the structural and chemical changes such as de-cohesion at the grain boundaries, and nucleation and growth of internal cracks and voids. From the material and component design perspectives, the steady-state creep rate ($d\varepsilon_s/dt$) is of considerable importance. The steady-state creep rate depends on the magnitude of applied stress, σ, and temperature, T, according to

$$\frac{d\varepsilon_s}{dt} = K'\sigma^n \exp\left(-\frac{Q}{RT}\right),\qquad(1\text{-}12)$$

where Q is the activation energy for creep, K' and n are empirical constants, and R is the gas constant. The exponent n depends on the dominant mechanisms of creep under a given set of experimental conditions, which include vacancy diffusion in a stress field, grain boundary migration, dislocation motion, grain boundary sliding, and others. The materials used in modern aircraft engines provide a classic example of how high-temperature creep resistance is enhanced through microstructure design in Ni-base alloys by dispersing nanometer-size oxide particles; this is discussed in Chapter 6.

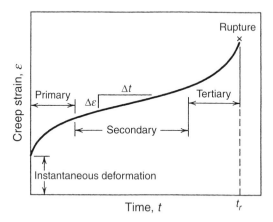

FIGURE 1-29 *Schematic creep curve showing strain versus time at constant stress and temperature. The minimum creep rate is the slope of the linear region in the secondary creep regime.* (W. D. Callister, Jr., Materials Science & Engineering: An Introduction, 5th ed., Wiley, New York, 2000, p. 226).

Deformation Processing

A large number of manufacturing processes employ solid-state deformation of hot or cold metals and alloys to shape parts. Hot-working is the mechanical shaping operation that is performed at temperatures high enough to cause the processes of recovery and recrystallization to keep pace with the work hardening due to deformation. In contrast, cold-working is performed below the recrystallization temperature of the metal; however, both hot- and cold-working are done at high strain rates. Cold-working strain hardens a metal, and excessive deformation without intermediate annealing causes fracture. It is a standard practice to anneal the metal between multiple passes during cold-working, to soften the metal and prevent fracture. In hot-working, the work hardening and distorted grain structure produced by mechanical deformation are rapidly erased, and new stress-free equiaxed grains recrystallize. In contrast, during cold-working the material hardens and its flow stress increases with continued deformation. In this section, we focus on the metallurgical changes that occur during hot working.

Hot-working erases compositional inhomogeneity of the cast structure (e.g., coring); creates a uniform, equiaxed grain structure; and welds any residual porosity. However, some structural inhomogeneity may be caused in hot-working because the component surface experiences greater deformation than the interior, resulting in finer recrystallized grains on the surface. Also, because the interior cools slower than the surface, some grain coarsening is possible in the interior, leading to low strength. In addition, surface contamination from scale formation and chemical reactions with gases in the furnace atmosphere, and the disintegration and entrapment of the scale in the surface and subsurface layers, can cause surface defects, poor finish, and part embrittlement. Hot-worked steels can decarburize at the surface, leading to strength loss and poor surface finish. Tighter dimensional tolerances are possible with cold-working than hot-working because of the absence of thermal expansion and contraction in the former.

In practice, hot-working is done in a temperature range bound by an upper and a lower limit. The upper temperature is limited by the melting temperature, oxidation kinetics, and the melting temperatures of grain boundary secondary phases (e.g., low-melting eutectics) that segregate at the grain boundaries during primary fabrication (e.g., casting). The melting of even a small amount of a low-melting-point secondary phase residing at the grain boundaries could cause material failure (hot-shortness). The lower hot-working temperature depends on the rate of recrystallization and the residence time at temperature. Because the recrystallization temperature decreases with increasing deformation, a heavily deformed metal will usually require a lower recrystallization temperature.

Temperature gradients develop in the workpiece during hot-working. This is because of the chilling action of the die. Because the flow stress is strongly temperature-dependent, a chilled region produces local hardening, and a non-deformable zone in the vicinity of a hotter, softer region. This could lead to the development of shear bands, and the localization of flow into these bands, resulting in very high shear strains and shear fracture.

Heat Treatment

Heat treatment is done to transform the metallurgical structure in alloys to develop the desired properties. Iron-carbon alloys serve as a classic example of how thermal treatments can be designed to transform the metallurgical structure for enhanced properties. Heat treatment practices make use of the thermodynamic data provided by the Fe-C (more often the Fe-Fe$_3$C) phase diagram, and the experimental measurements of the transformation kinetics.

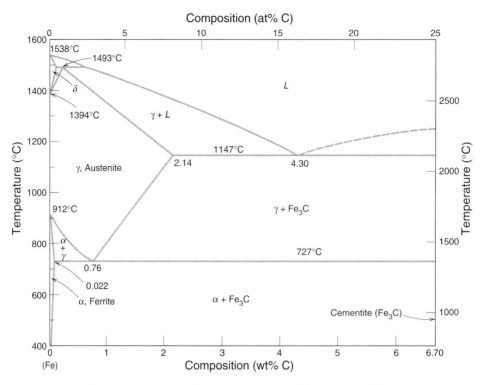

FIGURE 1-30 *The iron-iron carbide phase diagram.* (Adapted from Binary Alloy Phase Diagrams, 2nd ed., vol. 1, T. B. Massalski, editor-in-chief, 1990). Reprinted with permission from ASM International, Materials Park, OH (www.asminternational.org).

Many (but not all) phase transformations involve atomic diffusion and occur in a manner somewhat similar to the crystallization of a liquid via nucleation and growth (discussed in Chapter 2). In this section, we briefly review the basic elements of practical heat treatment approaches that are used to design the structure and properties of alloys.

Iron-carbon alloys are the most widely used structural materials, and the phase transformations in these alloys are of considerable importance in heat treatment. The Fe-Fe$_3$C diagram (Figure 1-30) exhibits three important phase reactions: (1) a eutectoid reaction at 727°C, consisting of decomposition of austenite (γ-phase, FCC) into ferrite (α-Fe, BCC) and cementite, i.e., $\gamma \leftrightarrow \alpha + \text{Fe}_3\text{C}$; (2) a peritectic reaction at 1493°C, involving reaction of the BCC δ-phase and the liquid to form austenite ($\delta + \text{L} \leftrightarrow \gamma$); and (3) a eutectic reaction at 1147°C, resulting in the formation of austenite and cementite (L $\leftrightarrow \gamma + \text{Fe}_3\text{C}$). Figure 1-31 depicts the development of microstructure in three representative steels at different stages of cooling: a hypoeutectoid steel, a steel of eutectoid composition (0.76% C), and a hypereutectoid steel.

Consider a eutectoid steel that is cooled from a high temperature where it is fully austenitic to a temperature below 727°C, and isothermally held for different times at the temperature. The isothermal hold will transform the austenite into pearlite, which is a mixture of ferrite (α-Fe) and cementite (iron carbide, Fe$_3$C). Experiments show that transformation (conversion) kinetics at a fixed temperature exhibit an S-shaped curve (Figure 1-32) that is mathematically described by the Avrami equation, $y = 1 - \exp(-kt^n)$, where y is the fraction transformed,

k is a temperature-sensitive rate constant, and t is transformation time. The Avrami equation applies to a variety of diffusion-driven phase transformations, including recrystallization of cold-worked metals and to many nonmetallurgical phenomena whose kinetics display a sigmoid behavior of Fig. 1-32. Usually the isothermal transformation data of Figure 1-32 is organized as temperature-versus-time map (with time on a logarithmic scale) to delineate the fraction (0 to 100%) transformed. The approach to construct an isothermal transformation map is shown in Figure 1-33. A complete isothermal-transformation (I-T) diagram for a eutectoid steel (0.77% C) is shown in Figure 1-34. The different iron-carbon phases (e.g., coarse and fine pearlite, upper and lower bainite, etc.) that form under different isothermal conditions are also displayed on this diagram. The physical appearance of some of the phases as revealed under a microscope is shown in the photomicrographs of Figure 1-35.

In practice, steels are rapidly cooled in a continuous fashion from the fully austenitic state to the ambient temperature, rather than held isothermally to trigger the transformation. In continuous cooling of steel, the time required for a transformation to start and end is delayed, and the

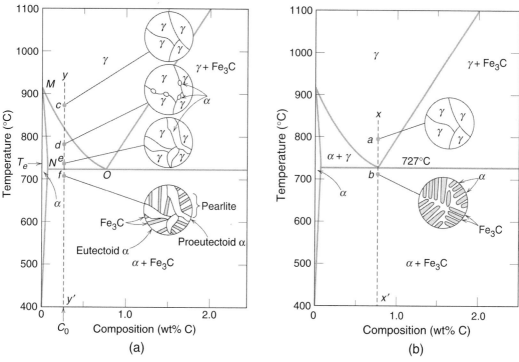

FIGURE 1-31 *(a) Microstructure evolution in a Fe-C alloy of hypoeutectoid composition C_0 during cooling from within the austenite region to below the eutectoid temperature of 727°C (the alloy contains less than 0.76 wt% C).* (W. D. Callister, Jr., Materials Science & Engineering: An Introduction, 5th ed., Wiley, New York, 2000, p. 279). *(b) Microstructure evolution in a Fe-C alloy of eutectoid composition (0.76% C) during cooling from within the austenite region to below 727°C.* (W. D. Callister, Jr., Materials Science & Engineering: An Introduction, 5th ed., Wiley, New York, 2000, p. 277). (c) Microstructure evolution in a Fe-C alloy of hypereutectoid composition C_1 during cooling from within the austenite region to below the eutectoid temperature of 727°C (the alloy contains more than 0.76 wt% C). (W. D. Callister, Jr., Materials Science & Engineering: An Introduction, 5th ed., Wiley, New York, 2000, p. 282).

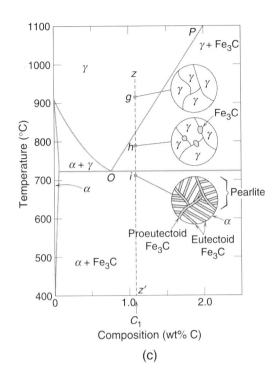

(c)

FIGURE 1-31 *Continued*

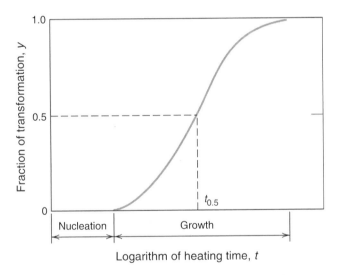

FIGURE 1-32 *Plot showing fraction reacted (y) versus the logarithm of time (t) typical of many isothermal solid-state transformations. The mathematical relationship between y and t is expressed by the Avrami equation, $y = 1 - \exp(-kt^n)$, where k and n are time-independent constants for a particular reaction.* (W. D. Callister, Jr., Materials Science & Engineering: An Introduction, 5th ed., Wiley, New York, 2000, p. 296).

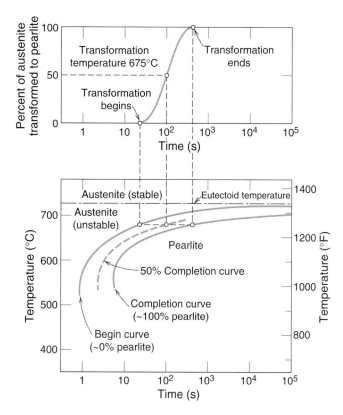

FIGURE 1-33 *Procedure to develop an isothermal transformation (I–T) diagram (bottom) from an experimental Avrami (fraction transformed versus log time) curve.* (Adapted from H. Boyer, ed., Atlas of Isothermal Transformation and Cooling Transformation Diagrams, 1977, p. 369). Reprinted with permission from ASM International, Materials Park, OH (www.asminternational.org).

isothermal transformation curves all shift to longer times. Figure 1-36 shows a continuous cooling transformation (CCT) diagram for a eutectoid steel with lines representing different cooling rates superimposed on the diagram (the constant cooling rate lines appear as curves rather than straight lines because time is plotted on a logarithmic scale).

Heat treated steels harden because of the formation of a hard and brittle phase, martensite, which forms only on rapid cooling below a certain temperature, called the martensite start temperature, M_s. which is shown in Figure 1-36. Martensite is a non-equilibrium phase and does not appear on the equilibrium Fe-Fe$_3$C phase diagram. Martensite has a body-centered tetragonal (BCT) structure (Figure 1-37), and a density lower than that of the FCC austenite. Thus, there is volume expansion on martensite formation that could cause thermal stresses and cracks in rapidly cooled parts. The situation is exacerbated for relatively large parts in which the center cools more slowly than the surface. Special surface-hardening treatments for steels such as nitriding do not, however, require martensite formation; instead, hard nitride compounds from the alloying elements in steel, and the solid solution hardening due to nitrogen dissolution, result in a hard and wear-resistant exterior in nitrided steels.

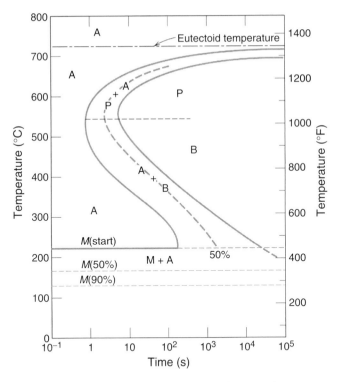

FIGURE 1-34 *The complete I–T diagram for a Fe-C alloy of eutectoid composition. A, austenite; B, bainite; M, martensite; P, pearlite.* (Adapted from H. Boyer, editor, Atlas of Isothermal Transformation and Cooling Transformation Diagrams, 1977, p. 28). Reprinted with permission from ASM International, Materials Park, OH (www.asminternational.org).

Steels hardened using martensitic transformation are frequently given a secondary treatment called tempering to impart some toughness and ductility. Hardened steels are tempered by heating and isothermal holding at temperatures in the range 250°C to 650°C; this allows for diffusional processes to form tempered martensite, which is composed of stable ferrite and cementite phases. Figure 1-38 shows the effect of tempering temperature and time on the hardness of a heat-treated eutectoid steel.

The ability of a steel to form martensite for a given quenching treatment is represented by the "hardenability" of the steel. A standardized test procedure to characterize the hardenability of steels of different compositions is the Jominy end-quench test (ASTM Standard A 255). A cylindrical test specimen (25.4 mm diameter and 100 mm length) is austenitized at a specified temperature for a fixed time, whereafter it is quickly mounted on a fixture and its one end is cooled with a water jet. After cooling, shallow flats are ground along the specimen length along which the hardness is measured, to develop a hardness distribution curve (hardenability curve). Because a distribution of cooling rates is achieved along the length of the specimen (whose one end is cooled), the hardness variation along the specimen length becomes a measure of cooling rate. The maximum hardness occurs at the quenched end where nearly 100% martensite forms. Away from the quenched end, cooling rate progressively decreases, thus allowing more time for

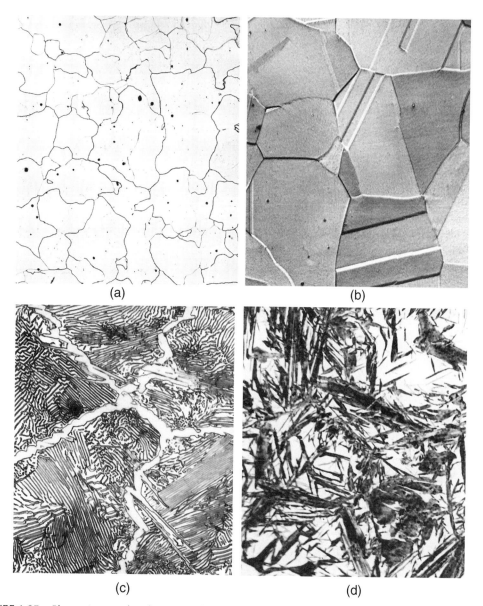

FIGURE 1-35 *Photomicrographs of (a) α-Fe (ferrite), and (b) γ-Fe (austenite). (Copyright 1971, United States Steel Corporation). (c) Photomicrograph of a 1-4wt% C steel having a microstructure consisting of a white proeutectoid cementite network surrounding the pearlite colonies (Copyright 1971, United States Steel Corporation). (d) Photomicrograph showing plate martensite. The needle-shaped dark regions are martensite, and the lighter regions are untransformed austenite (Copyright 1971, United States Steel Corporation).* Reprinted with permission from U.S. Steel Corporation, Pittsburgh, PA.

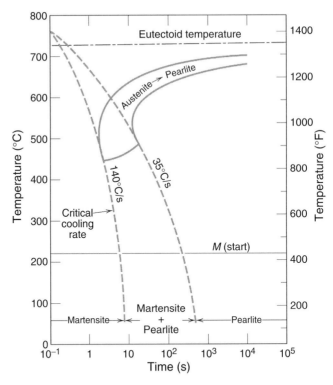

FIGURE 1-36 *Continuous cooling transformation (CCT) diagram for a eutectoid steel (0.76wt% C) and superimposed cooling curves, demonstrating the dependence of the final microstructure on the transformations that occur during cooling.* (W. D. Callister, Jr., Materials Science & Engineering: An Introduction, 5th ed., Wiley, New York, 2000, p. 313).

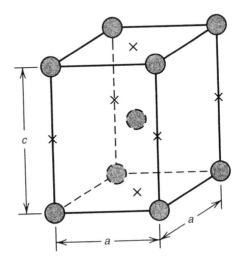

FIGURE 1-37 *The body-centered tetragonal (BCT) structure of martensite.* (W. D. Callister, Jr., Materials Science & Engineering: An Introduction, 5th ed., Wiley, New York, 2000, p. 306).

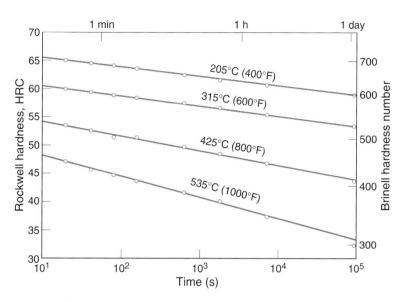

FIGURE 1-38 *Hardness versus tempering time for a water-quenched eutectoid plain carbon (1080) steel.* (Adapted from E. C. Bain, Functions of the Alloying Elements in Steels, 1939, p. 233). Reprinted with permission from ASM International, Materials Park, OH (www.asminternational.org).

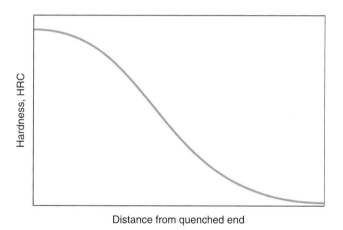

Distance from quenched end

FIGURE 1-39 *Typical hardenability curve showing Rockwell C hardness as a function of the distance from the quenched end in a Jominy end-quench test cooling.* (W. D. Callister, Jr., Materials Science & Engineering: An Introduction, 5th ed., Wiley, New York, 2000, p. 333).

diffusion of carbon to form softer phases (e.g., pearlite). A steel with a high hardenability will exhibit high hardness to relatively large distances from the quenched end. The hardenability of plain carbon steels increases with increasing carbon content. Likewise, alloying elements such as Ni, Cr, and Mo in alloy steels improve the hardenability by delaying the diffusion-limited formation of pearlite and bainite, thus aiding martensite formation. Figure 1-39 shows a schematic hardenability curve obtained from the Jominy end-quench test.

Precipitation Hardening

Some nonferrous alloys (e.g., Al alloys) can be strengthened through a special thermal treatment that forms a distribution of very fine and hard second-phase precipitates that act as barriers to dislocation motion. These alloys generally exhibit high solubility of the solute in the parent matrix, and a rather rapid change in the solute concentration with temperature. The alloy is first solutionized (or homogenized) by heating it to a temperature T_0 within the single-phase (α) region of the phase diagram (Figure 1-40) followed by quenching to near room temperature in the two-phase ($\alpha + \beta$) field. Quenching preserves the single-phase structure, and the alloy becomes thermodynamically unstable in the ($\alpha + \beta$) field because the α-phase becomes supersaturated with the solute. The atomic rearrangement via diffusion is sluggish because of the low temperatures after quenching. The supersaturated α-phase solid solution is then heated to a temperature T_2 in the ($\alpha + \beta$) field to enhance the atomic diffusion and allow finely dispersed β particles of composition, C_β, to form during isothermal hold, a treatment called "aging." An optimum aging time at a fixed temperature is required for optimum hardening. Overaging (prolonged aging) weakens the alloy through the coarsening (competitive dissolution) of fine precipitates in a distribution of coarse and fine precipitates. Higher temperatures enhance the aging kinetics, thus allowing maximum hardening to be achieved in a shorter time. The most widely studied precipitation-hardening alloy is Al-4%Cu, in which the α-phase is a substitutional solid solution of Cu in Al, and the second phase (θ-phase) is an intermetallic compound, $CuAl_2$. During aging of this alloy, intermediate non-equilibrium phases (denoted by θ'' and θ') form prior to the formation of the equilibrium θ-phase; the maximum hardness is associated with the θ'' phase, whereas overaging is caused by the continued precipitate coarsening and formation of θ' and θ phases. A schematic hardness (or strength) versus aging time profile after precipitation hardening for this alloy is shown in Figure 1-41.

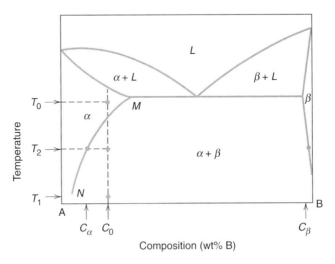

FIGURE 1-40 *Hypothetical phase diagram of a precipitation hardening alloy of composition C_0.* (W. D. Callister, Jr., Materials Science & Engineering: An Introduction, 5th ed., Wiley, New York, 2000, p. 342).

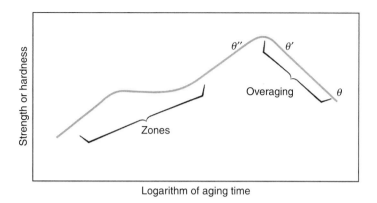

FIGURE 1-41 *Schematic diagram showing strength and hardness as a function of the logarithm of aging time at constant temperature during the precipitation hardening treatment.* (W. D. Callister, Jr., Materials Science & Engineering: An Introduction, 5th ed., Wiley, New York, 2000, p. 343).

Thermal Properties

Solid materials experience an increase in temperature upon heating due to energy absorption. The heat capacity, C, of a material specifies the quantity of heat energy, Q, absorbed to result in a unit temperature rise, i.e., $C = dQ/dT$. The quantity of heat absorbed by a unit mass to produce a unit increase in the solid's temperature is the specific heat, c, of the material. The heat capacity is expressed in J/mol·K or cal/mol·K, and the specific heat in J/kg·K or cal/g·K. At a fixed temperature, the atoms in a solid vibrate at high frequencies with a certain amplitude; the vibrations between atoms are coupled because of interatomic forces. These coupled vibrations produce elastic waves of various frequency distributions that propagate through the solid at the speed of sound. Wave propagation is possible only at certain allowed energy values or quantum of vibration energy, the "phonon" (which is analogous to a photon of electromagnetic radiation). Thus, the thermal energy of a solid can be represented in terms of the energy distribution of phonons.

For many solids, the heat capacity at constant volume, C_v, increases with increasing temperature according to $C_v = AT^3$, where A is a constant. Above a critical temperature, called the Debye temperature, θ_D, C_v levels off to a value $\sim 3R$ (where R is gas constant, 8.314 J/mol·K), or ~ 25 J/mol·K. Thus, the amount of energy required to produce a unit temperature change becomes constant at $T > \theta_D$.

The coefficient of linear thermal expansion (CTE), α, of a solid represents the change in length, Δl, per unit length, l_0, of a solid when its temperature changes by an amount, ΔT, due to heating or cooling, i.e., $\alpha = \frac{\Delta l}{(l_0 \cdot \Delta T)}$. The CTE represents the thermal strain per unit change in the temperature. The change in the volume, ΔV, of a solid on heating or cooling is related to the coefficient of volumetric expansion, α_v, by $\alpha_v = \frac{\Delta V}{(V_0 \cdot \Delta T)}$, where V_0 is the original volume. For uniform expansion along all directions upon heating, $\alpha_v = 3\alpha$; in general, however, α_v is anisotropic. Table 1-5 presents the average CTE data on some solid materials.

The genesis of thermal expansion of solids lies in the shape of the potential energy curve for atoms in solids (Figure 1-42). At $0°$K, the equilibrium separation between atoms is r_0;

TABLE 1-5 Coefficient of Thermal Expansion (CTE), Thermal Conductivity (k), and Specific Heat of Selected Materials

Material	$CTE, 10^{-6} \times C$	$k, W/m \cdot K$	$C_p, J/Kg \cdot K$
Al	23.6	247	900
Cu	17.0	398	386
Au	14.2	315	128
Fe	11.8	80	448
316 Stainless steel	16.0	15.9	502
Invar (64Fe-36Ni)	1.6	10.0	500
Al_2O_3	7.6	39	775
SiO_2	0.4	1.4	740
Pyrex (borosilicate glass)	3.3	1.4	850
Mullite ($3Al_2O_3.2SiO_2$)	5.3	5.9	–
Beo	9.0	219	–
ZrO_2(PSZ)	10.0	2.0–3.3	481
TiC	7.4	25	–
High-Density Polyethylene	106–198	0.46–0.50	1850
Polypropylene	145–180	0.12	1925
Polystyrene	90–150	0.13	1170
Teflon (polytetrafluoroethylene)	126–216	0.25	1050

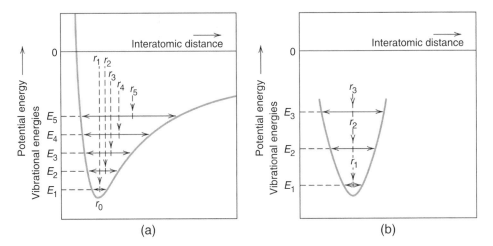

FIGURE 1-42 *(a) Plot of potential energy versus interatomic distance, demonstrating the increase in interatomic separation with rising temperature. On heating, the interatomic separation increases from r_0 to r_1 to r_2 and so on. (b) For a symmetric potential energy–interatomic separation curve, there is no net increase in interatomic separation with rising temperature ($r_1 = r_2 = r_3$, etc.), and therefore, no net expansion. (Adapted from R. M. Rose, L. A. Shepard, and J. Wulff, The Structure and Properties of Materials, vol. 4, Electronic Properties, 1966, Wiley, New York).*

as the solid's temperature is successively raised to T_1, $T_2 \ldots T_n$, the potential energy rises to E_1, E_2, $\ldots E_n$, and the mean position of an atom changes from r_0 to r_1, $r_2 \ldots r_n$. Because the shape of the potential energy trough in Figure 1-42 is nonsymmetric, the mean positions r_1, $r_2 \ldots r_n$ represent a net shift in the mean position of the atoms, i.e., a net expansion of the solid (for a symmetric potential energy trough, an increase in the temperature will lead to a greater vibrational amplitude but no increase in the mean atomic positions, i.e., zero expansion). For solids with strong atomic bonding, the mean positions will shift by only a small amount for a given temperature rise; as a result, such solids will exhibit a small *CTE*. For example, Al with a bond energy of 324 kJ·mol^{-1} has a *CTE* of 23.6×10^{-6} °K^{-1}, whereas W with a bond energy of 849 kJ·mol^{-1} has a *CTE* of 4.5×10^{-6} °K^{-1}.

Thermal conductivity, k, represents the ability of a solid to spatially distribute the thermal energy, i.e., k characterizes the rate at which energy is transported down a thermal gradient, dT/dx, between a region at a high temperature and a region at a low temperature within a solid. Experimental measurements of k for selected materials are summarized in Table 1-5. The thermal conductivity, k, is given from $k = -q/(dT/dx)$, where q is the heat flux (energy transferred per unit area per unit time) assumed to be constant, and the negative sign indicates that heat flows from the high- to low-temperature region. If q is not constant (non-steady thermal conduction), then the temperature changes with time, t, according to the Fourier equation,

$$\frac{\delta T}{\delta t} = \left(\frac{k}{c\rho} \right) \frac{\delta^2 T}{\delta x^2}, \tag{1-13}$$

where ρ is the solid's density, and the ratio, $\frac{k}{c\rho}$ is called the thermal diffusivity, α, of the solid. By analogy to the process of mass transport by diffusion in solids, to which Fick's equation (i.e., Equation 1-1) applies, (with C as the concentration), it is seen that D is analogous to thermal diffusivity, α (both α and D have the SI units, m$^2 \cdot s^{-1}$).

Thermal conduction in solids occurs via energy transmission through lattice waves (phonons) and also via free electrons, so that the conductivity, $k = k_l + k_e$, where the subscripts "l" and "e" denote the lattice and electronic contributions to conductivity, respectively. The lattice contribution is due to the migration of phonons (and transfer of thermal energy) from the high- to low-temperature regions. The electronic contribution is due to the migration of high-energy free electrons from the high-temperature regions to the low-temperature regions, where some of the kinetic energy of free electrons is transferred to atoms. However, electron collisions with crystal defects rapidly dissipate this energy. In very pure metals, free electron energy is not rapidly dissipated through collisions with defects, because of fewer crystal defects; as a result, the high-energy electrons can travel larger distances before their energy is transferred to atoms. In addition, a large number of free electrons exist in metals, which also aids heat conduction. Because electrons contribute to both thermal and electrical conduction in pure metals, the thermal and electrical conductivities are related. The fundamental law describing this relationship is called the Wiedemann-Franz law, according to which $L = k/\sigma T$, where σ is the electrical conductivity, and L is a constant, with an approximate value of $2.44 \times 10^{-8} \Omega \cdot W/K^2$. Alloying reduces both the thermal and electrical conductivity of metals because of the increased scattering of electron energy by the impurity atoms.

Electrical Properties

The conduction of electricity in solids follows a fundamental law, called Ohm's law, according to which $V = IR$, where V is the voltage (in volts), I is the electrical current (in amps or coulombs per s), and R is the electrical resistance of the material (in Ohms), respectively. The resistance, R, is influenced by the specimen geometry. In contrast, the electrical resistivity, ρ, is independent of specimen geometry and is related to the electrical resistance, R, by $\rho = RA/l$, where A is the cross-sectional area of the specimen perpendicular to the direction of current, and l is the distance between the points over which the voltage is applied. The units of ρ are Ohm·meters (Ω·m). The electrical conductivity, σ, is the inverse of the resistivity, i.e., $\sigma = \rho^{-1}$, and the units of σ are $(\Omega \cdot m)^{-1}$, or mho·m^{-1}. Solids exhibit an extremely wide range of σ (Table 1-6), with σ values varying from $10^7 (\Omega \cdot m)^{-1}$ for conductive metals to 10^{-20} $(\Omega \cdot m)^{-1}$ for insulators.

Electronic conduction is due to motion of charged particles (electrons and positively and negatively charged ions). However, not all electrons in atoms of a solid contribute to electronic conduction, and the number of electrons available to participate in conduction depends on the electron energy levels. Electrons in each atom may exist in certain allowed shells (1, 2, 3, and so on) and subshells (s, p, d, and f), with the electrons filling the states having the lowest energies. The electron configuration of an isolated atom (or an atom far removed from its neighbors in a solid) begin to be perturbed by the forces due to electrons and positive nuclei of its neighbors when the isolated atom is brought within relatively short distances of its neighbors. This leads to a splitting of closely spaced electron states in the solid's atoms, and the formation of electron energy bands. The energy gap between bands is large, and electron occupancy of the band gap region is normally prohibited. The electrons in the outermost energy band participate in the electron conduction process. The energy of the highest filled electron state at 0°K is called the Fermi energy, E_f. For metals such as Cu, the outermost energy band is only partially filled at 0°K, i.e., there are available electron states above and adjacent to filled states, within the

TABLE 1-6   Representative Room-Temperature Electrical Conductivity (σ) of Metals, Alloys, and Semiconductors

Material	$\sigma(\Omega\text{-m})^{-1}$
Al	3.8×10^7
Cu	6.0×10^7
Ag	6.8×10^7
Fe	1.0×10^7
Stainless steel	0.2×10^7
W	1.8×10^7
Nichrome (80%Ni-20%Cr)	0.09×10^7
60%Pb-40%Sn	0.66×10^7
Si	4×10^{-4}
Ge	2.2
GaAs	1×10^{-6}

same band. For metals such as Mg, an empty outer band overlaps a filled band. For both these types of metals, even small electrical excitation could cause electrons to be promoted to lower-level empty states (above E_f) within the same energy band; these "free" electrons will then contribute to electrical conduction. For insulators and semiconductors, one energy band (called the valence band) is completely filled with electrons but is separated by an energy gap from an empty band (conduction band). For insulating materials, the band gap is large ($E_g > 2$ eV), and for semiconducting materials, the band gap is smaller ($E_g < 2$ eV); the Fermi energy, E_f, lies near the center of the band gap. For electrical conduction to occur in such materials, electrons must be promoted (usually, by thermal or light energy) across the band gap into low-lying empty states of the conduction band. For materials with large E_g, only a few electrons can be promoted into the conduction band at a given temperature; increasing the temperature increases the number of thermally excited electrons into the conduction band, and the conductivity increases. Ionic insulators and strongly covalent insulators have tightly bound valence electrons, which makes it difficult for thermal excitation to increase the conductivity. In contrast, semiconductors are mainly covalently bonded but have relatively weak bonding. Thermal excitation of valence electrons into the conduction band across the relatively small E_g is, therefore, possible. As a result the conductivity rises faster with increasing temperature than for insulators. The electrical and electronic properties of semiconducting materials are discussed in Chapter 7.

Upon application of an electrical field to a solid, electrons are accelerated (in a direction opposite to the field), and the electrical current should increase with time as more and more electrons are excited. However, a saturation value of current is reached almost instantaneously upon the application of a field, because of frictional resistance offered to electron motion by the universally present crystal imperfections such as impurity atoms (substitutional and interstitial), dislocations, grain boundaries etc. The electron energy loss due to scattering by such defects leads to a drift velocity, v_d, which is the average electron velocity in the direction of the applied field. The drift velocity, $v_d = \mu_e E$, where E is the electrical field strength, and μ_e is called the electron mobility (m^2/V·s). Clearly, both the electron mobility and the number of electrons will influence the solid's conductivity. The electrical conductivity, σ, increases with both the number of free electrons, and the electron mobility. The conductivity $\sigma = ne\mu_e$, where n is the number per unit volume of free or conduction electrons, and e is the elementary charge (1.6×10^{-19} C).

In the case of metals, the electrical resistivity, ρ, is influenced by the temperature, impurities (alloying), and plastic deformation. The temperature dependence of ρ is given from the linear relationship, $\rho = \rho_0 + aT$, where ρ_0 and a are material constants. Thus increasing the temperature increases the thermal vibrations and lattice defects (e.g., vacancies), both of which scatter electron energy, thereby increasing the resistivity or decreasing the conductivity. The increased crystal defect content because of plastic deformation also increases the energy scattering centers in the crystal lattice, thus adversely affecting the conductivity. Alloying usually decreases the resistivity. The impurity content, C, of a single impurity in a metal decreases the resistivity according to $\rho = AC(1 - C)$, where A is constant, which is independent of the impurity content.

The electrical conductivity of a class of materials comprised of Si, Ge, GaAs, InSb, CdS, and others is extremely sensitive to minute concentration of impurities. These materials are called semiconducting material (see Chapter 7). In intrinsic semiconductors (e.g., Si and Ge), the electrical behavior is governed by the electronic structure of the pure elemental material, and in extrinsic semiconductors, the small concentration of the impurity (solute) atoms determines the electrical behavior. Both silicon and germanium are covalently bonded elements, with a small E_g (1.1 eV for Si and 0.7 for Ge); in addition, compounds such as gallium arsenide (GaAs), indium antimonide (InSb), cadmium sulfide (CdS), and zinc telluride (ZnTe) also exhibit intrinsic semiconduction. The wider the separation of the two elements forming the compound semiconductor in the periodic table, the stronger will be their bonding, and larger will be the E_g, making the compound more insulating than semiconducting.

In all semiconducting materials, every electron excited into the conduction band leaves behind a missing electron or "hole" in the valence band. Under an electric field, the motion of the electrons within the valence band is aided by the motion in an opposite direction of the electron hole, which is considered as a positively charged electron. The electrical conductivity of a semiconductor is, therefore, sum of the contributions made by electrons and holes, i.e.,

$$\sigma = ne\mu_e + pe\mu_h, \qquad (1\text{-}14)$$

where n and p are the number of electrons and holes per unit volume, respectively, and μ_e and μ_h are the electron and hole mobility, respectively. Because in intrinsic semiconductors every electron elevated to conduction band leaves behind a hole in the valence band, the numbers of electrons, n, and holes, p, are identical (n = p) and, therefore,

$$\sigma = ne(\mu_e + \mu_h). \qquad (1\text{-}15)$$

The electron and hole mobilities of common intrinsic semiconductors are in the range 0.01–8 m^2/V·s, and $\mu_h < \mu_e$. Both electrons and holes are scattered by lattice defects.

In the case of extrinsic semiconductors, the electrical behavior is dominated by the impurity atoms, and extremely minute impurity levels (e.g., one impurity atom in every 1000 billion atoms) can influence the electrical behavior. The intrinsic semiconductor, silicon, has four valence electrons, and it can be made into an extrinsic semiconductor by adding an element that has five valence electrons per atom. For example, P, As, or Sb from Group VA of the periodic table can be added to pure Si to leave one electron that is loosely bound to the impurity atom, and can take part in electronic conduction. Because of the extra negative electron, this type of material is called an n-type extrinsic semiconductor, with the electrons constituting the majority carriers and holes (positive charge) constituting the minority carriers. If Si is doped with a small amount of a Group IIIA element such as Al, B, or Ga, that have three valence electrons per atom, then each Si atom becomes deficient in one negative charge in so far as electron bonding is concerned. This electron deficiency is likened to availability of a positive hole, loosely bound to the impurity atom, and capable of taking part in conduction. Such materials containing an excess of positive charge are called p-type extrinsic semiconductors. Devices based on n- and p-type semiconductors are discussed in Chapter 7.

Dielectric and Magnetic Properties

Dielectric properties are exhibited by materials that are electrical insulators in which the centers of positive and negative charges do not coincide. These materials consist of electrical dipoles, which interact with an external electric field such as in a capacitor. The dielectric properties are in fact best understood in terms of the behavior of an electrical capacitor. In a parallel plate capacitor (plate separation, l, and area, A) the capacitance, C, is related to the charge stored on either plate, Q, by $Q = CV$, where V is the voltage applied across the capacitor. With a vacuum between the plates, the capacitance is given from $C = \varepsilon_0 A/l$, where ε_0 is the permittivity of vacuum ($\varepsilon_0 = 8.85 \times 10^{-12}$ Farads/m, and Farad = Coulombs per volt). With a dielectric material of permittivity ε residing in the gap between the plates, the capacitance, C, becomes, $C = \varepsilon A/l$, and $\varepsilon > \varepsilon_0$. The relative permittivity, $\varepsilon_r = \varepsilon/\varepsilon_0$ is called the material's dielectric constant, which is always greater than unity. In an electric field, the dipoles in the dielectric material experience a force that aligns (polarizes) them along the direction of electric field. As a result of polarization, the surface charge density on the plates of the capacitor's changes. The surface charge density D (in coulombs/m^2) on a capacitor with a dielectric material between the plates is defined from, $D = \epsilon E$, where E is the electric field strength. Alternatively, the surface charge density, $D = \varepsilon_0 E + P$, where P is the polarization defined as the increase in charge density relative to vacuum in the presence of a dielectric material in the capacitor.

The magnetic properties of solids have their origin in the magnetic moments of individual electrons. In essence, each electron is a tiny magnet whose magnetic properties originate from the orbital motion of the electron around the positive nucleus, and the spinning of each electron about its own axis. Each electron in an atom has a spin magnetic moment of absolute magnitude, μ, and an orbital magnetic moment of $m\mu$, where μ is called the Bohr magneton, and has a value of 9.27×10^{-24} A·m^2, and m is termed the magnetic quantum number of the electron. The net magnetic moment of an atom is the vector sum of magnetic moments of all electrons constituting the atom. Elements such as He, Ne, Ar, etc., have completely filled electron shells and cannot be permanently magnetized, because the orbital and spin moments of all electrons cancel out.

Different materials exhibit different types of magnetic behaviors, and some materials exhibit more than one type of magnetic behavior. Diamagnetic materials exhibit very weak magnetic properties when subjected to an external field, which perturbs the orbital motion of the electrons. Actually, all materials are diamagnetic, but the effect is usually too weak to be observed especially when the other types of magnetic behavior are stronger. In solids in which the electron spin or orbital magnetic moments do not completely cancel out, a net permanent dipole moment exists, giving rise to paramagnetic behavior. These randomly oriented dipoles rotate to align themselves when subjected to an external field, thus enhancing the field. The magnetic behavior is characterized in terms of the magnetic susceptibility, χ_m, which for a paramagnetic material, is a positive number in the range, 10^{-5} to 10^{-2}. The magnetic susceptibility, χ_m, is defined from $\chi_m = \mu_r - 1$, and μ_r is the permeability (μ) of the solid relative to that of the vacuum (μ_0), i.e., $\mu_r = \mu/\mu_0$. The permeability, μ, characterizes the magnetic induction (or the internal field strength), B, in a solid when subjected to an external field of strength H, and $\mu = B/H$.

Ferromagnetic materials such as Fe, Co, and Ni possess a permanent magnetic moment in the absence of an applied field. These materials are strongly magnetic, with very large positive ($\sim 10^6$) susceptibilities. In these materials, the electron spin moments do not cancel out, and the mutual interactions of these moments cause the dipoles to align themselves even in the absence of an external field. These materials reach a maximum possible magnetization (saturation magnetization) when all the magnetic dipoles are aligned with an external field. Many ceramics also exhibit ferromagnetic behavior; examples include cubic ferrites (e.g., Fe_3O_4), hexagonal ferrites (e.g., $BaFe_{12}O_{19}$), and garnets (e.g., $Y_3Fe_5O_{12}$).

The behavior called antiferromagnetism arises from the antiparallel alignment of electronic spin moments of neighboring atoms or ions in a solid. The oxide ceramic manganese oxide

(MnO) is antiferromagnetic by virtue of an antiparallel alignment of Mn^{2+} ions in the crystal lattice. However, because the opposing magnetic moments cancel out, there is no net magnetic moment in the solid. Finally, ferrimagnetic behavior arises from the incomplete cancellation of the electron spin moments, but the saturation magnetization for these materials is less than that of ferromagnetic materials.

The magnetic behavior is sensitive to temperature changes because enhanced thermal vibration of atoms perturbs the alignment of magnetic moments in a crystal. At the absolute zero temperature ($0°K$), the thermal vibrations are minimum and the saturation magnetization is a maximum for ferro- and ferrimagnetic materials. The saturation magnetization decreases with increasing temperature, and rapidly drops to zero at a critical temperature called the Curie temperature, T_C, because of complete destruction of coupling of spin moments. The value of the Curie temperature varies with the metal, but it is typically a few hundred degrees. At $T < T_C$, a ferromagnetic material consists of small regions or domains within each of which all magnetic dipole moments are aligned along the same direction; several domains could form within each grain of a polycrystalline solid. The adjacent domains are separated by domain walls or boundaries, across which the orientation of the dipole moments changes gradually. In a manner similar to the ferromagnetic behavior, antiferromagnetic behavior also diminishes with increasing temperature, and completely disappears at a temperature called the Neel temperature, T_N. Depending on their intrinsic magnetic property, different magnetic materials become paramagnetic above either T_C or T_N.

The internal field strength of materials is characterized in terms of the magnetic induction, B. For most materials, the magnetic induction, B, is not a simple function of the magnetic field strength, H, and the B–H plot (i.e., magnetization curve) usually forms a loop whose area represents the loss of magnetic energy per unit volume of the material during each magnetization–demagnetization cycle (Figure 1-43). For soft magnetic materials (99.95% pure Fe, Fe-3Si alloy, etc.), the area under the loop (also called the hysteresis curve) is small and the saturation magnetization is reached at small applied field strength. For a given material composition, structural defects such as voids and nonmagnetic inclusions limit the magnetic behavior of soft magnets. The hysteresis losses manifest themselves in heat generation. Another source of heating is the small-scale or eddy currents induced by an external magnetic field in the material (this is minimized by alloying to raise the electrical resistance of the material). Hard magnetic materials such as magnetic steels, Al-Ni-Co alloys, and Cu-Ni-Fe alloys are characterized by a higher hysteresis energy loss than soft magnetic materials. The relative ease with which magnetic boundaries move determines the hysteresis behavior, and microstructural features such as fine second-phase particles that inhibit the magnetic wall movement will lead to large energy losses. Judicious selection of alloying additions and heat treatment practices that lead to fine precipitates (e.g., WC and Cr_3C_2 in steels containing W and Cr) increase the resistance to the domain wall movement.

Optical Properties

The interaction of electromagnetic (EM) radiation (including visible radiation) with solids gives rise to physical processes such as absorption, scattering, and transmission. The range of wavelength of visible light is 0.4 μm to 0.7 μm, with each wavelength corresponding to a specific spectral color. The energy corresponding to each wavelength is expressed in terms of the energy, E, of a discrete packet or quantum (photon) from the Planck's equation, $E = h\nu = hc/\lambda$, where

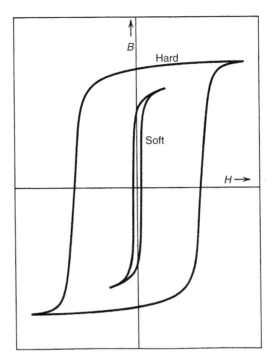

FIGURE 1-43 *Schematic magnetization curve for soft and hard magnetic materials.* (From K. M. Ralls, T. H. Courtney and J. Wulff, Introduction to Materials Science & Engineering, Wiley, New York, 1976).

h is Planck's constant ($h = 6.63 \times 10^{-34}$ J·s), and ν and λ are the frequency and wavelength of the EM radiation ($c = \nu\lambda$), and c is the velocity of light in vacuum. Thus, EM radiation of short wavelength has higher energy than radiation of large wavelength. When EM radiation crosses an interface between two different phases, the flux or intensity, I_0 (in units of W/m^2), of the incident EM radiation is partitioned between the intensities of the transmitted, reflected, and absorbed radiation. From the conservation of energy, $I_0 = I_T + I_R + I_A$, or $1 = (I_T/I_0 + I_R/I_0 + I_A/I_0)$. Materials with high transmissivity (i.e., large I_T/I_0) are transparent, materials that are impervious to visible light are opaque, and materials that cause internal scattering of light are translucent. Internal scattering of radiation occurs because of factors such as grain boundaries in polycrystalline materials (different refractive indices of grains having different crystallographic orientations), finely dispersed second-phase particles of different index of refraction, fabrication-related defects such as dispersed porosity, and in the case of polymers, the degree of crystallinity. In metals, the incident radiation is absorbed in a thin surface layer (typically less than 0.1 μm), which excites the electrons into unoccupied energy states. The reflectivity of the metals is a consequence of the reemitted radiation from transition of an electron from a higher energy level to a lower energy level, resulting in the emission of a photon. The surface color of a metal is determined by the wavelength of radiation that is reflected and not absorbed. For example, a bright, shiny metal surface will emit the photons in the same number and of same frequency as the incident light. Nonmetallic materials exhibit refraction and transmission in addition to absorption and reflection.

The velocity, v, of light in a transparent material is different from the velocity in vacuum, c, and the ratio $c/v = n$ is the refractive index of the material; the slower the velocity v in the medium, greater is the n. For nonmagnetic materials, $n \approx \sqrt{\varepsilon_r}$, where ε_r is the dielectric constant of the transparent material. The refraction phenomenon (i.e., bending of EM radiation due to velocity change at an interface) is related to electronic polarization, and large ions or atoms lead to greater polarization and larger n. For example, addition of large lead or barium ions to common soda-lime glass appreciably increases the n.

The reflectivity (I_R/I_0) represents the fraction of incident light scattered or reflected at an interface between media having different refractive indices. Generally, the higher the refractive index of a material, the greater is its reflectivity. Light reflection from lenses and other optical devices is minimized by applying a very thin coating of a dielectric material such as magnesium fluoride (MgF_2) to the reflecting surface.

The light absorption in nonmetals occurs by electronic polarization and by electron transfer between conduction band and valence band. A photon of energy $E = hc/\lambda$ may be absorbed by a material if it can excite an electron in the nearly filled valence band to the conduction band, i.e., if $E > E_g$, where E_g is the energy of the gap between the valence band and conduction band. Taking $c = 3 \times 10^8$ m/s, and $h = 4.13 \times 10^{-15}$ eV·s, and considering the minimum and maximum λ values in the visible spectrum to be 0.4 μm and 0.7 μm, respectively, the maximum and minimum values of band gap energy, E_g, for which absorption of visible light is possible, are 3.1 eV and 1.8 eV, respectively. Thus, nonmetallic materials with $E_g > 3.1$ eV will not absorb visible light, and will appear as transparent and colorless. In contrast, materials with $E_g < 1.8$ eV will absorb all visible radiation for electronic transitions from the valence band to conduction band, and will appear colored. Different materials could become opaque to some type of EM radiation depending on the wavelength of radiation and the band gap, E_g, of the material.

The intensity, I'_T, of radiation transmitted through a material depends also on the distance, x, traveled through the medium; longer distances cause a greater decrease in the intensity, which is given from $I'_T = I'_0 e^{-\beta x}$, where I'_0 is intensity of incident radiation, and β is called the absorption coefficient of the material (a high β indicates that the material is highly absorptive). When light incident on one face of a transparent solid of thickness, L, travels through the solid, the intensity of radiation transmitted from the opposite face is, $I_T = I_0(1 - R)^2 e^{-\beta x}$, where R is the reflectance (I_R/I_0). Thus, energy losses due to both absorption and reflection determine the intensity of transmitted radiation.

The color of solids arises out of the phenomena of absorption and transmission of radiation of specific wavelengths. As mentioned earlier, a colorless material such as single-crystal sapphire will uniformly absorb radiation of all wavelengths. In contrast, the semiconducting material cadmium sulfide ($E_g = 2.4$ eV) appears orange-yellow because it absorbs visible light photons of energy greater than 2.4 eV, which include blue and violet portions, the total energy range for visible light photons being 1.8 to 3.1 eV. Impurity atoms or ions influence the phenomenon of color. Thus, even though single-crystal sapphire is colorless, ruby is bright red in color because in the latter Cr^{3+} ions (from Cr_2O_3 added to sapphire) substitute Al^{3+} ions of Al_2O_3 crystal, and introduce impurity energy levels within the band gap of sapphire. This permits electronic transitions in multiple steps. Some of the light absorbed by ruby that contributes to the electron excitation from the valence band to the conduction band is reemitted when electrons transition between impurity levels within the band gap. Colored glasses incorporate different ions for color effects, such as Cu^{2+} for blue-green and Co^{2+} for blue-violet.

References

Ashby, M. F., and D. R. H. Jones. *Engineering Materials—An Introduction to their Properties and Applications,* 2nd ed. Boston: Butterworth-Heinemann, 1996.

Askeland, D. R., and P. P. Phule. *The Science & Engineering of Materials,* 4th ed. Pacific Grove, CA: Thompson Brooks/Cole, 2003.

Budinski, K. *Engineering Materials: Properties and Selection.* Englewood Cliffs, NJ: Prentice Hall, 1979.

Cahn, R. W. *Coming of Materials Science.* New York: Elsevier, 2001.

Callister, W. D., Jr. *Materials Science & Engineering: An Introduction,* 5th ed. New York: Wiley, 2000.

Darken, L. S., and W. R. Gurry. *Physical Chemistry of Metals.* New York: McGraw-Hill, 1953.

DeGarmo, E. P., J. T. Black, R. A. Kohser and B. E. Klanecki. *Materials and Processes in Manufacturing,* 9th ed. New York: Wiley, 2003.

Dieter, G. E. *Mechanical Metallurgy,* 3rd ed. New York: McGraw-Hill, 1986.

Epifanov, G. I. *Solid State Physics.* Moscow: Mir Publishers, 1974.

Flinn, R. A., and P. K. Trojan. *Engineering Materials and Their Applications,* 4th ed. New York: Wiley, 1990.

Gaskell, D. R. *Introduction to the Thermodynamics of Materials,* 3rd ed. Washington D.C.: Taylor & Francis, 1995.

Ghosh, A., and A. K. Mallik. *Manufacturing Science.* New York: Wiley.

Jastrzebski, Z. D. *The Nature and Properties of Engineering Materials.* New York: Wiley.

John, V. *Introduction to Engineering Materials.* New York: Industrial Press, Inc.

Kalpakjian, S., and S. R. Schmid. *Manufacturing Engineering & Technology.* Englewood Cliffs, NJ: Prentice Hall, 2001.

Kittle, C. *Introduction to Solid State Physics,* 6th ed. New York: Wiley, 1986.

Meyers, M. A., and K. K. Chawla. *Mechanical Metallurgy, Principles and Applications.* Englewood Cliffs, NJ: Prentice Hall, 1984.

Mitchell, B. S. *An Introduction to Materials Engineering and Science for Chemical and Materials Engineers.* New York: Wiley Interscience, 2004.

National Materials Advisory Board, National Research Council. *Materials Science & Engineering: Forging Stronger Links to Users.* NMAB-492, Washington, D.C.: National Academy Press.

Shackelford, J. F. *Introduction to Materials Science for Engineers,* 6th ed. Englewood Cliffs, NJ: Prentice Hall, 2005.

Shewmon, P. G. *Diffusion in Solids,* 2nd ed. Warrendale, PA: The Minerals, Metals & Materials Society (TMS), 1989.

Swalin, R. A. Thermodynamics of Solids. New York: Wiley, 1972.

Trivedi, R. *Materials in Art and Technology.* Ames: Taylor Knowlton, 1998.

2   Casting and Solidification

Casting Techniques

Metal-casting techniques constitute an important and widely used class of metal fabrication processes chiefly because of the low viscosity of liquid metals that facilitates flow into complex shapes. Different metals and alloys have widely different melting temperatures, which is a consequence of their different atomic bonding energies. As a result, the pouring temperature, T_p, i.e., the temperature at which a metal is introduced in a mold for shaping a part, varies over a wide range (Table 2-1). For example, T_p is $\sim505°$K for low-melting-point solder alloys and $\sim2172°$K for refractory zirconium alloys. Different metals and alloys must, therefore, be cast into shapes over a wide range of temperatures, and the material used for making a mold for casting a specific alloy must be stable at the pouring temperature of that alloy. This suggests that the thermal properties of both the metal being cast and the mold material play an important role in casting. The mold material must be able to absorb and dissipate the heat lost by solidifying metal fast enough to prevent the mold temperature from rising to its melting point. This depends on the thermal properties, as well as the relative size of the casting and the mold. In fact, it is possible to pour high-alloy steels ($T_p \sim 1860°$K) in an aluminum mold (melting point $= 933°$K) without causing melting of the mold material even in absence of external cooling of the mold.

General considerations in a casting operation include pattern design and pattern fabrication, selection of the mold and core materials and molding process (e.g., sand or permanent mold), selection of the melting technique (furnace type, fluxing, degassing, and inoculating practices), selection of pouring technique (gravity pouring, pressurized injection), control of the solidification process (solidification time, thermal gradients, grain structure), part separation from the mold (shakeout, part ejection), cleaning, finishing, and inspection (internal and external defects, dimensional tolerances, surface finish, metallurgical quality, strength property, etc.), and secondary treatment of cast components. Generally, the casting process starts out with the design and fabrication of pattern, which could be a simple flat-back pattern, a split-pattern, or a match plate, which is essentially the pattern together with its runner system laid out on a wood board or metal plate. Most patterns are made out of wood, metal, plastic, wax,

TABLE 2-1 Approximate Pouring Temperatures of Metals and Alloys

Alloy	Pouring Temperature, °K
Solder	505
Sn alloys	589
Zn alloys	616–727
Al alloys	895–1005
Cu alloys	1172–1450
Cast irons	1616–1755
Ni-base Superalloys	1700–1810
High-alloy steels	1755–1866
Low-alloy steels	1840–1980
Ti alloys	1977–2089
Zr alloys	2116–2172

or similar materials. The fabrication of the pattern involves decisions related to pattern draft, surface finish, corner radius, section thickness, and shrinkage and machining allowances.

The development of shrinkage is an important consideration in metal casting. There are three principal sources of shrinkage during casting: (1) volumetric contraction of the superheated liquid metal as it cools down to the solidification temperature, (2) the solidification shrinkage during liquid-to-solid transformation at the solidification temperature, and (3) patternmaker's shrinkage from contraction of the hot cast as it cools down to ambient temperatures. These different shrinkages correspond to the different regions of Figure 2-1a, which displays the volumetric contraction of the metal as it transforms from a superheated liquid into a solidified casting at room temperature. The first shrinkage (i.e., liquid shrinkage) seldom poses a problem in a casting operation because it occurs while the metal is being introduced in the mold, so fresh incoming metal compensates this shrinkage. The second type of shrinkage is eliminated by proper sizing and placement of the riser, and the third type of shrinkage is eliminated by oversizing the pattern (based on the thermal expansion of the metal). Table 2-2a gives the solidification shrinkage of some metals together with their crystal structure, melting temperature, and density in the liquid and solid states. When cores are used to make internal surfaces (e.g., holes) in a casting, they must also be oversized, because the metal and the hole will shrink during cooling. If machining allowances are needed, they should be subtracted from core dimensions, because machining will increase the hole size.

Expendable-Mold Casting
Green Sand Casting

Green sand casting is one of the simplest and most popular casting processes. Figure 2-1b shows the schematic diagram of a simple green sand mold together with all the components of the gating system. The gating system of a sand mold consists of pouring basin, sprue, sprue basin, runner, gates, and runner extension. Risers (e.g., the cylinder atop the plate casting of Figure 2-1b) used to feed the solidification shrinkage are generally not considered part of the gating system even though risers are often placed in the gating system of a mold. This is because riser design is based primarily on thermal considerations, whereas the design of the gating system mainly involves

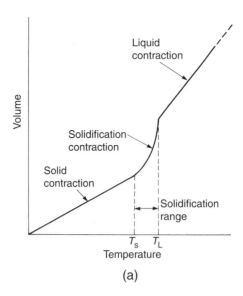

(a)

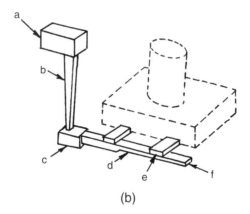

(b)

FIGURE 2-1 *(a) Schematic diagram of the three sources of volumetric contraction of a liquid alloy as a function of temperature during cooling: (1) contraction of superheated liquid from pouring temperature to liquidus temperature, T_L, (2) solidification contraction over the freezing range of the alloy (T_L to T_S), and (3) contraction of the solid casting as it cools down to ambient temperature.* (J. Campbell, Castings, Butterworth-Heinemann, Boston, 1999, p. 63). Reprinted with permission from Elsevier. *(b) Components of a simple sand mold: (a) pouring basin, (b) sprue or downsprue, (c) sprue basin, (d) runner, (e) gates, and (f) runner extension. Also shown is a cylindrical riser atop a plate casting.* (Adapted from Basic Principles of Gating & Risering, 1985). Reprinted with permission from the Cast Metals Institute, Schaumburg, IL (www.castmetals.com).

considerations of fluid flow (discussed in a later section). To avoid formation of casting defects, it is necessary to control the rate of mold filling. Too fast a metal flow causes air entrapment, porosity and dross formation, and erosion of sand, whereas too slow a flow causes the metal to solidify prematurely, yielding defects such as misrun (an incomplete casting) and cold laps

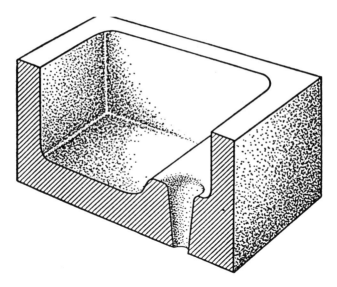

FIGURE 2-2 *Recommended design of the pouring basin: rectangular cross-section and flat base with a fillet at the base.* (H. F. Taylor, M. C. Flemings, and J. Wulff, Foundry Engineering, Wiley, 1959, p. 182). Reprinted with permission from Howard Taylor Trust, Boston, MA 02108.

(inhomogeneous or layered casting surface). A well-designed gating system evenly distributes the incoming metal to all parts of the mold without causing turbulence or sand erosion.

A pouring basin is either rectangular or square in cross-section, with a flat base and a fillet at the base near the sprue entrance (Figure 2-2). Spherical pouring basins with a curved base cause vortex flow and casting defects. A sprue connects the pouring basin to the runner. A sprue is usually tapered to allow for downward laminar flow (Figure 2-3), and the sprue basin (Figure 2-4) provides space for the metal to dissipate some energy before changing its direction as it flows into the runners. When multiple gates are placed on a single runner, the runner cross-section is made progressively smaller to enable uniform flow into each gate, as shown in Figure 2-5a. The runner extension shown in Figure 2-5b at the end of the runner traps the first metal, which may carry impurities such as eroded sand particles and dross. Another technique commonly employed to prevent unwanted impurity particles from entering the casting employs ceramic filters (e.g., honeycombs, drilled mica sheets, etc.) that are embedded in the runner or placed at the base of the sprue. Figure 2-5c-e shows some examples of filters used in the gating system of molds and basic principle of their operation.

Figure 2-6 shows the two basic types of gating designs widely used in sand casting practice: a pressurized gating system (Fig. 2-6a) and an unpressurized gating system (Fig. 2-6b). In an unpressurized system the choke (smallest cross-section) is located at or near the sprue base, whereas in a pressurized system the choke is at the gates. The metal velocity is higher at the gates in a pressurized system than in an unpressurized system. In the latter case, the lower velocity at the gate limits the erosion of sand during mold filling, but the runners may not be completely filled with the metal, leaving air pockets in the runner, which could lead to dross or porosity formation. In an unpressurized gating system, an open riser is sometimes used as a bubble trap provided the choke is located downstream past the riser. The choke restricts metal flow and lowers the metal velocity in the bubble trap, thereby allowing any entrapped air or gas

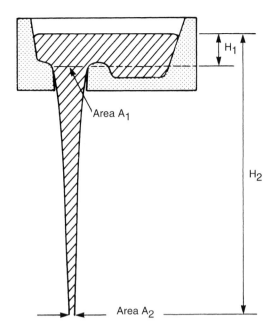

FIGURE 2-3 *A tapered sprue with the metal head H_1 in the pouring basin, and the total metal head, H_2. The cross-sectional areas at the top and the bottom of the sprue are A_1 and A_2. Note that the fluid mass crossing the areas A_1 and A_2 per unit time is constant. As the metal velocity at A_1 is $\sqrt{(2gH_1)}$, and the velocity at A_2 is $\sqrt{(2gH_2)}$, the product $A_1 \cdot \sqrt{(2gH_1)}$ must equal $A_2 \cdot \sqrt{(2gH_2)}$. This yields the fundamental relationship $A_1 \cdot \sqrt{H_1} = A_2 \cdot \sqrt{H_2}$.* (J. Campbell, Castings, Butterworth-Heinemann, Boston, 1999, p. 38). Reprinted with permission from Elsevier.

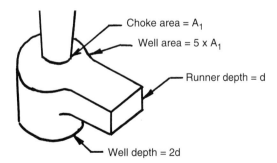

FIGURE 2-4 *Recommended design of the sprue basin: cross-sectional area of the basin is about five times the area at the base of the sprue, and the depth of the basin is twice that of the runner.* (Adapted from Basic Principles of Gating & Risering, 1985). Reprinted with permission from the Cast Metals Institute, Schaumburg, IL (www.castmetals.org).

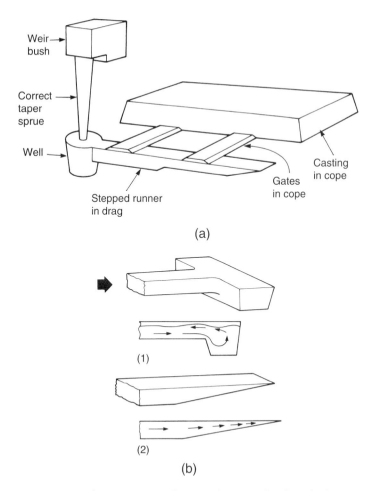

Weir bush

Correct taper sprue

Well

Stepped runner in drag

Casting in cope

Gates in cope

(a)

(1)

(2)

(b)

FIGURE 2-5 *(a) Diagram showing a stepped runner that is used with multiple gates. Keeping the runner cross-section constant would cause uneven flow from the gates.* (J. Campbell, Castings, Butterworth-Heinemann, Boston, 1999, p. 33) *(b) Design of the runner extension: (1) incorrect design that causes backwash and impurities in the casting, and (2) correct design that prevents backwash.* (J. Campbell, Castings, Butterworth-Heinemann, Boston, 1999, p. 54). Reprinted with permission from Elsevier.

bubbles to float out of the melt. Ceramic filters trap inclusions and dross particles, and are usually placed in the runner after the choke and before first gate. The gates are usually the bottommost part of the mold so that no further aspiration takes place due to free fall of metal.

In green sand casting the molding aggregate consists of a mixture of a base sand, clay, and water. Common varieties of sand such as silica, olivine, zircon, and chromite are used as base sand. The base sand is mixed with clay binders, water, and other additives in a fixed proportion to obtain the molding aggregate. Common sand additives include corn and wheat cereals to control sand expansion, cellulose to control the sand expansion and improve mold collapsibility and shakeout characteristics, iron oxide to improve mold rigidity, polymers and chemicals to lower the surface tension of water and improve the wettability of clay particles, and carbons to control

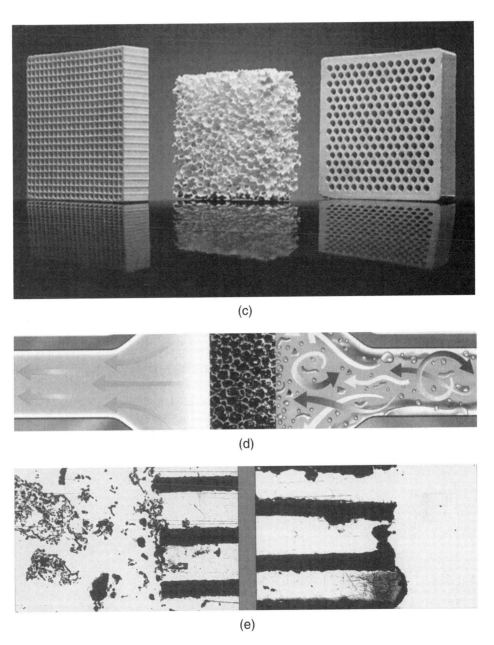

(c)

(d)

(e)

FIGURE 2-5 continued *(c) Examples of ceramic filters and honeycomb used to trap impurities in the gating system of a sand mold.* (M. Sahoo et al., Modern Castings, May 1995, p. 42). Reprinted with permission from Modern Casting, Schaumburg, IL (www.moderncastings.com). *(d) Diagram showing the effect of filter on flow pattern.* (B. Braun, Modern Castings, March 2004, American Foundry Society, Des Plaines, IL, p. 2). *(e) Photograph showing entrapment of impurity particles at the entrance to a filter.* (M. Sahoo et al., Modern Castings, May 1995, p. 42). Reprinted with permission from Modern Casting, Schaumburg, IL (www.moderncastings.com).

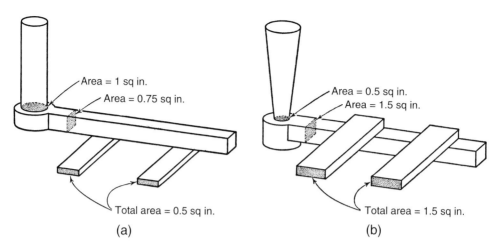

FIGURE 2-6 *(a) A pressurized gating system with choke (i.e., smallest cross-section) at the gate.* (H. F. Taylor, M. C. Flemings, and J. Wulff, Foundry Engineering, Wiley, 1959, p. 187) *(b) An unpressurized gating system with choke at the base of the sprue.* (H. F. Taylor, M. C. Flemings, and J. Wulff, Foundry Engineering, Wiley, 1959, p. 188). Reprinted with permission from H. F. Taylor Trust, Boston, MA.

expansion and reduce fusion between sand grains. The base sand is the major ingredient ($>85\%$) of the molding aggregate and is selected on the basis of considerations of cost, environmental and health factors, availability, type of metal being cast, and the properties desired in the molding aggregate. Some of the molding sand characteristics are green strength to ensure mold handleability and resistance to deformation from the weight of the pattern, dry strength to resist erosion and mold enlargement under metallostatic pressure, permeability to enable gases to escape and prevent internal pressure buildup in the mold, thermal stability to achieve dimensional accuracy and prevent sand expansion-related defects, refractoriness to withstand heat without melting or vitrifying of sand, and collapsibility to enable easy shakeout and prevent hot tearing (hairline cracking due to hindered contraction during cooling in a mold). These and other sand properties (e.g., compactibility and mold hardness) are characterized using standard foundry tests and are discussed in some of the textbooks on foundry processes listed at the end of this chapter. Many variables influence these properties, which include the size and shape of sand grains, the binder and water content, level of compaction, the melting point, and thermal expansion characteristics of sand. As an example, the effect of moisture content in the molding aggregate and grain shape on the green compression strength of ordinary molding sands is shown in the schematic of Figure 2-7. Both low and high water contents (and angular sand grains) yield low green strength; low water contents prevent adequate bond formation, and high contents cause excessive plastic flow and poor compactibility.

Commonly used foundry sands and their physical and thermal properties are listed in Table 2-2b. The effect of temperature on the thermal expansion of some common foundry sands is shown in Figure 2-8. All types of sand are essentially granular materials formed by natural attrition of rocks. Sand grains can be round, angular, or subangular, with an average size in the range 50 to 3360 μm. The grain shape, grain size, and size distribution determine the physical characteristics such as the density and openness of sand. Silica sand (essentially quartz or SiO_2) is cheap, abundant, and readily available in the preferred round shape, which improves the

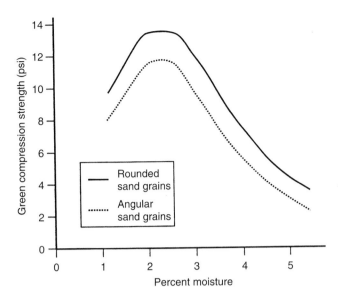

FIGURE 2-7 *Effect of moisture content and shape of sand grains on the green compression strength of molding sands.* (D. L. Zalensas, ed., Aluminum Casting Technology, 2nd ed., 1997, p. 191). Reprinted with permission from American Foundry Society, Schaumburg, IL (www.afsinc.org).

flowability of sand. The major disadvantages of silica are the health hazard (silicosis) that it poses, and its relatively high thermal expansion, which could result in mold wall dilation, and uneven expansion and cracking. Olivine sand (a silicate of Mg and Fe) contains no free silica (and is, therefore, less harmful than silica sand), and it has a lower thermal expansion than silica. It is refractory (i.e., high melting point, Table 2-2b) and has an angular grain structure. Zircon has a low thermal expansion (about one sixth of silica) and a relatively high thermal conductivity, which aids in faster cooling and shortens the production cycle. It also imparts better surface texture on the cast object due to rapid chilling and restricted metal penetration between packed sand grains. Zircon is a fine sand and is often used as a facing sand for better finish. However, it is relatively expensive and less abundant than the other two varieties discussed above.

Thermal Considerations

During cooling, thermal gradients are set up in the mold and the cooling rate depends on the thermophysical properties of both the metal and mold material. Materials with a greater capacity for heat extraction will exhibit steeper thermal gradients and faster heat dissipation. The relative chilling ability of different mold materials (relative to a value of unity for silica sand) can be used as a qualitative index for making a quick judgment about the cooling efficiency of different mold materials. Alumina, zircon, chromite, and silicon carbide have chilling abilities of 1.40, 1.23, 1.12, and 1.63, respectively. In contrast, the chilling ability of permanent mold materials are much higher; for example, copper, steel, and graphite have chilling abilities (relative to unity for silica) of 4.05, 3.95, and 3.34, respectively. Because the mold material will not only transfer the heat from the metal to the surroundings under steady-state but will also absorb the heat, the density and specific heat are also important material properties besides the thermal conductivity. The combined effect of these properties is called the thermal diffusivity

TABLE 2-2a Solidification Shrinkage, Density, Melting Point, and Crystal Structure of Common Metals

Metal	Crystal Structure	Melting Point °C	Liquid Density (Kg/m^3)	Solid Density (Kg/m^3)	Volume Change (%)
Al	fcc	660	2368	2550	7.14
Au	fcc	1063	17380	18280	5.47
Co	fcc	1495	7750	8180	5.26
Cu	fcc	1083	7938	8382	5.30
Ni	fcc	1453	7790	8210	5.11
Pb	fcc	327	10665	11020	3.22
Fe	bcc	1536	7035	7265	3.16
Li	bcc	181	528	—	2.74
Na	bcc	97	927	—	2.60
K	bcc	64	827	—	2.54
Rb	bcc	39	1437	—	2.30
Cs	bcc	29	1854	—	2.60
Tl	bcc	303	11200	—	2.20
Cd	hcp	321	7998	—	4.00
Mg	hcp	651	1590	1655	4.10
Zn	hcp	420	6577	—	4.08
Ce	hcp	787	6668	6646	−0.33
In	fct	156	7017	—	1.98
Sn	tetrag	232	6986	7166	2.51
Bi	rhomb	271	10034	9701	−3.32
Sb	rhomb	631	6493	6535	0.64
Si	diam	1410	2525	—	−2.90

From J. Campbell, *Castings*, 1999, Butterworth-Heinemann. Reprinted with permission from Elsevier.

TABLE 2-2b Thermophysical Properties of Common Foundry Sands

Sand	MP, °C	K, cal/cm · s · K	C, cal/g · K	ρ, g/cm³	α, cm²/s
Silica (SiO_2)	1750	0.0009–0.0013	0.269–0.281	1630	2.45×10^{-3}
Olivine (Fe, Mg silicate)	1872	0.0023–0.0025	—	2125	—
Zircon ($ZrSiO_4$)	2202	0.0025–0.0028	0.139–0.152	2960	6.15×10^{-3}

Note: MP, melting point; K, thermal conductivity; C, specific heat; ρ, density; α, thermal diffusivity.

(conductivity / [density × specific heat]) and determines the temperature gradients in the mold during solidification.[1] The thermal property data summarized in Table 2-2b show that thermal diffusivities of silica and zircon sands are 0.0021 and 0.0061 cm²·s^{-1} respectively. In contrast, the thermal diffusivities of cast iron and copper are over an order of magnitude greater (cast iron, 0.074, and copper, 0.932 cm²·s^{-1}, respectively). Further discussion of thermal effects during

---

[1]Thermal diffusivity is distinguished from heat diffusivity, which is defined as $(K.c.\rho)^{1/2}$, where K, c, and ρ are the thermal conductivity, specific heat, and density of the material respectively. Heat diffusivity is a measure of the heat-absorbing capacity, whereas thermal diffusivity is a measure of the heat mobility or heat dissipation rate.

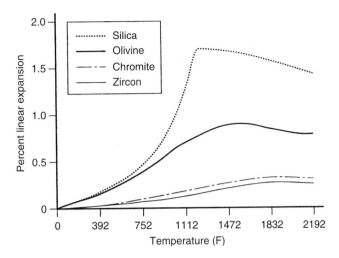

FIGURE 2-8 *Percent linear expansion of common foundry base sands as a function of temperature.* (D. L. Zalensas, ed., Aluminum Casting Technology, 2nd ed., 1997, p. 194). Reprinted with permission from American Foundry Society, Schaumburg, IL (www.afsinc.org).

solidification is presented in a later section. A factor that adversely influences the cooling rate is the formation of an air gap at the mold–metal interface because of differential contraction between the mold and the casting; as air is a thermal insulator, the air gap leads to an undesirable decrease in the cooling rate. Air gaps are eliminated and rapid cooling achieved when the metal solidifies under large external pressures (as in squeeze casting, discussed later).

A green sand mold can be constructed either manually via bench molding or by employing an automated process in an industrial setting (Figure 2-9). The industrial mass-production of parts via green sand casting uses match plate or production patterns for rapid fabrication of molds. A match plate is positioned at the parting plane of two matching halves of a metal container, as shown in Figure 2-9, and a mold release (or parting compound) is sprayed over the match plate. Premixed molding sand is then blown over one side of the match plate. A plunger squeezes and packs the sand on the match plate. The metal flask is inverted and the process of blowing and squeezing the sand is repeated on the other side of the match plate. A pouring basin is manually cut in the packed sand, the two flasks are separated, the match plate is removed, and cores are manually placed in the mold cavity. The mold is then reassembled and readied for pouring.

Although green sand molding is simple and cost effective, it has low process yield, which is usually 50 to 60%. Thus, only about 40 to 50% of the metal poured ends up in the actual part; the rest of the metal in the riser and gates is recycled and combined with fresh stock to make new castings. Care must, however, be taken in reusing foundry returns especially when fabricating premium quality castings. This is because repeated heating of recycled metal transforms a progressively increasing fraction of the metal into unreclaimable metal oxide, which both increases the metal consumption and causes dross and oxide defects in castings. In addition, some alloying elements may be lost during reheating, resulting in undesirable compositional changes. This may require on-line detection of melt quality and control of melt chemistry through judicious alloying. Any critical alloying elements lost, for example, due to

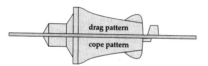

1. In this example, a cope pattern and a drag pattern are mounted onto a plate to form a matchplate.

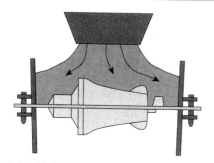

2. A flask is placed around the matchplate, and prepared sand is dumped or blown in on top of it.

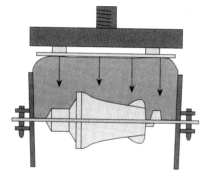

3. A squeeze board is placed over the sand, and the sand is compacted around the pattern.

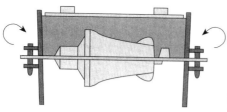

4. After compaction, the flask is turned over.

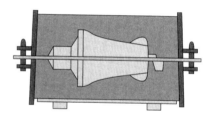

5. After rollover, steps 2 and 3 are repeated for the cope half of the mold.

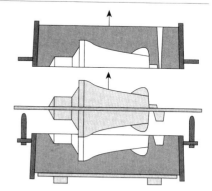

6. A sprue is cut into the cope. The cope then is lifted carefully off the drag, and the matchplate is withdrawn from the mold.

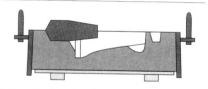

7. Cores (if any) are placed in core prints in the drag half of the mold.

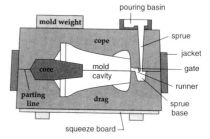

8. The cope is placed on top of the drag, and the flask is replaced with a mold jacket. Weights are placed on the assembled mold, and a pouring basin is added. The mold is read for pouring.

FIGURE 2-9 *Steps in making a green sand mold using a semi-automatic molding process.* (D. L. Zalensas, ed., Aluminum Casting Technology, 2nd ed., 1997, p. 187). Reprinted with permission from American Foundry Society, Schaumburg, IL (www.afsinc.org).

evaporation must be replenished. For example, fresh ingots of A356 have higher Mg content than foundry return (some Mg is lost due to evaporation on heating). Magnesium must, therefore, be added to the alloy to retain the age-hardening characteristics, which depend on the formation of the Mg_2Si precipitates.

Dry Sand and Skin-Dried Molds

The problems related to relatively low green strength (poor handle-ability) and high moisture contents (porosity defects) in a green sand mold are overcome with the use of dry sand molds. A dry sand mold is made in the same manner as a green sand mold, but excess water is added to the molding aggregate as compared to a green sand mold. The mold is baked at about 150 °C to remove the moisture. The baking step improves the mold strength and reduces moisture-related defects in cast parts, but it also adds to the process cost and increases the mold production time. A useful compromise between green and dry-sand molds is the skin-dried mold, in which roughly a 0.5- to 1-inch-thick layer of green sand is dried with a torch before the pour. This reduces the energy consumption and moisture-related defects while providing adequate mold strength for casting alloys such as cast iron. Frequently, a silica or zircon wash is applied to the mold cavity prior to drying to improve the casting finish.

Sodium Silicate-CO_2 Process

In the sodium silicate-CO_2 process, the mold is hardened through a chemical reaction at ambient temperatures. The process is used to make both molds and cores. A fine base sand is mixed with 3 to 4% water glass (sodium silicate) in a sand muller to obtain a soft molding aggregate. The mold is made in the same manner as a green sand mold. The completed sand mold is exposed at room temperature to carbon dioxide gas through vents that are made either in the match plate or in the sand mold itself (Figure 2-10). The exposure to CO_2 causes the sodium silicate in the sand aggregate to harden through a complex reaction that, in a simple form, can be written as $Na_2SiO_3(l) + CO_2(g) \rightarrow Na_2CO_3(s) + SiO_2(s)$. The ability to create strong molds without baking is an advantage over dry-sand molds and skin-dried molds. In addition, the curing reaction uses an inexpensive and nontoxic gas. The soft molding aggregate (water glass + sand) must be used immediately for molding after mixing is completed because prolonged exposure to atmospheric

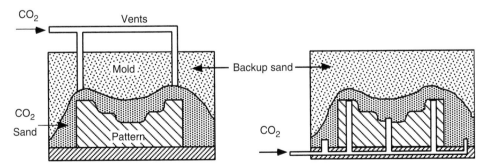

FIGURE 2-10 *Hardening of sand and sodium-silicate mixture by exposure to CO_2 gas at room temperature. The basic cure reaction is $Na_2SiO_3 + CO_2 \rightarrow Na_2CO_3 + SiO_2$* (R. Trivedi, Materials in Art and Technology, 1998, p. 117, Taylor Knowlton, Inc., Ames, IA).

CO_2 will partially harden the aggregate and affect its moldability. One disadvantage of this process is that when the reaction-formed silica comes in contact with hot metal, there is partial firing of the ceramic. This strengthens the ceramic and confers poor shakeout characteristics to the molding sand.

Vacuum Molding

Vacuum molding uses a vacuum to hold a foundry base sand in the desired mold shape while the metal is poured and allowed to solidify in the mold. Specially designed flasks, match plate patterns with vents, and a hollow carrier plate (with vents on its top surface) are used as the basic tooling. The flasks and the carrier plate are designed to permit the creation of a sand mold with vacuum assist. The basic steps of the process are illustrated in Figure 2-11. A match plate pattern with vent holes is placed on top of the hollow carrier plate that also has vent holes on its top surface. The pattern is then covered with a flexible Teflon film by pulling a vacuum through the carrier plate. A flask is placed over the film and filled with loose, unbonded foundry sand. A pouring cup and sprue are formed, and a second plastic film is placed to cover the

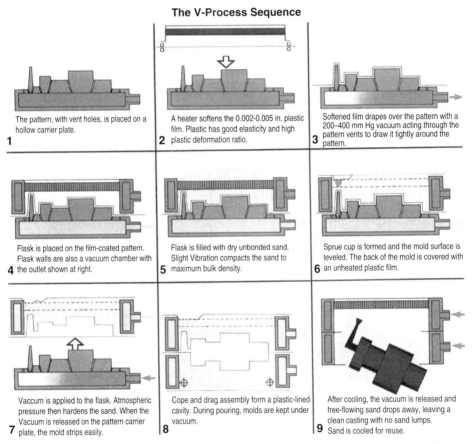

The V-Process Sequence

1. The pattern, with vent holes, is placed on a hollow carrier plate.

2. A heater softens the 0.002-0.005 in. plastic film. Plastic has good elasticity and high plastic deformation ratio.

3. Softened film drapes over the pattern with a 200–400 mm Hg vacuum acting through the pattern vents to draw it tightly around the pattern.

4. Flask is placed on the film-coated pattern. Flask walls are also a vacuum chamber with the outlet shown at right.

5. Flask is filled with dry unbonded sand. Slight Vibration compacts the sand to maximum bulk density.

6. Sprue cup is formed and the mold surface is leveled. The back of the mold is covered with an unheated plastic film.

7. Vaccum is applied to the flask. Atmospheric pressure then hardens the sand. When the Vacuum is released on the pattern carrier plate, the mold strips easily.

8. Cope and drag assembly form a plastic-lined cavity. During pouring, molds are kept under vacuum.

9. After cooling, the vacuum is released and free-flowing sand drops away, leaving a clean casting with no sand lumps. Sand is cooled for reuse.

FIGURE 2-11 *Sequence of steps involved in making sand castings using the vacuum-molding process.* (J. Pohlman, Modern Casting, May 1995, p. 33). Reprinted with permission from Modern Castings, Schaumburg, IL (www.moderncastings.com).

sand. A vacuum is pulled through the flask, and air is allowed to enter the bottom carrier plate. This allows the sand mold to be stripped from the match plate without causing the mold to disintegrate. The matching half of the mold is made in a similar fashion, and the two halves are assembled and readied for pouring. A vacuum is maintained on the flasks during pouring and subsequent solidification. As the plastic-lined mold cavity is filled, the film volatilizes, and the gases are removed by vacuum suction. The liquid metal in the sprue acts like a liquid seal that prevents air suction and mold collapse as the plastic film volatilizes. The major advantages of the process are that no special sand treatment or conditioning is needed (saves cost of binders, mixing equipment, shop floor space), there is complete mold venting (elimination of gas-related defects), and there is no shakeout required. Releasing the vacuum on the molding flasks allows the sand and the casting to drop and separate from each other. The sand is cooled and reused without any further special treatment.

Shell-Molding

The molding aggregate in shell-molding is a mixture of a fine base sand and a thermosetting binder. A metal match plate is used as a pattern and forms the cover of a dump box that is filled with the molding aggregate. The match plate is heated to about 150–230 °C, and the dump box is inverted to allow the resin-bonded sand to physically contact the hot pattern. The thermosetting plastic begins to cure and harden, forming a solid sand shell around the pattern. The dump box is then brought back to its normal upright position, and excess (uncured) sand mixture is removed. The partially cured shell is stripped from the match plate with the help of ejector pins, and the curing is completed in an oven. The steps are repeated to make the matching half of the shell mold. After curing, the two mold halves are assembled, clamped, and readied for the pour (metal shot or sand is used as a physical support for the mold halves). Some of the resin evaporates during the pour and represents an unreclaimable material loss. Very tight dimensional control (tolerances of 0.002–0.005 inch) and excellent surface finish are achieved using shell-molds. This reduces the need for machining, but very precise patterns are needed to start with. The process is amenable to automation for mass production of parts. Figure 2-12 shows the basic steps involved in shell molding.

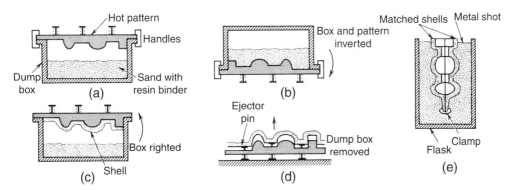

FIGURE 2-12 *Shell-molding process that uses a fine base sand mixed with a thermosetting binder. The mixture partially cures and forms a hard shell upon contact with a hot match plate (which forms the cover of the dump-box containing the sand mixture).* (E. P. DeGarmo, J. T. Black, R. A. Kohser, and B. E. Klanecki, Materials and Processes in Manufacturing, 9th ed., Wiley, New York, 2003, p. 308).

Investment Casting

Investment casting uses expendable wax patterns to make castings. Primitive versions of investment casting existed in ancient times, and archeometallurgists have excavated tools and implements made using this process dating back to 3000–2000 B.C. In its modern, refined incarnation, the process is extremely versatile. It is used to make turbine blades for aircraft engines, fuel system components, combustor chamber parts, turbine vanes, prosthetic devices, air frame parts, precision machine tools, dental implants, jewelry and sculpture, and a wide variety of other industrial and consumer items from ferrous and nonferrous alloys. Net-shape parts with complex geometry and thin sections (typically 0.015 inch) with tight dimensional tolerances (0.005 to 0.010 inch per inch) and excellent surface finish are routinely produced with this process. Parts weighing a few grams to over 200 kg are investment cast. The process is good for difficult-to-machine alloys such as Ni-base superalloys used in aircraft engine parts. The high design flexibility of investment casting and its ability to cast complex parts to near-net shape reduces the fabrication and assembly costs.

The basic steps of industrial investment casting are shown in Figure 2-13. Wax patterns are formed by injecting molten wax in a die. Cores for making internal surfaces can be incorporated in the wax pattern by placing preformed cores in the die and injecting molten wax around the core. Cores could be made out of a water-soluble material or ceramics; the latter types are removed only after the casting has solidified. Soluble cores are made from polyethylene glycol binder that contains fillers (mica, silica, or NaCl) and carbonates. Several identical wax patterns are injection molded and attached, usually manually, to a common sprue-and-riser assembly, also made out of wax. The wax cluster is submerged in a ceramic slurry containing ceramic fines suspended in a liquid vehicle. Continuous gentle agitation of the slurry prevents sedimentation. The wet cluster is either sprayed with coarse ceramic powder or submerged in a fluidized bed of coarse ceramic powders (stucco coat). The more open refractory stucco coat minimizes drying stresses and provides for frictional bond during subsequent coats. The clusters are hung on a conveyor and dried in a room with controlled humidity and temperature. The dried molds are then coated a second time with the liquid slurry and stucco coat, and the process is repeated until approximately a $\frac{1}{4}$-inch-thick shell is formed. The multiple coating and drying cycles may take a couple of days to a week to complete. Once a shell of desired thickness has formed, the wax cluster is melted out in an autoclave, leaving a hollow ceramic shell that is fired for additional strength before being readied for the pour. The remelted wax is reused for making more patterns, but the investment is usually lost. Pressure- or vacuum-assisted pour and gravity pour are commonly employed. After solidification, the shell is broken to recover the cast part.

Investment casting is a near-net-shape fabrication process applicable to a wide range of alloys (Al, Cu, steel, Co, Ti, and Ni-base superalloys). Although some intermediate steps (e.g., immersion, stucco coat) can be automated, the process generally involves high labor cost and is, therefore, suitable for relatively small production volumes of high-precision castings.

The investment casting technique is also used to make metallic foams, which can be regarded as gas–metal composites having excellent damping capacity and low thermal conductivity. A fluid refractory material is poured in voids of a spongy, foamed plastic and hardened. This is followed by heating to vaporize the plastic component and leave spongy lattice pores. Molten metal is poured in this spongy mold and solidified. After removing the refractory, a metallic foam casting that mimics the original spongy plastic is obtained. Metallic foams have also been prepared by adding a "blowing" agent, e.g., TiH_2 or ZrH_2, to a molten metal. The mixture is heated to decompose the blowing agent and form gas, which expands and produces foamed metal after solidification. Interconnected cellular pores are formed in foamed metals by casting the metal around granules, e.g., NaCl, introduced in the casting mold or by stirring the granules

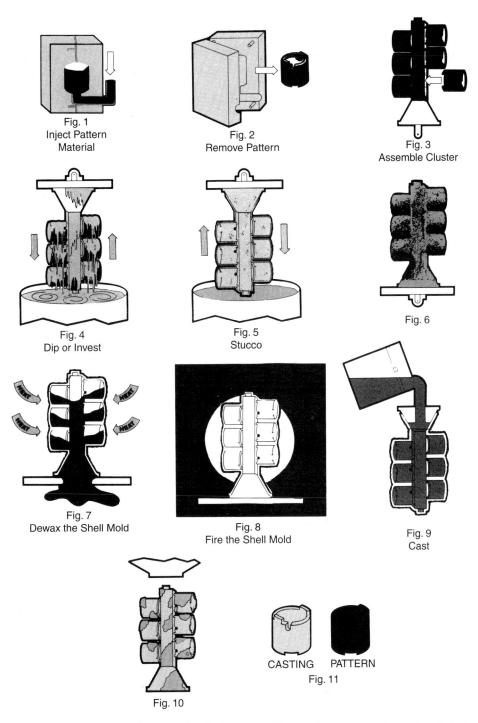

Fig. 1
Inject Pattern
Material

Fig. 2
Remove Pattern

Fig. 3
Assemble Cluster

Fig. 4
Dip or Invest

Fig. 5
Stucco

Fig. 6

Fig. 7
Dewax the Shell Mold

Fig. 8
Fire the Shell Mold

Fig. 9
Cast

Fig. 10

CASTING   PATTERN
Fig. 11

FIGURE 2-13 *Sequence of steps used in the investment (lost-wax) casting technique. These include wax injection and solidification in a prefabricated die, formation of a "tree" or "cluster" of wax patterns on a wax sprue, repeated immersion in a fine ceramic slurry, and a dry "stucco" coat. This is followed by melting the wax out, firing the shell, and pouring.* (Courtesy of Investment Casting Institute, Dallas, TX).

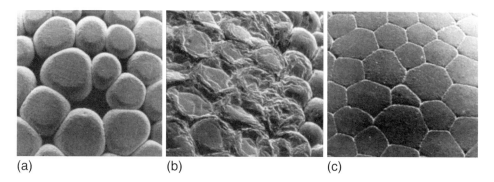

<table>
<tr><td>(a)</td><td>(b)</td><td>(c)</td></tr>
</table>

FIGURE 2-14 *Surface finishes on the foam used to make patterns in the lost-foam casting process: (a) and (b) are underfused pattern (susceptible to damage and rough surface finish), and overfused pattern (susceptible to warpage and wrinkled surface finish), respectively, which yield poor finish on the casting. The finish in (c) is the optimum surface texture of the pattern.* Reprinted with permission from Modern Casting, Schaumburg, IL (www.moderncasting.com).

into the melt followed by solidification in a die. These granules are later leached out with a chemical to leave a porous metal.

Lost-Foam Casting

The lost-foam casting process uses expendable polystyrene patterns. Different versions of the basic process are called the "full-mold" process, "evaporative pattern casting" (EPC), and "expanded polystyrene" process. Styrofoam beads are used to form the pattern. The beads are injected in a steam-jacketed die under low pressures and allowed to expand and fuse to form the pattern. The surface quality of the foam pattern determines the surface texture of the cast part, as illustrated in Figure 2-14. Both underaged foam patterns (incompletely fused beads) and overaged patterns (partially melted beads that create wrinkles on the surface) impair the casting's surface quality. After ejecting the foam pattern from the die, gates and risers made out of foam are glued to appropriate surfaces. A thin coat of a fine ceramic is applied via immersion in a slurry to cover all the surfaces of the foam pattern (except the pouring basin). The coating improves the casting surface finish by acting as a barrier between the supporting sand and the foam. After the coating has dried, the coated pattern assembly is either buried in loose, free-flowing sand or covered in lightly packed green sand. The coating also provides some stability to the mold and prevents sand from caving in the cavity created by evaporating foam, especially when the pattern is buried in loose (rather than packed) sand. The metal is poured, allowing the pattern to volatilize and progressively create the mold cavity to be continuously filled by the incoming metal. Pouring is usually assisted with a vacuum that removes gases from the burnt foam through the semipermeable coating, thus enabling uninterrupted metal ingress. Alternatively, pressurized lost-foam casting is used to eliminate gas porosity to a nearly undetectable level. After solidification, the casting is readily extracted from loose sand by robots, thus eliminating shakeout.

The process is readily automated with the coating application, metal pouring, part recovery, and final cooling (water quenching) steps handled by robots. The process features high design flexibility, and the light weight of the foam eliminates problems related to pattern weight such as mold deformation from the pattern weight (e.g., a mere 45-lb foam pattern will yield a nearly eight-ton iron casting!). Because the foam pattern is replaced progressively rather than instantaneously by the liquid metal, some of the defects resulting from turbulence are eliminated.

Grain refiners and modifiers are incorporated within the pattern to achieve in-mold modification and grain refinement of the alloy. The pattern draft is not needed as the pattern is not removed from the mold before the pour. The pattern is, however, wasted in each run. The process was patented in the 1950s but was used on a production scale starting in the 1980s by the automotive industry.

Selection of an appropriate pouring and gating technique is important. Even though venting may be adequate because of porous ceramic coating and free-flowing sand surrounding the foam pattern, the back pressure of gases from volatilizing foam tends to push against the liquid metal in the pouring cup and the sprue; this leads to an interrupted flow of metal in the mold that could lead to misrun in thin sections. High pouring temperatures (that depend on the density of the foam being used) are required because of the latent heat required to sublimate the expendable foam pattern. The casting yield in the lost-foam casting is usually less than 70%. The lost-foam casting is applicable both to monolithic alloys and metal-matrix composites and has been widely used by automotive manufacturers in making engine parts such as intake manifolds.

Other Expendable Mold Processes

Modifications of some of the preceding expendable mold processes are also used in industrial practice. Some examples are plaster molds, oil-bonded sand molds, and no-bake sand molds. Plaster molds use gypsum or hydrated plaster-of-Paris, often mixed with a small amount of talc and cereal binders, as the molding aggregate. The plaster is mixed with water and poured over a match plate, normally coated with a release agent for easy separation of the hardened shell from the pattern. The hardened plaster shell is stripped from the pattern, and the matching mold half is constructed in a similar fashion. The two halves are assembled, and heated in an oven before pouring. Plaster molds yield an excellent surface finish but are generally not recommended for high-melting-point alloys such as steels due to the poor thermal shock resistance of the plaster. Oil-bonded sand molds use a petroleum-based binder in sand. Because oil has a higher vaporization temperature than water in the green sand mold, low-permeability molds can be used with oil-bonded sands. Thus molds can be made from finer sands and compacted to greater densification levels, resulting in improved surface finish on the casting. Even a small amount of oil vapor, however, has an unpleasant odor, and the process is generally not used for mass production. Another expendable mold process uses sands that contain phenolic resins capable of hardening at room temperature (no-bake sand molds).

Multiple-Use Mold Casting
Permanent Mold Casting

Permanent molds are made out of metals such as iron, copper, and aluminum, or out of graphite. Metals normally cast in permanent molds include Al, Mg, and Cu alloys. The life of the die could vary from a few hundred parts to over 250,000 parts. Dies are made in multiple sections, capable of being readily assembled, and are generally given a refractory wash to reduce the extent of corrosive attack by molten metal. The die is preheated before the metal is poured. Ceramic cores can be inserted in the die to make internal surfaces and cavities. Production rate can be increased by using multiple-cavity molds and rotating platforms. Better texture and a finer grain structure are obtained in permanent molds than in green sand molds. The casting yield of conventional permanent mold casting (also called gravity die casting) is, however, only about 60%, which is comparable to the yield of traditional sand casting. Figure 2-15 shows a sketch of the conventional permanent mold (gravity die casting) process.

Modifications of the permanent mold casting include low-pressure permanent mold (LPPM) casting and vacuum permanent mold (VPM) casting. In LPPM (Figure 2-16), liquid metal is

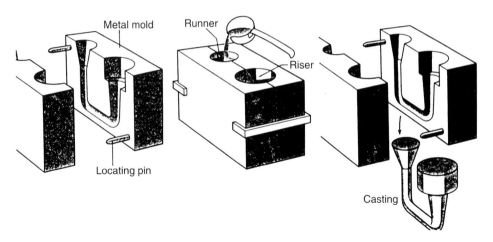

FIGURE 2-15 *Gravity die casting in a permanent mold.*

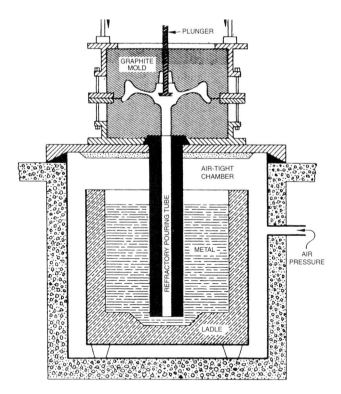

FIGURE 2-16 *Low-pressure permanent mold casting in which an inert gas is used to pressurize the molten metal counter to gravity through a feed tube and into the permanent mold that is placed on top of the pressure vessel. The solidification path is designed to enable the shrinkage to be fed by the pressurized molten metal.* (Courtesy of Amsted Industries).

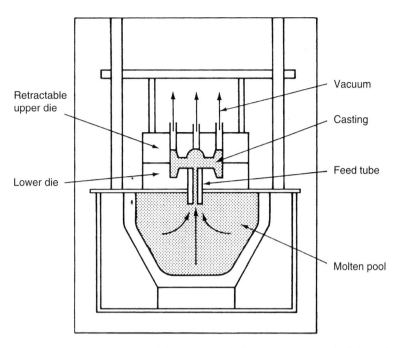

FIGURE 2-17 *Vacuum permanent mold casting in which a vacuum is applied through vents in the mold to raise the molten metal via a feed tube into the mold.* (E. P. DeGarmo, J. T. Black, R. A. Kohser, and B. E. Klanecki, Materials and Processes in Manufacturing, 9th ed., Wiley, 2003, p. 328).

introduced with the help of gas pressure in a permanent mold positioned atop a pressure chamber containing molten metal via a refractory feed tube. The flow of metal is countergravitational and modulated through control of gas pressure, which permits laminar flow. Air or an inert gas is introduced typically at 100 kPa in the pressure chamber that houses a furnace containing the molten metal. Because the pressurized metal is drawn from the bulk rather than the melt surface, only clean metal enters the mold. Risers are not needed as the pressurized metal in the feed tube acts to fill the solidification shrinkage. Casting yield of 80% or greater is achieved by LPPM.

In VPM, a vacuum is used to introduce molten metal in a permanent mold atop the vacuum chamber via a refractory tube (Figure 2-17). In a manner similar to LPPM, clean, dross-free metal flows counter to gravity, in a laminar fashion in the mold. The solidification is directional (toward the melt), and vacuum suction of the melt provides easy feeding of solidification shrinkage. Cleanliness of the metal is superior to that of LPPM because a pressurizing gas (e.g., air) is not used, and casting yield generally exceeds 80%.

Die Casting and Semisolid Casting

Die casting (also called pressure-die casting) is an important industrial process that is used to mass produce thin-walled, complex parts having good dimensional control and excellent surface finish. The process involves injecting a liquid (or partially solidified) alloy in a prefabricated die, and solidifying the alloy under pressure. Tool steel dies with multiple sections for ease of part ejection are generally used. The dies are water cooled for rapid heat extraction and coated with corrosion-resistant ceramic coatings. The tooling cost is high, and the process is justified in high-volume production of complex parts for which the cost can be spread over a large number of parts. Because thin sections are die cast and water-cooled dies are used, solidification times are short (on the order of a few seconds). The actual solidification time does not, therefore, limit the production rate. Risers to feed the solidification shrinkage are not needed because external pressure on the metal aids feeding. Pressures are typically in the range 15 to 60 MPa. The metals cast include Sn, Zn, Al, Mg, and Cu alloys, with zinc die castings representing the largest tonnage products.

Two basic type of die-casting machines are used: hot-chamber and cold-chamber die casting (Figure 2-18). Hot-chamber die casting, used for low-melting-point metals such as Sn and Zn, has a melting pot as a part of the die-casting machine, whereas cold-chamber die casting, used for Al and Mg alloys, has the melting furnace separate from the die-casting unit, and requires metal to be transported from the furnace to the die-casting unit. While hot chamber machines reduce the time to deliver the liquid metal, they exhibit greater molten metal contamination than cold chamber machines because of dissolution of the iron tooling in the metal. A casting yield of 90% or higher is possible with die casting, which is significantly greater than that of gravity die casting and most expendable mold processes. Die-casting machines vary considerably in capacity, from small units capable of injecting 1 to 2 lb of metal per shot to large units that can inject over 100 lb of Al per shot. Electrically heated, refractory-lined channels called molten metal launder systems deliver the exact amount of metal needed to die-casting machines from nearby holding furnaces. Launder systems minimize oxidation and turbulence problems, and eliminate the hazards associated with transporting exposed molten metal from the furnace to the die-casting machine.

A typical pressurization cycle for die casting is shown in Figure 2-19. During metal injection in the die with a hydraulically actuated plunger, some turbulence could occur (Figure 2-20), causing voids and porosity in the casting. Turbulence during die filling is minimized by (1) employing large gates and low injection pressures (followed by an increase in the pressure during the solidification stage), and (2) replacing fully molten alloys with pasty or mushy semisolid alloys with increased viscosity that reduces the turbulence. The lower operating temperatures with the use of semisolid alloys also reduce the thermal shock and hot corrosion of the die, and the amount of solidification shrinkage to be fed, but the abrading (grinding) action of the partially solidified slurry causes increased erosive wear, which necessitates special die surface treatments (e.g., abrasion-resistant ceramic coatings). Another method to reduce the porosity content in die cast parts, called the "pore-free" casting process, injects pure oxygen in the die before injecting the metal. Oxygen displaces air and water vapor (source of hydrogen porosity in castings) from the die, and reacts with injected metal to form oxide dispersions that bond with and strengthen the casting much like a composite casting. This should be contrasted to conventional sand casting, in which stray oxide particles essentially serve as strength-limiting defects because of the void space between the particle and the metal, and a lack of bonding.

Semisolid alloy slurries containing even 40–50% solidified crystals can be designed to have low deformation resistance and are used for low-energy, net-shape forming via mechanical deformation. The low deformation resistance is achieved through control of the solid crystallite shape (spherical rather than dendritic), which improves the flow behavior under low shear. The basic idea is implemented in a process called rheocasting in which the matrix alloy is held between its solidus and liquidus temperatures to create an equilibrium volume fraction of fine (50–100 μm) primary solid phase. A spheroidized rather than dendritic structure is created by mechanical (or electromagnetic) stirring, and by rapid dissolution of regions of sharp curvature on the particles via the process of "coarsening" or "Ostwald ripening." The shape and size of the

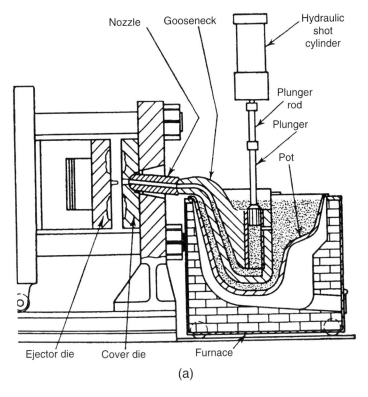

Nozzle　Gooseneck　Hydraulic shot cylinder

Plunger rod

Plunger

Pot

Ejector die　Cover die　Furnace

(a)

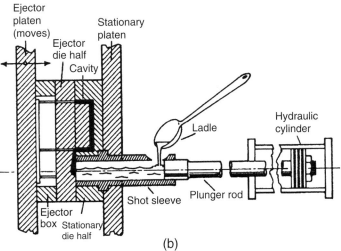

Ejector platen (moves)　Stationary platen

Ejector die half

Cavity

Hydraulic cylinder

Ladle

Shot sleeve　Plunger rod

Ejector box　Stationary die half

(b)

FIGURE 2-18 *(a) Hot-chamber die casting machine in which the melting furnace is a part of the die casting machine. The gooseneck design of the feed tube is used to inject the metal in the die cavity. (Courtesy of North American Die Casting Association) (b) Cold-chamber die-casting machine in which the molten metal is transferred from the melting furnace (separate from the die-casting machine) into the shot chamber of the machine. The metal is transported via an electrically heated channel (launder system). (Courtesy of North American Die Casting Association).*

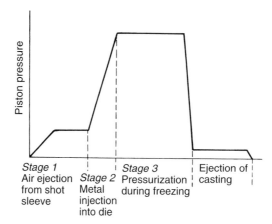

FIGURE 2-19 *Schematic diagram showing a typical pressurization cycle in die casting.* (J. Campbell, Castings, Butterworth-Heinemann, Boston, 1999, p. 64). Reprinted with permission from Elsevier.

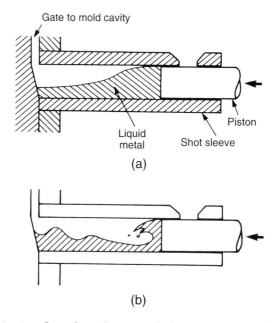

FIGURE 2-20 *(a) Laminar flow of metal in controlled pressurization, and (b) turbulent flow in uncontrolled pressurization, which results in porosity in the cast part.* (J. Campbell, Castings, Butterworth-Heinemann, Boston, 1999, p. 64). Reprinted with permission from Elsevier.

primary solid crystals depend on the cooling rate, degree of shear, and duration of stirring. The fine, spheroidized rheocast structure can also be produced by reheating a fine-grained, equiaxed dendritic feed material to a semisolid state followed by isothermal hold to allow coarsening and spheroidization to take place. Figure 2-21 shows typical microstructures of semisolid alloy slurries, and Figure 2-22 shows some semisolid formed components.

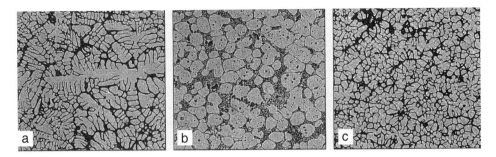

FIGURE 2-21 *Microstructure of an A357 Al-Si alloy under different casting conditions: (a) normal dendritic structure in a conventional casting, (b) globular structure with improved flow characteristics in a semisolid alloy slurry, and (c) fine globular structure of the semisolid alloy slurry when cast using a magneto-hydrodynamic casting process that causes vigorous shear in the slurry and reduction in globule size.* (J. Boylan, Advanced Materials & Processes, October 1997, p. 27). Reprinted with permission from ASM International, Materials Park, OH (www.asminternational.org).

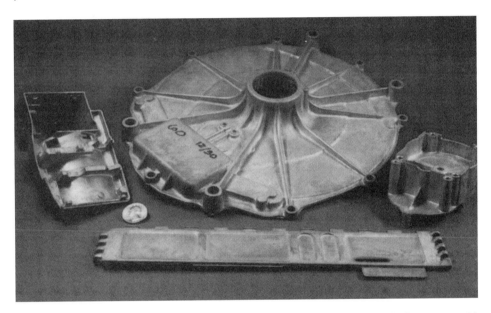

FIGURE 2-22 *Semisolid formed magnesium alloy components made by a modified injection molding process, called thixomolding.* (S. B. Brown and M. C. Flemings, Advanced Materials & Processes, January 1993, p. 36). Reprinted with permission from ASM International, Materials Park, OH (www.asminternational.org).

Stirring of the semi-solid alloy slurry to achieve better flow characteristics is accomplished either with the help of mechanical impellers or by use of electromagnetic (EM) induction, in which case physical contact between impeller and the metal is avoided. This prevents melt contamination from impeller erosion at high temperatures. EM stirring has been incorporated in continuous casting to produce billets for subsequent reheating into the semisolid state. Various modifications of the stirring techniques have been developed to create the rheocast structure, such as strain-induced melt activation, which involves reheating a heavily deformed fine-grained material, flow casting, which consists of solidifying the metal as it flows through a series of winding channels comprised of a series of small left- and right-hand helical elements, and super stir-casting, which uses very large (>1000 rpm) stirring speeds under vacuum to create fine-grained semisolid alloy slurries. Electric current pulses also have been used to produce fine-grained slurries by disintegrating dendrites into small primary crystallites.

Temperature and shear rate are the most critical parameters in rheocasting, because they control the relative amounts of liquid and solid phases in the slurry and the flow behavior of the slurry. The deformation stresses decrease as the temperature and the holding time in the semisolid state increase. This is profitably used in "thixomolding," which employs forging or extrusion of semisolid slurry to create net-shape, high-precision parts. Figure 2-23 shows some magnesium parts made using thixomolding, and Figure 2-24 shows a schematic diagram of a thixomolding machine. In thixomolding, the deformation rate must be relatively high to prevent separation of the liquid and solid phases, and surface tensile stresses should be minimized during deformation to prevent cracking. Special precautions are necessary to prevent surface oxides on semisolid billets to be trapped inside the component, and special die surface treatments are needed to reduce the abrasive (grinding) action of the solid crystallites in the semisolid slurry.

In semisolid forming, liquid handling is eliminated, and the feed material is cut into slugs, heated into a semisolid state, and shaped into final part. The semisolid slugs retain their shape

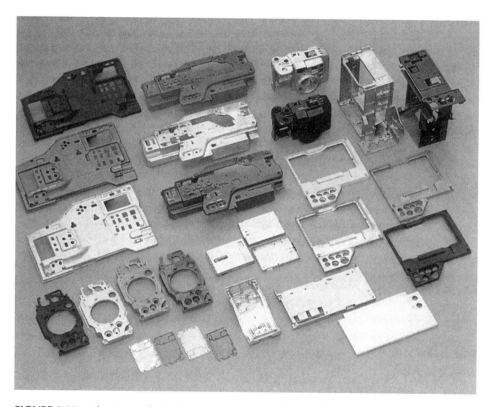

FIGURE 2-23 *Aluminum electronic components, stainless steel valve bodies, and brass components injection molded from reheated semi-solid billets.* (S. B. Brown and M. C. Flemings, Advanced Materials & Processes, January 1993, p. 36). Reprinted with permission from ASM International, Materials Park, OH (www.asminternation.org).

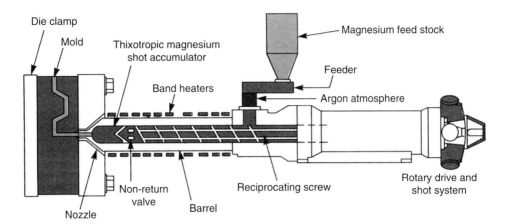

FIGURE 2-24 *Schematic of a semisolid casting technique (thixomolding) that combines features of die casting and injection molding for magnesium components. Magnesium feedstock is added from a hopper into a multizone, temperature-controlled barrel with a reciprocating screw. The screw rotation conveys the material through the heated barrel, producing a thixotropic (shear-thinning) semisolid Mg alloy slurry that is injected at high velocity in a die.* (R. F. Decker et al., Advanced Materials & Processes, February 1996, p. 41). Reprinted with permission from ASM International, Materials Park, OH (www.asminternational.org).

while they are transported to the shot sleeve of the die-casting machine. Because the slug deforms only under pressure, die closure can accompany placement of the semisolid slug in the shot chamber, resulting in reduced cycle times relative to the traditional die casting. Casting temperatures are low; e.g., for aluminum alloys, casting temperatures are typically 100 °C lower than traditional casting temperatures. Because of its rheological properties, the semisolid slurry fills the die with a laminar front whose physical location is precisely controllable from shot to shot by use of sensors installed in the machine. Laminar flow and gradual die filling eliminate the problem of gas entrapment. At the same time, the shear caused by the plunger motion reduces the viscosity of the semisolid slurry because of the latter's pseudoplastic or shear-thinning nature. As a result, relatively low pressures allow filling of complex cavities at high injection speeds. Figure 2-25 shows the effects of fraction solid, shear rate, and the cooling rate on the apparent viscosity of semisolid Pb-Sn alloy slurries. High shear rates lead to lower apparent viscosity at a fixed solid loading. Also, as the semisolid alloy is only about 40–50% melted at the time of injection, the problem of solidification shrinkage is reduced and metal loss is eliminated. The reduced amount of latent heat to be dissipated and a high solid fraction in a semisolid alloy permit quicker solidification and shorter production runs.

Squeeze Casting

Squeeze casting, or liquid metal forging, differs from die casting in two fundamental ways: (1) a metered quantity of metal is gravity-poured (not injected) in a preheated metal die, and (2) the metal is pressurized during solidification to higher pressures (~60–250 MPa). Figure 2-26 shows the basic steps of squeeze casting. The problem of turbulence during die filling is minimized as the metal is poured rather than injected, and high pressures during solidification eliminate all porosity, thereby yielding defect-free premium quality castings. Cycle times are short, and casting yield is close to 100% because runners, gates, and risers are not needed. The process is a high-precision, net-shape casting process that yields excellent finish and high-strength parts. Ceramic fiber-reinforced aluminum alloy pistons, Ni-base hard crusher wheel

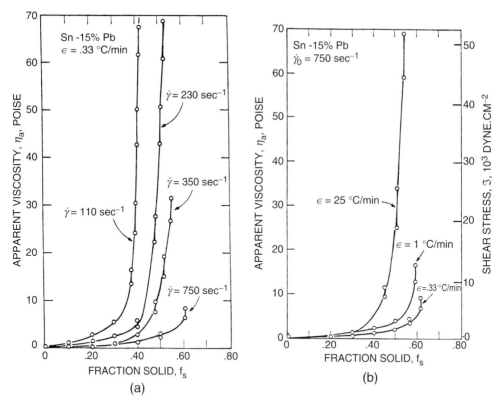

FIGURE 2-25 *(a) The effect of shear rate and solid fraction in the semi-solid slurry on the apparent viscosity of a Sn-15% Pb partially solid alloy slurry at a fixed cooling rate. (P. A. Joly and R. Mehrabian, J. Mater. Sci., 11, 1976, p. 1393, Chapman & Hall, London) (b) The effect of cooling rate and solid fraction in the semisolid Sn-15% Pb alloy slurry when sheared at a constant rate of 750 s⁻¹. (P. A. Joly and R. Mehrabian, J. Mater. Sci., 11, 1976, p. 1393, Chapman & Hall, London).*

inserts, steel missile components, and cast iron mortar shells are some of the parts cast using squeeze casting. Wrought alloys can be used because the application of pressure eliminates the need for good fluidity.

Because the external pressure eliminates the air gap that would normally form at the casting–die interface, the thermal contact is perfect and cooling is rapid. The application of a large hydrostatic pressure to a solidifying metal could lower its melting point, thus undercooling the metal to a greater degree and refining the grain structure by promoting copious nucleation. The magnitude of melting-point depression under a pressure, P, is expressed from the Clausius-Clapeyron equation

$$\left(\frac{dP}{dT}\right) = \frac{\Delta H}{T \Delta V} \tag{2-1}$$

where T is the temperature, P is the pressure, ΔH is the latent heat of solidification, and ΔV denotes the specific volume of the metal. Different metals will experience different levels of melting point depression. For example, it has been estimated that a pressure of 151 MPa will lower the melting point of pure tin by about 4.3 °C.

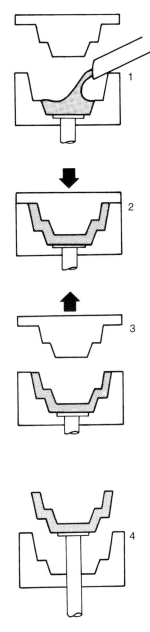

FIGURE 2-26 *Squeeze casting is analogous to forging and involves introduction of a metered quantity of metal in a die, followed by die closure and pressurization with a hydraulic press. The near-perfect thermal contact at the metal–die interface and increased nucleation under a large external pressure refine the grain structure, and eliminate the porosity.* (J. Campbell, Castings, Butterworth-Heinemann, Boston, 1999, p. 72). Reprinted with permission from Elsevier.

Centrifugal Casting

In centrifugal casting, molten metal is poured either in a permanent mold or a sand mold, rotated at 300 to 3000 rpm either in a vertical plane or a horizontal plane. Rotation is continued until solidification is complete. Three-dimensional hollow parts such as gear blanks, pressure vessels, and propeller hubs are made via the centrifugal casting technique without the use of cores that would be needed in a stationary casting process. Sequential pouring of different alloys in a rotating mold is used to create multilayer castings having different compositions at different locations. Gas bubbles, being lighter than the metal, segregate near the axis of rotation and float out, thereby yielding a sound casting. Risers are not needed as solidification is completed while the centrifugal pressure is acting on the metal, which aids the feeding of the shrinkage. Centrifugal casting is highly material efficient, with casting yields close to 100%. The most defective region of the casting is likely to be the inner circumference, where some gas bubbles and lighter inclusions might remain segregated. After machining out this region, however, premium-quality castings are obtained. Centrifugal casting is also used to make cast composites with selectively lubricated surfaces enriched with a solid-lubricant such as graphite, and to effect infiltration of fiber bundles to make fiber composites (Chapter 6).

Continuous Casting

In continuous casting, a water-cooled copper or aluminum mold, open at both ends, is continuously filled with metal at one end, and the solidifying metal is continually withdrawn from the other end. To initiate solidification, a dummy seed of the same cross-section as the casting is inserted in the mold near its exit. A tundish or reservoir with a bottom opening is positioned between the pouring ladle and the mold, and is used to regulate the flow into the mold. In continuous casting of thin sheets, two water-cooled rolls of steel are rotated and the molten metal is directly fed into the roll bite between water-cooled rolls. The metal freezes, undergoes some rolling reduction, and emerges as solid strip or sheet with good surface finish. The process, called the strip-casting technique, eliminates the two-step fabrication of thin sheets by first solidifying a thick sheet and then reducing its cross-section by hot rolling. Figure 2-27 shows schematic diagrams of the conventional continuous casting process and the improved Ohno continuous casting (OCC) process in which the metal is cooled in a way that orients the grains parallel to the die walls to reduce the interfacial friction at the die–metal interface. This is in contrast to the conventional process in which the grains are perpendicular to the die wall, which increases the friction at the die surface. Figure 2-28 shows the microstructure of an OCC cast Al-Cu alloy with a uniform and oriented grain structure.

Single-Crystal Casting and Directionally Solidified Structure

A polycrystalline material consisting of numerous grains is a highly disordered system in which atomic arrangement changes from one grain to another. The transition region between neighboring grains, or grain boundary, is a crystal defect that affects many physical and mechanical properties. Single crystals have ordered atomic arrangement and can be grown in sizes from very small to very large. For example, turbine blades of a modern aircraft engine are cast as single grain to achieve very high creep resistance and service life.

In the Czochralski technique (Figure 2-29) of growing single crystals, a single-crystal seed is brought in contact with a melt contained in a heated crucible. As the tip of the seed crystal melts, the temperature of the crucible is gradually decreased while the seed crystal is slowly withdrawn from the melt. This is frequently accompanied by rotation of the seed crystal to obtain a circular cross-section in the grown crystal. The crucible may be simultaneously rotated to control the

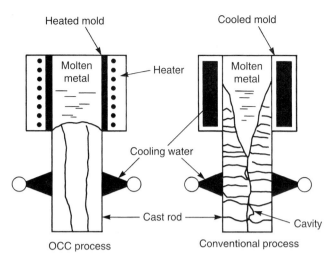

Heated mold   Cooled mold

Molten metal — Heater   Molten metal

Cooling water

OCC process   Conventional process

Cast rod   Cavity

FIGURE 2-27 *Two basic versions of the continuous casting process. In Ohno continuous casting (OCC) the molten metal is quenched by water at the exit of a heated die, resulting in large columnar grains that are parallel to the direction of strip movement. In the conventional continuous casting process, cooling commences at the walls of the water-cooled die, resulting in grains oriented normal to the die walls. This increases the interfacial friction during strip withdrawal.* (H. Soda et al., Advanced Materials & Processes, April 1995, p. 43). Reprinted with permission from ASM International, Materials Park, OH (www.asminternational.org).

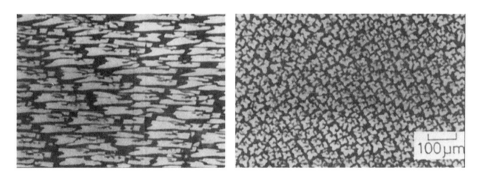

FIGURE 2-28 *Microstructure of an Al-Cu alloy rod produced by the OCC process showing grains oriented along the rod axis, and the uniform microstructure across the rod cross-section.* (H. Soda et al., Advanced Materials & Processes, April 1995, p. 43). Reprinted with permission from ASM International, Materials Park, OH (www.asminternational.org).

convection and achieve a uniform composition and homogeneous solidification. The method is widely used to grow highly pure and ordered semiconducting crystals of Si, Ge, and other materials. In the Bridgman crystal growth technique (Figure 2-30), an ampoule containing a seed crystal at one end and the material to be processed at the other end is placed in an electrically heated furnace. The temperature in the top zone of the furnace is maintained above the melting point and temperature in the lower zone is maintained below the melting point. The ampoule is initially positioned in such a way that only the small lower end of the single-crystal seed remains unmelted in the lower zone. After thermal stabilization has been achieved through an isothermal

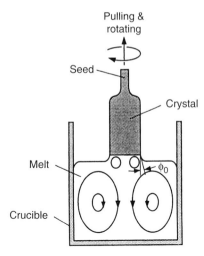

FIGURE 2-29 *Czochralski crystal growth technique that involves rotation and upward movement of a seed crystal in contact with the melt. (S. Kou, Transport Phenomena in Materials Processing, John Wiley and Sons, New York, 1996).*

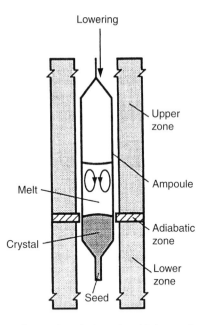

FIGURE 2-30 *Bridgman crystal growth technique in which a molten alloy contained in a glass ampoule is placed in the hot zone of a furnace, and gradually lowered into a cold zone to initiate crystal growth in contact with a seed crystal. (S. Kou, Transport Phenomena in Materials Processing, John Wiley and Sons, New York, 1996).*

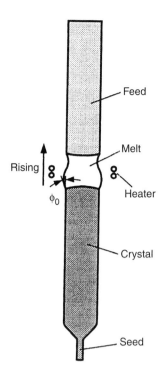

FIGURE 2-31 *The floating-zone technique of crystal growth in which a small molten zone is created at the contact region between a feed stock and a seed crystal via focused heating (induction, electron beam, or laser). The molten zone is held by surface tension forces. Very slow withdrawal of the feedstock out of the hot zone causes crystal growth.* (S. Kou, Transport Phenomena in Materials Processing, John Wiley and Sons, New York, 1996).

hold, the ampoule is slowly withdrawn into the lower (cooler) zone to allow oriented columnar crystals to grow. Some mechanism to filter out unwanted grains is needed so that a single grain emerges in the part. In one industrial version of the Bridgman process to grow single-crystal turbine blades of aircraft engines (discussed later in this section), a "pigtail" passage at the base of a ceramic investment mold is used as a filter for grain growth.

In yet another method to grow single crystals, called the floating-zone directional solidification technique (Figure 2-31), a small melt zone, created using induction power (or lasers and electron beams) and suitably designed current concentrators, is slowly traversed along the length of the feedstock specimen juxtaposed to a single crystal seed. The joint region is first melted, and the molten zone is slowly moved over the feed material to grow single crystals. The method can also be used to create large columnar grains oriented along the growth direction in the solidified region. The use of containers (crucibles) is avoided, and the molten zone is supported by surface tension forces between the solid and the liquid regions of the feed material. The natural convection in the zone (and forced convection from the electromagnetic field when induction heating is used), as well as the gravity, tend to destabilize the molten zone (Figure 2-32). A very careful control of the thermal gradient and growth (traverse) speed is needed to prevent destabilization of the molten zone. Mention must also be made of an important industrial technique (briefly discussed in a later section) for crystal purification, called zone refining, that is essentially a

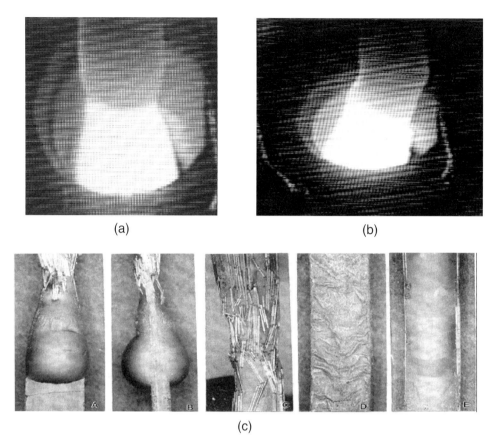

(a)

(b)

(c)

FIGURE 2-32 *(a) View of the molten zone produced in the floating-zone technique with the help of an induction heater and current concentrator. (b) An unstable molten zone in the floating-zone technique. Vigorous electromagnetic convection or a large size of the zone will cause instability. (c) Photographs showing unstable and stable zones produced in a fiber-reinforced composite specimen during floating-zone directional solidification. Vigorous convection led to fiber breakage and a large partial pressure of oxygen led to surface oxidation. The last photograph to the right is an optimally processed composite.* (S. N. Tewari, R. Asthana, R. Tiwari and R. Bowman, NASA Hightemp Review, 1993, 60-1-60-11, Advanced High-Temperature Engine Materials Technology Program, Vol. 2: Compressor/Turbine Materials-MMC[s], and CMC[s]. NASA Glenn Research Center, Cleveland, OH, 1993).

reversal of floating-zone solidification technique, and is widely used to produce ultra-high-purity (10^{-8} atom%) materials for the electronic industry.

The floating-zone technique yields a very pure material with low (<10 ppm) levels of common impurities such as N, O, C, and S. Computer-automated containerless levitation-zone melting techniques permit the growth rate, floating-zone diameter, interface position, and interface temperature to be continuously monitored during solidification and correlated with the structural characteristics of the grown material. The freezing interface can be purposely rotated (as in Czochralski method) to achieve good mixing in the floating zone and to increase the temperature gradient at the solid–liquid interface for increased stability of plane front solidification and coupled growth of the eutectic phases in alloys. The directionally solidified (DS) eutectic alloys lead to an aligned dual-phase microstructure; however, the structure is often noted to be interrupted at random intervals leading to discontinuities (banding) in the eutectic.

Banding is generally absent in composites grown using the Bridgman technique; however, the latter technique is less attractive than the floating-zone technique for high-temperature reactive alloys in which minutest contamination from the crucible material is not acceptable.

Structural and compositional changes accompany all solidification processes, and are discussed in later sections of this chapter. Segregation of solutes during crystal growth perturbs the chemical homogeneity of the grown crystal. High-purity electronic materials such as Ge and Si are often doped with a controlled amount of either a more electropositive or electronegative element to create doped semiconductors with controlled electrical conductivity (e.g., Si doped with P produces a p-type semiconductor). Because during solidification, the compositions at the solid–liquid interface (i.e., at the surface of the growing crystal) adjust according to the liquidus and solidus lines on the phase diagram, the compositions of the solid and liquid keep changing and an inhomogeneous composition might exist in the grown crystal. In addition, natural convection due to temperature gradients and solutal convection (from density changes in the liquid caused by solute segregation) can cause unwanted inhomogeneity in the composition at a microscopic scale. Such compositional inhomogeneity can be avoided during growth of very pure crystals using special techniques. For example, fluid convection is suppressed during crystal growth by applying a magnetic field across the solid–liquid interface; this results in a compositionally more homogeneous crystal.

The directional solidification (DS) of eutectic alloys is widely used to grow dual-phase composite microstructures composed of an in situ–grown unidirectionally aligned fiber (usually of a refractory metal) that acts as a reinforcement in a matrix phase. The reinforcement is either lamellar or fibrous in shape depending on the growth rate and temperature gradient during directional solidification. Figure 2-33 shows the microstructures of some low- and high-temperature directionally solidified alloys grown using different crystal growth techniques. These include a zone-directionally solidified bi-crystal of the ordered intermetallic compound, βNiAl (Fig. 2-33a), a Czochralski-grown specimen of NiAl(Cr) alloy (Fig. 2-33b), a Bridgman-grown Ni-base superalloy (PWA-1480) used in gas turbine engines (Fig. 2-33c-d), a Bridgman-grown Pb-Au eutectic alloy (Fig. 2-33e), and a floating-zone directionally solidified NiAl(Cr) alloy showing the dual-phase microstructures (Fig. 2-33f-g). The effect of growth

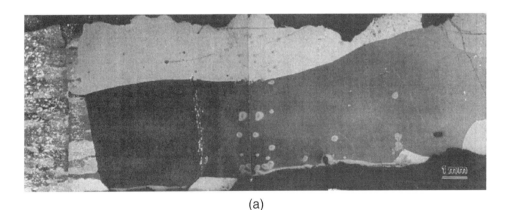

(a)

FIGURE 2-33 *(a) A bi-crystal of the ordered intermetallic β-NiAl produced by the floating-zone technique.* (R. Asthana, S. N. Tewari, and R. Bowman, unpublished work, 1992, NASA Glenn Research Center, Cleveland, OH).

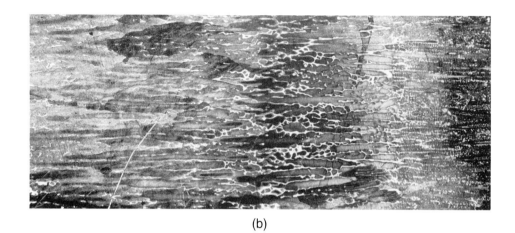

(b)

SOLID ··· MUSHY ZONE QUENCHED LIQUID 500 µm

(c)

500 µm

(d)

FIGURE 2-33 continued   *(b) Longitudinal microstructure of a NiAl(Cr) bar directionally solidified using the Czochralski (crystal-pulling) technique.* (S. N. Tewari, R. Asthana and R. Bowman, unpublished work, 1992, NASA Glenn Research Center, Cleveland, OH). *(c) Longitudinal microstructure of a Ni-base superalloy, PWA-1480, directionally solidified using the Bridgman crystal growth technique. PWA-1480 is used in single-crystal blades in gas turbine aeroengines. The nominal composition of the alloy is Ni-12Ta-10.4Cr-5Co-5Al-4W-1.5Ti (in wt%). The primary dendrites exhibit excellent alignment along the growth direction. The solidification initiated near the left end; the quenched liquid to the right end is the last region to solidify.* (M. Vijaykumar, S. N. Tewari, J. E. Lee, and P. A. Curreri, Materials Science & Engineering, A132, 1991, p. 195). Reprinted with permission from Elsevier. *(d) Transverse microstructure of the directionally solidified PWA-1480 superalloy. A very uniform distribution of the dendrites is noted across the specimen cross-section.* (M. Vijaykumar, S. N. Tewari, J. E. Lee, and P. A. Curreri, Materials Science & Engineering, A132, 1991, p. 195). Reprinted with permission from Elsevier.

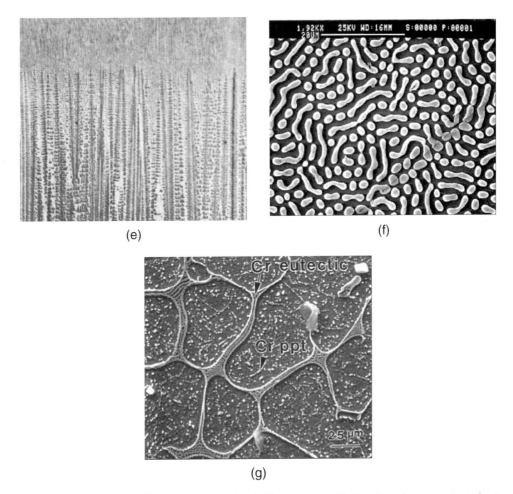

(e)

(f)

(g)

FIGURE 2-33 *(e) Longitudinal microstructure of a Bridgman-grown Pb-Au alloy showing oriented primary dendrites along the growth direction.* (S. N. Tewari, Materials Science & Engineering, A130, 1990, p. 219). Reprinted with permission from Elsevier. *(f) A eutectic colony in NiAl(Cr) directionally solidified using the floating-zone technique. (g) Transverse microstructure of a directionally solidified NiAl(Cr) bar showing eutectic colonies and primary Cr precipitates within the NiAl grains.* (R. Asthana, R. Tiwari and S. N. Tewari, Materials Science & Engineering, A336, 2002, 99–109).

speed on the microstructure of some directionally solidified Ni-base alloys (Ni-Al-Cr-Mo and Ni-Al-Cr) is shown in Figure 2-34. The oriented refractory fibers in these microstructures provide high-temperature creep strength for heat-resistant applications. Other examples of such in-situ fiber-reinforced eutectic composites that have been directionally solidified for high-temperature applications include NbC in Co, TaC in Co, and TiC in Ni. The DS of refractory alloys such as Mo-Mo$_5$Si$_3$ and Nb-Nb$_5$Si$_3$ has also been done to create creep-resistant in situ fibrous composites.

Modern aircraft engine gas turbine blades and vane components are cast by directionally solidifying Ni-base superalloys in investment casting molds using a modified Bridgman crystal growth technique. The DS is carried out in a special furnace with a ceramic (investment) mold

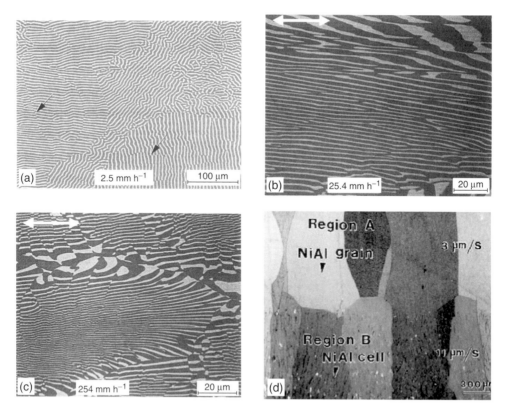

FIGURE 2-34 (a) Transverse microstructure of a Ni-33Al-31Cr-3Mo eutectic alloy directionally solidified using the Bridgman technique at a growth speed of 2.5 mm·h⁻¹. (S. V. Raj, I. E. Locci, J. A. Salem and R. J. Pawlik, Metallurgical & Materials Transactions, 33A, 2002, 597–612). (b) Transverse microstructure of a Ni-33Al-31Cr-3Mo eutectic alloy directionally solidified using the Bridgman technique at a growth speed of 25.4 mm·h⁻¹. (S. V. Raj, I. E. Locci, J. A. Salem and R. J. Pawlik, Metallurgical & Materials Transactions, 33A, 2002, 597–612). (c) Transverse microstructure of a Ni-33Al-31Cr-3Mo eutectic alloy directionally solidified using the Bridgman technique at a growth speed of 254 mm·h⁻¹. Note the considerable refinement that resulted from the higher growth speed. (S. V. Raj, I. E. Locci, J. A. Salem and R. J. Pawlik, Metallurgical & Materials Transactions, 33A, 2002, 597–612). (d) The effect of a change in the growth speed during directional solidification of NiAl by the floating-zone technique. At 11 mm·h⁻¹ (3 μm·s⁻¹), plane front solidification occurred (upper region), and at 40 mm·h⁻¹ (11 μm·s⁻¹) cellular solidification occurred. (R. Asthana, R. Tiwari and S. N. Tewari, Metallurgical & Materials Transactions, 26A, 1995, 2175–2184). Reprinted with permission from ASM International, Materials Park, OH (www.asminternational.org).

shaped like the turbine blade attached to a water-cooled chill via a "crystal selector" (pigtail), as shown in Figure 2-35. The tortuous pigtail opening allows only one grain eventually to emerge and consume the melt in the mold as the latter is gradually withdrawn out of the adiabatic zone of the furnace (a chill is used to initiate directional growth of grains at the entrance to the pigtail).

The gas turbine blade and vane components require high-temperature creep and rupture strengths, tensile strength, ductility, low density, and resistance to hot corrosion, and thermal and mechanical fatigue. Conventional investment cast superalloy engine components having a fine equiaxed grain structure fail prematurely due to rapid cracking along those grain boundaries (g.b.)

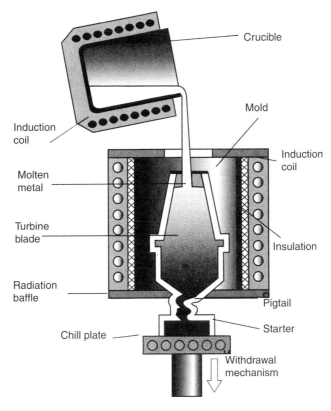

Crucible

Mold

Induction
coil

Induction
coil

Molten
metal

Turbine
blade

Insulation

Radiation
baffle

Pigtail

Chill plate

Starter

Withdrawal
mechanism

FIGURE 2-35 *Schematic of the technique to grow single-crystal turbine blades for gas turbine engines using investment casting and directional solidification. The investment shell is designed with a "pigtail"-shaped opening at the base that acts as a crystal selector. The mold is placed in the hot zone of the furnace and gradually withdrawn to initiate the solidification.* (Adapted from S. J. Mraz, Machine Design, July 24, 1997, p. 39). Reprinted with permission from Penton Media, Inc.

that are oriented perpendicular to the direction of centrifugal stress generated by blade rotation. DS allows growth of large oriented grains with g.b.'s parallel to the radius of rotation, and eliminates the crack-susceptible transversely oriented g.b.'s, which would lower the rupture life and fatigue resistance. Figure 2-36 shows the microstructure of a conventionally cast blade having randomly oriented grains, a directionally solidified blade having columnar grains, and a directionally solidified single-crystal blade.

Fluidity

Fluidity is an important characteristic of molten metals and alloys, and it directly affects the casting soundness and metallurgical quality. Fluidity is assessed by measuring the distance molten metal flows in a standard mold before the metal solidifies, so it involves both rheological and solidification factors. In a widely used test to characterize the fluidity of metals, called the spiral fluidity test, a mold with a spiral flow channel of a standard size and a semicircular cross-section is made out of iron or graphite. The total length (in inches) the metal travels in the spiral mold before solidifying is called its fluidity. Alloy composition, pouring temperature,

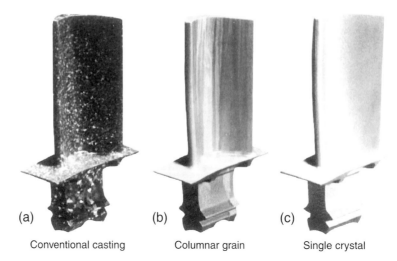

(a)	(b)	(c)
Conventional casting	Columnar grain	Single crystal

FIGURE 2-36 *Photographs showing a conventionally cast turbine blade with a random grain structure, a directionally solidified blade with columnar grain structure, and a single-crystal blade.* (Courtesy of Pratt & Whitney, adapted from W. D. Callister, Jr., Materials Science & Engineering: An Introduction, 5th ed., Wiley, New York, 2000).

solid impurities (e.g., oxides), and density and viscosity all influence the fluidity. Figure 2-37 presents the experimental fluidity data on some pure metals, and Al-Si and Al-Cu alloys, and displays the effect of temperature and alloy composition on spiral fluidity. In another fluidity test, the ability of the metal to fill strips of various thicknesses is characterized. The test evaluates the ability to produce near–net-shape castings, in particular thin-walled and complex parts.

A theoretical model developed by Flemings predicts the fluidity length of a metal in terms of the thermophysical properties of the metal and mold material. Consider a metal at its melting point, T_m, poured in a channel of radius, a, and flowing with an average velocity, V. The metal solidifies after a distance, L_f, by losing latent heat to the mold. The rate of heat dissipation by solidifying metal equals the rate at which heat is transferred across the mold–metal interface. The thermal resistance at the interface is specified in terms of an interface heat transfer coefficient, h, where the SI units of h are $J \cdot m^{-2} \cdot K^{-1} \cdot s^{-1}$. A large h indicates good thermal conductance or low thermal resistance at the interface.

Heat lost per unit time when a length, L_f, solidifies in time t is

$$\frac{\pi a^2 \rho_s L_f \cdot \Delta H}{t} = \pi a^2 \rho_s V \Delta H.$$

Heat transferred across the mold–metal interface in time t is

$$2\pi a L_f h(T_m - T_0).$$

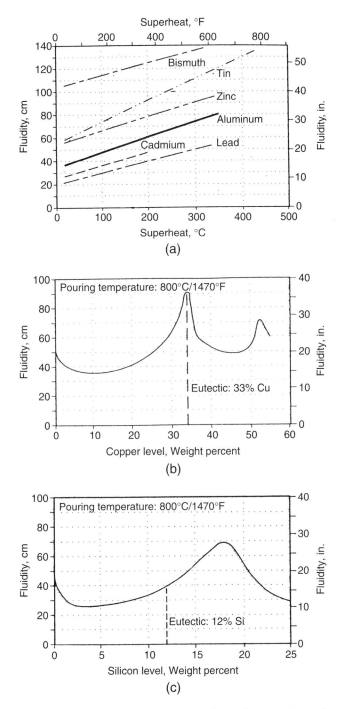

FIGURE 2-37 *(a) Fluidity (in inches) of some pure metals as a function of superheat temperature.*
(F. R. Mollard, M. C. Flemings, and E. F. Niyama, JOM, November 1987, p. 33) *(b) Fluidity of Al-Cu alloys as a function of the amount of Cu in the alloy.* (F. R. Mollard, M. C. Flemings, and E. F. Niyama, JOM, November 1987, p. 33) *(c) Fluidity of Al-Si alloys as a function of the amount of Si in the alloy.* (F.R. Mollard, M.C. Flemings and E.F. Niyama, JOM, November 1987, p. 33). Reprinted with permission from The Minerals, Metals and Materials Society, Warrendale, PA (www.tms.org).

On equating the preceding expressions, we obtain an expression for the fluidity length as

$$L_f = \frac{\rho_s V \Delta H a}{2h(T_m - T_0)}. \tag{2-2}$$

If the metal is poured in a superheated state (i.e., $T > T_m$), then besides the latent heat of solidification, the superheat also must be dissipated. This is accounted for by modifying the latent heat term as ($\Delta H + c \cdot \Delta T$), where $\Delta T = T_p - T_m$, and T_p is the pouring temperature, and c is the specific heat of the metal. This equation is based on the premise that flow of liquid metal stops when an element of the melt solidifies due to heat extraction in the channel. It further assumes that all resistance to heat flow is at the mold–metal interface, that there is no significant effect of surface tension on flow velocity, that flow channel is filled with liquid metal (fully developed flow), and that there is no decrease in velocity from friction effect. Despite these simplifications, the experimental measurements of fluidity of various metals are in good qualitative agreement with the preceding fluidity equation.

Alloying elements introduced in a metal to refine the grain structure or modify the eutectic morphology may also influence the fluidity. The modification of eutectic silicon in hypoeutectic and eutectic Al-Si alloys, and refinement of primary Si in hypereutectic Al-Si alloys by adding Na or Sr is known to alter the alloy fluidity. Under controlled atmospheric conditions (e.g., in the absence of metal oxidation), the fluidity of Al-Si alloys decreases because of modification of silicon; the extent of the fluidity reduction is, however, no more than about 10%. The decrease in alloy fluidity also correlates to the increase in viscosity through an inverse relationship between the two although viscosity measurements are done isothermally, whereas fluidity measurements involve continuous cooling of the liquid metal.

Melt Treatments

The interactions between molten metals and atmospheric gases cause defects in castings such as dross, slag, and porosity. Melt-cleaning processes aim to eliminate such defects in castings through special pouring and gating techniques, and prior treatment of the melt. For example, dross entrapment in castings is minimized by employing bottom pouring through ladles that keep the light dross floating on top of clean underlying metal. Similarly, use of filters (e.g., mica) embedded in the gating system can keep unwanted inclusions from entering the casting (the filters remain embedded in the gating system of the solidified metal after cooling, and are removed during remelting of foundry scrap).

Porosity in castings is caused by gases dissolved in the liquid metal (e.g., atomic hydrogen in Al). A reactive gas such as oxygen has very low solubility in Al and will not dissolve in the metal to any appreciable levels. In contrast, the solubility of atomic hydrogen in aluminum increases rapidly with increasing temperatures, especially in the molten state (Figure 2-38). Upon cooling, gas solubility in the metal decreases, and the rejected gas nucleates to form bubbles that may be trapped in the solidifying metal due to a progressive increase in the viscosity. Any impurity particles that are not wetted by the metal will act as preferential sites for gas nucleation and attachment. The adhesion of gas bubbles to inclusions decreases the float-up rate of bubbles in the melt as a result of the increased effective density of the cluster, and the increased drag force on the combined surface area of the cluster. The bubbles anchored to inclusions fail to float up, and are pushed into the last freezing boundaries by the nucleating crystals and form interfacial porosity. Many different approaches are used to minimize gas dissolution and porosity formation in castings. For example, metals can be melted under a protective atmosphere (a cover flux, cover

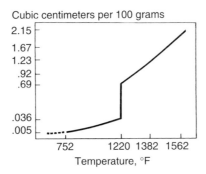

FIGURE 2-38 *The solubility of hydrogen in Al (in cubic centimeters per 100 g of metal) as a function of temperature* (D. L. Zalensas, ed., Aluminum Casting Technology, 2nd ed., 1997). Reprinted with permission from American Foundry Society, Schaumburg, IL (www.afsinc.org).

of an inert gas, or under vacuum). Careful pouring without turbulence and use of low superheat temperatures also reduce the gas content in metals. Dissolved gases can be effectively removed via vacuum degassing, which involves subjecting the gas-saturated melt to a vacuum with a low partial pressure of dissolved gas; the dissolved gas bubbles out as the metal establishes an equilibrium with the atmosphere. Alternatively, passing an inert gas (e.g., argon) through the melt or adding a degassing compound (e.g., hexachloroethane) to the melt will remove the dissolved gas. The compound hexachloroethane decomposes in the melt to form Cl_2 bubbles that flush the dissolved gas out of the metal; however, degassing with Ar is preferred because Cl_2 is toxic. Degassing must be done with adequate ventilation. Small quantities of metals can be degassed using a batch-type approach in which argon is introduced in the melt via a graphite lance (or a porous refractory plug or tile) almost touching the bottom of the holding crucible (to allow the argon to scan all of the liquid metal for most efficient degassing). In contrast, large industrial foundries use a continuous degassing process in which Ar is bubbled in the melt via a battery of gently rotating graphite impellers housed in a long chamber from which degassed metal is continuously withdrawn. The impellers have fine orifices to disperse Ar over a large melt volume. An important consideration in degassing is that it should preferably be done in the cooling mode; for example, furnace power supply should be turned off during degassing to prevent further gas absorption by the metal at higher temperatures that will defeat the purpose of degassing. At the conclusion of degassing, the metal is rapidly heated to its pouring temperature to prevent excessive gas absorption.

Another degassing method introduces an additive in the melt that causes the dissolved gas to form a low-density compound that will float to the top and can be skimmed off. Examples include addition of phosphorus to Cu to remove oxygen in the production of OFHC, or oxygen-free, high-conductivity Cu, and addition of Al to molten steel to form aluminum oxide inclusions (killed steel).

Rapid cooling of molten Al alloys results in more hydrogen porosity in aluminum castings because gas bubbles are readily trapped. Atmospheric conditions and alloying additions also affect gas content in the melt. For example, it is a common observation that more hydrogen porosity occurs in Al castings made during summer than winter; this is because cold air holds less moisture (source of atomic hydrogen) than warm air. Similarly, alloying elements such as Mg, Ti, Ni, and Li increase the solubility of atomic hydrogen in Al, whereas Si, Cu, Zn, and Mn decrease the gas solubility. Only the atomic form of hydrogen dissolves in Al, with the major source of H being water vapor in the atmosphere. The dissolution reaction is: $3H_2O + 2Al \rightarrow 6H + Al_2O_3$.

Many types of impurities can be removed and liquid metal protected by judicious use of foundry fluxes. Fluxes perform different functions, which include melt protection, melt cleaning, dross and inclusion removal, and degassing. Impurities may come from weathered or corroded charge, foundry returns (gates and risers), scrap metal (turning chips, etc.) and chemical reactions with the tooling. Common fluxes for Al include halogen compounds such as $AlCl_3$, AlF_3, CaF_2, etc. Cleaning fluxes are stirred in the melt using a special stirrer. After a few minutes' hold during which fluxes decompose and cause cleaning action, the impurities float to the top and are skimmed off. Cover fluxes for Al are mixtures of KCl, NaCl, cryolite (Na_3AlF_6), or NaF (in place of Na_3AlF_6). Usually, half the flux required for metal (based on the weight of metal) is added to cold (solid) metal in the crucible prior to the onset of heating. As the metal melts and forms a pool ("heel") at the base, the flux covers it and provides protection. The remaining quantity of flux is added after melting is complete. If all the flux were added after melting, then considerable oxidation would take place during heating to the pouring temperature, causing even greater melt contamination. One major drawback of fluxing is that certain fluxes attack the crucible, thus lowering its life; therefore, only the minimum quantity of flux needed for cleaning and protection of the metal must be used.

Metallic Foams and Gasars

Porosity in castings is usually considered a strength-limiting defect from the viewpoint of structural applications that demand maximum load-bearing capacity in a component. However, metals containing very large amounts of porosity often have unique combinations of properties that make them attractive for specific applications, including light-weight, low-stress structural applications where the specific strength (strength-to-weight ratio) is important. For example, metallic foams containing large amounts of porosity are light-weight metals with superior thermal insulation and vibration-damping properties, and high specific strength (i.e., strength-to-density ratio). Metallic foams are classified as monolithic foams and composite foams (i.e., particulate- and gas- "reinforced" metals). Particulate-reinforced composite foams are manufactured by stirring fine ($\sim$20 μm) ceramic particles in a gas-saturated molten metal, followed by solidification. The foam structure is stabilized by the presence of fine ceramic particles that anchor and stabilize gas bubbles in the structure. Another type of metallic foam, called syntactic metal foam, contains hollow particles in metals (e.g., fly ash microspheres in metals). The composite and syntactic foams can bridge properties between monolithic foams and solid materials.

Many different techniques have been developed to synthesize foamed metals and composites (see "Investment Casting" section). Some of the techniques to prepare gas-metal composites include (1) powder metallurgy (e.g., loose powder sintering), slip or slurry foaming, and slip pouring (sintering of a slurry), (2) chemical and electrochemical deposition, (3) physical vapor deposition, and (4) liquid-phase fabrication such as investment casting, lost-foam casting, infiltration, mixing, foaming, and the Gasar process. Of particular emerging interest is the Gasar process, a relatively novel way to produce ordered gas porosity in castings. The Gasar process generates cast parts with various pore morphologies, pore orientations, pore sizes (10–1000 μm), and porosity content (5–70%). Gasar materials have a monolithic matrix and pores with smooth walls. This structural feature imparts high durability, energy absorption, and plasticity to the material.

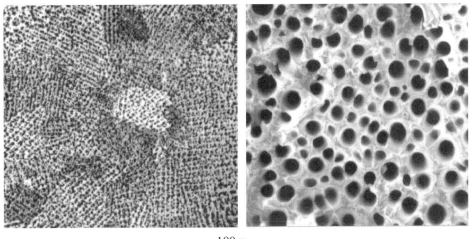

100 μm

FIGURE 2-39 *Photomicrographs showing the Gasar microstructure and Gasar castings.* (J. Sobczak, N. Sobczak, L. Boyko and R. Asthana, in Recent Research Developments in Materials Science, Vol. 3, 2002, 743–773).

In the Gasar process, a hydrogen-charged melt is directionally solidified, allowing the excess hydrogen to evolve in bubbles that may be frozen in the ingot interior via a eutectic solidification reaction, $L \rightarrow \alpha$ (gas) $+ \beta$ (solid). The resultant two-phase structure is similar to a eutectic structure, except that one of the phases is a gas. The dimensions, number, shape, and orientation of the pores may be controlled. The pore orientation inside the ingots depends on the direction of heat dissipation during solidification; for example, axially aligned or radially distributed pores of controlled size may be created. The kinetics of Gasar growth depends on gas pressure in furnace atmosphere, the partial pressure of hydrogen above the melt, and the melt temperature. Like other eutectic transformations, the gas eutectic reaction may result in the formation of either ordered or disordered structures depending on the thermodynamic conditions at the solidification front. Figure 2-39 shows Gasar casting and typical microstructure of the Gasar materials casting.

During directional solidification of a gas-saturated melt to form a Gasar material, either of the two phases (i.e., gas or solid) may lead the growth process in a manner similar to conventional eutectics. When the solid leads, dendrite centerlines penetrate deep into the melt while bubbles form in interdendritic spaces. Because cooling is controlled and directional, the micropits between dendrite branches make up an ordered system of depressions that can turn into bubble nucleation sites. The bubbles are mutually isolated and do not contact each other directly, so the terminal structure is determined by the shape of the dendrite skeleton. Bubbles may coalesce during growth. In metals with relatively low surface energies such as Cu and Mg, the bubbles are less prone to merger and coalescence than they are in molten Ni, Fe, or Co, which have a higher surface energy. For this reason, it is easier to engineer cylindrical pores in Gasars of the former group.

The industrial applications of Gasars are likely to lead to weight reduction, fuel-efficiency, and improved energy absorption. Potential Gasar applications include: recyclable filters for primary purification of oils, fuels, water, and other liquids, frictional components such as disks, shoes, drums, lining, and pads if the pores are filled with lubricants, sliding bearings, transpiration cooling elements of engine chambers, "sandwich" and other porous constructions, catalyst carriers for different chemical reactions, and energy damping elements.

Melting Furnaces

Furnaces used for the purpose of melting and shape casting in foundries are different from those that are used for primary fabrication of crude metal from mineral ores, such as a blast furnace. The common types of foundry furnaces include: reverberatory (or open-hearth) furnace, cupola, induction furnace, and electric arc furnace. Reverberatory furnaces are mostly used for nonferrous alloys and utilize oil or gas burners for energy source. The metal charge is brought in direct contact with blowing hot flue gases. This results in a direct and very efficient heat transfer, but cover fluxes are needed to minimize contamination from fuel gases. The furnace capacity is from a few thousand pounds to a few hundred thousand pounds. Cupolas are used for melting cast iron (both gray and nodular irons). Coke is used as the fuel and limestone as the primary flux. These materials and crude iron are arranged in layers in a vertical stack. On igniting the coke, a hot zone is created that causes the metal to melt and trickle down through the layers into a spout. The rather dusty process of layering involved in operating cupolas has limited their use in recent times in favor of electric melting. Induction furnaces are high-capacity (80–90 tons) electric furnaces that are fast and efficient, and used for melting both ferrous and nonferrous alloys. They cause less pollution than conventional gas or oil-fired furnaces. However, the vigorous convection from the presence of an electromagnetic field constantly exposes fresh metal to atmosphere, thus increasing the undesirable oxidation and gas absorption. For reactive (atmosphere-sensitive) metals, a vacuum cover is used in induction furnaces, and the melting and pouring operations are performed under vacuum. Arc furnaces are used mainly for melting iron and steel, and range in capacity from a few tons to nearly 200 tons. Up to 50 tons of metal can be melted in an hour. Cover fluxes are used to minimize atmospheric contamination. In the direct arc furnace, a powerful arc is struck between electrodes and the metal to be melted, and in the indirect arc furnace, the metal is placed in the vicinity of two electrodes between which an arc is established. An important aspect of all types of melting furnaces is the consideration related to refractory or furnace lining. This is discussed in Chapter 3, on ceramics.

Mold-Filling Time

In our discussion of green sand casting at the beginning of this chapter, it was stated that uncontrolled flow of metal leads to gas porosity, dross and slag, sand inclusion, misrun, cold shut,

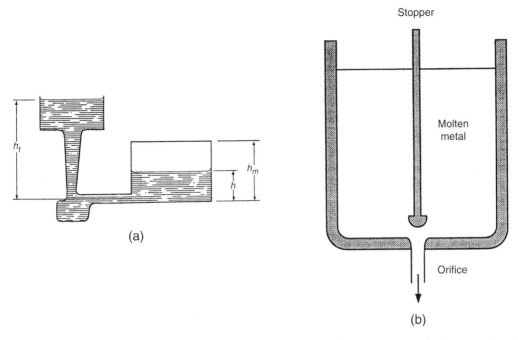

FIGURE 2-40 *(a) Schematic of a bottom-gated mold for filling time calculations. (b) Ladle discharge through an orifice at base of the ladle.* (R. A. Flinn, Fundamentals of Metal Casting, 1963, Addison Wesley Publishing Company, Reading, MA). Reprinted with permission from Addison-Wesley Publ. Co.

and many other types of defects. These defects can be eliminated by constructing a suitable gating system that is designed to evenly distribute the incoming metal to all parts of the mold without turbulence and sand erosion. Thus, in order to avoid casting defects, it is necessary to control the rate of mold filling. In this section, we make simplified calculations of mold filling time for two basic gating configurations: a top-gated mold and a bottom-gated mold. In a top-gated mold, the metal enters the mold cavity like a freely falling stream. By ignoring the fluid friction due to viscosity and drag due to changes in the direction of flow, mold-filling times can be readily calculated. The energy conservation for a liquid particle of mass, m, falling freely under gravity from an initial height, h, to a plane at $h = 0$, yields, $mgh = (1/2)mv^2$, where v is the particle velocity at $h = 0$ (location of the gate). The mold filling time (pouring time) will then be equal to the cavity volume divided by the product of gate cross-sectional area and velocity, v.

In a bottom-gated mold (Figure 2-40a), the incoming metal begins to experience a back pressure as soon as some metal has entered the mold, and the metal velocity progressively decreases. The velocity of the metal at the gate $= V_3 = \sqrt{2g(h_t - h)}$, where h is the instantaneous metal level in the mold, and h_t is total height of the metal in the sprue. This is the velocity of a jet discharging against a static head h, making the effective head as $(h_t - h)$. Now, for the instant shown in Figure 2-40a—that is, when instantaneous height is h—let the metal level in mold move up through an infinitesimal height, dh, in a time interval dt. If A_m and A_g denote the cross-sectional areas of the mold and gate, respectively, then mass balance considerations yield

$A_m dh = A_g\, V_3\, dt$. On substituting $\sqrt{2g(h_t - h)}$ for V_3, yields

$$A_m dh = \sqrt{2g(h_t - h)} \cdot dt \tag{2-3}$$

$$\frac{A_g}{A_m} \int_0^{t_f} dt = \frac{1}{\sqrt{2g}} \int_0^{h_m} \frac{dh}{\sqrt{h_t - h}} \tag{2-4}$$

where the integration limits stipulate that at time $t = 0$, $h = 0$ and at time $t = t_f$, $h = h_m$, where t_f is the time to completely fill the mold, and h_m is mold height. Integration of the preceding equation yields the mold-filling time as

$$t_f = \frac{A_g}{A_m} \frac{1}{\sqrt{2g}} (2\sqrt{h_t} - \sqrt{h_t - h_m}). \tag{2-5}$$

Because of a constantly increasing back-pressure on the incoming metal in a bottom-gated mold, the metal velocity progressively decreases in proportion to the instantaneous metal head $(h_t - h)$, and the time to fill the mold becomes greater than that for a top-gated mold.

Molds are poured with the help of a ladle, so it is of interest to determine the time to discharge the ladle. Figure 2-40b shows a ladle with an orifice at the bottom for discharging the metal. Let A_l and A_n denote the ladle and nozzle cross-sectional areas, respectively, and let h be the instantaneous metal level at some time t after pouring was initiated through the orifice. The average metal velocity in the nozzle (assuming no friction) is $V_n = \sqrt{2gh}$, and the mass flow rate can be written as

$$m = -\rho_m A_l \frac{dh}{dt} = \rho_m A_n V_n = \rho_m A_n \sqrt{2g\bar{h}} \tag{2-6}$$

where ρ_m is the metal density. Rearranging the preceding equation to separate the variables h and t yields

$$-\frac{dh}{dt} = \frac{A_n}{A_l} \sqrt{2gh} \tag{2-7}$$

Noting that at time $t = 0$, $h = h_i$, and at $t = t_f$, $h = h_f$, where h_i, and h_f are the initial and final heights of metal in the ladle, respectively, and integrating the preceding equation yields the following expression for the time for the metal level in the ladle to decrease from h_i, to h_f,

$$-\int_{h_i}^{h_f} \frac{dh}{\sqrt{h}} = \frac{A_n}{A_l} \sqrt{2g} \int_0^{t_f} dt \tag{2-8}$$

$$t_f = \frac{\sqrt{2}}{\sqrt{g}} \frac{A_l}{A_n} \left(\sqrt{h_i} - \sqrt{h_f} \right). \tag{2-9}$$

More rigorous fluid dynamic calculations can be made to account for non-steady flow with fluid friction effects and calculate the mold filling times for the purpose of mold design. Semi-empirical methods that combine the basic elements of fluid dynamics and the practical experience of foundrymen have also been developed to facilitate the design of gating system of a mold.

Gate and Runner Area Calculation

A simplified practical method to design the gating system has been developed by the American Foundry Society (AFS). The method permits calculation of gate and runner areas for sand molds. The method will be illustrated with the help of a simple example. Consider the gating system shown in Figure 2-41 for a plate-shaped Al casting measuring 20 cm × 20 cm × 4 cm. Assume that the casting will be poured in 4 seconds, using the unpressurized gating system of Figure 2-41 with a stepped runner. A tapered sprue of circular cross-section will be used, with an effective metal head of 15 cm. Let the cross-sectional areas of gates and stepped runners on each side of the sprue be denoted by A_{G1}, A_{G2}, and A_{G3}, and by A_{R1}, A_{R2}, A_{R3}, respectively. The AFS method begins with the calculation of the area at the base of the sprue (choke) using the following mass balance relationship:

$$\text{Choke Area} = \frac{\text{mass of the casting}}{\begin{bmatrix} \text{Metal} \\ \text{Density} \end{bmatrix} \begin{bmatrix} \text{Pouring} \\ \text{time} \end{bmatrix} \begin{bmatrix} \text{Spruce} \\ \text{Efficiency} \\ \text{Factor} \end{bmatrix} \begin{bmatrix} \sqrt{2gh} \end{bmatrix}} \tag{2-10}$$

where the sprue efficiency factor is a geometric correction factor and is taken to be 0.88 and 0.74 for circular and square cross-section of the sprue, respectively. Therefore, the choke area (A_B) becomes

$$A_B = \frac{1600}{4 \times 0.88 \times \sqrt{2} \times 981 \times 15}$$

or $A_B = 2.65 \text{ cm}^2 = \pi\, r_B^2$, where r_B is the radius at the base of the sprue, and $r_B = 0.919$ cm. At this point in the calculation, the concept of "gating ratios" is used to find the runner and gate areas. Gating ratio is defined as (choke area) : (total runner area) : (total gate area), and values of this ratio are specified. A widely used value of gating ratio is 1:4:4. Therefore, total runner area = $4 \times 2.65 = 10.6 \text{ cm}^2$, and total gate area = $4 \times 2.65 = 10.6 \text{ cm}^2$. As the gating system is

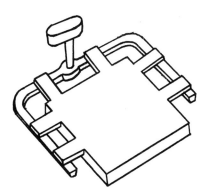

FIGURE 2-41 *A gating system that is symmetrical about the sprue (for the example problem in the text). Each arm has three gates.* (Adapted from, Basic Principles of Gating & Risering, 1985). Reprinted with permission from The Cast Metals Institute, Schaumberg, IL (www.castmetals.com).

symmetrical about the choke (i.e., same number of runners and gates on each side), the runner and gate areas will be same on both sides. Consider the left-side runner system.

Here

$$A_{R1} = A_{R5} = \frac{10.6}{2} = 5.3 \, \text{cm}^2$$

$$A_{G1} = \frac{5.3}{3} = 1.77 \, \text{cm}^2$$

$$A_{R2} = (A_{R1} - A_{G1}) + 5\% \, (A_{R1} - A_{G1})$$

$$= (5.3 - 1.77) + 0.05(5.3 - 1.77) = 3.71 \, \text{cm}^2$$

$$A_{G2} = A_{G1} + 5\% \, A_{G1} = 1.86$$

$$A_{R3} = (A_{R2} - A_{G2}) + 5\% \, (A_{R2} - A_{G2})$$

$$= (3.71 - 1.86) + 0.05(3.71 - 1.86)$$

$$= 1.94 \, \text{cm}^2.$$

Finally,

$$A_{G3} = A_{G6} = A_{G2} + 5\% \, A_{G2}$$

$$= 1.86 + 0.05 \times 1.86$$

$$= 1.95 \, \text{cm}^2.$$

Temperature Drop in Metal Flow

The gating calculations presented up to this point ignored the temperature changes that would occur during metal flow because of heat transfer to the surrounding mold. The temperature drop during flow can be estimated by making the simplifying assumption that dissipation of heat by the flowing metal is limited only by the thermal resistance of the mold–metal interface. This would be reasonable to assume for a mold made out of a high-conductivity material or a water-cooled die with unpolished interface and poor thermal contact. Consider a circular channel of radius, a, and length, L, shown in Figure 2-42a through which a metal of specific heat, C_L, and density, ρ, flows at an average velocity, v. The metal enters the channel at a temperature, T_i, and after traversing a distance, L, its temperature drops to T_f. The initial mold temperature is T_0, and the thermal resistance at the interface is represented by a heat transfer coefficient, h. We write a heat balance for a differential fluid element of thickness, dx, over a time interval, dt. The amount of heat lost by this element in time dt should equal the amount of heat that is transferred across the mold–metal interface in the same time, i.e.,

$$-\pi \, a^2 \, dx \, \rho \, C_L \frac{dT}{dt} = 2 \, \pi \, a \, dx \, h(T - T_0) \tag{2-11}$$

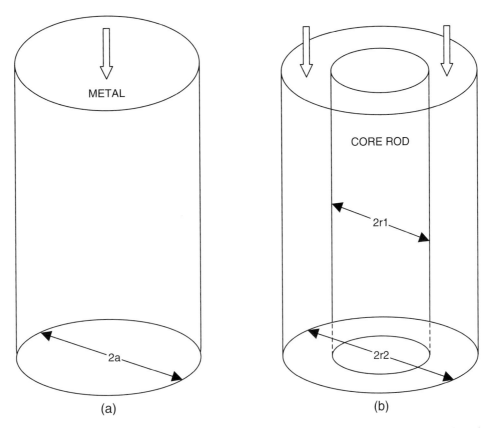

FIGURE 2-42 *Channel geometry for temperature drop calculations: (a) cylindrical channel, and (b) a cored channel with metal flowing through the annular region.*

where T is the instantaneous temperature of the fluid element.

$$\frac{dT}{dt} = \frac{2\,h\,(T - T_{\mathrm{o}})}{a\,\rho\,C_{\mathrm{L}}}.$$ (2-12)

Rearranging the equation and integrating it over the limits, $T = T_{\mathrm{i}}$ at $t = 0$, and $T = T_{\mathrm{f}}$ at $t = t$, we get

$$\int_{T_{\mathrm{i}}}^{T_{\mathrm{f}}} \frac{dT}{T - T_{\mathrm{o}}} = \frac{-2\,h}{a\,\rho\,C_{\mathrm{L}}} \int_{0}^{t} dt$$ (2-13)

$$\ln\,(T - T_{\mathrm{o}})\,\bigg|_{T_{\mathrm{i}}}^{T_{\mathrm{f}}} = \frac{-2\,h\,t}{a\,\rho\,C_{\mathrm{L}}}$$ (2-14)

or

$$\left[\frac{T_f - T_o}{T_i - T_o}\right] = \exp\left[\frac{-2\,h\,t}{a\,\rho\,C_L}\right] \tag{2-15}$$

Therefore,

$$T_f = T_o + (T_i - T_o)\exp\left[\frac{-2\,h\,t}{a\,\rho\,C_L}\right] \tag{2-16}$$

For small values of the argument of the exponential function, an approximate expression for $(T_i - T_f)$ can be obtained by truncating the series expansion of the exponential function,

$$\left[\frac{T_f - T_o}{T_i - T_o}\right] \approx 1 - \frac{2\,h\,t}{a\,\rho\,C_L} + 1\left[\frac{2\,h\,t}{a\,\rho\,C_L}\right]^2 + \frac{1}{2!} + \cdots. \tag{2-17}$$

or

$$\left[\frac{T_f - T_o}{T_i - T_o}\right] \approx 1 - \frac{2\,h\,t}{a\,\rho\,C_L}$$

or

$$1 - \left[\frac{T_f - T_o}{T_i - T_o}\right] \approx 1 - 1 + \left[\frac{2\,h\,t}{a\,\rho\,C_L}\right]$$

$$\frac{T_i - T_f}{T_i - T_o} \approx \frac{2\,h\,t}{a\,\rho\,C_L} \tag{2-18}$$

and

$$(T_i - T_f) \approx \frac{2\,h\,t\,(Ti - T_o)}{a\,\rho\,C_L} \tag{2-19}$$

This equation shows that the flowing metal will experience greater drop in temperature during flow in a channel of small radius, or when the flow velocity is small (i.e., time of flow is large). As an example, consider the temperature drop during flow of Al in a round channel 20 cm long and 5 cm in diameter. Assume that the initial mold temperature is 30 °C, and the metal enters the channel at 800 °C with an average velocity of 4 cm/s. The density and specific heat of Al are 2.4 g/cc and 0.28 cal/g·K, respectively, and the heat transfer coefficient at the mold–metal interface is 0.1 cal/cm^2 s·K. A direct application of the exponential form of temperature drop equation yields

$$T_f = 303 + (1073 - 303)\exp\left[\frac{-2 \times 0.1 \times 5}{2.5 \times 2.4 \times 0.28}\right]$$

$$= 727.6\ °K = 454.6\ °C$$

If, in contrast, the approximate expression for the temperature drop is used, we obtain, $T_i - T_f \approx 458.3$, and therefore, $T_f \approx T_i - 458.3 = 614.7\ °K = 341.7\ °C$. Note the large discrepancy

TABLE 2-3   Errors in Temperature Drop Approximation

A	B	$\dfrac{(A-B)\times 100}{B}$
$\left[\dfrac{2ht}{a\rho C_L}\right]$	$1 - exp\left[\dfrac{2ht}{a\rho C_L}\right]$	% Error
0.001	0.000995	0.05%
0.01	0.00995	0.50
0.05	0.04877	2.50
0.10	0.09516	5.09
0.50	0.39347	27.10
1.00	0.63212	92.07
1.50	0.77687	93.08
2.00	0.86466	131.30

between the exact and approximate solutions in this particular case. The percent error in using the approximate solution in place of the exact solution is tabulated in Table 2-3. For the values of the exponent ($2ht/a\rho C_L$) greater than 0.10, the error becomes significant.

The preceding analysis for temperature drop was oversimplified as only the interface thermal resistance was considered, and the flow velocity and mold temperatures were assumed to be constant. In addition, metal superheat and possible freezing of the metal due to latent heat dissipation were not considered. The solution illustrates the basic physics of liquid cooling during flow but yields only an estimate of the temperature drop. Note that in the preceding example, the final temperature of the metal, T_f, is less than the solidification temperature (660 °C) of Al, indicating that the metal would possibly solidify even before it travels the 20-cm-long channel (ignoring the dissipation of latent heat of solidification). A large value (0.1 cal/cm^2·s·K) of h was used in the calculation, which indicates excellent interfacial conductance and rapid heat transfer across the interface, which does not strictly match the assumption of poor thermal contact made in the analysis. If, however, a lower h value is assumed, say, 0.01 cal/cm^2·s·K, in the above example, then the exact solution yields,

$$\frac{T_i - T_f}{(1073 - 303)} = 1 - exp\left[\frac{-2 \times 0.01 \times 5}{2.5 \times 2.4 \times 0.28}\right]$$

Therefore, $T_i - T_f = 44.49$, and $T_f = 1028.5$ °K or 755.5 °C. From the approximate solution, we obtain, $T_i - T_f \approx 45.83$, or $T_f = 1027.2$ °K = 754.2 °C. The approximate and exact solutions yield similar values for the temperature drop. In both cases, the final temperature is above the solidification temperature, and metal solidification in the channel is not expected even though the latent heat effects are not considered.

Similar calculations of temperature drop during metal flow can be made for simple modifications of the channel geometry. For example, in a channel of semicircular cross-section, the surface area and the volume of the fluid element of thickness dx will be ($\pi a\, dx + 2\, a\, dx$) and $(1/2)\, \pi \cdot a^2 \cdot dx$, respectively. The thermal balance between the element dx and the surrounding mold will then yield

$$-\pi \cdot a^2 \cdot \frac{dx}{2} \cdot \rho \cdot C_L \cdot \frac{dT}{dt} = (\pi a\, dx + 2\, a\, dx)\left[h(T - T_o)\right] \tag{2-20}$$

Rearrangement and integrating over the limits $T = T_i$ at $t = 0$, and $T = T_f$ at $t = t_f$, will yield

$$\int_{T_i}^{T_f} \frac{dT}{T - T_o} = \frac{2(\pi + 2)h}{\pi \, a \, \rho \, C_L} \cdot \int_0^{t_f} dt \qquad (2\text{-}21)$$

$$\text{Ln}(T_f - T_o) - \text{Ln}(T_i - T_o) = \frac{-2(\pi + 2)h \, t_f}{\pi \, a \, \rho \, C_L} \qquad (2\text{-}22)$$

$$\frac{(T_f - T_o)}{(T_i - T_o)} = \exp\left[\frac{-2(\pi + 2)h \, t_f}{\pi \, a \, \rho \, C_L}\right] \qquad (2\text{-}23)$$

And an approximate expression for small argument of the exponential function will yield

$$\frac{(T_i - T_f)}{(T_i - T_o)} \approx \frac{2(\pi + 2)h \, t_f}{\pi \, a \, \rho \, C_L} \qquad (2\text{-}24)$$

Consider the approximate expressions derived above for $(T_i - T_f)$ in a circular and a semicircular channel. Assuming that metal velocity, interface thermal resistance, channel length, and mold temperature are identical in the two cases, the ratio of the expressions for temperature drop, $(T_i - T_f)$, leads to

$$\frac{\Delta T_{\text{semi}}}{(T_i - T_o)} \times \frac{(T_i - T_o)}{\Delta T_{\text{round}}} \approx \frac{2(\pi + 2)h \, t_f}{\pi \, a \, \rho \, C_L} \times \frac{a \, \rho \, C_L}{2 \, h \, t_f}$$

$$\frac{\Delta T_{\text{semi}}}{\Delta T_{\text{round}}} \approx \frac{(\pi + 2)}{\pi}$$

This indicates that temperature drop will be greater in a semicircular channel than in a round channel, an obvious consequence of the larger surface area-to-volume ratio in the former.

Finally, consider the temperature drop in the cored cylindrical channel shown in Figure 2-42b. Assume that the mold and the core rod are made of the same material. Note that because the metal flows through the annular region between the core and the surrounding mold, it will lose heat to both the mold and the core. Writing a thermal balance across a fluid element dx yields

$$-\pi \left(r_2^2 - r_1^2\right) dx \cdot \rho \, C_L \frac{dT}{dt} = (2 \, \pi \, r_2 \, dx + 2 \, \pi \, r_1 \, dx) \, h \, (T - T_o) \qquad (2\text{-}25)$$

where r_2 and r_1 are the outer and inner radii of the annular region, respectively. Rearranging and integrating over the limits $T = T_i$ at $t = 0$ and $T = T_f$ at $t = t_f$, yields

$$-\left(r_2^2 - r_1^2\right) \cdot \rho \, C_L \int_{T_i}^{T_f} \frac{dT}{T - T_o} = 2 \, (r_1 + r_2) \, h \int_0^{t_f} dt \qquad (2\text{-}26)$$

which on simplification yields

$$\frac{(T_f - T_o)}{(T_i - To)} = \exp\left[\frac{-2 \, h \, t_f}{(r_2 - r_1)\rho \, C_L}\right] \qquad (2\text{-}27)$$

The corresponding approximate solution based on truncated series expansion of the exponential term is

$$\frac{(T_f - T_o)}{(T_i - To)} \approx \frac{2\,h\,t_f}{(r_2 - r_1)\rho\,C_L}. \tag{2-28}$$

Riser Design

A riser is designed to feed the solidification shrinkage in a part. From the standpoint of heat transfer, an ideal riser will be spherical in shape because it has the smallest surface area for a fixed volume of metal. This will slow down the cooling of the riser, thus allowing a smaller riser to be used for feeding of the shrinkage. Spherical risers may, however, be somewhat difficult to use in sand molds because they present some challenges in retrieving the spherical riser pattern (usually a split-pattern) from a completed sand mold. A more practical and widely used type of riser shape is a cylinder with a hemispherical base. It should be evident that a riser should take longer to solidify than the casting that it is designed to feed; as a rule of thumb, a riser should take about 25% more time to solidify than the casting, and this observation is used as the basis of a practical method to design the risers discussed below.

Two types of risers are commonly used: a hot (or live) riser and a cold (or dead) riser (Figure 2-43). When a casting is gated through the riser, the latter is called a hot riser because it receives the hottest (last incoming) metal. A cold riser is positioned on a noncritical surface of the part, from where it can be readily cut off after casting with minimum damage to the part surface. In actual practice, hot risers are preferred to cold risers for their obvious advantage in promoting directional solidification and ease of feeding the solidification shrinkage. However, for reasons of economy of mold size, cold risers may be used (with the mold size increasing vertically rather than sideways). It is recommended that when using a dead riser some hot metal be actually back-poured in the riser to ensure that it also has sufficiently hot metal in it.

An open riser cools by losing heat to the surrounding mold and the atmosphere. The rate of heat dissipation from the riser into the mold can be decreased by inserting a cylindrical sleeve of an insulating material in the riser cavity. Similarly, combustible hot-topping compounds can

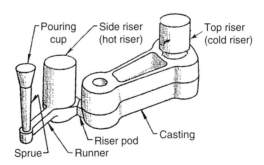

FIGURE 2-43 *The two basic types of risers used in sand casting practice: a hot (live or side riser) and a cold (dead or top riser). When the casting is gated through the riser, the latter is a hot riser. A dead riser is usually placed on a top surface of the casting.* (Adapted from AFS Cast Metals Institute, Basic Principles of Gating & Risering, American Foundry Society, Des Plaines, IL, 1985).

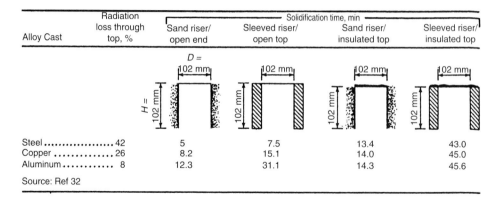

Alloy Cast	Radiation loss through top, %	Solidification time, min			
		Sand riser/ open end	Sleeved riser/ open top	Sand riser/ insulated top	Sleeved riser/ insulated top
Steel 42		5	7.5	13.4	43.0
Copper 26		8.2	15.1	14.0	45.0
Aluminum 8		12.3	31.1	14.3	45.6

Source: Ref 32

FIGURE 2-44 *Use of insulating sleeve and hot topping in a riser. The figure shows the solidification times of steel, Cu, and Al risers (102 mm high × 102 mm dia) in a sand mold. The relative effectiveness of hot topping and riser sleeve depends on the metal being cast* (Metals Handbook, Vol. 15: Casting, 9th ed.). Reprinted with permission from ASM International, Materials Park, OH (www.asminternational.org).

be added to the top surface of the molten metal in the riser cavity to slow down the atmospheric cooling from the surface of the riser. Slowing down the cooling of the riser metal enables a riser to stay molten for a longer time, and, conversely, a smaller riser to feed a given amount of shrinkage, thus increasing the casting yield. However, when using hot topping, the issue of contamination of the riser metal from the carbon, sulfur, and other impurities released from the combustion of the hot topping compound (which comes in direct physical contact with the riser metal) must be carefully assessed. In addition, hot topping must be dry and free of any absorbed atmospheric moisture to prevent steam explosion. Figure 2-44 shows an example of savings achieved in riser solidification time with the use of insulating sleeves and hot topping for 102 mm × 102 mm (4 in tall × 4 in dia) risers of steel, copper, and aluminum. This figure shows that in the case of steel for which radiation and convective heat losses are large, hot toppings are more effective than insulating sleeves, whereas in the case of Al, where thermal conduction through the mold rather than radiation and convection to atmosphere is the dominant cooling mode, insulating sleeves are more effective than hot topping. Using sleeves in conjunction with hot topping is the most effective way to slow down the cooling of the riser for both steel and Al as seen from an increase in the riser solidification time for steel and Al from 5 min and 12.3 min. without these riser aids to 43.0 min. and 45.6 min., respectively, when both these types of riser aids are used.

Many other factors must be considered in riser design. The riser must be tall enough to develop sufficient hydrostatic pressure for metal to flow into the shrinkage cavity in the part. Metals and alloys such as steels are called "skin-forming" because a solidified layer of metal forms on all the surfaces, including the top surface of an open riser, very early during cooling. This solid skin can prevent the atmospheric pressure from acting on the liquid metal in the riser and obstruct metal flow into the shrinkage void developing in the casting. This problem is overcome by using "pressure-risering" in which the solid skin on top of the riser is physically punctured with the help of a core rod to allow atmospheric pressure to aid the flow of metal. Other factors to consider in risering practice include feeding distance considerations and feeding problems in alloys with large mushy zones; these are discussed in a later section.

The design of the riser is based primarily on thermal considerations. For complex industrial castings, it is effective to use computer-based methods to design the riser. Riser design software programs calculate the size, number, and location of risers, and highlight potential hot spots and defect areas. Computer simulation of the casting process permits correction of shop-floor errors before they actually occur. It also allows for more efficient and cost-effective production of parts. Some casting simulation software incorporate both mold-filling behavior and solidification process. These software reveal the flow behavior on the basis of the solution to either the Navier-Stokes equations or Bernoulli's approximation. Fluid flow through various types of filters used in foundry practice can also be simulated and pressure drop or flow rate can be predicted. The thermal and physical property data for various alloys and mold materials needed for casting simulation are usually provided with the software. In the following section, the basic principles of riser design are illustrated through calculations for relatively simple geometries.

Naval Research Lab Method

The Naval Research Lab (NRL) method, also called the shape-factor method, makes use of standard NRL databases, developed primarily for ferrous alloys, to design top risers. Representative NRL graphs used in riser design are depicted in Figure 2-45. The first step in applying the NRL method is to estimate the shape factor (SF) of the casting, which is defined from

$$SF = \frac{\text{length} + \text{width}}{\text{thickness}}$$

Consider a plate-shaped ferrous casting measuring 20 in. $\times$ 20 in. $\times$ 2 in. For this casting, $SF = 20$. From the first NRL graph of Figure 2-45a, the ratio, R_v/C_v, for the SF value of 20 is ~0.25, where R_v is the riser volume and C_v is the casting volume. As $C_v = 20 \times 20 \times 2 = 800$ in^3, $R_v = 800 \times 0.25 = 200$ in^3. Once the riser volume is known, the riser height (H) and diameter (D) can be directly obtained from the second NRL graph (Figure 2-45b). For this casting, the following height and diameter combinations are possible: $H \times D = 7$ in. $\times$ 6 in., 5.25 in. $\times$ 7 in., and 4 in. $\times$ 8 in. Although in principle all these sizes will work, a judicious choice must be made based on factors such as the specific gravity of the alloy. Generally, an

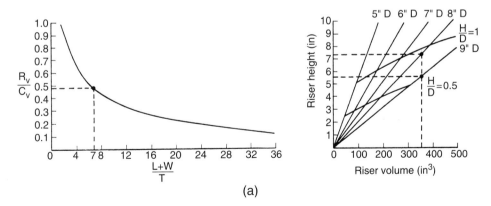

(a)

FIGURE 2-45 *(a) Naval Research Lab (NRL) graphs used to design the risers for ferrous alloys.* (Metals Handbook, Vol. 15: Casting, 9th ed.). Reprinted with permission from ASM International, Materials Park, OH (www.asminternational.org).

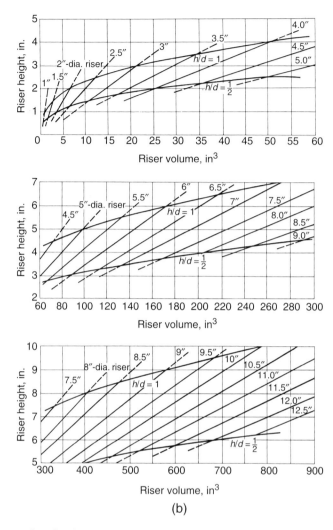

(b)

FIGURE 2-45 continued *(b) An expanded NRL graph to design risers having small volume.* (R. A. Flinn, Fundamentals of Metal Casting, 1963, Addison-Wesley Publishing Co., Reading, MA). Reprinted with permission from Addison-Wesley Publ. Co.

alloy of low specific gravity will need a tall riser for the metal in the riser to develop sufficient pressure to flow and feed the solidification shrinkage.

An important consideration in risering practice is the feeding distance of a riser, which is the distance in the casting over which molten metal can be delivered by the riser. The feeding distance is limited by the minimum section thickness of the part. This is because progressive solidification (i.e., solidification from sides toward the center) chokes the flow of metal, thus limiting the maximum distance the riser metal can reach. Feeding of shrinkage is made easier by judiciously placing the riser in the mold to take advantage of directional solidification toward the riser. For Al alloys, the feeding distance is roughly twice the minimum thickness of the casting. Figure 2-46a shows an example of how repositioning a riser from side to top can

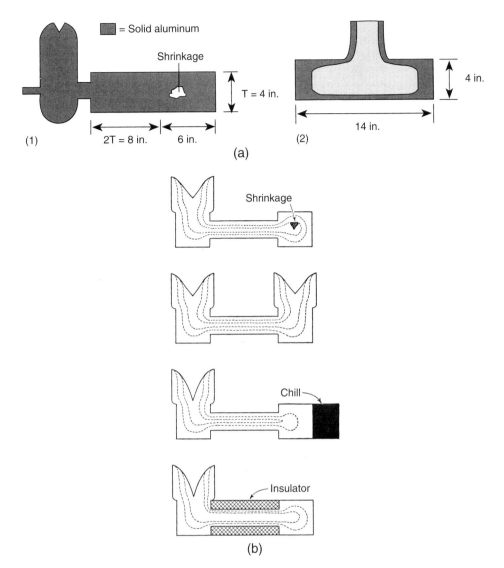

FIGURE 2-46 *(a) Sketch showing (1) incorrect and (2) correct placement of a riser to feed a 14-inch-long aluminum plate casting of 4-inch minimum section thickness. For Al, the feeding distance of the riser is twice the minimum section thickness (i.e., 8 inches in the figure). Repositioning the riser on the top allows the entire casting to be brought within the feeding distance of the riser.* (D. L. Zalensas, ed., Aluminum Casting Technology, 2nd edition, 1997). Reprinted with permission from American Foundry Society, Schaumburg, IL (www.afsinc.org). *(b) Use of chills and insulation to assist a riser feeding the solidification shrinkage. Chills placed in contact with a massive section enhance the cooling rate, and insulation wrapped around the thin section slows the cooling and prevents freeze choking, thus permitting a distant riser to feed the shrinkage. In the example shown, two risers will be needed in the absence of chills and insulation.* (H. F. Taylor, M. C. Flemings, and J. Wulff, Foundry Engineering, Wiley, New York, 1959, p. 135). Reprinted with permission from H. F. Taylor Trust, Boston, MA.

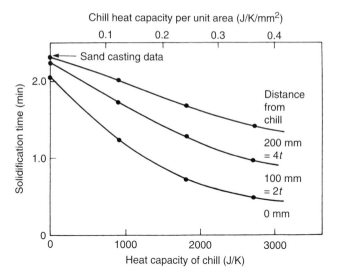

FIGURE 2-47 *The solidification time of a plate 225 mm × 150 mm × 50 mm of Al-5Si-3Cu alloy at various distances from the chilled end decreases as the chill is approached, and as the chill size (heat capacity) is increased.* (Rao and Panchanathan, 1973, adapted from J. Campbell, Castings, Butterworth-Heinemann, Boston, 1999). Reprinted with permission from Elsevier.

bring an entire aluminum casting within the feeding distance of the riser. Generally, shrinkage in a region is fed by metal from a neighboring region, which, in turn, is fed by another region and so forth. If the solidification path is properly designed then the last region to solidify will be fed by the riser. Chills and exothermic sleeves assist risers in feeding the shrinkage; internal and external chills enhance the cooling rate in thick sections, and insulation or exothermic sleeves slow down the cooling in thin sections, thus effectively increasing the feeding distance of the riser. Figure 2-46b shows an example of the use of chills and insulation to feed a casting. Chills made out of high heat capacity materials are more effective in decreasing the solidification time than chills of low heat capacity materials as shown in Figure 2-47. Both external and internal chills are used by foundrymen; external chills are reusable, but internal chills (pins, chaplets, coils etc) remain embedded in the casting and become a permanent part of the casting. Internal chills are, therefore, made out of the same alloy as the casting.

Riser Size Estimation Using Chvorinov's Rule

Chvorinov's rule provides a generic relationship between the solidification time, t, the volume (V), and the surface area (SA). The basic equation is $t = B\left(\frac{V}{SA}\right)^n$ where the exponent n is between 1 and 2, and B is a mold constant, which depends on the thermophysical properties of the metal and mold material, latent heat of the metal, pouring temperature, and mold geometry. In applying Chvorinov's rule to riser design, the foundrymen's rule-of-thumb alluded to earlier is used according to which the solidification time of the riser should be about 25% greater than that of the casting, i.e., $t_R \approx 1.25\ t_c$, where R and c refer to the riser and casting, respectively. Upon combining the Chvorinov's equation with this rule-of-thumb, one obtains the following relationship between the surface area-to-volume ratios of the riser and the casting

$$\left(\frac{V}{SA}\right)_r^2 = 1.25 \left(\frac{V}{SA}\right)_c^2 \tag{2-29}$$

where the subscripts "r" and "c" again denote the riser and casting, respectively (note that the mold constant B will be same for the riser and the casting). This equation provides a method to design risers. The following example illustrates the use of this method. Consider a 2 in. × 4 in. × 6 in. rectangular plate-shaped casting that is connected to a cylindrical side riser ($H/D = 1.5$) through a small neck. Here $V_{cast} = 2 \times 4 \times 6 = 48 \text{ in}^3$, $SA_{cast} = 48 + 24 + 16 = 88 \text{ in}^2$, and

$$V_{Riser} = \pi \left(\frac{D}{2}\right)^2 H = \frac{\pi (D)^2 H}{4} = \frac{\pi (D)^2 (1.5D)}{4} = \frac{1.5}{4} \pi D^3$$

$$SA_{Riser} = 2\pi \left(\frac{D}{2}\right) H = 2\pi \left(\frac{D}{2}\right)^2 = \pi D (1.5D) + \frac{\pi D^2}{2} = 2\pi D^2$$

On substituting these expressions for volumes and areas in

$$\left(\frac{V}{SA}\right)_r^2 = 1.25 \left(\frac{V}{SA}\right)_c^2$$

we obtain after simplification the result that $D = 3.25$ in., and therefore $H = 1.5 \times 3.25 = 4.88$ in.

If, in the previous example, the riser were placed directly on top of one of the large rectangular faces of the plate casting, with the area of the circular base of the riser being common to both the riser and the casting, then the base area must be subtracted from the heat-dissipating surfaces of both the riser and the casting. This is because the circular base of the riser forms an adiabatic surface across which there is not heat transfer. The common area of the riser base is $= 0.25 \pi (D)^2$, and the corrected areas will be $SA'_{cast} = 88 - 0.25 \pi D^2$, and $SA'_{Riser} = 2\pi (0.5D)^2 + \pi DH - 0.25 \pi D^2$. If, as before, we assume that ($H/D = 1.5$), then, $V_{Riser} = 0.25 \pi D^2 H = 0.375 \pi D^3$. The new surface area of the riser is $SA_{Riser} = \pi DH + \pi (0.5D)^2 = 1.75 \pi D^2$. Substituting these expressions for corrected areas of the casting and the riser in the expression

$$\left(\frac{V}{SA}\right)_r^2 = 1.25 \left(\frac{V}{SA}\right)_c^2$$

yields the following cubic equation in D: $D^3 - 112.14 D + 319.27 = 0$. This equation can be solved numerically (for example, by using the Newton-Raphson method) to yield $D = 3.11$ in, 8.65 in. and a negative (unrealistic) root. Note that both the positive roots are viable, but better casting yield will be achieved by using the smaller diameter riser. Let $D = 3.11$ in. and, therefore, $H = 1.5 D = 1.5 \times 3.11 = 4.67$ in.

As discussed earlier, the use of exothermic compounds (hot topping) and insulation will slow down the cooling of the riser, thereby permitting a smaller riser to be used for a given part. The curves of the type shown in Figure 2-48 permit the riser size to be estimated when using these riser aids. First, we need to estimate the following parameters—the volume ratio (VR) and the

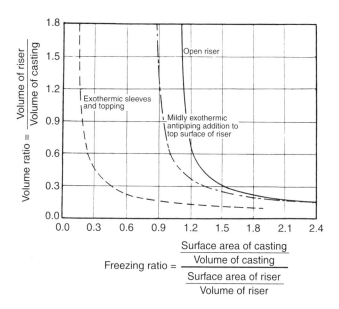

FIGURE 2-48 *Surface area-to-volume relationships for various types of risers (open riser, riser with hot topping, and riser with exothermic sleeves and hot topping).* (Adapted from Basic Principles of Gating & Risering, 1985). Reprinted with permission from The Cast Metals Institute, Schaumburg, IL (www.castmetals.com).

freezing ratio (FR). VR is the ratio of the riser volume to casting volume, and FR is the ratio of the riser modulus to casting modulus, where modulus denotes the volume-to-surface area ratio. If the VR and FR for a given combination of riser and casting yield a point that falls to the right of a curve in Figure 2-48, then the riser will be adequate for feeding the part. For example, if the point falls in the region between the two dashed curves, then the selected riser will be adequate when both hot topping and exothermic sleeves are used in the riser to slow down its cooling. However, the riser will be inadequate as an open riser, even when a mildly exothermic antipiping compound is used. The following example illustrates the use of this approach to design risers. Consider a cylindrical riser ($H/D = 0.5$) that is used to feed a circular plate-shaped casting whose thickness is half its radius. The riser diameter is two thirds the diameter of the casting. The riser is placed on top of the casting at its center. We want to find if this riser will be adequate as an open riser for the plate. Here,

$$V_R = \pi \left(\frac{2R}{3}\right)^2 \left(\frac{2R}{3}\right) = \frac{8\pi R^3}{27}$$

$$V_c = \pi R^2 \left(\frac{R}{2}\right) = \frac{\pi R^3}{2}$$

$$SA_R = 2\pi \left(\frac{2R}{3}\right)^2 = 2\pi \left(\frac{2R}{3}\right) = \left(\frac{2R}{3}\right)$$

$$SA_R = \frac{16\pi R^2}{9}$$

$$SA_c = 2\pi R^2 + 2\pi R \left(\frac{R}{2}\right) = 3\pi R^2$$

The corrected areas (after subtracting the common areas) are

$$SA_{R'} = \frac{16\pi R^2}{9} - \frac{4\pi R^2}{9} = \frac{4\pi R^2}{3}$$

$$SA_{c'} = 3\pi R^2 - \frac{4\pi R^2}{9} = \frac{23\pi R^2}{9}$$

Therefore,

$$VR = 0.593, \text{ and}$$

$$FR = \frac{\dfrac{23\pi R^2}{9} \Big/ \dfrac{\pi R^3}{2}}{\dfrac{4\pi R^2}{3} \Big/ \dfrac{8\pi R^3}{27}} = \frac{(46/9R)}{(36/8R)} = 1.14$$

For VR of 0.593 and FR of 1.14, Figure 2-48 shows that an open riser will be inadequate for feeding the shrinkage in the circular plate. However, this riser will be acceptable if a mildly exothermic topping were used to slow down the atmospheric cooling of the riser via convection and radiation.

Solidification Rate

The rate at which a casting solidifies depends on the rate of heat dissipation, which influences the casting production rate and the metallurgical structure of the part. Heat transfer during solidification is controlled by a number of factors, which include (1) the thermal conductivity of the liquid metal, which governs the diffusion of heat from the bulk of the liquid to the mold–casting interface, (2) radiation and convection to atmosphere, (3) the thermal conductivity of the solid metal, (4) the mold–metal interface resistance, characterized by heat transfer coefficient, which depends on mold coatings, surface finish, and air gap at the interface, (5) the thermal conductivity of the mold material (and whether the mold is externally cooled), and (6) convection by air currents at the outer surface of the mold. A generalized temperature distribution during solidification is schematically profiled in Figure 2-49.

In order to calculate the casting solidification rate, all the preceding thermal factors must be considered. For most practical situations, however, it is possible to identify the dominant thermal resistance, which will limit the rate of heat dissipation during solidification. In the following paragraphs, standard mathematical solutions and their application for calculating the solidification time under two limiting thermal conditions are presented without detailed derivation. Several assumptions are made in deriving the simplified mathematical solutions for the solidification rate given in the following paragraphs. These include one-dimensional heat transfer, temperature-independent mold and metal properties, and negligible fluid convection. These assumptions considerably simplify the mathematical analysis.

Sand Mold

If a casting solidifies in an insulating mold, the low thermal conductivity of the mold material dominates all other thermal resistances. For example, when a large casting solidifies in a sand mold or in an investment mold, almost all the thermal resistance is offered by the mold, and we

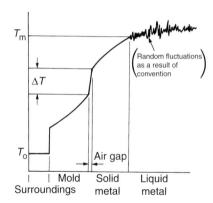

T_m

Random fluctuations as a result of convention

ΔT

T_o

Air gap

| | Mold
Surroundings

Solid metal

Liquid metal

FIGURE 2-49 *Schematic temperature distribution during solidification of a metal in a mold.* (J. Campbell, Castings, Butterworth-Heinemann, Boston, 1999, p. 125). Reprinted with permission from Elsevier.

can ignore other thermal resistances. Figure 2-50a shows a schematic temperature distribution in a large sand mold where the dominant thermal resistance is offered by the low thermal conductivity of the mold material. Under these conditions, heat conduction through the mold is governed by the non-steady heat conduction equation, whose one-dimensional form with constant thermophysical properties is

$$\frac{\partial T}{\partial t} = \alpha \frac{\partial^2 T}{\partial^2 x} \tag{2-30}$$

where T, t, and α denote the temperature, time, and thermal diffusivity, respectively (with $\alpha = k/\rho C$, where ρ is the density, C is the specific heat, and k is the thermal conductivity of the mold material). The solution of the heat conduction equation subject to appropriate initial and boundary conditions yields the temperature distribution in the mold. From a knowledge of the temperature distribution, and the energy balance at the solidification interface, the rate of solidification is calculated. The final solution can be expressed in terms of two nondimensional parameters, β and λ, defined from

$$\beta = \frac{(V/A)}{\sqrt{(\alpha t)}}, \text{ and } \lambda = \frac{T_f - T_0}{\rho_m L'}(\rho C), \text{ and } L' = L + C_m (T_p - T_f) \tag{2-31}$$

where V = volume of the casting, A = surface area of the casting, t = solidification time, α = thermal diffusivity of the mold material ($\alpha = k/C\rho$), k = thermal conductivity of the mold material, C = specific heat of the mold material, ρ = density of the mold material, T_f = freezing temperature of the metal, T_0 = initial mold temperature, ρ_m = density of the metal, L = latent heat of solidification of the metal, c_m = specific heat of the metal, and T_p = pouring temperature.

The following expressions give the solidification time, t, in terms of the parameters β and λ for some standard casting geometries

Long Plate:

$$\beta = \frac{2\lambda}{\sqrt{\pi}} \tag{2-32}$$

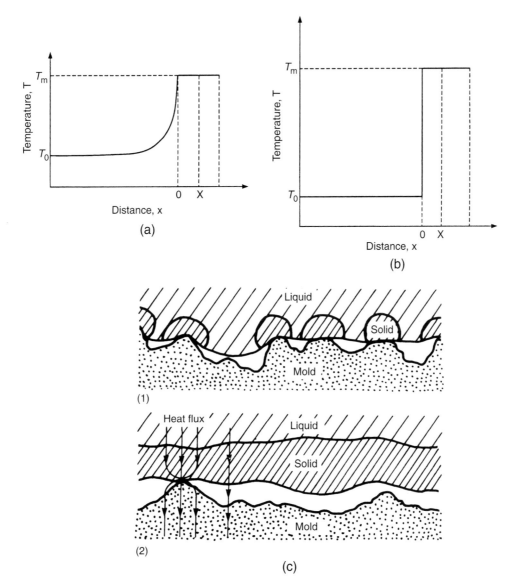

FIGURE 2-50 *(a) Idealized temperature distribution during solidification of a metal in an insulating sand mold (x = 0 denotes the mold–metal interface position). The low thermal conductivity of the sand is the dominant thermal resistance. (b) Temperature distribution in a water-cooled die (as in die casting) where the mold–metal interface resistance is the dominant thermal resistance. (c) Illustration of mold–metal interface at (1) an early stage during cooling when solid nucleates at points of good thermal contact between the mold and the liquid metal, and (2) a later stage during cooling when the contact area decreases due to localized shrinkage and deformation. This increases the air gap at the interface and lowers the dynamic interface heat transfer coefficient* (J. Campbell, Castings, Butterworth-Heinemann, Boston, 1999, p. 129). Reprinted with permission from Elsevier.

Long cylinder:

$$\beta = \lambda \left(\frac{2}{\sqrt{\pi}} + \frac{1}{4\beta} \right) \tag{2-33}$$

Sphere:

$$\beta = \lambda \left(\frac{2}{\sqrt{\pi}} + \frac{1}{3\beta} \right) \tag{2-34}$$

Part geometry influences heat dissipation; as a result, different expressions are obtained for the solidification times of different casting geometries. For example, in a long plate the cross-sectional area across which heat is dissipated remains constant during solidification, whereas in a sphere and a cylinder the area constantly changes as solidification progress. Note also that the expression

$$\beta = \frac{(V/A)}{\sqrt{(\alpha t)}}$$

can be rearranged as

$$t = \frac{1}{\alpha \beta^2} \left(\frac{V}{A} \right)^2 \tag{2-35}$$

which is analogous to Chvorinov's equation

$$t = B \left(\frac{V}{A} \right)^n \tag{2-36}$$

where $n = 2$, and the mold constant $B = (1/\alpha\beta^2)$.

Die Casting

In die casting (and to a limited extent, in permanent mold casting), the mold or the die is water cooled. This means that thermal resistance of the mold material can be ignored in a simplified analysis of solidification rate because cooling is continuous. In addition, section thickness of die-cast parts is usually small so that the thermal resistance in the liquid metal and the solid metal can be ignored. Under these conditions, the thermal resistance of the mold–metal interface dominates all other thermal resistances, and the temperature distribution can be schematically represented, as in Figure 2-50b. Table 2-4 provides approximate values of interface heat transfer coefficient (h) for different types of molds and interface conditions. Note that interface conditions during solidification are highly dynamic, as illustrated in Figure 2-50c, and the value of h used in the computations is only an approximation. For example, with continued solidification the contact area at the mold-metal interface decreases (Figure 2-50c) because of the volumetric shrinkage of the metal, and the interface conductance will likely decrease during the later stages of cooling.

Consider that a thickness dx of the metal solidifies in a time interval dt when it is introduced in a water-cooled die at its melting temperature (i.e., zero superheat). The rate of heat loss during

TABLE 2-4 Representative Values of Heat Transfer Coefficient (h)

Mold Type/Surface Condition	$h(cal \cdot cm^{-2} \cdot K^{-1} \cdot s^{-1})$
Metal mold/polished	0.0956
Metal mold/coated	0.0179
Metal mold/polished and water cooled	0.1195
Metal mold/coated and water cooled	0.0239
Die casting	1.195
Rapid solidification (melt spinning)	2.389

solidification of the element dx is ($A dx \rho \cdot L/dt$), where A is the die cross-section, and ρ and L are the density and latent heat of solidification of the metal, respectively. All the heat released upon solidification of the element dx must be transferred across the die-casting interface in time dt, so that from Newton's law of cooling and energy conservation, we can write,

$$A\rho L \left(\frac{dx}{dt} \right) = hA(T_f - T_0)$$

$$\frac{dx}{dt} = \frac{hA(T_f - T_0)}{\rho L} \tag{2-37}$$

where h is the die–metal interface heat transfer coefficient, T_0 is the die temperature, maintained constant by circulating water, and T_f is the freezing or solidification temperature. The initial conditions are $x = 0$ at $t = 0$. Integration of the preceding equation yields the thickness, x, solidified as a function of time, t

$$x = \frac{h(T_f - T_0) \cdot t}{\rho L} \tag{2-38}$$

Because the thickness solidified, x = (volume/surface area), this equation can be rearranged to yield

$$t = \frac{\rho L}{h(T_f - T_0)} \cdot \left(\frac{V}{A} \right) \tag{2-39}$$

which is essentially Chvorinov's rule, $t = B(V/A)^n$, where $n = 1$ and B is a mold constant with a value equal to the term $\rho L/h(T_f - T_0)$.

The data presented in Figure 2-51 show the effect of mold type, casting modulus (or, alternatively, thickness solidified), and the type of alloy cast on solidification time of different castings. As expected, conductive molds and externally cooled molds yield shorter solidification times, thus reducing the production time of castings. The linear plots of Figure 2-51 indicate that a simple parabolic equation of the type $x = a\sqrt{t}$ relates the thickness solidified and solidification time, t; this is consistent with Chvorinov's equation $t = B(V/A)^n$, with $n = 2$, because the volume-to-area ratio (V/A) is the thickness x for simple shapes such as plates.

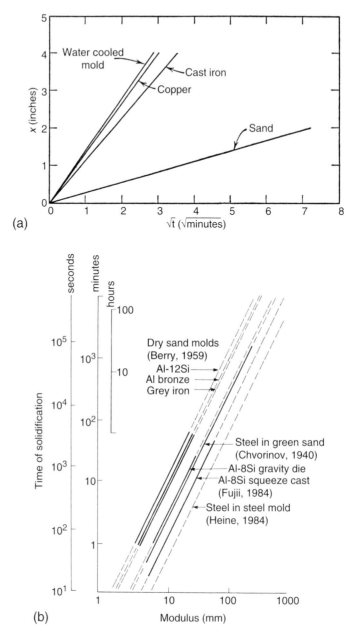

FIGURE 2-51 *(a) Solidification of steel in different mold materials. The plots show the thickness, x, solidified as a function of square root of solidification time. The linear behavior suggests parabolic solidification kinetics. (H. F. Taylor, M. C. Flemings, and J. Wulff, Foundry Engineering, Wiley, 1959, p. 113). Reprinted with permission from H. F. Taylor Trust, Boston, MA. (b) Solidification times for plate-shaped castings of different alloys in various types of molds as a function of casting modulus (volume-to-surface area ratio). (J. Campbell, Castings, Butterworth-Heinemann, Boston, 1999, p. 37). Reprinted with permission from Elsevier.*

As an example of solidification time calculations, consider the die casting of Mg, which is sometimes considered as an alternative to Al in view of similarities in their properties. Let us calculate the solidification time of a given die cast part made from Mg and Al. For Al, $T_{\mathrm{m}} = 933\,°K$, density $= 2.70$ g/cc, heat of fusion $= 1.08$ kJ/cm^3, and for Mg, $T_{\mathrm{m}} = 922\,°K$, density $= 1.74$ g/cc, and heat of fusion $= 0.64$ kJ/cm^3. Because the part thickness, x, is the same in both Al and Mg, we can write

$$x = \frac{h\left(T_{\mathrm{f}}^{\mathrm{Al}} - T_{\mathrm{o}}\right) \cdot t^{\mathrm{Al}}}{\rho_{\mathrm{m}}^{\mathrm{Al}} L^{\mathrm{Al}}} = \frac{h\left(T_{\mathrm{f}}^{\mathrm{Mg}} - T_{\mathrm{o}}\right) \cdot t^{\mathrm{Mg}}}{\rho_{\mathrm{m}}^{\mathrm{Mg}} L^{\mathrm{Mg}}}$$

Therefore,

$$\frac{t^{\mathrm{Mg}}}{t^{\mathrm{Al}}} = \frac{\left(T_{\mathrm{f}}^{\mathrm{Al}} - T_{\mathrm{o}}\right) \rho_{\mathrm{m}}^{\mathrm{Mg}} L^{\mathrm{Mg}}}{\left(T_{\mathrm{f}}^{\mathrm{Mg}} - T_{\mathrm{o}}\right) \rho_{\mathrm{m}}^{\mathrm{Al}} L^{\mathrm{Al}}}$$

$$\frac{t^{\mathrm{Mg}}}{t^{\mathrm{Al}}} = \frac{(933 - 303) \times 0.64 \times 10^3}{(922 - 303) \times 1.08 \times 10^3} = 0.60$$

Therefore, $t^{\mathrm{Mg}} = 0.60\, t^{\mathrm{Al}}$

In other words, magnesium die casting will save about 40% time in comparison to an identical Al casting.

Stages of Solidification—Nucleation and Growth

Nucleation

During solidification, the transformation of the random atomic structure of a liquid into an ordered atomic arrangement of a crystalline solid occurs via processes of nucleation and growth. In the earliest stages of cooling of a superheated melt, thermal vibrations of liquid atoms constantly bring groups of atoms sufficiently close together to form atomic clusters, which can be conceived of as the smallest crystals. As the melt temperature is still above the melting point at this time, liquid is the thermodynamically stable phase, and the atomic-size clusters remelt. These unstable clusters can be considered as subcritical solid nuclei or embryos that do not contribute to the solidification process. When the metal temperature has dropped only slightly below the equilibrium melting point (and the solid is the more stable phase), the atomic-size solid clusters continue to form and melt back. This is because the formation of a stable cluster or solid nuclei requires creation not only of the nucleus volume but also its surface (i.e., a new interface must form between the solid nuclei and the surrounding liquid). As all interfaces represent a discontinuity with respect to the bulk, the creation of a surface requires energy consumption. It turns out that only when the liquid temperature has dropped sufficiently below the normal solidification point will there be a large enough driving force to form the bulk volume *and* the new surface of the solid nuclei. Thermodynamically, the melting temperatures of particles depend on their radius; particles with a large radius melt at higher temperatures than particles with a small radius. Therefore, large atomic clusters will survive when the temperature has dropped only slightly below the normal melting point, but smaller clusters will readily melt back at these same temperatures. The surviving nuclei serve as precursors or seed crystals for the growth of the solid during subsequent cooling.

Homogeneous Nucleation

The processes of clustering and declustering (remelting) of nuclei continue with a drop in the temperature until a dynamic equilibrium is reached at which clustering and declustering rates become equal. If the cluster size exceeds a critical value, then thermodynamically stable nuclei begin to form in the liquid. For an atomic cluster to be thermodynamically stable against chance fluctuations favoring declustering, its radius must be greater than or equal to a critical radius, r^*. In terms of the free energy change accompanying the formation of a nucleus, we can write $\Delta G = \Delta G_s + \Delta G_v$, where ΔG_v is the volume free energy change per unit volume when liquid transforms into solid, and ΔG_s is the surface free energy change per unit volume. For a spherical nucleus of radius r, the *total* free energy change can be written as

$$\Delta G = 4\pi r^2 \cdot \gamma_{LN} + \frac{4}{3}\pi r^3 \cdot \Delta G_v \qquad (2\text{-}40)$$

where γ_{LN} is the interfacial energy of the liquid–nucleus interface. Figure 2-52 shows a schematic of the variation of the volume energy and surface energy to form a stable nucleus as a function of the particle radius. For $r < r^*$, the cluster is subcritical and will disappear, whereas if $r \geq r^*$, the cluster will become a stable nucleus and continue to grow. The critical nucleus size is obtained by noting that at $r = r^*$, $(d\Delta G/dr) = 0$. On differentiating Equation 2-40 with respect to r and equating the result to zero, we obtain

$$r^* = \frac{2\gamma_{LN}}{\Delta G_v} \qquad (2\text{-}41)$$

The Gibb's free energy change $\Delta G = \Delta H - T \cdot \Delta S$ where ΔH and ΔS denote the enthalpy and entropy changes, respectively, and by definition, $\Delta S = \Delta H/T$. At $T = T_m$ (melting point), solid and liquid phases have the same free energy so that $\Delta G = 0$, and hence $\Delta S = (\Delta H/T_m)$. Substituting this expression for ΔS into $\Delta G = \Delta H - T \cdot \Delta S$ yields the volume free energy change at any temperature T below T_m. This expression is

$$\Delta G_v = \frac{\Delta H}{T_m}(T_m - T) \qquad (2\text{-}42)$$

The term $(T_m - T)$ is the undercooling or supercooling, and is denoted by ΔT. Combining equations 2-41 and 2-42 yields the following expression for r^* in terms of ΔT

$$r^* = \frac{2\gamma_{LN} T_m}{\Delta H \cdot \Delta T} \qquad (2\text{-}43)$$

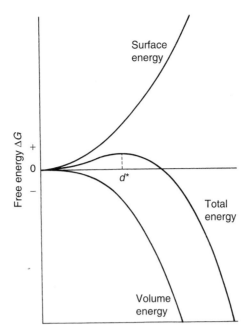

FIGURE 2-52 *Free energy change for nucleation as a function of the radius of the nucleus. The plot shows the surface free energy, volume free energy, and the total free energy change.* (J. Campbell, Castings, Butterworth-Heinemann, Boston, 1999, p. 139). Reprinted with permission from Elsevier.

The free energy change, ΔG^*, for the formation of a critical nucleus is obtained by substituting for r^* from Equation 2-43 in Equation 2-40. Upon simplification, this yields

$$\Delta G^* = \frac{16\pi \gamma_{LN}^3 T_m^2}{3(\Delta H \cdot \Delta T)^2} \tag{2-44}$$

This equation shows that at large undercooling, even a small nucleus will become supercritical or stable. The equilibrium number of embryos, n, of radius r, at a temperature T can be assumed to be given from the Boltzmann distribution function,

$$n = N \cdot \exp\left(-\frac{\Delta G}{kT}\right) \tag{2-45}$$

where N is the total number of atoms per unit volume in the crystal, ΔG is the free energy change given from Equation 2-44, and k is the Boltzmann constant. For the critical nucleus of radius, r^*,

$$n^* = N \cdot \exp\left(-\frac{\Delta G^*}{kT}\right)$$

Note that once an embryo reaches the critical radius r^*, it continues to grow and is no longer a part of the population of embryos at the temperature T. The nucleation rate, or the rate at which critical nuclei form, is then determined by the rate at which subcritical nuclei or embryos (with $r < r^*$) reach the critical size. The smaller embryos reach the critical size at a rate at which atoms of the liquid attach to a subcritical nucleus, making it supercritical. The nucleation rate, I, is given from

$$I = n^* \varepsilon \upsilon N \cdot \exp\left[-\frac{\Delta G^* + \Delta G_D}{kT}\right]$$
(2-46)

Where ΔG_D is the free energy of activation for diffusion in the melt ($\Delta G_D \sim kT$), υ is the atomic jump frequency, and ε is the probability of an atomic jump in a given direction ($\varepsilon \sim 1/6$). In essence, Equation 2-46 expresses the fact that the rate of nucleation is determined by the probability of an atom having sufficient energy to jump (diffuse) and the energy to attach itself to the solid (binding energy). The final expression for the nucleation rate can be written as

$$I = \frac{NkT}{h} \cdot \exp\left[-\frac{\Delta G_D}{kT}\right] \cdot \exp\left[-\frac{16\pi \gamma_{LN}^3 T_m^2}{3kT \cdot (\Delta H \cdot \Delta T)^2}\right]$$
(2-47)

where h is Planck's constant. Figure 2-53 shows the pronounced effect of undercooling, ΔT, on the nucleation rate. At a critical value ΔT^* equal to about 0.2 T_m (i.e., at a temperature of 0.8 T_m), the nucleation rate increases sharply. At this critical temperature, a sensible nucleation rate is achieved. Experiments in pure melts free of impurity particles have been performed to test these predictions. Table 2-5 presents experimental data on undercooling achieved for different metals.

The mathematical form of the nucleation rate equation (Equation 2-47) is such that at very large undercoolings (large ΔT and small T_m in Equation 2-47), nucleation rate should slow down. Physically, this is a consequence of a decreased mobility (diffusion) of atoms at lower temperatures.

The analysis of homogeneous nucleation previously presented applies to pure metals. For alloys, the situation becomes more complex, because the growing crystals will have a composition different from the surrounding liquid, per the phase diagram for the alloy under consideration. As a result, the alloying elements will be unevenly distributed, causing a composition gradient to develop, which, in turn, will drive the process of diffusion. Thus, nucleation in alloys could involve diffusion.

Homogeneous nucleation seldom occurs under normal conditions, because of the large amount of energy needed to create the surface of the nucleus. Nucleation becomes energetically feasible if a preexisting surface is present in the liquid (e.g., mold wall, inclusions etc.). This is called heterogeneous nucleation and will be discussed in the next section.

Figure 2-54 shows a schematic diagram of the cooling curve of a pure metal, which is essentially the temperature versus time map of the metal as it cools in a mold. The drop in

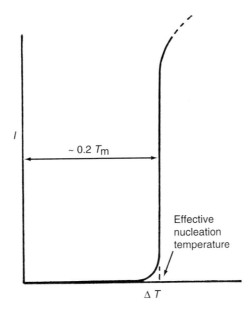

FIGURE 2-53 *Schematic of homogeneous nucleation rate as a function of undercooling,* ΔT. (G. J. Davis, Solidification and Casting, John Wiley & Sons, New York, 1973, p. 18). Reprinted with permission from Applied Science Publishers, Ltd., Barking Essex, U.K.

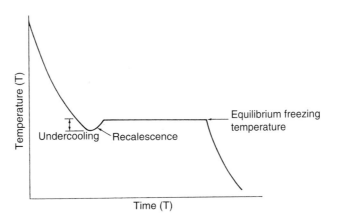

FIGURE 2-54 *The cooling curve of a pure metal showing regions of undercooling and recalescence prior to the onset of growth (solidification).* (E. P. DeGarmo, J. T. Black, R. A. Kohser, and B. E. Klanecki, Materials and Processes in Manufacturing, 9th ed., Wiley, 2003, p. 281).

TABLE 2-5 Undercooling of Metal Droplets (adapted from G. J. Davies, Solidification and Casting, John Wiley & Sons, New York, 1973)

Metal	Tm, K	DT, °C
Sn	505.7	105
Pb	600.7	80
Al	931.7	130
Ag	1233.7	227
Cu	1356	236
Ni	1725	319
Fe	1803	295

melt temperature below its equilibrium freezing point is also shown. This figure also shows that nucleation is followed by a rise in the temperature of the metal back to its equilibrium freezing point. This phenomenon is called recalescence, and it is followed by the isothermal growth of crystals.

Heterogeneous Nucleation

The presence of solid impurity particles in a solidifying melt could assist the process of crystal nucleation by providing a surface for the formation of the critical nucleus. It is found that nucleation often occurs at smaller values of undercooling than $\Delta T^*(= 0.2\, T_m)$ when the walls of the container or impurity particles in the melt catalyze the nucleation process. Nucleation promoted by such means is called heterogeneous nucleation. Because a surface already exists, the energy needed to form the nucleus surface (first term in Equation 2-40) is smaller than if there were no preexisting surface available on which to form the nucleus.

Consider a small, spherical, cap-shaped nucleus in contact with a particle in the melt (Figure 2-55). If, for simplicity, we assume that the nucleus contacts a flat surface on this particle (a reasonable assumption because nucleus size ≪ size of commonly found inclusions in metals), and if we further assume that the nucleus is in mechanical equilibrium on the particle, then from a balance of interfacial forces at the contact line between the substrate (s), cap (c), and the liquid (l), we can write

$$\gamma_{LC} \cos\theta + \gamma_{CS} = \gamma_{LS} \tag{2-48}$$

where γ_{CS}, γ_{CL}, and γ_{LS} are the interfacial energies of the interfaces between the subscripted phases (i.e., γ_{CS} is the interfacial energy of the spherical cap nucleus–substrate interface), and θ is the angle of contact between the nucleus and the substrate.

The critical radius for heterogeneous nucleation can be derived in a manner similar to that for homogeneous nucleation provided adjustments are made for the surface area and volume of the spherical-cap nucleus in contact with the substrate. The volume and surface area of the spherical cap nucleus of Figure 2-55 are

$$Volume = \frac{1}{3}\pi h^2(3r - h),$$

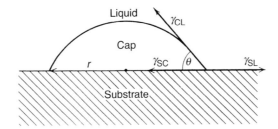

FIGURE 2-55 *A spherical cap nucleus in equilibrium on a substrate under the action of the three interfacial tensions γ_{CL}, γ_{SC}, and γ_{SL}, where C, L and S represent the spherical cap, liquid, and the substrate, respectively. The angle θ is the equilibrium contact angle the nucleus makes on the substrate.*

and

$$Area = 2\pi rh,$$

where r is the radius of the base of the spherical cap and h is its maximum height. In terms of angle, θ, the expressions for volume and surface area become

$$Volume = \frac{1}{3}\pi r^3 (2 - 3\cos\theta + \cos^3\theta)$$

and

$$Area = 2\pi r^2 (1 - \cos\theta)$$

In a manner similar to the derivation of equations for homogeneous nucleation, the energy minimization for heterogeneous nucleation yields the critical nucleus size as

$$r^*_{het} = \frac{2\gamma_{LC}}{\Delta G_v},$$

and the free energy change required for nucleation is

$$\Delta G^*_{het} = \frac{4\pi \gamma_{LC}^3 (2 - 3\cos\theta + \cos^3\theta)}{3\Delta G_v^2} = \Delta G^*_{hom} \cdot f(\theta) \qquad (2\text{-}49)$$

where $f(\theta) = (1/4)(2 - 3\cos\theta + \cos^3\theta)$, and ΔG^*_{het} and ΔG^*_{hom} are the free energy changes for critical nucleus formation by heterogeneous and homogeneous nucleation, respectively. The variation of the parameter $f(\theta)$ with the contact angle θ is shown in Figure 2-56.

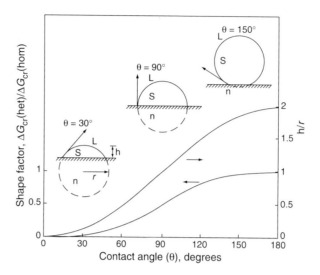

FIGURE 2-56 *The variation of the shape-factor f(θ) as a function of the angle θ for heterogeneous nucleation.* J. H. Perepezko, in Metals Handbook, Volume 15: Casting, 9th ed. Reprinted with permission from ASM International, Materials Park, OH (www.asminternational.org).

It is clear that for $\theta < 180°$, the energy barrier to nucleus formation will be smaller compared to homogenous nucleation; the smaller the value of θ, lower will be the undercooling needed to form the critical nucleus. Thus, heterogeneous nucleation of new crystals during solidification is facilitated by the presence of a wettable surface (i.e., small θ). The nucleation analysis for nonspherical (e.g., disc-shaped) nucleus and non-flat substrates yields similar conclusions. For example, calculations show that nucleation is easier at the root of sharp cracks in a substrate than on a flat surface because the volume of the material needed to obtain the critical radius is relatively small at the crack root.

For heterogeneous nucleation to occur, θ must be small and $\cos \theta$ must be large, i.e., γ_{CS} should be smaller than $\gamma_{LS} (\sigma_{CL}$ is the interfacial tension of the liquid, which for a pure liquid is constant at a fixed temperature). For $\gamma_{CS} < \gamma_{LS}$, the crystal planes of the nucleus and the substrate should match on an atomic scale at the interface, i.e., atomic disregistry between contacting atomic planes should be small. The atomic disregistry is specified in terms of a parameter, δ, where $\delta = (\Delta a/a)$, and a is the lattice parameter of the nucleating crystal, and Δa is the difference between the lattice parameters of the substrate and the nucleus at the interface. If the atomic disregistry (δ) is less than about 10% along two normal directions at the interfacing crystal planes, the nucleating ability of the substrate is relatively good and undercooling required to achieve nucleation is small. For larger values of δ, the nucleation potency of the substrate is poor. Figure 2-57 shows the effect of lattice disregistry between various ceramics and pure iron on the supercooling of molten iron.

Besides atomic disregistry, the nature and strength of the chemical bond between the nucleus and the substrate influences the heterogeneous nucleation. The crystallographic anisotropy of interfacial energies, γ, also plays a role; the solid phase must be suitably oriented with respect to the nucleating crystal to present a low-energy crystallographic plane for nucleation to occur. An example is that yttrium is more effective as a nucleant in metals when its prismatic plane rather than the basal plane is exposed to the nucleating phase. The nucleating crystal shape depends on the anisotropy of γ. For an isotropic interfacial energy, γ, the crystal shape is spherical,

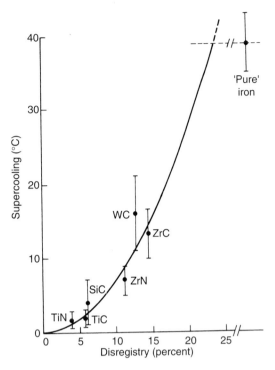

FIGURE 2-57 *Supercooling of molten iron in the presence of various nucleating agents. The six carbides and nitrides in this plot exhibit varying degrees of effectiveness in reducing the undercooling required for nucleation.* (J. Campbell, Castings, Butterworth-Heinemann, Boston, 1999, p. 140). Reprinted with permission from Elsevier.

but for an anisotropic γ, the crystal shape corresponding to the minimum free energy can be non-spherical.

Nucleation and Grain Refinement

Understanding nucleation during solidification is important for control of the grain size of industrial castings. For example, the addition of inoculants (heterogeneous nucleants) to molten alloys is an efficient method to refine the grain size. Inoculants that form a small contact angle with the nucleating phase are usually more effective than those that form large contact angles. Convection also affects nucleation. During solidification, nucleation continues while previously formed critical nuclei grow. These growing crystals often fracture because of convection, and the fragmented crystals act as preferred sites for heterogeneous nucleation of new crystals.

In many metal-matrix composites produced by solidification techniques, the reinforcing phase promotes nucleation and grain refinement. For example, carbon, Al_2O_3, silica, and SiC particles in hypereutectic Al-Si alloys promote heterogeneous nucleation of primary silicon, resulting in the refinement of Si. Heterogeneous nucleation is also observed on carbon dispersed in CuPbTi alloys, alumina dispersed in AlCuTi alloys, and titanium diboride (TiB_2) dispersed in TiAlMn alloys. Similarly, in Al containing TiC particles, TiC reacts with Al to form Al_4C_3 and complex carbides such as Ti_3AlC that promote heterogeneous nucleation of Al. Reactive or surface-active alloying elements can cause surface modification via chemical reactions or adsorption, which can promote nucleation provided a low-energy interface (even an intermediate transition layer) first forms.

Nucleation is often masked by the subsequent growth of crystals, and experimental observations of nucleation are quite difficult. However, certain glass-forming alloys such as Al-Ni-Y-Co alloys have been found to serve as slow-motion models of undercooled liquids, and enable high-resolution microscopic observations of the nucleation mechanisms to be made. For example, experiments show that TiB_2 particles dispersed in an amorphous Al-Ni-Y-Co alloy first form an adsorbed layer of $TiAl_3$ over which 2- to 5-nm-size Al crystals nucleate. However, the atomic disregistry at the TiB_2–$TiAl_3$–Al interface gives rise to strains in the nucleated Al.

Nucleation may be initiated by the application of external energy to molten metals; for example, exposing a liquid during cooling to high-frequency vibrations (ultrasound) promotes the nucleation and grain refinement. The propagation of ultrasound through liquids is accompanied by continuous formation and implosion of tiny gas-filled microcavities, a process called "cavitation." The positive pressure generated by implosion is large enough to increase the melting temperature of the liquid so that the liquid becomes supercooled with respect to its raised melting point, and homogenous nucleation can occur. Vibration is transmitted into the melt through a suitable guide tube; for example, a titanium horn with a TiB_2-coated tip to prevent erosion or dissolution in the metal. The very large ($\sim 10^3$ to 10^5 times the gravity) accelerations generated near the probe end immersed in the solidifying melt create large pressure gradients that could cause partial melting of dendrites, leading to refinement of the structure and to removal of air and dissolved gas. However, very intense vibration can also damage the tooling and erode the material from the probe assembly and the holding crucible.

In recent years, there has been much interest in developing materials with nanometer-size grains. Nanomaterials consist of grains with sizes on the order of a few nanometers, nearly a thousand times smaller than the micrometer-size grains found in common metals. The nanometric grain size leads to an enormous increase in the grain boundary area (barrier to dislocation slip), which provides greater strengthening than does micrometer-grained metal. Additional strengthening can be achieved by combining nanoscale ceramic particles with a nanocrystalline metal. The particles will also stabilize the nanometric grain structure by preventing grain coarsening during heat treatment.

Growth During Solidification

Nucleation is followed by the growth of the solid phase from the liquid as the latter cools in a mold. Casting soundness, compositional homogeneity, and grain size and morphology are directly influenced by the process of growth. Growth is controlled by heat, mass, and fluid transport. Important considerations in growth during solidification are structure of the interface, compositional changes accompanying growth, and rate of growth.

Atomic Structure at the Solidification Interface

The atomic structure of the solidification front that separates the solid from the liquid determines many important structural features of the casting. Much of our current understanding of the atomic processes in solidification, and their influence on industrial castings, has developed in the last sixty years. The following discussion closely follows the treatment of atomic processes during solidification presented in books by Davis, Flemings, and Kurz and Fisher (listed at the end of this chapter). The reader is referred to these sources for a more complete and authoritative coverage of this topic.

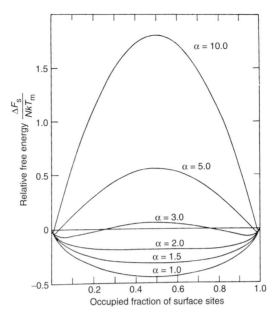

FIGURE 2-58 *Relative surface free energy change upon atom attachment to a solid–liquid interface as a function of the fraction of surface sites occupied on the interface. The parameter α depends on the crystal face, the type of crystal, and the phase from which the crystal grows.* (K. A. Jackson, Liquid Metals and Solidification, 1958, p. 174). Reprinted with permission from ASM International, Materials Park, OH (www.asminternational.org).

On an atomic scale, the solid–liquid interface could be either sharp (or smooth) or diffuse (or rough); in the latter case, the transition from the liquid to solid extends over several layers of atoms. The diffuseness of the solid–liquid interface is characterized in terms of a structure parameter, α defined from $\alpha = (L\zeta/kT_{\mathrm{m}})$, where L is the latent heat of solidification, k is the Boltzmann constant, and ζ is a crystallographic parameter, less than unity, that specifies the fraction of the total system energy that binds an atom in a crystal plane at the solid–liquid interface to other atoms in the same plane. The α factor determines the change in the free energy on randomly adding atoms of liquid to an initially planar and atomically smooth interface. The change in the free energy, ΔG, is given from

$$\Delta G = NkT_{\mathrm{m}}[\alpha \cdot x(1 - x) + x \ln x + (1 - x)\ln(1 - x)] \qquad (2\text{-}50)$$

where $x(= n/N)$ is the fraction of N possible sites of the interface that can be occupied by an atom at the equilibrium melting temperature, T_{m}. Equation 2-50 is depicted graphically in Figure 2-58, which shows the variation of the relative free energy change, $(\Delta G/NkT_{\mathrm{m}})$, as a function of x for a range of α values. The energy minima in this figure represent the thermodynamically stable interface configurations. For $\alpha \leq 2$, there is a single minimum in the curves at $x = 0.5$. Interfaces with $\alpha \geq 5$ exhibit an energy minimum corresponding to very small and very large values of x; in other words, the free energy is at a minimum when there are either only a few occupied sites (small x) or only a few unoccupied atomic sites (large x). Interfaces with $\alpha < 2$ are atomically rough (or non-faceted), whereas those with $\alpha > 5$ are atomically smooth (or faceted). The thickness (i.e., the degree of diffuseness) of the interface increases with

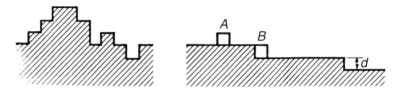

FIGURE 2-59 *Schematic illustration of an atomically rough and an atomically smooth solid–liquid interface. A smooth interface will grow faster sideways than normal to the surface because of energetically favorable attachment of liquid atoms to atomic steps.* (P G. Shewmon, Transformations in Metals, McGraw-Hill, New York, 1969).

decreasing α. Most metals have $\alpha < 2$, whereas most inorganic and organic liquids have $\alpha > 5$; a few metals such as Si and Bi have $5 > \alpha > 2$, and exhibit a complex mixed type of behavior. Elemental semiconductors such as Ge and compound semiconductors such as GaAs and GaSb solidify with faceted interfaces. The formation of facets is a consequence of the anisotropic growth of solid crystals along different crystallographic directions.

An atomically rough interface will grow in an isotropic manner; i.e., its growth rate will be independent of the crystallographic direction because such an interface does not offer any preferred sites for attachment to liquid atoms. As a result, liquid atoms attach themselves to the interface in a purely random manner, and there is no preferred growth direction. Atomically rough surfaces grow by a normal or continuous growth mechanism (i.e., completely random addition of atoms), and the mean interface growth rate is given from, $R = a \cdot \Delta T$, where ΔT is the undercooling and a is a constant ($\sim$1 cm/s·K). A smooth interface can grow only when liquid atoms attach themselves to low-energy positions on the smooth interface; i.e., at steps or ledges on the surface.[2] This is illustrated in Figure 2-59, where atom B is energetically more stable than atom A. Thus, existing steps will grow easily sideways, and a smooth or faceted interface will continue to grow. However, nucleation of new steps will be energetically difficult because this will require several atoms to spontaneously form a cluster on a flat interface, and the cluster itself will be weakly bonded to the underlying step because of small number of nearest neighbors in that plane. Because of this, interface growth is slowest along a direction normal to the faceted interface. The completely grown crystal will then be bounded by the slowest growing crystal planes, and will have a faceted appearance. The difficulty of nucleating a new step on an atomically smooth interface is reduced if a crystalline defect such as a screw dislocation intersects the free surface of the growing crystal; in this case liquid atoms can attach themselves to the step formed by the crystal imperfection, and growth perpendicular to the faceted surface will be possible.

In materials that solidify in a faceted manner, the atomic attachment on atomically flat solid–liquid interface requires a driving force in the form of melt undercooling, termed the kinetic undercooling. For faceted growth of smooth interfaces, the rate of growth, $R = b \cdot \exp(-c/\Delta T)$, where b and c are constants, and ΔT is the kinetic undercooling. This equation shows that at small undercoolings, the growth rate will be extremely small. When a screw dislocation intersects a faceted growth front and provides a step for atom attachment, the rate of growth is given from $R = d \cdot \Delta T^2$, where d is a constant. For faceted interfaces, foreign atoms (impurities),

---

[2]An atom with the largest number of nearest neighbors has the lowest energy; on a smooth interface, an incoming atom will encounter the largest number of neighbors at a step or a ledge.

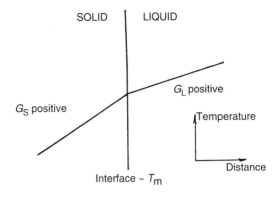

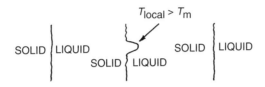

FIGURE 2-60 *Solidification in a positive temperature gradient. The temperature gradient in the melt stabilizes a planar front by melting any convex protuberance in the front.* (G. J. Davis, Solidification and Casting, Wiley, New York, 1979, p. 44).

crystal defects such as dislocations and twins, and fluid convection due to temperature- and solute-gradients (thermosolutal convection) can alter the growth rate.

Growth in Pure Metals

Consider a superheated pure liquid ($T > T_m$) being solidified in a directional manner (Figure 2-60) by heat extraction through the growing solid phase. The temperature at the S–L interface is close to the solidification temperature, T_m, and the temperature gradient, G_l, in the liquid and the gradient, G_s, in the solid are both positive, as shown in Figure 2-60. Under these conditions, the interface will grow in a stable planar form, and any perturbations (instabilities) in the growth front will melt back because of higher melt temperature ahead of the perturbed region of the interface. If, however, the liquid is undercooled so that the temperature gradient in the liquid, $G_l < 0$ (Figure 2-61), then any instabilities in the interface shape will survive and amplify during growth. For example, an initially planar front might break down into columnar cells, and these cells might further break down into dendrites (tree branch–like crystalline structures) by growth of lateral side branches. Further growth will cause higher-order branches to appear, resulting in a complex three-dimensional network of solid crystals, as shown in Figure 2-62. However, the conditions leading to cell and dendrite formation exist only during a short period (called recalescence) preceding the attainment of normal solidification temperature (see Figure 2-54). Once the normal solidification temperature is attained, growth proceeds by thickening of cells and dendrites that takes place by heat dissipation from the bulk of the liquid.

Dendritic growth is complicated by the process of coarsening that involves competitive dissolution and growth and leads to thickening of larger dendrite arms at the expense of smaller

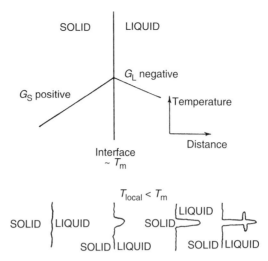

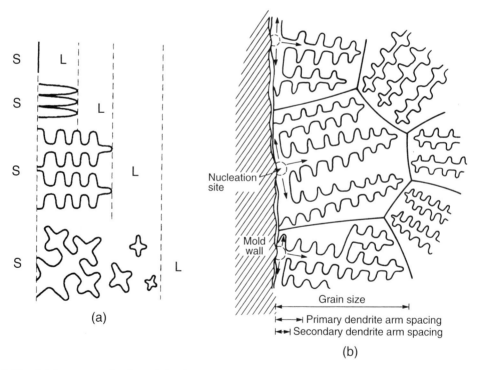

FIGURE 2-61 *Solidification in a negative temperature gradient. The temperature gradient in the melt destabilizes the planar front and amplifies any perturbation in the front, thereby forming cells and dendrites.* (G. J. Davis, Solidification and Casting, Wiley, New York, 1979, p. 45). Reprinted with permission from Applied Science Publishers, Ltd., Barking Essex, U.K.

FIGURE 2-62 *(a) Formation of cells and dendrites in an undercooled melt. S and L denote solid and liquid, respectively. (b) Illustration showing the relationship between dendrites and grains. Each grain consists of a raft of dendrites that evolve from a single nucleus and have similar crystallographic orientation.* (J. Campbell, Castings, Butterworth-Heinemann, Boston, 1999, p. 146). Reprinted with permission from Elsevier.

arms, which melt and disappear. This results in an increase in the average spacing between the secondary dendrite arms. The kinetics of coarsening are usually slower than the kinetics of growth (usually parabolic in time) and follow a one-third power law with respect to time, i.e., $R^3 = K \cdot t$, where R is the secondary dendrite arm spacing, K is the coarsening rate constant, and t is time. This competitive dissolution and growth process is driven by a need to minimize the total liquid–solid interface area; fine dendrite arms become unstable and disappear because they increase the total surface area. In the case of alloys, the composition at the solid–liquid interface varies with the interface curvature according to the Gibbs-Thompson relationship, $\Delta c = \sigma V_m / r \Delta S$, where Δc is the excess concentration at a curved interface of radius r relative to a plane interface, σ is the interfacial energy, ΔS is the entropy of fusion, and V_m is the molar volume of the solid. Fine dendrite arms present regions of high curvature, and high interface concentration, which create steeper concentration gradients and cause rapid coarsening via atomic diffusion.

Growth of Single-Phase Alloys

Growth in alloys involves not only heat transfer as in pure melts, but also solute redistribution. This has major practical consequences: solute redistribution results in undesirable segregation or compositional inhomogeneity in a casting. However, solute redistribution also permits production of ultrapure materials using an innovative alloy solidification process called zone refining (discussed next) that sweeps impurity atoms to a specific region of the casting, thus purifying the remainder of the casting.

Heat transfer during solidification occurs from bulk liquid by conduction through a thermal boundary layer at the interface and through the solid. The characteristic dimensions of the thermal boundary layer are on the order of (α/V), where α is the thermal diffusivity and V is the growth velocity. For alloy solidification, solute transport involves mass transfer by atomic diffusion through the solid phase, through a boundary layer in the liquid at the interface, and through bulk convection in the liquid ahead of the diffusion layer. The characteristic diffusion length is on the order of (D/V), where D is the diffusion coefficient.

Under slow growth rates where equilibrium exists at the interface, and the compositions of the solid and liquid phases are given from the phase diagram, an equilibrium distribution coefficient (or partition coefficient), k, is defined such that $k = (C_s/C_l)$, where C_s and C_l are the solute concentrations in the solid and the liquid, respectively, at a temperature T. Depending on the original alloy composition and the slope (positive or negative) of solidus and liquidus lines on the phase diagram, k could be either positive or negative (Figure 2-63).

Consider an alloy of composition, C_0, with a distribution coefficient, $k < 1$. The first solid to form will have a composition of kC_0, and as $k < 1$, the liquid surrounding the first-formed solid becomes enriched in solute. This occurs when the excess solute is rejected by the solid into the liquid ahead of the growth front. This excess solute can (1) migrate to low-solute areas by diffusion through the liquid, or (2) it can be homogeneously distributed throughout the liquid via convection (termed complete mixing), or (3) it can redistribute by both diffusion and convection (partial mixing). As solid phase continues to form, the surrounding liquid becomes progressively enriched in the solute. The new solid that forms becomes progressively enriched in solute in a manner consistent with the variation of solid composition along the solidus curve on the phase diagram. The solute concentration distribution and growth rates have been analyzed for the three physical situations alluded to above, that is, growth by solute diffusion in the liquid, growth by fluid convection, and growth by partial mixing (i.e., diffusion plus convection). The analyses assume that the composition of the solid does not change during growth, i.e., diffusion in the

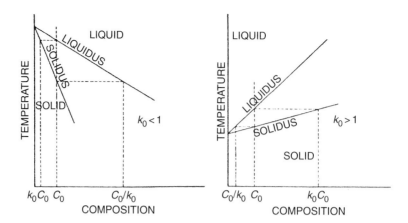

FIGURE 2-63 *Portions of hypothetical phase diagrams for which the partition coefficient, k_0, is less than or greater than unity ($k_0 = C_s/C_l$, where C_s and C_l are the solute concentrations in the solid and liquid, respectively).* (H. Biloni, in R. W. Cahn and P. Haasen, eds., Physical Metallurgy Principles, Elsevier, 1983, p. 478). Reprinted with permission from Elsevier.

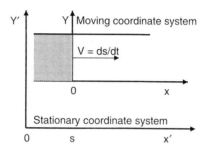

FIGURE 2-64 *Schematic illustration of a stationary coordinate system, and a coordinate system attached to a solidification front moving at a steady velocity, $V = ds/dt$.* (adapted from W. Kurz and D. J. Fisher, Fundamentals of Solidification, 4th ed., Trans Tech, Switzerland, 1998).

solid is negligible, and, therefore, solute concentration gradients exist in the growing solid.

1. *Diffusion in the liquid.* Because there is no liquid mixing in this case, a solute concentration gradient develops in the liquid ahead of the growing solid, and solute atoms diffuse down this gradient toward the bulk of the liquid. The solute distribution in the liquid is governed by Fick's second equation, which in one dimension, has the form

$$\frac{\partial C}{\partial t} = D\frac{\partial^2 C}{\partial x^2}, \tag{2-51}$$

where D is the diffusion coefficient of the solute in the liquid, assumed to be independent of solute concentration. Because diffusion during solidification takes place at a moving interface, it is convenient to describe the diffusion process with reference to a coordinate system (x, y) attached to the moving interface (Figure 2-64). The relationship between this moving coordinate system (x, y) and a stationary reference system (x', y') is $x = x' - vt$, where v is the constant interface velocity and t is the time. Differentiation of $x = x' - vt$ with respect to t yields $(dx/dt) = -v$. Note also that $(\delta x/\delta x') = 1, \delta C/\delta x' = (\delta C/\delta x)(\delta x/\delta x') = \delta C/\delta x$, and $(\delta^2 C/\delta x'^2) = (\delta^2 C/\delta x^2)$. The concentration distribution $C(x', t)$ in the stationary system is transformed to $C(x,t)$, the concentration distribution in the moving reference frame by using the chain rule of differentiation, i.e.,

$$\frac{dC}{dt} = \frac{\partial C}{\partial x} \cdot \frac{\partial x}{\partial t} + \frac{\partial C}{\partial t} \qquad (2\text{-}52)$$

Because

$$\frac{\partial x}{\partial t} = -v,$$

Fick's equation becomes

$$D\frac{\partial^2 C}{\partial x^2} = -v\frac{\partial C}{\partial x} + \frac{\partial C}{\partial t} \qquad (2\text{-}53)$$

For steady-state movement of the S–L interface,

$$\frac{\partial C}{\partial t} = 0,$$

so that

$$\frac{\partial^2 C}{\partial x^2} + \frac{v}{D} \cdot \frac{\partial C}{\partial x} = 0 \qquad (2\text{-}54)$$

To account for lateral solute diffusion, Equation 2-54 will need to be written in the generalized three-dimensional form in x, y, and z coordinates. For the one-dimensional form given by Equation 2-54, the general solution is $C = A + B\exp(-vx/D)$, where A and B are constants, and v is the growth velocity. For growth under steady-state, the boundary conditions are $C = C_0/k$ at $x = 0$ (i.e., at the growth front), and $C = C_0$ at $x = \infty$, where C_0 is the bulk liquid composition. The final solution to Equation 2-54 then becomes

$$C = C_0\left[1 + \frac{1-k}{k} \cdot \exp\left(-\frac{vx}{D}\right)\right] \qquad (2\text{-}55)$$

Equation 2-55 describes the solute composition ahead of the solidification front when diffusion in the liquid is the only mechanism of solute redistribution. This equation shows that high growth speeds, v, and small solute diffusivity, D, will both result in smaller concentration, C, at a fixed distance, x. In other words, a steeper concentration gradient will develop at high growth speeds and low solute diffusivity. Under these conditions, large-solute pile-ups will occur at the interface, which will destabilize a planar interface, and make the solution (Equation 2-55) inapplicable due to non-steady growth. The solution given by Equation 2-55 will, therefore, apply to the case of shallow gradients produced at small v and large D.

2. *Complete mixing in the liquid.* If there is complete solute mixing because of fluid convection, then the solute rejected into the liquid by the growing solid is homogeneously redistributed throughout the liquid. The solute composition, C_s, at a point where a fraction, f_s, of the sample has solidified during unidirectional solidification is

$$C_s = C_0 k (1 - f_s)^{(k-1)} \qquad (2\text{-}56)$$

where C_s is the concentration in the sample at a point where a fraction, f_s, of the material has solidified. This equation is called the Scheil equation or the nonequilibrium lever rule.

3. *Mixed control.* For growth controlled by both mixing and diffusion, it is assumed that the solute rich boundary layer of thickness, δ, ahead of the solidifying interface is gradually broken down by convection, induced either by vigorous stirring or by temperature gradients (natural convection). The solute concentration in the solid for this case has been derived and is given by

$$C_s = C_0 k_e (1 - f_s)^{(k_e - 1)} \qquad (2\text{-}57)$$

where k_e is an effective partition coefficient, defined from

$$k_e = \frac{k}{k + (1 - k) \cdot \exp\left[-\frac{v\delta}{D}\right]} \qquad (2\text{-}58)$$

To account for the different densities of the solid and liquid, the argument of the exponential term in the denominator of Equation 2-58 is multiplied by the density ratio. The composition of the solid obtained by this equation is intermediate between the two limiting situations represented by purely diffusive and purely convective growth considered above. The solute profiles during purely diffusive, purely convective, and mixed control situations are displayed in Figure 2-65.

The quantity of solute in the diffusion boundary layer increases as the interface growth velocity decreases. If a small perturbation in the growth velocity (due, for example, to some external disturbance) causes the velocity to increase momentarily, then the total quantity of solute in the diffusion layer in the liquid layer will need to decrease. Consequently, the solid will reject less solute in the liquid at that instant. As a result, that localized portion of solid will exhibit a higher solute content, leading to the phenomenon of banding that is often encountered in alloy solidification because of fluctuations of growth rate.

Constitutional Supercooling and Interface Instability

Consider an alloy of initial composition C_0 with $k < 1$ (Figure 2-63) solidifying under a steady-state condition. The solid and the liquid phases will be in equilibrium at all times during cooling, and their compositions will be given from the phase diagram. Because of equilibrium cooling, the temperature at the interface within the solid will be the solidus temperature, T_S, and the temperature at the interface within the liquid will be T_L, where T_S and T_L are given by the appropriate phase diagram. The equilibrium liquidus temperature of the alloy can be written with reference to a hypothetical binary phase diagram with liquidus and solidus lines assumed to be straight, as $T_L = T_m - m \cdot C$, where m is the slope of the liquidus line, and C is the liquid composition. The interface temperature, $T_i = T_m - m \cdot (C_0/k)$. If the solute rejected by the growing solid is removed only via diffusion through the liquid, then a solute gradient will develop in the liquid, and therefore, different regions of the liquid ahead of the interface in the

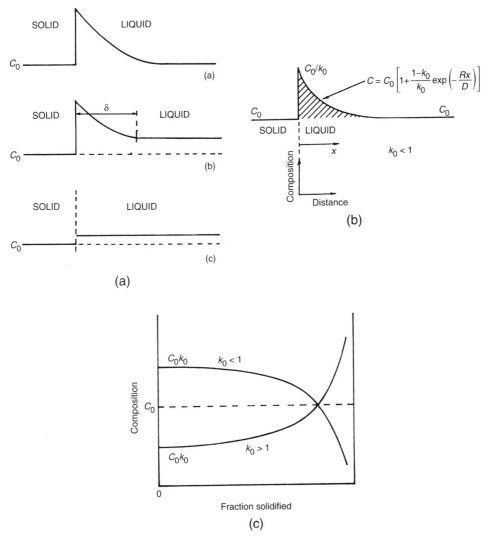

FIGURE 2-65 *(a) The effect of the mixing conditions on the nature of the solute layer at the solid–liquid interface: (1) no mixing, diffusion only, (2) partial mixing, and (3) complete mixing. (G. J. Davis, Solidification and Casting, Wiley, New York, 1979, p. 53) (b) The steady-state solute concentration profile in the liquid ahead of a moving solidification front with solute redistribution by diffusion only. (G. J. Davis, Solidification and Casting, Wiley, New York, 1979, p. 48) (c) Solute concentration profile of a bar solidified under conditions of complete solute mixing in the liquid as a function of fraction solid formed for alloys with $k_0 < 1$ and $k_0 > 1$. (G. J. Davis,* Solidification and Casting, *Wiley, New York, 1979, p. 52). Reprinted with permission from Applied Science Publishers, Ltd., Barking Essex, U.K.*

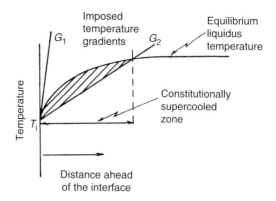

FIGURE 2-66 *Diagram showing the origin of constitutional supercooling resulting from solute rejection in the liquid ahead of the front. The constitutionally supercooled zone extends over the hatched region shown in the diagram, i.e., over the distance over which the equilibrium liquidus temperature is greater than the temperature due to imposed gradient.* (G. J. Davis, Solidification and Casting, Wiley, New York, 1979, p. 56). Reprinted with permission from Applied Science Publishers, Ltd., Barking Essex, U.K.

diffusion-affected zone will have different equilibrium liquidus temperatures corresponding to the solute concentration at any point. The liquidus temperature at different points in the liquid will be given from $T_L = T_m - m \cdot C$, where C is the liquid composition at a particular point in the liquid ahead of the interface. An expression for the liquidus temperature is obtained by substituting Equation 2-55 for liquid composition in the relationship $T_L = T_m - m \cdot C$. This yields

$$T_L = T_m - mC_0 \left[1 + \frac{1-k}{k} \cdot \exp\left(-\frac{vx}{D}\right) \right] \tag{2-59}$$

Because the interface temperature $T_i = T_m - m \cdot (C_0/k)$, the temperature T_m in the preceding equation can be replaced by the term $[T_S + m \cdot (C_0/k)]$. The resulting equation for the liquidus temperature, T_L, at different points in the liquid is

$$T_L = T_i + \frac{mC_0(1-k)}{k} \cdot \left[1 - \exp\left(-\frac{vx}{D}\right) \right] \tag{2-60}$$

A schematic plot of the liquidus temperature as a function of distance, x, ahead of the interface is shown in Figure 2-66. Note that because of heat extraction by an external heat sink (e.g., a mold), temperature gradients, G, exist in both the liquid and the solid. If, as shown in Figure 2-66, the temperature gradient in the liquid, G_L, is smaller than a critical value, then the liquid ahead of the interface is supercooled with respect to the actual temperature (determined by the imposed thermal gradient in the liquid). This phenomenon is called *constitutional* supercooling. The critical value of G at which constitutional supercooling is zero is the gradient that is tangent to the liquidus curve in Figure 2-66 at $x = 0$. Thus, constitutional supercooling will be absent when the imposed gradient $G > (dT_L/dx)$ at $x = 0$. Taking the first derivative of T_L with respect to x from Equation 2-60, and noting that $x = 0$ at the interface, we obtain the following condition for zero supercooling:

$$\frac{G}{v} > \frac{mC_0}{D}\left(\frac{1-k}{k}\right) \qquad (2\text{-}61)$$

This equation shows that constitutional supercooling will not occur when one or more of the following conditions is satisfied: there is a large thermal gradient in the liquid (so that the imposed gradient, G, is not greater than dT_L/dx at $x = 0$), the growth rate is low, the solute concentration, C_0, in the alloy is low (dilute alloy), the liquidus line on the phase diagram has a small slope (small m), the solute diffusion coefficient, D, in the liquid is large, and the partition coefficient, k, is large (for $k < 1$). If constitutional supercooling does occur, then the amount of undercooling due to this effect can be determined from $\Delta T = T_L - T$, where T is the actual temperature in the liquid at a distance x ahead of the interface.

The physical significance of the constitutional supercooling criterion is that it provides an initial assessment of the stability of an interface during phase change. In the presence of constitutional supercooling, a planar interface advancing into a liquid region of positive temperature gradient will become unstable and break down into cells and later into dendrites. This is because any small localized disturbance or perturbation in the shape of the front will amplify in the constitutionally supercooled liquid rather than melt back to a planar form. If, however, there was no constitutional supercooling, the interface will remain planar as in the case of a pure melt. Thus, Equation 2-61 can be used as a practical criterion to predict the conditions of interface breakdown. More rigorous theories for the stability of a planar solidification front have been developed that consider some important factors not considered by the simple constitutional supercooling criterion, such as the effect of capillary forces that tend to stabilize a planar front. The reader is referred to the books by Flemings and Kurz and Fisher for a discussion of the interface stability theories.

The preferential segregation of solutes due to their rejection by the solidifying interface has led to an important technique of purifying crystals, called zone refining. The technique is essentially a reversal of the floating-zone directional solidification technique (discussed earlier in this chapter), and is used to produce ultrahigh purity (10^{-8} atom%) materials for the electronic industry. In the zone-refining technique, a single heater or a battery of heaters is traversed along the length of a commercial purity ingot. For alloys with the partition coefficient $k<1$ (where $k = C_s/C_l$), the growth front rejects solutes into the liquid, which are swept to one end of the ingot (for $k > 1$, solutes will accumulate at the beginning portion). The molten zone is kept short to reduce mixing, and multiple passes are executed to obtain very high purity levels. In each pass, the concentration of solute, C_s, at a distance x from the first end of the bar is given from

$$C_s = C_0[1 - (1-k)e^{-kx/l}], \qquad (2\text{-}62)$$

where l is the length of the molten zone, and C_0 is the initial solute concentration in the ingot. After multiple passes in the same direction, a steady-state distribution of solutes is reached, and no further purification is possible. Under these conditions, the solute concentration at a distance x from the origin is, $C(x) = Ae^{Bx}$, where A and B are given from $A = C_0BL/(e^{BL} - 1)$, and $k = Bl/(e^{Bl} - 1)$, and L is the length of the bar. Zone refining is essentially the solidification analogue of the progressive fractionation process used in chemical industry, and it was instrumental in advancing the growth of semiconductor technology by providing ultrapure Si and Ge crystals for use in transistors.

Eutectic Solidification

In binary alloy systems such as Al-Si, Pb-Sn, Ni-Al, Al-Cu, and many others, a eutectic reaction occurs at a fixed composition. During this reaction, the liquid alloy isothermally transforms into a solidified structure consisting of alternating layers of two chemical phases α and β. Figure 2-67 shows the appearance of the eutectic phases in the microstructures of hypoeutectic and hypereutectic Al-Si alloys, and in the pseudobinary NiAl(Cr) alloys. For a binary alloy A-B of eutectic composition, solidification will commence when the eutectic temperature, T_e, is reached. At T_e, the liquid phase (L) decomposes into two solid phases—α (A-rich) and β (B-rich), via the reaction $L \rightarrow \alpha + \beta$. The two phases in the eutectic could grow with a variety of morphologies, such as lamellar (alternating plates of α and β), acicular (randomly oriented needles of α and β), and rodlike or globular (spheroidal α and β particles). Figure 2-68 shows common eutectic phase morphologies in metallurgical alloy systems. A number of alloys exhibit a lamellar eutectic morphology in which the α and β phases grow as lamellae in a direction that minimizes the total energy of the system. It is interesting to note that even though a globular shape will present the smallest surface for a given volume, it might not be the energetically most favorable shape. This is because of the role played by the magnitude and crystallographic anisotropy of the interfacial energy ($\gamma_{\alpha\beta}$) of the α/β interface. In many alloys, the magnitude of $\gamma_{\alpha\beta}$ depends on the crystallographic orientation of the interface; as a result, α and β phases (and hence, α–β interface) grow in a direction that minimizes the total surface free energy. Because of this orientation dependence of $\gamma_{\alpha\beta}$, non-globular eutectic morphologies could form. The morphology depends also on the volume fraction of the phases in the eutectic—a lamellar structure forms when α and β phases have roughly equal volume fraction. In contrast, a rod morphology is favored when one of the phases has a disproportionately lower volume fraction than the other phase.

Eutectic growth is also influenced by the atomic-scale roughness of the solid–liquid interfaces (i.e., α–L and β–L interfaces). If both interfaces are rough (i.e., the α-factor of the interface < 2), then eutectic growth is limited by the diffusion of atoms in the liquid ahead of the α–L and β–L interfaces rather than by the atom attachment kinetics. This is because the atomic diffusion through the liquid is slower than the rate at which liquid atoms arriving at the

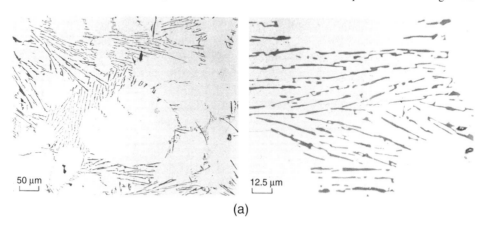

50 µm   12.5 µm

(a)

FIGURE 2-67 *(a) Low and high magnification views of the microstructure of a conventionally cast hypoeutectic Al-Si alloy showing primary Al dendrites (light phase), and acicular Al-Si eutectic.* (D. L. Zalensas, ed., Aluminum Casting Technology, 2nd ed., 1997, p. 41). Reprinted with permission from American Foundry Society, Schaumburg, IL (www.afsinc.org).

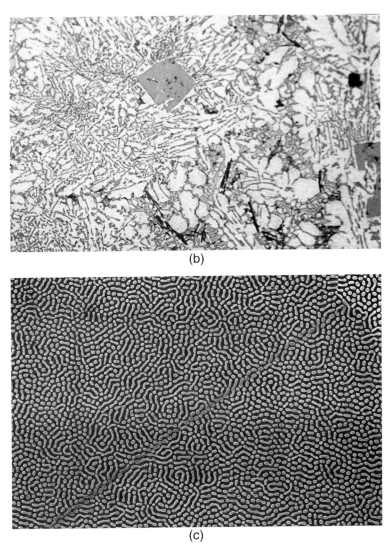

(b)

(c)

FIGURE 2-67 continued *(b) Microstructure of a conventionally cast hypereutectic Al-Si alloy showing primary Si and the eutectic* (Courtesy of S. Das, Regional Research Lab, Bhopal, India) *(c) Eutectic phase mixture in directionally solidified NiAl containing chromium.* (R. Asthana, S. N. Tewari and R. Bowman, unpublished work, 1992, NASA Glenn Research Center, Cleveland, OH).

growth front attach themselves to an atomically rough growth front (which provides several nearest neighbors and, therefore, presents a small energy barrier to the attachment process). As one of the phases in the eutectic is richer in element A and the other in element B, the α and β phases grow in a "coupled" manner, i.e., A-type atoms rejected by B-rich β diffuse through the liquid toward α–L interface and are incorporated in the A-rich α-phase. Similarly, the B-type atoms rejected by A-rich α-phase diffuse toward the β–L interface. Figure 2-69 shows the diffusion process occurring during eutectic growth. If in contrast, one of the S–L interfaces is rough and the other smooth (faceted), then the faceted interface will grow faster

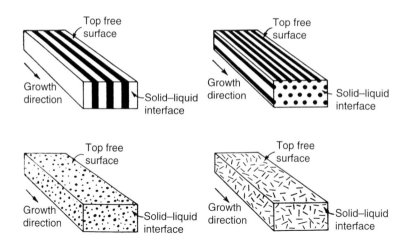

FIGURE 2-68 *Typical eutectic morphologies in alloys. These include plates (or lamellae), rods, globules, and fibers or needles. (*After W. C. Winegard, An Introduction to the Solidification of Metals, Institute of Materials, London, 1964).

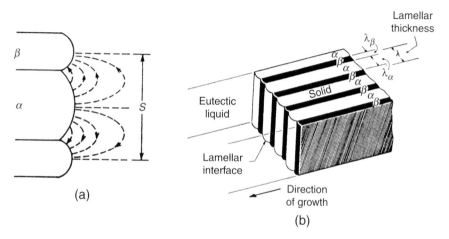

(a)

(b)

FIGURE 2-69 *(a) Coupled growth of the eutectic phases α (solute-impoverished) and β (solute-enriched). The solute atoms rejected by the α phase diffuse through the liquid ahead of the composite interface toward the β phase. (b) A three-dimensional view of the lamellar structure of eutectic phase mixture showing the lamellar thickness.* (W. A. Tiller, Liquid Metals and Solidification, 1958, p. 276). Reprinted with permission from American Society for Materials, Materials Park, OH (www.asminternational.org).

than the rough interface. This is a consequence of the fact that growth of the faceted interface will be faster along preferred directions due to availability of steps and ledges. In eutectic Al-Si alloys, the α-factor for Si is greater than 2, and Si grows in a faceted manner. Finally, when both phases grow with faceted S–L interfaces, the growth is not coupled and a random mixture of the two phases is produced. Metallic alloys usually do not show this type of eutectic growth, but certain organic and inorganic alloys do.

Studies on the directional solidification of eutectic alloys show that the spacing between the constituent phases growing in a coupled manner is related to the growth velocity R by

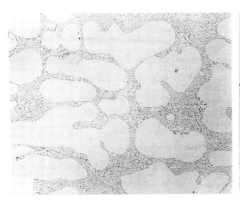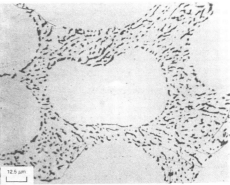

12.5 μm

FIGURE 2-70 *Low and high magnification views of the microstructure of a conventionally cast hypoeutectic Al-Si alloy in which modification treatment resulted in a fine, globular eutectic (compare this figure with the unmodified structure shown in Figure 2-67a). The modification treatment effected by adding a small amount of Na, Sr, or P to the alloy alters the acicular shape of the silicon, thus reducing the stress concentration and improving the strength.* (D. L. Zalensas, ed., Aluminum Casting Technology, 2nd ed., 1997, p. 42). Reprinted with permission from American Foundry Society, Schaumburg, IL (www.afsinc.org).

$\lambda = a \cdot R^{-1/2}$, where λ is the interlamellar spacing in the eutectic. This relationship shows that if R is increased, the spacing λ will decrease, i.e., a fine eutectic structure will form at high growth velocities.

In the preceding discussion, an alloy of eutectic composition was considered. For an off-eutectic (hypo- and hypereutectic) alloy, the formation of a proeutectic phase (e.g., α in a hypoeutectic alloy and β in a hypereutectic alloy) will precede the growth of the eutectic. For many alloys, however, it is difficult to microscopically distinguish the proeutectic phase from the same chemical phase formed in the subsequent eutectic reaction. For example, under a microscope, proeutectic α–Al in a hypoeutectic Al-Si alloy is indistinguishable from the eutectic Al. In contrast, in a hypereutectic Al-Si alloy (Si > 12%), the proeutectic Si is clearly distinguishable from the eutectic Si because the former grows as large and blocky faceted particles, whereas eutectic Si grows in an acicular fashion under normal growth conditions. An important alloy treatment used in casting practice modifies the normal acicular shape of the eutectic Si in Al-Si alloys to a globular morphology. This is accomplished by adding a small amount of Na, Sr, or P to the molten alloy. The atoms of these modifiers poison the growth of the acicular Si along its long axis during alloy solidification, thus causing the crystallizing Si to spheroidize. Modification treatment reduces the stress concentration resulting from the sharp ends of the acicular Si and improves the tensile strength and ductility of cast Al-Si alloys. Figure 2-70 shows the modified structure of the eutectic in a hypoeutectic Al-Si alloy; a comparison of this structure with that of an unmodified Al-Si alloy (Figure 2-67) reveals the significant change in the silicon morphology achieved through the modification treatment.

Solidification of Industrial Castings:

Grain Structure
The solidification process and the cast structure in real industrial castings is more complex than in binary alloys directionally solidified (DS) under controlled conditions. However, the scientific

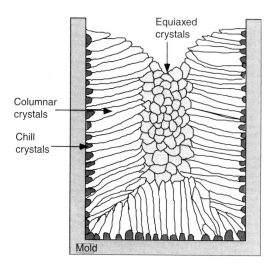

FIGURE 2-71 *Schematic illustration of the microstructure of a real casting solidified in a mold. Three distinct regions can be identified: an outer chill zone, an intermediate columnar zone, and an interior equiaxed zone.* (R. Trivedi, Materials in Art and Technology, Taylor Knowlton, Ames, IA, 1998, p. 177).

insights derived from the controlled DS studies provide a useful framework to understand the structure of industrial castings. Such studies have enabled unprecedented control and design of the cast microstructure for properties.

Figure 2-71 shows a schematic view of the microstructure that forms in a real casting that is cooled in a normal (multidirectional) fashion. Three geometrically distinct structural zones form in such a casting: a chill zone, a columnar zone, and an equiaxed zone. When a superheated liquid metal is introduced in a cold mold, the liquid layer in contact with the cold surface quickly supercools, usually even before the entire mold has filled. Copious heterogeneous nucleation in this supercooled liquid occurs due to the availability of a substrate (mold wall). This results in the formation of a very fine grain structure in a thin solidified layer at the mold wall, called the *chill zone*. Chill zone is, however, rarely of technological significance because it is too thin to influence the bulk properties and is removed during subsequent machining and surface preparation. Heat released from the solidified chill zone raises the mold wall temperature, decreases the temperature gradient through the solidified layer, and slows the growth of new solid crystals in the supercooled melt. Furthermore, heat flowing from the bulk liquid toward the growing solid continues to decrease the supercooling at the solid–liquid interface, and the solidification rate of dendritic crystals emanating from the chill zone progressively decreases. These dendritic crystals grow fastest normal to the mold wall, in a direction opposite to the direction of heat extraction by the mold leading to the formation of oriented columnar crystals (columnar zone). Primary dendrites (i.e., the central stem of growing crystals) whose preferred crystallographic growth direction is not normal to the mold wall will grow slower and be physically eliminated by neighboring faster growing dendrites that are oriented perpendicular to the mold wall. Experiments show that for relatively small-size castings made from an alloy of a fixed composition, the thickness of the columnar zone increases as the pouring temperature increases. In contrast, at a fixed pouring temperature more concentrated alloys (with higher solute content) exhibit a thinner columnar zone.

The liquid in the center of the mold solidifies with randomly distributed, roughly spherical (or equiaxed) grains. This region of a casting displays an *equiaxed zone*; the grain size in this zone is coarser than that in the chill zone. High pouring temperatures reduce the size of the equiaxed zone. The equiaxed grain structure could form because of several reasons. For example, many fine crystals (pre-dendritic nuclei) formed near the chill zone during pouring may be transported by fluid convection to the center of the casting where they act as preferred sites for grain growth (the Big Bang mechanism). This mechanism will operate only at low to moderate superheats that will not remelt the preformed nuclei. Another mechanism of equiaxed zone formation involves remelting and mechanical detachment of small dendrite arms under fluid flow, and their migration toward the center of the casting, where they provide a surface for the growth of equiaxed grains.

Thus, when heat flow is directional through the solid (i.e., both solid and liquid have a positive temperature gradient), columnar growth will occur in the casting. If, on the other hand, liquid is undercooled (negative temperature gradient at the interface), then an equiaxed grain structure will form. It is possible to control the solidification conditions to make castings with either columnar or equiaxed grain structure. Grain boundaries form when dendritic crystals growing out of neighboring nuclei impinge; thus each grain evolves from a single nucleus.

Segregation

Large-scale segregation of the alloying elements, spanning distances that scale with the size of the casting, can develop during solidification. Such segregation, called macrosegregation, results from the bulk movement of liquid and solid during solidification process. This movement may be caused by gravitational forces due to density differences arising out of changes in temperature or composition, solidification contraction, or capillary forces such as those caused by surface tension gradients at the free surface of the melt (where cooling causes surface tension to increase).

Small-scale solute segregation at length scales comparable to cell and dendrite size could also occur during solidification. This type of segregation—called microsegregation or coring—results from the constitutional supercooling. The solidification microstructures of a commercial Al-Cu (2014Al) alloy displayed in the photomicrographs of Figure 2-72 reveal the microsegregation within the dendrites. In an earlier section, it was mentioned that constitutional supercooling occurs when the solute diffusion is relatively slow, and microsegregation develops at micrometer-length scales. This is because of solute rejection and its limited diffusion in the liquid, which lead to a difference in the composition at the center of a cell (formed first), the outer regions (formed later), and the liquid in the gap between neighboring cells (intercellular region). For alloys with the partition coefficient, $k < 1$, the liquid in the intercellular regions becomes progressively rich in solute, and may form secondary phases. In contrast, alloys with $k > 1$ show an opposite behavior, and intercellular regions become depleted in solute. For single-phase alloys, microsegregation is expressed in terms of a segregation ratio, which is the ratio of the maximum solute concentration (at cell root) to the minimum solute concentration (at cell tip).

Both macro- and microsegregation are undesirable phenomena because of the compositional inhomogeneity they introduce in a cast part. Segregation can be erased through a secondary heat treatment called *homogenization*. Because microsegregation occurs over small distances (on the order of micrometers), it is eliminated at short times and relatively low heat treatment temperatures. The kinetics of homogenization depend on the cell spacing and the diffusion coefficient in the solid. In contrast, macrosegregation usually requires longer times because solute atoms must diffuse over larger distances (comparable to the dimensions of a casting). The

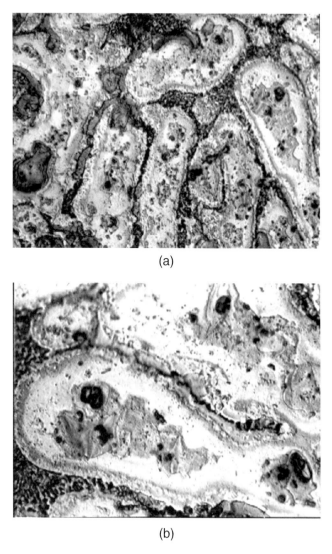

(a)

(b)

FIGURE 2-72 *(a) Microstructure of a cast Al-Cu (2014) alloy showing the dendritic structure and coring patterns revealed by etching with a strong oxidizing reagent (regions of different shades in this black-and-white photograph appear as color patterns on a color photomicrograph and display the solute distribution across the dendrites). (b) A higher magnification view of a single cored dendrite of Figure 2-72 (a) showing segregation patterns revealed via color etching.* R. Asthana and P. K. Rohatgi, Zeitschrift für Metallkunde, 84, 1993, 44–47.

time required for homogenization can be estimated from $t \sim (x^2/4D)$, where D is the diffusion coefficient of the solute in the solid at the heat treatment temperature, and x is the diffusion distance.

Homogenization treatment can also dissolve any secondary phases in the microstructure. A simple analytical model for a platelike dendrite morphology with an initial sinusoidal composition distribution in the primary phase gives the following equation for the volume fraction of the secondary phases as a function of the dissolution (homogenization) time

$$\frac{g+a}{g_0+a} = \exp\left[-\frac{\pi^2}{4}\frac{Dt}{l_0^2}\right] \qquad (2\text{-}63)$$

where g is the volume fraction of the secondary phase at time t, g_0 is the initial volume fraction (at $t = 0$), l_0 is one-half of the spacing between secondary dendrite arms, a is the concentration parameter defined as $a = (C_M - C_m)/C_p$ and C_M is the solute concentration in the matrix far from the second phase, C_m is the initial matrix concentration, and C_p is the concentration of the secondary phase (usually constant during dissolution). The critical time, t_c, for complete dissolution is obtained by setting $g = 0$ in Equation 2-63. This yields

$$t_c = \frac{4l_0^2}{\pi^2 D} \cdot \ln\left(\frac{g_0+a}{a}\right) \qquad (2\text{-}64)$$

This equation shows that a large diffusion coefficient and fine secondary dendrite arm spacing will cause the secondary phases to be erased rapidly.

Constrained Solidification in Small Regions

When a liquid solidifies in a small, confined region whose size is on the order of a few micrometers or less, the transport processes responsible for solute redistribution and solidification morphology are physically restricted. Such a situation is encountered when, for example, a liquid solidifies within a fine-bore capillary or within pores of a liquid-saturated porous solid. Alloy solidification within the interstices of a fiber bundle that is infiltrated with a molten metal to make metal-matrix composites is another example of solidification in small regions. In all such cases, the impervious walls of the zone will act as a barrier to solute diffusion and fluid flow, restrict the thermal diffusion during solidification (if the barrier is thermally insulating), and alter the undercooling at the growth front.

Let us first consider a small thermally insulating physical barrier ahead of an advancing planar solidification front in an undercooled melt at slow growth rates. When the planar front approaches the stationary obstruction, the motion of the interface segment nearest the barrier is retarded relative to the rest of the interface, causing a concavity of finite curvature to develop in an initially planar front. Introduction of concavity in the front will induce lateral solute diffusion currents, which will further enhance the curvature and develop steeper solute gradient in the liquid at the leading interface region. As a result, a planar front could break down into a cellular or a dendritic interface.

Now consider solidification within small channels. In very small channels, the interaction distances for solute and thermal fields could become comparable to the zone size. The order of magnitude of diffusional and thermal interaction distances are (D/V) and (α/V), respectively, where D is the diffusion coefficient, V is the front velocity, and α is the thermal diffusivity. For metallic alloys solidifying under normal cooling conditions, the diffusional and thermal interaction distances are on the order of a few μm. Pores and channels of comparable sizes will modulate the solutal and thermal fields. This in turn will alter the growth morphology and composition. Figure 2-73 illustrates the influence of finite zone size on solidification morphology and solute segregation in micrometer-size channels formed when a ceramic fiber bundle was infiltrated by an Al-Cu alloy. When the secondary dendrite arm spacing (DAS) in freely solidifying alloy is less than the zone size (Case 1, Figure 2-73a), solidification morphology and microsegregation are not affected. When DAS is comparable to zone size, secondary phases

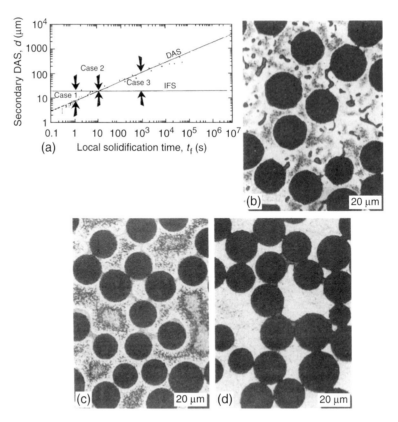

FIGURE 2-73 *Secondary dendrite arm spacing (DAS) as a function of local solidification time in an alumina fiber-reinforced Al-Cu alloy in which the solidification occurred in small regions between aligned alumina fibers. Different coring patterns and solidification morphologies result depending on whether the secondary DAS is larger than, comparable to, or smaller than the interfiber spacing (IFS).* (A. Mortensen, M. N. Gungor, J. A. Cornie, and M. C. Flemings, JOM, March 1986, p. 30). Reprinted with permission from the Minerals, Metals and Materials Society, Warrendale, PA (www.tms.org).

(e.g., eutectic) deposit onto the wall where the solute content is highest (minimum solute content occurs in the center of the zone). Finally, when the zone size is smaller than the DAS (which can be altered by the rate of external heat extraction), microsegregation is reduced, less secondary phases precipitate, and the solute content in the matrix is higher than that in the unconstrained matrix (Fig. 2-73d). Constrained solidification is also observed within the pores between ceramic particulates (Fig. 2-74) when the latter are infiltrated with a molten alloy to synthesize discontinuously-reinforced metal-matrix composites. The solidification zone morphology in the pores between ceramic particulates of arbitrary shapes is less ordered than that between fibers in a composite perform. In addition, some rearrangement of the ceramics could occur during exposure to melt, causing unpredictable changes in the pore size and shape. Nevertheless, the physical constraint offered by the pore alters the crystal growth morphology, and in sufficiently small zones, microsegregation may be completely obliterated.

Figure 2-75a shows some experimentally observed dendrite morphologies in narrow channels in the organic compound, succinonitrile, which is used to simulate the metal

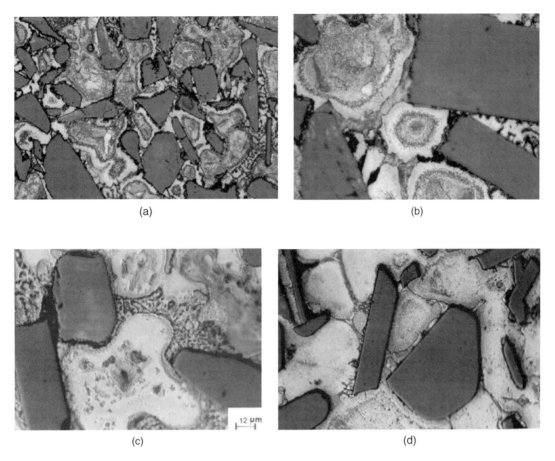

FIGURE 2-74 *(a) Coring patterns in an Al-Cu matrix solidified between SiC platelets. (b) A higher magnification view of a cored region of Figure 2-74(a). (c) Dendritic growth constrained by SiC platelets in a composite made by pressure infiltration of SiC platelets by an Al-Cu alloy. (d) Another example of constrained dendritic growth between closely spaced SiC platelets.* R. Asthana and P. K. Rohatgi, Zeitschrift für Metallkunde, 84, 1993, 44–47.

solidification behavior, and Figure 2-75b presents a schematic map for morphological transformations during solidification in finite zones. In Figure 2-75b, the growth velocity is plotted as a function of the ratio (d/λ), where d is the channel width and λ is the secondary DAS in normal (unconstrained) solidification of an alloy. For very narrow channels (small d/λ), a dendritic structure will tend to become cellular as the ratio (d/λ) is decreased. If, however, (d/λ) is close to or slightly larger than unity, then a cellular structure in a large channel will become dendritic as $(d/\lambda) > 1$. At very small (d/λ), cellular regime is expanded as both planar and dendritic structures tend to become cellular. When (d/λ) is only slightly greater than unity, the dendritic regime will be expanded to lower velocities. In addition, half-dendrites and half-cells could also form in finite zones as is illustrated in Figure 2-75a.

The shape of the solidification zone (i.e., channel shape) will also influence the growth morphology. Experiments show that larger cells form in capillaries of circular cross-section

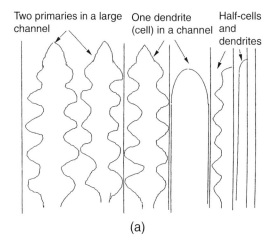

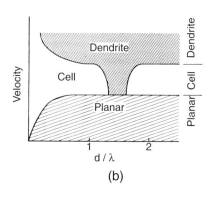

FIGURE 2-75 *(a) Possible morphologies of the solid phase crystallizing in a constrained space. (J. A. Sekhar, R. Trivedi, and S. H. Han, in Solidification of Metal-Matrix Composites, ed. P. Rohatgi, 1990, p. 21, The Minerals, Metals and Materials Society, Warrendale, PA). (b) Schematic microstructure map showing regions of different solidification morphologies for a range of solidification front velocity and the parameter (d/λ), where λ is the interfiber spacing, and d is the secondary dendrite arm spacing. (J. A. Sekhar, R. Trivedi, and S. H. Han, in Solidification of Metal-Matrix Composites, ed. P. Rohatgi, 1990, p. 21, The Minerals, Metals and Materials Society, Warrendale, PA). Reprinted with permission from The Minerals, Metals & Materials Society, Warrendale, PA (www.tms.org).*

than cells growing between closely spaced flat parallel plates. If the cross-section of the zone varies (as in a diverging or converging channel), transport processes will not be unidirectional, and interface velocity will constantly change. When the changed cross-section is larger than the primary dendrite arm spacing in unconstrained growth, there is no effect on dendrite spacing. However, large changes in cross-section could significantly change the growth morphology; for example, a planar front might break into cells and dendrites. Also, as the growth front approaches a progressively narrowing region, fluid flow will occur to compensate for the volume contraction. This can erase or reduce the solute gradient at the interface.

In small solidification zones, coarsening of secondary dendrite arms is accelerated and the extent of microsegregation is reduced. The solute concentration at the surface of a dendrite will vary spatially because of local variations in the surface curvature. Solute gradients will develop along the surface, and solute diffusion will occur to minimize the curvature via coarsening. When the solid volume fraction is low (during early stages of solidification), small dendrite arms melt and resolidify onto larger arms. In contrast, at high solid volume fractions, small dendrite arms merge or join into larger dendrites. The coarsening of secondary arms by diffusion within solid is enhanced because the finite size of the zone limits the diffusion distance. This reduces the extent of microsegregation. As microsegregation is reduced in a constricted space, an economical way to erase coring without the need for homogenization is to cool the material at a moderate rate to the solidus temperature, and hold it there to enable dendrite coalescence and erasure of microsegregation to occur.

The accelerated coarsening of dendrite arms in small zones will completely eliminate the dendritic structure after some critical time, t_c. Experiments show that t_c increases roughly linearly with the total solidification time, t_f. At $t < t_c$, the structure is dendritic and at $t_c < t < t_f$, a nondendritic, featureless structure forms; thus (t_c/t_f) determines the degree of dendritic character of the solidified microstructure. If t_c is reached early during solidification (i.e., small t_c/t_f), then a nondendritic structure is favored, and if (t_c/t_f) is large, then a dendritic structure forms. The condition $(t_c/t_f) = 1$ gives the *minimum* solidification time t_f, which would yield a fully coalesced (featureless) structure. Experiments show that minimum solute concentration in the zone occurs at the zone center for alloys with a partition coefficient k < 1, and is greater than that in free (unconstrained) solidification; this concentration increases with a decrease in the zone size and increase in t_f. Thus, within the smallest zones and at long t_f, the minimum solute concentration (a measure of microsegregation) would increase significantly.

Rapid Solidification and Metallic Glass

Rapid solidification involves quenching an initially liquid metal at extremely fast rates (10^5 to 10^6 degrees per second) to produce ultrafine crystalline or noncrystalline (glassy) metals. During solidification at such fast rates, the mobility of atoms responsible for interface motion (i.e., solid's growth) is much greater than the rate of redistribution of solute atoms. As a result, solute diffusion and microsegregation are suppressed. Rapidly solidified structures can be achieved using a variety of techniques such as laser surface melting, atomization of molten metal droplets, melt spinning, and quenching in twin-rollers. In laser surface melting, a thin film of molten metal is rapidly formed on the surface of a solid metal with the help of a laser gun. The liquid self-quenches via rapid heat conduction through the cooler solid metal. Rapid solidification is also encountered in weld solidification as discussed in the next section.

Melt atomization is used to form fine alloy powders that are hot-consolidated into net shapes. Atomization under a helium atmosphere transforms the molten metal into rapidly quenched fine (10–100 μm) powders that are collected and hot-pressed or hot-extruded in dies to form net-shape objects. The temperatures during pressing and extrusion are kept relatively low to avoid grain coarsening (or partial crystallization in the case of glassy metal powders). Rapidly solidified powders can also be obtained by crushing and milling of brittle melt-spun ribbons.

For solidification of gas-atomized metal droplets, the surface-to-volume ratio is very large. Droplets will cool by Newtonian cooling (i.e., by heat transfer across the gas–droplet interface via convection) and by heat conduction through the droplet. The relative contributions of these heat transfer processes is expressed in terms of a dimensionless group, Biot number, $Bi = hR/K$, where h is the heat transfer coefficient at the gas–droplet interface, R is the droplet radius, and K is the thermal conductivity of the droplet.

In melt spinning, molten alloy is forced out of a ceramic crucible via a nozzle at the crucible base with the help of a steady gas pressure. The molten metal comes in contact with a rapidly spinning copper wheel, usually cooled by circulating water, and rapidly quenches into ribbons 20–50 μm thick and up to 15 cm wide. Process variables such as the wheel material and surface finish, wheel speed, nozzle size and shape, ejection pressure, melt flow rate, angle of jet impingement, metal temperature, and gaseous atmosphere all influence the quenched material's structure and properties. A range of microstructures—plane front, cellular, dendritic, equiaxed— are observed across the melt-spun crystalline ribbons due to varying solidification conditions. Melt spinning is also known to lead to elongated grains and preferred texture.

In melt spinning, a thin atmospheric gas film constantly travels with the rotating wheel, and affects the heat transfer and ribbon texture. Turbulence in the gas film is deleterious to the ribbon quality, and the gas film should not exceed a critical Reynold's number, Re, where Re $= vd\rho/\eta$, and v is the gas velocity (i.e., surface speed of the rotating wheel), d is the ribbon width, and ρ and η are gas density and static viscosity, respectively. Slow rotation and light and thermally conductive gases (e.g., helium) yield low Re, and better quality ribbons. Finally, in the twin-roller method, a stream of molten metal is constantly introduced under gravity at the pinch point between two rotating metal rollers. The advantage of using twin rollers is that both surfaces of the liquid are quenched (unlike melt spinning where only one surface is quenched), but the contact time between the melt and the rollers is very short, which yields a slower quench.

For solidification against a metal substrate (as in melt spinning), heat transfer and temperature distribution are determined by Biot number ($Bi = h'l/K'$) and Fourier number ($Fo = \alpha t/L^2$), where h' is the mold–metal interface heat transfer coefficient, l is the thickness solidified in time t, K' is the conductivity of the liquid, and α is the thermal diffusivity of the liquid metal ($\alpha = K'/C\rho$, where C is the specific heat and ρ is the density). Extremely rapid cooling rates can be achieved most readily in thin sections (e.g., splats or droplets, i.e., at small R and l). This is because for thin sections, thermal conduction through the liquid and solidified layers is negligible compared to the surface heat transfer (via convection, conduction, and/or radiation), which can be enhanced by employing an external cooling medium.

Rapid solidification of alloys leads to several compositional and structural modifications, which include extension of solid solubility, grain refinement, formation of metastable phases, reduced microsegregation, and formation of glassy metals. Grain refinement enhances the tensile strength, and in the case of steels, lowers the martensitic transformation temperature. Rapidly solidified light-weight Al-Li alloys have very high elastic modulus and are used in aircraft structural parts. Net-shape forming by hot pressing of rapidly solidified powders made from difficult-to-machine Ni-base superalloys that are used in the aircraft industry also shows promise, although the fine grain structure (and large grain boundary area) of rapidly quenched superalloys is a disadvantage for high-temperature creep resistance. Iron-based metallic glasses (e.g., $Fe_{82}B_{12}Si_4C_2$, $Fe_{40}Ni_{40}P_{14}B_6$, and others) have superior magnetic permeability, low hysteresis losses and high resistance to eddy currents, and are used in electrical transformers.

Weld Solidification

Welding involves joining of metals by fusion. Welding techniques differ primarily in the method employed for heating. In gas tungsten arc welding (GTAW), an arc is struck between a nonconsumable tungsten electrode and the workpiece under a cover of an inert gas, usually argon or helium. Arcing melts a filler rod or wire that is introduced in the fusion zone, causing it to deposit at the joint region. The method is suitable for reactive metals such as titanium that form refractory oxides in air. However, high welding currents used in GTAW could cause partial melting of the tungsten electrode and formation of brittle tungsten inclusions in the deposit. In gas metal arc welding (GMAW), an arc is established between a consumable filler metal electrode and the work under an argon or helium cover. In electron beam welding, a high-energy beam of electrons is focused to a narrow diameter (< 1 mm) over the work, and in laser beam welding, a laser beam is used as the fusion source. Electron beam welding requires use of vacuum to prevent electron scattering by air molecules, but laser welding is preferred under normal ambient atmosphere.

Three distinct zones form in a fusion weld (Figure 2-76): fusion zone (FZ) or weld metal, the heat-affected zone (HAZ), and the unaffected base metal (BM). From a solidification standpoint, FZ is the most important region. FZ can be likened to a tiny casting in which the evolution of the solidification microstructure is determined by the alloy composition, growth rate (R), temperature

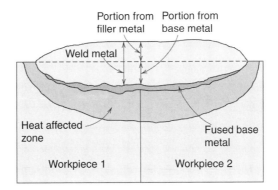

Portion from
filler metal

Portion from
base metal

Weld metal

Heat affected
zone

Fused base
metal

Workpiece 1

Workpiece 2

FIGURE 2-76 *Microstructurally distinct zones in the vicinity of a typical fusion weld.* (W. D. Callister, Jr., Materials Science & Engineering: An Introduction, 5th ed., Wiley, New York, 2000, p. 356).

gradient (G) and degree of undercooling (ΔT). In the heat-affected zone (HAZ), the metal is subjected to an intense thermal cycle, which leads to solid-state phase transformations that play an important role in influencing the joint properties. The HAZ structure is characterized by a mixture of different grain sizes, with grain size being the largest near the fusion line.

The basic solidification behavior of fusion welds is quite similar to that of an ordinary casting. However, solidification rates and temperature gradients in the liquid are much greater than those in normal castings. This has significant ramifications in terms of the solidification response. For example, in fusion welds, a chill zone does not form, and partially melted grains at the molten zone boundary promote columnar solidification in the weld pool. A nucleation barrier is not encountered for columnar growth. Another difference between ordinary cast structure and weld microstructure is the absence of equiaxed zone in the latter. Even though fluid flow in the molten pool is extensive, the dendrite fragments (from previously formed dendrites) could remelt due to the very high melt temperature and may not promote equiaxed grain formation. However, the solidification microstructure in welding is influenced by convection in other ways. Convection is induced by (1) buoyancy (density of liquid decreasing with increasing temperature), (2) electromagnetic (EM) forces due to nonuniform electric current field in the pool and the induced magnetic field, and (3) surface tension forces.

Rapid solidification in welding causes nonequilibrium phases to form in the structure. In addition, interaction of a heat source with the metal causes evaporation, remelting, vigorous liquid convection, solid-state transformations, and thermal stresses. Unlike the fixed shape of a casting, the weld pool geometry is dynamic, and the thermal gradient, G, and the growth rate, R, vary within the weld pool. As mentioned earlier, solidification in the weld pool can occur without an energy barrier to heterogeneous nucleation because wetting is perfect (i.e., contact angle = 0), and in some situations, partially melted base metal grains may provide the substrate for nucleation and crystallization. Growth could proceed in an epitaxial manner (i.e., each solidified layer bears a specific crystallographic orientation relationship with the previous layer), or it could lead to growth texture (e.g., columnar grains).

Compared to a normal casting, the microstructure length scales (i.e., cell and dendrite size) are much finer in fusion zone (FZ). Furthermore, rapid solidification causes the solute distribution between the liquid and solid phases to depart from the equilibrium value, causing extension of the solid solubility, and formation of nonequilibrium (metastable) phases and modified growth morphology at very high growth rates. For example, it is found that at very high growth rates in laser welding of low-alloy steels, only planar solidification occurs and dendritic solidification is suppressed.

Solidification Under Reduced Gravity

Solidification processes in the terrestrial environment invariably occur under Earth's gravity unless special techniques are devised to counter the influence of gravity for short intervals. Short periods of low gravity are produced using special drop towers that allow a brief span of testing in free-fall. Electromagnetic (EM) forces are also used to levitate small pools of conductive molten metals for special purposes. In the low-gravity environment of space, many physical processes caused by Earth's gravitational field are either completely eliminated or partially suppressed. For example, the buoyancy forces and fluid convection are reduced and sedimentation or flotation effects are suppressed. The thermal and mass transport processes driven by fluid convection will, therefore, be suppressed in a microgravity environment.

Materials processing studies, including controlled solidification experiments, have been conducted in space for well over two decades. In addition, containerless processing of alloys has been done in space to accurately determine the thermophysical and transport properties of reactive liquids, such as viscosity, diffusion coefficient, and surface tension. Accurate values of these properties are needed for use in mathematical and numerical modeling of the solidification process and microstructure formation in industrial castings. Many of the solidification experiments have involved study of crystal growth in metals and alloys (e.g., Bi-Sn and Sn-Pb), semiconducting materials (InSb, Si, Ge, GaSb, and GaAs), and nonmetallic liquids that simulate alloy solidification in metallic systems. For example, binary succinonitrile (SCN)-acetone alloys have been used to simulate the columnar growth during alloy solidification with non-faceted interfaces. In Earth-grown SCN-acetone samples, the solid–liquid interface is deformed and becomes irregular because of fluid convection. But SCN-acetone alloys grown in space exhibit an interface that is unperturbed by fluid flow and is more regular. This enables direct observations of columnar solidification unperturbed by fluid convection, which is difficult to accomplish in the terrerstrial environment.

Because fluid convection is suppressed in space, crystallization occurs purely by heat flow and solute diffusion unperturbed by fluid flow. This results in a highly regular pattern of cellular and dendritic crystals. It is also found that dendrites are usually larger in space-grown samples. The formation of equiaxed zone in castings is also influenced by the microgravity of space. The absence of sedimentation of crystal fragments and nuclei in space-grown samples due to suppression of fluid convection leads to a more homogeneous equiaxed grain structure as compared to Earth-grown samples where a large dispersion in the size of equiaxed grains is noted in the center of the casting.

Crystals of semiconducting materials such as InSb, Si, Ge, GaSb, and GaAs grown in space are defect free and chemically homogeneous. It has been possible to grow larger crystals of these materials in space using both Bridgman crystal growth technique and floating-zone technique (in a containerless manner). Whereas gravity-driven convection could be largely suppressed in space, not all convection can be eliminated. For example, surface tension variations due to temperature differences in the solidifying liquid could give rise to Marangoni convection, which is driven by surface tension gradients. This type of flow will limit the chemical purity and homogeneity even in space-grown crystals. Furthermore, booster rockets that keep the spacecraft on its course result in "g-jitters" or small random accelerations that could perturb the crystal growth process unless the test unit was positioned exactly at the center of mass of the spacecraft.

The phenomena of particle agglomeration and particle-solidification front interactions are important for inclusion control in castings and crystal purification. These are influenced by the microgravity environment of space. Microgravity experiments on melting and solidification of particulate composites, Ostwald ripening (microstructure coarsening) in the solid–liquid mixtures of alloys (e.g., Sn-Pb), and liquid-phase sintering in binary systems such as Ag-W, Cu-SiO_2, and Mo-Cu show that the tendency for phase segregation and clustering is reduced, and a more uniform phase distribution is achieved in the low-gravity environment. In one study on sedimentation and agglomeration, hot-pressed composite specimens of Ag containing W particles were melted and isothermally held in the molten state for prolonged periods aboard a spacecraft. No large-scale sedimentation of W in Ag melt occurred, but clustering of W particles because of their poor wetting with Ag melt was noticed. This clustering was, however, eliminated and a more uniform distribution achieved when nickel, which is a wettability-promoting element, was added to the composite specimen. The reference sample, processed in Earth's gravity, exhibited pronounced sedimentation, stratification of W particles, and agglomeration. In another space experiment, the distribution of SiO_2 particles in SiO_2-Cu (a poor wetting system) and of Mo particles in Mo-Cu (a good wetting system) was investigated. In SiO_2-Cu, the particles separated from molten Cu in space but in Mo-Cu, a uniform and stable suspension was achieved. The Mo-Cu specimens retained their uniform distribution when directionally solidified under microgravity conditions.

Levitation melting is another method to counter the effect of gravity in small samples. The method involves melting small quantities of electrically conductive materials that are suspended with the help of EM forces without the need of a crucible to contain the molten material. Molten alloys contained in crucibles at high temperatures could become contaminated by minute quantities of chemical species from reaction of the melt with the crucible material. Because a container is not needed in levitation melting, ultra-high-purity liquids can be obtained. Atmospheric contamination is eliminated by conducting levitation melting in a specially designed, leak-tight pressure vessel in either a vacuum or inert gas atmosphere. A specially designed induction coil is used to levitate and melt a small quantity of the material, and solidification is done by allowing the melt to fall into a mold. Usually, only small (<100 g) samples are processed.

When an electrical conductor is placed in an alternating magnetic field, a voltage is induced in it and an alternating current of the frequency of the inducing field but $180°$ out of phase with it, flows in the conductor. The alternating magnetic field is produced with an induction coil. The induced current heats the conductor. It also generates an opposing magnetic flux, which tends to push the object into a region of lower field strength; i.e., out of the coil. This force is given from the Lorentz equation $F = qvB$, where F is the force, B is the magnetic flux density, q is charge on an electron, and v is the electron velocity. The product qvB is the force that pushes against the specimen, and tends to push it into a region of lower magnetic flux density. If the magnetic field were uniform in space, there will be no net force tending to push the objective. It is, therefore, necessary to provide a field gradient to create a lift force to levitate the material. This is done by using a conical inductor coil. A buckling plate is located above the induction coil. Current induced in this plate causes an opposing magnetic field, distorting the magnetic flux, and creating a stable pocket of balanced magnetic forces for the levitated material.

Interactions of Solidification Front with Insoluble Particles

Control of solid inclusions and entrapped gas bubbles in the melt during solidification of metals, and growth of impurity-free single crystals demand a purification role for the solidification front, i.e., the solidification front must selectively reject the impurities. When a liquid containing insoluble foreign particles is solidified, the growing solid interacts with the particles in one of three distinct ways: the solid engulfs a particle instantaneously on contact, it pushes the particle indefinitely, segregating it in the last freezing liquid, or the solid engulfs the particle after pushing it over some distance (Fig. 2-77a). Experimental measurements on the effect of particle diameter on the solidification front velocity required to engulf the particles in some organic matrices are displayed in Figure 2-77b. These data show that greater velocity is needed to capture fine particles than coarse particles; this is a general behavior regardless of the materials used, and is observed in all metallic, and non-metallic liquids.

The particle–solidification front interactions are encountered during inclusion control in casting, solidification of discontinuously reinforced cast composites, growth of high-purity crystals, and cryopreservation. Such interactions can also occur in biological systems such as in phagocytosis, which involves ingestion of microorganisms by single white cells. In phagocytosis, the bacterium (particle) is transferred from the plasma (liquid) into the interior of the phagocytic cell (solid).

A particle is engulfed when particle–liquid interface and liquid–solid interface are replaced by a single solid–particle interface. For interface substitution to be energetically favorable, the free energy change, ΔF_{net}, of the process must be negative. Thus, particle engulfment is spontaneous when $\Delta F_{net} < 0$, and it is unfavorable when $\Delta F_{net} > 0$, where $\Delta F_{net} = \sigma_{sp} - \sigma_{pl}$, and σ's are surface energies of subscripted interfaces (s = solid, p = particle, l = liquid).

A particle will be pushed by growing solid as long as a liquid film occupies the gap between particle and solid, and prevents physical contact. The stability of a thin film supported between two solids depends on its energy, which is influenced by the local intermolecular forces at the supporting solid surfaces. The surface energy of the liquid film confined between two solids becomes a function of its thickness because the structure of the liquid changes over a few molecular diameters near the interface (Figure 2-77c and d).

The basic mechanism of pushing involves a balance of repulsive forces arising from the need to maintain a stable liquid film (which prevents contact and particle engulfment) and the attractive forces (e.g., fluid drag) (which compress the particle toward the front during growth and therefore favor engulfment). Because the particle interferes with the mass transport processes, the local growth velocity of the front behind the particle is altered. The front, therefore, acquires a net curvature, which in turn leads to a change in the melting point or a change in the free energy of fusion. The shape of the perturbation in the front that develops under the particle can be determined from the relationship between the interface temperature and kinetic undercooling, Gibbs-Thompson curvature factor, and temperature changes associated with external forces (e.g., gravity) and fluid drag. Because the repulsive forces arise from the need to maintain a stable film, the surface energy term $\Delta\sigma_0$ provides the driving force for repulsion, and a larger value of $\Delta\sigma_0$ makes engulfment more difficult and hence a higher front velocity is required to engulf the particles in systems having a large value of $\Delta\sigma_0$ (Figure 2-78).

During the growth of a solid from a pure melt under a positive temperature gradient, foreign matter with thermal properties different from the melt can distort the gradient locally, by serving as local thermal resistance. In the case of relatively coarse particles (>500 μm) in a positive temperature gradient, if $K_p < K_l$, the particle shields the local segment of the interface underneath

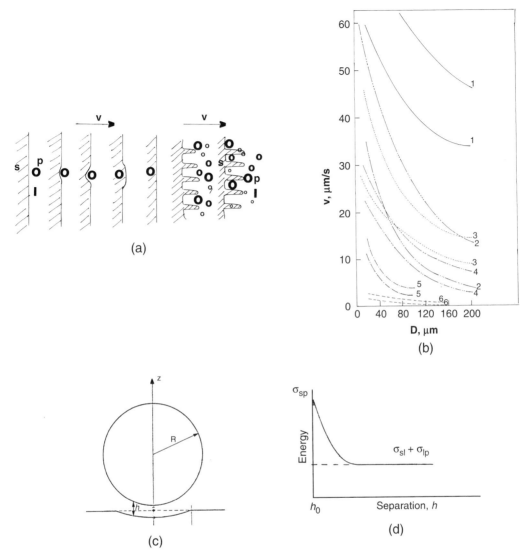

FIGURE 2-77 *(a) Diagram illustrating the interactions between an advancing solidification front and particles dispersed in the melt ahead of the front. (b) Upper and lower bounds on the measured critical solidification front velocity in various organic liquids for the engulfment of fine particles. Curve 1-1: acetal-naphthalene, 2-2: acetal-biphenyl, 3-3: nylon-naphthalene, 4-4: nylon-biphenyl, 5-5: PMMA-naphthalene, and 6-6: acetal-salol. (Data are from S. N. Omenyi and A. W. Neumann, J. Appl. Phys., 47 (9), 1976, p. 3956; and S. N. Omenyi, R. P. Smith, and A. W. Neumann, J. Colloid Interface Sci., 75, 1980, p. 117). R. Asthana and S. N. Tewari, Processing of Advanced Materials, 3, 1993, 163–180. (c) A thin liquid film of molecular dimensions between the solidification front and the particle ahead of the front. The film should remain unfrozen for pushing to occur. (d) Schematic variation of the surface free energy as a function of the liquid film thickness in the gap between the front and the particle. Physical contact may be thermodynamically forbidden if the solid–particle interfacial energy, σ_{sp}, is greater than the total free energy $(\sigma_{sl} + \sigma_{pl})$ prior to contact.*

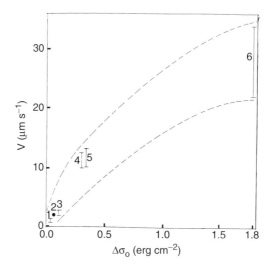

FIGURE 2-78 *Measured critical engulfment velocity as a function of the surface energy difference,* $\Delta\sigma_0$, *where* $\Delta\sigma_0 = \sigma_{sp} - (\sigma_{sl} + \sigma_{pl})$ *in the following systems: (1) Salol-PMMA, (2) Salol-Nylon, (3) Salol-Acetal, (4) Biphenyl-PMMA, (5) Naphthalene-PMMA, and (6) Biphenyl-Nylon.* R. Asthana and S. N. Tewari, J. Materials Science, 28, 1993, 5414–5425.

the particle, resulting in a cooler spot and hence a faster growth velocity. This causes a convex protuberance to appear on the front underneath the particle. Because the hydrodynamic force favoring engulfment is expected to be lower in front of a convex surface compared to a concave or planar surface, the particle tends to be pushed by the convex protuberance. For $K_p > K_l$, heat flow is preferentially through the particle that reduces the temperature gradients, and the particle is engulfed by a concave depression formed in the solidification front. The ratio, K_p/K_l, affects the depth and the curvature of the depression; a large ratio leads to a deeper depression, which promotes particle engulfment even at low velocities. Thus, the ratio K_p/K_l is a measure of the pushing–engulfment tendency. A more general criterion to predict the pushing–engulfment tendency is the ratio of the heat diffusivities of particle and liquid ($\sqrt{K_p \cdot C_p \cdot \rho_p} / \sqrt{K_l \cdot C_l \cdot \rho_l}$); engulfment is favored when this ratio is greater than unity. Large critical velocities are required to capture particles when a large temperature gradient exists at the solid–liquid interface. Experimentally it is found that larger the value of K_p/K_l, the smaller is the critical velocity for particle capture, i.e., capture is facilitated at progressively increasing values of the thermal conductivity ratio.

For growth from alloys, both solute and heat transport are involved. The obstruction of diffusion field by the particles in front of the phase change interface tends to reduce the concentration gradient G_c, at the interface. As a result, the local growth velocity is reduced and a depression appears on the front, which favors particle engulfment because of increased drag force on the particle in front of a concave interface. Fine particles provide less obstruction to solute diffusion than coarse particles.

As mentioned earlier in this section, the viscous drag on a particle ahead of a solidification front is an attractive force favoring engulfment, and large particles (which experience greater fluid drag) are more easily engulfed than fine particles. In other words, the critical velocity for particle engulfment in a given system will be larger for fine particles than for coarse particles, an observation that is confirmed in experiments (Fig. 2-77b). Similarly, the attractive drag forces are larger on a particle in melts of higher viscosity; hence, smaller capture velocities are required in melts of high viscosity. For a sphere in front of a planar solid–liquid interface, the drag force is a function of particle radius, viscosity, and front velocity, and is given from the expression

$$F_d = 6\pi \cdot \mu \cdot V \cdot R^2/h, \tag{2-65}$$

where μ is the viscosity, V is the growth velocity, R is the particle radius, and h is gap width between the particle and the front. Once the front begins to bend locally, the nature of liquid flow as well as drag force changes. The drag force compressing the particle toward a curved front depends on the curvature of the solid–liquid interface behind the particle, and is given by

$$F_d = 6\pi \cdot \mu \cdot V \cdot R^2/(1 - \alpha)^2 \cdot h, \tag{2-66}$$

where the constant α characterizes the curvature of the front such that $\alpha = 0$ for a planar interface and $\alpha = 1$ for a hemispherical front. Engulfment occurs when fluid flow in the gap between the curved front and the particle becomes insufficient to keep the thin liquid film from solidifying. Gas bubbles require a higher velocity for their capture compared to solid particles of the same size, because the hydrodynamic forces compressing the second phase toward the front are larger on the bubble compared to a solid particle (the bubble–melt interface is in fact the free surface of the liquid).

The buoyancy force may assist or impede particle engulfment, depending on the differences in the densities of particle and the melt, and the direction (parallel or antiparallel to gravity) of movement of the solidification front. Thus, in countergravitational growth (melt at top and solid at bottom) buoyancy forces will favor engulfment when $\rho_p > \rho_l$, and oppose engulfment when $\rho_p < \rho_l$.

The geometric entrapment or capture of particles by nonplanar, converging growth fronts (e.g., cellular and dendritic interfaces) is of interest in real castings. With nonplanar interfaces, the particles may be geometrically entrapped between two or more converging growth fronts (secondary dendrite arms); in such cases, lines of particles decorate cell or dendrite boundaries and separation between particles is on the order of cell dimensions. Also, when the diffusive interactions are strong, the particles may introduce morphological transitions (such as dendrite tip splitting or healing of an initially cellular interface) during growth.

References

Campbell, J. *Castings*. Boston: Butterworth-Heinemann, 1999.

Chalmers, B. *Principles of Solidification*. New York: Wiley.

Dantzig, J. A., and C. L. Tucker, III. *Modeling of Materials Processing*. Cambridge University Press, 2001.

Davies, G. J. *Solidification and Casting*. New York: Wiley, 1973.

Flemings, M. C. *Solidification Processing*. New York: McGraw Hill, 1974.

Flinn, R. A. *Fundamentals of Metal Casting*. Reading, MA: Addison Wesley, 1963.

Kou, S. *Transport Phenomena in Materials Processing*. New York: Wiley, 1996.

Kurz, W., and D. J. Fisher, *Fundamentals of Solidification,* 4th ed. Switzerland: Trans Tech, 1998.

Metals Handbook, Casting, Vol. 15, 9th ed. Materials Park, OH: American Soc. for Materials, 1989.

Taylor, H. F., M. C. Flemings, and J. Wulff, *Foundry Engineering*. New York: Wiley, 1959.

Zalensas, D. L., ed. *Aluminum Casting Technology,* 2nd edition. Des Plaines, IL: American Foundry Society, 1977.

3   Powder Metallurgy and Ceramic Forming

Crystalline Ceramics and Glasses

Ceramics are inorganic materials composed of both metallic and nonmetallic constituents. For example, Al_2O_3 (Al, metal; O, nonmetal), TiC (Ti, metal; C, nonmetal), and TiO_2 (Ti, metal; O, nonmetal) are all ceramics. This class of materials includes both traditional ceramics such as clay, tile, porcelain, and glass, as well as modern technical ceramics such as carbides, borides, oxides, and nitrides of various elements, which are used in high-technology applications. Examples of technical ceramics include aluminum nitride, boron carbide, boron nitride, silicon carbide, titanium diboride, silicon nitride, sialons, zirconium dioxide, barium titanate, and ceramic superconductors. The modern ceramic era started around World War II, and many key innovations in ceramic materials and processes of making them were developed in response to defense needs. For example, tape-casting for ceramic tape manufacture, discussed later, started because the supply of capacitor-grade mica was cut off during World War II.

Compared to metals, ceramics are hard and brittle, and their mechanical properties are more sensitive to flaws. As monolithic polycrystalline materials, they exhibit low toughness but high stiffness. Generally stronger under compression than under tension, these materials are also usually poor conductors of heat and electricity. They are predominantly ionic compounds and have high melting points.

Ceramics can be either crystalline or amorphous. Crystalline ceramics can be classified into different groups based on their crystal structure. For example, NaCl, MgO, and LiF form a rock salt structure (Figure 3-1a) in which anions (negatively charged nonmetallic ions) and cations (positively charged metallic ions) form two interpenetrating FCC lattices, with a coordination number of 6 (six nearest neighbors). In the cesium chloride (CsCl) structure (Figure 3-1b), the coordination number for both anions and cations is 8, with eight anions located at the corners of a cube and a single cation residing at the cube center. In the zinc blende or ZnS crystal structure (Figure 3-1c), all corner and face centers are occupied by sulfur atoms, and the interior tetrahedral positions are filled by Zn atoms. Other ceramics that exhibit a zinc blende structure are zinc telluride (ZnTe) and silicon carbide (SiC). In the crystal structure of CaF_2 (Figure 3-1d),

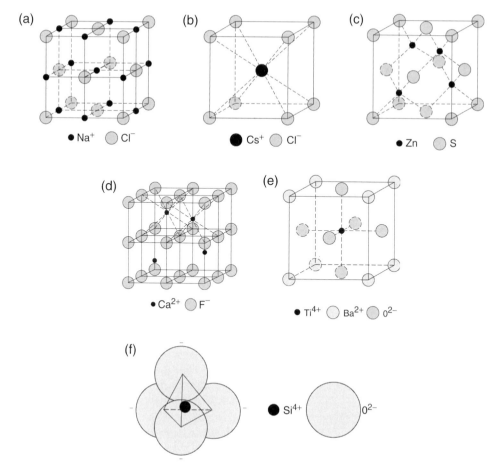

FIGURE 3-1 *(a) A unit cell of rock salt, or NaCl crystal structure, formed as two interpenetrating FCC lattices, one composed of the Na^+ ions and the other Cl^- ions. (W. D. Callister, Jr., Materials Science & Engineering: An Introduction, 5th ed., Wiley, New York, 2000, p. 386) (b) A unit cell of cesium chloride (CsCl) crystal structure with the Cs^+ ion at the center of the cell and Cl^- ions at the corners of the cube. (W. D. Callister, Jr., Materials Science & Engineering: An Introduction, 5th ed., Wiley, New York, 2000, p. 387) (c) A unit cell of zinc blende (ZnS) crystal structure. Each Zn atom is bonded to four S atoms and vice versa. All corner and face center positions are occupied by S atoms, and the tetrahedral positions are occupied by Zn atoms. (W. D. Callister, Jr., Materials Science & Engineering: An Introduction, 5th ed., Wiley, New York, 2000, p. 387) (d) A unit cell of CaF_2 crystal structure. Calcium ions are at the centers of cubes, and fluoride ions are at the corners. (W. D. Callister, Jr., Materials Science & Engineering: An Introduction, 5th ed., Wiley, New York, 2000, p. 388) (e) A unit cell of perovskite crystal structure displayed by barium titanate ($BaTiO_3$). Ba^{2+} ions are positioned at all eight corners of the cell, and a single Ti^{4+} ion is at the center of the cube. The O^{2-} ions are located at the center of each of the six faces of the cube. (W. D. Callister, Jr., Materials Science & Engineering: An Introduction, 5th ed., Wiley, New York, 2000, p. 388) (f) A silicon-oxygen tetrahedron in which each Si atom is bonded to four O atoms, which are located at the four corners of a tetrahedron, with Si atom at the center of the tetrahedron. The SiO tetrahedron is the repeating unit in silicate minerals. (W. D. Callister, Jr., Materials Science & Engineering: An Introduction, 5th ed., Wiley, New York, 2000, p. 393)*

Ca^{2+} ions are positioned at the center of cubes, and fluorine ions occupy the corner positions. Because there are unequal charges on the cation and anion (Ca^{2+} and F$^-$), there are twice as many F$^-$ ions as there are Ca^{2+} ions. When there are more than one type of cations in a ceramic compound (e.g., two in barium titanate, BaTiO$_3$), a perovskite structure (Figure 3-1e) forms. In this structure, Ba^{2+} ions are located at cube corners, Ti^{4+} ions at cube center, and O^{2-} ions at the center of cube faces. A crystal structure similar to perovskites is found in spinel compounds (e.g., magnesium aluminate or MgAl$_2$O$_4$). Here Mg^{2+} and Al^{3+} ions occupy tetrahedral and octahedral positions respectively, whereas O^{2-} ions form an FCC lattice. Ferrite crystals are ceramic magnets with a spinel-like structure. A variety of ceramic materials have a silicon-oxygen tetrahedron (SiO$_4^{4-}$) structure as a repeating unit, with each unit having a -4 charge on it because each Si^{4+} is bonded to four O^{2-} ions. In each tetrahedron, Si is covalently bonded to four O atoms located at the corners of a tetrahedron, with the center of the tetrahedron occupied by the Si atom (Figure 3-1f). Silica (SiO$_2$) is a simple silicate mineral that has SiO$_4^{4-}$ tetrahedron as a repeating unit. In vitreous or fused silica, the fundamental crystal unit is SiO$_4^{4-}$ tetrahedron, but there is no long-range order in the arrangement of this unit. Common glasses are silica-based ceramics to which oxides of Al, Ca, Mg, B, Na, and K, etc. have been added; these oxides lower the melting point, and the viscosity of glass, thus considerably easing the shaping of glass into complex objects. Glass is an amorphous solid solution of these various oxides, with silica (>50%) as the principal constituent. Glasses have a disordered or liquid-like atomic structure. As a solid-solution alloy of ceramic oxides, glasses do not have a single melting temperature and are characterized by a melting range. Various other naturally occurring ceramic minerals also contain silica, such as kaolinite clay, talc, and mica. Besides SiO$_4$, tetrahedral structures are found in other chemical species such as AlO$_4$, SiN$_4$, and AlN$_4$. These tetrahedrons are the same size as SiO$_4$, so it is possible to replace one with the other in silicate structures (provided charge neutrality can be maintained). It has been found that two-thirds of the Si in β-Si$_3$N$_4$ can be replaced by Al without changing the silicon nitride crystalline structure, provided an equivalent amount of nitrogen is replaced by oxygen. This type of compound is called a sialon, which is essentially a solid solution of Si$_3$N$_4$ and Al$_2$O$_3$. Whereas the physical and mechanical properties of sialons are similar to those of β-Si$_3$N$_4$, sialons have a lower vapor pressure, and they form more liquid phase at lower sintering temperatures with sintering additives than does β-Si$_3$N$_4$. This permits pressureless sintering to be used for densification of the ceramic. Other types of sialons have also been synthesized, in particular those based on α-Si$_3$N$_4$, and Si$_2$N$_2$O structure.

Carbon is sometimes classified with ceramics, although it is not exactly a ceramic; some of its polymorphs have a crystal structure similar to ceramics. For example, diamond has a zinc blende–type structure in which each carbon atom is covalently bonded to four other C atoms (Figure 3-2). Diamond is the hardest known substance, with excellent thermal conductivity, and is used in grinding and machining other materials. Thin, polycrystalline diamond films deposited on metals using vapor deposition techniques combine the benefits of high toughness of a backing metal with the exceptional hardness of the surface diamond film. Graphite is another polymorph of carbon, with a layered crystal structure of hexagonal sheets of C atoms bonded to one another with weak van der Waals secondary forces. Within each sheet, carbon atoms are strongly covalently bonded to three other C atoms. Figure 3-3 shows the layered crystal structure of graphite. The weakly bonded sheets permit shearing of graphite at low stresses, which imparts excellent lubricating property to graphite. Other minerals with a layered structure similar to graphite include talc, tungsten disulfide (WS$_2$), and molybdenum disulfide (MoS$_2$).

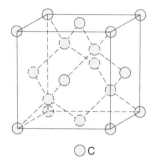

FIGURE 3-2 *A unit cell of the diamond cubic crystal structure, which is similar to the zinc blende structure (Figure 3-1c) in which C atoms occupy all the Zn and S positions. Each C is bonded to four other C atoms. The structure also forms in Si and Ge.* (W. D. Callister, Jr., Materials Science & Engineering: An Introduction, 5th ed., Wiley, New York, 2000, p. 397)

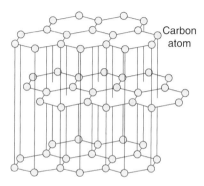

FIGURE 3-3 *The crystal structure of graphite, which consists of layers of hexagonally arranged C atoms. Each C atom is covalently bonded to three other C atoms in the same layer. The bonding between adjacent hexagonal sheets is of weak van der Waals type, which provides for easy shear and excellent lubricating properties of graphite.* (W. D. Callister, Jr., Materials Science & Engineering: An Introduction, 5th ed., Wiley, New York, 2000, p. 399)

These layered compounds are also good solid lubricants. Graphite is also used as a heating element, thermal insulation, furnace electrode, and electrical contact material. In a fibrous form, it is used as a reinforcement for strengthening metals and plastics. Many physical, mechanical, and electrical properties of graphite exhibit anisotropy; for example, the electrical conductivity and strength are higher in a direction parallel to the hexagonal sheets than perpendicular to them. This has permitted high-strength, high-conductivity graphite fibers to be manufactured for special applications.

Another polymorph of carbon is fullerene, a relatively new polymorph discovered in 1985. It consists of 60 carbon atoms arranged in the form of a hollow geodesic dome, or "buckyball" (in honor of Buckminster Fuller, the inventor of the geodesic dome). In addition to this spherical structure, other molecular shapes of carbon, such as carbon nanotubes, have been recently

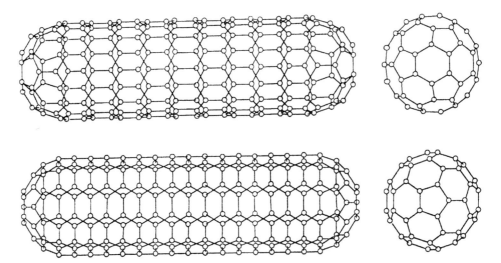

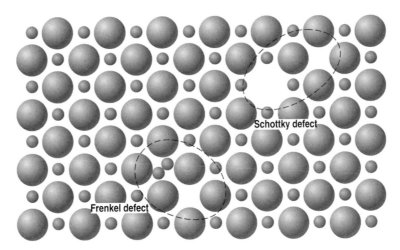

FIGURE 3-4 *Crystal structure of a fullerene molecule and carbon nanotubes.* (R. W. Cahn, The Coming of Materials Science, Pergamon, an imprint of Elsevier, New York, 2001). Reprinted with permission from Elsevier.

FIGURE 3-5 *Frenkel and Schottky defects in ionic solids.* (W. D. Callister, Jr., Materials Science & Engineering: An Introduction, 5th ed., Wiley, New York, 2000, p. 401)

discovered (Figure 3-4). Many of these materials have interesting physical and mechanical properties and are being investigated for their technological applications.

Atomic defects form in ceramics in a manner similar to metals. However, as ceramics are composed of anions and cations, the formation of crystal defects must not destroy the overall charge neutrality. A Frenkel defect in a crystalline ceramic forms when a cation (positive ion) leaves its normal position and moves into an interstitial position; thus a Frenkel defect involves a cation interstitial–cation vacancy pair. A Schottky defect forms when a cation and an anion are removed from the crystal interior and placed on the outer surface; this results in a cation vacancy–anion vacancy pair, or Schottky defect. Figure 3-5 illustrates the Schottky and Frenkel defects. Formation of these defects maintains charge neutrality because the number of cations and anions remains unchanged. These crystal defects play an important role in the physical and mechanical properties of ceramics.

Various techniques are used to manufacture ceramic parts. Some of these techniques apply also to powdered metals. For example, both crystalline ceramics and powdered metals can be shaped by employing solid-state powder metallurgy techniques, which involve pressing and sintering (heating) of fine powders. In addition, special slurry-based techniques such as extrusion, tape-casting, slip-casting, and injection-molding are also used to shape ceramics. In contrast to powdered ceramics, amorphous ceramics such as glasses are fabricated using techniques that also apply to molten polymers, such as blowing, pressing, and rolling.

Powder Metallurgy

The basic powder metallurgy (PM) technique to fabricate ceramic and metal parts involves the following steps: (1) making powders from metals or ceramics, (2) mixing or blending, (3) pressing or consolidation, and (4) sintering or firing. In addition, a variety of secondary treatments such as coining (or sizing) and liquid infiltration are applied to sintered parts to create either fully dense parts, or special components such as oil-impregnated bearings. The major advantages of PM are that difficult-to-melt refractory materials can be shaped into the final component without a need for melting of the raw materials, and parts with controlled porosity can be produced (e.g., filters, porous bearings, honeycomb structures, etc.). PM processes are usually very material efficient (high yield), with material utilization levels of nearly 95% or higher, and allow mass production of complex parts such as gears. At present, nearly 70% of PM parts are used in the automotive industry (e.g., in bearing caps, connecting rods, etc.). The current world-wide PM market is roughly constituted by 25% ceramics, 60% metals, and 15% carbides (cutting tools, drill bits, etc.). There are, however, some limitations of PM such as relatively high tooling cost, high cost of powders, porosity variation within a part, and limitations on part design (part must be ejectable from the die after compaction). In spite of these limitations, powder metallurgy competes against fabrication processes such as machining and casting in terms of part precision and part complexity, and has an expanding niche market for specialized parts.

Powder Production

The starting raw materials for powder metallurgy are powders, typically in the size range of 0.1 to 200 μm. Melt atomization, electrodeposition, chemical synthesis, and crushing and milling are commonly used methods to manufacture powders. *Atomization*, used to make metal rather than ceramic powders, involves disintegration of a molten metal stream into fine droplets under impingement from either an inert gas or water (Figure 3-6a). The droplets solidify to yield powders that are collected in a chamber filled with an inert gas to minimize atmospheric contamination. Figure 3-6b shows the appearance of atomized metal powders. Water-atomized powders are relatively rough, irregular in shape, and contain more oxide impurities than gas-atomized powders, which are more spherical. The collection surface (e.g., tank bottom) is cooled to prevent agglomeration and caking of collected hot powders. Frequently, the powders are cooled by passage of a cold inert gas that forms a fluidized bed. High-surface-tension metals require high atomization pressures (typically 0.3–3 MPa) to make fine powders that are needed for part manufacture.

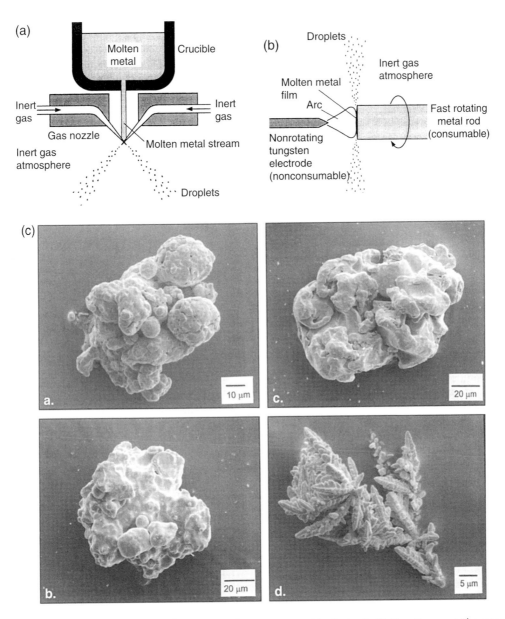

FIGURE 3-6 *(a) Melt atomization and atomization from a consumable electrode.* (S. Kou, Transport Phenomena in Materials Processing, John Wiley and Sons, New York, 1996) *(c) Photographs of atomized metal powders: a. atomized Cu, b. sponge Fe, c. water atomized Fe, and d. electrolytic Cu.* (Powder Metallurgy Design Manual, 3rd ed., 1998, p. 84. Reprinted with permission from Metal Powder Industries Federation, Princeton, NJ)

Solidification of Atomized Droplets

The control of powder size and powder morphology requires an understanding of the solidification behavior of fine molten droplets. During its flight, a droplet of initial radius, R, loses heat to the surrounding gas via convection (Newtonian cooling) and radiation. According to Newton's law of cooling, the heat flux, q, being transferred from the droplet surface to the gas is given from $q = h(T - T_g)$, where h is the heat transfer coefficient at the gas–droplet interface and T_g is the surrounding gas temperature. A thermal analysis of the problem of cooling of atomized droplets is given below. It is assumed that the droplet material has high thermal conductivity so there are no temperature gradients within the droplet. The total heat flux from the droplet to the gas can be written as the sum of convective and radiative heat losses from the droplet,

$$q = h(T - T_g) + \sigma\varepsilon(T^4 - T_g^4) \tag{3-1}$$

where σ and ε are Stefan-Boltzmann constant (56.69×10^{-9} W·m^{-2}·deg^{-4}), and emissivity of the droplet, respectively. A heat balance at the droplet surface yields

$$\frac{4}{3}\pi R^3 \rho C \frac{dT}{dt} = -[h(T - T_g) + \sigma\varepsilon(T^4 - T_g^4)] \cdot (4\pi R^2) \tag{3-2}$$

Noting that at $t = 0$, $T = T_i$ (initial droplet temperature), the time required to cool the droplet in flight to a temperature T_s can be obtained from numerical integration of the preceding equation. On separating the time and temperature variables, we obtain

$$\int_{T_i}^{T_s} \frac{dT}{[h(T - T_g) + \sigma\varepsilon(T^4 - T_g^4)]} = -\frac{3t}{R\rho C} \tag{3-3}$$

If surface radiation is small, the radiative term can be ignored, and the preceding equation simplifies to the case of heat transfer controlled by the interface heat transfer coefficient, h. The resulting equation is

$$\int_{T_i}^{T_s} \frac{dT}{h(T - T_g)} = -\frac{3t}{R\rho C} \tag{3-4}$$

Upon integration, Equation 3-4 yields

$$\ln(T_s - T_g) - \ln(T_i - T_g) = -\frac{3ht}{R\rho C} \tag{3-5}$$

which can be expressed as

$$\frac{T_s - T_g}{T_i - T_g} = \exp\left(-\frac{3ht}{R\rho C}\right) \tag{3-6}$$

Equation 3-6 shows that small droplets of low-specific-heat materials will be cooled to lower temperatures in a given flight time than large droplets of high-specific heat materials. Similarly, longer flight times and low interface thermal resistance will cause greater cooling.

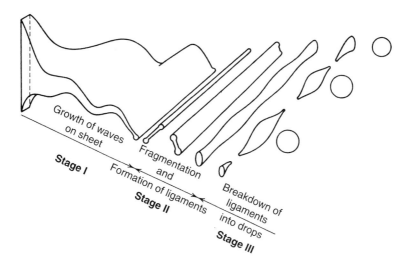

FIGURE 3-7 *Disintegration of a liquid sheet into droplets due to impact from a gas jet during atomization.* (N. Dombrowski and W. R. Johns, Chemical Engineering Science, 18, 1963, p. 203). Reprinted with permission from Elsevier.

The physical process of disintegration of a liquid stream into fine droplets during atomization occurs in several stages (Figure 3-7). Upon impact from a fluid jet (water or an inert gas), the surface of the metal stream experiences a mechanical stress pulse. Because liquids cannot support shear, the stream progressively thins down into a sheet. Below a critical sheet thickness, a ligament detaches from the stream, and this ligament then fragments into fine droplets at a distance sufficiently far from the nozzle, where the surface tension forces override the force because of gas impact. These surface tension forces allow the droplet to minimize its surface area by spheroidizing. Spherical droplets form, provided the droplets have not solidified or liquid viscosity has not increased because of cooling to a level where spheroidization might become difficult. The scheme of droplet formation is shown in Figure 3-7. Droplets of atomized metals of high surface tension, cooling relatively slowly form spherical powders, whereas rapidly cooled low-surface-tension liquid metal droplets yield irregular powders. This is because high surface tension favors minimization of surface area, but the relaxation time for shape adjustment is dictated by the cooling rate. Relationships have been derived for spheroidization time, τ_{sph}, and solidification time, τ_{solid}, the latter assuming convective heat transfer (i.e., the thermal resistance at the droplet–gas interface as the main resistance to heat transfer) as was done in the preceding analysis of droplet cooling. Depending on the operating atomization parameters and liquid metal properties, a range of the ratio τ_{sph}/τ_{solid} is obtained. If spheroidization time is significantly less than solidification time ($\tau_{sph}/\tau_{solid} \ll 1$), spherical powders are obtained, and if spheroidization time is much greater than solidification time ($\tau_{sph}/\tau_{solid} \gg 1$), highly irregular particles are formed. These predictions have been verified through experiments. Surface active solutes (e.g., Mg or Ca in Al and Cu) are added to metals to lower the melt surface tension, with the result that spheroidization becomes sluggish (i.e., the spheroidization time increases). This results in an irregular powder morphology. In water atomization, there is more effective heat transfer between the droplet and the atomizing medium than in gas atomization; as a result, irregularly shaped particles form. However, some loss in cooling efficiency occurs because of formation of steam, which interferes with heat transfer due to film boiling. High-melting-point metals yield more nearly spherical particles in water atomization than low-melting-point metals because steam production slows droplet cooling, thereby increasing the solidification time, and providing more time for droplet spheroidization. The microstructure of atomized powders varies from nanocrystalline and dendritic to fully amorphous, with the amorphous structure being promoted by glass-forming solutes such as phosphorus and boron in metals.

Gas or steam entrapment because of droplet collisions during atomization can cause contamination and porosity in powders. Droplet collisions during flight also alter particle size, shape, and size distribution. The collision frequency increases with metal flow rate and is highest in the center of atomized stream where the mass flux of liquid metal is greatest. Droplet collisions are minimized by using special techniques such as centrifugal atomization in which the droplet mass flux (mass per unit area) rapidly decreases with the radial distance from the center of rotation.

Theoretical relationships have been developed to predict the size of atomized particles. A widely used equation for the average diameter, d_{av}, of gas-atomized powders is

$$d_{av} = KD \left[\frac{v_m}{v_g} \frac{1}{We} \left(1 + \frac{M}{A} \right) \right]^{1/2} \tag{3-7}$$

where K is a constant and We is the dimensionless Weber number, which is the ratio of the inertial to surface tension forces and is given by $We = DV^2/\rho\gamma$. The other parameters in Equation 3-7 are D, the nozzle diameter (or metal stream diameter); v_m and v_g, the kinematic viscosities of the liquid metal and atomizing gas, respectively; V, the velocity of atomizing gas; γ and ρ, the surface tension and density of liquid metal, respectively; and M and A, the mass flow rates (mass/time) of metal and atomizing gas, respectively. Equation 3-7 shows that as the Weber number increases, particle size decreases; i.e., as surface tension decreases or atomizing gas velocity increases, the size of the atomized powders decreases. High temperatures decrease the surface tension and viscosity, and yield finer atomized powders. Particle size depends strongly also on the pressure of atomizing gas and nozzle design, which control the velocity and mass flow rate, respectively.

Other Methods of Powder Manufacture

A mechanical method of making powders from ceramic materials is milling. Milling involves continuous collisions between hardened balls that impact the coarse powders entrapped between the balls via a process called microforging. This refines the powder size. Surface active agents are added to solid powder mixtures to prevent particle welding and agglomeration. The mechanical process of milling deforms, fractures, and cold welds the particles through impact, abrasion, shear, and compression. Generally, metals do not respond well to milling because of their high ductility and tendency to cold weld. Also, because milling action generates heat, partial recrystallization could occur in metal powders. Milling is more useful to make fine powders from brittle ceramics than metals. Individual particles of crushed powders used as feed material for milling almost always contain minute cracks and microscopic notches that weaken the particles. Following the Griffith theory of brittle fracture, the stress, τ, to fracture a particle containing a preexisting crack of length a is $\tau = \sqrt{(2E\gamma_s/\pi a)}$, where E is the Young's modulus and γ_s is the solid's surface energy. Cracks of different sizes are distributed within the particle and on its surface, and the largest crack in the powder determines the fracture stress during milling. An optimum rotational speed is needed for effective milling; both very low and very high rotational speeds decrease the milling efficiency (e.g., fast rotation increases the centrifuging tendency on balls and reduces the intensity of the grinding action in milling). Milling is also used to synthesize prealloyed powders in a special high-efficiency ball mill called an attritor mill. It consists of a rotating impeller in a tank filled with hardened balls and the feed material.

Ceramic powders are usually granulated to obtain more consistent feed material for powder metallurgy-based manufacture. Granulation involves deliberate agglomeration of fine particles into larger particle clusters or agglomerates that result in improved powder flow and handling characteristics. Ceramics are commonly granulated using the spray-drying process, which typically produces spherical granules averaging 100–200 μm in diameter (the unagglomerated powder size is in the range of 0.1 to 10 μm). In spray drying, the powder is mixed with organic compounds to form a slurry. The slurry is sprayed into a heated chamber, where the organic component volatilizes during free fall of powders, and the powders agglomerate.

High-purity powders of metals such as Cu, Pd, and Ti are manufactured using an electrolytic deposition process. The technique makes use of an anode (which dissolves during electrolysis), a cathode (which serves as the deposition surface), and an electrolyte (usually a sulfate salt). The metal deposited at the cathode is ground into fine, spongelike powders that are highly porous. Problems of powder contamination from bath impurities and low deposition efficiency are some of the drawbacks of this method. A chemical method to manufacture metal powders uses gaseous reduction of fine powders of metal oxides by H_2 or CO. Temperature and gas pressure are the key process variables; high temperatures increase the rate of oxide reduction, and low temperatures prevent powders from sintering into agglomerates, thus yielding fine powders. The reduction process must be thermodynamically favorable (i.e., the free energy change, ΔG, for the reduction reaction at the process temperature should be negative). For the production of iron powder, gaseous reduction of FeO by H_2 occurs according to $FeO(s) + H_2(g) \rightarrow Fe(s) + H_2O(g)$. Because $\Delta G = -RT \ln K$, where K is the equilibrium constant ($K = p_{H_2O}/p_{H_2}$, p = partial pressure), the free energy change for the reaction can be converted into a working gas pressure ratio. The kinetics of reduction are temperature sensitive because thermally activated chemical and mass transport processes are involved in the conversion reactions. These processes include diffusion of reactants inward (into the initial oxide), adsorption and surface chemical reaction, nucleation and precipitation, and diffusion of the products outward.

Particle Size and Shape

Particles used in PM are quite fine, and their size, shape, and texture vary greatly with the powder production process and operating parameters. For example, water-atomized powders are rough and irregular, whereas gas-atomized powders are more regular and spherical. Various methods are needed to characterize the shape, size, and texture of powders; these include microscopy, sieve analysis, sedimentation analysis, diffraction techniques, and various other methods. Table 3-1 compares the approximate size detection capability of some particle size measurement techniques.

Microscopy techniques use computer-based quantitative image analysis techniques on several hundred powder particles viewed under an optical or scanning electron microscope (SEM). The diameter, width, length, and surface area are recorded. For particles of arbitrary shape, a characteristic size is defined in terms of maximum chord length between opposing edges. The concept of an aspect ratio is used to characterize nonspherical particles and is defined as the ratio of the longest and shortest chord lengths in a two-dimensional view. The aspect ratio characterizes the deviation of a particle shape from perfect sphericity (for a sphere, aspect ratio is unity). Alternatively, for nonspherical particles, the diameter of a circle with an equal

TABLE 3-1 Powder Size Analysis Instruments and Their Approximate Detection Capability

Instrument/Technique	Size Range (μm)
Ultrasonic attenuation spectroscopy	0.05–10
Centrifugal sedimentation (optical)	0.01–30
Centrifugal sedimentation (x-ray)	0.01–100
Coulter counter (electrical resistance zone sensing)	0.4–1200
Laser light diffraction	0.004–1000
Light microscopy	>1.0
Scanning electron microscopy (SEM)	>0.1
Sieving	5–100,000

projected area to that of the particle is used as a measure of the size. Although microscopy techniques are fast, they require considerable operator judgment in interpretation, because the imaging software often views agglomerated particles as a single particle.

Sieve analysis uses a vertical stack of sieves with the finest aperture sieve at the bottom and the coarsest aperture sieve at the top (a diaphragm collects the finest particles below the finest sieve). A weighed quantity of powders is placed on the top sieve, and the sieves are agitated with either a high-frequency air pulse or sonic vibrations to allow the powders to fall through the sieves. The powders residing on each sieve are weighed after screening for a fixed time, and a frequency distribution curve of weight residing on different sieves is generated. Possible errors include small (usually less than 10%) variations in the size of opening in manufactured sieves. Special sieves with less than 2-μm openings are sometimes used, although fine powders show a strong tendency to agglomerate because of large surface forces; this can cause errors in particle size measurements.

In the sedimentation analysis for powder size distribution, the time required for a particle to settle through a fixed distance in a liquid of known density and viscosity is measured. Ideally, sedimentation techniques require nearly perfect dispersion of spherical powders in the carrier liquid, with no mutual interference and no agglomeration. These conditions are easily satisfied in very dilute suspensions (<1% solid). The formal basis of the method is the well-known Stokes's equation, according to which the velocity, u_0, of a particle of radius R and density ρ_p settling in a fluid of density ρ_l and viscosity μ is given by

$$u_0 = \frac{2(\rho_p - \rho_l)gR^2}{9\mu}, \tag{3-8}$$

where g is the acceleration due to gravity. As an example, consider the settling of spherical alumina (Al_2O_3) particles in water at room temperature. Using the handbook data for material properties, it is found from Stokes' equation that a 10-μm diameter alumina sphere will take 1 min to settle through 1 cm distance under gravity, whereas a 1-μm-diameter alumina sphere will take 2 h to settle through the same distance. Because of slow settling of fine powders used in PM processes, very long observation times might be needed. The observation time is reduced by employing centrifugal force, which enhances the rate of settling (with g in the Stokes's equation replaced with the centrifugal acceleration). It is usually very difficult to observe the settling of a single particle, and commercial sedimentation units use a finite volume fraction of powders rather than a single particle. They measure the particle concentration versus time at a fixed plane in the fluid. The concentration is determined optically from relative intensity of transmitted light or x-rays. The relationship between the initial intensity, I_0, and transmitted intensity, I, and the particle radius is

$$-\ln(\frac{I}{I_0}) = K \sum N_i R_i, \tag{3-9}$$

where N_i and R_i are the concentration of particles of radius R_i. At the onset of settling, all sizes are uniformly dispersed in the liquid. With the progression of settling, large particles settle faster than fine ones. After a given time, t_x, all particles whose diameter is larger than x units would have settled below the level of light beam. Therefore, at time $t > t_x$, the particle concentration at the level of light beam equals the original concentration minus all particles whose diameter is equal to greater than x. The intensity of the transmitted light is now greater than the original suspension; these changes in light intensity are calibrated to yield particle size distribution.

Measurements of particle size by sedimentation analysis are influenced by the finite volume fraction of the powders that causes hindered settling (or flotation), and corrections are needed to account for this. The volume fraction effects are accounted for by noting that the hindered settling velocity is given by

$$u = u_0(1 - \phi)^{4.65} \tag{3-10}$$

where u_0 is the Stokes settling velocity of a single spherical particle, u is the hindered settling velocity, and φ is the particulate volume fraction. The settling rate is influenced also by the shape and orientation (in the case of non-spherical particles) during settling of the powders because viscous drag depends on the exposed surface, and corrections are introduced to account for these. Other errors in the sedimentation analysis could be caused by formation of agglomerates during settling and "wake" effect. Interestingly, sedimentation could occur even in a dry powder and lead to stratification. For example, dry powders in a container can rearrange by settling even during handling (or walking) by the operator. Each footstep can send a stress wave or pressure pulse through the powders that will rearrange the particles.

Light-scattering techniques to measure the particle size use the diffraction phenomenon to determine the size distribution of particles varying from $\sim$0.1 to 300 μm. The angle at which incident light is diffracted (scattered) by particles depends on particle size; the angle of diffraction is inversely proportional to the particle size, which means that fine particles yield greater scattering angles. Measurement of angular distribution of diffracted light is used to obtain data on particle size and size distribution.

Powder Mixing

Powder mixing is done to accomplish several objectives. Different size fractions of powders are blended to control the part porosity, aid sintering (e.g., with use of fine fractions), and facilitate powder compaction (e.g., with use of coarse fractions). New alloy compositions are obtained with greater ease by mixing different elemental powders than by mixing prealloyed powders. This is because prealloyed powders are often harder than elemental powders, have poorer compaction characteristics, and cause greater tool wear than elemental powders. Mixing also permits addition of binders to enhance the green strength of pressed part, and lubricants to assist pressure transmission during compaction and during part ejection after compaction. Stearate compounds (e.g., zinc stearate and calcium stearate) are commonly used lubricants for powders although other types of lubricants are also used. These compounds have low vaporization temperatures and are easily removed during the early stages of sintering. Mixing is done using rotating containers, screw mixers, and blade (impeller) mixers, which create different mixing patterns. Baffles are often added inside the mixing vessel to promote better mixing. Rotational speed and the amount of powders and other additives are important variables in mixing.

Powder Compaction

Powder compaction allows the basic part shape to be created through partial densification of powders under an external pressure. Figure 3-8 shows two common compaction configurations—single-action compaction and double-action compaction. Single-action compaction is used for relatively simple shapes and involves uniaxial pressing of powders in a suitable die with the help of a single punch. More uniform densification is achieved with the use of double-action compaction in which two movable punches press powders from opposite directions. Depending upon the complexity of the part, additional punches may be used to achieve a uniform density.

In double-action compaction, minimum pressure and minimum densification occur at the midplane or neutral axis in the powder bed (Figure 3-8) and maximum pressure and densification at the ends. However, even in double-action compaction the compact might develop an uneven density along the pressure axis because of interparticle friction and friction at the die wall. Because the porous central portion of the pressed part shrinks more than the end regions during sintering, the part may become uneven. For example, the part diameter at the ends could be a few hundred micrometers larger than the diameter at the center. Such variations can be reduced

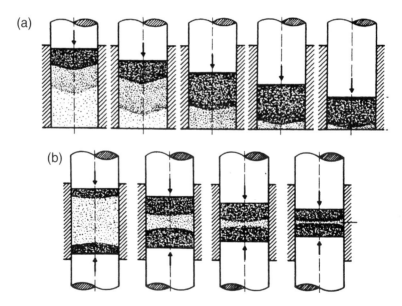

FIGURE 3-8 *(a) Powder compaction with a single punch.* (G. Zapf and K. Dalal, EPMA Educational Aid, The European Powder Metallurgy Association, Shrewsbury, England) *(b) Powder compaction with a double-action press.* (G. Zapf and K. Dalal, EPMA Educational Aid, The European Powder Metallurgy Association, Shrewsbury, England)

by use of advanced net-shape powder-forming processes such as powder injection-molding, gel-casting, hot-pressing, and shock consolidation.

The compaction sequence in powder compaction involves: die filling, initial pressing stroke and de-airing, second pressing stroke, and part ejection. De-airing removes the air that could interfere with interparticle bonding during second stroke, which is the actual compaction stroke. Entrapped air increases the spring-back upon part ejection, and the probability of defects such as lamination and cracking in green parts. Some spring-back is, however, inevitable because the elastic energy stored in compressed powders is released upon part ejection, leading to a slight increase in the compact dimensions. In reality, some differential spring-back between the punch and the part is needed to effectively separate the part from the punch during ejection. A linear spring-back of about 0.75% is typical; too large a spring-back could cause defects. In addition, rapid load application and rapid decompression (punch removal) also increase the chance of defects in the green compact. However, longer dwell at peak compaction pressure reduces the tendency for defects such as laminations because more extensive plastic flow and mechanical bond formation are facilitated.

Dies and punches used for powder compaction are made out of hardened tool steel, and ceramic (carbide) inserts are used to further reduce the wear of dies and punches. The clearance between the die and the punch is usually small ($10–100 \mu m$), and the die wall is tapered for ease of part ejection. Pressing times vary from a fraction of a second for small parts to a few minutes for larger parts. Compaction rates of nearly 5000 parts per minute are possible by use of multi-station rotating powder compaction units. Compaction pressures vary widely, with 20–150 MPa being a common range, although pressures on the order of 700 MPa are sometimes used. Die life depends on the hardness of powder, compaction pressure, and presence of lubrication, and can be several hundred thousand pieces for low-pressure compaction. Powder metallurgy parts generally vary from approximately 0.006 to 25 in^2 in cross-sectional area (perpendicular to the pressing direction), and 0.03 to 6 in. in length (parallel to pressing direction). For most applications, however, a practical limiting length is around 3 inches.

The PM route to manufacture parts has some advantages over competing processes such as stamping, die-casting, and investment casting. Powder metal parts can have varying metal thickness in contrast to stamping, where the sheet metal thickness is fixed. This gives the designer of PM parts an opportunity to reduce part weight without compromising part functionality. In addition, a PM press usually requires less shop-floor space, because of vertical stroke, than do die-casting and -stamping machines. Although, in general, casting processes cover a wider range of part sizes than does PM, the more mass production-friendly casting processes such as die-casting—which compete with PM in part precision and complexity—are suitable for relatively small, thin-walled parts of primarily nonferrous alloys. Likewise, investment (lost-wax) casting competes with PM (in particular, with metal injection-molding, discussed later in this chapter) in terms of part precision and detail, and is applicable to a wide range of ferrous and nonferrous alloys, but it is less amenable to automation and large production volumes than PM.

Dynamics of Powder Densification

Powder mixtures subjected to an external pressure become more dense via particle deformation and fracture. Compacted part density is, however, less than the theoretical density of the pore-free material, and some porosity persists in the structure. This residual porosity diffuses out during sintering, thus yielding a nearly fully dense material. Experimental data on the effect of compaction pressure on green density of alumina, tile, and potassium bromide powders are shown in Figure 3-9a. This figure shows that a relatively rapid initial densification is followed

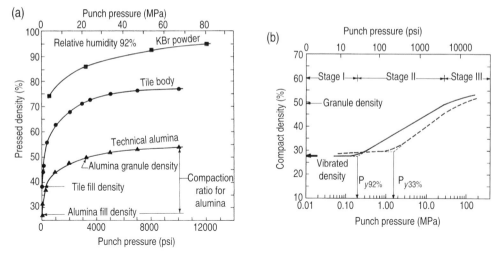

FIGURE 3-9 *(a) Pressed density (as percentage of theoretical density) as a function of punch pressure during uniaxial compaction of KBr, tile, and alumina bodies. (J. S. Reed, Principles of Ceramic Processing, 3rd ed., Wiley, New York, 1995) (b) Semilogarithmic plot of percent compact density as a function of punch pressure during uniaxial compaction of alumina showing three distinct stages of powder compaction. (J. S. Reed, Principles of Ceramic Processing, 3rd ed., Wiley, New York, 1995)*

by a subsequent sluggish rise in the density at larger punch pressure. If the data of Figure 3-9a are replotted with pressure on a logarithmic axis as shown in Figure 3-9b, then three stages of compaction are readily distinguished. In stage I, the pressure is small, and very little densification occurs because of sliding and rearrangement of particles. In stage II, the pressures are greater, and deformation and fracture of powder particles occurs, causing a decrease in the porosity (and an increase in the density). In this stage, the compact density, ρ_c, increases roughly linearly with the logarithm of the pressure ratio, P_a/P_y, according to

$$\rho_c = \rho_f + n \ln\left(\frac{P_a}{P_y}\right), \tag{3-11}$$

where ρ_f is the fill density (i.e., density in the die prior to compaction), P_y is the apparent yield strength of the powder material (or fracture strength of brittle powders), P_a is the applied pressure (punch pressure), and n is a compaction constant. The fill density characterizes the volume that a given mass of powder occupies in a die prior to compaction. A high fill density yields a high green density, a defect-free green compact, and a high sintered density. The ratio (ρ_f/ρ_c) of fill density to the compact density is called the compaction ratio. The preceding logarithmic relationship yields a linear plot of the compact density, ρ_c, as a function of $\ln(P_a)$, which is characteristic of stage II compaction. In the last stage of compaction (stage III), where the applied pressures are very large, most porosity disappears, but a small amount persists. Further increase in the applied pressure in stage III does not increase the compact density, which becomes roughly constant. Figure 3-10 shows a schematic illustration of the behavior of powders in these three compaction stages.

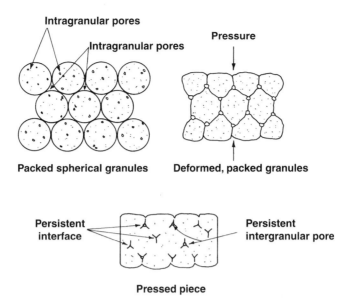

Intragranular pores

Intragranular pores

Pressure

Packed spherical granules   Deformed, packed granules

Persistent interface —— Persistent intergranular pore

Pressed piece

FIGURE 3-10 *Behavior of powders during compaction in the three stages displayed in Figure 3-9b. At low pressures (stage I), powders rearrange themselves, resulting in negligible increase in the density. At intermediate pressures (stage II), powders deform and fracture, resulting in a linear increase in the density with pressure that corresponds to a decrease in the interparticle porosity and porosity within individual particles. At high pressures (stage III), no further decrease in porosity occurs, and very fine pores persist in the structure. Most of this residual porosity is annihilated during sintering.* (J. S. Reed, Principles of Ceramic Processing, 3rd ed., Wiley, New York, 1995)

At a fixed punch pressure, the densification depends upon the hardness of the material being compacted. The effect of powder hardness on the green density and porosity is shown in Figure 3-9a; at a constant compaction pressure, softer potassium bromide (KBr) powders are pressed to significantly greater densities than harder alumina powders. Various empirical equations have been developed to predict the effect of compaction pressure on characteristic properties of powder compacts. For example, the pressure dependence of green density, green strength, and porosity is described by the following empirical equations

$$\ln \varepsilon = B - CP, \ \sigma = B'\sigma_0 P = KP, \ \text{and} \ \sigma = \sigma_0 \rho^m, \tag{3-12}$$

where ε is fraction porosity, ρ is fraction density, P is compaction pressure, σ is green strength of compact, σ_0 is the strength of wrought material, and B, B', C, K, and m are empirical constants. These equations can be cast in a linear form and readily fitted to experimental data to determine the empirical constants. For example, one can write the equation $\ln \varepsilon = B - CP$ as $\ln(1 - \rho) = B - CP$, which is a linear relationship between $\ln(1 - \rho)$ and punch pressure, P. Similarly, the power-law equation $\sigma = \sigma_0 \rho^m$ can be expressed as $\ln \sigma = \ln \sigma_0 + m \ln \rho$. The experimental data on powder compaction for a wide variety of metal and ceramic powders have been found to be in broad agreement with the preceding relationships.

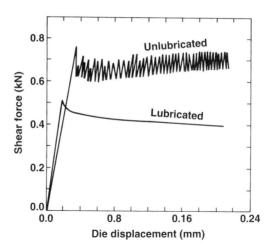

FIGURE 3-11 *The magnitude of shear force as a function of die displacement during ejection of pressed alumina compacts from lubricated and unlubricated dies.* (J. S. Reed, Principles of Ceramic Processing, 3rd ed., Wiley, New York, 1995)

The distribution of transmitted pressure through packed powder beds has been studied using both computer simulations and experiments with photoelastic materials and glass beads. Photoelastic materials are used in the form of disks, and they show internal stress distributions as color changes when viewed under polarized light. Studies show that applied force is transmitted through a network of contacting granules; however, not all granules experience the same pressure. Some granules experience higher pressure than others. Model studies using compaction of glass beads have demonstrated that small granules experience a higher level of stress during compaction and break down at a lower stress than large granules. As a result, large granules tend to persist in the compact even at high applied pressures.

Interparticle friction and friction at the die wall determine the magnitude of pressure transmitted within the compact. They also influence the pressure required to eject the compacted part from the die. The die-wall friction and internal powder friction can be determined accurately using specialized equipment. Higher compaction pressures produce stronger green compacts but also increase the pressure required for part ejection. The ejection pressure is related to the die wall friction and compaction pressure, and these can be controlled with the help of lubricants. The ejection pressure can be directly measured by attaching a sensor to the ejection punch. Figure 3-11 shows the shear force for part ejection in lubricated and unlubricated dies. Besides lowering the pressures needed for part ejection, die lubrication also overcomes the stick-slip motion, which causes large ($\sim$100 kN in Figure 3-11) force fluctuations and propensity toward surface cracking.

The sliding and redistribution of powders during compaction are resisted by the friction from the surface roughness on particles and by interparticle (surface) forces. Sliding between individual particles occurs when the applied load exceeds the (Coulombic) frictional resistance because of microscopic surface roughness, which is a measurable parameter. However, rolling and collective movement of particles reduce the resistance to compaction. Scanning electron microscopy (SEM) and profilometry are used to characterize the roughness on particle surface. In the latter technique, a fine-tipped stylus moves in contact with the surface and generates an electrical signal that is amplified and measured. The fluctuation of the mean amplitude of the signal is converted to a centerline-average (CLA) roughness. Powder surface roughness depends on the material type and powder fabrication method. For example, layered silicate minerals exhibit low surface roughness (<100 nm), whereas crushed ceramic minerals have steplike surface features because of cleavage fracture, with large roughness (>1 μm). Similarly, atomized metal droplets undergo rapid cooling and solidification, and exhibit surface dendritic morphology, which increases the surface roughness.

The presence of agglomerated powder particles reduces the transmitted pressure and inhibits the densification. Fine powders readily agglomerate because of electrostatic forces from adsorbed surface ions and because of universally present weak van der Waals forces. In addition, strong interparticle forces originating in the capillary forces of a wetting liquid (e.g., a liquid binder) also contribute to powder agglomeration.

Isostatic Compaction and Hot Isostatic Compaction (HIP)

When dies and punches are used for powder compaction, the transmitted pressure is directional. More homogeneous compaction can be achieved by pressing the powders uniformly in all directions. This is accomplished with the use of isostatic compaction, which employs flexible molds for containing the powders, and oil as a pressure transmission medium. In isostatic powder compaction, flexible synthetic rubber or sheet metal molds are filled with powders, de-aired, and sealed. The sealed molds are then immersed in oil and uniformly pressurized under large hydrostatic pressures. For reactive or atmosphere-sensitive metals such as Ni, Be, Zr, Ti, and V and their alloys, powder compaction and sintering are done concurrently using hot isostatic pressing (HIP). This minimizes powder contamination from exposure to atmosphere in the time interval between compaction and sintering in a two-step PM process. In HIP, flexible sheet metal molds containing powders are suspended in a pressure vessel containing argon gas rather than oil because sintering temperatures are usually very high. The temperature and pressure of the gas are raised to preset values, which depend on the powder material. For example, ferrous parts are normally hot-isostatically pressed at 70–100 MPa pressures and at temperatures of about 1250°C, whereas Ni-base alloys require nearly 300 MPa pressures and 1500°C sintering temperatures. HIP is a slow process and takes 6–24 hours for completion, and the step involving "canning" of powders is relatively costly. The technique is therefore used mostly for low production volumes of costly alloys and advanced composites that are susceptible to atmospheric contamination or undesirable chemical reactions during sintering. The sintering kinetics in HIP are given from the empirical relationship,

$$\ln\left(\frac{p}{p_0}\right) = -kt, \tag{3-13}$$

where p is the porosity at time t, p_0 is the initial porosity, and k is a temperature-dependent hot-pressing constant, determined from the analysis of experimental data. This equation shows that porosity decreases exponentially with increasing time [p = $p_0 \exp(-kt)$], which is a consequence of the rapid densification under the combined influence of pressure and temperature.

Analysis of Pressure Distribution in Uniaxial Compaction

The degree of densification within compacted powders depends on the extent to which the applied pressure can be transmitted within the powders against the force of friction. A simplified analysis of pressure distribution during uniaxial compaction in a cylindrical die, with friction

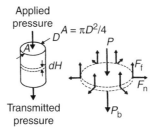

Applied
pressure
$A = \pi D^2/4$
dH
P
F_f
F_n
Transmitted
pressure
P_b

FIGURE 3-12 *A differential element of compressed powders in the form of a disc of thickness, dx, for analyzing the pressure distribution during uniaxial compaction.* (R. M. German, Powder Metallurgy Science, 1st ed., 1984, p. 120. Reprinted with permission from Metal Powder Industries Federation, Princeton, NJ).

at the die wall as the only resisting force, is presented next. Consider the forces acting on an infinitesimal disc of compressed powders (Figure 3-12) under an applied pressure, P_a. The pressure difference across the disc of thickness dH located at a depth h is $dP = P - P_b$, where P and P_b are the pressures on the top and bottom faces of the disc, respectively. The force of friction $F_f = \mu F_n$, where μ is the coefficient of friction between the die and powders, and F_n is the force normal to the die wall. Because a static equilibrium is achieved when the powders are neither accelerating nor decelerating under the applied pressure, P_a, a balance of forces along the cylinder axis yields $\Sigma F = 0 = A dP + \mu F_n$, or $dP = -\mu F_n/A$, where A is the cross-sectional area of the die ($A = \pi D^2/4$, with D as the diameter of the die). The pressure normal to the die walls is experimentally found to be a constant fraction of the axial pressure at any given plane within the powders, and a parameter, Z, is defined as the ratio of the radial-to-axial pressure.

Note that $F_n = \pi D \, dH \, Z \, P$, and $F_f = \mu F_n = \mu \pi D \, dH \, Z \, P$. Substituting this expression for F_f in the equation $dP = -\mu F_n/A$ yields

$$dP = -\frac{4\mu\pi DZPdH}{\pi D^2} = -\frac{4\mu ZPdH}{D} \tag{3-14}$$

This equation is integrated over the limits, $P = P_a$ at $h = 0$, and $P = P_x$ at $h = x$, so that

$$\int_{P_a}^{P_x} \frac{dP}{P} = -\frac{4\mu Z}{D} \int_0^x dH \tag{3-15}$$

and therefore,

$$\ln \frac{P_x}{P_a} = -\frac{4\mu Zx}{D} \tag{3-16}$$

Or,

$$P_x = P_a \exp\left(-\frac{4\mu Zx}{D}\right) \tag{3-17}$$

Thus, the transmitted pressure within the powder bed decreases exponentially with increasing depth. Note also that $P_x \to 0$ as $x \to \infty$, which is a consequence of the fact that the preceding derivation has implicitly assumed an infinite cylinder with negligible end effects.

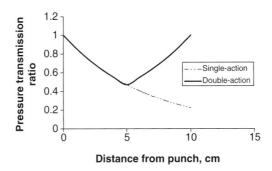

FIGURE 3-13 *Calculated pressure transmission ratio (P_x/P_a) during uniaxial powder compaction using single- and double-action press. The calculations are based on the friction coefficient, $\mu = 0.2$, and the radial-to-axial pressure ratio, $Z = 0.4$.*

As an application of Equation 3-17, consider a uniaxially pressed cylindrical compact 10 mm in diameter with the radial-to-axial stress ratio as 0.5 and friction coefficient as 0.3. To find the depth at which the pressure in the compact will become one-half the applied pressure, we note that (P_x/P_a) = 0.5, so that on substituting $\mu = 0.3$, $Z = 0.5$, and $D = 4$ cm in Equation 3-17, we obtain

$$\frac{P_x}{P_a} = 0.5 = \exp\left(-\frac{4 \times 0.3 \times 0.5 \times x}{10}\right)$$

where $0.5 = -0.06x$, or $x = 11.55$ mm. For double-action compaction of this part, Equation 3-17 will still apply provided the distance, x, is measured from the nearest punch across the midplane or neutral axis. Figure 3-13 shows the variation of transmitted pressure into the compact for both single-action compaction and double-action compaction.

Pressure Distribution in an Annular Cylinder

Now consider uniaxial compaction of an annular cylinder that is formed with an outer diameter, D, and an inner diameter, d. In a manner similar to the right circular cylinder analyzed in the preceding section, an axial force balance is written between the die wall friction and the force because of axial pressure at a depth h in the powders. Note that there are two curved surfaces over which die wall friction must be considered for this configuration. The force normal to the die walls is $F_N = \pi D\, dH\, Z\, P + \pi D\, dH\, Z\, P = (D+d)\pi\, Z\, P\, dH$. The force because of die wall friction is, therefore, $F_f = \mu F_n = \mu(D+d)\pi Z\, P\, dH$. The equilibrium of axial forces yields $dP = -F_f/A$, where A is the cross-sectional area. Substituting the expression for F_f in this force balance yields

$$dP = -\frac{F_f}{A} = -\frac{4\mu\pi(D+d)ZPdH}{\pi D^2 - \pi d^2} \tag{3-18}$$

$$dP = -\frac{4\mu ZPdH}{(D-d)} \tag{3-19}$$

Specifying the integration limits as $P = P_a$ at $x = 0$ and $P = P_x$ at $x = x$ yields

$$\int_{P_a}^{P_x} \frac{dP}{P} = -\frac{4\mu Z}{D - d} \int_0^x dH \tag{3-20}$$

On integration and rearrangement, we obtain an exponentially decaying pressure distribution,

$$P_x = P_a \exp\left(-\frac{4\mu Z x}{D - d}\right) \tag{3-21}$$

Note that the pressure decays faster in a thin powder ring than in a thick powder ring, i.e., as the compact wall thickness $(D - d)$ decreases, the transmitted pressure also decreases. For double-action compaction in an annular die, Equation 3-21 will apply with the distance x measured from the nearest punch across the midplane in the powder bed.

Powder Injection-Molding (PIM)

The discussion up to this point focused on mechanical compaction of dry powders to shape parts. An alternative approach to part manufacture, called powder injection molding, utilizes pressurized injection of suitably designed liquid slurries containing fine powders into prefabricated water-cooled dies. Powder injection-molding (PIM), used to make both ceramic and metal parts, combines the knowledge and experience gained in plastics injection-molding with that in sintering of ceramic and metal powders. Ceramic parts have been injection-molded for over 70 years, although injection-molding of metal parts is by comparison recent. The feed material in PIM consists of nearly 40% polymer binders and 60% metal or ceramic powders. The binders and the powders are mixed in a hot extruder to create the feed material for injection molding. The feed material is heated to about 150°C above the glass transition temperature of the plastic binders, injected into a water-cooled die under 30–100 MPa pressure, and allowed to solidify. The part is ejected and transferred to a debinding system, where the major portion of the binder is removed.

The binder system used in powder injection molding is actually a combination of a major binder (polystyrene, paraffin, cellulose, etc.) and a minor binder (e.g., liquid epoxy). The major binder should burn out at a lower temperature and provide pore channels for escape of gas produced on the decomposition of the minor binder. The minor binder provides strength while gaseous products from the major binder diffuse through the low-permeability structure. Ash content and carbon residue after burnout are important considerations in selecting the binder systems. Frequently, plasticizers (petroleum oil, stearic acid) are added to control the glass transition temperature and flow behavior of the binders, and surfactants are added to improve the wetting and spreading of the liquid binders on the powders to prevent interfacial void formation. Table 3-2 gives a summary of polymer binders used in PIM feedstock. Binders must be chemically inert to the powders and easy to remove during debinding. They should have good thermal conductivity to facilitate the solidification of the slurry in water-cooled dies (conductivities in the range 2.3–2.7 W/m·°K are normally acceptable).

Debinding creates a highly porous part in which powder particles are held together only with the aid of the minor binder. Debinding is done by thermal treatment (slow bake), dissolution in a solvent (e.g., heptane), or catalytic removal in which nitric acid (or oxalic acid or formaldehyde)

TABLE 3-2   Binders Commonly Used in Powder Injection-Molding Feedstock

---

77% Paraffin wax, 22.2% low-molecular-weight polypropylene, 0.8% stearic acid

2% Polyvinyl alcohol, 98% water

20% High-density polyethylene, 10% carnauba wax, 69% paraffin wax, 1% stearic acid

60% Epoxy resin, 30% paraffin wax, 10% butyl stearate

60% Water, 25% methylcellulose, 15% glycerine

55% Vegetable oil, 40% high-density polyethylene, 5% polystyrene

---

Source: Adapted from J.S. Reed, *Principles of Ceramic Processing*, 3rd ed., John Wiley & Sons, New York, 1995.

vapors leach out the binders without causing the part distortion, warping, or cracking that frequently occur in thermal debinding. A sweeper or carrier gas is circulated through the debinding cell to prevent binder deposit on cold chamber walls. The vapors are condensed by passing them through a distillation column and reused. Solvent and thermal debinding are slow (10 to over 30 h) processes, but catalytic debinding is faster and usually takes 2–4 hours for completion. Large, bulky parts present considerable difficulty in binder removal and are not good candidates for PIM.

Either a thermosetting or a thermoplastic resin-based binder system may be used, but a thermoplastic system is usually preferred. The consistency of powder size distribution and consistency of powder loading must be maintained. Important process variables in powder injection molding include temperature, injection pressure, flow rate, and cooling rate. Temperature control must be precise in order to control the flow and deformation of the slurry. In addition, flow behavior also depends on the injection rate, injection pressure, and die design. Flow fronts can join together into a defect-free monolithic part only if the cavity is rapidly filled and air is displaced through numerous small vents in the die. For ceramics, special abrasion-resistant dies and inserts are used in high-wear areas.

Metal injection molding is a relatively recent offshoot of the powder injection molding process. Metal injection molding (MIM) uses polymer binder-coated fine ($<20\ \mu m$) metal powders for injection-molding. MIM is used to make small, highly complex parts that would require extensive finish machining or assembly operations if manufactured by any other forming process. Figure 3-14 shows the sequence of steps involved in metal injection-molding. MIM has been used to manufacture a variety of industrial and consumer items, such as computer disk drive and printer components, connectors for fiber-optic cables, firearms, medical and surgical tools, dental braces, hair clippers, lock parts, bicycle parts, wrist watch cases, cell phone clips, pump bodies, turbocharger injectors, etc. The most common MIM materials are 316 and 304 stainless steels, 17-4 PH precipitation-hardened steels, and Fe-Ni alloys, such as the controlled expansion alloys (Fe-36Ni and Fe-29Ni-17Co) used in glass–metal seals and electronic packaging.

MIM tooling is expensive, and the process is justified at large production quantities. In general, as the part complexity increases, processing and tooling costs also increase. The design of the mold for MIM must account for shrinkage, which is typically 20% in all directions. All design details are retained after sintering of the injection molded part. The feed material (i.e., binder and powder) is fed into the mold cavity through gates, which are critical to proper filling of the die. Gates are normally located at the parting line of the die on less critical surfaces of the part because they leave a visible mark on the part. For parts with varying wall thickness, gates are located so that material flows from thicker section to thinner. The molded part is ejected

Mixing and pelletization

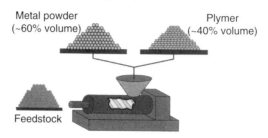

Metal powder (~60% volume)

Plymer (~40% volume)

Feedstock

Molding

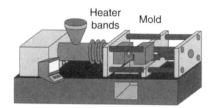

Heater bands

Mold

Debinding

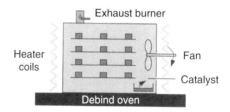

Exhaust burner

Heater coils

Fan

Catalyst

Debind oven

Catalytic debinding

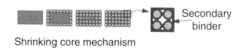

Secondary binder

Shrinking core mechanism

Sintering

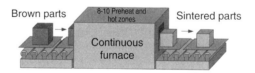

Brown parts

8-10 Preheat and hot zones

Sintered parts

Continuous furnace

FIGURE 3-14 *Sequence of steps involved in metal injection molding (MIM).* (Reprinted with permission from Phillips Plastics Corporation, Phillips Metals Division, Menomonie, WI, 2002)

from the die using ejector or knockout pins. The ejector pins must be sufficient in number to release the part without distorting it and are located near areas that require the highest ejection force and will require some finish machining. Single or multiple cavity molds can be used for injection molding depending on the production rate needed. With multiple-cavity molds, the volume of the part cavity plus the sprue and gates should be less than the shot capacity of the injection-molding machine.

Rheological Considerations in Powder Injection-Molding

The viscosity of powder injection-molding slurries is nearly 100 to 1000 times greater than the viscosity of the plastics used in conventional plastic injection-molding. Higher pressures are, therefore, needed to create defect-free parts by PIM. High shear rates on the slurry lower the viscosity and aid flow but may also lead to segregation or separation of powders and liquid binders. Injection speed is, therefore, a critical parameter in powder injection-molding. Figure 3-15 shows the effect of shear rate and temperature on the effective viscosity of injection-molding slurries of ceramics and metals.

The amount of powders (solid loading) in the slurry affects part shrinkage, distortion, and void content. Too low a loading causes extensive shrinkage and slumping of the part, whereas too high a loading causes uneven coating of powder particles, voids between powders, and formation of weld lines in molded part. At very high solid loadings, the slurry becomes very viscous and stiff, and difficult to injection-mold. The optimum loading is in a range of solid volume fraction that is slightly lower than the critical loading at which the slurry viscosity increases asymptotically. Figure 3-16 shows experimental data on normalized slurry viscosity versus solid loading for fine (BASF) and coarse (IC-218) iron powders dispersed in polyethylene. The steeper increase in normalized viscosity for the finer powders causes its critical loading to be smaller than that of the coarser powders. Near critical loading, minute errors in weighing the constituents can lead to marked rise in the viscosity and create a poor feed material. As a

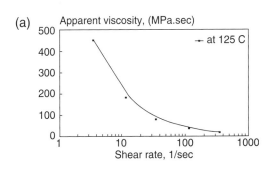

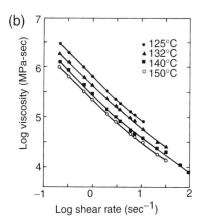

FIGURE 3-15 *(a) Apparent viscosity of an injection molding slurry as a function of shear rate.* (A. Bose, Advances in Particulate Materials, Butterworth-Heinemann, Boston, 1995). Reprinted with permission from Elsevier. *(b) A log–log plot of the viscosity of an injection molding slurry as a function of shear rate at different temperatures.* (J. S. Reed, Principles of Ceramic Processing, 3rd ed., Wiley, New York, 1995)

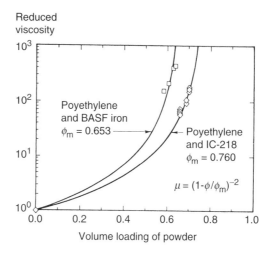

Reduced viscosity

$\mu = (1-\phi/\phi_m)^{-2}$

Poyethylene and BASF iron $\phi_m = 0.653$

Poyethylene and IC-218 $\phi_m = 0.760$

Volume loading of powder

FIGURE 3-16 *Reduced viscosity (i.e., slurry viscosity normalized with respect to the base polymer viscosity) as a function of the volume fraction of two types of iron powders in polyethylene: coarse IC-218 iron and fine BASF iron powders.* (A. Bose, Advances in Particulate Materials, Butterworth-Heinemann, Boston,1995). Reprinted with permission from Elsevier.

result, the PIM feedstock is optimally designed to have solid loading slightly lower than the value corresponding to the critical value at which the viscosity sharply rises to infinity.

Viscosity is a fundamental hydrodynamic property that characterizes the flow behavior of liquids and slurries. For pure liquids, the viscosity, μ, decreases with a rise in the temperature according to

$$\mu = \mu_0 \exp\left(-\frac{Q}{RT}\right) \tag{3-22}$$

where μ_0, and Q refer to viscosity at the melting point and the activation energy, respectively. At a fixed temperature, alloying can affect the viscosity, and theoretical models have been proposed to predict the effect of liquid composition on viscosity. For example, the composition dependence of viscosity is expressed as

$$\mu = (\mu_A x_A + \mu_B x_B)\left(1 - \frac{2x_A x_B \Omega}{RT}\right) \tag{3-23}$$

where x_A and x_B are the mole fractions of the solute and the solvent, respectively, and μ is the regular solution interaction parameter that accounts for the clustering tendency between unlike atom pairs relative to like atom pairs in the solution.

The viscosity of suspensions of fine solid particles in liquids such as those used in PIM does not follow Equation 3-23, which is applicable to true solutions rather than finely dispersed slurries. The PIM slurries containing fine particles are non-Newtonian fluids, that is, fluids whose viscosity at a fixed temperature and composition is not fixed but varies with the degree of shear. For a Newtonian liquid, the shear stress, τ, is proportional to the shear gradient, γ, and in terms of a constant of proportionality, μ, the shear stress is expressed as $\tau = -\mu\gamma$,

An equation for the relative viscosity of slurries containing acicular particles is

$$\frac{\mu}{\mu_0} = \left[1 - \frac{\phi}{0.54 - 0.0125p} \right]^2 \tag{3-31}$$

In practice, the applicability of these equations to real slurries is tested against the actual measurements of viscosity because there is little theoretical basis to adopt one equation in preference to the other. Other factors such as clustering or agglomeration can markedly influence the slurry viscosity. Agglomeration increases the viscosity because the liquid entrapped between agglomerated particles is immobilized and does not contribute to the shear gradients in the fluid. This may be likened to an effective increase in the solid's volume fraction.

Solid–liquid slurries often exhibit thixotropy, which is the continuous decrease of viscosity with time under shear, and subsequent recovery of viscosity when shearing is discontinued. In a thixotropic material, the viscosity at a given shear rate is function of time of shearing. When a slurry is sheared after a rest period, its viscosity gradually decreases to a steady-state value. The rate of decrease of the viscosity is a function of the initial shear rate, the duration of rest period, and the volume fraction and the nature of the solid phase (e.g., whether deformable). The decrease in viscosity with increasing shear rate is caused by a progressive reorientation of particles in the direction of flow, and the processes of agglomeration and de-agglomeration that accompany shearing.

In an agglomerate, the outer particles are loosely bonded and experience directly the hydrodynamic forces that lead to attachment, rearrangement, and detachment of particles. In the core region of an agglomerate, fluid shear is less severe because of protection by the outer layers. The inner particles have a larger number of neighbors (coordination number) and may form necks and welds by diffusional processes if the temperature is high.

Settling and Segregation in Powder Injection-Molding Slurries

Settling and segregation behavior of particles even in high-viscosity PIM slurries is an important consideration because the stability of the slurry is determined by both agglomeration and settling tendencies. The effect of the solid's volume fraction on the settling velocity is expressed from

$$V = V_0 (1 - \phi)^n,$$

where V is hindered settling velocity, V_0 is Stokes' velocity of a single spherical particle, ϕ is the solid's volume fraction, and the constant n is 4.6 to 5. The preceding relationship is valid at a small Reynold's number, Re, where $Re = \rho V d / \mu$, and ρ, V, and D are particle's density, velocity, and diameter, respectively, and μ is the viscosity of the melt. In a binary mixture with spherical particles of two sizes, under certain conditions the smaller particles do not settle but move upward, and large and small particles separate completely. Flocculation also affects the settling of suspensions. For coarse monosized particles, flocculation effects on sedimentation are small, but fine particles cluster appreciably and alter the settling rate. Flocculation is due to the presence of velocity gradients in a liquid (orthokinetic flocculation) and due to thermal diffusion (perikinesis). The degree of flocculation depends on the collision frequency between particles due to shear gradients and thermal diffusion, and the nature of interparticle forces in the liquid. The dependence of the rate of sedimentation (i.e., mass settled per unit area and unit time) on solid's volume fraction is different for flocculated and nonflocculated suspensions. For nonflocculated suspensions, the sedimentation rate decreases continuously with increasing particle concentration, whereas for flocculated suspensions, the sedimentation rate goes through a maximum followed by a progressive decrease in the sedimentation rate that eventually attains a constant value above a critical solid fraction. In dilute suspensions of nonagglomerated particles, the rate of settling is approximated by the first two terms in the series expansion of

$$V = V_0(1 - \phi)^n$$

$$\text{i.e., } \frac{V}{V_0} = (1 - \phi)^n \approx 1 - k\phi + \cdots , \tag{3-32}$$

where k is an empirical constant roughly equal to 5.0.

Because of settling, the initial particle distribution in the slurry is not preserved, and the settling rate is altered. In a highly concentrated suspension of monodispersed (i.e., equisized) spherical particles, settling is completely suppressed when the solid's volume fraction approaches close-packing limit. In a non-Newtonian liquid, the settling particles will generate a shear stress and will locally alter the viscosity. The maximum shear stress from settling of an isolated sphere of radius, R, in a liquid of viscosity, μ, is $(1.5 \ \mu V/R)$, where V is the settling velocity. Note that a non-Newtonian fluid may behave as a Newtonian fluid at low shear stresses, and Stokes' equation for settling velocity would apply if the viscosity is replaced with the viscosity at zero shear rate. In concentrated non-Newtonian suspensions, flocculation decreases the viscosity, and the maximum shear stress because of settling is small. Under certain conditions, a non-Newtonian slurry containing flocculated particles may exhibit good storage stability. In contrast, stirring action may accelerate the settling in a non-Newtonian suspension.

Real slurries usually contain a distribution of particle sizes (polydispersed suspension), and this could influence the settling rate. First, the smaller particles are dragged by the motion of large particles and are accelerated (the "wake effect"). Second, large particles settle through a suspension of fine particles that has high effective density and effective viscosity. Under these conditions, steeper velocity gradients develop because of the restricted space between fine particles, and because of increased solid–liquid contact area. The buoyancy force on the settling particles is determined by the effective density of the suspension rather than the density of the fluid, and the fluid drag is determined by the effective suspension viscosity. Thus, for settling of large particles in a polydispersed slurry, the density term in Stokes' equation is modified as $(\rho_p - \rho_s) = \rho_p - [\rho_p \cdot \phi + \rho(1 - \phi)]$ or $(\rho_p - \rho)(1 - \phi)$, where ρ_p, ρ_s, and ρ are the densities of the particle, suspension, and the liquid matrix, respectively.

Fine suspensions of aqueous solutions containing ionic species and charged particles will exhibit settling rates that depend on the flocculation tendency of particles. However, flocculation is rapidly completed in fine suspensions, and it is difficult to experimentally observe the settling rate in nonflocculated suspensions of very fine particles.

When a slurry contains sedimented particles, vigorous shear (e.g., stirring) is needed to lift and uniformly suspend the particles in the liquid phase. Two essentially different mechanisms of dispersion of sedimented particles may operate depending on the stirring conditions. At low shear rates, the initial particle distribution is not changed by stirring, and the dispersion of sedimented particles proceeds upward at very slow rates. At higher shear rates, the particles are lifted to the top of the melt in the initial stages of stirring, followed by settling of particles (assumed to be heavier than the melt). The rate of particle dispersion in the axial direction increases with both increasing speed and diameter of the stirrer; i.e., sedimented particles are lifted and uniformly suspended in the liquid medium. The lifting of particles to the top at the beginning of stirring is widely observed in rotating flows and is associated with the development of the Ekman boundary layer, which causes a secondary flow in the axial direction. Further discussion of this phenomenon is beyond the scope of this book.

Sintering

Sintering involves heating the compacted (or de-bound PIM) part to a temperature below the melting point of the powders, followed by isothermal soak at temperature to allow powders to bond without melting, and then slow cooling to ambient temperature. Table 3-3 gives approximate sintering temperatures for some alloys.

Both batch-type and continuous sintering furnaces are used in industrial practice; the latter uses a conveyor and pusher mechanisms to control the rate of translation through the furnace and the residence time of the "green" part in the sintering zone of the furnace. Powders of atmosphere-sensitive metals are sintered under inert (e.g., Ar or N_2) atmosphere or inert and reducing (Ar+H_2) atmosphere (the use of pure N_2 is not recommended for ferrous alloys in which the formation of harmful nitride compounds could impair the ductility of the part).

During sintering, parts shrink and densify without losing their pressed or molded shape. The sintering temperature and time are important process variables. The densification behavior of compacted high-purity Al_2O_3 powders doped with MgO as a function of sintering temperature at four different sintering times is shown in Figure 3-18, based on over 300 sintered specimens. The densification is accompanied by a decrease in the total porosity content with increasing temperature. Experiments also show that densification is rapid at higher temperatures, a consequence of the thermally activated mass transport mechanisms of sintering, which are discussed in the next section.

The role of thermal activation during sintering is revealed when the data of Figure 3-18 are plotted as an Arrhenius plot (i.e., natural log of density versus inverse temperature). Figure 3-19 shows an Arrhenius plot of the data of Figure 3-18. As shown in Figure 3-19, two distinct linear regimes are noted, each of which is consistent with the Arrhenius relationship $\rho = \rho_0 \cdot \exp(-Q/RT)$, where ρ and ρ_0 are the densities at the sintering temperature, T, and at room temperature, respectively. A somewhat abrupt transition (slope change) near $\sim 1400°C$ (i.e., T^{-1} of 6.0×10^{-4} $°K^{-1}$) is observed in Figure 3-19. Such transitions in Arrhenius plots indicate a transition in the dominant mechanisms driving the process of sintering. For example, Figure 3-19 shows that a low activation energy mechanism operates at high temperatures, and a high-activation energy mechanism operates at lower temperatures. Whereas Figure 3-19 does not identify the operative sintering mechanisms, it shows that sintering is a thermally activated process.

TABLE 3-3 Approximate Sintering Temperatures of Common Metals and Alloys

Material	$T, °K$
Al alloys	863–893
Brass	1163–1183
Iron	1393–1553
Fe-C	1393
Fe-Cr	1473–1553
316 SS (Fe-Cr-Ni)	1473–1553
430L SS (Fe-Cr)	1473–1553
W alloys	1673–1773

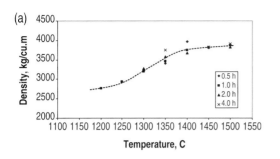 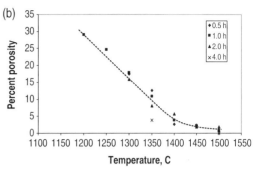

FIGURE 3-18 *(a) Density of sintered MgO-doped alumina (nominal size: 380 nm) as a function of sintering temperature for different times. (b) Percent total porosity in MgO-doped sintered alumina as a function of sintering temperature at different times.* R. Asthana, D. J. Bee and R. Rothaupt, Annual Meeting of the American Society for Engineering Education (ASEE), Salt Lake City, UT, 2004.

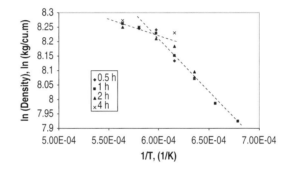

FIGURE 3-19 *The data of Figure 3-18(a) plotted as Arrhenius plot showing natural log of sintered density as a function of inverse absolute temperature for different sintering times. A change in the slope of the linear segments represents a transition in the dominant mass transport mechanism driving the densification process during sintering.* R. Asthana, D. J. Bee and R. Rothaupt, Annual Meeting of the American Society for Engineering Education (ASEE), Salt Lake City, UT, 2004.

The effect of particle size on the densification behavior of alumina compacts during sintering is shown in Figure 3-20. This figure compares the densification data on coarse (1300 and 800 nm) and fine (380 nm) Al_2O_3 powders and shows that fine powders sinter faster than coarse powders. Coarse powders require higher temperatures to attain full densification. This indicates the importance of using fine ceramic powders for rapid or low-temperature sintering, thus allowing harder and stronger ceramics to be produced at a lower processing temperature and with less energy consumption.

Mechanism of Sintering

In sintering, particles bond with one another by atomic diffusion. The driving force for sintering is the minimization of the solid–vapor interface area (i.e., the total area of the powders in contact with the surrounding vapor) and elimination of the regions of sharp curvature at powder contacts. In the initial stages of sintering, small necks form and grow between contacting particles by mass transfer via atomic diffusion. Fine powders increase the driving force for

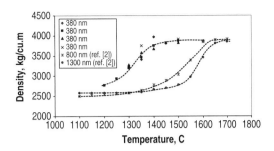

FIGURE 3-20 *The effect of alumina powder size on the density of sintered alumina as a function of sintering temperature.* (Data for 800-nm and 1300-nm powders are from J. S. Reed, Principles of Ceramic Processing, 3rd ed., Wiley, New York, 1995). R. Asthana, D. J. Bee and R. Rothaupt, Annual Meeting of the American Society for Engineering Education (ASEE), Salt Lake City, UT, 2004.

sintering because of a larger surface area per unit volume, which increases the total solid–vapor interfacial energy. However, only a portion of the solid–vapor interfacial energy drives the sintering because some energy is associated with the new grain boundaries that form at every contact region between particles. The net driving force for sintering depends on both surface and grain boundary (g.b.) energies, and can in fact be quite small, and sintering can be quite sluggish. But the emergence of grain boundaries at contacting regions also provides a short-circuit path for atomic diffusion that facilitates sintering. As a result, g.b.'s could be important in maintaining a high sintering rate.

The actual mechanisms of mass transport via atomic diffusion depend upon the type of material and the temperature. These mechanisms consist of different paths of atomic motion that produce mass transfer between powder particles. The important mechanisms of mass transfer during sintering include: 1) diffusion through the vapor, 2) diffusion along the grain boundaries, 3) diffusion along the particle surface, 4) diffusion through the crystal lattice (volume or bulk diffusion), and 5) viscoplastic flow of the material. Theoretical models have been developed to predict the contact area between particles as a function of time, temperature, and various material properties such as interfacial energies and diffusion coefficients. For example, the following equation predicts how the radius, X, of the neck between neighboring spherical particles of radius, R, (Figure 3-21), varies with sintering time,

$$\frac{X}{R} = \left(\frac{40\gamma a^3 D}{kT}\right)^{1/5} R^{-3/5} t^{1/5} \tag{3-33}$$

Here, γ is the surface energy of the solid (i.e., solid–vapor interfacial energy), D is the self diffusion coefficient at the temperature, T, of sintering, a is the lattice constant of the crystal, and k is the Boltzmann's constant. This equation shows that the rate of neck growth, dX/dt (or, equivalently, rate of part densification), decreases with increasing time but increases with increasing temperature. Higher temperatures increase the neck growth because the diffusion coefficient, D, increases exponentially with T according to $D = D_0 e^{-Q/RT}$, where D_0 is the frequency factor, which depends on the atomic jump processes, Q is an activation energy (J/mol·K) for diffusion, and R is the universal gas constant.

During the initial stages of sintering, pore structure is interconnected and pore shape is random. The growth of neck is given from a generic equation analogous to Equation 3-33, which has the form $(X/R)^n = Bt/R^m$, where the exponents n and m depend on the mechanisms of mass

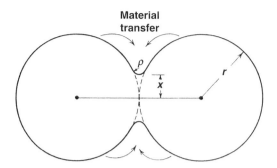

FIGURE 3-21 *Neck formation between particles during the early stages of sintering. Arrows indicate the direction of mass transfer from regions of positive curvature to regions of negative curvature, and ρ and R denote the radius of the neck region and particle, respectively.*

transport, X and R are the radii of the neck region and particle, respectively (Figure 3-21), and B is a constant that depends on material properties, particle geometry, and atomic diffusion coefficient, D. The values of constants n and m have been deduced for various transport mechanisms operative during sintering. For example, the values of n and m for mass transfer via evaporation-condensation are 3 and 1, for volume diffusion, 5 and 3, for grain boundary diffusion, 6 and 4, and for surface diffusion, 7 and 4, respectively. The ratio X/R is the neck size ratio and is used to follow the progress of sintering. The initial stage occurs when $(X/R) < 0.3$. Later, the necks merge and the neck size ratio may become ill-defined. As a result, other parameters, such as the density, are used to follow the progress of sintering at long times. For example, a densification parameter, ϕ, is defined as the change in density from the green state divided by the maximum possible density change, i.e.,

$$\phi = \frac{(\rho_s - \rho_g)}{(\rho_{th} - \rho_g)}, \qquad (3\text{-}34)$$

where ρ_{th} is the theoretical density (i.e., density of the fully dense material), ρ_s is the sintered density, and ρ_g is the green density (density of pressed part). One advantage of using this parameter to characterize part densification during sintering is that it normalizes the effect of a variable green density. Shrinkage is another important parameter, closely related to the densification parameter ϕ, that is used to monitor sintering. Shrinkage occurs because with the progress of neck growth, the center-to-center spacing between particles decreases. Thus neck size can be related to the approach of the particle centers or shrinkage. Shrinkage is a useful practical parameter because it eliminates the need for measurements of individual neck sizes. From a practical point of view, the occurrence of shrinkage requires oversizing of the tooling (e.g., die used for powder compaction) in order to maintain acceptable tolerances on part dimensions. It is essential to minimize green density variations in compacted parts because density gradients translate into differential shrinkage during sintering, leading to warpage and nonuniform shapes.

In addition to the large initial solid–vapor interfacial area, the change in net curvature over a relatively small region (Figure 3-21), i.e., sharp curvature gradients at interparticle contact regions is also responsible for the rapid sintering associated with the initial stage. Plastic flow via dislocation motion could also make a transient contribution in the early stages of sintering (e.g., during heating). Similarly, external mass transport via surface diffusion and vapor transport are more important in the early stages. Vapor transport occurs because of vapor pressure gradients. The pressure over the neck region is lower than equilibrium because it has a net concave curvature. However, the bulk of the material (i.e., a powder particle) is emitting vapor at a pressure above the equilibrium pressure because of the convex curvature. As a consequence, there is net mass flow into the neck region. This mass flow is accompanied by a reverse flux of vacancies from the neck to the particle surface. This is because for a concave surface the vacancy concentration is higher than the equilibrium value, whereas for a convex surface it is lower than the equilibrium value. In the later stages of sintering, bulk processes of mass transport become more active. These processes provide for neck growth by use of internal mass sources (e.g., plastic flow, grain boundary diffusion, volume diffusion etc.). These concepts are used in calculating the neck size as a function of the material properties and process variables during sintering.

The neck size ratio equation presented earlier, i.e.,

$$\left(\frac{X}{R}\right)^n = \frac{Bt}{R^m},$$ (3-35)

shows that smaller particle sizes result in more rapid sintering. Also, because the temperature effects are included in the parameter B via the diffusion coefficient (compare this equation with Equation 3-33), which increases exponentially with increasing temperature, small temperature changes can have a large effect on sintering rate. Time has a relatively small effect in comparison to temperature and particle size on the initial rate of sintering.

The models for sintering assume a uniform geometry for particles. However, in real powder systems there is a distribution in particle size, number of contacts per particle, and contact flattening (large radius of curvature) because of compaction. This complicates the mass transfer processes, which are sensitive to particle size and shape. With a change in particle size, the dominant sintering mechanisms are altered. For example, surface and grain boundary diffusion are enhanced relative to the other mass transport processes by a decrease of particle size. In contrast, lattice diffusion is not as sensitive to particle size as are these two diffusion processes. In general, finer particles cause faster neck growth and need less sintering time than coarser particles, or alternatively, need a lower sintering temperature to achieve an equivalent degree of sintering.

Following neck growth in the early stages of sintering, densification is accompanied by growth of new grains. This is the intermediate stage of sintering, and is the most important in determining the properties of sintered part. In the intermediate stage, grain boundaries (g.b.'s) interact with the residual porosity in the material. Grain growth requires migration of grain boundaries. These g.b.'s either drag the pores along with them as the grains continue to grow, or the g.b.'s break away from pores, leaving isolated pores within the interior of the grains. Pores that remain anchored at moving grain boundaries lower the total energy by decreasing the total grain boundary area. In contrast, when the pore and g.b.'s become separated, the total energy increases because of the newly created surface. The pore can be thought of as having a binding energy in relation to the grain boundary; this binding energy increases as the porosity increases. At the beginning of the intermediate stage of sintering, there is negligible separation between g.b.'s and pores, but with time, breakaway occurs because of the slower mobility of the pores relative to g.b.'s and the diminishing pinning force on g'b's. Temperature has a marked effect on the breakaway; at low temperatures, grain growth is retarded by the pores because the slower mobility of the pores increases the pinning force on g.b.'s, but at high temperatures, the driving force for grain growth and grain boundary mobility increase and breakaway becomes possible. The separation of pores from g.b.'s limits the density possible in the sintered material; in order to obtain high density, grain growth should be minimized. This is done by controlling the temperature, adding second-phase inclusions such as oxide particles (which pin the grain boundaries and restrain grain growth), and using a narrow particle size distribution in the starting powder materials. The isolation of the pores at grain interior results in a decrease in the densification rate.

Theoretical models have been developed to predict the densification rate in the intermediate stage on the basis of the dominant mechanism of elimination of pores. For example, if the pores are eliminated by the diffusion of vacancies through the bulk toward a sink (grain boundaries), then the density is given from $\rho = \rho_i + B_i \ln(t/t_i)$, where ρ and ρ_i represent the density at time t, and at the beginning of the intermediate stage, respectively; t_i is the time corresponding to the onset of the intermediate stage; and B_i is a material constant that varies inversely with the rate of grain growth. Therefore, a retarded grain growth aids densification. If, however, diffusion of vacancies along the grain boundaries controls the kinetics, then the density is given from the equation $(1 - \rho) = C_i + B_b/(t^{1/3})$, where C_i represents the conditions at the beginning of the intermediate stage, and B_b contains several geometric and materials properties, including diffusivity, surface energy, atomic volume, and grain boundary width.

The appearance of spherical, isolated pores signals the inception of the last stages of sintering in which the driving force is strictly the elimination of the pore–solid interfacial area. Figure 3-22 shows the pore–grain boundary configurations during the initial, intermediate, and final stages of sintering. In the final stage, individual spherical pores left after boundary breakaway shrink by losing vacancies to distant grain boundaries that have moved farther away because of grain growth. These vacancies have to diffuse through the bulk to reach the distant g.b.'s; hence volume diffusion is the dominant mechanism of densification in the final stage. The increased diffusion distance and the inherently slower rate of volume diffusion make the densification in the final stage sluggish. Another physical process that occurs in the final stage of sintering is pore coarsening, which increases the mean pore size while the number of pores (and percent porosity) decreases. Coarsening, or Ostwald ripening, is the process of competitive growth of larger pores at the expense of smaller pores, which lose vacancies to the larger pores and disappear. Coarsening occurs because of the differences in the radius of curvature of differently sized pores; local vacancy concentration varies inversely with the radius of the pore, and smaller pores emit more vacancies (and are more quickly annihilated)

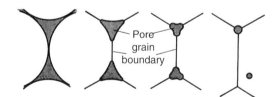

FIGURE 3-22 *Annihilation of porosity and evolution of a grain boundary between powders during sintering.* (R .M. German, Powder Metallurgy Science, 1st ed., 1984, p. 150). Reprinted with permission from Metal Powder Industries Federation, Princeton, NJ.

than larger pores that serve as vacancy sinks. The rate of pore elimination depends also on the solubility and internal pressure of any gas trapped within the pores. The rate of pore elimination in the last stages of sintering is expressed from $(d\varepsilon/dt) = \varepsilon_f - C_f\ln(t/t_f)$, where ε is the porosity, and ε_f and t_f are the porosity and time corresponding to the point at which pores become closed (i.e., the end of the intermediate stage), and C_f is a material constant. It has been observed that a homogenous grain size and a soluble gas aid densification in the final stage.

Other Considerations in Sintering

Frequently, mixtures of different metal powders are used in powder compaction to form alloys after sintering. Mixed-powder sintering works best for fine particles for which the diffusion distances are small and, therefore, compositional gradients are rapidly eliminated and homogeneity rapidly achieved, especially at high temperatures. Generally, fine particles, high sintering temperatures, and long sintering times promote better homogenization of the composition. For rapid part production with low energy consumption, however, sintering temperature and time must be decreased. In order to achieve these goals or to obtain better properties in the sintered part, specific chemicals are added to the powders. This is called activated sintering. An example of activated sintering is tungsten powder treated with Ni, Pd, or Pt as surface coatings or activators. The activator must stay preferentially segregated at the particle surface and form a low melting temperature phase, which could provide a high diffusivity path for rapid sintering and lower the activation energy barrier for diffusion to take place. In addition, the activator must have a low solubility in the base metal (so the base metal contamination can be minimized), and the base metal must have a large solubility in the activator (so the liquidus temperature is lowered). Therefore, an activator that decreases the liquidus or solidus temperatures of the base metal is preferred because it also remains segregated at the interface.

During sintering of mixed powders, a liquid phase may form because of different melting ranges of the components. In such a case, the liquid phase may enhance mass transport and sintering kinetics provided the following conditions are met: 1) liquid wets the solid phase and uniformly coats it, 2) the diffusivity of the solid's atoms in the liquid is large, and 3) the solid is soluble in the liquid. A wettable liquid film enables the surface tension forces to draw the powders together, thus aiding densification and pore elimination. The liquid film also lubricates the solid's surface and facilitates particle rearrangement. All these changes contribute to a rapid decrease in the compact volume. Other metallurgical changes may also occur in the presence of a liquid film. For example, the solid phase may partially dissolve in the liquid film and either promote the solidification (transient liquid-phase sintering), or cause a solution–reprecipitation process to occur, in which small particles dissolve and reprecipitate onto larger particles. Even though a liquid phase forms in both activated sintering and liquid-phase sintering, the latter leads to more secondary phases in the microstructure. Liquid-phase sintering is observed in many metallurgical systems such as Cu-Co, W-Cu, W-Ni-Fe, W-Ag, Cu-Sn, Fe-Cu, W-Co, and Cu-P. One of the unwanted effects of liquid phase sintering is swelling. Swelling occurs because the melt penetrates the grain boundaries in the solid phase, and the solid phase disintegrates into smaller particles and separates. Swelling can be controlled by selecting fine particles, low compaction pressures, and slow heating rate.

Homogenization

In multiphase alloys produced using the PM process, a homogenous structure is achieved through a heat treatment called homogenization, which involves the dissolution of unwanted

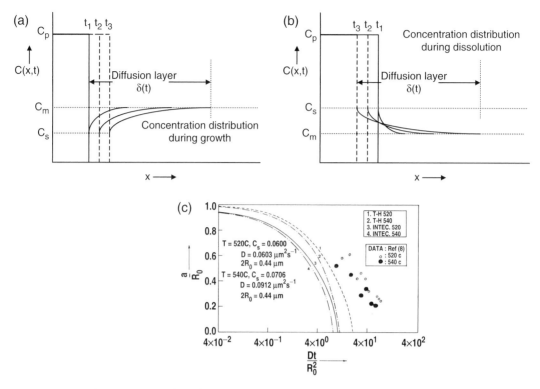

FIGURE 3-23 *Schematic temporal modulation of concentration profiles during (a) growth and (b) dissolution of a solid phase of constant concentration via diffusion in the matrix. (c) Experimental and theoretical dissolution kinetics of CuAl$_2$ precipitates in an Al-4.5Cu alloy at 520°C and 540°C. The theoretical analyses are based on Tanzilli and Heckel's finite difference solution (T-H) and an approximate analytic solution based on the heat balance integral technique (INTEG). R. Asthana and S. K. Pabi, Materials Science & Engineering, A128, 1990, 253–258.*

second-phase particles at elevated temperatures. The elimination of secondary phases that are thermodynamically unstable at high temperatures improves the mechanical properties of the PM alloy. Many other types of material processes also require control on the size, morphology, and distribution of the second-phase particles in a matrix. For example, in dispersion-strengthened alloys, in situ eutectic composites, and metal-matrix composites, the secondary phase must actually be stable at elevated temperatures over prolonged periods in order for it to serve as an effective strengthening agent. The kinetics of second-phase dissolution and growth are determined by the diffusion processes and can be predicted from the theory of diffusion. In dissolution, the second phase with a finite, nonzero initial radius decreases in size by rejecting solute in the matrix behind a receding interface. At any position, R, in the matrix phase far from the interface [$r \gg R(t)$], the solute concentration increases with time; close to the interface [$r > R(t)$] the solute concentration decreases with time; and at intermediate positions, the solute concentration will increase, decrease, or remain constant. Figure 3-23 shows the time modulation of schematic concentration profiles during dissolution by solid-state diffusion. These concentration profiles are compared with and contrasted to the concentration profiles that develop during the growth of a secondary phase in a two-phase alloy, which are also shown in Figure 3-23. During growth from a supersaturated matrix, the latter is depleted of solute immediately ahead of the moving interphase boundary. A comparison of the concentration profiles in dissolution and growth shows that time modulation of solute concentration distribution during dissolution is more complex than during growth.

The prediction and control of the rate of dissolution of secondary phases is important in the design of the processing schedule. Analytical and numerical models have been developed to predict the size, size distribution, volume fraction, and shape of secondary phases. The analytical models of diffusion-controlled dissolution invoke one or more of the following simplifying assumptions: stationary interfaces (slow interface motion), steady-state, linearized concentration gradient, dissolution of an isolated particle in an infinite (unbound) matrix (i.e., zero volume fraction), and constant (concentration-independent) diffusion coefficient. In general, the solute concentration can be obtained from the solution of Fick's second equation with a concentration dependent diffusion coefficient $D(U)$

$$\frac{\partial U}{\partial t} = \frac{1}{r^n} \frac{\partial}{\partial r} \left[r^n D(U) \frac{\partial U}{\partial r} \right] \tag{3-36}$$

where $n = 0, 1$, and 2 for plates, cylinders, and spheres, respectively, and U is the concentration distribution in the matrix. The mass balance at the moving interphase boundary is expressed from

$$(C_p - C_s)\frac{dR}{dt} = D(C_s) \left(\frac{\partial U_1}{\partial r} \right)_{R_0} \tag{3-37}$$

where R is the instantaneous position of the dissolving interface, which is initially located at $R_0(t)$. The solution to the preceding equations for appropriate boundary and initial conditions will yield the kinetics of dissolution. Table 3-4 summarizes some of the analytical models of diffusion-controlled dissolution. Figure 3-23c shows a comparison of the experimental data on the dissolution kinetics of the $CuAl_2$ precipitates in binary Al-Cu alloys in terms of a dimensionless time, $\tau = Dt/R_0^2$, and a dimensional radius (R/R_0) of the dissolving particle (R_0 is the initial radius of the precipitate). These data are compared with the predictions of some analytical and numerical models of dissolution in Figure 3-23c. A reasonable agreement is noted between the numerical model and the experimental data. Among the analytical solutions, models based on the assumption of a stationary interface have been found to offer the best approximation at slow interface movement and constant diffusion coefficient.

Coarsening

Coarsening (also called Ostwald ripening) consists of diffusion-controlled growth of second-phase particles or droplets of low interface curvature at the expense of second-phase particles or droplets of high curvature. Particles of low curvature grow by adding the atoms released by the dissolution of the particles of high curvature. Thus, coarsening originates from the thermodynamic need of a system to minimize the nonuniformity in the interfacial curvature of the secondary phase. The driving force for coarsening is the reduction of interfacial free energy. A particle with a radius of curvature, a, has increased solubility in a matrix relative to a particle with a planar interface ($a \rightarrow \infty$). The increased solubility, C_a, at a temperature, T, is related to the particle radius, a, by the Thomson-Freundlich equation given from $RT \ln \frac{C_a}{C_P} = \frac{2\gamma M}{a\rho}$, where C_P is the solubility of a planar interface, γ is the interfacial energy, M is the molecular weight, ρ is the density, and R is the universal gas constant. Fine particles with a large surface

TABLE 3-4   Analytical Models of Diffusion-Controlled Dissolution of Secondary Phases During Homogenization of Alloys

Model	Assumptions
$r(t) = r_0 - k\sqrt{Dt}$; where $k = \dfrac{C_s - C_m}{\sqrt{(C_p - C_m)(C_p - C_s)}}$	Linearized concentration gradient, isolated particle, constant D, valid at short times when $C_s \gg C_m$ (Aaron, 1968)
(short times): $r(t) = r_0 - \dfrac{k'Dt}{2r_0} - \dfrac{k'}{\sqrt{\pi}}\sqrt{Dt}, k' = \dfrac{2(C_s - C_m)}{(C_p - C_s)}$ (long times): $\ln(y^2 + 2py\sqrt{\tau} + \tau) = -\dfrac{2p}{\sqrt{1-p^2}}\tan^{-1}\left[\dfrac{\sqrt{1-p^2}}{(y/\sqrt{\tau}) + p}\right]$, where $p = \sqrt{\dfrac{C_p - C_m}{2\pi(C_p - C_s)}}, y = \dfrac{r(t)}{r_0}, \tau = \dfrac{2Dt}{r_0^2}\dfrac{(C_s - C_m)}{(C_p - C_s)}$	Quasi-stationary interface, isolated particle, constant D. The quasi-stationary interface approximation means that the concentration distribution in the matrix ahead of the moving phase boundary is the same as that which would exist if the interface had been fixed at r(t) from the start. The parameters y and τ are dimensionless particle radius and dimensionless time, respectively (Whelan, 1969).
$C(r,t) = \sum_{n=0}^{\infty} \dfrac{A_n \exp(-t/\tau_n)C_n \sin(\lambda_n r(t) - \delta_n)}{r(t)}$, where $\tau_n = (\lambda_n^2 D)^{-1}, \lambda_n r_0 = \beta\alpha_n, \lambda_n r_s = \alpha_n, \beta = r_0/r_s,$ $\alpha_n(1-\beta) = \tan^{-1}\left[\dfrac{\alpha_n(1-\sigma\beta)}{1+\sigma\beta\alpha_n^2}\right] + n\pi, n = 0, 1, 2, 3,$ and $\tan(\alpha_n - \delta_n) = \alpha_n, \sigma = (\dfrac{k''r_0}{D} + 1)^{-1}; k''$ is the interface reaction rate constant and r_s is the radius of an equivalent sphere around each particle. The coefficients A_n and C_n in the series solution for the concentration distribution C(r,t) are given in terms of concentration parameters and α, β and δ.	Uniform distribution of equal-sized spheres (diffusion field impingement), constant D, stationary interface, dissolution is limited by either diffusion rate or interface attachment rate, or by both (mixed control). The dissolution rate is obtained by using the expression for C(r,t) given in the first column in conjunction with the interface mass balance via numerical integration (Nolfi et al, 1969)
(short times): $y(\tau) = y_1(x) = 1 + 2\beta_0 x + \beta_1 x^2 + \beta_2 x^3 + \cdots; x = \sqrt{\tau}$, and $2\beta_0 = \dfrac{4\lambda}{\sqrt{\pi}} + \dfrac{8\lambda^2}{\pi^{3/2}} + \dfrac{4\lambda^3[(8/\pi) - 1]}{\pi^{5/2}} + \cdots; \beta_1 = 2\lambda + 3\lambda^2 + \cdots$ (long times): $y(\tau) = 1 + \lambda\phi_0(\tau) + \lambda^2\phi_1(\tau) + \cdots;$ $\phi_0(\tau) = 2\tau + 4\sqrt{\tau/\pi}; \phi_1(\tau) = 2.67\tau\sqrt{\tau/\pi} + 3\tau + 2.55\sqrt{\tau/\pi};$ and $\lambda = \dfrac{(C_m - C_s)}{(C_p - C_s)}$	Isolated spherical particle, dissolution with moving boundary, constant D, obtains series solutions to the dissolution kinetics (Luybov and Shebelev, 1973)
(isolated planar particle): $x(t) = x_0 - k_1\sqrt{t}$	Planar or spherical particles, dissolution with moving interface, zero or finite volume fraction (soft impingement), obtains closed-form solution for isolated plate. Based on an approximate solution to Fick's second equation with a cubic concentration profile (Pabi, 1979; Asthana and Pabi, 1990; Asthana, 1993)

C_p = solute concentration in the dissolving precipitate (assumed to be constant)
C_m = solute concentration in the matrix far from the interface (assumed to be constant)
C_s = solute concentration at the particle-matrix interface (assumed to be constant)

Aaron, H.B., Met. Sci. J., 2, 1968, p.192.
Whelan, M.J., Met. Sci. J., 3, 1969, p. 95.
F.V. Nolfi, Jr., P.G. Shewmon, and J.S. Foster, Trans. TMS-AIME, 245, 1969, p. 1427.
B. Ya. Luybov and V.V. Shebelev, Phys. Met. Metallog., 35(2), 1973, p. 95.
Pabi, S.K., Acta Metall., 27, 1979, p. 1693.
R. Asthana and S. K. Pabi, Mater. Sci. Eng., A128, 1990, 253–260.
R. Asthana, J. Colloid Interface Sci., 158, 1993, 146–151.

area-to-volume ratio have greater solubility than coarser particles. As a result, fine particles will disappear by coarsening and the coarser particles will grow at the expense of the smaller particles.

The coarsening phenomenon has been theoretically modeled in order to predict important structural characteristics of multiphase systems, such as average particle size and size distribution. Theory shows that at long times, the cube of average particle radius increases linearly with time according to $r^3(t) = K \cdot t$, where r is the average particle radius at time t, and K is the coarsening rate constant. Theory further predicts that any arbitrary distribution of particle radii will tend to a unique time-independent form when the particle radii are scaled by the average particle radius. A major limitation of most theories of coarsening is that they are rigorous only at vanishingly small-volume fraction of the coarsening phase (i.e., two isolated particles in an infinite matrix). In real systems, coarsening occurs at a finite-volume fraction of the secondary phase, which leads to mutual diffusional interactions among a large number of particles of different curvatures.

The kinetics of coarsening have been experimentally measured in a variety of solid–solid, liquid–solid, liquid–liquid, and vapor–liquid systems. Solid–solid systems (i.e., solid particles in a solid matrix) generally exhibit slow coarsening rates due to slow solid-state diffusion and misfit elastic strains of a solid embedded in another solid (matrix). In liquid–liquid systems, droplet coalescence takes precedence over coarsening because of high droplet mobility and strong surface forces. Coalescence is less important in solid–liquid systems (i.e., solid particles in a liquid matrix) in which the liquid phase wets the solid particles. In both liquid–liquid and liquid–solid systems, however, second-phase segregation because of settling (or flotation) could interfere with coarsening. The particles can migrate because of density differences or liquid convection under the influence of gravity. To avoid large-scale sedimentation of second-phase particles in the terrestrial environment, it has been necessary to use highly concentrated systems in which the large-volume fraction of the solid phase forms a skeletal network, and inhibits fluid convection and large-scale sedimentation. Gravity aids skeleton formation by inducing particle contacts and particle motion. However, it may also tend to destabilize the skeleton. For example, a group of particles may extricate itself from the skeleton and float or sink via interstices of the skeleton. The critical volume fraction at which a stable skeleton forms is a function of the density of the system; for example, the value of critical volume fraction is 0.55 for the Pb-Sn system, whereas it is 0.33 in the more closely density-matched Co-Cu system. Coarsening experiments have also been done in the microgravity of space to eliminate sedimentation in dilute solid–liquid mixtures, and in the presence of electromagnetic fields that levitate the particles in the liquid medium by countering the force of gravity. Microgravity removes most convection and sedimentation, and allows coarsening kinetics to be measured at low solid-volume fractions. This allows fundamental theories of coarsening developed for dilute systems to be tested against experiments. Finally, it must be noted that besides particle sedimentation and fluid convection, other factors can perturb the coarsening kinetics. For example, in concentrated alloys, the simultaneous diffusion of mass and heat affects the coarsening kinetics, and the coarsening rate becomes a function of both heat and mass diffusivities.

Ceramic Forming

Slip-Casting

Slip-casting is used to make ceramic objects by consolidating fine ceramics suspended in a liquid over a porous mold. Liquid drainage causes the solid to deposit on the mold (Figure 3-24).

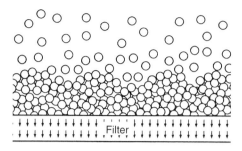

FIGURE 3-24 *Schematic illustration of the principle of slip casting, showing deposition of fine ceramics on a porous ceramic mold and capillary extraction of carrier fluid through the mold.*

The process employs homogeneous dispersion of fine ceramics in a liquid vehicle to which are added binders, surfactants (wettability promoters), and deflocculants for controlling the slurry properties such as viscosity and density. The mixture (slip) is poured into a porous gypsum mold having fine, micrometer-size pores and large (40–50%) amounts of porosity. With time, the liquid is drained by capillary suction, leaving a leather-like solid layer of the ceramic. The layer is stripped from the mold and is dried, fired, and glazed to improve the surface finish. Bathroom fixtures, china and dinnerware, porous thermal insulators, and other simple parts are made using slip-casting.

The viscosity and specific gravity of the slip are critical parameters, and must be carefully controlled to obtain homogeneous deposits. Both low- and high-viscosity slips are detrimental to the quality of the final product. Slips are pseudoplastic, so their viscosity decreases with increasing shear (mixing) rate. Slip viscosity is influenced by the agglomeration of fine ceramics dispersed in the slip, and deflocculants or stabilizers are added to control the extent of agglomeration. However, both over-deflocculated (i.e., low-viscosity) slips and under-deflocculated (i.e., high-viscosity) slips yield poor quality deposits. Common sintering aids, such as boron and carbon, that are added to aqueous slips are hydrophobic (liquid repellants) and hinder the formation of a homogeneously dispersed slip. The specific gravity of the slip is also important and must be high in order to increase the casting rate. However, mechanical properties such as the modulus of rupture (MOR) of cast and fired ceramics decrease at a greater rate at high specific gravity of the slip. Thus, slip composition and consistency must be judiciously selected to obtain the optimum viscosity and specific gravity so that a reasonable casting rate and the desired final properties of the cast object are achieved.

Factors important to slip-casting as a production process include casting rate, strength of the deposit, drainage behavior through the porous mold, and shrinkage and release from the mold. The kinetics of slip-casting are derived by likening the liquid drainage process to a filtration process and applying Darcy's equation for fluid flow through a porous body. According to Darcy's equation, the apparent flow rate, J (volume V per unit area A and time t), is $J = \frac{1}{A}\frac{dV}{dt} = -\frac{k}{\eta}\frac{\Delta p}{\Delta x}$, where k is the average permeability of the consolidated layer, η is the fluid viscosity, and Δp is the effective pressure difference over the cake (deposit) thickness, Δx. A mass balance condition in conjunction with Darcy's law has been shown to lead to parabolic kinetics for the evolution of the cake thickness with time, t, i.e., $x^2 = a \cdot t$, where a is a constant. Both the experimental measurements of casting rate and the theoretical predictions of parabolic kinetics indicate that a limiting thickness of the deposit is reached, and longer casting times do not produce greater

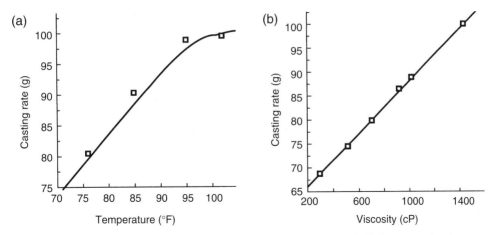

FIGURE 3-25 *(a) Relationship between temperature and casting rate of 32% ball clay + 18% kaolin. (B. Leach, H. Wheeler, and B. Lynn, American Ceramic Society Bulletin, 75 (8), 1996, 49–51) (b) Relationship between viscosity and casting rate of 32% ball clay + 18% kaolin. (B. Leach, H. Wheeler, and B. Lynn, American Ceramic Society Bulletin, 75 (8), 1996, 49–51)*

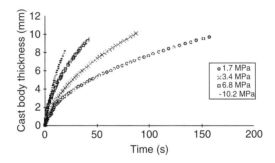

FIGURE 3-26 *Wall thickness as a function of casting time directly measured for pressure casting performed at different applied pressures from a SiC slip containing 3% carbon.* (www.ceramicbulletin.org, November 1998, 65)

thickness. Temperature, viscosity, and specific gravity influence the casting rate. Figure 3-25 shows the experimental measurements of the effect of slurry viscosity and temperature on the casting rate in deposits consisting of ball clay and kaolin.

Traditional slip casting is relatively inexpensive and material efficient (uses close to 100% of the material). However, it is slow and time consuming and may take several hours to complete. Casting rates of slips can be increased by employing pressure or vacuum to increase the liquid drainage through the porous gypsum mold. This permits rapid mass production of parts with thickness greater than that achievable using conventional slip-casting. In the case of an external pressure acting on the slip, the pressure term in Darcy's equation can be modified as follows, $\Delta p = \Delta p(\text{suction}) + \Delta p(\text{appld.})$, where $\Delta p(\text{suction})$ is the suction pressure of the porous mold, and $\Delta p(\text{appld.})$ is the external pressure applied to the slip. At high applied pressures, greater thickness is achieved in a given time, as shown in Figure 3-26 for pressure-cast SiC slip containing

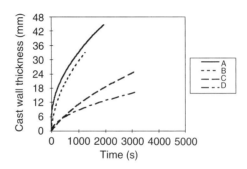

FIGURE 3-27 *Pressure slip casting curves showing dependence of green body thickness on time for different ceramic slips cast at 3.4 MPa. A through D denote the different composition of the slip; A, 4.5% clay – 95.5% nonclay; B, 17% clay – 83% nonclay; C, 35% clay – 65% nonclay; and D, 56% clay – 44% nonclay. Clay includes kaolinite, illite, and montmorillonite, and nonclay includes alumina, quartz, and feldspar.* (A. Salomoni and I. Stamenkovic, www.ceramicbulltein.org, November 2000, pp. 49–53)

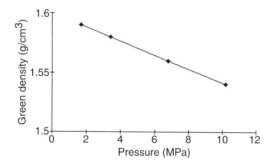

FIGURE 3-28 *Green density of pressure cast ceramic samples as a function of the applied pressure.* (www.ceramicbulletin.org, November 1998, p. 65)

3% carbon. Figure 3-27 shows the effect of casting time on cast wall thickness under a fixed pressure of 3.4 MPa in clay green bodies containing different percentages of clay (kaolinite, illite, and montmorillonite) and nonclay additives (alumina, quartz, and feldspar); the different curves marked A, B, C, and D correspond to different compositions. This figure shows that slip composition strongly affects the casting rate, and the thickness of the deposit does not increase at a constant rate with longer process time at a constant pressure. A saturation cake thickness is attained at long casting times in agreement with the theoretical parabolic kinetics.

Although the high casting rates achieved in pressure casting lead to faster production rates, they also result in lower packing levels because particles do not have enough time to rearrange themselves into a high-density (closely packed) mass. The migration of fine particles is accelerated under pressure, which causes clogging of the pores. As a result, the higher the applied pressure, the lower the deposit (green) density, as shown in Figure 3-28, for SiC slip containing 3% carbon. The densification pattern of the deposit reveals the differences in the packing mechanisms in slip-casting and pressure-casting. For example, in pressure-casting the extrapolation

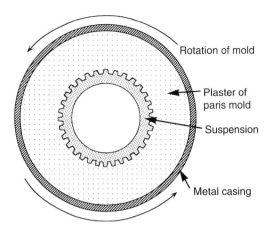

Rotation of mold

Plaster of paris mold

Suspension

Metal casing

FIGURE 3-29 *Cross-section of a centrifugally slip cast gear and plaster of Paris mold.* (G. A. Steinlage, R. K. Roeder, K. P. Trumble, K. J. Bowman, American Ceramic Society Bulletin, 75(5), 1996, 92–94)

of the green deposit density versus pressure plot to zero pressure (i.e., atmospheric pressure) does not lead to the same deposit density that is obtained by conventional slip-casting in the absence of external pressure.

A variation of pressure slip-casting is the centrifugal slip-casting in which powder consolidation takes place on a porous mold wall under large (20–60 g) centrifugal accelerations. The process is used to make tubular parts from monolithic and reinforced ceramics, and gradient or layered structures. Figure 3-29 shows a centrifugally slip-cast ceramic gear and the mold. The microstructural features of the deposit, such as the ratio of different phases, layer thickness, and preferred orientation, can be designed through control of mold shape, rotational speed, casting radius, particle size and distribution, and suspension viscosity. For example, deposits made from slips containing both platelets and nodules of alumina particles exhibit platelet alignment because of consolidation under centrifugal force. Figure 3-30 shows a centrifugally slip-cast composite gear with gradient layers of Al_2O_3 platelets in a $Ce-ZrO_2/Al_2O_3$ matrix; the layers are seen to be aligned parallel to the outer surfaces of the gear.

Well-dispersed colloidal suspensions can be centrifugally cast to make dense compacts; however, a narrow size distribution and the addition of a stabilizing agent (e.g., nitric acid and polyacrylate-based compounds) prevents flocculation. A narrow size distribution also prevents differential settling during centrifuging. The stabilizer prevents flocculation, but its concentration must be carefully controlled. At low stabilizer concentration, flocculation occurs, leading to a low-density deposit with rough surface. At high stabilizer concentration, the slurry can become too stable and well dispersed, and the deposit could remain fluid-like; i.e., as soon as rotation stops, consolidated particles redisperse. The commercial use of the centrifugal slip-casting technique has been somewhat limited owing to the need for high rpm (typically 20,000 or higher). One interesting application is in improving the toughness of SiC by positioning interlayers of graphite between SiC layers. Graphite is able to deflect cracks and improve the toughness of SiC in the final part. The layering is achieved by sequential pouring and centrifuging of different slurries in the rotating mold.

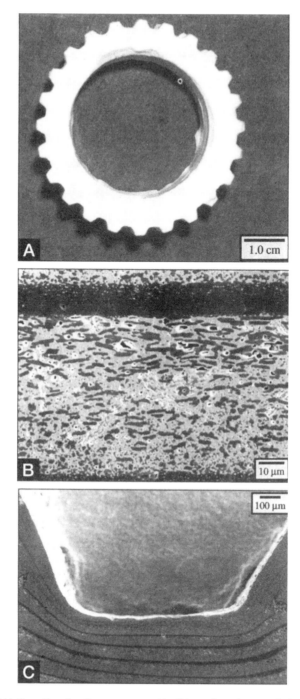

FIGURE 3-30 *(A) Centrifugally slip cast gear with (B) gradient layers of alumina platelet reinforcement in a Ce-ZrO$_2$/Al$_2$O$_3$ matrix. (C) Layers follow parallel to the outer surfaces of the gear.* (G. A. Steinlage, R. K. Roeder, K. P. Trumble, K. J. Bowman, American Ceramic Society Bulletin, 75(5), 1996, 92–94)

Tape-Casting

Tape-casting is widely used to fabricate ceramic substrates for high-technology applications. It is used to make components such as inductors and varistors, piezoelectric devices, thick film insulators, ceramic fuel cells, and multilayer capacitors (Figure 3-31). In tape-casting, a specially formulated ceramic slurry is allowed to form a wet coating of controlled thickness on a moving plastic film (usually Teflon^{TM} or cellulose acetate, or Mylar^{TM}). The plastic film must be clean and smooth. A precision doctor blade controls the deposit thickness (Figure 3-32), and a drier dries the deposit into a leather-like tape. After casting and drying, the tape is stripped from the carrier film. Ease of release is important; the carrier surface should be free of residual tape and the carrier side of the tape should be free of pullout, etc. The dried tapes are cut to desired length for subsequent firing. Cutting is done with a punching motion rather than a slicing motion, because the latter causes distortion and defects during firing. In addition, the drying temperature should be kept below the boiling point of the solvents to prevent formation of bubbles and defects during drying of the tape.

FIGURE 3-31 *Schematic diagram of a multilayer ceramic capacitor.* (M. L. Korwin, www.ceramicbulletin.org, December 1997, 47–50)

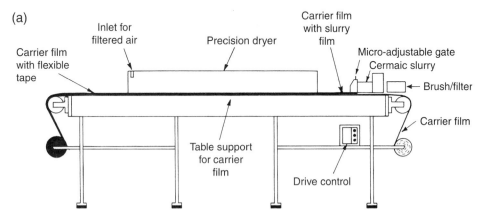

FIGURE 3-32 *(a) Schematic illustration of the doctor-blade tape casting process.* (J. S. Reed, Principles of Ceramic Processing, 3rd ed., Wiley, New York, 1995)

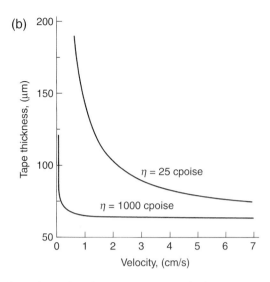

FIGURE 3-32 *(b) Effect of speed and viscosity on tape thickness in tape casting of ceramics.* (J. S. Reed, Principles of Ceramic Processing, 3rd ed., Wiley, New York, 1995)

The chief advantage of tape-casting is that it produces long, thin, and flat films of ceramics with controlled thickness, which are extremely difficult to produce using powder compaction, especially when holes and slots of various shapes and sizes are required in the tape. Such features are easily created into unfired flexible ceramic tapes made using tape-casting. The tape thickness (typically 3 µm to 1 mm) depends on the height of the doctor blade, viscosity, film speed, and drying and firing shrinkages. Figure 3.32b shows the effect of velocity and viscosity on the thickness of the tape; higher velocity and greater viscosity yield thinner tapes. The thickness becomes constant above a critical velocity, and this velocity increases with decreasing viscosity.

Tape-casting slurries contain several additives (deflocculants, surfactants, etc.) in a solvent such as xylene. The correct quantity of deflocculant is based on the surface area and particle size distribution of the powders to be used. Deflocculation is important because it improves particle packing during drying, thus increasing the green density of the tape-cast product. Fish oil is an inexpensive high-viscosity deflocculant commonly used in polyvinyl butaryl (PVB) binder systems for tape-casting, and it readily dissolves in xylene. PVB is used as a resin binder for tape-casting ceramics that are to be sintered in oxidizing atmosphere. Other additives may be added to lower the glass transition temperature of the resin and to permit increased room temperature flexibility of the tape. Tape-casting slurries are pseudoplastic and slightly thixotropic and exhibit shear-thinning behavior under the doctor blade during processing.

Variations of ceramic tape-casting exist in the paper, plastics, and paint industries. For example, paint manufacturers use the basic approach of removal of excess material from a moving surface being coated, using a doctor blade, to test the covering power of paint formulations. For this purpose, thin (<50 µm) films of paint are uniformly deposited on a standard black-and-white background using a doctor blade, and the degree to which the background is hidden by the deposit is measured optically.

Tape-casting is somewhat similar to slip-casting, although important differences exist between the two. For example, the deposition process is evaporative in tape-casting, whereas it is absorptive in slip-casting (i.e., filtration through a porous mold), and necessitates use of different types of solvents (aqueous for slip-casting and nonaqueous for tape-casting). The porous permeable surface in slip-casting is replaced with a nonpermeable, flexible carrier surface in tape-casting (earlier approaches used less flexible steel strip in place of flexible plastics). But slurry filtration (similar to slip-casting) just prior to tape-casting has been used to eliminate certain defects in tape-cast parts.

The emerging applications of tape-casting are in the creation of 3D objects using iterative laser cutting and lamination processes for rapid prototyping of parts, and fabrication of ceramic parts for applications in polymer-based lithium-ion battery membranes for cell phones and laptop computers. Tape-casting is also used by the aerospace industry to make thin sheets of brittle intermetallic materials such as iron-aluminide and nickel-aluminides that are used as the feedstock for fabrication of fiber-reinforced composites. These composites are made by stacking alternating layers of prefired tapes of the aluminides and reinforcing fibers, followed by hot consolidation. Similarly, tough SiC-C ceramic composites have been made by stacking alternating layers of tape-cast SiC and graphite, followed by hot-pressing.

Ceramic Extrusion

Clay and other types of plastic ceramic bodies are extruded to form parts such as ceramic filters and honeycombs that are used in automobile catalytic converters and in the filtration industry. Both piston (batch extrusion) and auger (continuous extrusion) machines are used. The important variables in extrusion include: entrance angle (α), reduction ratio (i.e., ratio of barrel diameter, R_0, to finishing tube diameter, R_F), yield strength of the extruded material, friction at the barrel surface and die wall, and extrusion pressure and velocity. Figure 3-33 shows the effect of extrusion pressure on extrusion velocity for various values of the reduction ratio (R_0/R_F). A large reduction ratio requires greater extrusion pressure to achieve a fixed extrusion velocity.

The yield strength of the extruded materials should be less than their adhesion strength to the barrel, and there should be no slippage at the walls (ribs are added to the walls to increase the friction). Figure 3-34 displays the schematic flow behavior of typical extrusion bodies through

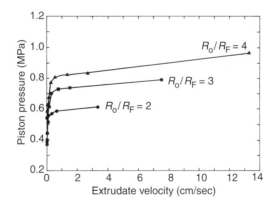

FIGURE 3-33 *Dependence of pressure for extrusion of ceramics on the extrusion velocity and reduction ratio of the die (R_0/R_F), where R_0 and R_F are the radii of the barrel and finishing tube, respectively.* (J. S. Reed, Principles of Ceramic Processing, 3rd ed., Wiley, New York, 1995)

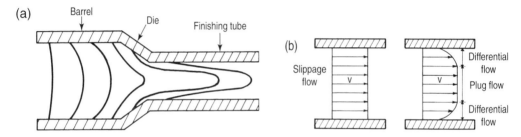

FIGURE 3-34 *(a) Schematic flow pattern of ceramics through barrel, die, and finishing tube. (J. S. Reed, Principles of Ceramic Processing, 3rd ed., Wiley, New York, 1995) (b) Flow patterns of ceramic extrusion bodies: slippage flow (negligible wall friction), and differential and plug flow. (J. S. Reed, Principles of Ceramic Processing, 3rd ed., Wiley, New York, 1995)*

the barrel, die, and finishing tube. The flow could involve the undesirable slippage flow (negligible wall friction), and differential and plug flow. The extrusion pressures vary from less than 1 MPa to about 15 MPa; extrusion capacity could be as high as ∼100 tons/h, and extrusion velocity is about 1m/min. Extrusion bodies are specially formulated mixtures of ceramic powders, polyvinyl alcohol (PVA), and a liquid vehicle (usually wax, water, or an oil). The permeability of the mixture must be low to avoid liquid extrusion through the ceramic powders and segregation problems.

Electrolytic and Electrophoretic Deposition

Electrical deposition of ceramics is used to create monolayers and multilayers, powders, composites and nanostructured materials. Two types of electrical deposition techniques are used: electrolytic deposition and electrophoretic deposition. In the electrolytic deposition process, solutions of metal salts are used. In electrophoretic deposition, fine ceramics suspended in a liquid vehicle are electrically charged and driven toward an oppositely charged electrode to form the deposit. The electrolytic deposition (ED) process can be performed from relatively dilute aqueous suspensions of metallic salts. Because a film or coating forms from ionic species, coating formation takes place at the atomic level, and subsequent sintering can be done at relatively low temperatures because of fine particle size. The amount of deposited material is controlled by variation of deposition time and current density. Both conductive and nonconductive materials can be deposited using ED; however, insulating ceramics form thinner deposits than more conductive materials. Some tendency toward microcracking could persist in the deposit because of drying shrinkage, as with most other wet ceramic-forming processes. Both thin (<1 μm) and thick (>10 μm) deposits are possible. The ED of oxide ceramics, carbon (fibers and felt), platinized silicon wafers, and a variety of composites is done for industrial applications. Table 3-5 gives examples of the ED ceramic coatings with proven applications.

Cathodic electrodeposition is the most widely used form of the basic ceramic electrodeposition technique. The deposition is achieved through hydrolysis of metal ions or complexes to form oxide, hydroxide, or peroxide deposits on cathodic substrates. The hydroxide and peroxide deposits can be converted to oxide films through a thermal treatment. The principal chemical reactions in cathodic ED involve reduction of water, nitrate ions, and dissolved oxygen according to

$$2H_2O + 2e^- \rightarrow H_2 + 2OH^-$$

$$NO_3^- + H_2O + 2e^- \rightarrow NO_2^- + 2OH^-$$

$$O_2 + 2H_2O + 4e^- \rightarrow 4OH^-$$

TABLE 3-5   Proven Applications of Ceramic Electrodeposition (ED)

ED Coating Material	Field of Proven Applications
Nickel hydroxide	Electrodes for rechargeable batteries
Titania (TiO_2)	Biomedical implants, gas sensors, capacitors, photodetectors, solar cells
Niobium oxide (Nb_2O_5), Ruthenium oxide (RuO_2)	Electrodes, catalysts, integrated circuits
Lead zirconate titanate (PZT), $BaTiO_3$	Multilayer capacitors, piezoelectric transducers, memory devices, and IR detectors
Zinc oxide (ZnO)	Piezoelectric transducers, solar cells, chemical sensors
RuO_2-TiO_2	Dimensionally stable anodes in the chloro-alkali industry
Hydroxyapatite and other calcium phosphate materials	Coatings on biomedical implants
Alumina (Al_2O_3), ceria (CeO_2), zirconia (ZrO_2), chromia (Cr_2O_3), and composites (Al_2O_3-TiO_2 and Al_2O_3-ZrO_2)	Protective films on metals

These reactions generate hydroxyl ions and increase the pH at the electrode. The metal ions, Me^{n+}, in the vicinity of the cathode are hydrolyzed by the electrically generated base (OH^- ions) to form colloidal (metal hydroxide) particles. The surface charge on colloidal particles depends on the pH; at low pH, particles are positively charged, but as pH increases because of OH^- ions formation according to the above reactions, the surface positive charge on particles decreases, the repulsion between them decreases, and particles form a dense and coherent deposit on the cathode surface. The rates of hydroxyl ion and metal ion Me^{n+} generation are important; when the rate of OH^- ion generation is faster than their consumption by Me^{n+} ions to form metal hydroxide particles, some OH^- ions are transported via diffusion and electrical current away from the cathode. As a result, the high pH boundary moves away from the cathode surface, resulting in weaker adhesion of the electrodeposited colloidal particles.

The colloidal metal hydroxide particles experience mutual repulsion and attraction. The interaction energy between particles consists of an attractive energy because of London–van der Waals forces, and a repulsive energy because of charge on particles. When the repulsion is greater than the London–van der Waals attraction, the total energy of the particles displays a maximum that essentially is the energy barrier to coagulation and deposition of colloidal particles on the cathode. The energy maximum decreases as the electrolyte concentration increases so that barrier to film deposition disappears at high concentrations. Besides the dispersion forces due to the attractive London–van der Waals interactions, there is some long-range attraction (spanning several particle diameters) between similarly charged colloidal particles in the vicinity of the cathode, which aids clustering and flocculation. The origin of the attractive forces is not well understood but is believed to result from the fluid flow arising from ionic migration through the solution. The ionic current stems from electromigration of hydroxyl ions generated in the above reactions. This gives rise to current gradients that generate localized flow, which in turn results in segregation and clustering of particles. In dilute solutions, where ionic species migrate over large distances, the deposition rate is limited by diffusion, and the deposit thickness (or weight) is directly proportional to the square root of deposition time. Figure 3-35 is a plot of deposit weight versus time for electrolytic deposition of zirconia (curves S1 and S2) from $ZrOCl_2$ electrolyte, titania (S3 and S4) from $TiCl_4$ electrolyte, and alumina (curve S5) from $Al(NO_3)_3$ electrolyte. The deposition was done at a current density of 20 mA·cm^{-2} for all systems except S5, for which the current density was 5 mA·cm^{-2}.

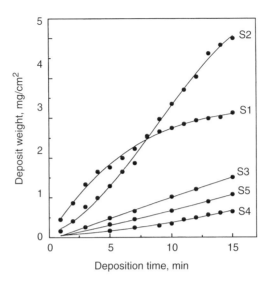

FIGURE 3-35 *Electrolytic deposition of coatings of ZrO$_2$ (S1 and S2) from ZrOCl$_2$ electrolyte, TiO$_2$ (S3 and S4) from TiCl$_4$ electrolyte, and Al$_2$O$_3$ (S5) from Al(NO$_3$)$_3$ electrolyte. The current density was 20 mA.cm^{-2} for all except S5 (5 mA.cm^{-2}). (I. Zhitomirski, www.ceramicbulletin.org, September 2000, 57–63)*

Electrophoretic deposition (EPD) involves migration of charged macroscopic ceramic particles toward an oppositely charged electrode (as opposed to migration of ionic species that combine with hydrolyzed metal ions to form the deposit in the ED process). Two basic steps are involved in EPD: electrophoresis, which is the migration of charged particles in an electric field, and their deposition and film formation. Both these processes are independent of the nature of the solid particles in the colloidal suspension. As a result, a wide range of materials can be deposited, including multilayer ceramic composites. EPD permits relatively high deposition rates to be achieved, with rates on the order of 200 μm/min being easily attained. The shape of the electrode determines the shape of the deposit. Ceramic laminates can be made by changing the suspensions to various ceramic compounds. EPD also permits simultaneous deposition of two or more different types of particles to synthesize ceramic–ceramic composites. For example, SiC-C composites having high fracture toughness have been produced by the EPD of alternating layers of SiC and graphite by changing the type of colloidal solids in the suspension. Figure 3-36 shows some SiC-graphite–laminated tubes produced by EPD, and the microstructure of an alumina deposit onto a graphite substrate. Figure 3-37 shows crack deflection and potential toughening in laminated SiC-graphite composite after bend test. EPD has been used to produce a variety of laminated configurations, such as multilayers, functionally gradient materials (FGM), and continuous FGM shown in Figure 3-38. Figure 3-39 shows the scanning electron micrograph of Al$_2$O$_3$/Y-TZP multilayers deposited onto Zn from aqueous slurries containing deflocculants, and EPD yttria-stabilized zirconia (YSZ) on graphite cloth after sintering or hot-pressing at 1500°C. Fully dense composites are obtained with a uniform distribution of YSZ in the matrix.

EPD is usually performed in organic solvents because the decomposition of water in aqueous solutions limits the voltage that can be attained and the rate of deposition is low. Stirring the suspension during deposition leads to smoother deposits. Because the voltage is usually

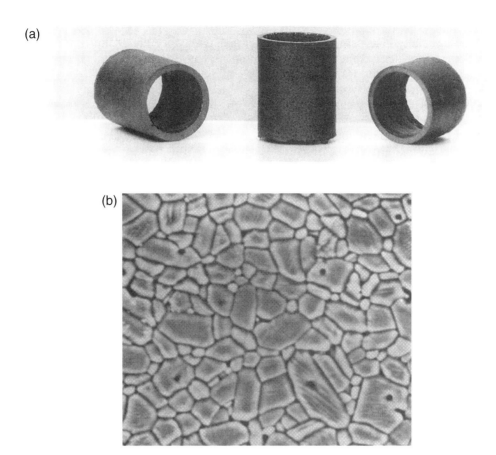

FIGURE 3-36 *(a) Some SiC-carbon laminated tubes produced by electrophoretic deposition.* (L. Vandeperre, O. Van der Biest, F. Bouyer, and A. Foissy, www.ceramicbulletin.org, January 1998, 53–58) *(b) Scanning electron photomicrograph of an Al_2O_3 self-supported deposit, obtained onto graphite by applying 6.46.4 mA.cm^{-2} current density for 10 min.* (R. Moreno and B. Ferrari, www.ceramicbulletin.org, January 2000, 44–48)

maintained constant and the quantity of the colloidal suspension is limited, the deposition rate decreases with time. The deposition rate is proportional to the solids concentration in the solution; as the suspension is depleted of the solids, the deposition rate decreases. Another reason for the slowing of the deposition rate is an increase in the electrical resistance because of the formation of an insulating ceramic layer. Figure 3-40 shows a plot of electrophoretic yield as a function of deposition time for suspensions of boron-doped SiC containing different amounts of *n*-butylamine and isopropyl alcohol.

The deposition rate in ED is usually faster and the deposit more homogeneous than in EPD. In EPD, the uniformity of the deposit and the minimum deposited thickness are limited by the size of the ceramic powders used. Some tendency toward settling and agglomeration will usually persist with ceramic suspensions used in electrophoretic deposition. Better control is possible

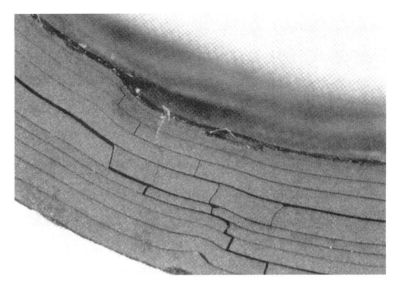

FIGURE 3-37 *Details of a ring cut from an electrophoretically deposited SiC-graphite composite laminated tube after mechanical testing, showing crack deflection between laminates.* (L. Vandeperre, O. Van der Biest, F. Bouyer, and A. Foissy, www.ceramicbulletin.org, January 1998, 53–58)

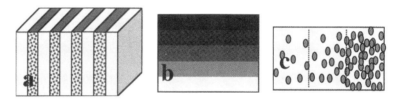

FIGURE 3-38 *Schematic illustration of some types of laminated materials that have been shaped by electrophoretic deposition in aqueous media: (a) multi-layers, (b) functionally gradient materials, and (c) continuous functionally gradient materials.* (R. Moreno and B. Ferrari, www.ceramicbulletin.org, January 2000, 44–48)

on the uniformity and thickness of the coatings by the ED process because ionic species rather than macroscopic particles are involved. Table 3-6 provides a comparison between the ED and EPD processes of depositing ceramics.

Glass Forming

Glass is formed by heating a mixture of quartz sand and soda ash, limestone, potash, and other chemicals to 1500–1600°C for 24 to 48 hours. Usually up to 30% recycled scrap glass is also added to the raw materials for economic reasons and also for better heat conduction and rapid fusion (glass has lower melting point and better thermal conductivity than the other raw materials used). The melting process involves many complex chemical reactions, which include formation of a eutectic melt of Na_2CO_3 and $CaCO_3$, decomposition of carbonates (and other compounds) into oxides together with the evolution of gases (e.g., CO_2 from carbonates), and

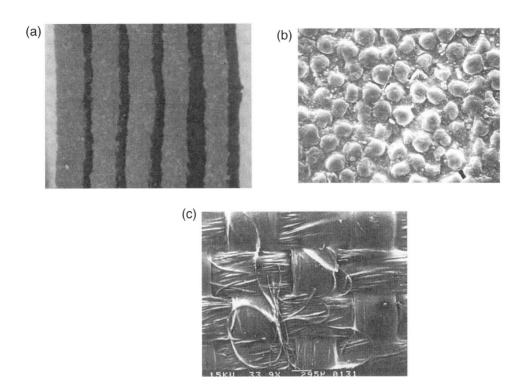

FIGURE 3-39 *(a) Scanning electron photomicrograph of Al₂O₃/Y-TZP multilayers deposited via electrophoretic deposition onto Zn from aqueous Al₂O₃ and Y-TZP slurries.* (R. Moreno and B. Ferrari, www.ceramicbulletin.org, January 2000, 44–48) *(b) Scanning electron photomicrograph of electrophoretically deposited yttria-stabilized zirconia (YSZ) on graphite cloth, hot pressed at 1500°C for 1 h. YSZ is distributed uniformly, and porosity is absent.* (P. S. Nicholson, P. Sarkar, and S. Datta, American Ceramic Society Bulletin, 75(11), November 1996, 48–51) *(c) Scanning electron photomicrograph of electrophoretically deposited YSZ on graphite cloth sintered at 1550°C.* (P. S. Nicholson, P. Sarkar, and S. Datta, American Ceramic Society Bulletin, 75(11), November 1996, 48–51)

gas removal by addition of chemical refining agents. Molten glass is tapped from the furnace and shaped using a variety of fabrication techniques. Common fabrication techniques for shaping glass objects include pressing, blow molding (inflating a blob of molten glass in a mold using compressed air), spinning, sheet forming using float-glass technique, and glass fiber formation. Figure 3-41 shows a schematic diagram of the float-glass technique of making sheet glass in which molten glass is allowed to spread over molten tin in an atmosphere of nitrogen (to prevent oxidation of Sn), followed by cooling and drawing. The viscosity of the glass is an important process variable affecting component fabrication and varies over a wide range depending on the temperature, as shown in Figure 3-42, for a common type of soda-lime glass. This figure also shows the range of viscosity needed for annealing, working (shaping), and melting of glass. The fabricated glass objects are usually annealed (or tempered) to relieve thermal stresses introduced during fabrication and prevent fracture. Tempering of glass involves heating the glass object to above its glass transition temperature but below its softening temperatures, followed by cooling (in air or oil). During cooling, the surface cools faster than the interior and becomes rigid

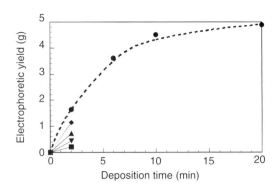

FIGURE 3-40 *Electrophoretic yield as a function of deposition time for suspensions of boron-doped SiC from mixtures of different concentrations. The different curves are for different amounts (0 to 90%) of isopropyl alcohol (an acid solvent) in a basic medium. The addition of increasing amounts of the acid solvent to the basic medium decreases the electrophoretic mobility and lowers the deposition rate.* (L. Vandeperre, O. Van der Biest, F. Bouyer and A. Foissy, www.ceramicbulletin.org, January 1998, 53–58)

TABLE 3-6 Comparison of Electrolytic and Electrophoretic Deposition of Ceramics

	Electrophoretic Deposition	*Electrolytic Deposition*
Medium	Suspension	Solution
Moving species	Particles	Ions or complexes
Electrode reactions	None	Electrogeneration of OH^- and neutralization of cations
Preferred liquid medium	Organic solvent	Mixed solvent (water-organic)
Required liquid conductivity	Low	High
Deposition rate	$1–10^3\ \mu m\cdot min^{-1}$	$10^{-3}–1\ \mu m\cdot min^{-1}$
Deposit thickness	$1–10^3\ \mu m$	$10^{-3}–10\ \mu m$
Deposit uniformity	Limited by particle size	On nanometer scale
Deposit stoichiometry	Controlled by stoichiometry of powders used	Can be controlled by use of precursors

(Source: I. Zhitomirsky, J. Materials, JOM-e, 52(1), 2000, http://www.tms.org./pubs/journals/JOM/0001/Zhitomirsky/Zhitomirsky-00)

before the interior has cooled. Later, when the interior regions cool, the rigid exterior prevents their contraction. This introduces tensile stresses in the interior and compressive stresses on the surface. The surface compressive stresses counteract external tensile stresses and prevent fracture. The thermal shock resistance of glass is enhanced by forming glass ceramics, which are partially crystallized glasses with superior mechanical and thermal shock resistance. These materials are formed by adding fine TiO_2 powders to molten glass to promote nucleation and recrystallization on cooling. Typical glass ceramics contain as much as 90% crystalline glass, and the rest is amorphous glass that fills the grain boundaries of partially crystallized grains, thus creating a pore-free structure. Grain sizes are $0.1–1.0\ \mu m$. Because glass fills grain boundaries, a pore-free structure forms, and stress-concentrating pores are eliminated.

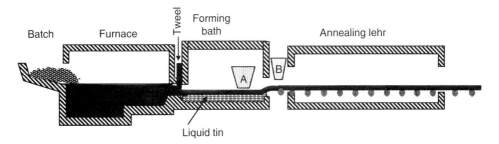

FIGURE 3-41 *Schematic diagram of a float-glass manufacture system. Locations for chemical vapor deposition of coatings (A), and spray pyrolysis coatings (B) are also shown.* (M. Arbab, L. J. Shelestak, and C. S. Harris, www.ceramicbulletin.org, January 2005, 30–35)

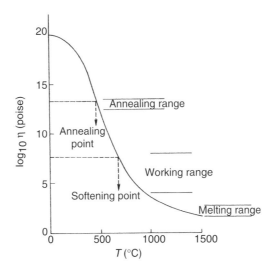

FIGURE 3-42 *The viscosity of a typical soda–lime–silica glass from room temperature to 1500°C. Above the glass transition temperature (450°C in this case), the viscosity decreases in an Arrhenius fashion.* (J. F. Shackelford, Introduction to Materials Science for Engineers, 1985, p. 330, Prentice Hall, Upper Saddle River, NJ)

The glassy state is a metastable state, but the kinetics of transformation to crystalline form is extremely slow at normal ambient temperatures, which is why ancient glass objects retain their amorphous state over millennia. The structure of glass is very open, and it can incorporate various types of chemical species that either aid or resist the tendency toward glass formation. For example, in soda-silica glass, controlled amounts (<20%) of Na_2O are added to glass as a source of sodium ions that are said to be network modifiers because they disrupt the silicon-oxygen network of glass. The reason for adding Na_2O is that Na^+ ions reduce the viscosity of the glass at elevated temperatures, thereby facilitating component fabrication (very large additions of Na_2O can, however, completely obliterate the SiO network and suppress glass formation). In contrast to soda-silica glass, in Pyrex glass, a glass former (or network former) such as B_2O_3

is added. This increases the viscosity and chemical inertness and decreases the coefficient of thermal expansion, thus making Pyrex glass more resistant to chemical corrosion and thermal shock than ordinary glass.

Pore Characterization

The shape, size, and volume fraction of pores change as a powdered metal or ceramic component progresses from the green state through dried state to sintered state. When a compact is sintered, pores shrink in size and become more spherical in shape. The total porosity content consists of open porosity (accessible to an external fluid) and closed porosity (inaccessible to an external fluid). Many physical and mechanical properties are affected by the presence of porosity, and it is important to know the amounts of open and closed porosity, average pore size and size distribution, and pore surface area. In this section, we review the most commonly used techniques to characterize these properties of the porous materials.

The bulk volume of a sintered part, based on its external dimensions, measured using a calipers or similar device, equals the true volume of the solid material in the part plus the open and closed pore volumes. The apparent volume of a part is defined as the true volume of the part plus the closed pore volume, and it does not include the volumetric contribution of the open pores. The apparent volume is conveniently measured using an immersion technique that allows all the open pores to be filled with a wetting liquid of known specific gravity. From the definitions of the bulk volume and apparent volume, it follows that the bulk volume of a part equals its apparent volume plus the open pore volume. These basic definitions are useful in estimating the extent of open and closed pore volume in a sintered part, and in characterizing the part density.

The pore surface area and porosity content are characterized using mercury porosimetry, gas adsorption, pycnometry, and permeametry. In *mercury porosimetry* (or more generally, liquid intrusion porosimetry), porosity is characterized by forcing a liquid, usually mercury, through a porous material under an external gas pressure, and measuring the volume of mercury needed for penetration as a function of applied pressure. As mercury has high surface tension and does not wet most solids (contact angle $\sim130–150°$), an external pressure is needed to initiate and sustain its flow through the porous material. At low pressures, only large pores are filled, but at high pressures, smaller pores are also filled. Increasing pressures are required for penetration of progressively finer pores in the solid. When a plot of pressure (P) versus volume (V) is made, usually a limiting or saturation value of liquid volume V_{max} is reached such that further increase in the pressure does not cause any appreciable increase in volume of liquid in the porous body. The total pore volume (or volume of pores of a given size) is equal to the volume of mercury that fills the pores. The method has also been used to measure the contact angle of different materials in contact with mercury, and this requires knowledge of the specific surface area, A, of the powders in the packed bed. The specific surface area (i.e., surface area per unit mass) is conveniently measured using a gas adsorption technique. *Gas adsorption* methods are based on measurements of the amount of a gas adsorbed on the surface of a powder to a monomolecular depth on the particle's surface. The amount of gas adsorbed in a monolayer is obtained from an adsorption isotherm, which is essentially a series of measurements of the volume of gas adsorbed as a function of pressure of gas. Gas adsorption methods yield precise measurements of the total surface area of porous materials and loose powders.

Mercury porosimetry offers high resolution in measurement of pore size, with measurable pore size range being 500 to 0.003 μm. The technique is sensitive to the contact angle of mercury and powdered samples being tested. Usually the contact angle, θ, of solids with Hg is taken as 130–150°; however, different solids exhibit different contact angles with Hg. The pressure, P_c, to force mercury into pores varies with cosine of the contact angle, θ, according to Young-Laplace equation,

$$P_c = \frac{2\sigma_{lv} \cos\theta}{r},\tag{3-38}$$

where σ_{lv} is the surface tension, and r is the pore radius. Errors in measurement are likely because of an assumed value of θ. In addition, the surface tension, σ_{lv}, varies with temperature, atmosphere and chemical purity, all of which influence the measurements. Errors can also arise from the fact that large pores may be connected through smaller necks; this may give rise to the erroneous finding that there are a large number of fine pores.

Pycnometry, based on Archimedes' principle, measures the volume of a liquid displaced on submersion of a porous body. This is a popular and inexpensive method to characterize the porosity. It consists of determining the quantity of water that fills the open pores after immersion of the sample for 2 hours in boiling water. Data are usually expressed as the percentage of water absorbed in pores with respect to the total mass of the dry sample. For solids that absorb the liquid, gas pycnometry is used, which measures the pressure changes on insertion of a porous body in a gas chamber. However, the gas should not adsorb on the powdered material and should be able to penetrate extremely fine pores to the limit of 0.1 nm. *Helium pycnometry* measures the total porosity content in a material by calculating the sample volume from the observed pressure changes between two chambers when a porous sample is introduced in one of the chambers. With no sample present in the chamber, the gas pressure in each chamber is the same, but when a sample is introduced in a chamber, a difference in pressure is observed. The pore volume is obtained from the gas law relationship between pressure and volume of a gas.

Permeametry measures the resistance to flow of a liquid or a gas through a compacted bed of powders and is based on models of laminar flow through porous solids. The pressure drop, ΔP, in a liquid flowing across a porous bed of thickness, l, is expressed from Darcy's equation, which accounts for the pressure drop because of fluid drag. Darcy's equation is

$$\frac{\Delta P}{l} = \frac{\mu}{k_1} V,\tag{3-39}$$

where V is the volumetric flow rate per unit cross-section perpendicular to flow, μ is the fluid viscosity, and k_1 is called the Darcian permeability constant. Darcy's equation is valid at small flow rates (laminar flow); at high flow rates, energy losses because of inertial forces become important as well as energy losses because of fluid viscosity. At high flow rates, the following more general equation, called the Forscheimer equation, is used

$$\frac{\Delta P}{l} = \frac{\mu}{k_1} V + \frac{\rho}{k_2} V^2,\tag{3-40}$$

where ρ is the fluid's density and k_2 is called the non-Darcian or inertial permeability. At high flow rates, the second term with square of the velocity, V, which represents the energy losses because of inertial forces, overrides the viscous dissipation. The deviation between the Darcy and Forscheimer equations is characterized in terms of Forscheimer number, Fo, where

$$Fo = \frac{\rho V}{\mu} \left(\frac{k_1}{k_2} \right)$$ (3-41)

so that the Forscheimer equation becomes

$$\frac{\Delta P}{l} = \frac{\mu}{k_1} V(1 + Fo).$$ (3-42)

If $Fo \ll 1$, viscous losses dominate, and Darcy's equation is obtained.

The permeability constants k_1 and k_2 are related to the structure of the porous medium. Several models have been developed to relate these constants to the void content and particle diameter. For example, the following relationships developed by Ergun are widely used:

$$k_1 = \frac{\varepsilon^2 d^2}{150(1 - \varepsilon)^2}$$ (3-43)

and

$$k_2 = \frac{\varepsilon^2 d}{1.75(1 - \varepsilon)},$$ (3-44)

where ε is the porosity and D is mean particle diameter in the porous material. These relationships have also been applied to complex cellular solids such as ceramic foams and filters, which may contain 75–95% porosity. However, it is not easy to define particle diameter, D, for a cellular solid. The diameter D has a rather clear meaning for powder-based materials, whether compacted or loose, but it is difficult to define D for the weblike structure of solid filaments connected in three dimensions. Two approaches are used in dealing with cellular ceramics: assuming particle diameter to be same as pore diameter, and using a hydraulic diameter derived from measurements of foam specific surface area. Pore diameter can be related to pore count on the surface of the cellular solid; typical linear pore density on the surface is 1–100 pores per square inch. Thus, with certain assumptions it is possible to use Ergun equations to predict pressure drop in cellular solids.

For cellular solids, the Darcian permeability, k_1, increases as pore diameter increases, and it approaches zero when pore diameter decreases. When the solid's surface area is large (i.e., pore size is fine), viscous losses are high, and k_1 is low. With a small pore size, more area is exposed to fluid. In contrast, non-Darcian permeability, k_2, represents inertial energy losses, and depends on the kinetic energy (ρV^2). These inertial losses could arise from the disturbances to fluid flow caused by turbulence and flow obstruction by solid. At high fluid velocities, energy losses will depend on the kinetic energy and curvature. In very fine pores, the pore size becomes comparable to the mean free path of gas molecules, and molecular flow (or diffusion or Knudsen flow) takes over, which is governed by the kinetic theory of gases. Considerable scientific judgment is needed in identifying and applying the theoretical models in order to estimate the porosity in cellular solids.

Microscopy techniques are also used to estimate the porosity. Microscopy is combined with the image analysis techniques and uses computer processing of images collected using optical or scanning electron microscopy, and determines the porosity from the ratio of pore area to total analyzed area. Images must have high contrast and high resolution to distinguish objects (i.e., porosity) to be analyzed from the rest of the material. Image analysis of several images of the sample yields not only percent porosity but also interparticle distance, aspect ratio, and other statistical parameters characterizing the sample.

Properties of Ceramics

Mechanical Properties

Ceramics and glasses are hard and brittle solids whose strength properties are extremely sensitive to the presence of minute flaws. Compared to metals, the number of slip systems for dislocation movement through polycrystalline ceramics are limited. For both ceramics and metals, however, the slip direction is the closest packed direction and the slip plane is the closest packed crystal plane. In the case of ceramics, the presence of charged ions (rather than neutral atoms as in metals) imposes certain restrictions on the number of slip systems that is responsible for their brittle behavior. In the case of amorphous ceramics (e.g., glasses), slip by dislocation motion does not exist as there is no crystalline order in the material. As a result, plastic flow is possible only through thermally activated motion of atoms, which is restricted at temperatures below the glass transition temperature. This leads to very strong brittle behavior of glass.

The sensitivity of the fracture strength of ceramics to minute flaws leads to considerable scatter in strength, which is because of the spatial distribution of flaws, the range of flaw sizes, and the orientation of the flaws with respect to the direction of applied stress. Usually, large ceramic objects show greater dispersion in strength, and lower strength values than smaller objects. This is due to a greater probability of finding strength-limiting flaws in large objects. A mathematical function, called the Weibull distribution function, simulates the strength distribution of brittle ceramics, and is expressed from

$$ P = \exp\left[\left(\frac{-V}{V_0}\right)\left(\frac{\sigma}{\sigma_0}\right)^m\right], \tag{3-45}$$

where P is the probability of survival of a specimen of volume V, σ is the fracture stress, V_0 is the unit volume, and σ_0 and m are constants. A key parameter in the Weibull distribution function is the Weibull modulus, m. A large value of m is associated with a narrow strength distribution, that is, smaller scatter in the distribution of measured fracture stress. For modern technical ceramics, the Weibull modulus, m, is usually in the range 5 to 10 in comparison to 50 to 100 for metals.

The Weibull function can be written in a linear form by taking the double logarithm of the preceding equation. This yields

$$ \ln \ln P = \ln\left(\frac{-V}{V_0}\right) + m \ln\left(\frac{\sigma}{\sigma_0}\right) \tag{3-46}$$

which is the equation of a straight line between ln lnP and ln(σ/σ_0), where P is the survival probability at a stress value, σ. The Weibull function is sometimes expressed in terms of the probability of failure (1-P). In a statistical population of size, N, the Weibull probability distribution function will be followed if a plot of ln ln(N+1/N+1−i) versus lnσ is linear ("i" denotes the ith sample in the population when the fracture stress data are arranged in an ascending order). Figure 3-43 shows a typical log–log Weibull plot for the fracture stress of brittle alumina fibers.

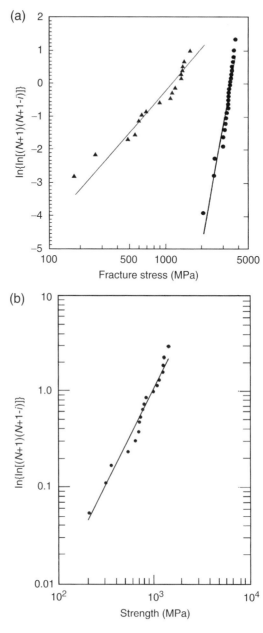

FIGURE 3-43 *Weibull plot of the strength distribution of single crystal sapphire fibers extracted from Ni-base superalloy matrix composites. (a) Haste alloy matrix (• denotes as-received fiber)* (R. Asthana, S. N. Tewari and S. L. Draper, Metallurgical and Materials Transactions, 29A, 1998, pp. 1527–1530) *and (b) Ni₃Al matrix.* (S. Nourbakhsh et al., Metallurgical & Materials Transactions, 25A, 1994, p. 1259)

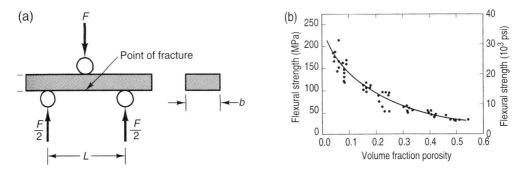

FIGURE 3-44 *(a) Schematic diagram showing the three-point bend test configuration. (b) Room-temperature flexural strength of aluminum oxide as a function of volume fraction porosity.* (R. L. Coble and W. D. Kingery, Journal of the American Ceramic Society, 39 (11), 1956, 382)

This figure shows the percentage of cumulative population of fibers that failed at progressively increasing stress levels. The slope of the straight lines is the Weibull modulus, m. The Weibull distribution function is the most widely used mathematical function to represent the strength distribution of brittle ceramics.

The mechanical strength of brittle ceramics is experimentally characterized using a three- or four-point loading technique. A particularly simple technique is the three-point bend test (ASTM Standard C1161). The basic test configuration is shown in Figure 3-44a. A standard bar of circular or rectangular cross section is placed over two supports that are separated by a distance, L. A normal load is applied at the center of the top face of the bar until the bar fractures. The fracture stress is related to the specimen thickness, bending moment, and the moment of inertia of the cross section. The test yields the modulus of rupture (MOR), also called the transverse rupture strength or flexural strength, and is given for rectangular bars from $MOR = 3F \cdot L/2b \cdot h^2$, where F is the breaking force, L is the gap between the supports, b is the width of the specimen, and h is the specimen thickness. For samples of circular cross section of radius, R, the MOR is given from $MOR = F \cdot L/\pi R^3$. MOR is a measure of the tensile strength of brittle materials; actually, the ultimate tensile strength (UTS) of a brittle solid is ~0.6MOR.

The mechanical properties (MOR, hardness, elastic modulus) of ceramics are sensitive to the amount of porosity. Generally, MOR and hardness exponentially decrease with increasing porosity content. For example, the flexural strength or MOR exponentially decreases with the porosity content according to $\sigma = \sigma_0 \cdot \exp(-nP)$, where σ and σ_0 are the flexural strength of the porous and nonporous ceramic, respectively, n is an empirical constant, and P is the porosity. Data on the flexural strength of alumina as a function of the porosity content are presented in Figure 3-44b; the flexural strength decreases with increasing porosity content, and the decrease is rather precipitous at low levels of porosity. A similar drop is observed in the hardness of ceramics as a function of the porosity content. The elastic modulus of ceramics also decreases with increasing levels of porosity, with the modulus of many ceramics following the empirical relationship, $E = E_0(1 - aP + bP^2)$, where E_0 is the Young's modulus of the pore-free ceramics, a and b are constants, and P is the porosity. Table 3-7 gives representative room-temperature values of the elastic modulus, MOR, and hardness of selected ceramics.

The theoretical fracture strength of brittle ceramics can be predicted from an analysis of the strength of the ionic and covalent bonds. Calculations show that the theoretical strength of brittle solids is about 0.1 E, where E is the Young's modulus. The actual measured strength of ceramics is, however, significantly lower than this value because of the presence of a population of

TABLE 3-7 Representative Mechanical Properties of Ceramics

Ceramic	E, GPa	MOR, MPa	Hardness, kg/mm^2
Diamond	1000	—	7000
WC	650	—	2100
SiC	450	170–820	2500
TiC	379		
Zirconia (PSZ)[a]	205	800–1500	1300
Graphite	27	—	—
Al$_2$O$_3$	375	210–340[b]	2100
Al$_2$O$_3$ single crystals	380	340–1000	—
Si$_3$N$_4$	310	250–1000	1650
Quartz (SiO$_2$)	110	107	800
B$_4$C[c]	290	340	2800

[a] Partially stabilized zirconia (3 mol% yttria).
[b] Sintered alumina (5% porosity).
[c] Hot-pressed (5% porosity).

flaws or minute cracks that raise the stress locally at the tip of cracks, thus weakening the material. Cracks, pores, and grain boundary corners can all act as stress raisers in brittle ceramics. Table 1-4 of Chapter 1 listed the plain-strain fracture toughness, K_{IC}, of some materials, including ceramics. This table showed that monolithic ceramics have considerably lower toughness than metals because of a lack of mechanisms to dissipate the energy through plastic deformation. The low fracture toughness of polycrystalline ceramics is a major deterrent to their widespread use as structural materials in spite of their many attractive properties. Innovative materials design approaches have, however, been developed to overcome the poor toughness of ceramics, and some of these are discussed in the following paragraphs. Several methods have been developed to determine the fracture toughness of ceramics. For example, toughness can be estimated from the total length of all cracks in a material. For this purpose, a ceramic is subjected to an indentation test in which a pyramidal diamond indenter makes an indent under an external load at which cracks form near the corners of the indent. The total length of all cracks is used as a measure of the material's toughness. Alternatively, a standard beam specimen containing a chevron-shaped notch is subjected to a four-point bend test to fracture, and the mode I fracture toughness, K_{IC}, is calculated from the fracture force, specimen dimensions, Poisson's ratio, and Young's modulus of the material.

Because of their many outstanding properties (low density, high melting point, high oxidation resistance, chemical inertness, and low thermal conductivity), ceramics are used in numerous applications, but their low toughness has limited their potential for even wider use, especially in structural applications at high temperatures where they can outperform other material classes. In the last few decades, many novel approaches to materials design were developed and applied to ceramics to overcome their inherent low toughness. One method to overcome the low toughness of ceramics is to increase the misorientation between grains to hinder crack propagation across the grain boundaries. Another method introduces tiny microcracks in noncubic crystalline ceramics (Al$_2$O$_3$, TiO$_2$, etc.) in order to increase the fracture energy, and therefore the toughness. These subcritical microscopic cracks provide an additional mechanism to dissipate the energy although they decrease the strength. Microcracks form during cooling from the processing temperature because of anisotropic thermal expansion of noncubic crystals. Yet another method to toughen ceramics is based on the beneficial role of sintering aids that are frequently added to ceramics to stimulate or expedite sintering. Sintering aids often form compounds that constantly deflect cracks during the latter's growth, thus requiring increased energy consumption for crack propagation. Toughness can be enhanced and cracks can be blunted by distributing a soft, plastically deformable phase in a ceramic matrix, which is the basis of a class of materials called, cermets (e.g., Co binder in WC). Cermets contain fine (0.5–10 μm) grains of a hard carbide (WC or TaC) bonded with a thin (0.5–1 μm) layer of a metallic binder such as cobalt, which partially dissolves the carbide grains and forms a strong chemical bond to it. Cermets may also contain fine grains of two carbides; for example, WC intermixed with TaC and TiC grains. These latter carbides stabilize the tungsten carbide and reduce the erosion during machining. Cermets are manufactured using a powder metallurgy process in which fine W powders are mixed with the oxides (TiO_2, Ta_2O_3, etc.) and carbon black and heated under an H_2-rich reducing atmosphere to 1400–2700°C. The powders are then mixed with Co powder to create a uniform Co film on carbide grains. Hot or cold compaction of powder mixture, and liquid-phase sintering above the melting temperature of the Co binder (1320°C) under vacuum or a reducing atmosphere, yield a strong, hard, and wear-resistant cermet.

Combining two or more ceramics in a composite such as SiC whisker–reinforced Al_2O_3 is yet another method to toughen ceramics. Considerable toughness gains are possible with this approach; for example, Al_2O_3 containing SiC whiskers nearly doubles the fracture toughness of alumina. The whiskers inhibit crack advance and absorb energy when they are pulled out or fractured by a propagating crack. The interface strength between different ceramics in such materials must be carefully tailored; too high a bond strength will impart poor toughness because of limited fiber pullout, and too low a strength will consume little energy during pullout, with negligible gains in toughness. A judicious selection of the ceramic constituents of the composite is important; whiskers should have higher elastic modulus than the matrix, and nearly identical coefficient of thermal expansion (CTE). In addition, these whiskers should be thermally and chemically stable at processing and service temperatures.

Ceramic-matrix composites based on silicon nitride (Si_3N_4), silicon carbide (SiC), zirconia (ZrO_2), alumina (Al_2O_3), and titanium carbide (TiC) are excellent cutting tool materials for high-speed machining of hard materials. These composites have high hardness at room and elevated temperatures, high compression strength, and excellent chemical and thermal stability. These composites can substitute tungsten carbide-cobalt cermets in cutting tools, and allow higher cutting speeds and superior finish to be achieved while saving strategic materials such as W and Co. Monolithic ceramics are not suitable as cutting tool materials because of their low toughness, poor thermal shock resistance, low transverse rupture strength, and tendency to fail somewhat abruptly. Ceramic-matrix composites have overcome some of these deficiencies, in particular the low toughness of monolithic ceramics. As mentioned in the preceding paragraph, in SiC whisker–reinforced Al_2O_3, the whiskers retard crack propagation by deflecting and/or bridging the crack. Thus, additional energy is needed to continue crack propagation, and fracture toughness improves. Self-reinforced silicon nitride ceramics have better toughness than conventional monolithic Si_3N_4; the reinforced material is designed to have highly acicular, oriented grains of β-Si_3N_4, which interlock and create tortuous paths that deflect cracks and contribute to toughness. Hybrid ceramic composites such as SiC whisker–toughened alumina that also contain dispersions of fine SiC particles achieve high strength and toughness. The processing and properties of some ceramic-matrix composites are discussed in Chapter 6.

An interesting approach to enhance the toughness of ceramics is to use the energy of propagating cracks to trigger a solid-state phase change in the vicinity of the crack in such a way

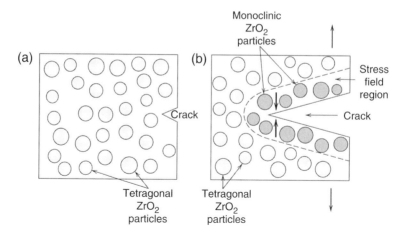

FIGURE 3-45 *Schematic diagram illustrating the operating principle of transformation toughening of zirconia. (a) A preexisting crack and (b) crack arrest due to the volume expansion accompanying the stress-induced transformation of zirconia.* (W. D. Callister, Jr., Materials Science & Engineering: An Introduction, 5th ed., p. 545, Wiley)

that volume changes on transformation accommodate the stresses to inhibit the growth of crack. This approach is called transformation toughening, and the classic example of this mode of toughening is yttria-stabilized zirconia. In transformation toughened zirconia ceramics (Figure 3-45), fine particles of partially stabilized ZrO_2 are dispersed within an Al_2O_3 or a ZrO_2 matrix. Small quantities (2–4%) of oxides such as MgO, Y_2O_3, and CaO are added to ZrO_2 and equilibrated at ~1100°C to form a phase with a tetragonal crystal structure. The oxide phase stabilizes the tetragonal structure. The material is then cooled rapidly to room temperature to prevent the transformation of the tetragonal phase into the more stable monoclinic form (the transformation kinetics for this reaction are rather sluggish). This stabilizes the metastable tetragonal ZrO_2 phase at room temperature rather than the more stable monoclinic phase. The stress field at the tip of an advancing crack causes the tetragonal ZrO_2 to transform into the stable monoclinic phase accompanied by a slight (~2%) volume expansion of transformed particles. This expansion of the dispersed ZrO_2 particles results in a compressive stress on the crack tip that arrests the latter's growth, thus imparting toughness to the ceramic.

Thermal Properties

Thermal expansion, thermal conductivity, and thermal shock resistance are important properties of ceramics. In Chapter 1, the thermal expansion of solids was seen to originate in the asymmetric shape of the potential energy versus atomic separation curve. It was also mentioned that strong interatomic bonds lead to low thermal expansion. Many ceramics are either ionic or strong-covalently bonded solids and exhibit a low (0.5×10^{-6} to 12×10^{-6} °C^{-1}) linear coefficient of thermal expansion (CTE). Generally, crystalline ceramics (other than those with a cubic structure) exhibit anisotropy in thermal expansion, and expansion is greater along certain crystallographic directions than others. In many applications, ceramics must be joined to metals with very different expansion characteristics. When the joint is exposed to temperature excursions, considerable thermal stresses may develop because of the different CTE of ceramics and metals. As an example, consider the thermal behavior of a brazed joint between yttria-stabilized zirconia (YSZ) and steel, which is of interest to solid oxide fuel cells (SOFC). SOFC utilize ionic oxide electrolytes for electrochemical generation of electrical energy, and require chemically and thermally stable interconnect materials because of the high operating temperatures (700–1000 C) that are needed to achieve electrical conductivity in the oxide electrolyte. For YSZ (3% yttria), the room-temperature CTE is $8.9 - 10.6 \times 10^{-6}/°K$, and the CTE of stainless steel $\sim 11 \times 10^{-6}/°K$. The CTE mismatch of YSZ with respect to stainless steel is, therefore, relatively small. Direct chemical bonding of YSZ with steel without a wettable interlayer is, however, not feasible, and a thin (~ 100 μm) interlayer of a braze alloy based on a noble metal (Au, Ag, or Cu) may be used. With a Cu interlayer (CTE$\sim 17.0 \times 10^{-6}/°K$) at the steel/YSZ interface, thermal stresses will be relatively large due to a large CTE mismatch. In contrast to Cu, gold has a lower CTE ($\sim 14.2 \times 10^{-6}/°K$), which means lower thermal stresses due to mismatch between CTEs. In addition, gold has excellent thermal conductivity (315 W/m·°K), resistance to oxidation, and very high ductility, which is beneficial to the accommodation of thermal stresses via plastic flow. Frequently, a noble metal braze also contains an active metal like Ti or V that promotes wetting and bonding between steel and zirconia. A comparison of the CTE ($\sim 17 \times 10^{-6}/K$) of a commercial Ti-containing gold braze with the CTE of YSZ (CTE ~ 8.9 to $10.6 \times 10^{-6}/K$) shows that the CTE mismatch, $\Delta\alpha$, is relatively large. The thermal strain, $\Delta\alpha \, \Delta T$, due to CTE mismatch at the joint during cooling from the braze application temperature ($\sim 1300°K$) is approximately 0.007. Using a modulus value of 90 GPa for a gold braze, the elastic thermal stress will be on the order of 630 MPa, which exceeds the yield strength of the braze. This will lead to plastic yielding in the braze.

Similar situations can be envisioned in other important applications. For example, carbon-carbon composites are extensively used for the nose cone and leading edges of the space shuttle, and in aircraft braking system where the extremely high frictional torque generates intense heat, raising the disk temperature to 500°C and the interface temperature to 2000°C. Most such applications require joining the composite to metals or other materials. Active metal brazing has been used for this purpose. For example, Ag-Cu braze alloys containing small amounts of Ti may be used to join C-C to other metals such as commercial purity titanium. During cooling of such a joint from the braze application temperature ($\sim 1125°K$), the CTE mismatch, $\Delta\alpha$, between C-C ($\sim 2.0 - 4.0 \times 10^{-6}/°K$ over $20 - 2500°C$) and a commercial Ag-Cu-Ti braze (typical CTE $\sim 19 - 20 \times 10^{-6}/°K$) is large, and will lead to large thermal stresses. The thermal strain, $\Delta\alpha\Delta T$, during cooling from 1125°K to room temperature will be 1.3×10^{-2}, a value that exceeds the yield strain (on the order of 10^{-3}) of the braze alloy. Again, plastic yielding at the C-C/Ti joint may be likely.

The thermal conductivity of ceramics is quite low (~ 2 to 50 W/m·°K) because of the small number of free electrons that can contribute to thermal energy transport. As a result, thermal conduction in ceramics is due mainly to the relatively less efficient mode of lattice vibrations or phonons that are readily scattered by crystal defects. In the case of amorphous ceramics (glasses), the disordered atomic structure causes even greater phonon scattering than crystalline ceramics. This leads to even lower thermal conductivity in glasses.

Because of their low thermal conductivity and high melting temperatures, ceramics are widely used as heat-resistant materials (e.g., thermal insulation) in a number of applications. In particular, porous ceramics are excellent thermal insulators. Because pores contain still air, the transfer of heat across a pore is very inefficient. This is because physical confinement of entrapped air eliminates convective heat transfer, and the low (0.02 W/m·°K) thermal conductivity of air precludes heat transfer via conduction within the pores.

The importance of thermal conductivity and thermal expansion of ceramics can be illustrated by the example of materials used in integrated circuits. In the microelectronic industry, the complexity of semiconductor chips has continuously increased and the size of integrated circuits decreased. Integrated circuits with a greater spatial density (i.e., number per unit area) of components consume more power per chip and generate more heat. This heat must be dissipated via a high thermal conductivity heat sink such as aluminum. However, use of a metallic heat sink leads to a large thermal expansion mismatch between the metal and the semiconductor. The coefficients of thermal expansion (CTE) of Al and intrinsic semiconductors Si and GaAs are $22 - 27 \times 10^{-6}/°K$ and $3 - 6 \times 10^{-6}/°K^{-1}$, respectively; this large CTE mismatch gives rise to thermal stresses during chip operation, leading to thermal fatigue and device failure. Aluminum nitride (AlN) ceramics have more closely matched CTE ($4.5 \times 10^{-6}°K^{-1}$) to semiconductors than do metals, and an acceptable thermal conductivity. AlN ceramics are used for fabricating heat sinks and packages for microelectronic devices.

Rapid fluctuations in temperature can induce substantial thermal stresses in both crystalline and amorphous ceramics, and lead to failure. This is a consequence of the poor thermal shock resistance of ceramics. The failure of brittle ceramics because of thermal shock is controlled by the elastic stress distribution, the breaking stress, and the thermal expansion and thermal conductivity. At high heat transfer rates from (or to) the body, as in rapid quenching of a hot ceramic over a temperature range, ΔT, the maximum surface stress, σ_m, is $\sigma_m = \frac{E\alpha\Delta T}{1-\nu}$, where E is the Young's modulus, ν is the Poisson's ratio, and α is the coefficient of thermal expansion. If the surface stress, σ_m, exceeds the fracture strength of the material, σ_f, then the ceramic will crack due to quenching stresses, but if $\sigma_m < \sigma_f$, then cracking will not occur. A critical temperature range, ΔT_c, can be specified for which a rapid (theoretically, infinitely fast) quench will lead to the condition, $\sigma_m = \sigma_f$. This temperature range defines the thermal shock resistance, which is the maximum tolerable temperature at rapid quenching rate of a material. Therefore,

$$\Delta T_c = \frac{\sigma_f(1-\nu)}{E\alpha}, \qquad (3\text{-}47)$$

where σ_f is the fracture strength. In many situations of practical interest, the quench rate could be moderate rather than rapid as was assumed in the preceding equation. For moderate rates of heat transfer, an approximate value of the thermal shock resistance is obtained from

$$\Delta T_c* = \frac{\sigma_c(1-\nu)K}{E\alpha}, \qquad (3\text{-}48)$$

where K is the thermal conductivity of the material.

The notion of "rapid" and "moderate" heat transfer rates can be understood with reference to the nondimensional Biot number, where $Bi = R \cdot h/K$, and R is a length scale (usually half-thickness of a plate or radius of a sphere), and h is the interface heat transfer coefficient (i.e., heat transferred per unit area and unit temperature difference between the body and the surroundings). If either R or h is very large or if K is very low, Biot number is high, which corresponds to very high heat transfer rates. For very large values of the Biot number, the resistance to thermal shock is essentially independent of h and K, and ΔT_c is estimated from the first of the two expressions given above (i.e., from Equation 3-47).

In reality, the value of h varies over a very wide range, being sensitive to the manner in which a body is cooled or heated, and the surface conditions (roughness, coatings, etc.). Approximate values of h for a variety of thermal configurations encountered in practical situations can be obtained from a number of semi-empirical equations developed in the theory of heat transfer. When the rate of heat transfer is moderate, and the value of Biot number is small, the thermal shock resistance is given from Equation 3-48.

Equations 3-47 and 3-48 show that high fracture strength and high thermal conductivity, and low modulus and low thermal expansion result, in better thermal shock resistance. These equations also provide a basis for selection of materials for applications in which the thermal shock resistance is of paramount importance. For example, calculations based on the preceding expressions show that Pyrex glass is better than ordinary bottle glass, and silicon nitride ceramics are better than alumina ceramics in their thermal shock resistance. Metals have better thermal shock resistance than ceramics because of their higher thermal conductivity and ductility (which helps in accommodating thermal stresses without fracture). A low thermal expansion is desirable for high thermal shock resistance because it involves low thermal strains and low thermal stresses. Compared to metals, ceramics have lower conductivity and ductility, both of which impair the thermal shock resistance. However, the lower CTE of ceramics is beneficial to the thermal shock resistance.

The low thermal conductivity and low thermal expansion of ceramics makes them excellent heat-resistant materials for use as thermal insulation in furnaces and tiles in space shuttles. Whereas carbon-carbon composites are used for nose tip and airfoil edges of space shuttles where the temperatures during reentry are highest, in other areas, thermally insulating ceramic tiles are used. Carbon-carbon composite ablates during reentry, but ceramic tiles do not undergo material loss. Therefore, barring mechanical damage to the delicate tiles, they survive several flights without needing replacement. The basic material for the insulating tiles is very fine, glassy fiber of silica. Two types of tiles are used for thermal insulation in the shuttle: HRSI (high-temperature reusable) tile designed for temperatures to 1530°K, and LRSI (low-temperature reusable tile) for use around 400–650°K. The HRSI and LRSI tiles differ chiefly in the surface coatings applied to tiles; HRSI has 15 mil thick high-emissivity black coating of reaction–cured borosilicate glass (high emissivity aids in radiating heat during reentry), and LRSI has a white SiO_2-Al_2O_3 coating to reflect the sun's radiation during the orbiting phase of the mission. The tiles protect the light-weight aluminum structure of the vehicle so the temperatures do not exceed 450°K. A small portion of the orbiter is covered with a stronger material, especially wear-prone areas near doors, and landing gear areas are covered with a hard and more wear-resistant material (FRCI-12), which is 80% silica fibers mixed with 20% aluminosilicate fiber doped with boron. Boron causes the silica and the aluminosilicate fibers to fuse during thermal treatment, strongly bonding the fibers and providing high strength, hardness, and wear resistance at high temperatures.

Electrical and Electronic Properties

Many crystalline ceramics possess interesting electrical and electronic properties and are used in a wide variety of modern devices. Crystalline ceramics such as Rochelle salt [$NaK(C_4H_4O_6).4H_2O$], barium titanate ($BaTiO_3$) and strontium titanate ($SrTiO_3$) exhibit *ferroelectric* behavior. These crystals have a dipole moment even in the absence of an electric field, i.e., the centers of positive and negative charges in their crystal structure do not coincide. In an electric field, a mechanical moment is induced on the crystal because of spatial separation of positive and negative charges (provided the electric field is not exactly aligned with charge centers). Ferroelectric ceramics are used in transducers to convert one type of excitation into another (e.g., optical to electrical). The ferroelectric behavior disappears above a temperature called the Curie temperature (see Chapter 1).

Ceramics such as quartz, tourmaline (an aluminosilicate that contains boron), Rochelle salt (potassium sodium tartrate), barium titanate, and lead zirconate titanate (PZT) have a capacity to react to an imposed electrical excitation by changing their dimensions, and conversely, electrically respond to mechanical excitation. These crystals are used to convert voltage to motion and vice versa. For example, a quartz crystal, ground and polished to a precise thickness, will oscillate at its natural frequency when an oscillating voltage is imposed across the crystal. This phenomenon, called *piezoelectric* behavior, is the basis of using the crystals in radio transmitters and receivers to tune specific broadcast frequencies. Piezoelectric ceramics are also used in ordinary electrical watches, ultrasound generators, phonograph pickups, microphones, and in SONAR (Sound Navigation and Ranging) in underwater sound detection. PZT is the primary material for production of piezoelectric ceramics used in these devices.

Ceramic crystals based on the magnetic iron oxide, Fe_2O_3, exhibit magnetic induction even in the absence of a magnetic field, and constitute the material class *ferrites*. Ferrites are used in signal-processing and information-recording applications. Other ferrites are based on MnZn, NiZn, and yttrium iron garnet ($Y_3Fe_5O_{12}$). Magnetic induction of ferrites depends strongly on the processing history of the material, grain size, density, and impurity content. Ceramics based on the semiconducting zinc oxide (ZnO) and containing Bi, Co, and Mn at the ZnO grain boundaries are used in circuit overload protection and are classified as *varistors*. The overload protection is achieved because of the unique nonlinear current (I) – voltage (V) response of varistors. Above a critical voltage, the current rises rapidly, and the circuit tries to draw a very large current, which causes the circuit to break. During normal operation, varistors act like "high-value" resistors. Ceramics having *ferromagnetic* properties are added in powder form to stealth coatings that absorb radar and other electromagnetic radiation and prevent detection. The electrically conductive ceramic tin oxide is added to carpet and floor tile to eliminate the effect of static electricity; tin oxide is white, so it works well with light-colored flooring and plastics.

Finely divided ceramics are used to make electrical conductors, capacitors, resistors, and dielectrics for use in microelectronic devices, and as substrates for mounting (packaging) these components. *Conductors* are based on silver powder mixed with ~10% lead-bearing borosilicate glass powder, which acts as a binder for the conductive Ag powder and the ceramic substrate to which it is applied. The powder mixture is dispersed in an organic liquid to create "ink" that is applied to the ceramic substrate by "screen-printing." The printed circuit is then dried and pyrolyzed to remove the organic base, and fired to fuse the ceramic (glass) binder and bond Ag to the ceramic substrate. The resulting film must retain definition, be adherent to and compatible with the substrate, and possess good electrical conductivity. *Resistors* are made in a manner similar to conductors, using "ink" composed of palladium-silver (Pd-Ag) or ruthenium-oxide powders mixed with 60–98% glass in an organic liquid. Glass controls the electrical resistance of the ink. This ink is applied to a ceramic oxide substrate and treated in a manner similar to conductors. Resistors are made oversize and then trimmed using lasers. *Dielectrics* allow conductor lines to be printed over each other (crossover) and multilayer capacitors to be produced. A suitable paste or ink is used in a manner similar to conductors and resistors. Crossover dielectrics (glasses) must survive multiple firings (to about 800°C) without shorting the conductors.

Oxidation and Corrosion Resistance

Ceramics can withstand harsh corrosive environments at high temperatures better than most materials. In particular, oxides, carbides, silicides, and borides of refractory metals Zr, Ti, Hf, Mo, and Ta have outstanding resistance to oxidation and corrosion at elevated temperatures. Figure 3-46 shows the elevated temperature oxidation kinetics of several ceramics as a function of reciprocal of oxidation temperature. In this figure, the kinetics are plotted as logarithm of the parabolic rate constant for diffusion-controlled oxidation of the ceramics. The linear plots indicate an Arrhenius-type thermally activated oxidation kinetics.

Many oxides, carbides, and borides can also be grown into small diameter fibers in order to reduce the number of flaws in their cross-section and increase their strength and modulus while retaining their superior corrosion resistance. Yttria-stabilized zirconia (YSZ) fibers have outstanding resistance to corrosive environments rich in alkali-metal chlorides and carbonates up to 700°C and can withstand short-term exposure to mineral acids at their boiling point. These fibers also have excellent resistance to oxidizing and reducing atmospheres at elevated temperatures. Similarly, high-purity alumina fibers have excellent resistance to chemical attack and allow long service life in corrosive environments. These fibers usually contain about 2–4% silica to inhibit grain growth that would weaken the fiber at elevated temperatures. Chemical

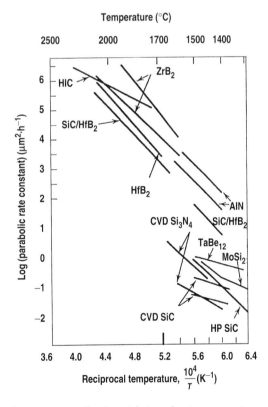

FIGURE 3-46 *Parabolic rate constant for the oxidation of various ceramics and ceramic composites as a function of the reciprocal of absolute temperature of oxidation.* (K. Upadhya, J. Yang, and W.P. Hoffman, American Ceramic Society Bulletin, December 1997, p. 51)

additives and surface coatings are incorporated to enhance the oxidation resistance and limit grain coarsening.

Most thermal protection materials are required to have excellent resistance to high-temperature oxidation and corrosion in various types of atmosphere. Silica or its derivatives are frequently used to increase the oxidation resistance of thermal protection materials. For example, space shuttle tiles contain silica fibers, and the C-C composites at the shuttle's leading edge and nose cap have an SiC coating for oxidation resistance to $1600°C$. Above this temperature, the protective influence of silica and SiC is lost in oxidizing atmospheres. Besides silica, many other oxide-based ceramics (MgO, CaO, BeO, HfO_2, ThO_2, ZrO_2, and Cr_2O_3) are chemically inert to very high temperatures and may be used either in bulk form or as coatings. Many of these are, however, very brittle, susceptible to thermal shock, or have other limitations. For example, MgO, CaO, and Cr_2O_3 all have high melting points but also high evaporation rates. In addition, MgO and CaO are hygroscopic and degrade through moisture absorption. Oxides of beryllium and thorium (BeO, ThO_2) are toxic and radioactive, respectively. Oxides of hafnium and zirconium (HfO_2 and ZrO_2) have high melting points ($2900°C$ and $2770°C$, respectively) and low volatility (i.e., tendency to vaporize), but both suffer from poor thermal shock resistance. Both these oxides undergo a solid-state phase transformation that leads to large-volume changes, residual stresses, cracking, and failure. Frequently, additives such as CaO, MgO, and Y_2O_3 are incorporated in ZrO_2 and HfO_2 to stabilize their crystal structure and prevent transformations and volume changes, but these additives also tend to lower their melting and softening temperatures, thus limiting their high-temperature use.

Carbides of refractory metals Hf, Zr, and Ta have higher melting points than their corresponding oxides, have fair resistance to thermal shock, and are more stable against stresses induced by phase transformations. However, these carbides have high brittle-to-ductile transformation temperature (BDTT) ($1725–1980°C$) depending on their chemical purity. In addition, in oxidizing atmospheres, these carbides form a multilayer oxide scale with a more porous outer layer that degrades the oxidation resistance of the carbides. The oxidation resistance is improved through addition of mixed carbides, e.g., addition of TaC to HfC improves the oxidation resistance of HfC.

In a manner similar to carbides, borides of refractory metals Ti, Zr, Hf, and Ta also have high melting temperatures, high hardness, low volatility, and good thermal shock resistance and thermal conductivity. These compounds generally have good oxidation resistance, and further improvements are achieved through the use of additives such as SiC, which improves the oxidation resistance of both ZrB_2 and HfB_2. For example, with SiC in ZrB_2, a thin, glassy layer of SiO_2 forms when the boride is exposed to an oxidizing atmosphere; the silica layer covers an inner layer of ZrO_2 on the surface of ZrB_2 base material. The outer glass layer provides good wettability, surface coverage, and oxidation resistance.

Among the thermal protection materials, carbon-carbon composites have very high specific strength (strength-to-density ratio), excellent thermal shock resistance, outstanding creep resistance and low thermal expansion coefficient. However, these composites have poor oxidation resistance above $350°C$. In order to profitably use their excellent high-temperature properties, their oxidation resistance is enhanced by (1) applying oxidation-resistant carbide coatings (e.g., SiC, mentioned previously), (2) adding oxidation inhibitors, (3) minimizing deleterious impurities in carbon, and (4) increasing the extent of graphitization. Carbide and boride inhibitors such as monolithic HfC and HfB_2, and their composites (HfC-SiC, HfC-TaC, and HfC-HfB_2), are either applied as surface coatings or impregnated in the porous carbon matrix to enhance the oxidation resistance. Mechanical incompatibility of the coating relative to carbon because of thermal expansion mismatch could, however, reduce the thermal shock resistance of the C-C composite. In addition, most such coatings and additives provide only relatively short-term protection. Considerable challenges exist in designing and developing advanced thermal-protection and corrosion-resistant materials capable of functioning in severe thermal and corrosive environments.

Another type of corrosion to resist which refractory ceramics are used involves molten metals. The corrosion and degradation of refractories and engineered ceramics by liquid metals is encountered in many manufacturing processes. Examples include melting and holding crucibles used in foundries, ceramic coatings on die-casting dies, electrode linings used in extraction and refining of metals, submerged refractory nozzles, and impellers. The selection of proper materials for such applications affects the melt and casting quality, scrap rates, and costs associated with frequent refractory replacement. Several factors must be considered in the selection of refractories such as wall thickness, porosity, corrosion/erosion resistance, outgassing and binder evaporation tendency, thermal expansion, thermal shock resistance, rate of melting, maximum operating temperature, and mechanical considerations such as physical abuse during cleaning, charging, heating, and handling. In addition, the refractory must be poorly wet by the molten metal to limit the contact area and the extent of corrosive attack.

A large number of proprietary formulations of refractories for specific types of melts have been developed. Zircon and silica refractories are recommended by refractory manufacturers for highly acidic melts, magnesite for basic melts, and alumina and spinel refractories for neutral melts. An acidic refractory in contact with a basic melt or slag will cause reaction, erosion, and inclusions in the melt, and should not be used. Most refractories are brittle at room temperature, but they become plastic at elevated temperatures, often because of liquid-phase formation on heating. On cooling, the liquid phase turns glassy and becomes more prone to thermal shock and cracking, thus degrading the refractory performance.

Refractories most commonly used by the cast metals industry to contain molten metals are alumina, mullite, and aluminosilicates (with varying $Al_2O_3.SiO_2$ ratios). Generally, these refractories also contain other compounds such as calcium oxide, sodium oxide, aluminum borate, aluminum fluoride, barium sulfate, and calcium fluoride. These additives inhibit the wetting of refractory by molten metals and improve their corrosion resistance.

A variety of engineered ceramic coatings have been developed for use in die-casting and injection-molding dies. These include erosion- and wear-resistant ceramics such as titanium nitride (TiN), vanadium carbide (VC), titanium carbide (TiC), and titanium diboride (TiB_2). The resistance to hot corrosion of these coatings in contact with corrosive liquids depends on the melt composition, porosity content in the coating, and processing conditions (temperature, pressure, injection rate). The depth of attack in the coating depends on the capillary penetration of open pores in the refractory coating by the corrosive liquid. However, whereas high porosity levels are detrimental to the corrosion resistance and mechanical strength, they are beneficial to the thermal shock resistance of the coating. Some aspects of processing and properties of ceramic coatings are discussed in Chapter 5.

One type of high-temperature process that is different from corrosion and oxidation, but that also leads to significant changes in the surface characteristics of the ceramics, is the basis of a technique of microstructure examination called thermal etching. Thermal etching is widely used in ceramic science and technology to reveal the microstructure of ceramics such as alumina that are inert to chemicals at ambient temperatures and present difficulty in observing microstructural features such as grain boundaries. Thermal etching involves heating a polished alumina specimen for 10 to 30 minutes at a temperature slightly below the sintering temperature. This allows surface atoms to be redistributed via thermal diffusion, thereby revealing grain boundaries. In addition, any minor surface imperfections (scratches and microcracks) get healed because of surface material transport. Bulk diffusion of atoms also occurs, which causes grain growth; as a result, the microstructure of the ceramic after thermal etching could be different from the original material. Because of this, thermal etching temperatures and time are kept as low as is necessary to reveal the grain structure through microscopy.

Bioceramics and Porous Ceramic Foams

Microporous bioceramics are special ceramics used for the repair or replacement of damaged body parts. Bioceramics could be crystalline such as hydroxyapatite, semicrystalline such as bioactive glass-ceramic, or amorphous such as bioactive glass. An example of bioactive glass ceramic is the three-phase silica-phosphate material consisting of apatite, wollastonite, and a CaO-SiO_2–rich glassy matrix. Bioceramics should be able to form a stable interface with living tissue in order to prevent interface failure. Certain glass compositions and glass ceramics form a strong mechanical bond to bone via a surface chemical reaction that forms a biocompatible compound interlayer that serves as the bonding interface between the glass implant and the tissue. Microporous bioceramics permit growth of living tissue into pores of the implant and form an interface within the pores. A porous bioceramic serves as a bridge or scaffold for bone growth. The large interfacial contact between the microporous ceramic and the tissue reduces movement of the tissue. However, very fine (<100–$150\,\mu m$) pores in the implant could limit the blood supply to the connective tissue, causing its degeneration and death. Porous metal implants have also been in use; however, these are coated with bioceramics to reduce the metal corrosion and discharge of metal ions in the tissue. Common bioceramic coatings for metal implants such as hydroxyapatite (HA) are plasma-sprayed over porous metal implants produced by sintering wires or meshes. High-purity alumina was one of the first bioceramics used in clinical work involving hip and knee prostheses, jawbone, bone screw, and dental implants because of its chemical stability, excellent biocompatibility, high strength, and high resistance to wear and impact fatigue. Likewise, tetragonal zirconia containing Mg or Y as stabilizers has also been used as a bioceramic for joint prostheses because of its high fracture toughness and strength. Table 3-8 lists some clinical uses and examples of bioceramics.

Ceramic foams are widely used in molten metal filtration and purification, catalytic combustion, and in fluid mixing and heat transfer applications. They also are used in support structures of high-temperature furnaces because the air trapped within the pores is immobilized, which

TABLE 3-8   Clinical Uses of Bioceramics

Material	Application
Al_2O_3, ZrO_2 (PSZ), hydroxyapatite, bioactive glass	Orthopedic
Hydroxyapatite, bioactive glass ceramic	Coatings for bioactive bonding
Calcium phosphate salts, tricalcium phosphate	Bone space fillers
Al_2O_3, hydroxyapatite, bioactive glasses	Dental implants
Hydroxyapatite, bioactive glass ceramic	Spinal surgery
Rare-earth–doped aluminosilicate, glasses	Treatment of tumors
Pyrolytic carbon coating	Artificial heart valve

reduces convection and conduction through the foam. The tortuous, weblike structure of ceramic foams enables fluid mixing at a microscopic scale and large interfacial area of contact between the fluid and the porous ceramic. Large pores improve the permeability but are inefficient for removal of small impurity particles. In contrast, fine pores permit good collection efficiency with fine particles, but the filter pressure drop is increased. Porous ceramic foams are made by first coating flexible, open-cell polymer foams with colloidal slurries of ceramics. The polymer is then burned out, and the ceramic skeleton is sintered to reproduce the open cell structure of the original polymer foam. Alternatively, ceramic powders can be added to a liquid thermosetting polymer such as polyurethane, which is then foamed. The porosity content and pore size are determined by these characteristics in the original foam; cell sizes ranging from 100 μm to over 1 mm, and cell densities of 0.1–0.3 are common. Models to predict the physical and mechanical properties of foamed ceramics have been developed. The failure process is controlled by the collapse of cell walls due to buckling and crushing under an external load. The fracture strength is sensitive to the porosity content; the strength decreases with increasing porosity. The generic relationship between ceramic foam properties and porosity is

$$\frac{P_f}{P_s} = c(1 - \varepsilon)^m \tag{3-49}$$

where P_f and P_s are the property (e.g., fracture strength) of the foam and fully dense solid, respectively, and c and m are empirical constants, which are determined from experimental measurements.

Joining of Ceramics

Joining of ceramics is a technologically important area of manufacturing. This is because even though net-shape manufacture of ceramic and ceramic composite parts is preferred because of the inherent brittleness of ceramics and related machining problems, many advanced applications require assembly and integration of smaller ceramic parts. For example, fabrication of complex structural components requires robust integration technologies capable of assembling smaller, geometrically simple parts into larger, more complex systems. This requires joining of ceramics to themselves and to dissimilar materials. Ceramic joining has been done using a variety of processes that include diffusion bonding; fusion welding; active metal brazing; brazing with oxides, glasses, and oxynitrides; reaction forming; and many others.

Brazing is probably the most widely used joining method for ceramics. It is a simple and cost-effective method applicable to a wide range of ceramics. Two somewhat different approaches are used for brazing ceramics depending upon the type of "glue" material. Metal brazing utilizes an intermediate layer of a metal or alloy that melts, flows, and bonds with the ceramic surfaces, whereas in ceramic brazing, a ceramic material (e.g., a glass) is used as the glue. Metal brazes used to join ceramics (oxide ceramics, glasses, carbon) are alloys based on noble metals Ag, Pt, Au, or Pd that form oxides that bond with the substrates. A more widely used type of metal braze contains an active metal, generally Ti (or Zr, Cr, Nb, and Y) as an ingredient in an alloy; the active metal reacts with the ceramic to form compounds that permit braze wetting and spreading, and formation of a strong joint upon braze solidification. Alternatively, the ceramic surface may be pre-metallized or coated with the active metal (or with a compound that decomposes to form the metal, such as TiH_2 for Ti). The surface film of active

TABLE 3-9 Examples of Active Metal Brazing of Ceramics

Joint Materials	Filler Metal	Joint Strength, MPa
Si_3N_4–Si_3N_4	Al, Al-Cu, Al-Si,Al-Mg	20–600
Si_3N_4–Si_3N_4	Ag-Cu-Ti	50–820
Si_3N_4–Si_3N_4	Au-Ni-Pd, Pd-Ni-Ti, Au-Cu-Hf, Cu-Si-Al-Ti, Ni-Cr-Si	0–510
Si_3N_4–Si_3N_4	Ag-Cu-In-Ti	270
Steel–Si_3N_4	Ag-Cu (Ti interlayer)	210
Steel–Si_3N_4	Ag-Cu-Ti (Cu interlayer)	180–350
Steel–Si_3N_4	Ag-Cu-Ti-Al	200
Steel–Si_3N_4	Ag-Cu-Ti (Mo interlayer)	390
Steel–Si_3N_4	Ag-Cu-Ti (Ni interlayer)	70–150
Inco 909 (superalloy)–Si_3N_4	Ag-Cu-Pd-Ti (Ni interlayer)	151
Si_3N_4–Si_3N_4	70 SiO_2–27Mg–3Al_2O_3	450
BN–BN	Al	6
Sialon–Sialon	Al	61
Al_2O_3–Al_2O_3	Cu-44Ag-4Sn-4Ti	80–120
PSZ ZrO_2–PSZ ZrO_2	Cu-44Ag-4Sn-4Ti	>400
Steel–TZP ZrO_2	Ag-4Ti	151

PSZ: partially stabilized zirconia ; TZP : tetragonal zirconia polycrystals

metal is deposited using sputtering, vapor deposition, or thermal decomposition (e.g., Ti-coated Si_3N_4, Ti-coated partially stabilized zirconia, Cr-coated carbon, and Si_3N_4 coated with Hf, Ta or Zr). The active metal (either as an alloying element or as a surface film) overcomes the poor flow characteristics of the filler metal with reaction-induced wettability. Titanium is the most commonly used active metal in braze fillers because it chemically reacts with ceramics to form compounds that provide strong bonding. In the case of oxide ceramics, Ti forms TiO, TiO_2, and Ti_2O_3; in the case of silicon carbide, silicon nitride, and sialons, Ti forms titanium silicides, titanium carbides, and titanium nitrides. Active metal brazing is generally done in vacuum furnaces under high-purity inert atmosphere, but noble metal brazes are used in air in order to form the oxide binders. Table 3-9 lists some examples of active metal brazing of ceramics to metals and to ceramics, and the joint strength achieved for each couple.

Braze fillers are often used as powders or pastes containing organic binders, although braze foils and wires are also used to accommodate complex joint configurations. Pastes and powders may leave unwanted residue from organic binders at the joint, whereas foils or ribbons may be difficult to produce in some braze alloys, especially those that contain metalloids like B, Si, and P, which make the alloy inherently brittle and lead to edge cracking and difficulty in producing continuous sheets. In recent years, rapidly solidified braze foils of amorphous Ni, Co, Cu, Ti, and Zr-base alloys have been successfully produced. These amorphous alloys are ductile and readily produced in foil form. Metallic glass brazes also provide greater strength, leak tightness, and resistance to shock and vibration compared to mechanically fastened joints. They also possess superior spreading and wetting properties than atomized powders or paste formulations that usually require wide gaps for filling and yield weak joints due to powder contamination from oxides that hinder fusion and bonding. Metallic glass braze fillers based on Ni possess excellent corrosion resistance, and are used in porous metal seals for rotating blades in jet turbine engines, compressor vane and shroud assemblies, and honeycomb structural panels joined to perforated face sheets for applications in exhaust plugs, cones, nozzles, turbine tailpipes, and fusion reactors. Most such applications have involved stainless steels and superalloys rather than ceramics, but some Ni, Ti-, and Cu-base metallic glass braze fillers have been used to join graphite to Mo, Cu, and V, and ceramic matrix composites (e.g., C/C, SiC/SiC, and C/SiC to titanium and Ni-base superalloys).

Ceramic braze materials include glasses (or mixture of glass with crystalline materials) such as $55SiO_2$-$35MgO$-$10Al_2O_3$ and CTS glass (CaO-TiO_2-$SiO2$). The reason for using glass as a ceramic filler is that many sintered ceramics have amorphous phases at their grain boundaries, and good wetting and bonding is expected between the glass filler and these grain boundary phases. A very widely used commercial method to join alumina ceramics is the molybdenum-manganese (Mo-Mn) process, in which a specially formulated paint (or slurry) containing powdered Mo (or MoO_3), Mn (or MnO_2), and a glass-forming compound is applied to the ceramic surfaces. The painted ceramic is fired in wet hydrogen at 1500°C, which causes the glass forming constituents from the ceramic to diffuse into Mo layer and form a strong bond. However, the poor toughness, low Young's modulus, and tendency to stress corrosion are the main challenges in using glass as a filler to join ceramics.

Ceramics are also joined using welding. Two basic methods are used: solid-state welding and fusion welding. Solid-state welding (also called diffusion bonding) is done by heating prefabricated (and usually, polished) ceramics under very high pressures to allow solid-state diffusion between the ceramics to occur and form metallurgical bond. Usually, an intermediate layer of a material (either a powdered ceramic or a ductile metal foil) is placed at the joint to enable faster interdiffusion and bond formation. A wide variety of oxides, carbides, nitrides, and borides have been joined to themselves and to their mutual combinations. However, the very large pressures needed in solid-state diffusion bonding can introduce mechanical damage to the ceramic. Fusion welding has been used to join Al_2O_3 to Al_2O_3, ZrB_2 to ZrB_2, TaC to TaC, and ZrB_2 to graphite. Arc welding, laser welding, and electron beam welding have been used. Fusion welding is more difficult with ceramics because of one or more the following problems: high melting temperatures, tendency to sublimate (evaporate without melting), excessive evaporation upon melting (due to very high vapor pressures), excessive thermal shock and fracture upon rapid heating and cooling, and undesirable phase transformations in the heat-affected zone due to very high temperatures needed for fusion welding.

Another method of joining ceramics, mainly for microelectronic and micro-electro-mechanical systems (MEMS) is the direct bonding (or wafer bonding) technique, in which atomically smooth and chemically clean surfaces are brought in physical contact to form physical (van der Waals) bonds. The bond strength is increased by heating the joint to a higher temperature. For example, atomically clean and smooth Si wafers are bonded to Au-plated Cu interconnects using this method. Upon heating, a small amount of Au-Si eutectic liquid forms at the interface, which solidifies to form a strong intermetallic bond layer. Another method to join ceramics makes use of the process of sintering (sinter bonding) in which atomic diffusion causes particles to bond either with the presence of an intermediate layer (e.g., glass frit) or without it.

Adhesive bonding is used for ceramics such as Si-Al-O-N that are difficult to be brazed, welded, or diffusion bonded because they tend to decompose at elevated temperatures rather than melt and bond. The presence of Y in Si-Al-O-N ceramics results in an intergranular glassy phase with a melting point of about 1350°C, which permits liquid-phase joining. Another high-temperature adhesive containing ~45% α-Si_3N_4 and oxides such as Y_2O_3, Al_2O_3 and SiO_2 is used to join Y-Si-Al-O-N ceramics at 1600°C. The adhesive has a composition similar to the ceramic, thus permitting joining without significant change in the composition and properties of the joint. The α-Si_3N_4 in the adhesive reacts to form fine acicular β-SiAlON, which reinforces and strengthens the joint. Such joints are stronger than those formed with pure glass-forming adhesives because of the latter's relatively poor resistance to crack growth. Another adhesive used to join ceramics employs phosphate binder systems. These have been used in a variety of alumina refractories. Some of these binders (e.g., aluminum phosphate) have been used as low-temperature binders for Si_3N_4 in applications requiring low-to-moderate bond strength.

An important subset of the field of joining of ceramics is the joining of ceramic-matrix composites. Ceramic composites such as SiC/SiC, C/C/SiC, C/C and others have been developed for a variety of advanced applications. For example, carbon-carbon composites containing SiC are promising for lightweight automotive and aerospace applications. In the automotive industry, C/C/SiC brake disks, made either by CVI or by hot pressing C/C composites followed by Si infiltration to form SiC, are already being used in some models. These composites are also being developed for hypersonic aircraft thermal structure and advanced rocket propulsion thrust chambers. C/C/SiC composites containing diamond, c-BN, B_4C, and similar hard particles are being developed for cutting tools for higher use temperatures and cutting speeds. Similarly, heat- and wear-resistant SiC/SiC ceramic-matrix composites have potential applications in combustor liners, exhaust nozzles, reentry thermal protection systems, radiant burners, hot gas filters, and high-pressure heat exchangers, as well as in fusion reactors, owing to their resistance to neutron flux. Carbon-carbon composites are used in the nose cone and leading edges of the space shuttle, solid propellant rocket nozzles and exit cones, ablative nose tips and heat shield for ballistic missiles, aircraft braking system, and first wall tiles of fusion reactors. Carbon-carbon composites have been brazed using Ag-, Au-, and Cu-base filler metals containing active metals Ti, Cr, and Zr (all strong carbide formers) for moderate use temperatures, and HfB_2 and $MoSi_2$ powders for very high use temperatures. Joining of ceramic composites is an active area of research in the field of joining science and technology.

Various factors must be considered in selecting a joining process for ceramics. Organic adhesives and mechanical fasteners are relatively easy to use, and these may be able to provide joint strengths comparable to brazing and soldering. The high-temperature strength is, however, poor, especially with organic adhesives, and hermetic seals may be difficult to form with fasteners. Active metal brazing is the most widely used joining technique because of its simplicity, cost-effectiveness, and ability to form strong and hermetically sealed joints capable of withstanding high temperatures. In a joint, a number of dissimilar materials may have to be brought together, all of which must function in the desired manner under the operating conditions. Thus, a number of factors must be considered in selecting a braze, which include: oxidation and corrosion resistance, stability at high temperatures, mismatch in the coefficients of thermal expansion (CTE), and retention of useful properties (e.g., electrical and thermal conductivity). As an example, consider the complexity of selecting a joining method and materials for use in the interconnects of solid oxide fuel cells (SOFC) based on ionic oxide electrolytes. These devices are becoming increasingly attractive for the electrochemical generation of electrical energy. Manufacturing technologically and commercially viable SOFC's requires creation of oxide/metal joints that are chemically and thermally stable to the operating temperatures (700–1000°C). Among the current joining technologies, glass seals and active metal brazing are the most important approaches for the SOFC joints. Glass seals have the limitations of softening close to the SOFC operating temperatures, and partial devitrification, which changes the CTE, causing greater CTE mismatch, thermal stresses and fatigue, and potential loss of joint hermeticity due to seal failure. Glass-ceramics (partially crystallized glass), cements, and mica also have been used for SOFC sealing applications with varying degree of success. The filler material must show good adherence to the substrates, prevent grain growth and long-term thermomechanical degradation due to creep and oxidation, have closely matched CTE with the joined materials, and possess liquidus temperature greater than the operating temperature of the joint but lower than the substrate's melting temperature. In the case of brazing, the major challenge is the poor wettability of the ceramic by molten alloys that may hinder flow and capillary penetration at the joint region. The wetting and spreading phenomena are discussed in Chapter 4.

References

Bose, A. *Advances in Particulate Materials*. Boston: Butterworth Heinemann, 1995.

Chiang, Y. M., D. P. Birnie, III, and W. D. Kingery. *Physical Ceramics: Principles for Ceramic Science & Engineering*. New York: John Wiley & Sons, 1997.

German, R. M. *Powder Metallurgy Science*. Princeton, NJ: Metal Powder Industries Federation (MPIF), 1984.

Kingery, W. D., H. K. Bowen, and D. R. Uhlmann. *Introduction to Ceramics*, 3rd ed. New York: John Wiley & Sons, 1993.

Kou, S. *Transport Phenomena in Materials Processing*. New York: Wiley, 1996.

Metals Handbook, Powder Metallurgy, vol. 7, 9th ed. Materials Park, OH: American Society for Materials, 1984.

Reed, J. S. *Principles of Ceramic Processing*, 3rd ed. New York: John Wiley & Sons, 1995.

Van Vlack, L. H. *Physical Ceramics for Engineers*. Reading, MA: Addison-Wesley, 1964.

4 Surface, Subsurface, and Interface Phenomena

Surface and interface phenomena play an important role in materials design and processing. These phenomena pervade joining, adhesion, sintering, impregnation, coatings, mixing, dispersion, detergency, compositing, and a large number of other processes of importance to manufacturing. Closely related to these are near-surface phenomena that extend from the surface to subsurface regions and profoundly influence material behavior and performance. These include wear, friction, lubrication, and corrosion. This chapter provides a brief introduction to selected surface, subsurface, and interface phenomena that are manifested in the synthesis and processing of various classes of materials.

All surfaces and interfaces have surface (or interface) energy associated with them, which is the excess energy that an atom has at (or near) a surface compared to an atom in the bulk. For example, surface tension of a liquid has its origin in the atomic interactions in the first few layers of surface atoms. Surface atoms experience a net attractive force from adjacent atoms directed inward (toward the bulk) and perpendicular to the surface. This attractive force reduces the number of atoms at the surface and increases the interatomic distance. The interatomic forces that create surface tension include both chemical forces (covalent, ionic, metallic, and hydrogen bonds) and physical and dispersion forces, collectively called van der Waals forces. In theory, therefore, surface tension of any substance can be considered as an additive contribution of these two types of forces. For example, in the case of liquid metals, surface tension can be divided into contributions from metallic bond and the dispersion forces.

The study of atomic and molecular origin of surface forces and their influence on macroscopic surface properties is a highly developed field, which is beyond the scope of this book. In this chapter, we are chiefly interested in macroscopic surface and interface phenomena from the viewpoint of materials processing and manufacturing. We shall, therefore, only briefly review the foundational concepts in the physics and chemistry of surface interactions, and devote more effort at understanding their practical manifestation in materials synthesis and processing.

Surface Forces

The genesis of all surface phenomena lies in chemical and physical interactions between atoms and molecules. These interactions could be either highly localized (short-range interactions) or span physically measurable distances (long-range interactions). The long-range interactions between atoms and molecules are "physical" in the sense that they do not involve any chemical reactions or charge transfer (as in ionic bonding) or charge sharing (as in covalent bonding). The two fundamental physical forces among atoms or molecules are (a) Coulomb forces, and (b) van der Waals forces. The former are electrostatic forces and owe their origin to simple physical interactions (attraction or repulsion) between point charges, and the latter forces involve interactions between permanent or induced dipole moments and other dispersion forces. Besides Coulombic and van der Waals interactions, charged species can also interact by mechanisms that are intermediate between these two types. These include polarization of polar or nonpolar molecules, and ion–dipole interactions. A dipole forms because of a nonsymmetrical charge distribution or charge separation. Even molecules without permanent dipole can undergo polarization and acquire induced dipole moment because of deformation of their electron cloud by an electric field. This is because the electric field causes a net displacement of the center of negative electron cloud relative to the positive nucleus. In fact, any atom or molecule can be polarized, and the ease of its polarization is defined in terms of its polarizability, α, which specifies the strength of the induced dipole moment when the atom or molecule is placed in an electric field of intensity, E. The polarizability is defined from $\mu = \alpha \cdot E$, where μ is the induced dipole moment (units: Coulomb.meter).

Electrostatic Forces

Coulombic forces drop off as the inverse square of the distance between two charges (Coulomb's law), decay slowly, and can be appreciable to several tens of nanometer separation. In reality, however, these forces drop off faster (e.g., with the inverse of the fourth power of distance) because of the electrostatic interactions between all (rather than two isolated) point charges that cause an electrostatic "screening" effect. In an equilibrium distribution of charges, the probability of a particle residing at a specified energy level, ΔG, with respect to a reference state, can be expressed from the Boltzmann distribution function, $c = c_0 \exp(-\Delta G / kT)$, where c and c_0 are the concentrations of the particle at a point x and at some reference point.

The energy, ΔG, for electrostatic interactions between point charges can be readily determined by invoking Coulomb's law of electrostatic forces. An isolated charge, q, in space will produce an electric field at a point x such that the energy needed to bring a unit charge from infinity to a distance x from q will be the electrical potential, ψ, at x. This potential can be calculated from the fundamental relationship between energy, $E(x)$, and force, F: $E(x) = -\int F \cdot dx$, where F is the electrostatic force between q and the unit charge, and integration is performed from infinity to x. In a system containing both positive and negative charges (ions), the term ΔG in Boltzmann's equation becomes $\Delta G = \pm z^+ e \psi$, where z^+ is the valence of the positive ions, e is the electronic charge, and the positive sign applies for the concentration distribution of negative ions, and the negative sign applies to the concentration distribution of positive ions. In other words, the Boltzmann function for the positive and negative ions in a system consisting of both positive and negative ions can be written as $c^+ = c_0 \exp(-z^+ e \psi / kT)$, and $c^- = c_0 \exp(+z^+ e \psi / kT)$, where c^+ and c^- are the concentrations of the positive and negative ions, respectively. Even though the system as a whole may be electrically neutral, locally, a charge imbalance may occur and lead to regions of excess positive or negative charge. The sign

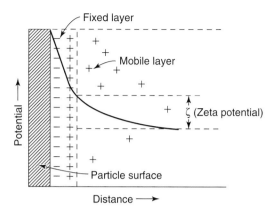

FIGURE 4-1 *Electrical double layer and potential drop in double layers. The potential drop in the diffuse or mobile layer is the zeta potential* (Z. D. Jastrzebski, The nature and properties of engineering materials, 2nd ed., 1977, p. 173, John Wiley & Sons).

of the excess charge will depend on the sign of the potential in a given region; in a region of negative electric potential, ψ, more positive than negative ions will be localized, and in a region of positive ψ, more negative than positive ions will be localized. Such segregation of charged ions is most clearly revealed in liquid solutions that contain an electrolyte. The excess charge around a potential is called the electrical double layer. The electrical potential, ψ, is called the zeta (ζ) potential (Figure 4-1) and can be experimentally measured to characterize the forces of interaction between charged particles in a solution.

In reality, the double layer is "diffuse" rather than sharply demarcated because of the thermal motion of the ions in the solution. Furthermore, because ions occupy a finite volume in space, a thin layer (a few nanometers thick), called the Stern layer, forms between the diffuse double layer and the charged region near a surface. The ions in the Stern layer are effectively fixed or immobilized relative to the ions in the double layer, and they partially neutralize the surface charge through electrical "shielding." Calculations show that the electrical potential, ψ, in an electrolyte near a surface decays exponentially with the distance from the surface, and the thickness of the electrical double layer decreases rapidly as the concentration of the electrolyte in the solution is increased. The charge density on a surface will change with the electrolyte concentration because of adsorption of additional ions on the surface or because of a decrease in the separation between charged surfaces (for example, with a decrease in the distance between charged particles dispersed in a liquid). This reduces the thickness of the electrical double layer as well as the electrical repulsion between surfaces. The Coulombic electrical interaction phenomena between charged ions, alluded to previously, has important ramifications in materials and manufacturing processes that involve dispersion or stabilization of solid-liquid suspensions. Examples of such phenomena are froth floatation technique of separating minerals; stabilization of slurries, emulsions, and paints; electrophoretic deposition of coatings and powders; slip-casting; tape-casting; and numerous other processes.

The preceding discussion dealt with the electrostatic interactions and surface forces due to charged species that were assumed to reside on the solid's surface. In fact, the existence of some type of electrical charge on the surface of a solid is widely observed in water or an aqueous solution (an exception being nonpolar solids in a nonpolar liquid). A surface may acquire an electrical charge from preferential dissolution of surface ions, ionization of surface groups, adsorption of ions from the solution, or other means. A classic example of surface charging due to ionic dissolution is the silver halide emulsions used in the photography industry. Silver halide emulsions develop charged surface ions on silver halide solid crystals from preferential dissolution of either Ag+ ions (in which case the solid crystal develops a negative charge) or halide ions (in which case the crystal develops a net positive charge). In this case, the magnitude and sign of the surface charge can be controlled by adding the oppositely charged ionic species (in the form of a salt) to the solution to suppress the dissolution of like ions. Polymers containing carboxylic ions develop surface charge because of direct ionization in a solution of a given pH. Some materials may adsorb ions to develop surface charge; for example, anionic and cationic surfactants used in industry produce negative and positive charges on the surface, respectively.

Van der Waals Forces

At the beginning of this chapter, it was stated that many different types of "physical" forces operate between atoms and molecules, and these are collectively called the van der Waals forces. One type of physical force that contributes to the van der Waals forces originates from interactions between induced or permanent atomic dipoles. Another contribution to the van der Waals forces is from a weak physical force called the London force. The London force is a universal force and makes the largest contribution to the van der Waals forces. The London force can be attractive or repulsive. It operates over relatively large (10 nm or higher range) distances, and it originates in complex quantum-mechanical interaction phenomena. Because of its universal nature, the London (or dispersion) force plays an important role in a number of surface and interface phenomena.

The total energy due to van der Waals interaction in vacuum is dominated by the contribution of the dispersion (London) forces, which vary with the inverse sixth power of the distance. In a non-vacuum medium, the magnitude of the van der Waals interactions is reduced because of the dielectric characteristics of the medium. Furthermore, the dispersion interaction falls off faster than the inverse sixth power of the distance in a non-vacuum medium, and it becomes important at shorter distances in a non-vacuum medium than in vacuum.

The preceding discussion applies to interactions between isolated atoms or molecules. In the macroscopic world of surfaces, numerous charged species reside on each surface, and their cumulative effect determines the forces between surfaces. For the simple case involving two flat surfaces separated by a distance, H, in a vacuum, the total interaction energy per unit area can be calculated. This energy is given from: $\Delta G = -A/(12\pi H^2)$, where A is called the Hamaker constant, which can be estimated from the polarizability of atoms and other properties. Typical values of Hamaker constant for some materials are listed in Table 4-1.

TABLE 4-1 Hamaker Constant for Some Materials

Materials	Intermediate Phase	Hamaker Constant, $\times 10^{20}$ J^{-1}
Fused quartz-fused quartz	Air	6.5
Fused quartz-fused quartz	Water	0.8
Polystyrene-polystyrene	Air	6.6
Polystyrene-polystyrene	Water	1.0
PTFE-PTFE	Air	3.8
PTFE-PTFE	Water	0.1

The cumulative effect of interactions between charged ions on surfaces is an attractive force between surfaces (i.e., negative ΔG). The energy due to this attractive interaction between surfaces decays more slowly than the energy between individual charged species (ions), which drops with the inverse sixth power of separation. As a result, macroscopic surfaces interact over distances significantly larger than those between atoms and molecules. If the interacting surfaces are located in a non-vacuum medium (e.g., an intervening liquid film), then the attractive interaction is reduced due to "screening" by the medium. This effect is accounted for by modifying the Hamaker constant, A, in the interaction energy equation as follows. If two surfaces of a material, 1, are separated by a medium, 2, the effective (or composite) Hamaker constant is $A_{\text{eff}} = (\sqrt{A_{10}} - \sqrt{A_{20}})^2$, where A_{10} and A_{20} are the Hamaker constants of the materials 1 and 2 in a vacuum, respectively. Note that if A_{10} and A_{20} become similar in magnitude, A_{eff} approaches zero and the interaction energy between the surfaces of phase 1 also approaches zero. This has important practical implications in the dispersions of fine powders in liquids where a judicious choice of medium 2 can be used to stabilize the dispersion and prevent agglomeration of particles.

Contact Angle, Surface Tension, and Young's Equation

When a liquid contacts a solid surface, the local intermolecular interactions may cause the liquid to either spread into a thin film or shrink to minimize its contact area with the solid. The spreading behavior of a liquid on a solid is most conveniently described in terms of the interfacial energies and the angle of contact, θ, that the liquid makes on the solid. The angle θ is measured through the liquid at the solid–liquid–vapor contact region after the liquid has attained an equilibrium configuration on the solid under the action of the forces acting at the contact region (Figure 4-2a). These forces are the interfacial tensions at the interfaces between solid (s), liquid (l), and surrounding gas or vapor (g or v). Thus, σ_{ls} is the interfacial tension of the solid–liquid interface, σ_{lv} is the interfacial tension of the liquid–vapor interface (also called the surface tension of the liquid, whose value for some liquids is given in Table 4-2), and σ_{sv} is

TABLE 4-2 Surface Tension of Liquids

Liquid	Surface Tension (σ_{lv}), $J \cdot m^{-2}$
Water	0.073
Epoxy resin	0.047
Glycerol	0.063
Petroleum oil	0.029
Silicone oil	0.021
Polyvinyl chloride	0.042
Polystyrene	0.043
Ge	0.621
Si	0.874
GaAs	0.401
GaSb	0.440

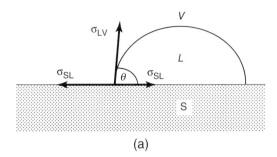

(a)

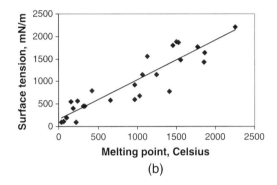

(b)

FIGURE 4-2 *(a) A sessile drop in mechanical equilibrium under the action of the interfacial tensions σ_{LV}, σ_{SV}, and σ_{SL} that correspond to the liquid–vapor, solid–vapor, and solid–liquid interfaces, respectively. The equilibrium condition is expressed from the Young-Dupre equation: $\sigma_{LV} \cos\theta + \sigma_{SL} = \sigma_{SV}$. (b) Variation of surface tension of pure liquid metals with melting tempera-ture.* (Surface tension data are taken from B. J. Keene, Review of data for the surface tension of pure metals', International Materials Reviews, 38(4), 1993, 157-192).

the interfacial tension of the solid–vapor interface (also called the surface energy of the solid). Under mechanical equilibrium of a stationary (sessile) drop, the interfacial tensions are related to the contact angle by the Young-Dupre equation, which is

$$\sigma_{lv} \cos\theta + \sigma_{ls} = \sigma_{sv} \qquad (4\text{-}1)$$

This equation essentially expresses the balance of interfacial forces parallel to the solid's surface in Figure 4-2a (the term $\sigma_{lv} \cos\theta$ is the resolved component of the surface tension force parallel to the solid's surface). The equation has also been rigorously derived using reversible thermodynamics and is used as a foundational relationship in practically all fields of enquiry that deal with the contact between liquids and solids. Note that if $\theta < 90°$ in Equation 4-1, then the solid is wet by the liquid, and if $\theta > 90°$, the solid is not wet by the liquid. The limits $\theta = 0°$, and $\theta = 180°$ define complete wetting and complete non-wetting, respectively. Gravity also acts to perturb the equilibrated shape of the droplet but its influence is negligible on thin films and small droplets.

Upon rearranging the Young-Dupre equation to write an expression for the angle, θ, we obtain $\theta = \cos^{-1}\frac{(\sigma_{sv} - \sigma_{ls})}{\sigma_{lv}}$. This expression shows that low values of θ will be obtained if the liquid surface tension is low, the surface energy of the solid is high, and the liquid–solid interfacial tension is low. In many liquid–solid systems, wettability is improved (i.e., θ is decreased) by lowering the surface tension of the liquid phase (e.g., by raising the temperature or adding another substance to the liquid), and by raising the surface energy of the solid (pre-cleaning and preheating the solid). In the case of reactive systems, all three interfacial energies depend on the nature of the chemical reaction and cannot be manipulated independently.

Wetting is also characterized in terms of a "Work of Adhesion," denoted by W_{Ad}, and defined from $W_{Ad} = \sigma_{lv}(1 + \cos\theta)$. A high value of W_{Ad} indicates good wetting, and a low value of W_{Ad} indicates poor wetting. Generally, good wetting indicates that strong adhesion or bonding will develop between the liquid (e.g., a coating material) and the solid.

In the Young-Dupre equation, σ_{sv} is the surface energy of the solid, i.e., the interfacial energy of the solid–vapor interface. This energy depends on the nature of the vapor phase. If the solid is in equilibrium with its own vapor or if it is in vacuum, the solid's surface energy is denoted by σ_s, whereas if the solid is in equilibrium with the saturated vapor of the liquid phase the solid's surface energy is denoted by σ_{sv}. The difference $(\sigma_s - \sigma_{sv})$ is called the spreading pressure, Π, of the liquid ($\Pi = \sigma_s - \sigma_{sv}$), which is the decrease in the solid's surface tension because of adsorption of the liquid vapor on the solid's surface. The surface structure of solids could be altered by exposure to the liquid, and therefore, the solid's surface energy could change upon contact with the liquid. For $\Pi = 0$, $\sigma_s = \sigma_{sv}$, which means that if the spreading pressure is negligible, then it is possible to express the Young-Dupre equation in terms of a true or intrinsic surface tension of the solid, σ_s, regardless of the liquid in contact with the solid. It is found that for most systems, if the contact angle is greater than about 10 degrees, the spreading pressure is negligible. If the spreading pressure is negligible, then the surface energy of the solid obtained from contact angle measurements with different liquids must be true surface energy of the solid, σ_s, independent of the liquids used in the measurements.

The surface tension, σ_{lv}, of most pure liquids decreases with increasing temperature and approaches zero at some critical temperature, T_c. Close to T_c, surface atoms in the liquid become very weakly bonded. For pure liquids, σ_{lv} decreases linearly with temperature according to $\sigma_{lv} = \sigma_m(\frac{T_c - T}{T_c - T_m})$, where σ_m and T_m are the surface tension at the melting point, and the melting point, respectively. Metals with a high melting point generally have high surface tension, as shown in Figure 4-2b. The preceding equation can be written in a differential form in terms of a temperature coefficient of surface tension, $(\frac{d\sigma_{lv}}{dT}) = (\frac{\sigma_m}{T_c - T_m})$. The magnitude of the temperature coefficient, $(d\sigma_{lv}/dT)$, is negative for most pure liquids, including metals (except for Cd and Zn and some alloys), and increases with increasing σ_{lv}. The magnitude and sign of the temperature coefficient $(d\sigma_{lv}/dT)$ is sensitive to alloying; for example, sulfur in molten steel decreases σ_{lv} and changes $(d\sigma_{lv}/dT)$ from negative to positive. This has ramifications in high-temperature materials processes where surface tension variations induce fluid convection as discussed later. When $(d\sigma_{lv}/dT)$ is positive, it indicates a tendency for surface adsorption and surface ordering. Table 4-3 gives the temperature coefficients of surface tension of some pure metals.

The surface properties of metals and alloys are strongly influenced by the electronic structure and interatomic bonding. For example, high liquid surface tension is associated with the electron bonding and decreases with a prominence of s-electron bonding. Likewise, the surface energy of solids is related to their interatomic bond strength. The surface energy of FCC metals increases with their mechanical strength, and this has been verified by plotting the surface energy of solids,

TABLE 4-3   Temperature Coefficients of Surface Tension of Some Pure Metals[1]

Metal	$(d\sigma_{lv}/dT)$, $J \cdot m^{-2}$
Al	-0.155×10^{-3}
Ga	-0.070×10^{-3}
Ge	-0.12×10^{-3}
Au	-0.20×10^{-3}
Fe	-0.39×10^{-3}
Mg	-0.26×10^{-3}
Ni	-0.35×10^{-3}
Si	-0.145×10^{-3}
Ag	-0.21×10^{-3}
Sn	-0.103×10^{-3}
Zn	-0.21×10^{-3}

[1]Data are from B. J. Keene, "Review of data for the surface tension of pure metals," *International Materials Reviews 38*(4), 1993, 157–192.

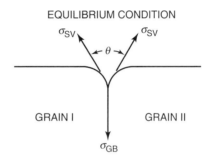

FIGURE 4-3   *Equilibrium of interfacial tensions at a grain boundary intersecting the free surface of a solid. The interfacial tensions σ_{SV} and σ_{GB} correspond to the solid–vapor interface, and the grain boundary between the neighboring grains, respectively. The angle θ is the dihedral angle.*

σ_{sv}, as a function of their Young's modulus at a constant temperature. In a manner similar to the surface energy of the solid (σ_{sv}) and that of the liquid (σ_{lv}), the solid–liquid interfacial energy (σ_{sl}) is influenced by electron configuration and valence electron concentration. In addition, σ_{sl} is also related to the Gibb's free energy of formation (ΔG) of the solid, stoichiometry, interfacial reactions, temperature, and atmosphere. The free energy of formation (ΔG) is a measure of the chemical stability of the solid, and a large negative value of ΔG of metal oxides and carbides correlates with a large contact angle. Thus, wettability of solid oxides and carbides by liquid metals decreases as the stability of these compounds increases.

Grain Boundaries in Polycrystals

Grain boundaries in polycrystalline materials represent a discontinuity between neighboring grains, and are associated with an interfacial energy. Consider the grain boundary intersecting the free surface of the bi-crystal shown in Figure 4-3. A mechanical equilibrium between the two grains is attained when a groove forms at the junction to satisfy the force balance

$$\sigma_{gb} = 2\sigma_{sv} \cos \frac{\theta}{2},$$ (4-2)

where σ_{gb} is the grain boundary energy, and θ and σ_{sv} are the dihedral angle and surface free energy of the solid crystals, respectively. If the free surface of the bi-crystal of Figure 4-3 is in contact with a liquid instead of a vapor, then

$$\sigma_{gb} = 2\sigma_{sl} \cos \frac{\theta'}{2},$$ (4-3)

where θ' is the dihedral angle in the presence of a liquid. If the liquid penetrates the grain boundary, then a single solid–solid interface is replaced by two solid–liquid interfaces. The change in the surface free energy, $\Delta\sigma$, when a solid–solid interface (the original grain boundary) is replaced by two solid–liquid interfaces is negative; i.e.,

$$\Delta\sigma < 0, \text{ where } \Delta\sigma = 2\sigma_{sl} - \sigma_{gb}.$$ (4-4)

The critical condition for equilibrium is

$$\Delta\sigma = 0, \text{ or } \sigma_{sl} = 0.5 \, \sigma_{gb}.$$ (4-5)

In the presence of a liquid phase, the grain boundary therefore adjusts its shape to satisfy the preceding equilibrium condition. This shape adjustment takes place through mass transport processes such as evaporation and condensation, bulk diffusion in the solid, diffusion in the liquid, and surface diffusion.

Three-Phase Equilibrium

The force balance (see Figure 4-2a) that leads to the Young-Dupre equation involves co-planar forces and is applicable to spreading in a plane (two-dimensional spreading). If, however, the solid is either non-rigid or reacts with the liquid, causing dissolution and formation of a small ridge on the solid's surface at the three-phase contact line, then the contact line motion could take place in three space dimensions (e.g., horizontal and vertical). Under these conditions, the co-planarity of the forces in Figure 4-2a is destroyed, and the Young-Dupre equation is no longer valid. For non–co-planar interfacial tensions, the general equation for equilibrium of forces leads to the following relationship:

$$\frac{\sin \phi_s}{\sigma_{lv}} = \frac{\sin \phi_l}{\sigma_{sv}} = \frac{\sin \phi_v}{\sigma_{sl}}$$ (4-6)

where ϕ_s, ϕ_l, and ϕ_v are the equilibrium dihedral angles in the solid, liquid, and vapor phases, respectively, as shown in the schematic of Figure 4-4. For a non-reactive solid in contact with a liquid, this configuration could arise, for example, when the vertical component of the liquid's surface tension, $\sigma_{lv} \sin \theta$, is not balanced by the vertical forces due to the rigidity of the solid (because the force $\sigma_{lv} \sin \theta$ acts on the substrate along the line of contact, it is not balanced by the droplet weight, which acts at the center of mass of the droplet). This unbalanced vertical

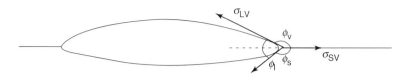

FIGURE 4-4   *The general three-phase equilibrium configuration.*

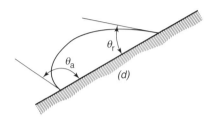

FIGURE 4-5   *Schematic of a drop on an inclined plane exhibiting advancing contact angle, θ_a, and receding contact angle, θ_r.*

force component could produce a small surface deformation in the solid. The deformation is generally negligible for most strong and stiff solids; however, highly deformable solids such as gels, rubber, and thin mica sheets in contact with certain liquids are known to undergo observable deformation leading to ridge formation at the line of contact. Once a ridge forms at the contact line, the co-planarity of the three interfacial tensions is destroyed, and Young-Dupre equation ceases to be valid. Continuum elasticity theory has been used to predict the deformation in plates and membranes due to capillary forces, and the results show that a ridge would form at the line of contact even in a weightless (small) drop. The extent of deformation of different solids could differ significantly. Furthermore, even in a deformable solid it may take considerable time to form a ridge and reach the three-vector equilibrium expressed by Equation 4-6. If the deformation time is considerably greater than the time in which an observation is made, then the Young-Dupre equation will appear to be valid even though true equilibrium might not have been attained. In other words, in the case of deformable solids with slow deformation rates, the Young-Dupre equation (Equation 4-1) will apply only to a metastable configuration rather than to a true equilibrium, yielding contact angle values that are not true equilibrium values. The more general Equation 4-6 has been used to study the morphological evolution of grain boundary grooves in crystalline metals in contact with liquids at high temperatures where atomic diffusion processes cause preferential dissolution, chemical reaction, and ridge formation on the solid.

Equation 4-1 suggests that the equilibrium contact angle is completely and uniquely defined by the three interfacial energies, which are thermodynamic quantities and are not influenced by gravity or other external forces. Common experience with liquid droplets on inclined planes (e.g., raindrops on glass windows), however, suggests otherwise. When a liquid droplet is in contact with an inclined plane in Earth's gravity (Figure 4-5), the front edge of the droplet begins to creep forward, whereas the rear edge remains anchored. As a result, the advancing angle, θ_a, increases and the receding angle, θ_r, decreases. If the receding angle is greater than zero, and the inclination of the solid plane is greater than a critical value, the droplet as a whole begins to move. The critical inclination manifests itself in a critical force that increases as the droplet size increases. This shows that droplet weight may influence the measured contact angle. Attempts have been made to test the Young-Dupre equation in microgravity experiments carried out aboard the space shuttle on liquid-phase sintered alloys. Generally, however, results of such studies have been inconclusive, owing to perturbations from solidification shrinkage and the solid's dissolution during sintering. In any case, it is generally agreed that if the droplet size is smaller than a critical length, called the capillary length, then gravity exerts a negligible influence on the droplet shape, and the contact angle measured on flat, horizontal substrates is the true angle, determined solely by the three interfacial tensions. The capillary length, L_c, is defined from: $\sqrt{(2\sigma_{lv}/\rho g)}$, where ρ is the liquid's density. The length, L_c, represents the maximum height of rise of perfectly wetting liquid on a vertical plate in Earth's gravity.

Microscopic Angles and Precursor Film

The Young-Dupre equation defines macroscopic contact angles. In many systems, the macroscopic wetting front is spearheaded by a thin "foot" or a "precursor film." The precursor film forms when the liquid has a finite, nonzero curvature near its contact with the solid because of local intermolecular interactions within a few nanometers distance. Thus, on a microscopic scale, the liquid droplet gradually thins in the vicinity of its peripheral contact with the solid. Because of its miniscule thickness, the precursor film may elude detection, and the observed contact angles (measured from an analysis of droplet images) may be "macroscopic" angles rather than true "microscopic" angles that form between the thin precursor film and the surface. The microscopic contact angles do not follow the Young-Dupre equation because there is no clear distinction between solid, liquid, and vapor phases at the foot of the precursor film. Alternative equations that strongly depend on the type of intermolecular forces between the liquid and solid atoms have been proposed. An example is the equation

$$\sigma_{sv} - \sigma_{lv}\cos\theta - \sigma_{sl} = 0.5\,\sigma_{lv}\cos\theta\sin^2\theta \qquad (4\text{-}7)$$

which has been derived by assuming a specific form of the potential energy function for intermolecular interactions (Lennard-Jones 6–12 potential). Other equations have been obtained by assuming other types of force distribution functions, such as an exponential repulsion term in place of the r^{-12} term of Lennard-Jones function. While all such models are theoretically interesting and insightful for the light they shed on the interrelationship of the fundamental forces between atoms and molecules and the observed spreading behavior of liquids, more practical application-oriented studies in virtually all disciplines make use of the concept of macroscopic contact angles defined from the Young-Dupre equation.

Roughness and Chemical Inhomogeneity

Measured contact angles are extremely sensitive to the nature of a solid's surface. In fact, contact angle measurement techniques serve as a high-resolution probe to characterize the surfaces of solids. Microscopic roughness and minute chemical inhomogeneity may cause significant changes in the measured contact angle. Strictly speaking, the equilibrium contact angles used in the Young-Dupre equation apply only to ideal surfaces; i.e., surfaces that are chemically and structurally homogeneous. Real surfaces, however, seldom conform to the conception of an ideal surface and contain chemical and structural irregularities. For example, real surfaces may have microscopic grooves, ridges, grain boundaries, dislocations, adsorbed films,

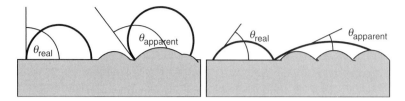

FIGURE 4-6 *Contact angle of a liquid droplet on a rough surface. The apparent contact angle on a rough surface increases if the real contact angle on a smooth surface is obtuse, and the apparent contact angle decreases on a rough surface if the real angle on a smooth surface is acute.* (D. Meyers, Surfaces, Interfaces and Colloids, 2nd ed., 1999, p. 115, Wiley-VCH, New York, NY).

and oxidized, corroded, or coated areas. In addition, crystallographic planes of different atomic packing density have different surface energies, and different crystal planes could expose different molecular groups to a liquid (only single-crystal substrates and freshly cleaved mica approach the consistency and cleanliness of an ideal surface). Such irregularities prohibit a liquid from tracing its path in a completely reversible fashion during forward and reverse motions. This leads to different values of the contact angle during forward motion (advancing angle) and reverse motion (receding angle), or the phenomenon of hysteresis of contact angle discussed below.

A rough surface can be conceived of as providing an additional contribution to the area of contact between liquid and solid. The additional surface effectively increases the solid–liquid interfacial free energy. The true area of contact is greater than the nominal or apparent area, and the ratio, r, of these areas (r = actual area/apparent area), is used to characterize the scale of surface roughness. The ratio is $r = 1$ for a smooth surface, and $r > 1$ for a rough surface. The apparent or measured contact angle (also called the Wenzel angle, θ_w) on a rough surface is related the equilibrium angle θ_y (defined from Equation 4-1) from the relationship: $\cos\theta_w = r\cos\theta_y$. This equation shows that Wenzel's angle will increase with roughness when the Young angle is greater than 90° and it will decrease when the Young angle is smaller than 90°. Figure 4-6 schematically illustrates the effect of surface roughness on the measured contact angles; an obtuse angle is further amplified by roughness, making the surfaces even more non-wettable, whereas an acute contact angle is decreased by surface roughness, thereby assisting liquid spreading. The spreading behavior on a rough surface is, however, highly dependent on the nature and distribution of roughness. A range of contact angle may be accessible to the liquid because on a rough surface, the liquid is in a metastable state rather than under true equilibrium. The contact line motion on a variety of roughness profiles, such as concentric circular grooves and saw-tooth undulations, leads to a multitude of thermodynamically allowed low-energy configurations. This yields a range of contact angle values depending on the configuration the system resides in during the period of observation.

Roughness gives rise to contact angle hysteresis, H, which is defined as the difference of the advancing and receding contact angles; i.e., $H = \theta_a - \theta_r$, where θ_a and θ_r are the advancing and receding contact angles, respectively. The advancing and receding angles are those that are obtained after the advancing or the receding motion has stopped. When the forward motion of a liquid over a solid is obstructed by a surface asperity, the contact angle undergoes a small transient change before attaining a constant value. Likewise, when the liquid is allowed to retreat across a solid, it shows a transient response before attaining a steady contact angle. The angles θ_a and θ_r denote the measured values of the contact angle after the transient phase.

For characterizing the spreading behavior of liquids on solids, it is a common practice to report the advancing angle rather than the receding angle; this is because the advancing angle is less sensitive to the surface defects. The extent of hysteresis increases with increasing surface roughness.

The origin of contact angle hysteresis lies in the existence of multiple thermodynamically allowed states in which the three-phase (solid/liquid/vapor) contact line can reside. Surface roughness (troughs and crests) hinders the contact line motion by creating energy barriers. The system can reside in any of the potential wells (minimum energy configurations) accessible to it. This is, however, possible only if the scale of the roughness is large enough to overcome the thermal loitering of the contact line. In fact, roughness may be less important at elevated temperatures, at which the thermal energy of droplet will enable the contact line to "jump" over the energy barriers due to roughness. At temperatures at which hysteretic losses occur, the contact line advances by a "stick-slip" motion over the surface asperities (especially at low velocities), and the extent of hysteresis depends on the spatial density of surface imperfections. As even minute surface irregularities can pin the contact line, only the most carefully prepared smooth and homogeneous surfaces will not hinder contact line motion.

The effect of roughness may be partially masked and hysteresis reduced when an interface separating a wetting fluid from a nonwetting fluid propagates on a rough surface. As the wetting fluid tends to fill the grooves, during reverse motion, the contact line propagates on partially wetted areas (grooves filled with wetting fluid). This partially masks the roughness effects. Furthermore, the motion of the contact line over a rough surface depends on the liquid velocity and the wavelength of surface asperities. For example, at large velocities (achieved in forced flow), the liquid meniscus virtually "slips" over the asperities without penetrating the wedges. In contrast, at low velocities the front is temporarily anchored to an asperity before breaking loose; the process repeats itself in the defect field. High velocity corresponds to large capillary numbers ($Ca > 0.01$, where $Ca = \mu U/\sigma_{lv}$, μ is the viscosity and U is the contact line velocity) and low velocity corresponds to small capillary numbers ($Ca < 0.01$). At large Ca, inertial forces in the liquid become important relative to viscous and surface forces.

Besides roughness, surface chemical inhomogeneity (adsorbed films, and oxidized, corroded, or coated regions) also alters the contact angle and spreading behavior. Most studies on spreading on heterogeneous solids use an effective contact angle as some kind of an average for the different chemical species. For example, the contact angle of a liquid on a two-phase surface constituted by two chemical phases covering area fractions f_1 and f_2 and having intrinsic contact angles, θ_1 and θ_2, can be expressed as an area fraction-weighted average according to

$$\cos\theta_c = f_1 \cos\theta_1 + f_2 \cos\theta_2 \tag{4-8}$$

where θ_c is an effective or composite contact angle. This equation has been modified to take into account surface inhomogeneities of molecular dimensions. Note that Equation 4-8 provides only an averaged behavior and does not shed light on flow fluctuations due to chemical inhomogeneity. Consider the rise of a liquid in a cylindrical capillary composed of alternating layers of two chemical phases, one with a wettable value of the contact angle for a given liquid ($\theta < 90°$), and the other with a non-wettable value of the contact angle ($\theta > 90°$). When the liquid crosses over from a wettable patch to a non-wettable patch, it experiences a retarding force; the liquid front will cease to advance only if the decelerating force is large enough to cause the fluid to lose all momentum prior to its contact with the next wettable patch. If the contact line reaches the next wettable patch before its momentum has dropped to zero, the contact line will again accelerate, and capillary rise will continue. In this example, the effective contact angle approximation (Equation 4-8) could also yield an angle smaller than 90°, just by selecting an appropriate combination of layer thicknesses, thus enabling the capillary rise to continue. However, local flow discontinuities (acceleration and deceleration) and associated energy losses due to the chemical inhomogeneity will not be revealed by the use of an effective contact angle approach.

The segregation of impurities to a surface or interface may change the surface energy, often dramatically, even at small impurity concentrations. Solute segregation at surfaces and grain boundaries is widely observed in a wide variety of polycrystalline materials. The effect of impurity segregation on the surface energy at low concentrations at a constant temperature is expressed from the Gibb's adsorption equation,

$$\Gamma = -\frac{1}{RT}\frac{d\sigma_{lv}}{d \ln \alpha_j},$$
(4-9)

where Γ is the excess surface concentration per unit area, $d\sigma_{lv}$ is the change in the surface energy, and α_j is the activity of component j in the solvent. This relationship shows that segregation occurs when the impurity causes a decrease in the surface energy of the solid. The greater the adsorption, the more the solute tends to lower the interfacial energy. Conversely, the component with the lower surface energy will preferentially segregate at the interface. For example, in Al-Mg alloys the surface tension of Mg is lower than that of Al at the latter's melting point; as a result, Mg preferentially segregates at the free surface of Al and weakens the surface oxide film on Al. Generally, elements that have a high surface activity exhibit low solubility, and surface activities can vary over orders of magnitude, especially because they are very sensitive to trace quantities of solutes. The surface effects of solutes could be pronounced, and even at low solute concentrations, the surface energies may be reduced appreciably.

Dynamic Contact Angles

The angle of contact between a liquid and a solid as given by the Young-Dupre equation is a static angle that is attained when the contact line is in mechanical equilibrium with respect to its surroundings. At high contact line velocities, the hydrodynamic forces distort the contact angle and lead to the phenomenon of dynamic contact angles. The dynamic effects due to fluid motion are unimportant at low velocities or in low-viscosity fluids, and the dynamic angle attains a quasi-static value; however, in high-viscosity fluids or at high speeds these effects can lead to departures from the thermodynamic equilibrium, and the contact angle begins to be influenced by the imposed fluid velocity (Figure 4-7a). For example, during spreading or capillary rise, most low-temperature organic liquids and water begin to exhibit velocity dependence of contact angle at $\sim 10^{-4}$ m·s^{-1}. The velocity dependence of contact angles has been characterized in experiments on forced flow of a large number of liquids (water, polyethylene, silicon oil, etc.) through tubes. Figure 4-7b shows the experimental measurements on dynamic contact angle in forced flow of liquids through tubes, plotted as the dynamic angle, θ, versus the logarithm of the fluid velocity.

Theoretical relationships have been proposed to predict the observed dependence of contact angle on velocity in different liquids. These relationships apply to ideal systems comprised of smooth, rigid, and homogeneous solids in contact with chemically inert and pure liquids of low vapor pressure. Generally, these relationships yield a power law dependence for the spreading

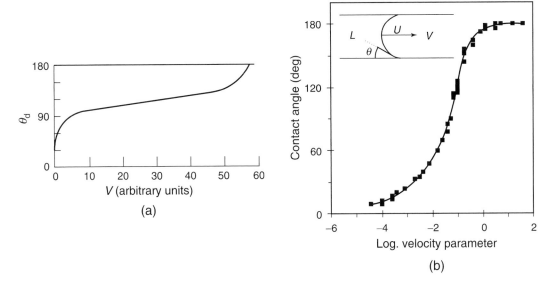

FIGURE 4-7 *(a) Schematic representation of the effect of meniscus velocity on the dynamic contact angle.* (D. Meyers, Surfaces, Interfaces and Colloids, 2nd ed., 1999, p. 436, Wiley-VCH, New York, NY) *(b) Experimental data on forced flow of liquids through glass tubes displaying dependence of contact angle on the logarithm of a velocity parameter.* (Adapted from N. Eustathopoulos, M. G. Nicholas and B. Drevet, Wettability at High Temperatures, Pergamon, 1999, p. 69. Data are from R. L. Hoffman, A study of the advancing interface I. Interface shape in liquid-gas systems, Journal of Colloid and Interface Science, 50, 1975, p. 228).

kinetics of small droplets on a solid's surface, i.e., $R^m = \text{const} \cdot t$, where R is the droplet base radius, t is the time, the exponent m is in the range 5–10, and the value of the constant depends on the specific model. We shall now consider a few models to obtain expressions for the constant in the power law relationship.

Perhaps the most widely used model for the dynamic contact angles is the De Gennes equation, according to which $Ca = \mu U / \sigma_{lv} = C' \theta^n$, where Ca is the capillary number, C' is a constant $(0.02\ \text{radian}^{-3})$, and the exponent n is approximately 3.0 ± 0.5 for velocities up to a few meters per second. This expression was developed from an analysis of the experimental measurements on the distortion of contact angle due to an imposed velocity in forced flow of liquids through tubes. This relationship also applies to velocity-dependence of contact angle of thin, hemispherical droplets of an inert liquid spreading on a solid surface. The De Gennes equation can be written as $U = C(\sigma_{lv}/\mu)\theta^n$, where σ_{lv} and μ are the surface tension and viscosity of the liquid, and θ is the dynamic contact angle distorted due to the imposed velocity, U. Noting that $U = (dR/dt)$, where R is the radius of the contact area between a small droplet and the underlying solid, and that $R^3\theta \approx (4V/\pi)$, where V is the droplet volume, the equation for the rate of spreading of the droplet is obtained as

$$U = \left(\frac{dR}{dt}\right) = C\left(\frac{\sigma_{lv}}{\mu}\right)\theta^3 = C\left(\frac{\sigma_{lv}}{\mu}\right)\left(\frac{64V^3}{\pi^3 R^9}\right) \qquad (4\text{-}10)$$

Upon integrating Equation 4-10 over the limits $R = R_0$ at $t = 0$ and $R = R$ at $t = t$, we obtain

$$R^{10} = \frac{64V^3}{\pi^3} C \left(\frac{\sigma_{lv}}{\mu} \right) \cdot t + R_0^{10} \tag{4-11}$$

which is a power-law expression of the form $R^{10} \sim t$ for the time dependence of radius of spreading droplets and is widely used to model the spreading of thin droplets of viscous polymers on solids. A relationship that has been applied to silicon oil on silane- and paraffin-coated glass and paraffin oil on paraffin-coated glass leads to a parabolic dependence of the cosine of dynamic angle, θ, on the spreading velocity. This relationship is

$$\cos \theta = \cos \theta_0 - 2\xi(1 + \cos \theta_0) \left(\frac{\mu}{\sigma_{lv}} \right)^{1/2} U^{1/2} \tag{4-12}$$

where θ_0 is the equilibrium contact angle, and ξ is an empirical constant. For thin droplets with small equilibrium angle, and with the dynamic angle θ not too different from the equilibrium angle, θ_0, one can write $\cos \theta_0 \approx 1$, and $\cos \theta \approx 1 - (\theta^2/2)$. Equation 4-12 then becomes

$$U = \left(\frac{dR}{dt} \right) = \frac{1}{16\xi} \left(\frac{\sigma_{lv}}{\mu} \right)^{1/2} \theta^2 = \frac{1}{16\xi} \left(\frac{\sigma_{lv}}{\mu} \right)^{1/2} \left[\frac{16V^2}{\pi^2} \right] R^{-6} \tag{4-13}$$

Upon integration together with the initial condition $R = R_0$ at $t = 0$, Equation 4-13 yields the following $R^7 \sim t$ relationship

$$R^7 = \frac{V^2}{\pi^2 \xi} \left(\frac{\sigma_{lv}}{\mu} \right)^{1/2} \cdot t + R_0^7 \tag{4-14}$$

Another model is based on a hydrodynamic analysis for the slug-flow of a liquid in a capillary. It assumes that the walls of the capillary are wet with a preexisting film of small thickness, δ. The dynamic contact angle is obtained as an asymptotic limit to the liquid–vapor profile in the region of contact with the solid. The resulting equation has the form

$$\tan \theta = 3.4 \left(\frac{\mu}{\sigma_{lv}} \right)^{1/3} U^{1/3} \tag{4-15}$$

For small θ, $\tan \theta = \theta$, and noting that for small hemispherical liquid caps, $\theta \approx (4V/\pi R^3)$, we obtain

$$U = \left(\frac{dR}{dt} \right) = \frac{1.63V^3 \sigma_{lv}}{\mu \pi^3} R^{-9} \tag{4-16}$$

which on integration together with the initial conditions $R = R_0$ at $t = 0$, yields a relationship similar to the de Gennes equation ($R^{10} \sim t$). This relationship is

$$R^{10} = \left[\frac{1.63V^3 \sigma_{lv}}{\mu \pi^3} \right] \cdot t + R_0^{10} \tag{4-17}$$

Models of the dynamic contact angle phenomenon that incorporate the effect of intermolecular interactions between the liquid and the solid have also been developed. For example, a model postulates that motion of the contact line occurs via sliding of the molecules along the interface, which is assumed to be a thermally activated process with a rate given from the theory of absolute reaction rates. The energy consumed in causing the flow is obtained as $E = \sigma_{lv}(\cos\theta_0 - \cos\theta)$, and this energy is used to raise or lower the activation energy for forward or reverse motion of the liquid molecules along the solid surface. The velocity dependence of the dynamic contact angle is given from

$$U = \left(\frac{dR}{dt}\right) = 2K\lambda \sinh\left[\frac{\sigma_{lv}}{\Delta nkT}(\cos\theta_0 - \cos\theta)\right] \tag{4-18}$$

where K is an oscillation frequency (number of molecular displacements occurring per unit time per unit length of the liquid–solid–vapor interface), λ is the average distance between sites on the solid's surface (related to planar density of surface atomic planes), Δn is the surface concentration of sites ($\sqrt{(\Delta n/2)} \approx \lambda^{-1}$), k is the Boltzmann constant, and T is the absolute temperature.

At high temperatures (or large Δn), $\sigma_{lv}(\cos\theta_0 - \cos\theta) \ll \Delta nkT$, and the hyperbolic sine is roughly equal to its argument, so that

$$\cos\theta = \cos\theta_0 - \left(\frac{kT}{K\lambda^3\sigma_{lv}}\right)\frac{dR}{dt}. \tag{4-19}$$

The parameter λ varies only slightly from system to system, and K is a rather specific characteristic of the system. Also, if θ_0 is small and $\theta \sim \theta_0$, then $\cos\theta_0 \approx 1$, and $\cos\theta \approx 1 - (\theta^2/2)$, and Equation 4-19 yields $U = \left(\frac{K\lambda^3\sigma_{lv}}{2kT}\right)\theta^2$, which is analogous to the De Gennes equation ($U = \text{constant}\cdot\theta^n$) with the exponent $n = 2$ instead of $n = 3$.

Note that Equations 4-18 and 4-19 do not include the bulk viscosity of the liquid; this is because the model postulates the existence of an interfacial viscosity that is the retarding force acting on the fluid.

The experimental validity of the preceding models of dynamic contact angles has been tested in a variety of low-temperature organic liquids. For example, dynamic angle measurements of ethanol, octane, hexane, silicone oil, and water on solid materials such as glass, stainless steel, aluminum, titanium, nylon, Teflon, PMMA, and other solids have been made and compared to the model predictions. It is found that different models apply to different solid/liquid systems, and to a given system at different velocities. For example, in many of the preceding liquid and solid combinations, Equation 4-18 applies at low and moderate velocities, whereas Equation 4-15 has been found to correctly describe the velocity dependence of the dynamic angle at high velocities. In the case of high-temperature metals in contact with solid metals or ceramics, considerable complexity is introduced because of chemical reactions, product layer formation, and dissolution. Figure 4-8 shows the experimental measurements of the instantaneous radius of droplets of some molten metals spreading on solid ceramics and metals as a function of time of contact. The data are plotted in a form consistent with the linear form of the power-law, $R^m = (\text{constant})\cdot t$ (i.e., log R versus log t). This figure shows that each system exhibits several different values of the power law exponent, m, indicating a more complex spreading behavior as compared to a simple, non-reactive liquid. Some of the values of m seem to agree with the standard models, whereas others depart from the known behavior. Thus, new models are needed to describe the complex spreading behavior of high-temperature liquids.

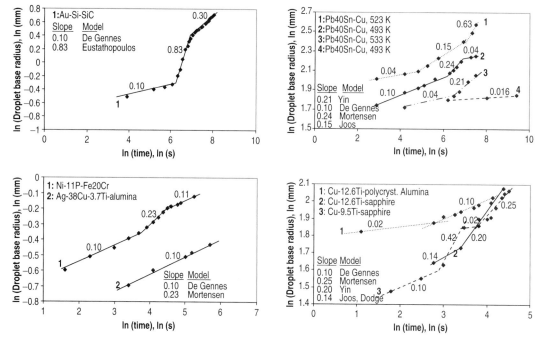

FIGURE 4-8 *Experimental spreading data plotted as the natural logarithm of the instantaneous radii of small metal and alloy droplets on solid substrates as a function of natural log of spreading time. Each system exhibits several different intermediate flow regimes that are consistent with the power law, $R^m = const. \, t$, where m is numerical exponent, R is the droplet base radius, and t is the time. (R. Asthana and N. Sobczak, JOM-e, 52(1), 2000, http://www.tms.org/pubs/journals/jom/0001Asthana/Asthana0001/html).*

Immersion of Solids in Liquids

Many materials processes involve immersion and dispersion of solids in liquids as an intermediate step in manufacturing. Immersion involves substitution of a solid–gas interface by an equivalent solid–liquid interface. Because the contact angle of different solids with liquids is either acute or obtuse, immersion is associated with either absorption or generation of energy; that is, the process is either spontaneous or requires external energy. The change in energy on a solid's submersion in (or withdrawal from) a stationary liquid can be determined from the surface properties of the liquids and solids. Likewise, fine particulate or fibrous solids can float (or remain partially submerged) on a liquid because the surface tension and gravitational forces (i.e., weight and buoyancy) balance out at an equilibrium position of the partially submerged solid. In other words, a solid approaching an interface between two fluids (e.g., liquid and air) may either pass through the interface and sink in the lower fluid under the influence of gravity, or it may take up an equilibrium position floating at the interface. The stability of solids at fluid interfaces plays an important role in processes such as mineral beneficiation; creation of slurries, pastes, and colloidal suspensions; paint dispersion; particulate composites; and others. Theory confirms that although incorporation of wettable particles in a liquid is energetically spontaneous, nonwettable particles need to surmount an energy barrier for immersion; in both cases, however, the energetics of particle transfer are path dependent, as discussed on the next page.

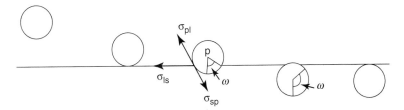

FIGURE 4-9 *Stages in transferring a spherical solid particle from a gas phase into a fluid across a planar gas–fluid interface. The semiapical angle ω represents the degree of immersion.* P. K. Rohatgi, R. Asthana, R. N. Yadav and S. Ray, Metallurgical & Materials Transactions, 21A, 1990, 2073–83.

Figure 4-9 shows the incremental steps in the transfer of a solid sphere into a liquid as a function of the angle of immersion, ω. For a reversible process of transfer in the absence of convection and fluid drag, the various interfacial energies, solid and liquid densities, and the dimensions of the solid will determine the energy for submersion, or conversely, the solid's equilibrium position at the fluid boundary. For a partially submerged solid particle, positioned at an interface at a semi-apical angle ω (Figure 4-9), the equilibrium position will correspond to the minimum in the total energy of the system comprised of surface, buoyancy, and potential energies. The net change in the surface energy during the process of immersion of a spherical solid particle of radius a is

$$\Delta E_s(\omega) = \pi a^2 \sigma_{lv} \left[- \sin^2 \omega + 2 \cos \theta (\cos \omega - 1) \right] \tag{4-20}$$

The change in the buoyancy energy (i.e., work done against the buoyancy forces) for the particle at the angular position ω is given from

$$\Delta E_B(\omega) = - \int_0^\omega V(\omega) \rho_m \, g \, dY_\omega \tag{4-21}$$

$$\Delta E_B(\omega) = - \frac{\rho_m \, g \, \pi a^4}{48} \left[108 \ln \frac{(2 + \cos \omega)}{3} + 39 - 40 \cos \omega + \cos^4 \omega \right] \tag{4-22}$$

where $V(\omega)$ is the volume of liquid displaced, $Y(\omega)$ is the distance along Y-axis the center of gravity of the particle is displaced during its immersion through an angle ω, and ρ_m is the density of the melt. If the planar liquid surface is taken as the reference level, the change in the potential energy in Earth's gravitation field is

$$\Delta E_P(\omega) = - \frac{4}{3} \pi a^4 \rho_p g (1 - \cos \omega) \tag{4-23}$$

where ρ_p is the density of the particle. As both buoyancy and potential energies vary with the fourth power of the particle radius, and the surface energy varies with the square of the radius, it is evident that for fine particles the total energy for immersion (and the solid's stability at the interface) will be determined mainly by the dominant surface energy. Similar calculations have been made for other shapes of solid; for example, change in surface energy for submersion of

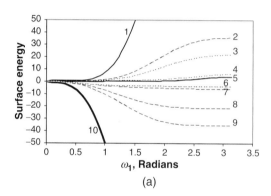

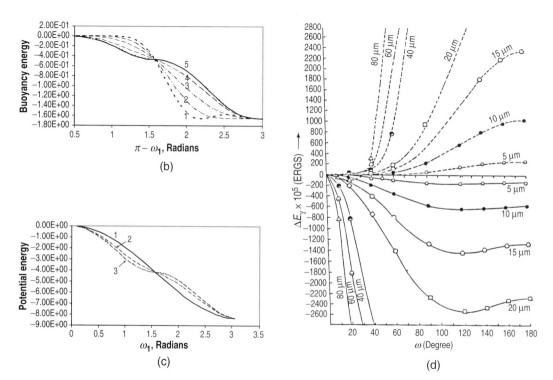

FIGURE 4-10 *(a) Change in dimensionless surface energy on submersion of a spheroid into a fluid under two wetting situations: wettable contact angles (bottom half), and nonwetting angles (top half). The change in energy is positive for nonwetting angles. The curves 1–10 are for different values of the contact angle θ and the aspect ratio of the spheroid α (= a/b), where 2a and 2b are the minor and major axes. The α and θ values are given here: (1) 5, 160°; (2) 2, 160°; (3) 1.5, 160°; (4) 0.5, 160°; (5) 0.01, 160°; (6) 0.01, 20°; (7) 0.5, 20°; (8) 1.5, 20°; (9) 2, 20°; and (10) 5, 20°. (b) Change in the nondimensional buoyancy energy as a function of the angle (π − ω₁) for the aspect ratio, α, given here: curve 1 (α = 2), curve 2 (α = 1.5), curve 3 (α = 1), curve 4 (α = 0.5), and curve 5 (α = 0.1). (c) Change in the nondimensional potential energy on submersion of a solid ellipsoid in a fluid. The different curves represent the different aspect ratio, α. Curve 1 (α = 1), curve 2 (α = 0.5), and curve 3 (α = 0.1). (R. Asthana, unpublished work, 2001) (d) The effect of particle radius on the total surface energy for particle submersion in a fluid for nonwetting (θ > 90°) and wetting (θ < 90°) values of the contact angle, θ. The energy change is plotted as a function of the semiapical angle of immersion, ω. The calculations in (d) are for immersion of graphite sphere in aluminum. (P. K. Rohatgi, R. Asthana, R. N. Yadav and S. Ray, Metallurgical & Materials Transactions, 21A, 1990, 2073–83).*

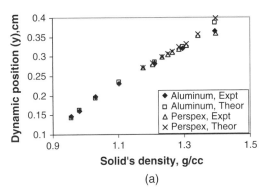

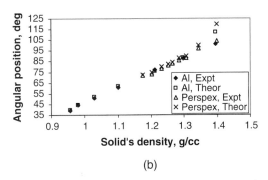

FIGURE 4-11 *Comparison of the experimental and theoretical equilibrium (a) vertical and (b) angular positions of aluminum and Perspex spheres of different densities on water. The vertical position, y, and the angular position, φ, are defined in Figure (c).* (Data are from S. Hartland and J. D. Robinson, Equilibrium position of particles at a deformable fluid-fluid interface, Journal of Colloid & Interface Science, 35(3), 1971, 372).

an ellipsoidal particle (with major and minor axes as 2b and 2a, respectively) is

$$\Delta E_s [\omega] = \pi \sigma_{lv} b^2 \left[-\frac{a^2}{b^2} \sin^2 \omega + \left(1 - \frac{a^2}{b^2}\right) \frac{(\cos^4 \omega)}{2} + \right.$$
$$\left. 2 \cos \theta \left\{ \frac{a^2}{b^2} (\cos \omega - 1) + \left(\frac{a^2}{b^2} - 1\right) \frac{(1 - \cos^2 \omega)}{3} \right\} \right]$$

(4-24)

Figure 4-10a-d shows the variation of surface, buoyancy, and potential energies for submersion of a solid in a liquid. The computed total surface energy for thermodynamically reversible submersion of graphite spheres of different sizes in molten Al as a function of the angular depth of immersion is shown in Figure 4-10d for two cases: wettable (Cu-coated) graphite and nonwettable (uncoated) graphite. Equations 4-20 through 4-24 are derived by assuming a flat, non-deformable fluid interface. In reality, however, the liquid surface is deformed in the region of contact with the solid, and the equations are modified. Theoretical analyses of the problem and their experimental validation have been presented in the surface science and chemical engineering literature. Representative examples are shown in Figure 4-11a and b, which compare the theoretical predictions with the experimental measurements of the equilibrium position (defined in Figure 4-11c) of spherical aluminum and Perspex spheres on liquid paraffin. The vertical displacement, y, with respect to the flat liquid surface (Figure 4-11a), and the angular position, φ (Figure 4-11b), are plotted as a function of the solid's density, which was artificially modified by injecting controlled amounts of lead through tiny holes drilled in hollow Al and Perspex spheres. The average agreement between the theory and experiment is better than 2% if the densest Al and Perspex spheres are disregarded.

All of the preceding discussion ignored the perturbations caused by the inertial forces and fluid drag. The relative contribution of these forces can be rationalized in terms of Reynolds number, $Re(Re = \rho_m V d^2/\mu)$ and Weber number $We(We = \rho_m V d^2/\sigma_{lv})$, where μ is the viscosity, σ_{lv} is the surface tension, ρ_m is liquid's density, and V and d represent the velocity and diameter of the particle, respectively. From the definitions of these dimensionless numbers, it is clear that Re is the ratio of inertial to viscous forces, and We is the ratio of inertial to surface tension forces, respectively. As an example of the use of these numbers, consider that a 100-μm-diameter graphite particle is introduced in pure molten Al devoid of surface oxide ($\rho_m = 2.7$ g·cm^{-3}, $\mu = 0.01$ cP, and $\sigma_{lv} = 800$ ergs·cm^{-2}) at an average velocity of 1 m/s. This will yield a very large Reynolds number ($Re = 27,000$) and a small Weber number ($We = 0.34$), which indicates that viscous drag is significantly smaller than both inertial forces (due to velocity) and capillary forces (due to surface tension), and that capillary forces are the dominant forces. Calculations based on detailed mathematical models show that extremely high velocities (typically > 1 m/s) are needed to transfer fine non-wetting particulates in the melt because of the very large magnitude of capillary forces opposing particle entrainment. The magnitude of the injection velocity or fluid shear needed to introduce fine, non-wetting particles in liquids can be decreased by chemical alloying to decrease the liquid surface tension, and by surface modification (e.g., coating) of the solid to reduce the contact angles and facilitate particle incorporation under melt shear. In liquid metals, surface oxide films provide additional resistance to particle immersion because of the formation of two additional interfaces (i.e., gas–oxide and oxide–melt interfaces).

Gas–Solid Attachment and Gas Stabilization

The attachment of bubbles and small droplets to solids dispersed within a liquid is encountered in a number of physical processes, such as mineral beneficiation, steel refining, agglomeration, emulsification, condensation, nucleation, detergency, and many others. The presence of gas within the melt may increase the instability of particle-melt suspensions. Nonwetting particles attach to gas bubbles, and the stability of bubble/particle cluster determines the energy required to detach and uniformly disperse the particles in the liquid. Frequently, fine particulates are purposely added to stabilize gas in a liquid as, for example, in the production of certain types of foamed metals by the addition of carbide particles to molten metal.

When a particle-bubble agglomerate forms within a stationary liquid, the particle attains an equilibrium position at the curved bubble–melt interface. Because the bubble has a finite volume of gas in it, it will slightly expand when it is partially penetrated by the particle as it attains an equilibrium position (the particle is either partially engulfed in the bubble or completely ejected out of it). The angular position of the solid at the surface of the bubble can be characterized by two angles, ω_1 and ω_2 (Figure 4-11d). Minimization of the total surface energy of the bubble-particle cluster shows that the particle will remain anchored to the bubble as long as $|(\sigma_{sl} - \sigma_{sg})/\sigma_{lg}| < 1$ regardless of the radius of curvature of the bubble. Particles poorly wetted by liquids are therefore more likely to be anchored to gas bubbles than are wetting particles. These may increase the apparent density of bubbles, prolong their residence time in the liquid, and make gas and particle removal difficult.

Agglomeration

A non-wetting fluid is spontaneously expelled from the gap between contacting particles. This results in the formation of an agglomerate because the force of adhesion between particles in the non-wetting liquid is greater than the force between them in air. The difference of these forces is the attractive capillary force (sometimes called the capillary contracting force) that causes agglomeration and destabilizes a slurry of fine particles. The existence of particle agglomerates with entrapped gases is well known in both metallic and nonmetallic liquids. Capillary forces also cause fiber bunching and segregation when fibers are brought in contact with a non-wetting liquid matrix during the fabrication of fiber-reinforced composites. If the fibers are not packed close to their theoretical packing density, then the fibers separate into regions of high packing density and low packing density upon exposure to a non-wetting matrix.

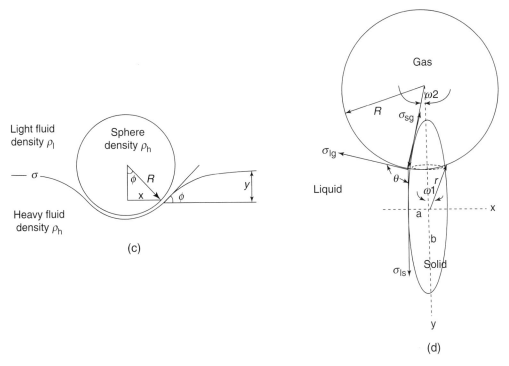

FIGURE 4-11 continued *(c) The vertical position, y, and the angular position, ϕ, of a sphere at a deformable fluid interface, and (d) schematic diagram showing the equilibrium position of a solid particle anchored at the surface of a gas bubble in a liquid.*

Many studies have analyzed the capillary forces between solids, the fluid profile in the contact region, and the size of the cavity created by the expulsion of the non-wetting fluid. The capillary-driven expulsion of a non-wetting fluid from the region of contact (e.g., Hg between a glass sphere and a glass plate) yields a vacuum or gas-filled micro-cavity. This cavity provides for interaction between particles at distances varying from a few nanometers for micrometer-size particles to several micrometers for millimeter-size particles. Calculations show that attractive capillary forces between fine solids in a non-wetting liquid decrease with increasing separation; the decrease is rapid at small separations, followed by a sluggish decrease and finally an abrupt drop to zero as a critical gap width is exceeded and the cavity disappears.

A wetting liquid in the contact region between two solids in air is not expelled from the gap and gives rise to attractive capillary forces that tend to draw the solids together. The force between particles in contact with a wetting liquid comprises two contributions—a force due to liquid surface tension ($2\pi r\sigma_{lv}\cos\theta$, where r is the radius of a pendular liquid ring between touching spheres), and a force due to the pressure difference, ΔP, across the curved liquid–vapor boundary, where $\Delta P = \sigma_{lv}[1/R_1 + 1/R_2]$ and R_1 and R_2 are the two principal radii of curvature of the liquid–vapor boundary. In the limit of small liquid volume between touching spheres, the ΔP term dominates the total pressure, and the maximum interparticle force due to capillary pressure between touching spheres can be calculated. This force is $F = 2(2)^{1/2}\sigma_{lv}\cos\theta/R$, where R is the radius of the sphere. The force increases with increasing liquid surface tension, and decreasing contact angle and particle radius. Non-spherical (e.g., jagged) particles produce not only normal attractive forces due to capillary interactions but also shear forces and torques that cause rotation and particle rearrangement. Many of these effects have been observed in liquid-phase sintering of compacted powders, in which a liquid film of a wettable phase coats the particles and promotes strong capillary attraction between them.

Fine particles dispersed in non-wetting liquids exhibit significant agglomeration tendency. Attractive capillary forces also develop between partially immersed solid particles at the interface of gas and liquid. Such interactions aid rapid agglomeration and clustering. The kinetics of agglomeration influences the suspension stability and the quality of the product being formed using that suspension. In a quiescent liquid, thermal diffusion (Brownian motion) causes particle collisions, and the diffusion coefficient of particles in thermal motion at a temperature T is given from the Stokes-Einstein equation: $D = kT/(6\pi a\mu)$, where k is the Boltzmann constant, a is the particle radius, and μ is the fluid viscosity. If collision is followed by recoil, then agglomeration will be feeble or non-existent. The formation of agglomerates depends, therefore, not only on the collision frequency but also on "sticking" probability; the sticking probability is unity when each collision leads to clustering. For a sticking probability of unity in an initially mono-dispersed suspension, each collision will reduce the number of particles by one (i.e., one aggregate forms when two particles are lost). The rate of decrease of total particle concentration, n, is given from

$$-\frac{dn}{dt} = \left(\frac{4kT}{3\mu}\right) n^2 = k_F n^2 \qquad (4\text{-}25)$$

where $k_F (= 4kT/3\mu)$ is called the agglomeration (or flocculation) rate constant and has a value of 6.13×10^{-18} m^3/s for aqueous suspensions at room temperature. As the particle radius, a, increases, the diffusion coefficient given by the Stokes-Einstein equation decreases, but the effective collision radius increases, so these effects exactly offset each other. Upon integration, Equation 4-25 yields the concentration $n(t)$ of particles in the suspension as a function of time, t, as

$$n(t) = \frac{n_0}{(1 + k_F n_0 t)} \qquad (4\text{-}26)$$

where n_0 is the initial concentration of particles in the liquid. A characteristic agglomeration time, t_F, is defined as the time in which the number of particles is reduced to half of the initial value [i.e., $n(t) = 0.5 \cdot n_0$]. As an example, for an initial concentration of 10^{16} particles per cubic meter in the liquid, and with an agglomeration rate constant of aqueous suspension as $k_F = 6.13 \times 10^{-18}$ m^3/s, the agglomeration time will be about 16 s (in actual practice, the times are usually greater than this value).

Particle collisions cause different types of aggregates to form. For example, even in a mono-dispersed suspension, continued collisions would form not only singlets and doublets but also triplets, and higher-order agglomerates. For m-fold agglomeration (i.e., m single particles in an agglomerate), the concentration of agglomerates is

$$n_m = \frac{n_0(t/t_F)^{m-1}}{(1 + t/t_F)^{m+1}} \qquad (4\text{-}27)$$

where t_F is the characteristic agglomeration time. Calculations based on this equation show that for all types of agglomerates (i.e., singlets, doublets, etc.), the concentration rises to a maximum at a characteristic time and then slowly decreases. The concentration of singlets is the largest of all other types of agglomerates at any given time. These predictions are in good agreement with the experimental results on rapid agglomeration in aqueous and non-aqueous systems.

The preceding discussion assumed strong attractive forces between particles so that each collision led to agglomeration and the possibility of de-agglomeration was ignored. In reality, collisions will cause not only agglomeration but also de-agglomeration; for example, doublets will disintegrate into singlets, and so forth. To account for deflocculation, a second term is introduced in Equation 4-25, so that the rate of concentration change becomes

$$\frac{dn_1}{dt} = -an_1^2 + bn_2, \tag{4-28}$$

where a and b are coefficients characterizing the probability of the formation and rupture of agglomerates per unit time, and n_1 and n_2 are the concentrations of singlets and doublets, respectively. The coefficients a and b are expressed in terms of fundamental parameters characterizing interparticle forces, such as the Hamaker constant.

If some energy barrier to particle contact and adherence exists (i.e., sticking probability <1), then agglomeration can be considered as a chemical reaction with an activation energy that must be surmounted for agglomerates to form. Under these conditions, slow agglomeration (termed coagulation) will occur. Many refinements have been introduced in the basic agglomeration model, discussed previously, in order to account for particle collisions due to fluid shear, and the effect of disjoining pressure in the liquid film between closely-spaced particles. Gravity may also influence the agglomeration. Because particles of different size or density will settle in a fluid at different rates, the relative motion between them would cause particle collisions and agglomeration. The relative velocity, u_R, due to gravity for two widely separated particles (or droplets) is

$$u_R = \frac{2(\mu_r + 1)\Delta\rho a_1^2(1 - \lambda^2)g}{3(3\mu_r + 2)\mu} \tag{4-29}$$

where μ_r is the relative viscosity of the particle (or droplet) with respect to the liquid (i.e., $\mu_r = \mu_p/\mu_l$, with μ_p and μ_l representing particle and liquid viscosities, respectively), $\Delta\rho$ is density differential ($= \rho_p - \rho_l$), and λ is the size ratio of the two particles ($\lambda = a_1/a_2$, where a_1 and a_2 are radii of the two particles in relative motion prior to collision). For solid particles in a liquid, $\mu_r \to \infty$, and the relative velocity becomes

$$u_R = \frac{2\Delta\rho a_1^2(1 - \lambda^2)g}{9\mu}. \tag{4-30}$$

Thus, a large density mismatch, and a small mismatch of sizes (large λ) will lead to large relative velocity and greater probability for collisions. External shear and other local disturbances in the fluid could also induce collisions. For example, the propagation of ultrasound through a suspension can induce relative motion between differently sized particles. Finer particles respond better to the inducing frequency and vibrate with greater amplitude than larger particles; as a result, differently sized particles will collide during propagation of sonic wave.

Agglomeration due to capillary forces between partially submerged particles is important in steel refining and mineral beneficiation processes. It was stated in reference to Figure 4-11 that fluid meniscus around a partially submerged particle is deformed as the particle attains an

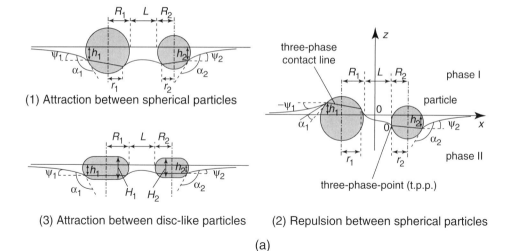

(1) Attraction between spherical particles

(3) Attraction between disc-like particles

(2) Repulsion between spherical particles

(a)

FIGURE 4-12 *(a) Schematic diagram showing capillary meniscus around two approaching particles and capillary interactions between them. The larger the meniscus deformation, the stronger will be the interaction between particles. If sin $\psi_1 \cdot sin \psi_2 > 0$, the particles will attract one another, and if sin$\psi_1 \cdot sin \psi_2 < 0$, the particles will repel one another. (From H. Yin, H. Shibata, T. Emi and J. S. Kim, Capillary attraction of solid particles at gas/steel melt interface, in Proceedings of International Conference on High Temperature Capillarity, 29 June – 2 July, 1997, Krakow, Poland, eds. N. Eustathopoulos and N. Sobczak, 380-387, Foundry Research Institute, Krakow, 1998).*

equilibrium position at the interface. The interface deformation from two closely-positioned particles creates an interaction force between the two; the larger the deformation, the stronger is the interaction force. The interaction force could be attractive or repulsive depending upon the sign (+ or −) of the slopes of deformed fluid interfaces (Figure 4-12a). For commonly encountered solid inclusions (e.g., alumina) in molten steel, the contact angle is obtuse ($\theta > 90°$), and the particle density is less than that of the steel. This leads to attractive capillary forces between the inclusions and a strong tendency to agglomerate.

Figure 4-12b shows the attractive force (estimated from in-situ observations in a confocal scanning laser microscope) between alumina particles in molten steel as a function of the factor $R_1^2 \times R_2^2$, where R_1 and R_2 are the radii of disc-shaped alumina particles. This figure shows a comparison of the experimental measurements and theoretical predictions based on capillary attraction between partially submerged alumina inclusions; the solid lines are the linear regression fit of the data. Whereas there is a large discrepancy in the magnitude of attractive force between theory and measurements, the slopes of the regressed theoretical and experimental lines are similar, which indicates that the observed attraction between alumina particles has common characteristics with the capillary interaction (which is assumed in the theory to be the fundamental cause of observed agglomeration of alumina inclusions in steels). The results also show that high particle density and large contact angles lead to stronger attractive capillary force. Similarly, attractive force is large when the surface tension of the liquid is small; a low surface tension can be obtained through alloying. Strong capillary attraction and agglomeration is also observed for other common impurities in molten steel, such as CaO and SiO_2 particles.

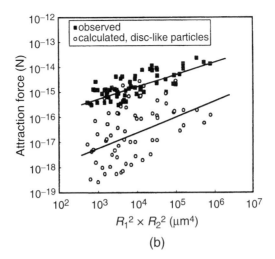

(b)

FIGURE 4-12 *(b) Comparison of experimentally measured and theoretically calculated attractive force between disc-like alumina inclusions in molten steel plotted as a function of the product* $R_1^2 \times R_2^2$, *where R denotes the particle radius.* (From H. Yin, H. Shibata, T. Emi and J. S. Kim, Capillary attraction of solid particles at gas/steel melt interface, in Proceedings of International Conference on High Temperature Capillarity, 29 June – 2 July, 1997, Krakow, Poland, eds. N. Eustathopoulos and N. Sobczak, 380-387, Foundry Research Institute, Krakow, 1998).

Capillary Flow

Capillary Pressure

The absorption of a liquid in fine capillaries and pores is encountered in numerous materials processing operations. A liquid is spontaneously wicked through the pores of a wettable solid. It is a common observation that if a glass capillary of fine bore is immersed in water, water spontaneously rises in the capillary to an equilibrium height. In contrast, if a glass capillary is immersed in liquid mercury, the level of Hg is depressed, and the meniscus acquires a convex curvature. In a wettable system (water/glass), interfacial tensions cause the liquid to spontaneously rise in the capillary, whereas in a non-wettable system (Hg/glass), an external pressure in excess of atmospheric pressure will be needed to initiate the rise. The pressure difference across the liquid–vapor interface is the capillary *pressure* P_c that drives (or opposes) the movement of the liquid in the capillary. This pressure depends on the liquid's surface tension, σ_{lv}, capillary radius, r, and the contact angle, θ, and is given from the Young-Laplace equation,

$$P_c = \frac{-2\,\sigma_{lv}\cos\theta}{r}. \tag{4-31}$$

For non-wettable solids, $\theta > 90°$, $\cos\theta < 0$, and $P_c > 0$, which indicates that a positive external pressure is needed to initiate the flow. Conversely, for wettable solids, $\theta < 90°$, $\cos\theta > 0$, and $P_c < 0$, and the liquid rises spontaneously because of a negative capillary pressure. In real solids with interconnected networks of irregular pores of different sizes, geometrical corrections

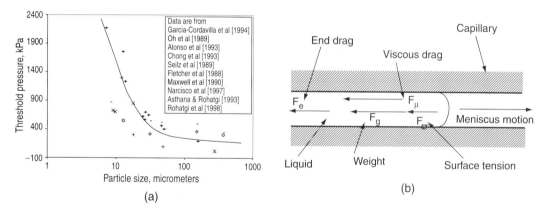

FIGURE 4-13 *(a) Literature data on the effect of particle size on the threshold pressure for infiltration of oxide and carbide ceramics by molten Al, Ag, and Zn. (R. Asthana, M. Singh and N. Sobczak, in Proc. 29th Int. Conf. on Adv. Ceramics and Composites, 2005, 249–261, American Ceramic Society). (b) Schematic representation of the forces acting on a fluid rising in a cylindrical capillary.*

are introduced to account for the pore shape, and the expression for the capillary pressure becomes

$$P_c = \frac{-(1-\phi)\rho A \sigma_{lv} \cos\theta}{\phi}, \qquad (4\text{-}32)$$

where ϕ is porosity, ρ is the solid's density, and A is the specific surface area of the solid (i.e., surface area per unit mass). In particulate beds, the effective pore size, r, in the Young-Laplace equation is taken to be the hydraulic mean diameter of the particles. Experimental measurements of capillary pressure in liquids are consistent with the inverse dependence of the pressure on the pore size given from the Young-Laplace equation. Figure 4-13a shows the magnitude of minimum external pressure, P_{th}, needed to initiate the flow of molten metals through packed beds of ceramic particulates as a function of the average particle diameter. The pressure P_{th} is called the threshold pressure and is a measure of the capillary pressure opposing metal ingress through the packed bed. Figure 4-13a shows that P_{th} increases with decreasing particle size and is related to the effective pore size.

Both the liquid surface tension, σ_{lv}, and contact angle, θ (and, therefore, P_{th}) depend upon temperature, atmosphere, and liquid composition. Experiments show that P_{th} for metal infiltration through porous ceramics decreases with increasing temperature, which is a consequence of the improved wettability at elevated temperatures. However, high temperatures are beneficial only under a "clean" atmosphere in which the wettability-inhibiting surface oxides on molten metals are unstable. In fact, the experimentally measured P_{th} is frequently found to be nearly the same for different types of ceramics such as SiC, TiC, and Al_2O_3 when they are infiltrated at a constant temperature by molten Al in air. This suggests that the contact angle of these ceramics with Al is the same, which is contrary to the sessile-drop measurements of contact angle at similar temperatures that show that these particulates form different contact angles with Al under controlled test environment. Inert and/or reducing atmospheres (Ar+H_2) yield low values of P_{th}. Thus, surface oxides obstruct metal ingress even at high temperatures under oxidizing atmospheres, and to realize the benefits of improved wettability at high temperatures, it is necessary to utilize an oxygen-deficient atmosphere. Many composite growth techniques utilize controlled atmosphere and wetting-promoters (e.g., Mg) to reduce the oxidation of molten metals and create large negative capillary pressures to cause self-infiltration ("wicking"). In most cases, a critical level of Mg is needed.

The equilibrium (or static) height, h_{eq}, attained by a liquid in a straight vertical capillary made out of a wettable solid is obtained from the balance of forces due to surface tension and the weight of the liquid column of height h_{eq}, i.e., $2\pi r \cdot \sigma_{lv} \cdot \cos\theta = \pi r^2 \rho \cdot g \cdot h_{eq}$, where ρ is the density of the metal. On rearranging the preceding equation, the equilibrium height is obtained as

$$h_{eq} = \frac{2\sigma_{lv}\cos\theta}{rg\rho}. \tag{4-33}$$

For non-wetting solids, $h_{eq} < 0$, indicating depression of the meniscus below the reference level ($h = 0$). For a non-wetting porous solid with liquid on top of the solid, an equilibrium penetration distance may be reached; in this case, flow is driven by gravity, and opposed by surface tension forces. In contrast, for a wetting solid with the liquid on top, both surface tension and gravity act in the same direction, and there will be no limiting length (either the solid will become fully saturated with the liquid, or the liquid will be exhausted).

Capillary Rise

Two fundamentally different approaches are used to model the rise of liquids in capillaries and pores: Darcy's equation, which yields an averaged flow behavior, and fluid physics-based models that solve the equation of fluid motion in a single pore or capillary. For steady-state unidirectional flow, Darcy's equation is $u = -\frac{\kappa}{\mu}\frac{dp}{dx}$, where u is the average velocity of the liquid, p is the pressure, μ is the dynamic viscosity of the fluid, x is the thickness of the porous body, and κ is the permeability of the porous medium that is intimately related to void fraction and pore size distribution. Empirical relationships to predict the permeability of packed beds of particulates and fiber bundles, including geometrical correction factors for shape deviation from perfectly spherical, have been derived. However, the complexity of real porous solids renders theoretical predictions somewhat difficult, with the result that κ is frequently determined with the aid of Darcy's law in conjunction with experimental measurements of fluid velocity and pressure gradient. This renders the entire approach somewhat circuitous. The second theoretical approach develops an equation of fluid motion within a single idealized pore in terms of the various forces acting on the fluid, and is discussed below. For counter-gravitational flow of a wettable liquid in a straight cylindrical capillary of radius R shown in Figure 4-13b, the fluid motion can be described by the following differential equation that includes surface tension, viscous drag, end drag, and gravitational forces

$$2\pi R\sigma_{lv}\cos\theta - \pi R^2\rho gh - 8\pi\mu h\frac{dh}{dt} - 0.25\pi R^2\rho\left(\frac{dh}{dt}\right)^2 = \frac{d}{dt}\left[\pi R^2 h\frac{dh}{dt}\right] \tag{4-34}$$

where σ_{lv} is surface tension, ρ is density of melt, θ is (equilibrium) contact angle, μ is viscosity, g is the acceleration due to gravity and h is the penetration distance. On the left-hand side of Equation 4-34, the first term is the force due to surface tension, the second term is the instantaneous weight of the liquid column, the third term is the fluid (Poiseuille's fluid drag), and the fourth term is the end drag (fluid friction at the entrance due to change of flow direction).

The sum of these forces is the net rate of change of the momentum of the liquid, which is the term on the right-hand side of Equation 4-34. The preceding equation can be rewritten as

$$h\left(\frac{d^2h}{dt^2}\right) + 1.25\left(\frac{dh}{dt}\right)^2 + ah\left(\frac{dh}{dt}\right) + gh = b \tag{4-35}$$

where

$$a = \frac{8\mu}{\rho r^2}$$

and

$$b = \frac{2\sigma_{lv}\cos\theta}{\rho r}.$$

Equation 4-35 is a non-linear differential equation that may be solved numerically subject to the initial conditions

$$h = \left(\frac{dh}{dt}\right) = 0; \quad t = 0.$$

There are, however, a few limiting cases of practical interest that permit analytical solution. For example, if the acceleration term is small, then the following solution to Equation 4-35 is obtained

$$h^2 = \left[\frac{r\sigma_{lv}\cos\theta}{2\mu}\right]t - \left[\frac{\rho ghr^2}{4\mu}\right]t - \frac{\rho^2 r^4}{32\mu^2}\left[\exp\left(\frac{-8\mu t}{\rho r^2}\right) - 1\right] \tag{4-36}$$

At slow rate of rise and negligible end drag, an asymptotic solution, or the Washburn equation, is obtained, which represents the long-time, steady-state solution. This equation has the form

$$h^2 = \left(\frac{\rho r^2}{4\mu}\right)\left[\frac{2\sigma_{lv}\cos\theta}{\rho r} - gh\right]t \tag{4-37}$$

If the total penetration length, h, is much smaller than the height, h_{eq}, to attain hydrostatic equilibrium (i.e., $h \ll h_{eq}$, where $h_{eq} = 2\,\sigma_{lv}\cos\theta/\rho gr$), then

$$h\frac{dh}{dt} = \frac{r^2}{4\mu}\left(\frac{\sigma_{lv}\cos\theta}{r}\right) \tag{4-38}$$

and the solution becomes a particularly simple parabolic expression

$$h^2 = \left(\frac{r\sigma_{lv}\cos\theta}{2\mu}\right)t \tag{4-39}$$

Equation 4-39 shows that penetration length depends mainly on the parameter $(r\sigma_{lv}\cos\theta/2\mu)$; the larger the value of this parameter, the greater the penetration distance is in a given time. Thus, the simplest capillary-rise model yields a parabolic increase in penetration distance h

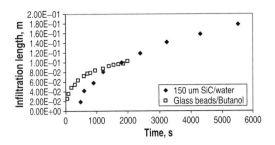

FIGURE 4-14 *Infiltration length as a function of time for countergravitational ascent of water in SiC and of butanol in glass beads at room temperature.*

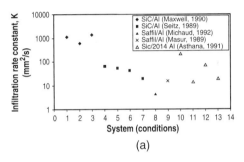

(a)

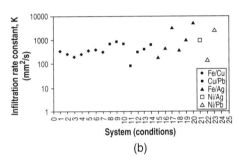

(b)

FIGURE 4-15 *(a) Estimated infiltration rate constant ($r\sigma_{lv}\cos\theta/2\mu$) for SiC-Al and saffil-Al systems based on the literature data. (b) Estimated infiltration rate constant ($r\sigma_{lv}\cos\theta/2\mu$) for some powder metallurgy systems based on the literature data. Numbers on x-axis indicate that data are for different test conditions. These conditions are, however, not identified for visual clarity.*

with time, t. The parabolic solution has been applied to infiltration of a wide variety of aqueous, organic, and metallic systems. Figure 4-14 presents examples of systems that exhibit a roughly parabolic dependence of penetration length on time: flow of water through packed SiC beds and flow of butanol through glass beads. In addition, experimental data on the kinetics of molten metal penetration in porous sintered metals (Cu, Ag, and Pb in Fe; Ag, Sn, and Pb in Cu; and Ag in Ni), and in porous ceramics (e.g., Al in TiC) have been fitted to a parabolic equation. The experimental capillary rise kinetics of molten metals in porous ceramics and metals are measured using interrupted infiltration tests, weight gain using thermogravimetry, dynamic methods that utilize implanted electrodes to track the liquid front in the porous solid, and non-invasive methods that measure a change in a physical property (e.g., capacitance) with the progression of the liquid front in the pores. Experiments also show that flow kinetics of metals in porous solids are influenced by a large number of variables, which include pore size distribution, pore shape, surface tension, contact angle, viscosity, density, magnitude of applied pressure (if used), rate of pressurization, the temperatures of the liquid and the solid, compositions, and gas atmosphere.

Equation 4-39 for parabolic infiltration kinetics is of the form $h^2 = kt$, which suggests that the parameter ($r\sigma_{lv}\cos\theta/2\mu$) can be considered as an infiltration rate constant, k. Figure 4-15 shows approximate range of measured infiltration rate constants for some metal–metal and ceramic–metal systems (in nonwetting ceramics, external pressure is needed to drive the flow,

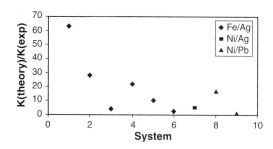

FIGURE 4-16 *Ratio of theoretical and experimental infiltration rate constants for Fe-Ag, Ni-Ag, and Ni-Pb systems.* (Data are from V. N. Eremenko and L. D. Lesnik, The Role of Surface Phenomena in Metallurgy, ed. V. N. Eremenko, Consultant Bureau, New York, English Translation, p. 102).

and the parabolic infiltration rate equation is modified to account for the pressure). For SiC-Al and Al_2O_3 (Saffil)-Al composites, k, is roughly 4.4–1428 mm^2·s^{-1}. Wettable coatings increase the k; for example, Cu coatings on SiC increase the rate constant for infiltration by Al from 72.3 to 213.2 mm^2·s^{-1}. The value of k for powder metal parts is generally in the range 174–812 mm^2·s^{-1}.

In a large number of high-temperature solid–liquid systems (e.g., molten metals in porous metals or ceramics), the rate constant, k, varies with temperature according to the Arrhenius relationship $k = A \exp(-Q/RT)$, where Q is the activation energy for capillary rise. The flow is thus thermally activated, and the value of Q depends on the dominant transport mechanisms, which are often difficult to identify. A representative value of Q for SiC-Al is 58.8 kJ·mol^{-1}, for TiC-Al it is 105.0–445 kJ·mol^{-1} (for 1–25 μm pores), and for dense and porous mullite infiltrated with Al, the values of Q are 70 and 185 kJ·mol^{-1}, respectively. Transitions in mechanisms that drive the flow occur with changes in temperature and/or pore radius (e.g., due to exothermic reactions, pore shrinkage, or expansion due to product phase deposition or ceramic dissolution). For example, the counter-gravitational rise of molten Al in porous TiC is limited by atomic diffusion at small capillary radius, r, and by interface reaction kinetics at large r. This is because with increasing r, the liquid volume in the capillary increases faster than the solid's surface area, which triggers reaction control and slower infiltration due to reduced supply of reactive species by the solid. However, at small r shallow solute gradients develop and flow begins to be limited by diffusion. These situations may exist during infiltration in reactive (e.g., ceramic-metal) systems, and may cause discrepancies with respect to the simplified theoretical model for inert liquids presented above. The flow behavior in reactive systems is discussed in Chapter 6 in the context of reactive synthesis of ceramic and metal composites.

Figure 4-16 shows the ratio (k_{th}/k_{exp}) of the theoretical and experimental infiltration rate constants for molten metal infiltration of sintered powder metal compacts, which is done to close the residual porosity with a low-melting point liquid metal. Figure 4-16 shows that the ratio (k_{th}/k_{exp}) varies from less than 10 to over 60. Quantitative agreement between theory and experiment is, therefore, not achieved in many cases of practical interest. Discrepancies may arise because of both hydrodynamic and interfacial factors, which include irreversible energy losses due to sudden expansion and contraction of pores in a powder bed, unsteady multidirectional flow in contrast to the unidirectional flow assumed in the theoretical models, uncertainties in representing the pore size by an averaged or effective pore radius, and transient capillary forces due to temporal evolution of the contact angle.

Refinements of the basic capillary rise model have been developed to overcome some of these deficiencies. For example, capillary penetration problem has been analyzed by relaxing the assumption of steady-state. The outcomes indicate that the Washburn equation (Equation 4-37) is not applicable to penetration at very rapid rates. Other analyses for unsteady flow of an incompressible liquid through porous solids start with the appropriate form of the Navier-Stokes equation and account for the skin friction effect. For example, the following differential equation for one-dimensional penetration of a vertical capillary has been derived for the unsteady laminar flow with skin friction

$$1 = X \left[1 + 8 \left(\frac{dX}{dt} + \frac{4\Omega}{3} \frac{d^2X}{dt^2} - \frac{\Omega}{144} \frac{d^3X}{dt^3} + \cdots \right) \right] \tag{4-40}$$

where X is the dimensionless penetration length and the parameter Ω is a temporal Reynolds number given from

$$\Omega = \frac{r^5 \rho_m^3 g^2}{2\mu^2 \sigma_{lv} \cos\theta}.$$

When the inertial effects are ignored, Equation 4-40 reduces to the Washburn equation in dimensionless form.

Another refinement of the capillary penetration model considers nonequilibrium or time-dependent contact angles. If the contact angle does not stabilize at a constant value during the time period of capillary rise, then θ varies with time, and the capillary rise kinetics can no longer be expressed by Equations 4-36 through 4-39. This will be true for liquids that exhibit rapid penetration and slow approach to equilibrium wetting. For example, for some polymeric liquids,

$$\cos\theta = \cos\theta_0 \left[1 - \exp\left(-\frac{t}{t_0} \right) \right], \tag{4-41}$$

where θ_0 is the equilibrium value of θ, and t_0 is a reference time. Analyses similar to the one presented previously have been developed to deduce the capillary penetration kinetics for an unstable contact angle during capillary rise. Figure 4-17 schematically illustrates the situation. Penetration is retarded when the contact angle decays from an initially large (acute) value to a smaller value during the time of rise in the capillary. Figure 4-18 compares the experimental

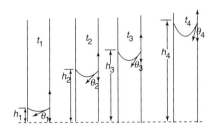

FIGURE 4-17 *Schematic diagram showing capillary rise in a wettable system with a time-dependent contact angle.*

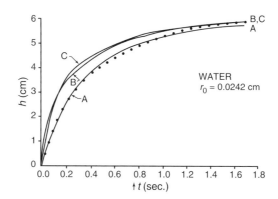

FIGURE 4-18 *Comparison of the experimental and theoretical capillary rise kinetics of water in a glass capillary (radius: 242 μm). Curve A incorporates unsteady contact angle given by Equation 4-4 in the calculations.* (From G. L. Batten, Jr., Liquid imbibition in capillaries and packed beds Journal of Colloid and Interface Science, 102(2), 1984, 513–518). Reprinted with permission from Elsevier. The theoretical models compared with the data are: curve A—Batten, curve B—Szekeley et al., and curve C—Letelier et al. (J. Szekeley, A. W. Neumann and Y. K. Chuang, 'The rate of capillary penetration and the applicability of the Washburn equation,' Journal of Colloid and Interface Science, 35(2), 1971, 273–278; M. F. Letelier, H. J. Leutheusser, and C. Rosas Z., Journal of Colloid and Interface Science, 72(3), 1979, 465–470).

kinetics of capillary rise of water in glass capillaries ($r = 0.0242$ cm) with the theoretical models based on constant as well as unsteady contact angles. This figure shows that a better agreement between theory and experiments is achieved when the effect of transient contact angles is considered in the capillary rise model. Another case of practical interest, sometimes encountered in reactive systems, is capillary rise with concurrently changing contact angle and capillary radius. This is discussed in a later section.

Wettability and Capillary Rise at High Temperatures

Effect of Oxide on Liquid

Most metal and ceramic systems are processed and/or used at temperatures at which chemical reactions begin to influence the spreading and flow behaviors. For example, most metals at high temperatures show a strong tendency to oxidize, and therefore oxygen strongly influences the infiltration and spreading behavior of metals. Solids may dissolve in metals, release oxygen, form reaction layers, and modify the metal and substrate chemistry. Thus, attainment of the contact angle equilibrium between the solid and the liquid could become difficult in high-temperature systems if the liquid–vapor boundary is in contact with a reactive atmosphere and becomes covered with a solid chemical reaction product (e.g., oxide film) that opposes the forces driving the spreading process toward equilibrium. If the instantaneous angle of contact of liquid droplets far from the Young equilibrium is θ, then the driving force causing the system to approach equilibrium is $F = \sigma_{sv} - \sigma_{sl} - \sigma_{lv} \cos \theta = \sigma_{lv}(\cos \theta_e - \cos \theta)$, where θ_e is the equilibrium angle. If the equilibrium angle is $0°$, then the driving force $F = \sigma_{lv}(1 - \cos \theta)$; this suggests that the driving force for spreading and approach to equilibrium decreases as the contact angle decreases. If a reaction product film envelops the liquid–vapor interface, the driving force will be opposed by the film, resulting in a large value of the contact angle. Residual oxygen in

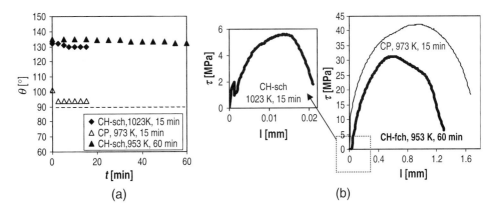

FIGURE 4-19 *Effect of test procedure on (a) contact angle, and (b) shear behavior of Al-polycrystalline Al$_2$O$_3$ couples. CH, contact heating (sch, slow contact heating; fch, fast contact heating); CP, capillary purification.* (N. Sobczak and R. Asthana, Ceramics Transactions, 158, 2005, 3–17, American Ceramic Society).

the atmosphere will inhibit the spreading by forming an oxide skin on the liquid front, which will hinder true contact. This will influence not only the θ, but also the structure and adhesion at the interface.

The effect of oxide films on wettability has been demonstrated with the help of a number of techniques. For example, the effect of oxide is revealed in sessile-drop tests that use a capillary purification (CP) technique to erode the oxides by extruding tiny droplets of metals through a graphite syringe just prior to their contact with the substrate under high vacuum (or inert atmosphere). Figure 4-19 compares the contact angle, θ, and interfacial shear stress, τ, of Al/Al$_2$O$_3$ sessile-drop couples made using two different wettability test procedures: contact heating (CH), in which the metal and substrate are jointly heated to the test temperature, and the capillary purification (CP) technique described previously. The data presented in Figure 4-19 clearly demonstrate that oxide removal even at low temperature decreases the θ and increases the τ. When the oxide is eroded using CP (or when very low oxygen partial pressures, p_{O2}, are used), acute θ ($\leq 90°$) is obtained even at low temperatures, thus permitting unhindered spreading. At high temperatures, the destruction of the oxide under vacuum is known to lower the θ. It has been shown that oxide destruction is due to (1) the formation of the volatile suboxide, Al$_2$O via the reaction: 4Al(l) + Al$_2$O$_3$(s) $\rightarrow$ 3Al$_2$O(g), and (2) partial dissolution of Al$_2$O$_3$ skin in metal drop. Figure 4-20 shows the oxygen partial pressure versus temperature relationship for the dissociation of the suboxide, Al$_2$O; high temperatures and low oxygen partial pressures favor the dissociation of this suboxide. The positive effects of oxide removal (or low p_{O2}) could mask the negative effects of a low wettability test temperature; thus, wettability may significantly improve in the absence of surface oxides even at temperatures at which complete non-wetting is observed in a given system (in most systems, contact angle decreases and wettability improves with increasing temperature). Experimental observations show that surface oxides can diminish the wettability (and joint strength) by obstructing the wettability-enhancing reactions in systems that are inherently reactive, but the oxide removal *per se* does not improve the wettability in non-reactive systems that are non-wettable.

At high temperatures, the incubation period for oxide dissociation is small, so very low contact angles may be achieved even in a short time. The equilibrium oxygen partial pressures

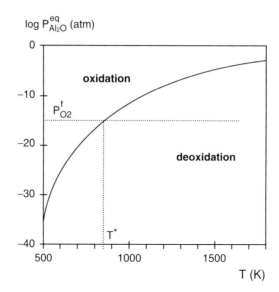

FIGURE 4-20 *Equilibrium pressure–temperature relationship for the formation of Al_2O.* (N. Eustathopoulos, M. G. Nicholas, and B. Drevet, Wettability at High Temperatures, Pergamon, New York, 1998). Reprinted with permission from Elsevier.

(e.g., $p(O_2)^{eq} < 10^{-32}$ atm at 1100°C) for Al oxidation are much lower than the physically realizable pressures in a wetting test or during capillary rise under controlled atmosphere. Therefore, the only way to establish a physical contact between the metal and the solid is through the dynamic processes involving unavoidable transport of oxygen atoms toward the metal and continual removal of oxygen as volatile AlO_x species. Calculations show that the oxide film on Al will be eroded at $T > 1100$°C under partial pressures of oxygen $< 10^{-10}$ atm.

Alloying and Surface Coatings

Alloying elements may alter the interfacial properties and the capillary forces driving the flow, and either accelerate or retard the flow. For example, wettable surface coatings and surface active alloying elements reduce the pressure needed to overcome the capillary forces resisting the flow in a non-wettable system. Mg in Al decreases the θ due to oxide disruption and lowering of σ_{lv}, and it also reduces the threshold pressure for infiltration by Al. In contrast, Cu as a solute in Al slightly increases the P_{th} but Cu in the form of coating lowers the P_{th} and permits greater infiltration lengths to be achieved in the SiC/Al composite, as shown in Figure 4-21. A similar behavior is observed with Ti in Al/Al_2O_3; Ti alloying does not improve the wetting by Al because no wettable reaction products form, but Ti coatings lower the θ of Al with alumina because of metal-to-metal contact during flow. Unlike Ti and Cu in Al, which do not improve the wettability as alloying additives, Cr is beneficial as both a coating and alloying element in Ni/Al_2O_3 and in Cu/C. With Si in Al, the effect of oxide removal is less dramatic, but Si slightly improves the wettability and reduces the P_{th}. Si aids the dissolution of Al_2O_3, causing interfacial roughening; during cooling, Si precipitates (along with reaction-formed Al_2O_3 crystallites) strengthens the ceramic/metal joint. Other wettable coatings (Ni, Cr, K_2ZrF_6, etc.) also generally lower the θ and P_{th} in ceramic/metal systems but the mechanisms of wettability improvement may be varied. Surface coatings of the compound K_2ZrF_6 disrupt the oxide, which is partially dissolved

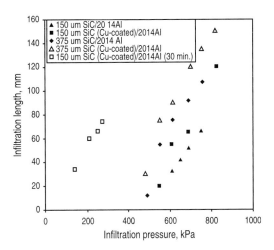

FIGURE 4-21 *The effect of copper coating on the infiltration-length-versus-pressure relationship for the infiltration of SiC platelets with molten 2014 Al alloy.* (Data are from R. Asthana and P. K. Rohatgi, Zeitschrift für Metallkunde, 83(12), 1992, 887–892).

by the fluoride species in the coating. The P_{th} decreases when K_2ZrF_6 is used but does not drop to zero. Larger infiltration lengths are attained with K_2ZrF_6 coatings on ceramics than with uncoated ceramics. In the case of molten Ni in contact with Al_2O_3, vapor-deposited Cr films on Al_2O_3 slightly decrease the contact angle, θ, of both polycrystalline (PC) and single-crystal (SC) Al_2O_3 with Ni. The θ decreases from 108° to 94° for $Al_2O_3^{PC}$ and from 100° to 98° for $Al_2O_3^{SC}$; in neither case, however, is good wetting ($\theta < 90°$) achieved. The sessile-drop wettability tests on Ni/Cr-coated Al_2O_3 (Figure 4-22) show three distinct regions around the Ni-droplet: an outer region of discontinuous Cr-film, an intermediate region composed of tiny droplets that are Cr-rich near the outer regions and Cr-depleted near the inner region, and a narrow region free of Cr nearest the drop. In this example, Ni was also present far from the sessile drop because of its evaporation and deposition and its surface diffusion. Thus, significant chemical and structural changes may occur near the liquid front because of coating or alloying and modify the solid's surface. These changes alter the spreading behavior of a liquid and the capillary rise phenomenon.

Reactive Infiltration

In reactive infiltration, the reinforcing phase forms via a chemical reaction between a liquid propagating through a porous solid. The reaction is usually exothermic (e.g., infiltration of silicon in porous carbon, and infiltration of aluminum in TiO_2, mullite [$3Al_2O_3 \cdot 2SiO_2$] and in Ni-coated alumina). The infiltration conditions can be controlled to achieve the desired level of conversion and structure. The reactive infiltration of Si in porous C forms reacted SiC phase and unreacted C in a Si matrix. Usually, preexisting SiC grains are employed as inert filler in the porous C preform to permit heterogeneous nucleation and bonding of SiC. The porous perform is made by pyrolyzing a high-char polymer precursor material, and some control on pore size, volume fraction, and morphology is possible. Reactive infiltration can also produce two or more phases via chemical reactions with a suitable choice of the precursor materials. The rate of capillary rise in reactive infiltration has been measured using thermocouples embedded in the porous solid. For example, the flow of Si through porous carbon has been tracked by recording

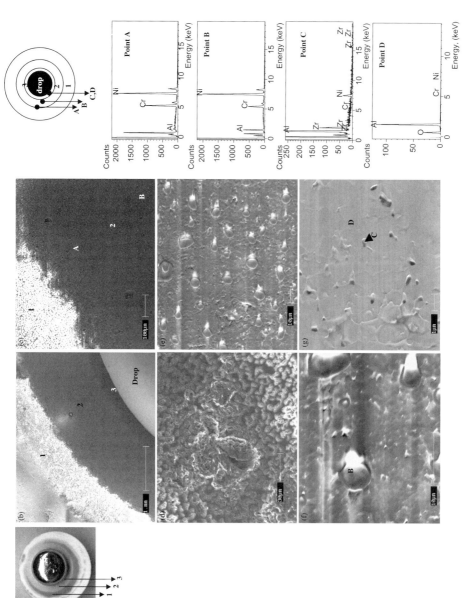

FIGURE 4-22 A composite figure showing the three concentric ring-shaped regions that formed around Ni in contact with a Cr-coated polycrystalline alumina substrate. The EDS spectra at points A, B, C, and D in these regions and the SEM views of the regions are also shown. The outer region 1 is Cr-film-covered substrate; the film is discontinuous due to heating. Region 2 (~1.5–2 mm wide) contains numerous tiny droplets; the droplets are Cr-rich (point A) near the outer areas of region 2, and are depleted in Cr (point B) near the inner area. Region 3 (~0.2 mm. wide) is free of Cr film; very small amounts of Ni and Cr exist in this region (points C and D). The small holes in the intergranular areas in region 3 are rich in Zr (from the sintering aid ZrO$_2$). (N. Sobczak, K. Nogi, H. Fuji, T. Matsumoto, Joining of Advanced & specialty Materials V, J. E. Indacochea et al (eds.), 2003, 108–115, ASM International).

the temperature rise via embedded thermocouples; a strong exothermic reaction between carbon and silicon causes temperature spikes. The peak temperatures decrease with increasing height of capillary rise. This indicates a progressive decrease in the infiltration front velocity with increasing distance, which is consistent with the behavior predicted by the parabolic solution to the capillary rise model (Equation 4-39). The numerical discrepancy between the capillary rise model and the measured velocity of Si in porous carbon is, however, nearly two orders of magnitude. This clearly suggests that the capillary flow model must be modified to include the reaction kinetics that appear to be driving the flow in a reactive system such as Si-C. It is thus necessary to first understand the mechanisms and kinetics of reactions in the solid–liquid system under consideration.

In the case of carbon-silicon, two basic reaction mechanisms have been proposed. In the first, a thin film of SiC forms on the carbon surface, through which Si and C diffuse (possibly along the grain boundaries in SiC), and cause continued reaction and conversion to SiC. The very large (nearly 58%) volumetric expansion upon conversion of C to SiC causes the freshly formed SiC layer to spall off, thereby exposing fresh carbon to Si. This exposure is believed to result in the dissolution of C in Si, saturation of Si with C, and reprecipitation of SiC, followed by SiC grain coarsening via a competitive dissolution and growth process (Ostwald ripening). Thus, in the first mechanism, the reaction kinetics are either controlled by the inward diffusion of Si through the reaction-formed SiC layer, or by the outward diffusion of carbon through the SiC layer. In the second mechanism proposed for the reaction of C and Si (i.e., the dissolution-reprecipitation mechanism), the dissolution of C in Si, and diffusion to cooler regions in the melt is believed to be responsible for the precipitation of SiC, which may subsequently coarsen with time. Analyses of the Si/C reaction kinetics data reveal that Si diffusion through the SiC layer is the more likely reaction mechanism operative during the infiltration of porous carbon by molten Si.

Reactive Penetration

Reactive penetration is distinguished from reactive infiltration, and it refers to metal ingress in a dense solid (e.g., single crystals, fully dense polycrystals, amorphous ceramics, etc.) which are devoid of pre-existing porosity. In such solids, the liquid advance is controlled by the movement of the reaction front, which reconstructs the contacting interface and permits penetration by the metal. Chemical reactions and phase transformations cause volumetric changes, which may lead to microcracks in the product phase followed by melt permeation and continued reaction. In both infiltration and penetration, chemical reactions form an interpenetrating network of metal and ceramic crystals (e.g., Al and Al_2O_3) reminiscent of a eutectic-type microstructure. For example, such structures form in single crystal ZnO (ZnO^{SC}) where there are no grain boundaries, and in amorphous SiO_2 and fully-dense mullite and fly ash when they are in contact with molten Al. It is interesting to note that reactive penetration occurs (and an interpenetrating Al-Al_2O_3 structure forms) in dense ceramics like ZnO, which are not wet by Al, but Al does not spontaneously infiltrate porous TiO_2 and ZrO_2 even though these ceramics are wetted by Al. Similar anomalies have also been observed in porous ceramics. For example, when porous carbon is in contact with Cu-Ti or Cu-Cr alloys, TiC and Cr_3C_2 form, respectively. It is observed that the TiC reaction layer in Cu-Ti/C does not prevent melt penetration of porous carbon (even though $\theta > 90°$), but Cr_3C_2 reaction layer in Cu-Cr/C hinders melt impregnation even though the measured contact angle, $\theta < 90°$. Most observations of reactive penetration are based on the sessile-drop wettability tests in which metal penetration of the underlying substrate is observed. Substrate cracking due to chemical transformations and secondary oxidation of the metal front may increase the resistance to penetration, although the behavior of each solid-liquid system is highly unique.

Reactive penetration of dense (and amorphous) fly ash by Al forms fine Al_2O_3 crystals while concurrently enriching molten Al with Si and Fe released from the dissolution of fly ash in Al. In a broader sense, the solid/liquid interactions may not always lead to the creation of a new product phase, and new crystals of a preexisting phase may nucleate and grow by a dissolution-re-precipitation process. For example, fine Al_2O_3 crystals form in non-reactive Al/Al_2O_3 system at the solid–liquid interface underneath sessile drops. The new Al_2O_3 crystals at the solid–liquid interface form by the dissolution of the Al_2O_3 substrate, saturation of the melt with O, and re-precipitation of new Al_2O_3 crystals that grow at the drop-side of the interface in an epitaxial manner (i.e., in specific crystallographic orientation relationship with respect to the underlying solid). In addition, O diffusion along the S/L interface may also aid the formation of these crystals. These micro-crystalline alumina precipitates strengthen the joint.

Reactive penetration in Al/TiO_2 and Al/mullite occurs in a manner similar to Al/fly ash. Detailed microstructural examination shows that in sessile drops of Al on TiO_2 at T ≥ 1173 K, new Al_2O_3 crystals form at the solid–liquid interface surrounded by Ti-rich Al. On the drop-side of the interface, large Al_2O_3 crystals form by dissolution-precipitation, and on the substrate side, very fine Al_2O_3 crystals, surrounded by an Al impregnated region, form. At even higher temperatures (≥ 1373 K), Al wets, reacts, and penetrates the TiO_2, forming a continuous network of Al_2O_3 crystals interpenetrated by a continuous Al network. The governing reaction is $4Al(l) + 3TiO_2 \rightarrow 2Al_2O_3 + 3Ti$. The penetration layer in Al/TiO_2 attains a thickness of a few micrometers (e.g., 2–8 μm at 1273 K in 30 min. contact). TiO_2 reacts with Al to form Ti aluminides via the reaction $TiO_2 + Al \rightarrow Al_2O_3 + Al_xTi_y$, where the type of aluminide (e.g., Ti_3Al, TiAl, $TiAl_3$) formed depends upon the TiO_2/Al ratio. Thus, wettability improvement in Al/TiO_2 is because of the formation of a favorable interface structure from these chemical reactions.

The penetration behavior in Al/SiO_2 is also reaction-assisted, promoted by the reaction $4Al(l) + 3SiO_2 \rightarrow 2Al_2O_3 + 3Si$. This reaction is more intense than the reaction of Al with TiO_2 and ZrO_2. The reaction product layer plays an important role in aiding or impeding the spreading and flow. Sessile-drop tests show that cracking of SiO_2 near the triple line (TL) (because of expansion accompanying $SiO_2 \rightarrow Al_2O_3$ transformation) does not hamper the propagation of the reaction product layer, which is able to advance beyond the TL. Alloying Al with Si does not change the preceding reaction between Al(l) and SiO_2 but reduces the reactivity and leads to poor wetting. In contrast, with Ti alloying of Al, the contact angle in the SiO_2/Al couples rapidly decreases, and Si (from the preceding chemical reaction) reacts with Ti to form titanium silicide ($TiSi_2$) via the reaction $Ti + 2Si \rightarrow TiSi_2$. Thus, spreading and penetration are accompanied by complex chemical reactions and microstructure changes.

Capillary Flow with Unsteady Contact Angle and Pore Size

In reactive infiltration of a porous solid, the pore size may continually change because of specific volume changes caused by the chemical reactions and product phase deposition within the pores (Figure 4-23). In addition, the contact angle, θ, may exhibit a protracted time dependence and stabilize at its equilibrium value, θ_∞, in times that scale with the time of infiltration. Both time-dependent pore size [$R(t)$] and contact angle [$\theta(t)$] will modulate the capillary forces that drive the flow. In a large number of solid-liquid systems, the experimentally measured contact angle, θ, decays with time according to the relationship $\theta(t) - \theta_\infty = (\theta_0 - \theta_\infty) \exp(-t/\tau)$, where θ_0 is the contact angle at zero time, and τ is a characteristic time of the system. In the context of solid-liquid reactions during capillary flow, two limiting situations can be envisioned: reaction limited by interface processes (attachment kinetics) and by diffusion processes. Under the simplifying assumption of a planar reaction front, it can be shown that for diffusion- and interface-controlled reactions, respectively, parabolic [$R(t) = R_0 - m\sqrt{t}$] and linear [$R(t) = R_0 - kt$] reaction

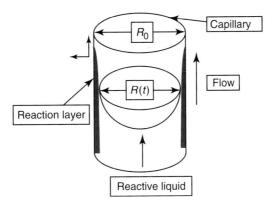

FIGURE 4-23 *Schematic of reactive infiltration of a cylindrical capillary with shrinking radius due to product-phase deposition within the capillary.*

kinetics will be obtained, where m and k are the respective reaction rate constants for diffusion- and interface control and R_0 is the initial pore radius. The preceding mathematical functions describing the temporal evolution of the contact angle and the pore radius can be incorporated in a model of capillary rise (e.g., Washburn equation) to derive the isothermal kinetics of capillary rise under the limiting cases of diffusion- and interface-control. Such a calculation has been performed, and representative computational results for the reactive carbon-silicon system are presented in the following paragraph.

During infiltration of porous carbon by molten Si, SiC formation causes volumetric expansion, and results in pore shrinkage. The experimental $\theta - t$ data for Si-C can be fitted to an exponential form, and the baseline values of the reaction rate constants may be taken approximately to be $k \approx 4 \times 10^{-8}$ m·s^{-1} and $m \approx 2 \times 10^{-7}$ m·s$^{-1/2}$ (these are only representative values, as they strongly depend upon the reaction conditions and temperature). The computational results presented in Figure 4-24 show that for Si in porous C ($R_0 = 10$ µm), the Washburn equation overestimates the flow kinetics. This is in agreement with the experiments in which the measured velocity of Si in C is found to be smaller than the Washburn velocity. Calculations also show that for both interface- and diffusion-limited flow, penetration lengths are larger when both θ and R vary with time than when R is allowed to decrease but θ (and the capillary pressure at the liquid

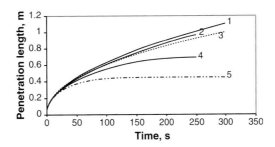

FIGURE 4-24 *Theoretical infiltration profiles for Si infiltration of a carbon capillary ($R_0 = 10$ µm). Curve 1, Washburn equation; curve 2, diffusion-limited rise with decreasing R and θ; curve 3, diffusion-limited rise with decreasing R but constant θ; curve 4, interface-limited rise with decreasing R and θ; and curve 5, interface-limited rise with decreasing R but constant θ. The values of parabolic reaction rate constant, m (linear rate constant, k), is taken to be 4×10^{-8} m·s$^{-1/2}$ (m·s^{-1}).* (R. Asthana, Metallurgical & Materials Transactions, 33A, 2002, 2119–2128).

front) is constrained to remain constant. The reaction-induced shrinking of pores will cause pore closure, flow cessation, and attainment of a limiting length by the liquid. The limiting lengths (typically, a few mm in the case of Si in porous carbon) increase with R_0 and are larger when both θ and R vary than when only R decreases but θ remains constant (Figure 4.25a). In both

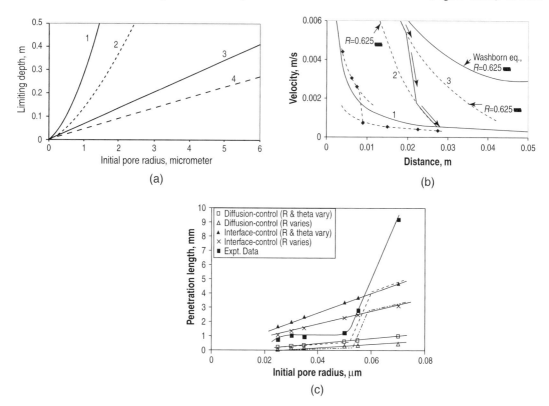

(a)

(b)

(c)

FIGURE 4-25 (a) Theoretical limiting length versus initial capillary radius (R_0) for the Si infiltration of porous carbon. Curve 1, diffusion-limited rise with decreasing R and θ, curve 2, diffusion-limited rise with decreasing R but constant θ, curve 3, interface-limited rise with decreasing R and θ, and curve 4, interface-limited rise with decreasing R but constant θ. The values of parabolic reaction rate constant, m (linear rate constant, k) are taken to be 4×10^{-8} $m \cdot s^{-1/2}$ ($m \cdot s^{-1}$). (R. Asthana, Metallurgical & Materials Transactions, 33A, 2002, 2119–2128). (b) Infiltration velocity of molten Si in porous carbon performs as a function of the infiltration distance, h. Also shown are theoretical curves based on various models: curve 1: De Gennes model, curve 2: diffusion-control with a variable θ, and curve 3: interface-control with a variable θ. The dashed curves show a possible path for transition from diffusion-driven flow to reaction-driven flow with increasing infiltration distance. (The experimental data are from P. Sangsuwan, S. N. Tewari, J. E. Gatica, M. Singh and R. Dickerson, 'Reactive infiltration of silicon melt through microporous amorphous carbon preforms,' Metallurgical and Materials Transactions, 30B, 1999, 933–944). (R. Asthana, J. Materials Processing & Manufacturing Science, 10(4), 2003, 187–197). (c) Infiltration distance of molten Si in porous carbon performs as a function of the initial pore radius. Also shown are the theoretical curves based on various approximations. (The experimental data are from P. Sangsuwan, S. N. Tewari, J. E. Gatica, M. Singh and R. Dickerson, 'Reactive infiltration of silicon melt through microporous amorphous carbon preforms,' Metallurgical and Materials Transactions, 30B, 1999, 933-944, and from J. G. Li and H. Hausner, Infiltration of carbon substrates by molten silicon, Scripta Metall. Mater., 32(3), 1995, 377). (R. Asthana, M. Singh and N. Sobczak, in Proc. of the 29th International Conference on Advanced Ceramics and Composites, 2005, 249–261, American Ceramic Society).

cases, diffusion-limited flow is faster than interface-limited flow, as would be expected from the slower product-phase growth under diffusion control. Figure 4-25b compares the theoretical predictions of infiltration velocity as a function of infiltration distance with the experimental measurements obtained from embedded thermocouples in Si-C system. The measured velocity of Si in porous C (median pore diameter $\sim$1.25 μm) exhibits an abrupt drop of $\sim$73 pct. over an infiltration distance, h, of 0.007–0.009 m. This drop is presumably a result of the transition in the reaction mechanism driving the flow; at short distances, flow is limited by interface reactions and at large distances, flow is limited by the diffusion of Si and C through the SiC product phase. The theoretical curves for diffusion-controlled and interface-controlled infiltration in Figure 4-25b qualitatively mimic the observed drop in the velocity even though the numerical magnitude of the discrepancy is quite large. Figure 4-25c shows the maximum penetration distance of molten Si in porous carbon as a function of the initial effective pore radius. This figure also includes the Si penetration data of stationary (sessile) drops of Si over carbon substrates. The data for pore radii less than 0.05 μm are for sessile drops of Si in contact with carbon. These measurements represent the effect of a finite liquid reservoir (droplet) on the maximum penetration distance, a situation that is quite different from conventional infiltration (represented by the rest of the experimental data in Figure 4-25c), in which virtually an inexhaustible supply of the liquid is in contact with the porous substrate.

Many factors render quantitative predictions uncertain and cause the large discrepancy noted from Figure 4-25b between theory and experiments. For example, accurate values of the reaction rate constants, m and k, are sometimes difficult to obtain for the infiltration conditions realized in a particular experiment. In addition, the exothermic effects of reactions, ignored in the calculations, can strongly influence the infiltration. In the case of Si in carbon, the temperature is observed to rise by 390° to 740° above the melting point of Si, which will alter the melt properties and the infiltration behavior. The contact angle, θ, can deviate from an exponential function that is used to describe its time dependence in the calculations, and it may actually depend on the test conditions (oxygen partial pressure, temperature, alloying, etc.). The compositional changes due to reactions during capillary flow will alter the melt properties and diffusion coefficients, and the morphological features and defect structure (micro-voids, grain boundaries, etc.) of the product phase layer may influence the extent of chemical attack via short circuit paths. A complete model to describe the complex capillary flow phenomena in reactive systems seems to be lacking at present.

Joining

A large number of manufacturing operations involve joining of materials using liquid phases as an intermediate step. For example, brazing and soldering are widely used joining techniques that distribute a liquid filler metal in the joint region by capillary forces. Upon solidification, the selected filler metal must meet the design requirements related to the strength, corrosion resistance, electrical conductivity, etc. Table 4-4 summarizes the compositions and physical and mechanical properties of some commercial brazes used in joining industry.

Metals and alloys are brazed using a wide variety of fillers. Ceramics are also joined using brazes, although diffusion bonding of ceramics is also used (see Chapter 3). Whereas brazing temperatures are higher than those used in diffusion bonding (which involves plastic flow and solid-state diffusion of the filler metal), the higher joining pressures in the latter may cause mechanical degradation of the ceramic assembly. Brazing is, therefore, more widely used than diffusion bonding. Ceramics (and ceramic-to-metal joints) are brazed by melting a thin metal

TABLE 4-4 Properties of Some Commercial Braze Alloys

Braze	Comp., %	T_L, K	T_S, K	E, GPa	Yield Strength, MPa	Ultimate Tensile Strength, MPa	Coefficient of Thermal Expansion, $\times 10^{-6}$ K^{-1}	Thermal Cond., W/m·°K	%Elong.	Elect. Cond., $\times 10^6 \Omega^{-1} \cdot m^{-1}$
TiCuSil	68.8Ag-26.7Cu-4.5Ti	1173	1053	85	292	339	18.5	219	28	29
TiCuNi	15Cu-15Ni-70Ti	1233	1183	144	—	—	20.3	—	—	—
Cu-ABA	92.8Cu-3Si-2Al-2.25Ti	1297	1231	96	279	520	19.5	38	42	5.1
PalCo	65Pd-35Co	1492	1492	—	341	661	—	35	43	4.6
PalNi	60Pd-40Ni	1511	1511	—	772	978	15.0	42	23	5.4
PalCuSil10	59Ag-31Cu-10Pd	1125	1097	—	327	374	18.5	145	18	18.9
PalCuSil15	65Ag-20Cu-15Pd	1173	1123	—	379	448	—	98	23	13.0
Gold ABA	97.5 Au-0.8Ni-1.75 V	1303	1276		209		16.1	25	29	
Gold ABA-V	96.4Au-3Ni-0.6Ti	1363	1318		143		17.3	8	31	

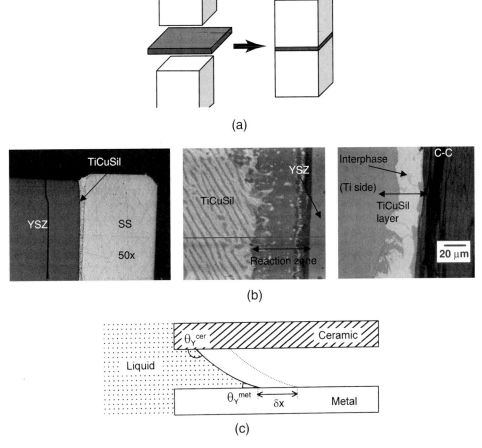

(a)

(b)

(c)

FIGURE 4-26 *(a) Schematic of ceramic joining using metallic braze foil, and (b) actual joint microstructures in stainless steel joined to yttria-stabilized zirconia (YSZ), and carbon-carbon composite joined to Ti. In both joints, a commercial Ag-Cu-Ti braze (TiCuSil) was used* (M. Singh, T. P. Shpargel, and R. Asthana, Proc. of the 29th International Conference on Advanced Ceramics and Composites, 2005, 383–390, American Ceramic Society. M. Singh, T. P. Shpargel, G. N. Morscher and R. Asthana, Materials Science & Engineering, 2005). *(c) Braze penetration in the gap between a ceramic and a metal substrate. The curvature of the fluid meniscus and the spreading rate are controlled by the braze contact angle on the ceramic and the metal.* (N. Eustathopoulos, M. G. Nicholas, and B. Drevet, Wettability at High Temperatures, Pergamon, New York, 1998).

interlayer and allowing it to uniformly spread under capillary forces (Figure 4-26a). Unfortunately, the high surface tension of liquid metals and their large contact angles on ceramics hinder braze spreading, thus reducing the contact area and leaving unhealed interfacial imperfections that weaken the joint by concentrating the stress. For example, for the common braze base metals such as Ni, Ag, and Cu, the surface tension, σ_{lv}, values are large; σ_{lv} for Ni is 1796 mN/m (1455°C), for Ag it is 925 mN/m (960°C), and for Cu it is 1330 mN/m (1085°C), respectively. Similarly, the large contact angles between the braze metal and non-wetting ceramic substrates reduce the joint contact area. Figure 4-26b shows that the different contact angles between the

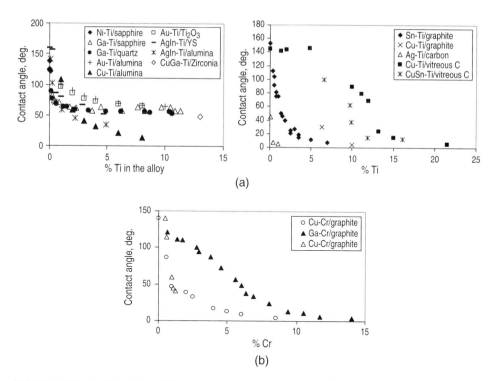

FIGURE 4-27 *(a) The effect of Ti content on the contact angle of various Ti-containing alloys on oxide ceramics and carbon. (b) The effect of Cr content in Cr-containing alloys on their contact angle with carbon.*

liquid filler and the ceramic and the metal substrates at the joint region cause uneven flow of the filler in the gap, increasing the likelihood of unhealed interfacial defects. Including a reaction between the braze and the ceramic by suitable alloying improves the adhesion as shown in the photomicrographs of Figure 4-26b.

Reactive additives in the braze (e.g., Ti in Cu-Ag braze) cause the liquid braze to flow over the ceramic surface (usually a reconstructed surface composed of wettable chemical phases, e.g., Ti_xO_y in the case of oxide ceramics, and TiC and TiN in the case of carbides and nitrides). Figure 4-27 shows the effect of selected reactive solutes (Ti and Cr) in molten metals on their contact angles with oxide and carbide ceramics. Even a small amount of the reactive additive can significantly decrease the contact angle because of chemical interactions that lead to beneficial interfacial changes. For example, Ti additions lead to the formation of various titanium oxides at an oxide/metal interface; usually, a complex, multi-phase interface forms. Thus, TiO, Ti_2O, and Ti_2O_3 reaction layers form at the interface in yttria-stabilized zirconia (YSZ) joined to itself using an Ag-Cu-Ti braze alloy, and TiO, Ti_3O, and Ti_6O layers form at the interface in CuGa-13 at %Ti/zirconia joints. Some of these titanium oxides are better wetted by metals than others; for example, TiO is better wetted than Ti_2O_3, which is strongly ionic. The extent of wetting and adhesion is characterized by the Work of Adhesion, W_{ad}, defined from $W_{ad} = \sigma_{lv}(1 + \cos\theta)$, where σ_{lv} is the surface tension of the liquid metal, and θ is the contact angle. Generally, a high value of W_{ad} indicates chemical bonding, which results in good wetting and adhesion (provided interfacial defects, thermal stresses, and interfacial plastic flow are negligible). The values of W_{ad} for joints between Cu (base metal in a Cu-braze) and Ti oxides are 740 mJ·m^{-2} (Cu/Ti_2O_3), 1460 mJ·m^{-2} ($Cu/TiO_{1.14}$), and 1650 mJ·m^{-2} ($Cu/TiO_{0.86}$), respectively; these values suggest a progressive improvement in wettability, with the sub-stoichiometric oxide, $TiO_{0.86}$, yielding the best wetting. This is a general behavior; i.e., metals that form chemical bonds exhibit higher values of W_{ad} than metals that form purely physical bonds with a substrate. For example, wetting of zirconia by Ag, Cu, Ni, and Co involves both physical (London-van der Waals) forces and chemical bonds. The wetting of zirconia by these metals is better than that by the alloying additions such as In, Sn, Bi, and Pb that are used in low-temperature brazes. All of these solutes (i.e., In, Sn, Bi, and Pb) form weak physical bonds and yield low Work of Adhesion (100–200 mJ·m^{-2}), whereas Ag, Cu, Ni, and Co form stronger chemical bonds. The values of the Work of Adhesion (in mJ·m^{-2}) for zirconia in contact with Ag, Cu, Ni, and Co are 430, 650, 900, and 950, respectively. These metals form chemical bonds between atoms of the liquid metal, Me, and the ion O^{2-} of the surface oxide; however, a new compound (oxide) is not formed.

Titanium as an additive in braze is also known to partially scavenge O atoms from oxide ceramics such as YSZ, leaving behind O-vacancies in the sub-stoichiometric zirconia, which becomes more surface-active and wettable by metals. Thus, moderate-to-strong improvements in wettability may occur just by the dissolution of an oxide ceramic in molten braze even if reaction product layers do not form at the joining temperature. Other systems that exhibit improvement in wetting without a product layer formation include Cu-Ti/Y_2O_3, and Sn-Ti/Al_2O_3. The practical benefit of some type of chemical interaction between the substrate and the liquid braze, and the resulting improvements in wetting and flow characteristics, is the formation of a strong joint.

The wetting of brazes containing reactive additives on ceramics is generally excellent, although the kinetics of wetting are usually sluggish (as is actually the case with most metallic liquids in contact with ceramics). For example, hot stage microscopy of the spreading of Ag-38Cu-3.7Ti drops on Al_2O_3 at $950°$ C has shown that the average spreading rate is $\sim 6 \times 10^{-4}$ mm per second. The radius, $R(t)$, of the contact region between the oxide and the molten braze increases with time, t, according to the empirical relationship $R(t) = R_0 \cdot \exp{(t/3\tau)}$, where R_0 is the initial contact radius (e.g., the contact radius of the solid metal in contact with the substrate), and τ is a characteristic time constant (~ 10 min. for the Ag-38Cu-3.7Ti braze). This relationship shows that after 5 min contact at $950°C$, an initial contact radius of 5 mm will increase by 18% and the contact area by 39%, whereas after 10 min, the radius will increase by 40% and the contact area by 95%. The joining conditions (e.g., brazing time and temperature) can be selected to achieved the desired level of coverage (or penetration) and reaction.

In the brazing literature, a Wettability Index (WI) defined from WI = (area covered by the braze metal) x cos θ is often used as a measure of the spreading and flow behavior. Higher the value of WI, the better is the wetting and spreading of the braze alloy. WI is used for a relative assessment of braze spreading behavior, and depends upon the volume of the braze metal at the joint. Generally, WI > 0.05 indicates good spreading, and WI > 0.10 indicates excellent spreading characteristics in a braze. The WI of Au, Ag, and Pd braze alloys in contact with stainless steels (316 and 304 grades) are given in Table 4-5 at different joining temperatures. These data show that the wetting of gold on steels ranges from good to excellent, even in the absence of reactive additives. The WI of Ag-Cu-Ni brazes on 316 and 304 stainless steels indicates that their wetting is moderate to good. Palladium is another braze base metal (and sometimes a constituent in Ag braze alloys) for high use-temperatures. It is a noble metal with good oxidation resistance and good ductility. These characteristics aid in preserving the integrity and stability of the joint under harsh conditions, and in accommodating thermal stresses via interfacial plastic flow in joints. Commercial Pd brazes include Palcusil, Palco, and Palni, and their WI with steels are also given in Table 4-5.

TABLE 4-5   Wetting Index of Some Commercial Braze Alloys in Contact with Stainless Steels

Braze	Temperature, K	Wettability Index (WI)	
		316 SS	304 SS
Gold	1343	–	0.088
Gold	1348	0.087	–
Gold	1373	0.358	0.238
Gold	1423	–	0.355
Palni (60Pd-40Ni)	1548	0.078	–
Palco (65Pd-35Co)	1548	0.073	–
Palcusil-10 (59Ag-31Cu-10Pd)	1123	0.035	0.015
62.5Ag-32.5Cu-5Ni	1123	0.024	–
62.5Ag-32.5Cu-5Ni	1173	0.038	0.026
62.5Ag-32.5Cu-5Ni	1223	0.062	0.057
62.5Ag-32.5Cu-5Ni	1273	–	–
75Ag-24.5Cu-0.5Ni	1073	0.007	–
75Ag-24.5Cu-0.5Ni	1123	0.029	–
75Ag-24.5Cu-0.5Ni	1173	0.039	–
75Ag-24.5Cu-0.5Ni	1223	–	0.029
75Ag-24.5Cu-0.5Ni	1273	–	0.082
77Ag-21Cu-2Ni	1123	0.027	0.001
77Ag-21Cu-2Ni	1173	0.045	0.014
77Ag-21Cu-2Ni	1223	0.063	0.045
77Ag-21Cu-2Ni	1273	–	0.09

Soldering is identical to brazing except the melting temperature of the filler metal in soldering is less than 450°C. Soldering is widely used in the electronics field. For example, low-temperature solders, such as the Sn-Pb eutectic solders in contact with Cu, Ag, and other conductive substrates, are used in electronic packaging applications. Similar to brazing, the spreading and wetting of solders on metals determines the integrity of the joint. Most metal–metal couples used in soldering applications exhibit reaction-limited wetting phenomenon in which chemical reactions control the liquid's rate of spreading. For example, in the case of solid Cu in contact with molten Sn-Pb solder, a reaction band composed of intermetallic compounds Cu_6Sn_5 and Cu_3Sn forms at the contact line. This band acts as a diffusion barrier between Cu and the solder cap and limits the growth of the compound layer and spreading of the filler metal. The spreading of the Sn-Pb solder on Cu consists of distinct stages. The first stage is the very rapid spreading under the surface tension forces at the solid–liquid, liquid–vapor, and solid–vapor interfaces and is described by $\sigma_{lv}\cos\theta + \sigma_{ls} < \sigma_{sv}$. No reaction products form in this stage because of the incubation time needed to initiate the reaction. In the second stage of spreading, the underlying solid substrate dissolves in the solder cap, forming intermetallic compounds, and replacing a single solder–substrate interface by at least two interfaces. In the third and final stage, a reaction band forms around the droplet. The spreading rate is now limited by the diffusion-controlled growth of the reaction band and its wettability with the eutectic solder. Growth could occur by surface, bulk, and grain boundary diffusion as well as by vapor transport; however, the dominant mechanism will be temperature and system dependent.

Another example that illustrates the role of capillarity in joining is the reactive infiltration process that is used to join technical ceramics and ceramic-matrix composites; in particular, a variety of silicon carbide based ceramics (e.g., reaction-bonded and sintered). Figure 4-28a and b shows some joint configurations in SiC-based ceramic-matrix composites produced by reactive infiltration. The basic joining steps include the application of a carbonaceous mixture in the joint area and curing at 110–120°C for 10–20 min. Silicon in paste form is applied in the joint region and heated to 1425°C for 5–10 min. The molten Si reacts with carbon to form silicon carbide with controllable amounts of residual silicon at the joint. Figure 4-28c and d shows the interface regions of reaction-bonded (RB)-SiC and sintered SiC joints produced by Si infiltration of a carbonaceous mixture that was applied at the mating surface of the joints. The thermomechanical properties of the joint interlayer can be tailored close to those of the silicon

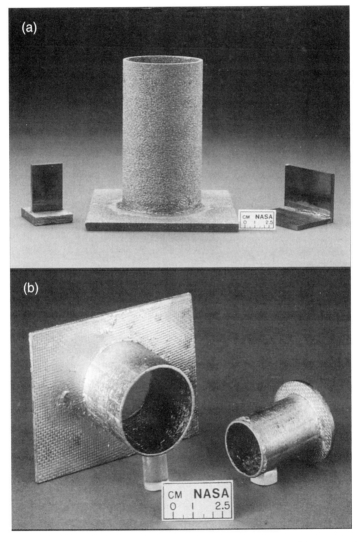

FIGURE 4-28 *(a) and (b) Photographs showing components fabricated from joined silicon carbide subelements.* Photo Courtesy of M. Singh, QSS Group Inc. NASA Glenn Research Center, Cleveland, OH.

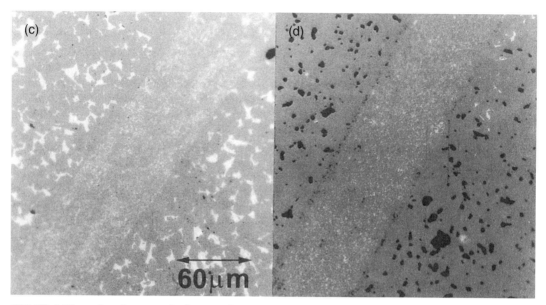

FIGURE 4-28 continued *(c) and (d) Optical micrographs of a SiC–SiC joint formed via a silicon infiltration process: (left) reaction-bonded SiC, and (right) sintered SiC. The white and gray areas are Si and SiC, respectively.* (J. M. Fernandez, A. Munoz, F. M. Varela-Feria, and M. Singh, Journal of the European Ceramic Society, 20, 2000, p. 2641). Reprinted with permission from Elsevier. Photo Courtesy of M. Singh, QSS Group, Inc., NASA Glenn Research Center, Cleveland, OH.

carbide-based matrix materials, and high-temperature fixturing is not needed to hold the parts at the high temperatures required for infiltration. Self-infiltration of the porous carbon interlayer in the joint region by molten Si occurs because of a rapid drop in the contact angle between Si and C to near zero. This decrease in the contact angle results from a strong carbide-forming exothermic reaction, which gives rise to a negative capillary pressure and wicking of Si through the carbonaceous interlayer. Upon solidification, a strong joint with little structural discontinuity is created.

Adhesion

Adhesion between materials is caused by interactions at surfaces, which may be purely physical (dispersion forces), chemical (adsorption, dissolution, compound layer formation), mechanical (residual stresses), or frictional (mechanical "keying" or interlocking of surface asperities). Purely physical adhesion because of, for example, electrostatic forces may develop in insulators or dielectrics, such as plastics, ceramics, and glasses. The electrostatic force is caused by electron transfer and polarization of surfaces and usually leads to relatively weak adhesion. In many systems, adhesion is caused by the solubility of chemical species, which forms a solution (or an alloy) via atomic diffusion. In joining of polymers, the long chain molecules of the adhesive interdiffuse and entangle with one another, leading to adhesion. In the case of metal–oxide joints created by combining liquid metals and non-reactive solid oxides, the joint strength will depend upon the extent of dissolution of O atoms in the liquid metal; adhesion will be low (and wetting will be poor) if very low levels of O (from the oxide substrate) dissolve in the metal. For non-reactive metal-oxide couples with more dissolved O in the metal from the substrate, the Me-O clusters are adsorbed at the interface, and yield low θ and stronger adhesion. The lowest

θ and strongest adhesion are achieved in reactive Me–oxide couples that form reaction products at the interface with a metallic character. Systems that are wettable generally show stronger solid–solid adhesion than non-wettable systems.

Adhesion also depends on the universally present residual (or "frozen") stresses in solids. When a piece of metal is fractured, these stresses are partially relaxed; as a result, the fracture surface profiles slightly change and the broken halves will not exactly match and adhere if brought together. Furthermore, the fractured surfaces exposed to atmosphere will immediately adsorb impurities (e.g., water vapor, organic contaminants, etc.), and the adhesion strength between the two surfaces will further deteriorate. The influence of residual stresses on joint strength depends strongly on the joint configuration; the strength may either increase or decrease because of residual stresses. Finally, mechanical interlocking between microscopic asperities on mating surfaces may provide for purely frictional adhesion, and surface roughening by mechanical abrasion or chemical etching may produce strong joints. Roughening will increase the number and size of asperities, as well as the actual area of contact between surfaces.

The adhesive bond strength between two chemically inert solid surfaces can be estimated from the Stefan equation, according to which,

$$ft = \left(\frac{3}{4}\right) \mu a^2 \left(\frac{1}{h_1^2} - \frac{1}{h_2^2}\right), \tag{4-42}$$

where f is the stress to separate the surfaces, t is the time, μ is the viscosity of the fluid between the surfaces, a is the length of contact, h_1 is the initial gap width between the two surfaces when pressed together, and h_2 is final separation during which a fluid (e.g., an adhesive) has filled the gap. Thus, if two solid plates are mechanically pressed together, and only the top plate is held, the bottom plate does not fall off under its own weight because time is needed for the ambient air to enter and fill the narrow gap between the two plates. Also, because the final gap width, h_2, after separation will be much larger than the initial width, the Stefan equation can be simplified by neglecting the second term. The preceding equation then becomes

$$ft = \left(\frac{3\mu a^2}{4h_1^2}\right). \tag{4-43}$$

In the example of two juxtaposed plates, this equation shows that the time it would take for the lower plate to fall off (or, equivalently, the time for air to fill the slit) will be very large if the initial gap and the applied stress, f, are small (here, f is the weight of the lower plate divided by its nominal—not real—contact area). Furthermore, a large radius of the disc, a, and high fluid viscosity will also increase the separation time. In other words, if conditions are right, the plates will adhere effectively for a very long time (several hours to several months), or conversely, a very large stress will be needed to separate the plates in a short time (a few seconds). The adhesion between plates in the preceding example may be further increased by a film of a wetting liquid whose volume is less than the volume of the slit between the plates. In this case, a liquid meniscus will form across the liquid–air interface within the gap, and an (attractive) capillary pressure will develop between the plates. The magnitude of this pressure is given from the Young-Laplace equation

$$P_{\mathrm{c}} = \frac{-2 \, \sigma_{\mathrm{lv}} \cos \theta}{r},$$

where r is the radius of curvature of the liquid meniscus between plates. For a perfectly wetting liquid, $\theta = 0°$, and we obtain

$$P_c = -(2\,\sigma_{lv}/r).\tag{4-44}$$

Because r is usually extremely small (typically <0.1 μm), a very large negative capillary pressure will develop between the plates, which will adhere permanently.

The preceding discussion focused on physical processes responsible for adhesion between solids. In a wide variety of joining processes, chemical changes at the joint promote adhesion. For example, surface coatings and alloying additions can improve the wetting and bonding. Figure 4-29a shows an example of the effect of a reactive additive (titanium) in braze alloys on the fracture strength of joints between yttria-stabilized zirconia (YSZ) and steel, Ti and YSZ; larger Ti contents lead to stronger adhesion and higher fracture strength in the joints. Similar beneficial effects are achieved in nickel-alumina couples through chromium alloying of Ni. Alternatively, thin Cr films (<1 μm) lower the contact angle of Ni on single-crystal and polycrystalline Al_2O_3 substrates, and improve the adhesion compared to uncoated Al_2O_3 in contact with Ni. Chromium interlayers also help in reducing interfacial cracks in Ni–Al_2O_3 by acting as stress-absorbing compliant layers because their coefficient of thermal expansion (CTE) is intermediate between the CTE of Ni and Al_2O_3. Figure 4-29b and c display the contact angle versus shear strength data for joints between Ni and Al_2O_3, and between Al and Al_2O_3 under a variety of joining conditions. Generally, a high interfacial shear strength correlates with a low contact angle as noted from these data, although considerable dispersion exists because of different testing conditions and joining procedures used in different studies, differences in the alloy compositions, surface preparation, and the test atmosphere. Figure 4-29c includes data on solid-state (diffusion) bonding as well as liquid-phase (braze) bonding, both of which yield excellent adhesion in Ni/Al_2O_3. An approximate value of W_{ad} between solid Ni and solid Al_2O_3 is 645 mJ·m^{-2} at 1273 K under H_2 and 518 mJ·m^{-2} at 1673 K under Ar. In contrast, in liquid-phase processed Ni–sapphire joints, a significantly higher W_{ad} of 1100 mJ·m^{-2} is obtained, which seems to suggest that liquid-phase processing will likely yield a stronger interface than a solid-state joining process in this particular system. This may not, however, always be the case, because a myriad of extraneous factors (structural defects, etc.) could mask purely thermodynamic effects. Correlations between measured fracture strength of the joint and W_{ad} also have been established for metal/oxide couples, which confirm that a high work of adhesion correlates with a high fracture strength in both diffusion-bonded and liquid-phase processed joints.

Capillarity in Miscellaneous Other Processes

Until this point, practical manifestations of the capillary phenomena in some of the physical processes important to materials processing were described. There are numerous other situations of practical interest where capillarity is of central importance. Fabrics, woven yarn, and paper represent complex open capillary systems with interconnected and tortuous pores. Fabrics are treated with liquids in dyeing and waterproofing and the success of such treatments depends on the flow of liquid through the pores. For dyestuff impregnation, the liquid should wet and penetrate the fabric via "wicking" action, whereas in waterproofing a high contact angle (poor wetting) is needed to repel the liquid. Similar situations are encountered in the paper and pulp industry. For dye impregnation, low contact angles and high liquid surface tension are desired; however, high-surface-tension liquids also form large contact angles on solids so that the quantity

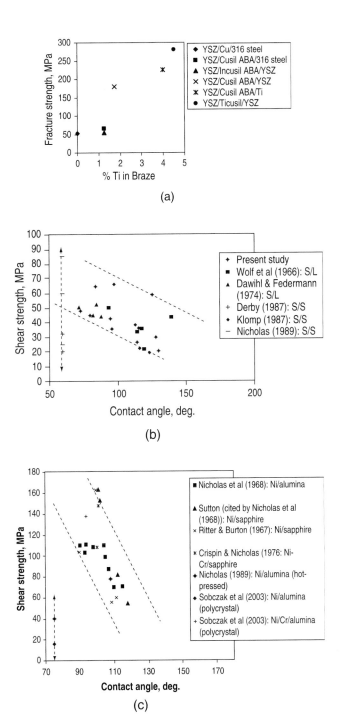

FIGURE 4-29 *(a) The fracture stress of joints between yttria-stabilized-zirconia (YSZ) and YSZ, steel and Ti as a function of percent Ti in the braze interlayer. (b) Data on shear strength of the Ni-alumina joint as a function of the contact angle from published studies.* (R. Asthana and N. Sobczak, Ceramics Transactions, 158, 2005, 3–17, American Ceramic Society) *(c) Data on shear strength of the Al-alumina joint as a function of the contact angle from published studies.* (N. Sobczak, K. Nogi, H. Fuji, T. Matsumoto, K. Tamaga and R. Asthana Joining of Advanced & Specialty Materials, J. E. Indacochea et al (eds.), 2003, 108–115, ASM International, Materials Park, OH)

$\sigma_{lv} \cos \theta$ remains essentially constant, and frequently it is difficult to independently change σ_{lv} and θ through addition of wetting agents or surfactants. Many surfactants are adsorbed on the surface of the textile fiber, and their concentration in the liquid is depleted, with the result that the σ_{lv} and θ are not favorably altered. In addition, small air bubbles could remain trapped in wedges and crevices on the fiber surface when a non-wetting liquid invades the fabric; extremely large pressures might be needed to expel these bubbles. The bubbles might, therefore, remain in the liquid and cause defects after the curing process is completed. Waterproofing is frequently done by applying a layer of a non-wetting material to the fabric or by applying a better wetting second material that would wick the liquid (water) from the fabric. For example, polyester-cotton clothing allows moisture to pass from the skin to the air but prevents the reverse process (atmospheric moisture, or water passing from the outside to the inner layers). This is done by using an inner layer of polyester (high contact angle with water) contacting a layer of cotton (low contact angle).

Detergency is used to remove a liquid (oil) or a solid (soil, grease) contaminant from a surface by using a second liquid (detergent). The liquid detergent should be able to penetrate between the impurity and the substrate via capillary action, and should cause the former to lift off and emulsify. Reattachment of the impurity is obstructed because the detergent molecules preferentially adsorb on the substrate surface. The capillary penetration of the cleaning solution requires that its contact angle with the substrate be smaller than that of the impurity–substrate interface. This is accomplished by formulating the composition of the detergent to include surfactants and other chemical cleaners.

Capillary and interfacial phenomena play an important role in the degradation of refractory in contact with corrosive liquids and hot metals. This aspect is important in crucible manufacture, coatings for die-casting dies, electrode linings in extraction and refining of metals, submerged refractory nozzles, and impellers. Refractory selection for such applications has a direct effect on product quality, scrap rates, and costs associated with frequent refractory replacement. Refractory and coating life can be increased by selecting compositions that are poorly wet by the liquid metals; this will minimize the contact area and corrosive attack. Refractory formulations include several non-wetting additives to reduce the spreading of liquid metals on the refractory.

Another capillary phenomenon important to a wide variety of materials processes is the flow of thin (<5 mm) liquid layers driven by surface tension gradients at the liquid surface. Such flows are called Marangoni flows and can result from temperature or composition changes in the liquid and other factors (e.g., electrical potential difference along the free surface of a liquid in an electric field). The liquid flow is always from a region of low surface tension to a region of high surface tension. As an example of Marangoni effect, consider the effect of sulfur on the surface tension of molten Fe. Minute amounts (40–50 ppm) of sulfur in Fe reduce the surface tension at 1650°C by about 25% and change the temperature coefficient of surface tension ($d\sigma_{lv}/dT$) from a negative value to a positive value. This affects the depth of weld penetration in welded steels; low values of surface tension, σ_{lv}, and a positive ($d\sigma_{lv}/dT$) cause deep welds, whereas large values of σ_{lv}, and a negative ($d\sigma_{lv}/dT$) cause shallow welds in steels. In a weld pool, the temperature in the center is higher than near the periphery, and because at high sulfur contents ($d\sigma_{lv}/dT$) > 0, σ_{lv} is greater in the center of the pool than near the periphery. Because Marangoni flow occurs from a region of low σ_{lv} to a region of high σ_{lv}, the liquid will move from the periphery to the center, where it will be driven downwards, thereby causing deeper melt back and a deeper weld fusion zone. The reverse will be true in weld pools of low S content in steels, and a shallow fusion zone will form. Marangoni flows also cause fluid convection in directional solidification of single crystals by the Czochralski method and the floating zone method that were described in Chapter 2. Such flows also enhance the degradation of refractories in furnaces where refractory dissolution causes local changes in the composition and surface tension near the slag line (i.e., near metal/slag or gas/slag interfaces), and these changes in σ_{lv} drive recirculatory flow, which intensifies refractory erosion.

Wear, Friction, and Lubrication

Up to this point, our discussion was centered on surface and interface phenomena in materials processes. Closely related to these are surface and near-surface phenomena, such as wear, friction, and lubrication, whose influence extends from the surface to subsurface regions in materials. Wear is defined as material loss under rolling or sliding motion between two contacting surfaces as, for example, in bearings, gears, piston rings, and brakes. Friction is the force that resists motion between two contacting surfaces. Most industrial applications require low wear and low friction, although low wear and high friction is required in brake pads, wrapping applications, and control of tension on belts. Failures that take place by wear and friction are gradual rather than catastrophic.

Figure 4-30a shows a plot of measured coefficient of friction as a function of surface energy of Cu sliding against Cu with or without very thin films of different lubricants (identified in the figure). Lubrication decreases the friction and the surface energy, which is essentially the adhesion energy of Cu to itself. Figure 4-30a shows that the higher the adhesion energy, the greater the measured coefficient of friction is. Similar relationships have been suggested between wear (e.g., wear particle size) of materials and surface energies. All such relationships establish the fact that surface energies and wear and friction behaviors are related in a very fundamental way.

Many different types of wear processes are encountered in real applications. These include adhesive wear (also called scouring, galling, seizure, or scuffing), abrasive wear, erosive wear, and fretting wear. Adhesive wear takes place by transference of material from one surface to another as a result of solid-phase welding during sliding contact. As all real surfaces have microscopic asperities (troughs and crests), an external load compressing two solid surfaces together is accommodated initially by elastic deformation of contact points. At large loads, plastic deformation occurs and leads to cold-welding at contact points. Under relative motion, the softer of the two surfaces transfers material to the harder surface, which leads to adhesive wear. Adhesive wear is characterized in terms of volume loss, V_{ad}, where $V_{ad} = ksL/H$, and s is the sliding distance, L is load, H is hardness of softer material, and k is the wear coefficient, which varies from 10^{-3} to 10^{-8} for common industrial materials The volume loss, V_{ad}, may be converted to linear penetration, e.g., increase in diameter of a bearing, or reduction in shaft diameter, which are directly measurable parameters.

As virgin metal-to-metal contact is rare in industrial practice, and all surfaces are covered by native oxide films as well as oxide films produced by frictional heating, the values of wear coefficient are essentially representative of oxide-to-oxide contact. Oxide films covering the surfaces must be disrupted before wear of underlying metal can begin. If the two metals are similar, the oxide films on both are disrupted. If the materials are different, then softer metal deforms more, i.e., oxide on softer metal breaks. If the oxide debris gets embedded in the softer metal, wear rate is decreased. Certain metals and alloys such as cast irons are less susceptible to seizure than other metals because cast irons contain graphite flakes that provide lubrication. Another example of in situ lubrication behavior is smearing of lead metal in a bearing alloy. Some

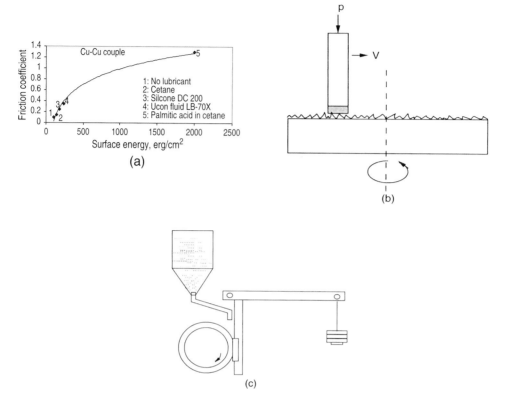

FIGURE 4-30 *(a) Plot showing measured coefficient of friction as a function of surface energy of Cu sliding against Cu with or without very thin films of different lubricants.* (Data are from E. Rabinowicz, in Fundamental Phenomena in the Materials Sciences, vol. 2, Surface Phenomena, L. J. Bonis and H. H. Hausner, eds., Plenum Press, New York, 1966). *(b) Schematic of the pin-on-disc wear test. (c) Schematic of the rubber-wheel-abrasion test (RWAT).*

metals tend to work harden during sliding motion (for example, austenitic manganese steels used in rock crushers), and although they would gall or seize in their soft condition, as they begin sliding together, they begin to harden on the surface and reduce tendency to cold-weld.

Abrasive wear is caused by plowing or scratching of a surface by motion of hard particles or protuberances on the surface. Two forms of abrasive wear are two-body wear and three-body wear. In two-body abrasive wear, abrasive particles are fixed (as in grinding wheels and sand paper), whereas in three-body wear, abrasive particles are loose (as in ball milling and lapping). The two-body wear leads to higher wear than three-body wear because in three-body wear, abrasive particles can roll instead of plowing through the softer material, thereby causing less wear. Figure 4-30b and c show standard tests for characterizing two-body wear (pin-on-disc test), and three-body wear (rubber wheel, abrasion test or RWAT). In a pin-on-disc test, the abrasive is attached to a rotating wheel, whereas in RWAT, abrasive is continuously introduced at the contact region between the test specimen and the rotating rubber wheel. Abrasive wear is characterized in terms of volume loss, V_{ab}, where $V_{ab} = K'P/H$, and K' is an empirical constant, P is the applied pressure, and H is the hardness of the surface being abraded. The most common mechanism of abrasive wear is delamination, which involves nucleation of subsurface cracks that propagate parallel to the surface for a short distance and then open up at the mating surface, thereby leading to formation of flaky wear debris.

Fretting wear is caused at two tight-fitting surfaces that are subjected to small-amplitude oscillations. Vibration amplitudes as small as one microinch can cause fretting wear; smaller amplitudes cause stresses that are absorbed by elastic deformation at contacting asperities and wear does not take place.

Erosive wear takes place on surfaces that are subjected to a flow of particles and gases that impinge on the metal at high velocities. Sand-blasting and abrasive water-jet cutting involve erosive wear. Other examples include airfoils, shrouds in fans, compressors and turbines, helicopter blades, and centrifugal pumps. The extent of material removal due to erosive wear depends on the abrasive type, base metal type, impact velocity, and angle of impact. Figure 4-31 shows the effect of impact angle on erosive wear of aluminum (ductile solid) and glass (brittle solid) caused by iron micro-spheres. The erosive wear of the ductile material reaches a maximum at an angle of about 20°, whereas erosive wear of brittle material reaches a maximum at normal incidence (90°).

Wear is reduced by use of liquid or semisolid lubricants (oils, grease) and solid lubricants (PTFE, graphite, MoS_2, WS_2, etc.). Lubricating action involves various mechanisms. In the case of liquid lubricants, the most widely encountered modes of lubrication are hydrodynamic, hydrostatic, and boundary lubrication, although other types (e.g., elastohydrodynamic) may be important in special situations. Hydrodynamic lubrication depends on the shape and relative motion of solid surfaces, which leads to formation of lubricant films having sufficient thickness and internal pressure to separate the two surfaces. In hydrostatic lubrication, a high external pressure is applied to fluid, which forces the two surfaces apart. Because fluid pressure is uniform

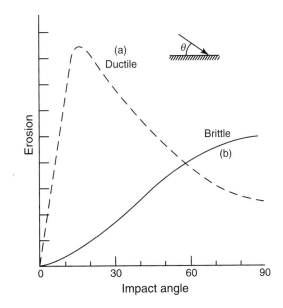

FIGURE 4-31 *Erosive wear of ductile and brittle materials as a function of the angle of impact.* (I. M. Hutchings, Tribology: Friction and Wear of Engineering Materials, CRC Press, Boca Raton, FL, 1999).

under external pressure, the film thickness is constant. Therefore, in a bearing shaft supported on a thick film of oil, the high friction and wear at startup and shutdown that characterize hydrodynamic lubrication are avoided. In boundary lubrication, very thin films at contacting asperities prevent cold-welding. For example, a single monolayer of stearic acid at asperities prevents adhesion. As the fluid film is extremely thin, frictional heat is not dissipated efficiently in boundary lubrication. Chemical additives in fluid film react with the metal to form soft solid products that prevent metal damage and adhesion. Because of high heat concentration at asperities, reaction rate is enhanced at contact points. Thus, in boundary lubrication, severe wear is replaced by mild corrosive wear by a judicious addition of chemical additives to lubricant.

The effect of material type on wear and friction can be illustrated by considering a classic solid lubricant, graphite. The hexagonal close-packed structure of carbon atoms in graphite is such that the basal planes are held together by weak dispersion forces; as a result, graphite is easily sheared parallel to its basal planes and serves as a good lubricant. The wear rate of highly oriented natural graphite rubbing on copper is shown in Figure 4-32. This figure shows that resistance to wear is very low (i.e., wear rate is high) when basal planes of graphite lattice are perpendicular to the copper surface. When basal planes are parallel to copper (i.e., along the direction of rubbing), wear resistance is dramatically decreased. The lubricating property of graphite depends on the presence of moisture. As the partial pressure of water in atmosphere increases, the wear rate rapidly decreases. Figure 4-33 shows friction coefficient of graphite as a function of test temperature; pure graphite becomes a poor solid lubricant at elevated temperatures because of loss of moisture. However, graphite doped with cadmium oxide has excellent antifriction properties even at high temperatures, as seen in Figure 4-33. The effect of temperature and atmosphere on the friction behavior of two other solid lubricants with layered structure (MoS_2 and WS_2) similar to graphite are shown in Figure 4-34. These materials do not depend on the presence of water vapor in atmosphere to attain good antifriction properties;

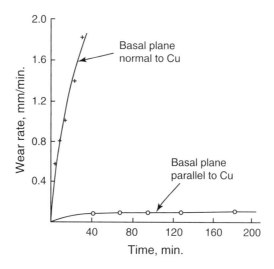

FIGURE 4-32 *Wear rate of oriented graphite rubbing against copper.* (Adapted from H. E. Sliney, Solid lubricant materials for high temperatures—a review, Tribology International, 1982, Butterworth, p. 303). Reprinted with permission from Elsevier.

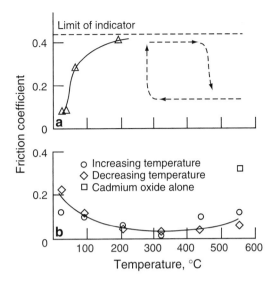

FIGURE 4-33 *Coefficient of friction of graphite as a function of test temperature: (a) graphite, and (b) graphite containing cadmium oxide.* (Adapted from H. E. Sliney, Solid lubricant materials for high temperatures—a review, Tribology International, 1982, Butterworth, p. 303). Reprinted with permission from Elsevier.

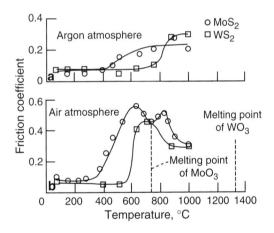

FIGURE 4-34 *Coefficient of friction of molybdenum disulfide and tungsten disulfide as a function of temperature under air and argon.* (Adapted from H. E. Sliney, Solid lubricant materials for high temperatures—a review, Tribology International, Butterworth, 1982, p. 303). Reprinted with permission from Elsevier.

however, the type of gas present in the atmosphere does affect their lubricity (Figure 4-34). Both materials retain their antifriction properties until higher temperatures in an inert argon atmosphere than in an air atmosphere. The rapid formation of surface oxides in air increases the friction coefficient even at relatively lower temperatures. Other layered materials such as mica also have a low friction coefficient because of low shear stress needed to slide the weakly bonded crystal planes; additionally, surface-adsorbed water molecules act as lubricant and further lower the friction coefficient of mica because the adsorbed H_2O is not extruded even at contact pressures of a few hundred megapascals.

Polyimide coatings are used in self-lubricating varnishes and as binders for inorganic solid lubricants. Polyimide coatings are good lubricants at temperatures in the range of 100–500°C, but they do not exhibit good lubricating property at lower temperatures (25–100°C). The low-temperature tribological properties of polyimide are improved by additives. For example, carbon fluoride coatings and MoS_2-bonded polyimide coatings have better wear and friction properties than pure polyimide; the inferior antifriction properties of polyimide at temperatures less than 100°C are totally masked by the addition of these additives to polyimide.

The wear and friction properties of solid lubricants depend on test conditions such as the load, velocity, temperature, atmosphere, and also film thickness. As the applied load increases at fixed velocity and film thickness, the coefficient of friction decreases; i.e., the film serves as a better lubricant at high loads. Also, under fixed test conditions a greater film thickness improves the antifriction properties (i.e., lowers the friction coefficient).

Hard materials wear less than softer materials, and the wear rate decreases as the hardness increases. This is also borne out by the fact that wear and friction properties can be actually improved by suitable alloying and microstructure control through processing and heat treatment. Thus, in the case of Al-Si alloys, as the amount of either silicon or a harder additive, SiC, increases, the wear decreases. In Fe-C alloys, as the structure changes from soft pearlitic and ferritic to martensitic, coefficient of friction increases because of a progressive increase in the hardness. The wear and friction properties depend also on the material pretreatment process, as well as on the process of depositing surface coatings. The different mechanical and chemical means of pretreating surfaces lead to different shapes, sizes, and architecture of surface asperities, which lead to different wear response.

The type and size of abrasive being used in a wear test also influences the wear rate. Figure 4-35 shows a schematic diagram for the wear rate of cemented carbide materials tested against a hard ceramic (e.g., SiC) abrasive of varying grit size. At a fixed grit size, the abrasive wear rate increases as the amount of soft ductile binder phase in the cermet increases (hard WC powders are bonded with FeNi alloys to produce cemented carbides, a very common cutting-tool material). This is because as the amount of soft binder increases, the spacing between the hard WC particles increases and SiC abrasive can remove material by plowing. Also, at a fixed binder

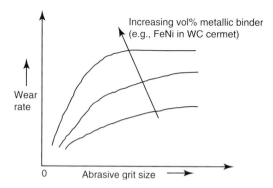

FIGURE 4-35 *Schematic diagram showing the effect of abrasive size and amount of metallic binder in a cermet on the wear rate.*

content in the cemented carbide, the wear rate increases as the abrasive size increases. This is because the large SiC particles begin to serve as indenters over a large region (which includes binder and WC particles) and deform and remove an "averaged" material, rather than individual constituents.

Corrosion

Material degradation by corrosion is a surface and near-surface phenomenon. Corrosion could be caused by electrochemical reactions, electromagnetic radiation, dissolution in liquids, oxidation, and high-temperature surface reactions. The most common corrosion processes (galvanic corrosion, rust formation, etc.) involve electrochemical reactions. It is customary to describe electrochemical corrosion in terms of electrochemical cells composed of electrodes, an electrolyte, and a semi-permeable membrane that is permeable to ionic species (cations and anions) but impervious to bulk convection of the electrolyte. As an example of a galvanic cell, consider an electrode of Cu immersed in an aqueous solution of 1.0 M $CuSO_4$ and a Zn electrode immersed in an aqueous solution of 1.0 M $ZnSO_4$, with a semi-permeable membrane dividing the two electrolytes (where M denotes the molarity of the solution and is the number of moles of solute per unit volume of the solution). A metal electrode immersed in a 1.0 M solution of its ions at room temperature is called a standard half-cell. If the electrodes are connected via an outer electric circuit, the following electrochemical reactions (half-cell reactions) will occur spontaneously at the anode and the cathode, respectively: $Zn \rightarrow Zn^{2+} + 2e$, and $Cu^{2+} + 2e \rightarrow$ Cu. Thus, a Zn electrode will dissolve (oxidize) in the electrolyte, releasing 2 moles of electrons per mole of Zn dissolved, and the electrons will travel via the outer circuit and reduce the Cu^{2+} ions in the $CuSO_4$ to form Cu deposit on a Cu electrode. The sulfate ions will travel from the $CuSO_4$ side across the membrane toward the $ZnSO_4$ side to neutralize the Zn^{2+} ions being produced in the solution. The overall reaction is the sum of the two half-cell reactions occurring at the anode and cathode; i.e., $Zn + Cu^{2+} \rightarrow Zn^{2+} + Cu$. The voltage measured in the circuit for this reaction will be 1.1 V and is a characteristic of the Cu-Zn galvanic couple. The electromotive force (EMF) for half-cell reactions for metals have been calculated with reference to an arbitrary standard reference cell that is called the standard hydrogen electrode. The standard hydrogen electrode consists of a Pt electrode in a 1.0 M solution of H^+ ions at 1 atmosphere pressure and room temperature. The noble metal Pt only acts as a surface over which H^+ may be reduced or H atoms may be oxidized. A standard EMF series has been constructed in which different metals are ranked according to their half-cell potentials with reference to the standard hydrogen electrode (by convention, the half-cell EMFs are given for the reduction reactions). Table 4-6 gives some electrode reactions and corresponding half-cell EMFs based on the standard EMF series. Metals near the top of the series are increasingly noble (resist corrosion), and metals near the bottom are increasingly anodic (prone to corrosion). The EMF generated by any combination of metals in the series can be estimated. In a Cu-Cr galvanic couple, the anodic and cathodic reactions are: $Cr \rightarrow Cr^{3+} + 3e$, and $Cu^{2+} + 2e \rightarrow$ Cu, respectively. The measured EMF for a Cu-Cr galvanic couple will be $E_{cathode} - E_{anode}$, where $E_{cathode}$ and E_{anode} are the standard half-cell potentials. This gives the EMF as $+0.34 - (-0.74) = 1.08$ V.

Besides galvanic corrosion, there are other types of corrosion processes that take place because of an imbalance of the solute concentration on the two sides of an electrochemical cell in which both electrodes are made of the same material. The driving force for corrosion is the difference in the ionic concentration in the electrolyte. Another corrosion process is triggered

TABLE 4-6 The Standard Electromotive Force Series (selected metals only)

Corrosion Tendency Potential (V)	Electrode Reaction	Standard Electrode
	$Au^{3+} + 3e \rightarrow Au$	+1.420
	$Ag^+ + e \rightarrow Ag$	+0.800
↑	$Fe^{3+} + e \rightarrow Fe^{2+}$	+0.771
	$Cu^{2+} + 2e \rightarrow Cu$	+0.340
Increasingly	$2H^+ + 2e \rightarrow H_2$	0.000
Noble (cathodic)	$Ni^{2+} + 2e \rightarrow Ni$	−0.250
	$Co^{2+} + 2e \rightarrow Co$	−0.277
	$Cd^{2+} + 2e \rightarrow Cd$	−0.403
	$Fe^{2+} + 2e \rightarrow Fe$	−0.440
Increasingly	$Cr^{3+} + 3e \rightarrow Cr$	−0.744
Active (anodic)	$Zn^{2+} + 2e \rightarrow Zn$	−0.763
	$Al^{3+} + 3e \rightarrow Al$	−1.662
↓	$Mg^{2+} + 2e \rightarrow Mg$	−2.363
	$Na^+ + e \rightarrow Na$	−2.714
	$K^+ + e \rightarrow K$	−2.924

by an imbalance of dissolved oxygen on the two sides of a semi-permeable membrane. Rusting of iron is an example of such a corrosion process. The electrochemical reactions responsible for the formation of rust, which is the compound, $Fe(OH)_3$, are: 1) dissolution of iron in the anodic regions via $Fe \rightarrow Fe^{3+} + 3e$, 2) formation of hydroxyl ions in the O-rich regions via $O_2 + 2H_2O + 4e \rightarrow 4OH^-$, and 3) formation of rust by the reaction $Fe^{3+} + 3OH^- \rightarrow Fe(OH)_3$. The extent of corrosion is characterized in terms of a corrosion penetration rate (CPR), which is given from: $CPR = KW/\rho \, At$, where W is the weight loss per unit time, t, ρ is density of metal, A is the exposed surface area of metal, and K is a constant such that $K = 534$ for CPR in mils per year (mpy), or $K = 87.6$ for CPR in mm per year.

Material degradation can also occur in a dry gaseous environment at high temperatures at which a material reacts with the gas to form a scale (e.g., an oxide layer). The experimental measurement of oxidation tendency of some metals and alloys in dried, flowing air at 1475 K is shown in Table 4-7. The comparative factor in this table denotes the degree of oxidation resistance; the higher the comparative factor, the better the resistance of the metal to oxidation is. Oxidation is an electrochemical corrosion process that may be represented by the generic chemical equations: $Me \rightarrow Me^{2+} + 2e$ and $(1/2)O_2 + 2e \rightarrow O^{2-}$. The overall reaction can be written as the sum of the two preceding reactions to yield $Me + (1/2)O_2 \rightarrow MeO$, where Me represents the metal, and MeO represents the metal oxide. The thermodynamic stability of oxides at different temperatures can be predicted from the Gibb's free energy, ΔG, of oxide formation. Figure 4-36 shows the calculated ΔG values for some metal oxides at different temperatures; an oxide with the largest (negative) ΔG value will be the most stable oxide at a given temperature.

In order for an oxidation reaction to continue, electrons released from metal ionization (i.e., from the reaction $Me \rightarrow Me^{2+} + 2e$) should be conducted to oxide/gas interface where they reduce the oxygen, the Me^{2+} ions must diffuse away from the metal/scale interface, and O^{2-} ions must diffuse toward this interface. The reaction front may be located at the metal–scale interface, oxide–gas interface, or within the oxide itself. The rate limiting step in oxide growth may either be the ionic diffusion or the kinetics of interface reaction. For diffusion-limited

TABLE 4-7 Oxidation of Various Materials at 1475 K in dry, flowing air (adapted from N. J. Grant, Refractory Metals & Alloys, Proceedings of the Metallurgical Society Conference, Detroit, MI, May 1960, vol 11, 1961, pp 119-168)

Material	Exposure Time, h	Metal Loss, in./side	Comparative Factor
Mo	0.5	0.018	1
Ta	1.25	0.032	1.4
Hf	18.25	0.047	14
W	4.0	0.010	14
Zr	18.25	0.024	27
Ti	66.25	0.044	54
Cr	66.25	0.0075	320
67Cr-33Ni	66.25	0.0028	850
310 Stainless steel	66.25	0.0015	1600

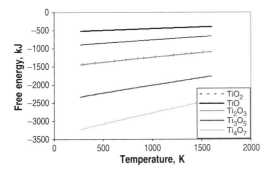 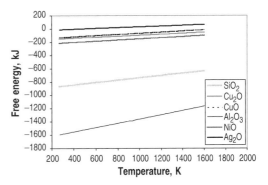

FIGURE 4-36 *Standard Gibb's free energy of formation of various oxides as a function of temperature.*

growth of a non-porous scale, a concentration gradient of diffusing species will develop across the scale; this gradient will progressively become shallower with increasing oxide thickness, resulting in sluggish diffusion and retarded growth. The kinetics of diffusion-limited growth can be expressed as $(dy/dt) = C_1/y(t)$, where $y(t)$ is the instantaneous scale thickness, t is time, and C_1 is a temperature-dependent constant. For $y = y_0$ at $t = 0$, the oxide thickness at time t is obtained from integration of the preceding equation. This yields

$$y^2(t) - y_0^2 = 2C_1t, \tag{4-45}$$

and for a negligible initial oxide thickness ($y_0 = 0$),

$$y^2(t) = 2C_1t,$$

which is a simple parabolic relationship characteristic of many diffusion processes. In contrast, for a porous oxide scale, bulk transport processes will maintain the oxygen concentration constant at the reaction front, and the oxide growth rate will be constant at a fixed temperature. Thus,

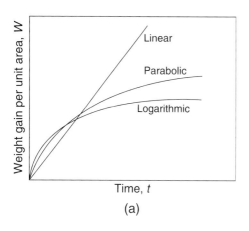

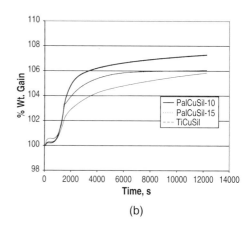

FIGURE 4-37 *(a) Schematic diagram showing oxide layer thickness as a function of oxidation time for linear, parabolic and logarithmic kinetics.* (From W. D. Callister, Jr., Materials Science & Engineering: An Introduction, 6th ed., John Wiley, New York, 2003, p. 593). *(b) Oxidation kinetics of some commercial braze foils in air measured with the thermogravimetric method.* (M. Singh, T. P. Shpargel, and R. Asthana, Proc. of the 29th International Conference on Advanced Ceramics and Composites, 2005, 383–390, American Ceramic Society, Westerville, OH)

$(dy/dt) = C_2$, where C_2 is a temperature-dependent constant, and with $y = y_0$ at $t = 0$, we obtain a linear growth kinetics for oxide growth; i.e., $y(t) = C_2 \cdot t + y_0$, where y_0 is the initial oxide thickness. The linear and parabolic kinetics are schematically represented in Figure 4-37a and represent just two limiting situations. Other growth laws (e.g., cubic and logarithmic) are also observed in different systems, or in the same system at different temperatures. If the density of the oxide does not change during growth, then the weight of the oxide can be substituted for thickness in the preceding growth laws. Figure 4-37b presents the oxidation data as a function of time for some Pd and Ti-base braze alloys oxidized in air; these data do not follow a simple oxide growth law, which is typical of many commercial alloys.

The tendency of the oxide scale to protect the metal from further oxidation depends upon whether the scale is non-porous and completely covers the underlying metal. Thus, the relative volumes of oxide and metal can be used to predict whether the oxide will form a protective or non-protective scale. The Pilling-Bedworth (PB) ratio is a parameter based on the relative volume criterion and is defined as PB $= (A_{MeO}\rho_{Me})/(A_{Me}\rho_{MeO})$, where A_{MeO} is the molecular weight of the oxide, A_{Me} is the atomic weight of the metal, and ρ_{Me} and ρ_{MeO} are the densities of the metal and oxide, respectively. For metals with PB < 1, the oxide volume is less than the volume of metal consumed in the oxidation reaction, and the oxide will be porous and non-protective. For metals with PB > 1, compressive stresses occur in the scale, and at PB > 2–3, the oxide will likely crack and flake off, and will become non-protective. The PB ratios for some metals are given in Table 4-8; for most metals the PB values generally agree with the predicted tendency for an oxide to form a protective scale, although there are some exceptions. For example, Si with PB > 2 forms protective scale, whereas alkali metals with PB < 1 form non-protective scale.

Another type of corrosion process involves the dissolution of solid metals and refractories in molten metals, which is encountered in some manufacturing processes. The extent of dissolution increases in the presence of recirulatory flows that cause erosion and sweep away chemical solutes

TABLE 4-8 The Pilling-Bedworth Ratios of Common Metals (adapted from B. Chalmers, Physical Metallurgy, John Wiley, 1959)

Metal	PB Ratio	Oxide Characteristic
Cu	1.68	Protective
Al	1.28	Protective
Si	2.27	Protective
Fe	1.77	Protective
Ni	1.52	Protective
Pd	1.60	Protective
Pb	1.40	Protective
Ag	1.59	Non-protective
Ti	1.95	Non-protective
Mo	3.40	Non-protective
V	3.18	Non-protective
W	3.40	Non-protective
Na	0.57	Non-protective

from the reaction front. Several theoretical and semi-empirical models to predict the dissolution kinetics in liquids have been developed. A widely used model is the Nernst-Shchukarev model (also known as the Berthoud equation), according to which the dissolution kinetics of a solid metal in molten metal is expressed from

$$\frac{dC}{dt} = \frac{kA}{\omega(C_S - C)} \tag{4-46}$$

where C is the instantaneous concentration of the dissolved species in the liquid, t is time, A is the surface area of the solid in contact with the liquid, C_s is the saturation concentration (solubility limit) in the liquid, ω is the liquid volume, and k is the dissolution rate constant, which is related to the diffusion coefficient, D, in the liquid from $k = D/\delta$, where δ is the diffusion layer thickness (typically a few tens to a few hundred micrometers). Integrating Equation 4-28 together with the initial conditions, $C = 0$ at $t = 0$, yields

$$C = C_S \left[1 - \exp\left(-\frac{kAt}{\omega} \right) \right] \tag{4-47}$$

Noting that $k = D/\delta$, the normalized concentration (C/C_s) of the dissolved species is

$$\frac{C}{C_S} = 1 - \exp\left(-\frac{DAt}{\delta\omega} \right) \tag{4-48}$$

The ratio (C/C_S) decreases as the diffusion layer thickness, δ, increases because a large δ causes a shallow solute gradient at the interface and a smaller driving force for the dissolution. The solubility of a metal depends upon the temperature and usually follows an Arrhenius-type equation,

$$C_S = C_S^0 \exp\left(-\frac{Q}{RT} \right), \tag{4-49}$$

where C_S^0 is a frequency factor, Q is an activation energy for solubility, and R and T are the gas constant (8.314 J/mol·K) and absolute temperature, respectively. The dissolution rate constant, k, in Equation 4-46 also displays an Arrhenius-type temperature dependency and is influenced by fluid convection. In the presence of recirculatory flows, the dissolution rate constant, k, can be estimated from semi-empirical equations that have been developed for different ranges of the Schmidt number (Sc), where Sc $= \upsilon/D$, and υ is the kinematic viscosity of the liquid. For liquid metals, Sc is usually less than 1000, and for diffusion-controlled dissolution, the dissolution rate constant of solid metals in liquid metals can be estimated from

$$k = 0.554\,I^{-1}D^{2/3}v^{-1/6}\sqrt{w}, \tag{4-50}$$

where the parameter I is a function of Schmidt number, and w is the angular velocity. The preceding equations have been used to analyze the dissolution kinetics of metals such as Mo, Nb, Cr, Y, and others in molten Al. The dissolution process usually leads to formation of complex intermetallic layers at the solid–liquid interface. Analytical models that consider both diffusion and chemical reactions during the growth of these interlayers have also been developed.

Many other types of material degradation processes could be classified as corrosion processes. For example, radiation of neutrons, electrons, protons, cosmic rays, and various types of electromagnetic radiation may break interatomic bonds and dislodge and eject atoms, thereby causing gradual loss of material. Because the energy of a photon of radiation of wavelength, λ, is $E = hc/\lambda$, where c is the velocity of light (0.2998×10^9 m/s) and h is Planck's constant (0.6626×10^{-33} J·s), radiation of short wavelength will be more damaging to a material than radiation of long wavelength. Radiation environments are encountered in fusion reactors, space exploration, medical treatment, and a variety of other areas.

References

Berg, J. C., ed. *Wettability.* New York: Marcell Dekker, Inc.

Boettinger, W. J., C. A. Handwerker, and U. R. Kattner. *The Mechanics of Solder Alloy Wetting and Spreading,* ed. F. G. Yost, et al. New York: Van Nostrand Reinhold, 1997, p. 103.

Bonis, L. J., and H. H. Hausner, eds. *Fundamental Phenomena in the Materials Sciences,* vol. 2, *Surface Phenomena.* New York: Plenum Press, 1966.

Eremenko, V. N., and L. D. Lesnik. *The Role of Surface Phenomena in Metallurgy,* ed. V. N. Eremenko. New York: Consultant Bureau, English Translation, p. 102.

Eustathopoulos, N., M. G. Nicholas, and B. Drevet. *Wettability at High Temperatures, Pergamon Materials Series,* ed., R. W. Cahn. New York: Pergamon, 1999.

Hutchings, I. M. *Tribology: Friction and Wear of Engineering Materials.* Boca Raton, FL: CRC Press, 1999.

Matijevic, E., ed. *Surface & Colloid Science.* New York: Wiley Interscience, 1969.

Murr, L. E. *Interfacial Phenomena in Metals and Alloys.* Reading, MA: Addison Wesley, 1975.

Myers, B. Surfaces. *Interfaces and Colloids.* New York: Wiley-VCH.

Naidich, Y. Ju. *Progress in Surface and Membrane Science,* ed. D. A. Cadenhead and J. F. Danielli. New York: Academic Press, 1981.

Neumann, A. W., and R. J. Good. *Surface and Colloid Science,* vol. II, ed. R. J. Good and R. R. Stromberg. New York: Plenum Press, 1979.

Rabinowicz, E. *Friction and Wear of Materials.* New York: John Wiley & Sons, Inc., 1995.

5   Coatings and Surface Engineering

Coatings and surface engineering are used to protect manufactured components from thermal or corrosive degradation, impart wear resistance and hardness to the surface while retaining the toughness and ductility of the bulk component, and enhance the aesthetic and decorative appeal. For example, modern heat engines use thermal barrier coatings to reduce heat losses and increase engine efficiency. Even thin, thermally insulating coatings are able to reduce heat loss in an engine, unlike conventional furnaces in which several layers of a thick refractory lining are needed to reduce the heat loss. Unlike a furnace, where heating is continuous and temperature increases monotonically with time, combustion is cyclic in an engine, and even a thin coating of an insulating material will provide an adequate thermal barrier because of the slow response of the coating to temperature changes.

Many factors must be considered in selecting coatings for a given application. It is important to remember that 1) there is no such thing as a universal coating that can be used under all conditions, 2) coatings are not inherently good or bad, and 3) all coating selections represent compromises made to satisfy all the variables involved. The process of selecting coatings must consider the operating conditions, material compatibility issues, nature and surface preparation of substrate, time or speed of application, cost, safety, environmental effects, coating properties, and structural design. Frequently the task of selecting the right coating is really a process of elimination rather than selection. Regardless of the coating process used, selecting the right coating for a given application usually requires considerable experience and judgment.

Design and Development of Coatings

The process of designing and developing coatings for modern technological applications is usually very complex. It combines experimentation and analysis. Consider the complexity of developing coatings for electrodes used in high-temperature glass melting furnaces. Electrodes used in glass-melting furnaces are usually made out of molybdenum (Mo). At high temperatures ($\sim$1873 K) normally used in such furnaces, Mo oxidizes and becomes an electrical insulator, and the furnace efficiency decreases. One popular approach to resist the oxidation of Mo electrodes is to apply $MoSi_2$ coatings to Mo electrodes. This, however, introduces problems of mismatch of coefficient of thermal expansion (CTE) between $MoSi_2$ and Mo, which may lead to interfacial cracking and failure of the coating. The CTE of the coating and Mo can be closely matched by adding controlled amounts of SiC to $MoSi_2$, thus reducing the cracking problem. However, a new problem is introduced by SiC additions; i.e., at high temperatures C and Si from SiC diffuse into Mo electrodes, which causes rapid contamination and degradation of the electrodes. This problem is overcome by applying a diffusion barrier layer. A very thin layer of a compound, $MoSi_xN_yC_z$, is sputter-deposited between the Mo electrode and the $MoSi_2$-SiC coating. This provides some protection against diffusion and contamination. However, the intermediate layers should be compatible with both the coating and the underlying substrate and should adhere to both.

Generally, coatings used in complex high-technology areas are application specific and cannot be transferred to other applications. One challenge for the designer of the coatings is to create multilayer, multifunctional coatings in which the synergy of the properties of different layers determines the combination of functions that can be performed.

Surface Pretreatments

Before a coating can be applied to a surface, the latter must be cleaned to remove physical and chemical impurities and structural defects such as surface micro-voids and cracks, sand, scale, flux residues, oil, grease, soil, and other chemical impurities. A variety of mechanical and chemical cleaning and pre-treatment techniques are used prior to coating. Abrasive cleaning is used to remove sand inclusions and oxide scale. Sand, steel grit, and steel shot are used as abrasives. Pressurized air and abrasive particles from a centrifugal pump impact the work piece at a high velocity. In shot-blasting, the air pressure is 60–100 psi (0.4–0.7 MPa) for hard ferrous parts and 10–60 psi (0.07–0.4 MPa) for softer non-ferrous parts. Cleaning is done in an enclosed hood; however, large parts are manually cleaned in rooms using masks and protective clothing.

Barrel finishing or tumbling is used to descale and derust parts. Tumbling is a low-cost, mass finishing operation, usually done in dry condition. Sand, granite chips, slag, and alumina pellets are used as abrasives. A barrel of mild steel plate lined with hard wood, or natural or synthetic rubber, is filled with the abrasive and the parts to be cleaned. The lining material increases tumbler life, improves surface finish, and reduces metal contamination. Rotary motion of the tumbler leads to "landsliding" action, which causes cutting, scrubbing, and polishing, and removes fins, scales, and sand particles. Only a small fraction of parts is finished at any time because most of the cleaning action is confined to a surface layer where agitation and landsliding are most intense. Higher speeds lead to faster processing but also part damage by impingement. Tumbling is used mainly on rugged, hard parts and is used only to a limited extent on softer parts, which are attached to racks within the barrel to reduce the damage from impact. Various shapes and sizes of abrasive media are mixed so that abrasive will reach all areas to be cleaned (corners, crevices, holes). Finishing action depends on the ratio of volume of work to volume of abrasive, speed of rotation, part geometry, and shape and size of abrasive. Several modifications of the basic tumbling process are used in industry. For example, in spindle-finishing, parts are attached to a rotating shaft inside the barrel, with abrasive media rotating in a direction opposite to the direction of rotation of parts. This provides for rapid cleaning with little part damage from impingement.

Another mass finishing operation is vibratory finishing. Unlike tumbling, which uses closed drums, vibratory finishing is done in open containers, which allows in-process inspection. Circular or rectangular tubs are vibrated at 900–3600 cycles per minute. Cleaning times are shorter in vibratory finishing than in tumbling because the entire load (not just the surface) is under agitation. Vibration frequency and amplitude depend on part size, shape, weight, material, and abrasive media. Both natural and synthetic abrasives are used, which provide both cleaning action and cushioning; synthetic media permit better control of abrasive shape and size. Polyester and resin are common matrices, and alumina, emery, SiC, and diamond are abrasives (50–70% by weight in media). Media particles must fit in holes and crevices without lodging, which highlights the importance of part geometry in the selection of abrasive shape and size. Chemical compounds (either acidic or alkaline) may be added for finer finish and for removing soils and oil.

Other common surface cleaning and finishing processes to remove physical impurities include ultrasonic cleaning, grinding, wire brushing, and polishing. Ultrasonic cleaning employs the cavitation phenomenon for surface cleaning. Cavitation is the formation and implosion of gas-filled microcavities during propagation of a sonic wave through a liquid, which generates very large local pressures and provides scrubbing action. Cleaning is done below the boiling point of the solvent, because boiling interferes with the propagation of the sonic wave and lowers the cleaning efficiency. Grinding and belt-sanding are traditional finishing operations in which work is held against an abrasive belt or wheel; abrasives of varying fineness are used. Common abrasives are emery, alumina, silica, and diamond, bonded by resin, shellac, or clay. These finishing processes are labor-intensive, amenable to automation only to a limited extent, and difficult to use on parts with recesses and internal corners. Wire-brushing uses high-speed rotary brushes that remove very little metal except high spots. Wire-brushing uniformly scratches the surface, and these scratches are removed by buffing and lapping. Buffing and lapping are polishing operations to produce scratch-free surfaces. The part is brought in contact with a revolving cloth (linen or cotton) charged with a fine abrasive (ferric oxide, corundum, silica, or diamond). The polishing operation can be either manual or semiautomatic, and generally produces an optically flat and mirror-like finish.

Chemical pretreatments such as alkaline cleaning, solvent cleaning, vapor degreasing, pickling, and salt bath cleaning remove oils, soil, grease, flux residue, and other chemical impurities. Alkaline cleaning uses caustic soda or trisodium phosphate to aid in emulsification of dirt. The part is rinsed after alkali treatment to wash off the impurity. The solution pH must be controlled; for example, for cleaning steel parts, a pH of 9–14 is commonly used. In solvent-cleaning, water- or petroleum-based solvents are used to dissolve oil, grease, and fats. Petroleum-based solvents include kerosene and naphtha, whereas water-based solvents include sulfonated castor oil and water. Cleaning is done either by part immersion in dip tanks at room temperature or by spraying and wiping. Petroleum-based solvents are relatively inexpensive but pose a fire hazard. In contrast, water-based solvents do not pose a fire hazard and are less toxic than petroleum-based solvents. They are suitable for cleaning metals (Al, Zn, and Pb) that are readily attacked by alkaline solvents.

Vapor-degreasing employs vapors of a chlorinated solvent (e.g., trichloroethylene) to clean the part. The solvent boils at a low temperature ($\sim$188 °F), and the part to be cleaned is suspended in a closed chamber to expose it to vapors. The vapor condenses on the part and causes degreasing action. The residue is washed off by spraying the part with the chlorinated solvent. With time, the bath becomes dirtier because of accumulation of the residue in the solvent, but the vapors remain clean. The hung part must not be warm or else the vapors will not condense on the part and no cleaning action will occur. Vapor-degreasing is a rapid cleaning process, but vapors do not remove all types of chemical impurity. Therefore, vapor-degreasing is usually followed by alkali-cleaning. In addition, vapor emissions into the atmosphere must be controlled to prevent environmental pollution. This is accomplished by the use of cooling coils to condense the vapor for reuse. Another chemical cleaning process is pickling, which is an acid-cleaning treatment that removes oxide scale and flux residues. Pickling involves cleaning of metal parts in dilute acids by spraying or immersion. Common acids used in pickling are 10% H_2SO_4 at 150–185 °F or HCl acid at room temperature. The part is first cleaned with an alkali to allow acid to reach all surfaces. Pickling solution does not attack certain types of oils and grease, and alternative cleaning treatments may be needed. After pickling, the part is rinsed and neutralized in an alkaline solution to prevent rusting. Pickling is widely used to clean large-tonnage products such as billets, sheet, strip, and tubing. Over-pickling can roughen and degrade the part, and pickling inhibitors are used to reduce the extent of chemical attack. Some of the problems with pickling are dilution of acid because of carryover of pre-cleaning medium to acid bath, hydrogen embrittlement because of evolution and attachment of gas bubbles to the part, and splashing and vapors from acid, which can cause corrosion of equipment and atmospheric contamination.

Heavy oxide scale on metals such as Fe, Cu, and Ni is removed by salt bath cleaning. The metal part is dipped in a large bath of NaOH containing 1.5–2% sodium hydride maintained at 370–470 °C. Immersion is followed by water quenching. Salt bath cleaning is usually not recommended for parts that were previously heat-treated or tempered below 370 °C in order to prevent undesirable metallurgical changes during cleaning. Another finishing process called electrocleaning (also called electropolishing, deplating, or reverse electroplating) makes use of electricity to produce a very high-quality surface finish. The part is made into an anode (positive terminal) in an electrolyte solution that also contains a metal cathode. The passage of an electric current (typical current density: 10–150 A per ft^2) forms gas bubbles on the anode surface. These bubbles burst, scrub and scour the scale, and lift soils. Material is removed fastest from high spots where the current density is very large, which yields a very smooth and mirror-like finish.

Coating Techniques

Vapor-Phase Deposition

Vapor-phase techniques deposit thin or thick films and coatings of a variety of materials from the vapor phase. Among the widely used vapor-phase deposition methods are vacuum-metallizing, sputtering, chemical vapor deposition, and ion-nitriding. In vacuum-metallizing, the coating material is evaporated from a source and condensed on work piece under vacuum (<0.001 Torr). Figure 5-1 shows the basic approach of batch type and continuous vacuum-metallizing processes. Thin metal coatings of Zn, Al, Cd, and other metals are readily deposited from the vapor phase on a variety of substrates. Parts can be rotated inside the chamber to expose different surfaces, fastened to racks, or masked for selective exposure. Multiple vapor sources in the chamber allow successive coatings of different metals to be deposited on the same work piece. The work piece is initially at room temperature but heats up due to vapor condensation; however, the part should not vaporize, excessively outgas, or undergo structural changes. Vaporization is difficult for metals with low vapor pressure, in which case better vacuum and higher temperatures are needed for metallizing. The change in vapor pressure with temperature

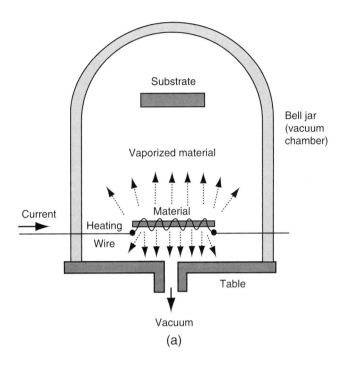

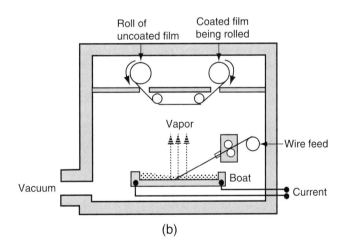

FIGURE 5-1 *(a) Vacuum metallizing of a substrate via evaporation and condensation of the coating material.* (R. Trivedi, Materials in Art and Technology, Taylor Knowlton, Ames, IA, 1998, p. 351). *(b) Continuous vacuum metallizing of a roll of film.* (R. Trivedi, Materials in Art and Technology, Taylor Knowlton, Ames, IA, 1998, p. 352).

is given by the Clausius-Claperyon equation, according to which,

$$\frac{d \ln p}{dT} = \frac{\Delta H}{RT^2},$$

where p is the vapor pressure, T is absolute temperature, R is the universal gas constant, and ΔH is the enthalpy of evaporation. This equation can be used to estimate the energy needed for evaporation during metallization.

Consider that molten gold at its boiling point is used to deposit a 5-μm thick coating on all the surfaces of a cylindrical ring of 8 cm outer diameter (OD), 4 cm inner diameter (ID), and 10 cm length. The atomic weight of gold is 196.9 amu, and the boiling point and density of gold are 2857 °C and 19.3 g/cc, respectively. The vapor pressure–temperature relationship for gold is

$$\ln p(\text{atm}) = -\frac{44,400}{T} - 1.01 \ln T + 21.88 \tag{5-1}$$

To estimate the energy needed to form the coating, we take the first derivative of Equation 5-1 with respect to T. This yields

$$\frac{d \ln p}{dT} = \frac{\Delta H}{RT^2} = \frac{44,400}{T^2} - \frac{1.01}{T} \tag{5-2}$$

At the boiling point of Au ($T = 3130$ K), Equation 5-2 yields, $\Delta H = 342.86$ kJ/mol. Now, with a 5-μm-thick gold film covering all the surfaces of the ring, the new ring diameter becomes $4 + 2(5 \times 10^{-4})$ cm, or 4.001 cm, and new height becomes $2 + 2 \times 5 \times 10^{-4}$, or 2.001 cm. Therefore, the coating volume $= \pi(4.001)^2 2.001 - \pi(2)^2 2 = 0.02513$ cm^3, and the weight of deposited film $= 0.02513 \times 19.3$ g/cm^3, or 0.4849 g. This weight of Au corresponds to (0.4849/196.9) or 0.00246 moles. Therefore, the energy needed to evaporate this amount of gold is $0.00246 \times 342.86 = 0.845$ kJ $= 845$ J. This is clearly an underestimate because the energy needed for melting was not considered in the calculation, and no energy losses were accounted for.

Coatings of metals with a high vapor pressure are readily deposited by vacuum metallizing. Aluminum has a high vapor pressure and is easy to deposit, followed by Se, Cd, Ag, Cu, Au, Cr, Pd, Ti, Pt, and other metals. The deposits are generally pure, with ordered structure, and can be mono-crystalline or polycrystalline. The coated surface may be protected from wear by applying an overcoat of a resin. The equipment for vacuum-metallizing includes vacuum chamber, vapor source, vacuum pumps, control valves, etc. Usually, a tungsten filament is heated by passing an electric current, and chips or wires of coating material are fastened to the filament. Alternatively, a ceramic crucible containing the metal to be evaporated is heated either by induction power or an electron beam. Pre-cleaning of the substrate is necessary to form adherent coatings. Generally, parts to be metallized are first cleaned by ionic bombardment using the glow-discharge technique. Glow-discharge cleaning is done in a vacuum chamber prior to coating deposition; ionic bombardment of work piece during glow discharge dislodges surface impurities. A wide variety of coatings are deposited using vacuum-metallizing. These include decorative coatings of Al or Au on molded plastics, metal parts, and paper rolls; optical coatings of Al, Cu, Cr, and Pt over-coated with thin SiO and MgF films for optical systems such as mirrors, headlamps, microscope and telescope lenses, sunglasses, filters, and beam splitters; corrosion-resistant coatings of Cd and Al on steels to prevent hydrogen embrittlement;

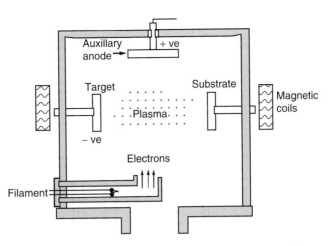

FIGURE 5-2 *Diagram showing the basic setup used for sputtering.* (R. Trivedi, Materials in Art and Technology, Taylor Knowlton, Ames, IA, 1998, p. 356).

and electrical and electronic thin film components such as resistors (NiCr, PtAu, CrAl, etc., on ceramics), capacitors (metal films on dielectric ribbons of plastic or glass), and magnetic memory devices (FeNi coatings on nonmagnetic plastic tapes for computer storage memory). Hundreds of diodes, triodes, transistors, resistors, and capacitors are deposited on small ($\sim$1 cm^2) Si wafers using this process.

In vapor-phase deposition of coatings by sputtering, plasma of ions and electrons is first created under vacuum (0.001–0.1 Torr) using electrical heating. The plasma energy ejects atoms from target metal, which strike the work piece at high velocities and provide good bonding. A small amount of a working gas (e.g., Ar) in the chamber serves as a source of ions for plasma. Figure 5-2 shows the basic principle of sputtering. Two mechanisms of ejection of target atoms upon collision from the working gas ions have been established: direct collision with surface atoms, and penetration into solid, which produces internal collisions with recoils that reach the surface (cascade collisions). Impact velocity, ionic mass, and angle of incidence are important variables in sputtering. The maximum yield occurs when the angle of incidence of impinging ions to the surface normal is about 60–70°. At larger angles, surface electric field deflects the impinging ions and lowers the yield. The sputtering yield, S, is defined as

$$S = \frac{N_a}{N_i},$$

where N_a and N_i are the number of atoms ejected and number of ions impinging, respectively. S is inversely proportional to the cosine of angle of incidence, θ, with respect to surface normal, and reaches a maximum at $\theta = 60$–70°.

The atomic impact of target atoms also dislodges adsorbed impurity atoms and disrupts any native oxide films. Because the coating material is not vaporized in sputtering, very hard and refractory coatings with controlled composition and structure can be deposited. In addition, functionally graded coatings such as wear-resistant Cr-CrN-Cr$_2$N coatings for forming applications have been deposited. For applications in cutting tools, sputtering has been used to deposit hard B$_4$C, WC, and TiC coatings on steels to achieve high surface hardness and bulk toughness.

Sputtering has also been used to deposit nanocrystalline Ti-Al-N coatings that have increased wear and oxidation resistance.

Nucleation and Growth in Vapor Condensation

The nucleation of crystals of a coating on a substrate from the vapor phase is analogous to the heterogeneous nucleation of solid crystals from a melt. Initially, vapor atoms must attach themselves to a suitable site to form a stable nucleus. In a manner similar to crystal growth from melt (Chapter 2), the total energy per unit volume to form a coating nucleus can be written as the sum of the volume free energy change (ΔF_v) and surface energy change (ΔF_s), i.e.,

$$\Delta F = \Delta F_v + \Delta F_s = \frac{RT}{V_m} \ln \frac{p}{p_e} \cdot \frac{4}{3} \pi R^3 \xi(\theta) + 4\pi R^2 \sigma_{cv} \xi(\theta) \tag{5-3}$$

Here V_m is the molar volume, p is pressure of vapor, p_e is the equilibrium vapor pressure, R is the radius of nucleated embryo, σ_{cv} is the crystal–vapor interfacial energy, $\xi(\theta)$ is a geometric factor given from

$$\xi(\theta) = \frac{(2 + \cos\theta)(1 - \cos\theta)^2}{4},$$

and θ is the angle of contact that the nucleus makes with the substrate; $\xi(\theta) = 0$ at $\theta = 0°$, and $\xi(\theta) = 1$ at $\theta = 180°$. A stable nucleus forms when ΔF is a maximum. On differentiating ΔF with respect to R, and setting the result equal to zero, the radius of critical nucleus is obtained as

$$R^* = -\frac{2\,\sigma_{cv}V_m}{RT \ln \dfrac{p}{p_e}}.$$

The total free energy corresponding to the critical nucleus is

$$\Delta F^* = \frac{16\,\pi(\sigma_{cv})^3 \xi(\theta)(V_m)^2}{3R^2 T^2 \left(\ln \dfrac{p}{p_e}\right)^2}.$$

Note that Helmholtz function, F, rather than Gibb's function, G, has been used for the total free energy because, unlike crystallization from a liquid, vapor-phase deposition is not a constant pressure process. The above relationships show that nucleation from a vapor will be easier at high temperatures, and nucleation will be difficult when σ_{cv} is large or when the pressure is very low (i.e., when there are only a few atoms in the vapor phase). Note that evaporation (and not condensation) becomes easier at low temperatures when pressure is low. An approximate expression for the rate of condensation from vapor, obtained from the kinetic theory of gases, is

$$rate = 5.85 \times 10^{-5} \alpha p \sqrt{\frac{M}{T}},$$

where M is the mole of substance to be evaporated, p is the pressure of the saturated vapor, and $\alpha(<1)$ is a condensation coefficient that takes into account the fact that only a fraction of

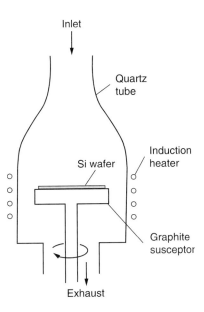

Inlet

Quartz
tube

Induction
heater

Si wafer

Graphite
susceptor

Exhaust

FIGURE 5-3 *Basic outline of chemical vapor deposition of a coating on a silicon substrate.*

the atoms impinging on the substrate actually condense to form the nucleus. At low levels of super-saturation in the vapor, i.e., when p is small, deposition of the coating may be limited by nucleation rather than growth.

In chemical vapor deposition (CVD), coating forms by reaction between vapors over a heated surface (not with the surface). The method consists of vaporizing a chemical compound, mixing it with other gases, and introducing the mixture in a chamber containing heated work piece (Figure 5-3). Chemical reactions between gases either at atmospheric pressure or under low vacuum (0.1–10 Torr) form the coating. Borides, carbides, oxides, nitrides, composites, and thin films of semiconducting and superconducting materials are deposited using CVD. CVD is used in the fabrication of semiconductor chips (e.g., GaAs coatings for CD players, laser printers, radon detectors, cellular telephones) and for depositing hard and wear-resistant coatings such as diamond thin films. Common precursor gases in the CVD process are volatile halides of metals such as Nb, Ta, Zr, Hf, Cr, Sn, and others. Volatile precursors and effluents (nickel carbonyl, arsine, phosphine, etc.) from the CVD process are generally flammable and toxic, and these are flushed out to a disposal system to neutralize toxic byproducts and un-reacted precursors before releasing them into atmosphere. A typical CVD exhaust manifold consists of particle traps, pressure indicators and sensors, chemical and mechanical scrubbers, and an incinerator system to decompose toxic gases. Leak-tight piping and ducts are an obvious requirement.

The ability to deposit extremely hard and wear-resistant coatings such as diamond thin films on metals makes the CVD process attractive in a wide variety of manufacturing applications. The CVD process parameters can be controlled to deposit smooth and graded nanocrystalline diamond thin films, which are required to reduce the cutting force and obtain high feed rates in machine tools. Similarly, smooth films are needed in optical applications to reduce radiation scatter due to roughness. Diamond film growth is improved by surface pretreatment (e.g., polishing) and by deposition of a buffer layer, such as AlN, TiN, CN, and WC on silicon; TiN and Si_3N_4 on steels; and Ti, TiN, and SiC on cermets. These buffer layers absorb thermal stresses and act as diffusion barrier to impurity atoms. To deposit diamond thin films on Ti to improve its wear resistance for advanced structural applications, a carbon-bearing gaseous precursor such as CH_4 is used. However, carbon from CH_4 also dissolves in Ti at the CVD temperatures, forming brittle carbide compounds and causing cracking. To overcome this problem, buffer layers of TiN, carbonitrides, or diamond-like carbon (DLC) are deposited on Ti. TiN buffer layers prevent hydrogen and carbon diffusion in Ti and DLC layers permit smooth diamond films to form on Ti by nucleation and growth because preexisting carbon from DLC aids diamond nucleation. However, both TiN and DLC cause poor adhesion; with TiN, an additional problem is the large CTE mismatch with respect to diamond. A composite nitrided layer (TiN) and a carbonitrided layer (TiCN) give the best adhesion and expansion compatibility. Similarly, in the deposition of oxidation-resistant CVD coatings of NiAl on Ni-base superalloys, a bond coat of Ni or Pt is needed for good adherence. First, hafnium is deposited on the superalloy, and then the surface of the superalloy is aluminized using CVD, which forms a NiAl coating doped with Hf for increased thermal barrier and oxidation resistance.

Chemical thermodynamics provides a basis to select the CVD precursors and the reaction conditions favorable for coating deposition (however, chemical kinetics must be considered for the rate processes involved). In a fundamental sense, the reaction with the more negative free energy change, ΔG, at a given temperature will be thermodynamically preferred. Thus, changes in entropy and enthalpy between products and reactants must be calculated from a knowledge of the temperature-dependence of the specific heat in order to determine the thermodynamic feasibility of the coating deposition reaction. Alternatively, standard free energy versus temperature diagrams, called Ellingham diagrams, can be used to evaluate the thermodynamic feasibility of a reaction and thus allow choices to be made of suitable precursor materials for the CVD reaction deposition process. Consider the deposition of wear-resistant titanium diboride coatings on steel using different precursors according to the following reactions:

1. $TiCl_4 + 2BCl_3 + 5H_2 \rightarrow TiB_2 + 10HCl$
2. $TiCl_4 + B_2H_6 \rightarrow TiB_2 + H_2 + 4HCl$

Calculations show that reaction 2 has a more negative free energy change than reaction 1, and therefore, $TiCl_4$ and B_2H_6 can be used as precursors for the deposition of TiB_2 coatings (this is a rather simplified scenario because in reality, sub-halides, not considered above, may also form). The free energy change, ΔG_T, of a reaction at a temperature, T, can be written in terms of the enthalpy (ΔH) and entropy (ΔS) changes as $\Delta G_T = \Delta H - T\Delta S$. At a pressure of 1 atm, the standard Gibb's free energy change is $\Delta G_T{}^\circ = \Delta H^\circ - T\Delta S^\circ$, and $\Delta G_T = \Delta G_T{}^\circ + RT \ln K_P$, where K_P is the equilibrium constant and is a function of pressure ($K_P = 1$ at equilibrium). In terms of changes in heat capacities of the products and reactants, the standard free energy change, $\Delta G_T{}^\circ$, can be expressed from

$$\Delta G_T{}^\circ = \Delta H^\circ_{298} + \int_{298}^{T} \Delta C_P dT - T\Delta S^\circ_{298} - T\int_{298}^{T} \frac{\Delta C_P}{T} dT \qquad (5\text{-}4)$$

As $\Delta G_T{}^\circ$ becomes more negative, the stability of a compound increases, and it becomes difficult to reduce it. Consider the formation of Mo coatings via CVD by reduction of $MoCl_5$ with H_2.

Free energy calculations (Ellingham diagram) show that $MoCl_5$ will be a suitable precursor because it will be reduced by H_2 gas to form HCl via the reaction $MoCl_5(g) + (5/2)H_2(g) \rightarrow Mo(s) + 5HCl(g)$. HCl has more negative free energy of formation than $MoCl_5$, and the $\Delta G_T{}^\circ$ versus T curve for $MoCl_5$ lies above the curve for HCl on Ellingham diagram. When the $\Delta G_T{}^\circ$ curves cross over, reduction is possible only at temperatures greater than the temperature at which the curves intersect. For example, H_2 can reduce $SiCl_4$ above 1500 K at 1 atm. pressure but not at lower temperatures. If, however, pressure is reduced, reduction would be possible. Consider the deposition of Cr coatings by the reduction of $CrCl_2$ by H_2 at 1400 K. The free energy curve for $CrCl_2$ does not intersect the curve for HCl at 1400 K and 1 atmosphere and reduction is not possible; however, if partial pressures are adjusted, then reduction may be possible. This is because the free energy change for the reaction $CrCl_2(g) \rightarrow Cr(s) + Cl_2(g)$ can be written as $\Delta G_T = \Delta G_T{}^\circ + RT \ln K_P$, where $K_P = (p_{CrCl_2}/p_{Cl_2})$, and p_{CrCl_2} and p_{Cl_2} are the partial pressures of gaseous $CrCl_2$ and Cl_2, respectively. If $(p_{CrCl_2}/p_{Cl_2}) = 1000$ at 1400 K, then, $\Delta G_T = \Delta G_T{}^\circ + 8.314 \times 1400 \times \ln(1000)$, and, therefore, $\Delta G_T = \Delta G_T{}^\circ + 80$ (kJ). The free energy gap between the HCl and $CrCl_2$ curves at 1400 K and 1 atm. pressure is less than 80 kJ, and, therefore, reduction of $CrCl_2$ to Cr will be possible with the above choice of partial pressure ratio (convertible to molar concentration). It should be noted that in some cases, CVD coating deposition may be possible just by gaseous decomposition even without a reduction reaction. For example, above a critical temperature, Si coatings may be deposited by decomposition of SiI_4 according to $SiI_4 \rightarrow Si + 2I_2$.

Whereas the preceding chemical thermodynamic considerations permit evaluation of the feasibility of CVD reactions, the actual rate of CVD coating deposition is determined by chemical kinetics and rate-limiting mass transport processes. Two rate-limiting conditions can be envisioned for coating deposition: gaseous diffusion and surface reactions (Figure 5-4). At high reactant gas velocity in the CVD chamber (or at low gas pressures and low temperatures), the diffusion boundary layer at the substrate surface is thin, and the diffusion through this layer fast. Under these conditions, coating deposition is limited by the kinetics of chemical reactions at the substrate surface (i.e., adsorption of precursor atoms, nucleus formation, and desorption of product[s] and unreacted gases). In contrast, at slow gas velocity (or at high gas pressures and high temperatures), the diffusion boundary layer at the substrate surface is thick, and gaseous diffusion through this layer is slow relative to the kinetics of chemical reactions at the interface. Under these conditions, gaseous diffusion through the boundary layer rather than adsorption,

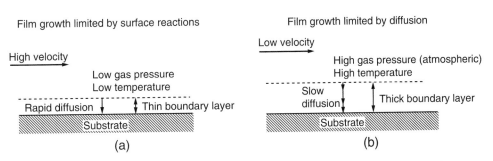

FIGURE 5-4 *(a) Physical transport processes involved in CVD film growth limited by surface reactions. (b) Physical transport processes involved in CVD film growth limited by diffusion.*

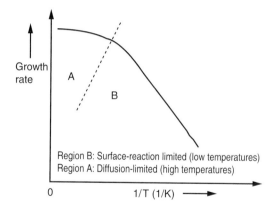

FIGURE 5-5 *A schematic Arrhenius diagram showing transition between mechanisms of CVD film growth due to a change in the operating temperature. Within each zone A and B, the kinetics of film growth follows the Arrhenius equation: $Rate = A \exp(-Q/RT)$, where A is a (temperature-independent) frequency factor, Q is an activation energy for diffusion, and R is the universal gas constant.*

reaction, nucleation, and desorption at the substrate surface, control the rate of coating deposition. Figure 5-5 shows a schematic of the natural logarithm of coating growth rate as a function of reciprocal of absolute temperature for a hypothetical CVD reaction. Experiments show that coating deposition is thermally activated, and consistent with an Arrhenius-type relationship, $rate = A \cdot \exp(-Q/RT)$, which leads to a linear variation of ln(rate) with inverse absolute temperature as shown by the linear segments in Figure 5-5. The change in the slope of the line in this figure indicates a change in the mechanism of coating deposition with changing deposition temperature; for example, at high temperatures (small $1/T$), gaseous diffusion becomes the slower (rate-controlling) step, whereas at low temperatures (large $1/T$), surface reactions become the slower step. In the intermediate temperature range, the deposition process is limited by both diffusion and surface reactions. These considerations are used to develop detailed models for designing the CVD process.

Ion-Nitriding

Ion-nitriding is a surface hardening technique from the broad group of ion implantation methods that are used to surface harden industrial parts. Ion-nitriding is a thermal-chemical process of surface hardening for production of nitrided surface layers that relies on the chemical diffusion of atoms into the sub-surface regions of the work piece. A pre-cleaned work piece is placed inside a vacuum furnace on a plate that is electrically isolated from the furnace. The furnace is evacuated, and the work piece is preheated under vacuum using either internal radiant heating, plasma heating, or both. A working gas is injected in the furnace, and a high voltage is applied between the plate (cathode) and the furnace (anode). The cathode starts to emit electrons, which ionize the surrounding gas molecules and cause ionic bombardment and heating of the work piece. On reaching a specified temperature (400–560 °C), nitrogen gas (in a predetermined ratio with respect to hydrogen) is added. This ionizes the gas mixture and creates a glow-discharge, and the gas ions bombard the surface. Nitrogen starts to diffuse into the surface of the parts, and creates a nitrided layer on the surface. As the temperature rises, the pressure of the gas mixture is increased to a level at which all holes, slots, and depressions are equally uniform in glow seam thickness; this assures a uniform case depth, regardless of part configuration. Then, by maintaining the temperature and gas pressure for a prescribed period of time, the desired case (i.e., nitrided layer) depth and hardness are obtained.

Ion-nitriding develops a compound layer at the surface that consists primarily of iron nitrides Fe_4N and Fe_3N. The relative amounts of these nitride phases can be controlled by proper selection of the processing conditions, which in turn determine the surface properties. For example, a carbon-free atmosphere coupled with a relatively low level of nitrogen will form Fe_4N, which is ductile and has excellent resistance to fatigue, impact loading, and wear. On the other hand, a carbon-rich atmosphere with a high level of nitrogen will form the hard Fe_3N phase, with low ductility but with excellent wear, erosion, and corrosion resistance. Ion-nitriding works on any ferrous alloy, allows for areas to be masked off from the treatment, and produces a highly repeatable layer composition. Unlike CVD, the process is pollution-free and environmentally safe.

Thermal Spray-Coating

The thermal spray techniques to deposit coatings consist of atomization and deposition of molten or semi-molten droplets of the coating material on substrates. Coating grows by rapid solidification of droplets that coarsen and sinter to cover the substrate. Multiple (up to a few hundred) passes may be needed to build up the desired deposit thickness. The feed material for thermal spray can be in the form of powder, wire, and rod. Industrial thermal spray techniques differ from one another in the type of energy source used for melting the feed material and associated hardware. The feed material is melted by using energy from fuel combustion, electric arc, or plasma. Flame spray, plasma spray (atmospheric plasma and vacuum plasma), HVOF (high-velocity oxy-fuel), and detonation gun are some of the most common thermal spray techniques for depositing a wide range of coatings. The basic approach of each of these techniques is illustrated in Figure 5-6.

In the flame spray process, temperatures are in the range of 3000–3350 °K and spray velocity is 80–100 m/s. Powder feed rate, working distance, and spray angle must all be controlled to achieve adherent coatings and homogeneous deposits. Flame spray coatings are usually deposited in air, and range in thickness from roughly 150 to 2500 μm. Porosity content is usually high (10–20%) and substrate–coating bond strength relatively low (15–30 MPa). In atmospheric plasma spray (APS) technique, electrodes (anode and cathode) are used to ionize a working gas, which expands and forms a high-velocity jet. Common working gases are argon, hydrogen, and mixtures of argon and hydrogen, argon and helium, and argon and nitrogen. Gaseous mixtures of diatomic and monatomic gases are usually preferred, because diatomic gases have better thermal conductivity than monatomic gases, whereas monatomic gas molecules can attain higher velocities. Plasma temperatures are nearly 14,000 °K, and gas velocity is about 800 m/s. The tensile bond strength is usually less than 40 MPa, porosity levels are 1–7%, and coating thickness is in the range of 50–500 μm. Vacuum plasma spray (VPS) is similar to APS, but the coating is deposited inside a vacuum chamber. This minimizes droplet contamination (e.g., reaction with gases in the atmosphere) and allows higher-impact velocities to be achieved because there is less atmospheric drag on droplets during flight. Plasma temperatures of 10,000–15,000 K are attained at velocities in the range of 1500–3500 m/s. Typically, 150–500-μm-thick coatings are deposited using VPS with high bond strength (>80 MPa) and low (<1%) porosity contents.

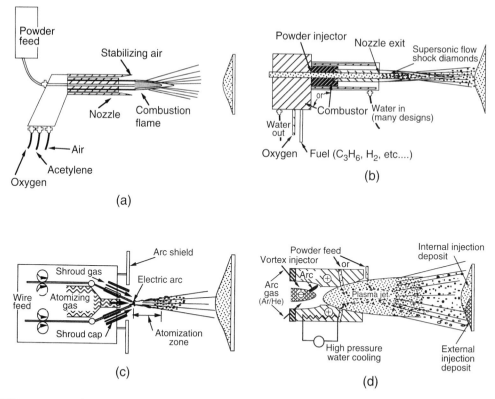

FIGURE 5-6 *(a) Schematic of the flame-spray process. (R. W. Smith and R. Knight, JOM, August 1995, p. 32). (b) Schematic of high-velocity oxy-fuel (HVOF) thermal spray process. (R. W. Smith and R. Knight, JOM, August 1995, p. 32). (c) Schematic of the wire-arc spray process. (R. W. Smith and R. Knight, JOM, August 1995, p. 32). (d) Schematic of the plasma spray process. (R. W. Smith and R. Knight, JOM, August 1995, p. 32).* Reprinted with permission from the Minerals, Metals & Materials Society, Warrendale, PA (www.tms.org).

In high-velocity oxy-fuel spraying (HVOF), a mixture of a fuel gas (e.g., acetylene, propane, or propylene) and oxygen is burnt at high pressures to form a high-velocity exhaust jet. Feed material introduced in the jet melts and is accelerated toward the work at high velocities (2000 m/s). Flame temperatures are, however, relatively low (about 3440 K). Coating thickness is in the range 100–300 µm, and high (>80 MPa) bond strength and low (<1%) porosity levels are achieved.

Physical Considerations in Thermal Spraying

The detonation gun (D-gun) process competes with the HVOF process in producing high-quality coatings with low-porosity content and excellent adhesion. The process consists of four steps: injection of oxygen and fuel mixture (e.g., acetylene) into the combustion chamber; injection of the powder and nitrogen gas (buffer), which separates the gas supply from the explosion area and prevents backfiring; ignition of the mixture and powder acceleration toward the substrate by the detonation wave; and ventilation of the barrel by nitrogen gas. Typically 8 to 10 detonation waves per second are possible. The droplet velocity can be as high as 1200 m/s, and the deposits have low (0–2%) porosity and excellent adhesion (up to 250 MPa).

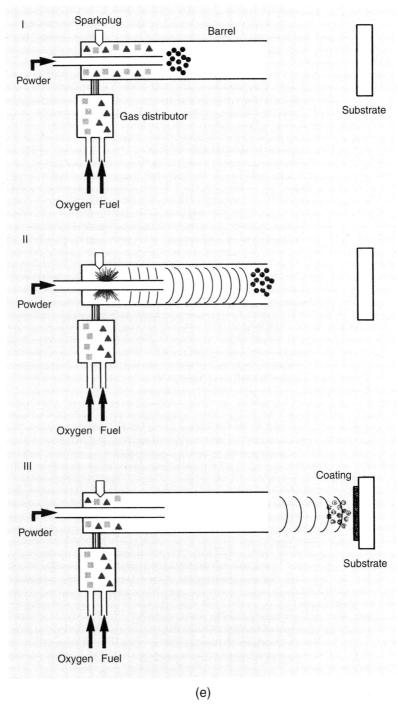

(e)

FIGURE 5-6 *(e) Schematic of the detonation gun thermal spray process.* (E. Kadyrov and V. Kadirov, Advanced Materials & Processes, August 1995, p. 21). Reprinted with permission from ASM International, Materials Park, OH (www.asminternational.org).

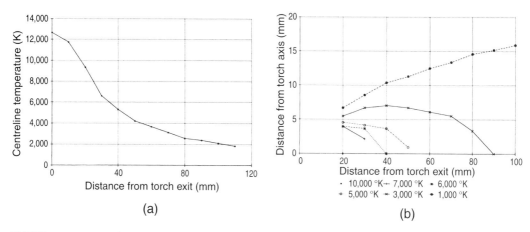

FIGURE 5-7 (a) Axial temperature distribution in the plasma spray process as a function of the distance from the gun nozzle. (Figure adapted from L. Pawlowski, The Science & Engineering of Thermal Spray Coating, Wiley, New York, 1995, p. 73). (b) Radial and axial temperature distribution in plasma spray. Each curve is an isotherm at the temperature shown. The temperature gradient is very steep; for example, the temperature drops from 10,000 °K to 1000 °K over a 4-mm radial distance at an axial distance of 20 mm from the gun exit. (Figure adapted from L. Pawlowski, The Science & Engineering of Thermal Spray Coating, Wiley, New York, 1995, p. 73).

The most important process parameters in thermal spray are velocity, temperature, working distance, powder feed rate, and gas atmosphere. In thermal spray, the work and/or spray gun traverse to build up the coating. Speed of traverse is important; high linear velocities lead to small coating thickness per pass but the gun returns to the same spot rapidly. This reduces the oxidation and cooling rate of the deposit; slow cooling allows incoming droplets to fuse and flow with the film to form a continuous coating, especially at high spray densities. If, in contrast, the speed of traverse is slow, then the deposit thickness per pass will be large, but appreciable temperature gradients may exist between the top and bottom surfaces of the deposited layer, resulting in thermal stresses. Also, more oxidation, splat boundaries, and porosity will result because more time is available for chemical reaction with the atmosphere.

The melted feed material exits the gun nozzle and spreads out in a cone-shaped region. The heat flux is high and heat loss low in the center of the spray cone. In contrast, heat loss to surrounding gases is greater around the periphery than in the center of the spray cone. As a result, droplets near the outer periphery of the cone may arrive at the substrate in partially solid or fully solid state, yielding greater porosity on the outer regions of deposits. Very steep temperature gradients exist both along the centerline of the spray cone and in the radial direction, as shown in Figure 5-7. For the example shown in Figure 5-7a, the temperature drops from 12,000 K to 2,000 K along gun axis over a distance of 10 cm.

The thermal properties of the feed material, together with the thermal conditions in the spray chamber, govern the extent of melting. When solid powders are injected into a combustion flame or plasma, the heating of particles depends on the magnitude of the Biot number, Bi, where $Bi = hd_p/2\lambda$, h is the heat transfer coefficient at the particle–gas interface, and λ is the thermal conductivity of the particle material. For $Bi < 0.01$, the temperature of the particle is uniform and there are no temperature gradients within the particle, whereas for $Bi > 0.01$, temperature gradients will exist within the particle, especially in low-conductivity materials. The temperature gradients within the particles can be calculated using the Fourier equation for

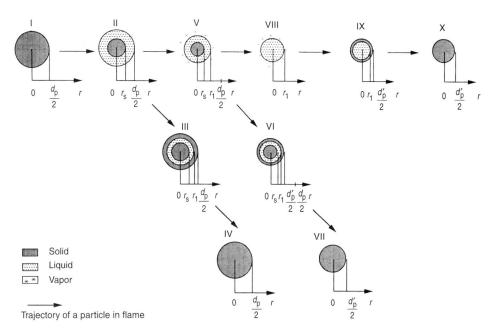

FIGURE 5-8 *Thermal response of nonconducting particles during flight in the flame spray process. Phase change could include partial or complete melting, boiling, and partial or complete resolidification of the melted powders.* (L. Pawlowski, The Science & Engineering of Thermal Spray Coating, Wiley, New York, 1995, p. 98).

heat conduction subject to appropriate initial and boundary conditions. The calculations are made for different thermal scenarios, some of which are shown in Figure 5-8 for thermally non-conductive particles. These include solid particles with a thin shell of liquid, partial evaporation of the liquid surrounding a solid core, and particles liquid inside and solid outside. Because of the existence of a size distribution in powders, several different thermal scenarios may operate concurrently. The thermal spray parameters are, however, optimized based on the largest powders in the distribution, and this may lead to overheating and boiling of fine particles. As illustrated in Figure 5-8, the droplet may completely melt and superheat, leading to surface boiling and evaporation. Further along its trajectory toward the substrate, the shrunk (i.e., partially evaporated) droplet may partially solidify on its surface prior to impact. Alternatively, there may be no evaporation but the liquid surface may form a solid crust while the droplet interior may remain molten. These various scenarios have been mathematically analyzed and modeled to predict the thermal behavior of droplets.

For thermally conductive particles with small values of Bi(i.e., $Bi < 0.01$), a heat balance between the particle and the flame yields

$$\pi d_p^2 h(T_g - T_p) = \frac{1}{6}\pi\, \rho_p C_p d_p^3 (\frac{dT_p}{dt}) \tag{5-5}$$

The solution of this equation, subject to appropriate boundary conditions, yields the droplet temperature as a function of residence time in the flame. For thermally conductive particles, the spray parameters can be optimized from a knowledge of the Difficulty-of-Melting Factor (DMF) of powders and the Ability-of-Heating Factor (AHF) of flame. The DMF of powders specifies the conditions under which the particle at the end of its trajectory is completely melted. The DMF is given from

$$\text{DMF} = \frac{H^2 d_p^2 \, \rho_p}{16},$$

where H is the heat content per unit mass of the particle, d_p is the particle diameter (taken to be the largest particle in the distribution), and ρ_p is the density of the particle. Because the DMF is calculated on the basis of the largest powders in the distribution to ensure complete melting of feed material, it is possible that smaller particles may overheat and vaporize. Particles of a given size will melt if the AHF of flame is greater than the DMF of particles. AHF is expressed in terms of arc current/input power parameters. Therefore, from a knowledge of DMF and from the condition that AHF > DMF for complete melting, power requirements for thermal spray can be estimated.

Upon reaching the substrate, the incoming droplets may impact a liquid film, a mushy layer, or a fully solid layer depending on the thermal conditions existing in the reactor (e.g., whether the substrate is preferentially cooled). The deformation of droplets impinging onto a cold surface leads to concurrent solidification and spreading, and this affects the shape and size of the grains in the deposit. If the thermal conditions in the reactor permit rapid heat extraction from the droplet via convection and radiation, then partial solidification of droplets may occur during flight, which will lead to high-porosity deposits. High-speed videography has been used to experimentally monitor the deformation processes occurring during high-velocity impact of droplets on solid surfaces.

The impingement, deformation, flow, and solidification behaviors of droplets determine the amount and distribution of porosity and other defects (e.g., unmelted grains, splat boundaries, oxide inclusions, etc.) in the deposit. In addition, the formation of porosity is influenced by the roughness of the substrate over which the impinging droplet spreads. Depending on the relative magnitudes of wavelength of substrate roughness λ and droplet diameter D, different mechanisms of spreading and porosity formation may take place. If $\lambda > D$, then spreading occurs by a repetitive acceleration–deceleration motion and ends with a catastrophic disintegration of the liquid phase. If, in contrast, $\lambda < D$, then the surface roughness plays a hindering role to spreading of droplet.

The spreading behavior of impacting droplets is mainly influenced by the inertial and viscous forces; surface tension forces are relatively unimportant at large impacting velocities even for small droplets (surface tension forces may, however, become important during co-deposition of composite coatings in which molten droplets and solid reinforcement particles come in contact with one another during flight). The viscous dissipation of a fraction of the kinetic energy of molten droplets occurs because of atmospheric drag. The residual kinetic energy transforms into thermal energy upon impact, which may raise the temperature by a few hundred degrees at impact velocities greater than about 400 m/s. The fluid drag experienced by the droplet as it moves through the stream of gas can be included in the equation of motion of the droplet. Assuming the viscous drag to be the major force experienced by the droplet during its flight, the equation of motion of the droplet can be written as

$$\frac{1}{6} \rho_p \, \pi \, d_p^3 \, \frac{dv}{dt} = \frac{1}{8} C_D \, \pi \, d_p^2 \, \rho'(u - v)^2 \tag{5-6}$$

where u is the gas velocity, v is the particle velocity, d_p is the particle diameter, C_D is the drag coefficient, which is a function of the relative velocity of particle with respect to gas via a Reynold's number, Re, where $Re = \rho' d(u - v)/\mu'$, and ρ' and μ' are the mass density and dynamic viscosity of the gas phase, respectively.

Upon impacting the substrate, the liquid droplet usually deforms into a lamellae with a pancake morphology (although other shapes such as flowers or rosettes may also form under certain conditions). A parameter ζ is used to describe the spreading behavior upon impact and is given from $\zeta = (2/d)\sqrt{(A/\pi)}$, where A is the surface area of the pancake-shaped splat and d is the diameter of the incoming droplet just before impact. The parameter ζ depends on Reynold's number, Re, and Weber number, We. These dimensionless groups are defined from $Re = \rho dv/\mu$ and $We = \rho dv/\sigma$, where ρ, μ, and σ are the density, viscosity, and surface tension of the liquid droplet, respectively. Expressions for the parameter ζ have been derived in terms of Re and We for different ranges of these dimensionless parameters. For example, at Re and We greater than 100, the expression for ζ has the following form

$$\frac{3\xi^2}{We} + \frac{1}{Re}\left(\frac{\xi}{1.2941}\right)^5 = 1.$$

The thickness of the lamellae is given from $h = 2d/3\zeta$ and the spreading time is given from $t = 2dRe/3v$, where v is particle velocity. The solidification time t of the splat is given from $t = (h/2a)(1/\alpha)$, where α is thermal diffusivity of the droplet material, h is the lamellae thickness, and a is the root of the equation:

$$K_\varepsilon + \phi(a) = \frac{k_1 \exp(-a^2)}{a},$$

where

$$k_1 = \frac{c_p(T_m - T_0)}{L\sqrt{\pi}},$$

and

$$K_\varepsilon = \frac{\lambda_p}{\lambda_s}\sqrt{\left(\frac{\alpha_s}{\alpha_p}\right)}.$$

Here T_m is the melt temperature, T_0 is the substrate temperature before impact, L is the latent heat of fusion of the droplet material, α and λ are the thermal diffusivity and thermal conductivity, respectively, $\phi(a)$ is the probability integral, and the subscript "s" refers to substrate. The preceding equation for solidification time assumes one-dimensional heat flow and a semi-infinite substrate with an ideal contact between the splat and the substrate. Under these assumptions, calculations show that for arc spraying of nickel powders on an iron substrate, a lamellae thickness, h, of 5 μm and solidification time of 1 μs are obtained at an atomizing gas pressure of 0.7 MPa, whereas at a pressure of 0.25 MPa the lamellae thickness and solidification times are 40 μm and 100 μs, respectively. During droplet impact, deformation, and solidification, large quenching stresses (as high as 100 MPa) are generated in the splat. Partial relaxation of these stresses takes place through micro-cracking and creep processes. However, most often a secondary heat treatment is required to remove quenching stresses and homogenize the metallurgical structure.

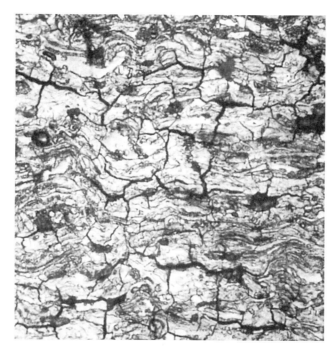

FIGURE 5-9 *Typical brick-wall structure of an as-deposited thermal spray coating. Evidence of internal cracking due to the thermal stresses is also seen.* (L. Pawlowski, The Science & Engineering of Thermal Spray Coating, Wiley, New York, 1995, p. 73).

Structure, Properties, and Applications of Thermal Spray Coatings

Structure. The structure of thermal spray coatings is either "columnar" (for metals such as Ni and NiCr deposited on high-conductivity substrates such as Cu) or "equiaxed" or "brickwall" type (for low-thermal-conductivity materials such as yttria and alumina ceramics). The photomicrograph of Figure 5-9 shows the brickwall structure of a thermally sprayed coating deposited on a low-conductivity substrate. Besides the actual coating material, the deposit may also contain unmelted grains, porosity, splat boundaries, and oxide inclusions (Figure 5-10). Due to rapid solidification, grain size in the deposited coating varies from a few nanometers (e.g., when substrate is cooled with liquid nitrogen) to a few hundred nanometers. Some variation in grain size occurs across the thickness of the deposited coating because of variation of cooling rate between the top (hotter) and the bottom (cooler) surfaces of the deposit. Adhesion between the deposit and the substrate develops because of mechanical gripping of surface asperities from the solidification shrinkage of splat, and because of chemical reactions with the substrate. For example, Mo and W metals sprayed onto a steel substrate form intermetallic compounds such as FeMo and Fe_2W, respectively. Generally, however, chemical interactions during thermal spray are limited owing to extremely short interaction times; subsequent heat treatment is usually done to enhance the chemical interactions (interdiffusion, reaction layer formation), which also alter the coating properties. Heat treatment also relieves the thermal stresses in as-deposited coating, which may be as high as 100 MPa. Heat treatment may also cause phase transformations, volumetric changes, and a decrease in the porosity. For example, normal thermally-sprayed Al_2O_3 coatings have a density of 3.67 $g \cdot cm^{-3}$. Upon heat treatment at 1200 K, the coating transforms into a phase of higher density (3.98 $g \cdot cm^{-3}$) with an accompanying shrinkage in

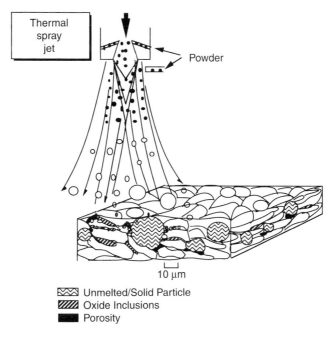

FIGURE 5-10 *Schematic illustration of the deposit structure in thermal spray. The deposit could include unmelted particles, porosity, and oxide inclusions.* (R. W. Smith and R. Knight, JOM, August 1995, p. 32). Reprinted with permission from The Minerals, Metals and Materials Society, Warrendale, PA (www.tms.org).

the volume. The as-deposited and heat-treated coatings are ground and polished using diamond, cubic boron nitride, or SiC, and any residual interconnected porosity is sealed by impregnating the pores with liquid sealants such as silica, Cr_2O_3, epoxy, silicone, or phenol.

The structure of the deposit may be inhomogeneous, with considerable spatial variation in the grain size and defect (e.g., porosity) content. This is because the liquid metal droplets undergo size changes as they are subjected to very large dynamic pressure forces from the gas stream velocity. The droplets, therefore, disintegrate into finer droplets because the dynamic forces far exceed the cohesive (surface tension) forces, which tend to preserve the integrity of the droplet. Droplets in the outer regions of the spray cone are smaller in size than droplets in the center of the spray cone, and droplet mass flux decreases with increasing radial distance from the spray axis (i.e., there are fewer droplets per unit area near the outer periphery of the cone). Because the core region of the spray cone has a higher droplet velocity and a higher volumetric concentration of droplets than the peripheral zones, the relative velocity between droplets and gas stream is small in the core region, and the thermal energy is large due to a higher droplet mass flux. These factors lead to a less effective convective heat transfer from the droplet to the gas during flight in the center of the spray cone than at the outer periphery. Cooling is, therefore, suppressed in the core region, and higher deposit densities are obtained in the center of the deposit. In contrast, microstructure at the outer surface of the deposit is dominated by droplets that arrive nearly fully solidified. This is because of smaller droplet mass flux and more effective heat transfer in the peripheral regions of the spray cone. The outer deposit generally shows a layered structure with deformed or fractured dendrites, porosity, and prior particle boundaries that persist even after heat treatment. The droplets arriving in the core region of the deposit show a fine spheroidal grain structure because of incremental solidification in which the rate of heat dissipation from an impacting droplet (by conduction through the deposit and the substrate, and by radiation and convection through the gas) is slow enough to keep a thin film of liquid on the surface of the deposit from solidifying. As a result, the impinging droplets meet the liquid film at the deposit surface, and the two liquids flow together before solidification occurs. This eliminates splat boundaries and yields a fine equiaxed grain structure at the center of the deposit. When solidification is not completed in flight and a thin liquid film is retained at the substrate surface, the mechanical force of impacting breaks any surface oxides on both the arriving droplet and the preexisting liquid film. The impact also disintegrates dendrite arms in partially solidifying liquid into smaller fragments. These fragmented dendrite arms aid the formation of an equiaxed grain structure via grain multiplication. In addition, turbulent convection in the melt caused by mechanical impact assists melting and detachment of dendrite roots. These broken dendrite fragments promote grain refinement in the deposited structure.

In the outer regions of the spray, solid nuclei may begin to form within the droplets during flight by heterogeneous nucleation, surface nucleation, surface oxidation, and droplet collision. The probability of achieving a high degree of undercooling before nucleation is high for small droplets. However, even for small droplets, the large undercoolings required for homogeneous nucleation may be impossible to achieve, and heterogeneous nucleation usually occurs at relatively low values of the undercooling. Collision between droplets in flight is a strong possibility and will catalyze the heterogeneous nucleation. The presence of impurity particles in the droplets will also catalyze the heterogeneous nucleation if the conditions are suitable; dendrites then appear to be emanating from the particle surface.

Properties. The most extensive use of thermal spray coatings is in applications requiring high hardness, wear resistance, and thermal protection. Hardness is a first approximation to wear resistance. The hardest thermal spray coatings are carbides and oxides, such as chromium oxide (Cr_2O_3) and tungsten carbide (WC). The hardest metal sprayed is molybdenum (Mo). Hardness depends on 1) thermal spray technique and operating parameters; 2) amount, size, and distribution of grains; 3) defect content; and 4) binders or additives. Different thermal spray methods and operating conditions yield different hardness in a given coating. For example, high-velocity oxy-fuel (HVOF) spray gives harder WC coatings than detonation gun (D-gun) technique, and both HVOF and D-gun give harder tungsten carbide coatings than atmospheric plasma spray (APS) technique. Hardness values are considered as a preliminary indicator of the quality of the deposited coating. Porosity, unmelted grains, and stray inclusions lower the coating hardness. The hardness, H, of as-deposited coating is very sensitive to the porosity content, P, and exponentially decreases with P according to $H = H_0 \exp(-aP)$, where H_0 is the hardness of nonporous bulk material, and a is an empirical constant. Additives and spray atmosphere also affect the hardness. For example, titanium dioxide (TiO_2) is added to aluminum oxide (Al_2O_3) and chromium oxide (Cr_2O_3) to lower their melting points and make the spray process energy efficient. In one study, it was found that coatings of pure Al_2O_3 led to 6.8% porosity and a Vickers hardness (HV) of 1080, whereas coatings of Al_2O_3 and 13% TiO_2 resulted in 3.7% porosity but the hardness also decreased to 1060. Heat treatment of as-sprayed coatings usually improves the hardness because porosity content is decreased. The hardness of the as-deposited coatings depends also on the gaseous atmosphere in the spray chamber. For example, oxygen content in the spray chamber can influence the hardness and porosity in the deposited material. In the case of molybdenum coatings, at low-oxygen partial pressures soft Mo coatings form, whereas at high-oxygen partial pressures, hard Mo coatings form because of partial oxidation of Mo in oxygen-rich atmosphere. Similarly, Ti coatings sprayed in pure Ar have lower hardness than Ti coatings sprayed in pure nitrogen that contain reaction-formed hard titanium nitride (TiN).

A variety of techniques have been developed to characterize the adhesion strength of thermal spray coatings. The tensile bond strength of thermal spray coatings is most widely characterized using a standard procedure that involves tensile pull on an adhesive joint created with epoxy between the coated test specimen and an identical uncoated base metal specimen. If failure occurs at the coating–substrate interface, then the failure stress will be a measure of the bond strength. If, however, failure occurs within the coating, then the bond strength is greater than the fracture strength of the coating material. Coating adhesion is influenced in a complex manner by the operating process parameters and the nature of the substrate and the deposit. Frequently a soft (ductile) alloy is used as a bond coat between the substrate and a hard topcoat in order to improve the adhesion.

Typically, the bond strength of thermally sprayed carbide coatings is greater than about 50 MPa, which indicates reasonably good adhesion. In contrast, bond strength of Ni-Al and Ni-Cr coatings is inferior ($\sim$30–40 MPa). Generally, the bond strength of thermally sprayed oxide coatings is less than that of carbide coatings. Also, if the substrate is oxidized, lower adhesion results. Adhesion is usually better on rough substrates, although in the case of brittle coatings, roughness also tends to increase the residual stresses and tendency for interfacial cracking because of notch formation. High-thermal-conductivity substrates (e.g., Cu and brass) yield low bond strength, and a bond coat of a low-conductivity material must be applied to enhance the adhesion strength. The gaseous atmosphere in the spray chamber also affects the adhesion. For example, titanium coatings sprayed in argon atmosphere show better adhesion to substrate than titanium coatings sprayed in nitrogen. This is because titanium nitride (TiN) that forms in nitrogen atmosphere deforms less than pure Ti on impact. This reduces the interfacial contact area and leads to poor adhesion.

Generally, the elastic modulus (E) and tensile strength (σ) of thermally sprayed coatings are less than the corresponding bulk values. Both E and σ depend on spray technique, operating parameters, and binder content. Usually carbide and oxide coatings attain only 20–40% of the bulk value of the elastic modulus of the coating material. The strength, σ, is typically 5–10% of bulk strength for most ceramic materials. The effect of the binder is system-specific; thus, both strength and failure strain of tungsten carbide coatings increase with the amount of low-melting Co binder in the coating. Increasing the spray distance reduces the coating modulus because large working distances lead to low-impact velocities, smaller temperature rise upon impact, and incomplete fusion of liquid or semi-solid droplets, all of which lead to more porosity and lower modulus. In the case of metallic coatings, coating modulus up to 85% of the bulk value has been achieved with vacuum plasma spray (VPS). The ultimate tensile strength of metal coatings such as Al, Ni, and Ti is 50–75% of the bulk value and about 20% for refractory metals such as W. Bond coat materials such as Ni-Al and Ni-Cr used for oxide ceramic overlays are relatively ductile and have modulus intermediate between coating and substrate materials.

A variety of wear-resistant coatings with low coefficient of friction (μ) can be deposited using thermal spray techniques. Hard and wear-resistant coatings of chromium oxide (Cr_2O_3) against steel have μ of 0.1–0.7 under dry unlubricated conditions, and 0.1–0.2 under lubricated conditions. The value of μ for dry sliding wear of Cr_2O_3 against graphite is 0.2–0.3. Typical range of μ for WC-Co coatings sliding against themselves is 0.4–0.6, and μ of Al_2O_3-ZrO_2 composite coatings and TiO_2 coatings rubbing against sapphire is 0.216–0.026 and 0.08–0.2, respectively. In a manner similar to other properties, wear and friction behaviors of thermal spray

TABLE 5-1 Coating Properties: Thermal Spray versus Hard Chrome

Coating	WC-12Co	CrC-25NiCr	Hard Chrome
Vickers hardness	1100–1400	900–1200	800–1000
Surface finish, RMS	140–170	180–200	40–50
Dry abrasive wear rate (relative)	0.2	0.4	1.0
Service temperature	1000	1500	800

Source: Adv. Mater. Proc., August 1997, p. 20.

coatings are also influenced by the coating technique; for example, abrasive wear of most oxide coatings against SiC is lower for vacuum plasma spray (VPS) coatings than for atmospheric plasma spray (APS) coatings because of lower porosity and higher hardness of VPS coatings. Thermally sprayed ceramic coatings have superior wear resistance than widely used electroplated hard chrome plating, as noted from the data of Table 5-1.

Unlike strength and modulus, the coefficient of thermal expansion (CTE) of thermal spray coatings is usually close to their bulk value. Porosity in the coating, and nature and roughness of substrate, minimally affect the CTE; however, chemical phases and crystal structure of the deposited coating significantly influence the CTE. The CTE mismatch between commonly deposited ceramic coatings and metal substrates is typically on the order of 20×10^{-6} to 22×10^{-6} per °K. Such a large CTE mismatch can give rise to large residual stresses in ceramic coatings deposited on metals during temperature excursions, and these stresses will decrease the bond strength of the coating and increase its propensity for fracture. Strain-tolerant compliant layers having CTE intermediate between the overcoat and the substrate are used as intermediate bond coat to improve the adhesion and decrease the thermal stresses. Another important property of the thermal shock resistance (TSR) of as-sprayed ceramic coatings depends on the CTE, modulus, and fracture strength of the coating materials and on the coating microstructure. TSR is the maximum allowable temperature difference (see Chapter 3) that a part can withstand without cracking. Generally, coatings of high elastic modulus and large coefficient of the thermal expansion have low thermal shock resistance, and high porosity content in the coating microstructure improves the TSR.

Applications. Thermal spray coatings are used principally for wear, erosion and corrosion resistance, and thermal protection. Selected applications of thermally sprayed ceramic, metallic and refractory metal coatings are described in this section. Chromium oxide (Cr_2O_3) and Ni-base superalloy coatings have been used for wear and erosion resistance in petroleum and gas search drilling installations. Molybdenum steel valve lifters in auto engines can be replaced with lightweight aluminum valve lifters having thermally sprayed coatings of an iron base alloy (Fe-C-Si-Mn); an overall weight reduction and reduction in fuel consumption are the major benefits. Landing gear components (e.g., inner cylinder, drag braces, and brake torque tubes) in the aircraft have traditionally been made from hard chrome-plated low-alloy steel. This material has been replaced by harder nickel aluminide (NiAl) thermal spray coatings that display superior wear performance. In textile machinery, textile fibers cause sliding wear of rotating steel drums. Wear-resistant thermal spray coatings of chromium oxide (Cr_2O_3) permit steel drums to be replaced by lighter and cheaper aluminum drums. Vacuum pumps and agro-alimentary pumps used in manufacture of liquid yogurt and chocolate use biocompatible ceramic coatings deposited using thermal spray processes. Rolls used for cold-rolling of metal sheets are coated with a ceramic alumina-zirconia coating to extend roll life over chrome-plated rolls that have been traditionally used in cold-rolling. Similarly, hot extrusion dies used for manufacturing seamless tubes of brass and Cu are made of an iron base alloy (Fe-5Cr-1C-V, Mo); with a thermal spray coating of alumina over a bond coat of nickel-chrome, die life increases dramatically.

Judiciously selected thermal spray coatings can reduce the corrosion and oxidation of industrial components and act as thermal barriers. In automobile alternators, aluminum mid-plates are thermal spray coated with Al_2O_3 to improve salt corrosion resistance, moisture resistance, and attain specified breakdown voltage at low cost. Thermal barrier coatings (TBCs) allow heat from combustion gases to do additional work via a turbocharger in an adiabatic engine; this eliminates the need for a cooling system. Multilayer coatings of zirconia (ZrO_2), yttria (Y_2O_3), and a cobalt-chrome alloy (CoCrAlY) are used to coat ductile iron, nickel base alloys, and aluminum composites for thermal protection. In aero-engines and marine gas turbine engines, turbine vanes and turbine blades are thermally sprayed with oxidation and corrosion-resistant coatings such as zirconia-yttria or zirconia-magnesia ($ZrO_2 + MgO$) with an intermediate bond coat of either NiAl or M-Cr-Al-X, where $M =$ Ni or Co, and $X =$ Y, Hf, Si, or Ta. Combustor parts are made of superalloys, often in single-crystal form, with melting temperatures in the range 1500–1600 K. Without thermal barrier coatings (TBCs), these superalloys will melt, oxidize, and creep at the operating temperatures approaching 1640 K. Similarly, thermal barrier coatings (TBCs) for combustion cans and vanes, WC-Co wear-resistant coatings for compressor fans, and Co-base and Cr_2C_3-NiC composite coatings for erosion resistance in compressor airfoil have all been applied using the thermal spray techniques.

Weld overlays of stainless steel and Ni-base coatings are used to protect joints from corrosive attack of combustion atmospheres in jet engines where H_2S, SO_2, and SO_3 form. These overlays have a relatively short service life. In contrast, Fe-Al overlays deposited using thermal spray techniques improve the service life by forming intermetallics (FeAl and Fe_3Al) that have high melting points, high strength, and excellent resistance to sulfidation. The Al content of Fe-Al coatings determines the corrosion resistance; high Al contents form a protective scale, but at low Al contents greater sulfide attack and corrosion occur.

Oxidation- and heat-resistant refractory silicide coatings such as titanium disilicide ($TiSi_2$) are also used in combustor applications. These coatings are deposited by plasma-spraying of $TiSi_2$ powders in vacuum (in air, $TiSi_2$ converts to Ti_5Si_3, which has a high melting point, 2130 °C, but very poor ductility at room temperature). The $TiSi_2$ coatings have a high melting point and retain strength to elevated temperatures. The compound $TiSi_2$ is highly resistant to oxidation, and $TiSi_2$ diffusion coatings are used to protect titanium from oxidation. In addition, among all transition metal silicides, $TiSi_2$ has the lowest electrical resistivity; it is used in integrated circuits as contacts and gate electrodes.

There are many other important applications of heat-, corrosion- and oxidation-resistant thermal spray coatings. These include TBCs on vane platforms in high-pressure turbines, plasma-sprayed coatings of zirconia-yttria or zirconia-calcia with a nickel-chrome bond coat in rocket thrust chamber liners, insulating ceramic overlayers for microelectronic substrates (e.g., high-conductivity Cu, Al, or steel), and Al, Ta, and Nb coatings on electrodes of double-layer capacitors made from activated carbon fibers. In addition, boilers in power generation plants receive thermal spray coatings of 316-stainless steel or a nickel-chrome alloy for resistance to carburizing and erosion. Boilers that burn pulverized coal have an inside tubing that is exposed to steam at high pressures and to burning coal, which leads to hot corrosion and erosion from coal particles. In the case of blast furnaces, shell-cooling steel pipes are submitted to carburizing and embrittlement; thermal spray composite coatings of alumina-titania reduce the carburization of cooling pipes. Other applications are in heater rolls in laser printers, fax machines, and copier machines. In these components, a steel pipe is bond-coated with a nickel-aluminum-moly (NiAlMo) alloy. Then a top coat of electrically insulating spinel compounds and heater coatings of copper-zinc ferrite or a cermet (e.g., titania-nickel-chrome and alumina-nickel chrome) are thermally sprayed. These coatings reduce warming time and power consumption compared to classical halogen lamp heating. Another application of thermal spray coatings is in melting crucibles made of oxide ceramics and graphite. To avoid contamination by carbon, duplex coatings of tungsten-tantalum and rhenium-tungsten are used on crucibles. In nuclear industry, parts of thermonuclear fusion devices require thermal fatigue resistance, reduced ion erosion, and high melting point. Titanium carbide, vanadium carbide, and titanium diboride coatings are thermally sprayed on Cu and steel for such applications. Fusion devices use thermal spray coatings of Cu on Cu-W tubes for brazing of graphite armor to tube.

Electroplating

Electrodeposition or electroplating deposits thin metal, alloy, or composite coatings on new parts for wear and corrosion resistance and aesthetic appeal, on worn or machined parts to restore dimensions within acceptable tolerances, and on parts that need to be precisely duplicated by electroforming. Electroforming consists of production of free-standing bodies by electrodeposition of metals onto shaped mandrels, which are capable of subsequent removal. This enables the shape and surface finish of the mandrel to be faithfully reproduced in a single operation.

Physical Considerations

The deposition rate and the structure and properties of electroplated coatings depend on the solution composition, pH, temperature, and electrical current density. Plating is done in tanks or vats of up to several-thousand liters' capacity. The solution in the tank is mildly stirred using air jets or work motion, and electric heaters or steam coils are used to maintain the desired temperature; plating temperatures are low ($<100\,°C$), and part distortion and undesirable metallurgical changes minimal. Work is mounted on a jig, which serves as the cathode, and suspended in the electrolyte. The time to deposit the required thickness is estimated from the applied current density and the known efficiency of deposition. The deposition efficiency is the ratio of the weight of metal actually deposited against the weight that should have been deposited theoretically by the current and time combinations used.

The foundational electrochemical principle in plating is the Faraday's law, which states that for any metal, 96,500 Coulombs of electric charge (or Faraday) will deposit one gram-equivalent (g-eq) of the metal, where gram equivalent is the atomic weight of the metal divided by the valence charge. Thus, theoretically (i.e., for 100% deposition efficiency) 1 F charge (i.e., 96,500 C) will deposit (63.54/2 or 31.77) g of Cu from a $CuSO_4$ solution (valence charge: 2, atomic wt. of Cu = 63.54 amu). If a current of 15 A is passed for 10 min. through the $CuSO_4$ solution, the mass of Cu deposited will be 15(A) × 600 (s) × 31.77 (g)/96,500 (C) = 2.964 g. If this amount of Cu is plated over a 1500 cm^2 area, then the thickness 'x' of the deposited film in μm will be 2.964 (g)/[1500 (cm^2) × 10^{-4} × 8.93 (g·cm^{-3})] = 2.2 μm, where the density of Cu is 8.93 g·cm^{-3}. The deposition efficiency is usually less than 100% because some of the electrical energy is consumed in side reactions (e.g., hydrolysis and gas evolution) that do not directly contribute to the plating process. Thus, if a 25 μm thick film of Zn (density: 7140 kg·m^{-3}, at. wt.: 65.38 amu) is electrodeposited on a 1.5 m^2 area from a $ZnSO_4$ solution (valence charge: 2) by passing a current of 1000 A for 15 min., then the actual weight of Zn deposited will be 1.5 (m^2) × 25 × 10^{-6} (m) × 7140 $(kg \cdot m^{-3})$ = 0.268 kg or 268 g. The theoretically expected weight from Faraday's law will be 1000 (A) × 15 (min.) × (60 s/min.) × (65.38/2) × (1/96,500 (C)) = 304.9 g. The efficiency of the deposition process (i.e., actual weight divided by the theoretical wt.) will be (268/304.9) = 0.879 or 87.9%.

In industrial practice, the electroplating of sheet and tubular products is usually a continuous process. For example, tin plating of continuous steel strip is done to enhance the corrosion resistance of steel. The Faraday's law permits estimation of plating process parameters (e.g., strip velocity through the plating tank, length of the tank, etc.) in a continuous plating operation. The following examples illustrate the basic approach in designing a continuous plating process. Suppose that it is required to deposit a 1 μm thick film of Sn in a continuous fashion on each side of a 1 m wide strip of steel at a current density of 7000 $(A \cdot m^{-2})$ by passing the strip through a 10 m long tank containing the electrolyte. It is required to estimate the velocity, V, of the strip through the plating bath. First, we note that the residence time of each meter of the strip in the tank will be (10/V) min. The mass of Sn deposited on both sides of a 1 m long segment of the strip will be 1 (m) × 1 (m) × 2 × 10^{-6} (m) × 7310 $(kg \cdot m^{-3})$ = 0.01462 kg or 14.62 g, where 7310 $kg \cdot m^{-3}$ is the density of Sn. The gram-equivalent of Sn is (118.7/2) = 59.35. Thus, to deposit 14.62 g of Sn will require (14.62/59.35) or 0.2643 F of charge. Because 1 F is 96,500 C, 0.2463 F will equal 23768 (A·s) or 396.19 A·min of electrical charge. This is the amount of charge that must be delivered to each linear meter of the strip in a time of (10/V) min. at a current density of 7000 $A \cdot m^{-2}$. Now because the total surface area of each linear meter segment of the strip (two sides) is 2 m^2, the current to this area will be 7000 $A \cdot m^{-2}$ × 2 m^2 = 14,000 A. The residence time of the segment in the plating tank will be (396.19 A·min/14,000 A) = 0.0283 min. = (10/V). Therefore, the strip velocity through the tank will be 353.4 m/min.

There is no technical limit to the thickness of the electrodeposited coating, although most surfacing applications do not require very thick coatings. Application of coating is not confined to the line-of-sight. Areas that do not need to be coated can be masked with non-conducting tapes. Part geometry affects the evenness of deposits; for example, thickness is more at projections than at recessions because current density is higher at projections. This can be overcome by making current density over the part as uniform as possible. Several approaches are used to accomplish this such as correct positioning and sizing of anodes, use of anodes that approximate the shape of surface to be plated, positioning 'burners' or 'robbers' (i.e., conducting metal plates that steal some of the deposit and prevent excessive deposition) near high current-density areas, and use of nonconducting shields to reduce current flow at sharp edges. The deposition rate in electroplating is seldom greater than 75 μm/h, but forced circulation can increase this rate. Because ionic species rather than macroscopic particles form the deposit, electroplating does not adequately fill holes, level rough material, or completely cover surface imperfections. Hard and inert nonmetallic particles (e.g., diamond, carbides, and oxides) can be co-deposited with the base metal coating from electrolytic suspensions containing these particles to form wear-resistant composite coatings. Plating conditions can be adjusted to alter the hardness, internal stresses, and metallurgical characteristics of the deposit.

Plating Practice

Practically any metal or alloy can be electroplated on a conductive substrate. Non-conductive substrates (e.g., ceramics or plastics) are first metallized, electroless-plated, or painted with a conductive paint to make them conductive. The most common plating metals include tin, nickel, copper, chromium, silver, gold, platinum, and their alloys. Tin-lead alloys are widely used in industrial applications; for example, alloys with 5–12% Sn are used in bearings, alloys with 10–60% Sn are used for soldering and corrosion resistance, and alloys with 60–98% Sn are deposited on printed circuit boards as a metal etch resist, to improve solderability of contact surfaces (the industrial usage of Pb in solders has, however, seen a steady decline because of its toxicity and the availability of alternative Pb-free solders). SnPb alloys are electroplated using solutions such as fluoborate, sulfonate, chloride, fluosilicate, and pyrophosphate baths. Anodes are high-purity lead bars that are either cast or extruded, and must contain less than 0.01 to 0.001% impurities such as Sb, Bi, Cu, Fe, Zn, and Al.

Electroplated nickel coatings reduce mechanical wear, corrosion, fretting, and scaling, and are extensively used in electroforming of tools, molds, and dies. Nickel plate is often applied over copper and under chrome to obtain a decorative finish. In addition, nickel tends to gall when rubbed against some metals even when lubricated, which requires chromium to be used as an overlay. The original process of electroplating of Ni was patented by Watts in 1916 (the baths are also called Watts baths). These baths use solutions of nickel sulfate, nickel chloride, and boric acid (a pH stabilizer). The major source of Ni in the bath is nickel sulfate and the minor source is nickel chloride. The chloride ions in the bath increase the electrical conductivity of the electroplating bath. Agitation and heat increase ionic diffusion and plating rate. Wetting agents are added to bath to prevent sticking of hydrogen gas that forms during plating to the workpiece, which could lead to surface imperfections such as pits. The bath is cleaned by circulation over activated carbon to trap impurities. Periodic cleaning is necessary because the adhesion of Ni coatings is impaired by contaminants such as Cu, Pb, Zn, and Cd.

Copper is mostly used as an under-plate because it rapidly tarnishes in contact with ambient atmosphere. Copper is the only metal that will readily plate on zinc die-cast parts, and is also widely used in printing rollers, surface lubricants in metal working, and electroforming. Copper is an excellent undercoat for other metal deposits because it covers holes and splinters efficiently, is relatively inexpensive, has high plating efficiency (good coverage), requires less hazardous waste treatment than Ni (recycling of bath possible), and has high electrical conductivity (good coat for PC bonds, and also as topcoat on steel wire to give high-strength, high-conductivity electrical cables). The deposition process uses one of the three types of baths—alkaline bath (cyanide and non-cyanide baths), acidic bath (sulfate and fluoborate baths), and mildly alkaline baths (pyrophosphate baths). Alkaline baths have better throwing power but are more difficult to control chemically. The bath contains copper cyanide (CuCN) and sodium cyanide (NaCN) as the major ingredients and may also contain sodium carbonate (pH stabilizer) and proprietary formulations. The bath temperature is 45–65 °C, and current densities are typically 100–300 amps per square decimeter. The maximum deposit thickness for most applications is 0.05–0.1 mil. Filtering of spent bath over activated carbon is required for removing the impurities. Acidic baths are cheaper than cyanide baths; use a mixture of copper sulfate, sulfuric acid, and chloride salts; and operate at 20–40 °C. High deposition rates are possible at high sulfuric acid concentrations and at low copper sulfate concentrations because of an increase in the electrical conductivity of the bath. Impurities such as Ni, Co, Cr, and Fe do not readily co-deposit with high-purity Cu films but reduce the bath conductivity at concentrations greater than 1000 ppm. In contrast, Sn and Sb co-deposit even at very low (~60 ppm) concentrations, and make the deposit dark and brittle. Plating of Cu-base alloys, particularly brasses, is widely used primarily because of their decorative appeal. Yellow brass (70Cu-30Zn) is more widely electroplated than red brass (80Cu-20Zn), and is used in brass-plated steel wire that promotes adhesion to rubber in steel-belted tires. Plating baths for electroplating of brass contain copper cyanide (CuCN), zinc cyanide (ZnCn), sodium hydroxide, soda ash, and wetting agents. The bath temperature is typically $\sim$40 °C, and the anode is a rolled or extruded bar with low ($<$0.03%) content of Pb and Sn impurities. The color and luster of the coatings are controlled by varying the relative ionic concentrations of Cu and Zn in the bath, and by controlling the cyanide ion concentration.

Electroplated chromium coatings are used on plastic molds, metal forming and drawing dies, cutting tools, gauges, cylinder liners, piston rings, crankshafts, hydraulic rams, and numerous other industrial components. In plastics-processing equipment, barrels, screws, nozzles, and runner plates are chrome plated. Metals most commonly chrome plated include carbon steel, cast iron, stainless steel, tool steel, copper, brass, aluminum, and zinc. Pure electroplated chromium is harder and more brittle than pure Ni. Frequently, electroplated Cr deposits are designed to contain a small amount ($\sim$1%) of molybdenum for improved resistance to both mechanical wear and corrosion in acidic environments. Chromium coatings have low coefficient of friction, resist sticking and corrosion, and retain room-temperature strength to 300 °C. Industrial Cr coatings are usually no thicker than about 0.5 mm; thicker deposits require overplating followed by finish machining by grinding. Chrome plating is energy sensitive because of the poor deposition efficiency of chromium, which is evident from the low values of the gram-equivalent of trivalent and hexavalent Cr ions (g-eq. is $52/3 = 17.3$ for Cr^{3+} and $52/6 = 8.7$ for Cr^{6+}). The plating efficiency of the standard hexavalent Cr deposition process is quite low ($\sim$14%), but smooth coatings having high hardness (950–1100 Vickers) are obtained. The plating efficiency for trivalent Cr deposits is higher than that for hexavalent Cr deposits. Both hexavalent Cr and trivalent Cr coatings are used in industrial practice, although hexavalent chrome is more toxic and provides poorer coverage in holes than trivalent Cr. Coating baths for hexavalent Cr contain chromic acid and sulphate ions in a fixed ionic concentration ratio; a ratio of 100:1 to 150:1 is typical for decorative chrome plating (e.g., in plumbing fixtures), whereas a ratio of 75:1 to 100:1 is used in hard (heavy) chrome-plating (e.g., in extrusion dies). Plating is carried out at a temperature of 35–65 °C in plating tanks, which have a lining of lead, acid brick, or synthetic resins. Surfaces not needing plating are "stopped-off" by masking with wax, plastic, vinyl wraps, or metallized tapes. Anodes are designed to be an even distance from the part; a number of anodes that conform to part geometry (e.g., curved surfaces) are used. Agitation of the bath is needed during plating because chromic acid solutions are viscous and tend to stratify. This leads to uneven temperature distribution in the bath and poor deposition efficiency. Anodes are made of lead or lead-antimony alloys with low levels of impurities, such as Fe, Cu, Ni, Pb, and their oxides, which reduce the electrical conductivity and plating efficiency. After plating, the part is rinsed, final finished (ground and polished), and annealed to remove hydrogen absorbed during plating to prevent stress cracking. Decorative Cr plating deposits thin layers and is usually plated over less expensive Cu and/or Ni, whereas hard chrome or functional plating is thicker and takes advantage of the inherent properties of Cr metal. Hard chromium deposits are brittle and show large (macro) cracks when stressed. Some stress relief is possible in chromium coatings that are designed to contain a high density of micro-pores. Microporous chromium coatings also improve the corrosion resistance because they spread the corrosion current over a wider surface area (i.e., pore area). Microporous chromium films are produced by first co-depositing a thin nickel plate

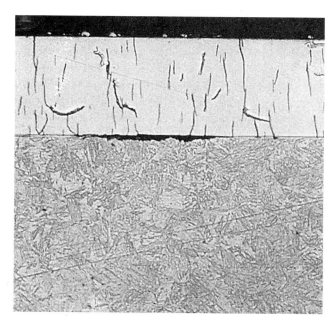

FIGURE 5-11 *Cross-section of an electroplated hard chrome layer showing internal cracks created in the deposit for stress relief.* (M. Schroeder and R. Unger, Advanced Materials & Processes, August 1997, p. 21). Reprinted with permission from ASM International, Materials Park, OH (www.asminternational.org).

containing a suspension of very fine inert particles. Chromium plated over this layer deposits preferentially around the particles, and this leads to micro-pores in the deposited film. Another method to create micro-pores in the film is to lightly spray the chromium deposit with fine sand or alumina powders. The brittle chromium deposit forms micro-pores and hairline cracks upon impact from the abrasives, which exposes the bright nickel underneath. Pore density is on the order of 10,000 pores per square centimeters; spray conditions must be judiciously controlled so that micro-cracks do not extend to the interface with the underplate, which may impair the adhesion. Figure 5-11 shows electroplated chromium with micro-cracks created for stress relief.

The trivalent Cr coatings are safer than hexavalent coatings, and the trivalent coating baths can tolerate current interruptions without passivation for long periods. Coating baths contain sulphate ions similar to Cr^{6+} baths, but have shielded anodes and a lower Cr-ions concentration than hexavalent Cr baths (5–6 g/liter for trivalent baths and 100–250 g/liter for hexavalent baths). Anodes are made of lead and are surrounded by boxes and sealed on one side with a selective ion membrane of perfluorinated sulfonic acid reinforced with inert Teflon fabric. The box surrounding the anode is filled with dilute sulfuric acid. The membrane prevents migration of trivalent Cr ions from reaching the anode and prevents their oxidation to the hexavalent state.

The electrodeposition of chromium differs from that of other metals in one important respect; deposition is controlled by one or more catalysts. The addition of a catalyst improves the plating efficiency of the otherwise sluggish chromium deposition processes. Several modifications based on catalyst addition have been developed for improved plating efficiency. These include 1) sulfate catalyst: self-regulated (add excess quantity of a single sulfate catalyst to bath. The excess amount initially settles to tank bottom but is periodically stirred using an air wand to make up for the spent amount). The current efficiency is, however, only about 14%; 2) self-regulated, mixed-catalyst bath (excess amount of a mixture of sulfate and fluoride catalysts is used and periodically stirred to reintroduce solid catalyst in bath). This process has an 18% efficiency and forms brighter and harder deposits than the first process; 3) completely self-regulated, mixed catalyst bath (proprietary mixture of sulfate and fluoride catalysts is added in excess amount; hoeing of solution reintroduces catalyst in solution). This method has 23% current efficiency, high plating speed, and smoother, brighter, and harder deposits; 4) partially self-regulated, high-speed, mixed-catalyst bath (requires more monitoring and adjustment of bath but allows higher deposition rates and higher efficiency to be achieved); and 5) self-regulated, crack-free process (produces a softer, more ductile, and crack-free matte finish. All other plating methods leave surface cracks due to internal stresses).

Pre-cleaning of parts is essential for obtaining high-quality adherent deposits on metals in all electroplating techniques. In the case of chrome plating, parts are ground and polished to desired smoothness followed by a four-step chemical cleaning process that consists of 1) dipping in an organic solvent (to remove heavy oils, wax, and grease) or vapor-degreasing; 2) dipping in water-soluble alkaline cleaners to remove light oils, soils, and flux residues; 3) anodic cleaning (i.e., electrolytic generation of oxygen gas bubbles on surface of part immersed in a solution that leads to scrubbing action); and 4) reverse etch with chromic acid, which leaves microscopic roughness on the surface for proper adhesion. A water rinse is used between each step.

The major disadvantage of chrome plating is the health hazards it poses (skin irritation and respiratory problems) and high process cost (expensive chromium compounds, low plating efficiency of only 12–24%, need for good ventilation, and sophisticated waste treatment). Chrome plating also reduces the fatigue strength of steels on which it is plated, and shot-peening and heat treatment are done to overcome this problem. The major advantages of chrome plating include high hardness, low friction, and excellent corrosion and thermal resistance.

Gold plating was originally used for decorative purposes. Recent non-decorative applications of gold plating in high-technology areas are primarily in the electronics industry, in which electrically conductive gold coatings are electroplated on electrical contacts and connectors, on semiconductors for soldering and bonding, and on printed circuit tabs. Several different plating bath compositions are used for gold plating; most of them use cobalt and nickel complexes in a solution of citric acid. These baths have long life (3 years or more), but citric acid in the bath leads to fungus growth, and regular filtering and cleaning of the bath is needed. High-purity gold deposits are soft (Knoop hardness number of 60–85), but functional gold plate has higher hardness (120–300 KHN). Common impurities such as Fe, Ni, Cu, and Pb must be avoided in the deposit to form high-quality, electrically conductive, and adherent deposits.

Electroless Plating

Electroless plating involves metal deposition without the use of electric current. Nickel is widely plated using electroless deposition method, and improves the hardness, wear resistance, corrosion resistance, and magnetic properties. Electroless Ni coatings are more uniform in thickness than electroplated Ni coatings. Plating baths for electroless Ni are composed mainly of nickel sulfate (source of Ni), sodium hypophsophate (reducing agent), and chemicals to prevent precipitation of insoluble nickel hydroxide. Bath pH is maintained at about 4.0, but generation of hydrogen lowers the pH; to offset this, alkaline salts of Na and K or ammonium ions are added. Good ventilation is needed to remove hydrogen gas during plating. Hydrogen gas bubbles may

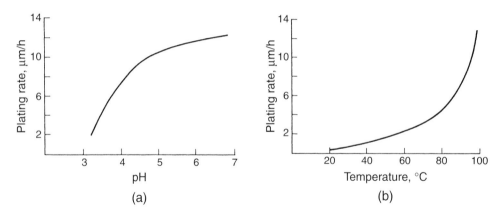

FIGURE 5-12 *(a) Effect of solution pH on the plating rate of electroless Ni.* (F. A. Lowenheim, Electroplating: Fundamentals of Surface Finishing, McGraw-Hill, New York, 1978, p. 395). *(b) Effect of solution temperature on plating rate of electroless Ni.* (F. A. Lowenheim, Electroplating: Fundamentals of Surface Finishing, McGraw-Hill, 1978, p. 396).

attach themselves to steel substrates during plating and cause embrittlement of the substrate. A post-plating anneal may be needed to remove hydrogen. Bath impurities such as phosphorus and boron must be minimized because they lower the deposit density, and co-deposit with nickel, forming amorphous alloys that disrupt the normal crystalline structure of the deposit. The bath is operated at about 85 °C in tanks made out of reinforced polypropylene or stainless steel. Figure 5-12 shows the effect of solution pH and bath temperature on the thickness of electroless nickel deposits.

During electroless plating, tanks also get plated; this is called "plate-out." Plate-out is removed once in 3–5 days for high-volume work. Waste treatment uses steel wool as substrate for Ni deposition in spent bath, and plated steel wool is rinsed, dried, and sold in the market. With plastic tanks, plate-out is dissolved in nitric acid and rinsed. With steel tanks, the tanks are passivated with nitric acid before plating, and tank body is maintained at a slightly anodic potential by application of a voltage between tank and a small cathode. Continuous plating depletes the bath of metal and other chemicals, and these must be replenished. For example, for a plating rate of 0.5 to 1.0 mil per hour, part loading of 0.5 square feet per gallon, and a bath Ni content of 6 g/liter, approximately 30% of total nickel will be consumed every hour. Additions may be needed every 30 minutes to an hour. Generally, plating bath needs to be replenished more frequently in electroless plating than in electroplating. The total amount of Ni replenished is measured in units of "cycles" or "metal turnover." A cycle is defined as the ratio of total quantity of Ni replenishment added to the total quantity of Ni present in bath. Automatic control of bath composition is achieved by the use of photometric cells that detect changes in Ni concentration by variation of solution light transmittance. Changes beyond preset limits activate pumps.

Pure electroless Ni deposits are harder (500–750 KHN) than electroplated Ni deposits. In addition, the abrasive wear resistance is excellent in Ni deposits that are low in phosphorus content. The hardness of the deposit is inversely proportional to the phosphorus content in the deposit, and high phosphorus deposits exhibit low hardness and high wear. However, high hardness can be restored in such deposits through a heat treatment at about 400 °C, which

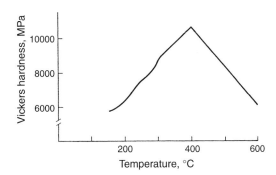

FIGURE 5-13 *Effect of heat treatment temperature on the hardness of electroless Ni deposit with one hour heat treatment time.* (F. A. Lowenheim, Electroplating: Fundamentals of Surface Finishing, McGraw-Hill, 1978, p. 399).

converts Ni-P amorphous alloy into crystalline Ni and a hard nickel phosphide phase. This increases the hardness of the deposit but also decreases the corrosion resistance. Figure 5-13 shows the effect of heat treatment temperature on the hardness of nickel-phosphorus coatings. High phosphorus contents also decrease the magnetic properties associated with Ni, and the single largest application of high-phosphorus electroless Ni deposits is as under-layers beneath the magnetic coatings in memory discs. In this particular application, 0.2–0.5 mil-thick high-P electroless Ni films are needed with a very smooth surface finish achieved through polishing (low hardness of high P coatings makes polishing easy). In addition, composite coatings of Ni containing either hard diamond and SiC particles, or soft polytetrafluoroethylene (PTFE) and fluorinated carbon powders, have also been deposited to increase the hardness and wear resistance. Fine powders of a variety of ceramics are also coated with metals such as Ni and Cu using the electroless plating process for better compatibility with metal matrices in particulate reinforced metal-matrix composites. Figure 5-14 shows a photomicrograph of metal plated ceramic particles.

Anodizing

Anodizing is an electrochemical process that forms a hard, chemically inert and durable oxide film on metals by passing an electric current in an electrolyte in which the metal to be anodized serves as an anode. Aluminum is the most widely anodized metal. The electrolyte serves as the oxidizing agent and acid solutions such as sulfuric, chromic, oxalic, boric, and phosphoric acids are common electrolytes. The cathode is either the side of the tank or a strip of commercial lead. Figure 5-15 shows the anodized film thickness and anodized film weight as a function of deposition time at different temperatures for an Al alloy in a sulfuric acid bath at a current density of 130 A/m^2. High temperatures cause coating dissolution and reduced anodizing efficiency. Similarly, very acidic and very alkaline electrolytes also cause the film to dissolve as it grows, and this will result in a porous anodic film. The reactions that occur at the anode and cathode during anodizing can be represented by the following equations:

$$\text{(anode)} \quad Me + nH_2O \rightarrow MeO_n + 2nH^+ + 2ne$$

$$\text{(cathode)} \quad 2ne + 2nH_2O \rightarrow nH_2 + 2nOH^-$$

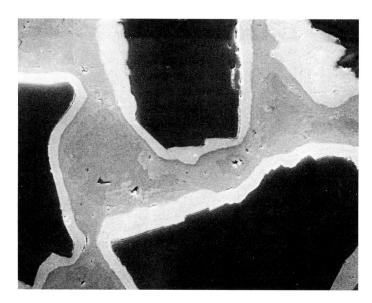

FIGURE 5-14 *SEM view of electroless Ni-coated SiC particles embedded in an Al alloy matrix. The composite was produced by infiltrating a bed of Ni-coated SiC powders with the molten alloy followed by solidification. Note the short contact times between the molten alloy and coated particles during infiltration prevented complete dissolution of the Ni coating.* (R. Asthana and P. K. Rohatgi, Journal of Materials Science Letters, 11, 1993, 442–445.)

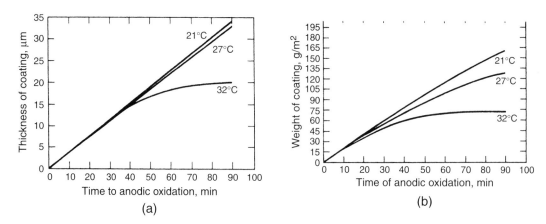

FIGURE 5-15 *(a) Anodized coating thickness as a function of anodizing time for Al alloy 1100 at three different bath temperatures (15% sulfuric acid bath, current density: 130 A·m^{-2}).* (F. A. Lowenheim, Electroplating: Fundamentals of Surface Finishing, McGraw-Hill, New York, 1978, p. 458) *(b) Effect of anodizing time on the weight of anodized film on an Al 1100 alloy sheet at different temperatures (15% sulfuric acid bath, current density: 130 A·m^{-2}).* (F. A. Lowenheim, Electroplating: Fundamentals of Surface Finishing, McGraw-Hill, 1978, p. 459).

TABLE 5-2 Relation Between Weight of Anodic Coating and Al Removed

Electrolyte	Current Density, $A \cdot M^{-2}$	Anodizing Time, min.	Coating Ratio[1]
Sulfuric acid (15% solution)	130	15	1.41
Sulfuric acid (15% solution)	130	30	1.33
Sulfuric acid (15% solution)	130	60	1.19
Oxalic acid (3% solution)	130	15	1.31
Oxalic acid (3% solution)	130	30	1.27
Oxalic acid (3% solution)	130	60	0.95
Chromic acid (3% solution)	130	60	0.87

[1] Coating Ratio = (weight of unscaled coating/weight of Al removed)
Source: Adapted from F. A. Lowenheim, *Electroplating* (New York: McGraw-Hill, 1978).

These reactions show that oxide will grow on the metal anode and hydrogen will be liberated at the cathode. Typical ranges of anodizing process parameters are summarized in Table 5-2.

Anodizing equipment includes power supply, tanks for cleaning and anodizing (steel or glass-reinforced polyester), heat source, storage and anodizing racks, thermometers, and ventilation system. Generally, the anodic film is porous and can be impregnated with dyes to give color before the pores are sealed. Anodized films on Al are typically 0.0004 to 0.0008 inches in thickness, although thicker (~0.005 inch) coatings are also produced. Parts are pre-cleaned using detergent or soap solutions, solvent spraying, vapor-degreasing, and ultrasonic cleaning. The most common method is immersion in a dip tank containing a hot solution of acidic or alkaline industrial detergents. After cleaning, rinsing, and drying, the part is cleaned either mechanically ("scratch brushing" with wire brushes and buffing) or chemically ("caustic etch" or "frosted finish" by using caustic soda solution, or acid etch using HCl, HF, and nitric acids). These treatments make the surface more receptive to the process of coating deposition. The part is again rinsed and dried before anodizing treatment.

During anodizing, an electric current is passed through an acidic electrolyte, which generates oxygen at the aluminum anode. Oxygen combines with Al to form a porous oxide film on the part. The oxidation reaction continues beneath the surface film via pores and the film formed last is next to metal, whereas the oldest part of the film is on the outer surface. When a desired oxide thickness has been achieved, the part is removed from the tank, rinsed, dried, and prepared for dyeing and/or sealing. The dyestuff impregnates porous film and produces brilliant colors with the luster of parent metal visible through color-impregnated oxide film. A thick oxide film is desirable for color stability as well as for corrosion and abrasion resistance. After dyeing, the part is rinsed and dried and is ready for sealing the porosity. Sealing is done by first immersing the part in a solution of nickel acetate at 95–100 °C for 3–5 minutes, followed by boiling the piece for 30–60 minutes in de-mineralized water. This converts the aluminum oxide film to a monohydrate with an attendant increase in volume and pore closure. The nickel acetate treatment prior to boiling plugs the pores and thus traps the dyes. This treatment is not needed on anodized parts that have not been impregnated with dyes.

The alloy composition and metallurgical structure of the part to be anodized are important factors because they affect the appearance, density, and hardness of the coating. When alloying elements dissolve during anodizing (e.g., Cu in Al-Cu alloys), density and corrosion resistance decrease and porosity increases. Alloying elements such as Si in Al-Si alloys do not dissolve but leave imperfections in the coating. Usually, wrought alloys yield bright and smooth anodic coatings, whereas sand- and permanent-mold cast alloys yield somewhat inferior coatings because of coarser grain structure in cast parts.

Coatings on Powders and Fibers

Ceramic powders and fibers are coated with metals and ceramics for use in a wide variety of composite materials. Carbon fibers are one of the most widely used reinforcement materials in metal matrices. These fibers have been coated with metals, and a variety of oxide, carbide, and boride phases for compatibility with different types of matrix metals. For example, ion-plating of carbon fibers with Ni, Ta, Ag, Co, and Cu-Co and Cu-Ni alloys has been done for use in magnesium alloys. Another surface modification technique, called the sodium process, has been successfully used to coat carbon fibers for use in Al matrices; however, the technique is unsuitable for Mg matrices because sodium severely embrittles the Mg alloy. In contrast, titanium coatings on carbon fibers are compatible with Mg and have been used in carbon fiber–reinforced magnesium-matrix composites synthesized using solidification techniques such as squeeze-casting. Gaseous chlorides of Ti, Si, Zr, Hf, Ta, V, Cr, Nb, and B are used in reactive environments to deposit metal borides, metal carbides, metal oxides, and metal nitrides on carbon fibers using the vapor-phase deposition techniques. In a typical vapor-phase coating process, a yarn or tow of C fibers is passed through a reactive mixture containing $TiCl_4(g)$, $BCl_3(g)$, and Zn vapor to deposit a thin coating of $Ti_xB_y/TiC/Ti_xB_yC_z$ on the fibers. The coated fibers are then drawn through a bath of liquid magnesium alloy to form composite wires that are then consolidated by hot-press diffusion-bonding techniques into near-net-shape components. Other common reinforcement materials such as silicon carbide, boron fibers, and alumina fibers are also surface-treated for chemical compatibility with metals. For example, boron fibers with a SiC coating (Borsic fiber) are especially suited for use in aluminum-matrix composites. Similarly, boron carbide–coated boron fibers are compatible with titanium alloys and are used in titanium-matrix composites.

Organic Coatings—Paints and Powders

The major constituents of liquid paints are a film former or binder (e.g., vegetable oil, varnish, epoxy, vinyl, etc.), a solvent or thinner if binder is not liquid at application temperature, fillers (e.g., finely divided mica and talc) to hide the surface, pigments (e.g., TiO_2, iron oxide, etc.) to add color, and extenders that add body to the liquid paint but do not provide color or hiding ability. To reduce the cost of pigments and modify their color intensity, white fillers such as barites, calcium silicate, mica, talc, silica, etc. are added, which also impart resistance to sunlight and corrosive liquids. Finely divided aluminum, bronze, lead, and zinc are added to enhance light reflectivity and reduce permeability of coatings to moisture and gases. The basic steps in the manufacture of liquid paints are disintegration of pigment aggregates into finer particles, transfer and mixing of particles in the liquid vehicle, displacement of air by the liquid vehicle, and stabilization of the resulting slurry.

Both thermoplastics and thermosets are used in liquid paints and powder coatings. Thermoplastics melt and flow with the application of heat but retain the same chemical composition when cooled. Common thermoplastics are PVC, nylon, polypropylene, and thermoplastic polyester. They are used as encapsulation materials for a variety of parts. In contrast to thermoplastics, thermosets undergo chemical change (polymerization) when heated and can no longer melt to plastic-like state. Common thermosets include acrylics, amino resins, epoxy, polyester, polyurethane, phenolics, silicones, and polyvinyl acetate. Acrylic resins are high-molecular-weight polymers (mol. wt. $\sim$80,000) and produce transparent coatings in unpigmented state. Plasticizers are added to acrylics to improve adhesion, crack resistance, and flexibility. Acrylic coatings are used for wood, concrete, paper, textiles, and metals. Amino resins are condensation products of formaldehyde and urea or melamine. Melamine coatings retain color better than urea coatings at elevated temperatures (300 °F) and are used for automotive topcoats and kitchen appliances. Urea coatings are used for paper, textiles, leather, furniture, and in automotive primer coats. Phenolic resins are brittle, have high melting points, and are used in water-resistant varnishes and as adhesives for plywood and for insulation. Polyvinyl acetate–based pigmented resins are used for house paints and adhesives; these give glossy coatings. Silicones are used as sealants in fabric coatings, as roofing coatings over polyurethane foams, and as release coatings for paper. Silicones with methyl groups cure rapidly and produce hard films, whereas those with phenyl groups are softer and have better adhesion properties.

As with most coating processes, paint application to an object is preceded by appropriate cleaning treatments. Pre-cleaning is done using volatile solvents, steam, and pickling, as well as mechanical processes such as wire-brushing, sanding, and sand-blasting. Common methods of paint application include use of brush, rollers, and spraying. Liquid paint viscosity and density are important properties. For example, variation of paint viscosity with shear rate is an indicator of the 'brushability' of the paint. Rollers with lambskin are used for manual work, whereas roller machines are used for automatic coating of both sides of flat sheets (e.g., venetian blinds and steel sheets that are fabricated into cans). The quality of the finish depends on the type of roller surface, type of paint, application speed, pressure, paint viscosity, density, humidity, and temperature. Spray-painting utilizes a spray gun to atomize the paint into very fine droplets. Compressed air, superheated steam, and liquid gases such as butane in aerosol spraying are used. Paint spraying is done using conventional spray gun and high-volume, low-pressure (HVLP) spray gun. The traditional siphon cup spray gun requires control of three factors: spray gun angle (usually 90 °), spray distance (6–10 in.), and travel speed. The HVLP spray technique has higher transfer efficiency (i.e., amount of paint a spray gun is capable of depositing) than traditional spray techniques. The higher the transfer efficiency, the lower the amount of paint wasted in overspray. Traditional spray guns atomize paints at 35–65 psi pressure, whereas HVLP uses a "soft spray" atomization, which employs a high volume of air but low (1–10 psi) pressures that limit overspray. Poor atomization will cause poor texture and mottled or blotchy colors. The variables in atomization include size of opening in the fluid tip, the air pressure at the air cap, paint supply source, and viscosity.

Dip-coating employs large tanks and part immersion in liquid paint, whereas flow-coating uses nozzles to flow paint over part, and requires smaller tanks than dip-coating. Dip-coating is used when part is too complex to be brush-coated or sprayed. Hollow objects are rotated when dipping and drawing. The equilibrium film thickness depends on the withdrawal speed, substrate geometry, melt viscosity, and surface tension, and can be theoretically predicted by solving the steady-state Navier-Stokes equation. When the part is being drawn, the paint may drain slowly; faster drainage is achieved by passing the coated work over a charged electrode. Another method of liquid paint application is called curtain-coating in which part moves beneath a curtain of liquid paint at a controlled speed, and coating thickness and uniformity are controlled by varying the speed, thickness of curtain, and coating flow rate. Curtain coating is used for flat surfaces, which may be transported along a conveyer belt. Very large parts are "flow-coated," in which paint is squirted on the work and the excess paint that runs off is re-circulated. Electro-coating (E-coating) uses the principle of electrophoresis or migration and deposition of charged paint particles in a liquid medium, and powder-coating techniques deposit dry paint powders, generally electrostatically charged, on a grounded object, followed by a thermal cure to form a uniform film. The elimination of solvents in powder coatings reduces or eliminates harmful volatile organic compounds (VOC) and makes powder coating an environmentally friendly coating technology. Owing to their industrial importance, the electrocoating and powder coating techniques are discussed in greater detail later. The coating or film forms either by evaporation of the solvent that could be aqueous or organic, or through polymerization. Film formation by evaporation requires that the binder be a high polymer. In contrast, film cure through poly-merization involves cross-linking of a monomer or low polymer, which combines to give larger molecules (e.g., alkyds, epoxies, polyurethanes, etc.). Cure ovens include direct-fired gas and oil ovens, electrically heated and steam-heated ovens, infrared ovens, and induction and radiation (UV or electron beam) cure ovens.

Automotive coatings constitute an important class of industrial organic coatings. Generally, these coatings consist of an undercoat or basecoat and a topcoat; both basecoat and topcoat may contain all the principal paint constituents, such as pigments, binders, and solvents. Three types of basecoats are commonly used: primers (prevent corrosion, improve adhesion), primer surfacers (act like primers but also fill defects), and sealers (prevent sinking of topcoat in imper-fections). The first step in coating application for automotive needs is substrate preparation for paint application. Pre-cleaning is done to remove rust using proprietary formulations fol-lowed by water rinse and drying. The cleaned substrate is primed using a primer, which is a bond coat used to provide maximum adhesion and corrosion resistance on a bare surface. Primer surfacers provide some of the same benefits as primers but contain a high level of pigment solids for filling sand scratches and other minor imperfections. For best results, primer surfacers are sanded before topcoating. Primer sealers provide same protection as primers with the advantage of giving good, uniform color holdout for the topcoat. When the sealers are used, the topcoat will soak into the primer sealer less and have a glossier appearance. Generally, topcoat involves a three-stage process called "tricoat." In a tricoat system, there is a highly pigmented basecoat, a translucent mid-coat containing mica, and a high solids clear-coat. The mica flake in the mid-coat is fine and has a coating of titanium dioxide or other materials for color effects. For example, iron oxide produces a gold pearl, and chromium oxide produces blue-green pearl. Clear-coats provide protection, gloss, and durability, and are composed of acrylic urethane, polyurethane, and acrylic lacquer.

Liquid paint application methods may introduce defects in the surface layers, most of which are caused by surface tension effects. Common paint defects include picture framing ('fat-edge' effect), orange peel, cratering, and Benard cells. Picture framing is due to surface tension effects; if the surface tension of the coating resin is greater than that of the solvent, the resin flows toward the edge. Orange peel results from local variations in surface tension, which cause localized fluid flow and mounds or dimples in coating. Surfactants are added to liquid paints to control the surface tension and increase the viscosity in order to lower the paint mobility. Cratering results from dust, oil, and foreign matter. These impurities alter the local surface tension of the liquid paint; therefore, resin flows away from low-surface-tension to high-surface-tension regions via Marangoni flows (see Chapter 4), and forms a crater. Benard cells appear during solvent evaporation. Small-scale eddy flows of liquid during solvent evaporation are caused by gradients of temperature, surface tension, or density, and these flows produce fine cellular morphologies in the paint film.

Powder Coatings

Powder-coating techniques deposit dry paint powders that may be electrostatically charged for better adhesion to a grounded object. This is followed by a thermal cure to melt the powders and form a uniform film. The principal advantages of powder coatings are improved quality, reduced finishing costs, elimination of harmful solvents, high material utilization, and reduced solid waste disposal costs. In addition, absence of solvents and mixing saves time, storage space, equipment, and labor, and pollution problems and fire hazard are minimized. The deposition process is cleaner, powder spills can be easily vacuum-cleaned, and safety regulations are readily implemented. The coating process can be fully automated.

Electrostatic Powder Spray. The electrostatic spray method is used to apply thin (1–5 mil) films, although heavier films (6–15 mils) are also possible. The main components of the coating unit are powder delivery system, powder charging system, reclaim system, and powder cure system. The powder delivery system uses a pump to draw powders from a delivery hopper (e.g., a fluidized bed), mixes it with controlled volume of air, and delivers the mixture to the part. A "soft" (low-velocity) powder cloud is used to gently cover the part in contrast to a paint spray gun that blows wet paint toward the part with great velocity. Because thermoplastic powders can soften and agglomerate if the air in the delivery system is warm (and clog the passageways in the pump and hose), the delivery air must be cold and delivery hoses short and free of bends. Minimum acceptable pressures and powder levels must be used to increase the space between the particles in the delivery stream. The powder charging system puts an electrostatic charge on each particle with the help of a needle-like wire (electrode) that is located at the gun exit, operating at 50 kV to 100 kV potential. The high voltage creates an ionizing field near the tip, and transfers static charge to the powders as they pass through this region. This causes the particle to be attracted to and held by the grounded work. Another charging technique, called tribocharging, uses high-velocity friction to generate static charge. Tribocharging is most effective with high-resistivity powders such as epoxies. The reclaim system separates unused powders from the air by passing the stream through fine screens, bag filters, or cyclones, and then returns the powders to the delivery hopper. Cleaned air is also recycled. This reclaim system allows close to 100% material utilization. Other application utility systems for electrostatic powder coating units include temperature and humidity controllers, and refrigerated drying and filtering of compressed air supply.

Powder coatings are cured in the same manner as liquid coatings, although there are important differences between cure ovens designed for powders and liquid coatings. Powder cure ovens should heat up the material fast to allow maximum flow-out before the curing reaction gels the film. In contrast, liquid cure ovens require slow heat up to avoid boiling of the solvent in the paint. Powder ovens can be shorter, and they consume less energy than liquid cure ovens because exhaust air to carry away solvents is not needed; heating the air that must be exhausted represents major energy consumption in liquid cure ovens. During curing, when the powder melt temperature is reached, the powder melts and flows, forming a continuous liquid film. If thermoset powders are used, an additional curing reaction takes place. When the part is maintained at the curing temperature for a proper length of time, the film becomes 100% cured and exhibits toughness and impact resistance.

Another method to apply powder coatings employs part immersion in a bed of dry thermoplastic or thermoset powders. Dipping methods are used for relatively thick coatings (>4–5 mil), and use a fluidized bed comprised of an upper and a lower compartment separated by a porous membrane. The powder is placed in the upper compartment, and dry air is blown into the lower compartment and passes through the membrane into the upper chamber, causing expansion of the powders into a fluid-like mass into which parts can be dipped. The degree of expansion or fluidization can be controlled by the volume and velocity of injected air. Parts are heated to a temperature above the melting point of the powders prior to dipping in the fluidized bed. Alternatively, electrostatically charged powders are used for some specialized applications. A bake cycle is required to cure the finish. The coating thickness depends on the part temperature and the residence time in the powder bed.

Powder coatings are used in a large number of modern industrial and consumer appliances. Applications include linings on the inside of oil-drilling pipe, reinforcing steel bars for cement on highway and bridge decks, wheels, bumpers, mirror frames, oil filters, battery trays, coil springs, washer tops and lids, range housing, freezer cabinets, dryer drums and microwave ovens, farm implements, outdoor furniture, light reflectors, lamp housing, window and door frames, bicycle frames, and air conditioner housing. Powder coatings provide special optical effects, thus enhancing the decorative appeal of the coating, and are now widely used on automotive bodies. The two–coat translucent effect is achieved by covering a brightly covered basecoat with a translucent topcoat that contains a pigment that gives it some color, yet the basecoat remains visible through it. This provides brilliance and depth. Wrinkle finishes are produced with two different powders, which melt and flow at different rates; peaks of wrinkles are formed by powder that flows at a slower rate. Veins are formed by combining different-colored powders that are chemically incompatible. On heating, multiple cure reactions cause the different types to separate and coalesce, creating lines or veins in the finished film. The chameleon effect yields changing color as the lighting or angle of view changes. Fade is a multicolor powder effect used on bicycle frames. An example is that a yellow powder is applied to the top half of the frame and a blue powder to the bottom half. When curing, the area where powders overlap fades to green.

Electrocoating

Electrocoating (E-coating) or electrodeposition is used mostly with epoxies and acrylics, and involves electrical deposition of paint film on a conductive part. The process is similar to electroplating, and uses anodes, cathodes, electrolytes, and electric current. E-coating offers numerous advantages such as automation (reduced labor cost), high line speed (improved productivity), material efficiency (eliminates overspray, recycles unspent bath constituents), uniform deposits (no sags or edge pull), and reduced emissions. The most significant uses of E-coating are in auto primers, appliances, electrical goods, toolboxes, lawn mowers, agricultural equipment, metal shelving, toys, and other industrial and consumer goods.

The deposition process involves electrophoretic migration of charged paint particles and their deposition on the work. When a voltage is applied to an electrolytic cell composed of anodes and cathodes immersed in an electrolytic suspension of paint particles, water decomposes at the electrodes (electrolysis), and paint particles acquire a charge. The charged particles move to the oppositely charged electrode, lose their charge, precipitate on the work, and become tightly packed (electro-endosmosis). With the thickening of the deposited layer, the work becomes electrically insulated, and deposition stops automatically. Two types of E-coat processes are employed: cathodic E-coat (the part is a grounded cathode) and anodic E-coat (the part is an anode). Cathodic process is more widely used as it gives better color and corrosion resistance than the anodic process. In the anodic E-coat process, paint particles are given a negative charge, and paint deposits at anode, whereas in cathodic E-coat paint particles are given a positive charge, and paint deposits at the cathode. Bath is cleaned by pressure filtration through a fine membrane that performs the function of kidneys in the human body. It concentrates resin and pigments, and returns paint solids to bath, removes soluble salts that might degrade paint performance, and provides rinsing medium (the liquid permeate removed in the filtration process is used in rinsing).

Electrocoating is a high-speed, highly automated finishing process, and permits complex shapes and recesses to be evenly painted. The thickness of the film is independent of bath solids, rheology, or shape of coated article. The process offers good paint transfer efficiency (>95%), a high degree of paint utilization, and yields high-density deposits having good corrosion resistance. There are, however, some limitations of the E-coat process, which include high capital cost (which is justified only in high-volume work), need for sophisticated process maintenance, and high electrical energy consumption. The process is not applicable to nonconductive materials and is less well suited for multiple color coatings than are powder coatings. In addition, it is difficult to deposit relatively thick films (>2 mils) because of the electrical resistance of the deposit. Whereas higher voltages can yield greater deposited thickness, the applied voltage should be less than the "rupture voltage" at which gases form under the film and cause it to "lift off."

Conversion Coatings

Conversion coatings are surface chemical treatments that provide benefits such as surface passivation or improved corrosion resistance, surface activation for improved paint receptivity, and increased surface hardness and abrasion resistance. They are frequently used as paint pretreatments to promote paint adhesion by either changing the surface texture or surface chemistry. They also provide corrosion resistance when paint is damaged during service. Common conversion coatings include zinc phosphate, iron phosphate, and chromate coatings. Zinc phosphating is done mostly on steel, Zn, and Al. The zinc phosphate microcrystals form on the surface by a chemical conversion process. When liquid paint is applied to such a surface, paint flows around these microcrystals and grips them, providing for better mechanical adhesion. Zinc phosphate is also a corrosion inhibitor. If Zn phosphate microcrystals on a steel substrate dissolve in a corrosive environment upon prolonged exposure, then phosphate ions react with (Fe^{3+}) ions from partially corroded steel to form insoluble ferric phosphide, which provides barrier to further corrosion. The part to be phosphated is first pre-cleaned, pickled in a mild acidic solution to remove rust, and neutralized in alkali, followed by zinc phosphating. Typical Zn phosphate coating weight is about 150–350 mg/ft^2 or 1.5 to 3.5 g/m^2. Zn phosphate crystal size is about 5–15 μm. Compared to zinc phosphating, iron phosphating is less expensive and is applied to a wide variety of iron and steel parts. In addition, iron phosphating is preferred to Zn phosphating in situations where the part must retain formability after the treatment; for example, iron phosphating is preferred to zinc phosphating of coil steel because the former treatment allows the part to retain formability without damage to the treated surface. Iron phosphating, however, yields relatively thin coatings (0.15–0.80 $g \cdot m^{-2}$), and provides less corrosion protection than zinc phosphate coatings. Chromating is a popular method of applying paint base to Zn or Al alloys but is rarely used for Fe and steel. Improved paint adhesion to a chromated surface is achieved because of chemical reaction of hexavalent chromium ions with the paint. Typical weight of chromate coating is in the range 0.1–5.0 $g \cdot m^{-2}$.

Vitreous Ceramic Coatings

Traditional ceramic items such as dinnerware, sanitary ware, electrical porcelain, and ceramic floor and wall tile are coated with a variety of vitreous ceramic coatings or glazes. Traditional ceramics are processed using the following steps: a ceramic body is formed by means of slip-casting, jiggering, or pressing. The body is then dried to remove bulk moisture content (drying) and trimmed via wet sponging, cutting, or grinding to remove rough edges (trimming). The dried ceramic body is heated to a temperature that will remove most chemically bound water and organics (bisque firing). The ceramic is then coated with a water-based slurry of powdered glass, fluxes, and colorants (glazing). The glaze-coated body is heated to a temperature that will vitrify the ceramic coating (glaze firing). After cooling, any sharp defects are ground off using abrasive grinding wheels, and the ceramic is decorated using hand painting, screen printing, or other techniques. Finally, the ceramic body with its applied decoration is heated to fuse the decoration to the ceramic.

Glaze coating is a vitreous coating, and each glaze formulation will function differently depending on the firing cycle chosen. A defect formed during the processing of the ceramic body may necessitate a change in the glaze firing cycle. Glaze raw materials consist of processed minerals whose relative amounts in the mixture can be controlled. A frit allows reactive elements to be incorporated into a glassy matrix. A mixture is first made from processed minerals and then sintered in a furnace until a fluid glassy melt is formed. The melt is then water-quenched to break up glass into particles that are ball-milled to fine sizes. A glaze also uses certain fluxes such as lead oxide and boron oxide to control the melting point and coefficient of thermal expansion (CTE). The CTE of glaze must be lower than that of ceramic body to resist cracking on cooling. If a glaze is to be fired to a high temperature, more silica can be used and less powerful fluxes are needed. Glaze properties are controlled by adding other oxides; e.g., B_2O_3 lowers the CTE of a glaze. Gums and binders are often added to aid in suspension of heavy glass particles, and these may age and degrade upon prolonged holding. Preservatives are added to the slurry to slow down the aging process. Because fine powders tend to severely agglomerate, a deflocculant (e.g., sodium silicate) is added to the glaze slurry. The extent of deflocculation achieved can be characterized by measuring the viscosity of the slurry. The specific gravity of a glaze slurry must also be periodically checked because it influences the sedimentation behavior of suspended powders. The processing and application of glaze must avoid contamination, and the ceramic body must be free of dirt and grease.

Traditionally, glaze is applied using spraying (may give uneven thickness), dipping (may leave run marks), and waterfall method (may cause improper drainage), and considerable practical skills are needed to produce defect-free parts. Frequent agitation of glaze is needed to prevent settling. During glaze firing, the temperature–time cycle must be carefully controlled to allow glaze to fuse and bond to the body and to allow time for gas to bubble through the surface of the molten glass. When firing several parts, they should be spaced evenly during firing because coloring oxides can "jump" from one piece to another if improper spacing is used. Overfiring must be avoided because glaze tends to dissolve the underlying ceramic body. Cooling at the conclusion of firing must be slow to avoid thermal stresses and cracking. Silica in the glaze undergoes a phase conversion during cooling, which lowers the CTE of glaze coating and increases the resistance against cracking. The time lag between glaze application and firing should be minimized to prevent glaze from gathering dust. The chemistry of the ceramic body must be as closely matched with that of the glaze as possible. Colors are obtained by adding metallic oxides to the glaze. Certain oxides favor crystal growth in a glaze; these crystals scatter light, which enhances the surface appearance and aesthetic appeal of the glazed object.

Surface Hardening

Surface hardening is a generic term, but it acquires a special and restricted meaning when applied to ferrous materials. Surface hardening allows a hard and wear-resistant surface to be produced through metallurgical phase transformations and/or compositional changes at the surface of a steel object. In surface hardening of steels, coatings are not deposited from external sources; instead various metallurgical treatments create a hard surface while allowing the part to retain a soft and tough interior.

Two fundamentally different approaches to surface hardening of steels are employed: selective hardening (e.g., flame, induction, laser, and electron beam hardening techniques) and diffusion hardening (carburizing, nitriding, carbonitriding, boriding, etc.). Selective hardening treatments do not introduce any chemical elements into the surface or subsurface regions from outside, and do not produce any compositional changes in a part. These treatments modify the microstructure in the surface and subsurface regions by producing hard, wear-resistant chemical phases through suitable heat treatments. In contrast, the diffusion methods introduce chemical species into the surface and subsurface regions of a steel part to create a hard, wear-resistant exterior. A basic difference between such surface treatments and the methods that deposit an external coating on a part (e.g., electroplating, vapor deposition, and thermal spray) is the absence of an abrupt chemical or physical discontinuity in the structure. Instead there is a gradient of chemical and/or structural change from the surface to the subsurface regions. This eliminates the problem of adhesion between dissimilar materials and reduces the thermal stresses caused by an abrupt change in the CTE of different materials. The range of hardness values that can be obtained through surface-hardening techniques is, however, limited by the chemistry of the part, and the thermodynamic (phase diagram–related) and kinetic (cooling rate) considerations. This is usually not the case with the externally deposited coatings where a wider range of hardness values can be achieved.

Selective Hardening

The basic principles of selective hardening of steels can be understood with reference to the Fe-Fe$_3$C phase diagram and the appropriate TTT (time–temperature–transformation) diagrams discussed in Chapter 1. The basic approach of hardening involves heating the steel part into the austenitic region (γ-iron), followed by quenching at a sufficiently fast rate to form hard and brittle martensite in the quenched region. If cooling curves for different types of quench (e.g., water quench, oil quench, etc.) are superimposed on the relevant TTT diagram, then it is possible to predict the kind of microstructure (and therefore properties) that will be obtained. At slow cooling rates, soft pearlite, which is a mixture of ferrite (α-Fe) and cementite (Fe$_3$C), will form and hardening will not be achieved. At moderately fast cooling rates, a mixture of bainite, martensite, and retained austenite may form with some improvement in hardness. At high quench rates (i.e., at rates where the cooling curve is through the austenite region and does not cut through the knee of the TTT curves), martensite forms via a diffusion-less process below the martensite start temperature on the TTT diagram. Alloying elements in iron-carbon alloys shift the curves on the TTT diagrams either to the right or to the left; as a result, different combinations of phases can form at a fixed quench rate.

Flame-hardening uses an oxyacetylene torch to heat the part to form austenite, followed by water quenching to form martensite on the surface. Although no compositional changes take place, the steel part should have adequate carbon content for hardening to take place. The method is used mostly for large parts such as gears that are difficult to heat in a furnace.

Induction-hardening employs induction-heating principles to heat the part and form austenite. A high-frequency alternating current is passed through an inductor (usually a water-cooled helical copper coil). A magnetic field is induced in the part, which generates eddy currents that lead to resistive heating of the part. The depth of heating increases as the frequency of the alternating current decreases. Localized heating of the part is possible with very little radiation losses. However, a small and precisely controlled gap between the part and the inductor is required.

Laser-hardening involves three types of hardening treatments. In laser surface transformation, the surface of the part is heated to form austenite, followed by quenching to form martensite. In contrast, in laser surface melting and in laser surface alloying, the surface of the part is melted and quenched either without any externally introduced alloying elements (laser surface melting) or with alloying elements introduced into the surface liquid pool (laser surface alloying). The quenching of austenitized surface of the part in laser surface transformation does not usually require any liquid or gaseous quench media for large objects; the part self quenches, i.e., the surface heat is rapidly conducted down very steep temperature gradients to the cooler bulk of the part, thereby quenching the surface and forming martensite. The requirement of self-quenching imposes certain restrictions on the part size; the working mass must be large enough to serve as a good heat sink. It has been estimated that a mass of eight times the mass to be hardened is needed around the area. In laser surface transformation, the energy densities typically vary in the range of 500 $W{\cdot}cm^{-2}$ to 5000 $W{\cdot}cm^{-2}$. It has been estimated that a laser energy input of 500 $W{\cdot}cm^{-2}$ will produce a temperature gradient of about 500 °C/mm. The effect of laser power and traverse speed on the depth of heating is shown in the schematic of Figure 5-16; low input power and low speeds yield greater depth of heating than high power and high traverse speeds. Unlike induction heating, no precise distance between the part and the laser source needs to be maintained. Also, the width of the laser beam or hot spot can be controlled through laser optics, thereby permitting selective heating of very small areas on a part.

Laser surface treatments have been used in many different ways to deposit coatings and surface engineer parts for a variety of applications. For example, lasers are used to deposit hard and corrosion-resistant vanadium carbide (VC) and titanium carbide (TiC) coatings on die casting dies to reduce the hot corrosion by molten metals, increase the thermal fatigue resistance, and reduce the wear. Thermal spray, PVD, and CVD are also used to deposit carbide coatings on die casting dies. Some of these processes are very slow, or require a bond coat for adequate adhesion. In contrast, lasers create a strong metallurgical bond between the carbide coating and steel, and eliminates the need of a bond coat. Typically, the base metal (steel or Al) is sprayed with a solution containing fine carbide powders and dried. The spray-coated object is then scanned with a high power laser beam to melt the surface of the underlying steel (or Al) part and allow the melt to wick through the interstices between the carbide powders, thus forming a strong bond upon cooling and solidification. Another application of laser surface hardening in general, and laser surface alloying in particular, is the deposition of a Cr-rich layer on the surface of steel tubes of Cr-Mo steels that are used in heat exchangers, boilers, and nuclear power reactors. High Cr content in the surface and subsurface layers improves the oxidation and corrosion resistance. In addition, hard particles such as CrB_2 can be added to the surface layers for increased wear resistance. For this purpose, the steel surface is melted using high power laser, and Cr and CrB_2 powders are injected in the molten surface pool.

Electron beam hardening is similar to laser hardening but uses an electron gun for surface heating. The gun accelerates, directs, and focuses a beam of electrons via electromagnetic (EM)

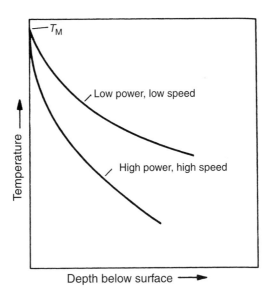

FIGURE 5-16 *Schematic temperature profiles during laser surface-melting showing the effect of laser power and scan speed.* (ASM Metals Handbook, Vol. 4, Heat Treating, 1991, p. 289). Reprinted with permission from American Society for Materials, Materials Park, OH (www.asminternational.org).

coils on the work. Unlike laser hardening, however, electron beam hardening must be done under a vacuum to prevent electron scattering and to minimize surface contamination. The exchange of parts in the vacuum system is difficult for large production volumes, and the technique is more suitable for low production volumes. Electron beams are also used to deposit hard, durable, and corrosion-resistant coatings, such as WC-Co–cerments.

Diffusion Hardening

Diffusion hardening involves thermally activated diffusion of solute atoms (e.g., carbon, nitrogen, or boron) into the surface and subsurface regions of a steel part. In carburizing, controlled amounts of carbon atoms are diffused in the surface layers of low-carbon steels at temperatures at which austenite is the stable phase. This is followed by quenching and tempering to form tempered martensite. Gas, vacuum, pack, and liquid carburizing methods are used. Gas and vacuum carburizing use a gaseous mixture as a source of carbon; the main constituents of the gaseous atmosphere are CO, CH_4, CO_2, N_2, H_2, and water vapor. Carburizing atmospheres are highly toxic and flammable and require use of leak-tight furnaces. Carbon monoxide released from the carburizing furnace is burned to carbon dioxide. The depth of hardened case is a function of the time, temperature, and atmosphere, and also depends upon the geometry of the part. For example, in a gear, case depth is smallest at concave surfaces (root of the tooth) and largest at convex surfaces (tooth tip).

In gas carburizing, a part is initially heated to carburizing temperature in an inert gas atmosphere. The inert gas is used to keep oxygen away, which would cause an explosion when methane (carbon carrier) is introduced. At the carburizing temperature, the inert gas supply is discontinued, and methane is introduced, which decomposes to release carbon. The parts are quenched either in an integral chamber or are moved through separate zones of the furnace on a wire mesh toward the quenching port. In vacuum-carburizing, part is heated in a vacuum furnace that is essentially a steel pressure vessel. The work chamber is a box inside the furnace that contains heating elements. Work is placed in the chamber in a way to allow for uniform heating. Furnace is then sealed, evacuated, and heated. When the part reaches the carburizing temperature, a carburizing gas (usually methane) is introduced in the furnace chamber. After a predetermined soak time, the temperature is lowered to reduce part distortion, and the part is quenched. Vacuum-heating is slower than gas and liquid bath heating but reduces undesirable surface reactions and scale formation. Parts stay clean and bright, and case depth is well controlled. The main disadvantage is the high cost of equipment and maintenance.

Pack carburizing uses solid compounds such as charcoal, barium carbonate, and calcium carbonate or sodium carbonate to cover the part and allow the carbon released from the decomposition of the solid charge at the carburizing temperatures to diffuse into the part. This is followed by quenching of carbon saturated austenite to form martensite. Pack carburizing is more labor intensive and dusty (needs packing of solid charge), and requires more physical handling (e.g., unearthing the part) for quenching than gas and liquid carburizing. It is also relatively slow, because more time is needed to heat and cool the thermal mass of solid compounds and container, and is generally used on relatively small parts. In the actual pack carburizing process, activated charcoal and the other ingredients are put in a welded sheet metal box with a gasket for sealing the top down (a good seal is needed to reduce the oxygen content, which, in excess, will burn charcoal). The sealed box is put in a furnace and heated to form carbon monoxide, which is the carburizing agent. After the part is soaked at the carburizing temperature to allow C from CO to diffuse in the part, the part is removed from the pack using a lifting device and transferred to a quench bath. Handling of charcoal and other solid compounds, loading, unloading and box maintenance, slow heating, discontinuous and dusty operation, and a lack of precise control on case depth are some limitations of pack carburizing.

In liquid-carburizing, the part is immersed and heated in a salt bath to form austenite and allow carbon to diffuse in the surface and subsurface regions, followed by quenching in oil, water or brine solution to form martensite. The process is carried out at 845–900 °C. Cycle times are faster in liquid carburizing than solid and gas carburizing because liquid salt baths heat up faster than both gaseous mixtures and solid charge. Salt baths contain a cyanide compound (usually sodium cyanide, NaCN) and a mixture of barium chloride, sodium carbonate, and chlorides of potassium and sodium. The dissociation of NaCN releases carbon that diffuses into the part. Because of the toxicity of cyanide baths, non-cyanide baths are also used for liquid carburizing. Here fine carbon particles are suspended in a salt bath using agitation with a mechanical stirrer, and serve as the source of carbon atoms for diffusion. Non-cyanide baths use slightly higher temperatures (900–955 °C) than cyanide baths, which could slightly increase the tendency for part distortion during subsequent quenching. Low-temperature salt baths produce a thin case (0.03 inch) and high-temperature salt baths produce a heavy case (up to 0.12 inch). Blind holes need extra care in processing, because salt can build up in holes and cause corrosive attack.

A major effect of carburizing is introduction of compressive residual stresses into the surface of the part, as shown in Figure 5-17. These stresses are because of both normal quenching and volumetric changes that take place during transformation of austenite to martensite; lattice distortion during transformation of FCC austenite to BCT martensite leads to volume expansion. The surface compressive stresses are beneficial because they counteract applied tensile stresses and because fatigue strength of the part increases. Peak compressive stress in properly carburized

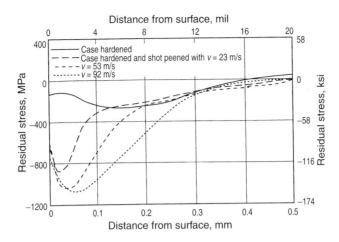

FIGURE 5-17 *Effect of shot-peening on surface compressive stresses of a carburized part. Shot-peening further enhances the in-built residual compressive stresses on the carburized part.* (ASM Metals Handbook, Vol. 4, Heat Treating, 1991, p. 371). Reprinted with permission from American Society for Materials, Materials Park, OH (www.asminternational.org).

parts vary in the range of 200 to 450 MPa. The magnitude of in-built compressive stresses is further increased by shot-peening of carburized parts.

Nitriding introduces nitrogen atoms in the surface layers of steel parts in the temperature range of 500–550 °C. In this temperature range, ferrite rather than austenite is the stable phase, so that nitrogen diffuses in ferrite. In addition, quenching is not required to develop surface hardness, which helps in reducing part distortion and permits relatively good dimensional control. The mechanisms of surface-hardening by nitriding are formation of iron nitrides, solid solution strengthening, and formation of nitrides of alloying elements (Al, Cr, V, and Mo) in steels. The major sources of hardening are the two iron nitride phases that form in the case: Fe_4N, which is a very hard and brittle phase, and Fe_3N, which is relatively tough and less brittle than Fe_4N. The relative thickness of these nitride layers is controlled by adjusting the nitriding process parameters to obtain the desired combination of surface hardness and toughness. Both gas and liquid media are used for nitriding; in addition, ion-nitriding is also used.

Another diffusion hardening process is carbonitriding, which introduces both carbon and nitrogen atoms in the surface layers when the part is heated into austenitic region (700–760 °C). The thermal treatment is followed by quenching. Carbonitriding makes possible use of low-carbon steels to achieve hardening, and also allows use of high-alloy carburized steel without the need of drastic quenching. Both carbon and nitrogen-rich gas atmospheres (e.g., CH_4 and NH_3) and liquid cyanide salt baths are used for carbonitriding. The cyanide salt (NaCN) used in liquid carbonitriding combines with oxygen at the surface of the bath to form sodium cyanate (NaCNO), which produces C and N for diffusion in the part. Because both carbon and nitrogen are austenite stabilizers, relatively large quantities of retained austenite may remain in the part after quenching, and reduce the hardness and wear resistance.

Gas carbonitriding is identical to gas carburizing with ammonia added to the carburizing gas mixture. This produces both N and C that diffuse in the part. Compared to nitriding, however, higher temperatures are needed in carbonitriding, and compared to carburizing, case depth is shallower. Carbonitriding has shorter cycle times as compared to nitriding, and permits higher hardness to be achieved at lower temperature compared to carburizing because of the diffusion of nitrogen.

Carburizing produces medium case depths, uses low- to medium-cost steels, and involves high capital investment. Nitriding produces shallow case depths, uses medium- to high-cost steels, produces little distortion, and requires medium capital investment. Carbonitriding produces shallow case depths, uses low-cost steels, and requires medium capital investment. Another process, called ferritic nitrocarburizing, involves diffusion of both C and N in the surface layers from a gas atmosphere at 560–670 °C, and produces a thin (<0.001 in.) and hard case on low-carbon steel. The process is similar to gas carbonitriding except that gas composition is slightly different and process temperatures are lower (100–150 °C lower as compared to carbonitriding). Methane and ammonia are used as the source of C and N in nitrocarburizing, but higher concentrations of ammonia are used than in gas carbonitriding. The lower process temperatures in nitrocarburizing make the C diffusion sluggish. Both high ammonia contents and low process temperatures lead to more nitrogen in the case.

Many other diffusion hardening treatments are available, such as aluminizing, siliconizing, chromizing, and boronizing (or boriding). Aluminizing involves diffusion of Al in the case to produce oxidation and corrosion resistance. Galvanized Zn coatings compete against aluminizing, which is more expensive but also more effective for higher-temperature applications; diffusion of Al in surface layers forms high melting point intermetallics, which provide better heat resistance (i.e., resistance to melting) than galvanized coatings. Aluminizing is used in jet engine parts and high-temperature fasteners. In siliconizing and boriding, silicon and boron, respectively, are allowed to diffuse in a part to provide surface hardening.

All of the diffusion hardening treatments involve non–steady-state diffusion of solute atoms at high temperatures in a concentration gradient. Thus, in the case of carburizing, a concentration gradient of carbon develops during isothermal soak at the carburizing temperature, which can be predicted using Fick's second diffusion equation, which for one-dimensional diffusion has the form

$$\frac{\partial C}{\partial t} = \frac{\partial}{\partial x}\left[D(C)\frac{\partial C}{\partial x} \right] \tag{5-7}$$

where C is the concentration, t is the time, and $D(C)$ is the diffusion coefficient, which is a function of concentration at a fixed temperature. The initial and boundary conditions for carburizing of a semi-infinite steel slab of initial carbon content, C_0, with a fixed carbon concentration, C_s, at the surface are, $C = C_0$ for $t = 0$ at $0 \le x \le \infty$, $C = C_s$ for $x = 0$ at $t > 0$, and $C = C_0$ for $x = \infty$ at $t > 0$. If for simplicity, the diffusion coefficient, D, is assumed to be independent of concentration, then the solution to the one-dimensional diffusion equation yields the carbon concentration, C, at a point x and time t as

$$\frac{C - C_0}{C_s - C_0} = 1 - erf\left(\frac{x}{2\sqrt{Dt}}\right) \tag{5-8}$$

where erf denotes the error function, and by definition

$$erf(z) = \frac{2}{\sqrt{\pi}} \int_0^z e^{-u^2}\, du.$$

The numerical values of the error function are tabulated in standard mathematical tables, and can be used in the above solution.

Laser Surface-Engineering

Lasers as a possible tool for surface engineering have attracted attention in the last few decades. Since its discovery, the laser has been accepted as a very powerful materials tool for variety of processing operations. Schawlow and Towns first established the theoretical concept of the laser in 1958, and Maiman invented a working ruby laser in 1960. Later, continuous CO_2 lasers with many kilowatts of power tremendously increased the possibility of commercial development of lasers in material processing. Recent developments in Nd:YAG, excimer, and fiber laser have enhanced such capability. The advent of the fiber laser has demanded the greatest change in our perception of the laser, with tremendous growth in this field. The primary factor of power density greater than 10^6 watts/cm^2 is the reason for the laser becoming an attractive tool in material engineering. Both the increasing demand for advanced materials and the availability of high-power lasers have stimulated interest in research and development related to laser applications. More obvious applications such as laser cutting, laser welding, and laser heat treatment are already present in industry. Lasers offer an easily controlled, chemically clean heat source that can produce a very narrow heat-affected zone and a very minute distortion. The application of lasers in surface alloying, cladding, glazing, and annealing of semiconductors offers the possibility of producing new materials with better quality and reproducibility. The very rapid rate of heating and cooling could also open up opportunities for producing novel surface properties. Laser-cladding technology has been applied in the automobile industry to produce erosion-resistant cylinder heads in lightweight aluminum alloy engines by depositing solid solution alloys.

Energy in the form of electromagnetic radiation, particularly in the visible light range, has certain advantages. It can be focused and transported relatively easily. A concentrated, collimated, and coherent light, as in a laser beam, can be used for several industrial processes. Laser-melting, -welding, -drilling, and -machining are well-established processes. This chapter is an overview of a relatively new technique: laser surface-engineering. As the name suggests, laser technology can be used to engineer the surface of structural components, thereby improving surface-related properties and performances. Most of the deteriorations of structural components initiate at the surface. Corrosion, oxidation, wear, erosion, and fatigue are but a few such surface-related deteriorations. Hence, engineering the surface holds potential for many technological advances.

Laser–Materials Interaction

Electromagnetic radiation propagates energy, which can be exchanged with matter only in elementary discrete quantities called radiation quanta or photons. The energy of photon is proportional to the frequency v of the associated electromagnetic radiation through the relation $E = hv$ where the Plank's constant $h = 6.62 \times 10^{34}$ Joule $\times$ sec. The energy flux per unit time can be expressed as a function of the photons per second per unit surface area. Therefore, $F = c\,n_f hv$ where n_f is the photon number per m^3 moving along the flux direction at the speed of the light, c, 3×10^8 m/sec.

The input of energy from pulsed/continuous wave lasers into the near-surface regions of solid involves electronic excitation and de-excitation within an extremely short period of time. The interest in these energy deposition techniques arises because lasers can be used to achieve extreme heating and cooling rates in the near-surface region (10^4–10^{10} K/s), while the total deposited energy (0.3–5 J/cm^2) is insufficient to affect, in a significant way, the temperature of the bulk material. This allows the near-surface region to be processed under extreme conditions, with little effect on bulk properties. The initial stage in all laser-metal processing applications involves the coupling of laser radiation to electron within the metal. This first occurs by the absorption of photons from the incident laser beam, promoting electrons within the metal to higher energy states. Electrons that have been excited in this manner can divest themselves of their excess energy in a variety of ways. For example, if the photon is energetic enough, excited electrons can be removed entirely from the metal. This is the photoelectric effect, and in most materials, requires photon energies greater than several electron volts. However, the laser photons have usually much less energy (CO_2 laser, 0.12 eV; Nd:YAG laser, 1.2 eV). Hence, the electrons that have absorbed energy from laser photons may not be energetic enough to leave the material, but they do have to lose energy to get back to ground state. This occurs when excited electrons are scattered by lattice defects. Such defects can be dislocation, grain boundary, and lattice distortion (due to photon, thermal, or other kinds of stress). Consequently, the photon energy is converted to thermal energy.

Photon interactions with matter occur usually through the excitation of valence and conduction band electrons throughout the wavelength band from infrared to ultraviolet regions. Since free carrier absorption by conduction-band electrons is the primary route of energy absorption on metals, beam energy is almost instantaneously transferred to the lattice by electron–photon interaction.

Types of Lasers

Certain atoms/molecules can take on various quantum energy states. When externally stimulated, such as by an electric discharge, the molecules or atoms jump to excited states. Atoms in high-energy states give up energy to go to a lower value, accompanied by release of photon with well-defined energy. Some of the photons travel along the optical axis and start oscillating between a pair of mirrors. The photons can be reabsorbed, diffracted, or can strike an excited molecule. If a photon strikes an excited molecule, it will stimulate that excited molecule to release its energy and fall to a lower energy state and thus emit another photon of identical wavelength, traveling in exactly the same direction and with the same phase. Similarly, multiple photons will be generated and oscillate in the cavity in phase, generating even more photons. The excited state becomes depleted, and so by the Boltzmann distribution, more and more of the energy is passed into that state, giving a satisfactory conversion of electrical energy into the upper state.

Active material has at least one such unit (atom, ion, or molecule) that can give rise to lasing. Rare earth elements are particularly popular as the lasing element. These elements are inserted in a glass or crystal to make it active. The energy states of these heavy atoms are not much affected by the crystal they are in. Neodymium (Nd) is an important transition element as the active atom in many crystals and glasses. Nd^{+3} in an yttrium aluminum garnet (YAG) crystal or a suitable glass is the appropriate medium for lasing. The Nd atoms partially substitute yttrium atoms in YAG crystal. The energy levels are broadened only slightly, preserving the multiple energy state essential for population inversion. Many materials show stimulated emission phenomena. However, only a few can achieve power levels useful industrially. Lasers suitable for industrial materials processing are CO_2, CO, Nd:YAG, Nd:Glass, excimer (e.g., KrF, ArF, XeCl), etc. A comparison of different industrial lasers being used for surface engineering is given in Table 5-3.

TABLE 5-3 Characteristics of Lasers Used for Surface Engineering

LASER	CO$_2$	Nd-YAG	Excimer
Wavelength	10.6 μm	1.06 μm	ArF: 0.193 μm KrF: 0.248 μm XeCl: 0.308 μm XeF: 0.351 μm
Energy/photon	0.117 eV	1.17 eV	3.53 – 6.42 eV
Continuous wave (CW) Power range	2 W – 25 kW	0.05 W – 3.5 kW	--------------
Pulse length	1 μs - CW	0.3 ms - CW	10 – 60 ns
Energy/pulse	Up to 1700 J at 0.05 Hz	100 J	Up to 4 J
Average power	Up to 25 kW	Up to 3 kW	Up to 150 W
Pump mechanism	Gas discharge, RF or DC excited, longitudinal or transverse (TEA)	Xe or Kr Flashlamps, diode	Transverse Discharge

Lasers as Surface-Engineering Tools

The first attempt at laser surface alloying was reported in 1964. However, the application of lasers in coating technology developed rapidly only after 1980. Primarily, the development in laser technology has been responsible for this leap. Availability of various kind of lasers such as CO$_2$, Nd:YAG, and excimer in adequate power levels and optical fiber delivery systems has extended the usability of laser in surface engineering and many other manufacturing processes. Manufacturers have recognized that by its very nature laser surface engineering has unique advantages over its conventional counterparts. Various fabrication processes that can be carried out during laser surface engineering are given in Figure 5-18.

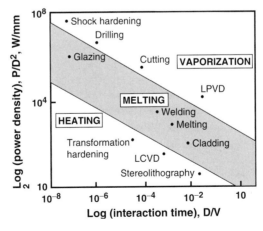

FIGURE 5-18 *Map of laser processes as a function of power density per unit time.* (From W. M. Steen, Laser Material Processing, Springer, New York, 1991, p. 173).

Characteristics of Laser Surface Engineering

1. Light is inertia-less, making it conducive to rapid processing.
2. Focused laser light provides high energy density.
3. Laser surface engineering can be carried out with and without atmosphere control.
4. Materials difficult to melt (such as ceramics and refractory metals) can be surface-treated.
5. Precise and selected surface area can be surface-treated.
6. Little or no contamination is involved.
7. Heat-affected zone (HAZ) is narrow or absent.
8. Nonflat surface (e.g., inner diameter, nonregular surface) can be surface-treated.
9. Coating is metallurgically bonded.
10. Laser equipment can be time-shared using fiber optics to process multiple components simultaneously.
11. Laser treatment is amenable to automation.

In recent years the laser has become an accepted tool in the field of materials processing, especially for cutting and welding, due to its versatility as a processing tool coupled with its flexibility, unmatched by any other conventional manufacturing technique. Furthermore, especially in the last decade, the lasers have been extensively studied and configured for the purpose of surface engineering of various finished components. Surface engineering is often the most efficient and inexpensive method of providing a component with surface properties different from the bulk properties of the material. Typical applications of surface engineering are to improve the wear and corrosion resistance.

The advantage lasers offer in materials processing is evident with its power density. With changes in power density, the laser can be made to perform a wide range of physical processes, ranging from just heating, to melting, to vaporization, and thereby corresponding fabrication/manufacturing processes as schematically depicted in Figure 5-18. In general terms, to fuse the laser's energy with a material, the laser beam must be focused on a small spot where the power density is sufficiently high enough to cause the material to absorb, rather than reflect, the laser beam. The efficiency of absorption is related to mainly the magnitude of power density and to some degree the wavelength of a laser beam. As shown in a schematic manner in Figure 5-19, absorption of power density is more for Nd:YAG laser beam ($\lambda = 1.06\ \mu$m) compared to CO_2 laser beam ($\lambda = 10.6\ \mu$m), and when material is vaporized, all the incident laser energy is absorbed. When the material is in a molten state, less laser energy is absorbed. The dramatic transition in absorption occurs when the material changes from solid to liquid. The efficiency of absorption is, therefore, linked to both the material's physical state and the laser power density at the surface. Higher power density is required to vaporize, compared to that required to melt the material.

Based on the power densities required to vaporize them or to melt them, materials can be put in three major classes. Within each class the power density values and laser process parameters are similar. Going from one class to the next, there are substantial differences in power density values and laser processing parameters. Examples of materials in each class and the corresponding power density values for vaporization and melting are listed in Table 5-4. In practice, however, one seldom deals with materials that are not dissimilar. Therefore, one should group the materials according to the compositions classified in Table 5-4. If all the components belong to the same class, the material is easy to laser-process; otherwise, the material is very difficult to process. Also remember that the power density within the beam is a function of not only the focal length of the lens and the wavelength but also of the beam divergence, which is directly

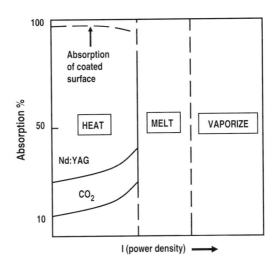

FIGURE 5-19 *Physical processes in laser materials interaction as a function of power and absorption.* (From S. L. Engel, Classification of materials, Manufacturing Engineering, February 1989, pp. 43–46).

TABLE 5-4 Classification of Materials Based on Power Required to Process

Class	Materials	Power Density, watts/mm^2	
		Vaporize	Melt
I	Good thermal/electrical conductors: Cu, Al, Au, Refractory materials: W, Va, Ta, Mo	10^6	10^4
II	Ferrous materials, Nickel alloys	10^4	10^2
III	Organic/plastic materials Some metals: Pb, Zn, Sn, Cd	10^2	2-5

Source: From S. L. Engel, "Classification of Materials," *Manufacturing Engineering*, pp. 43–46, February 1989.

related to the laser's transverse excitation mode (TEM). Thus, with collective consideration of absorption characteristics, materials, and type and mode of operation (pulse or continuous) of lasers, various surface-engineering operations listed in Figure 5-20 can be performed.

Lasers provide the tools for developing unique methods of treating the surfaces of materials. Initially, with development of continuous-wave CO_2 lasers of high power, and lately with evolution of Nd:YAG lasers into multikilowatt-power fiber-optic-beam delivery system, surface-engineering technology is rapidly growing in the identification of new and improved processing techniques.

Presently, however, at least a few percent of lasers are devoted to surface engineering–related technology compared to other materials processes in the manufacturing environment, as illustrated in Figure 5-21. In addition to the previously mentioned two types, excimer lasers are also employed in several laser surface-engineering processes. These three major types are

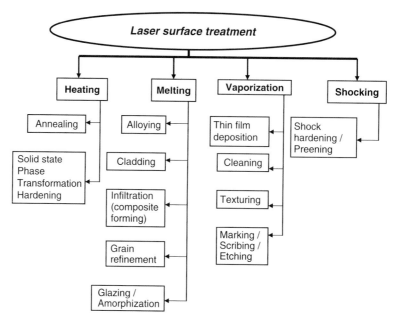

FIGURE 5-20   *Laser surface-engineering processes.*

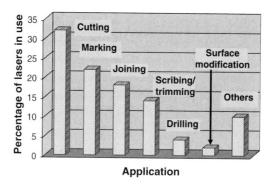

FIGURE 5-21   *Percentage of lasers in use for various processes of surface-engineering.*

suitable for surface engineering because of their operating characteristics listed in Table 5-3 and accordingly can perform a wide range of manufacturing processes on various classes of materials listed in Table 5-5. The recent developments of multikilowatt diode-pumped lasers even elevate the technology to a new level. Unlike Nd:YAG, CO_2, and excimer lasers, diode-pumped lasers remain a very compact machine that are extremely convenient and easy to integrate into a chain of manufacturing equipment with a relatively low cost for maintenance.

Laser energy can alter the surface properties of materials with great precision and low heat input. Lasers offer many benefits for surface engineering, including localized treatment, low thermal distortion, control of treatment depth, and the ability to address complex shapes. Typical applications of laser surfacing in the automotive industry include hardening and cladding engine

TABLE 5-5 Typical Potential and Current Applications for Laser Surface Modification

Component	Material	Process	Type of Laser
Diesel Engine Cylinder Liner	Cast Iron	Localized Transformation Hardening	CO_2
Auto Gas Engine Bore	Aluminum-Silicon Alloy	Ceramic Coating	Nd-YAG and CO_2
Turbine Blades	Superalloy	Cladding as repair operation	Nd-YAG and CO_2
Office Equipments: Electromechanical Parts	Plain Carbon Steel	Localized Transformation Hardening	Nd-YAG and CO_2
Automotive Steering Assembly	Cast Iron	Localized Transformation	Nd-YAG and CO_2
Automotive Valves / Valve Seats	Alloy Steel	Localized Cladding	Nd-YAG and CO_2
Pump Shafts	Alloy Steel	Localized Cladding	Nd-YAG and CO_2
Gears	Plain Carbon Steel	Localized Transformation Hardening	Nd-YAG and CO_2
Cam Shaft	Cast Iron	Localized Transformation Hardening	CO_2
Diesel Engine Valve Seat Inserts	Stainless Steel	Cladding	Nd-YAG and CO_2
Forming Dies	Tool Steels	Surface Alloying or Cladding	CO_2 and Nd-YAG
Bearings, Bearing Retainers, Slide Rails	Alloy Steels	Surface Alloying	CO_2 and Nd-YAG
Die Casting Dies	Die Steel	Surface Alloying or Cladding	CO_2 and Nd-YAG
Aircraft, Printing Machine, Microelectronic Components,Tools	Ferrous and Non-ferrous Alloys, Wood, Ceramics, Plastics	Scribing/Etching/Marking	Excimer, CO_2, and Nd-YAG

camshafts or valve seats; in the aerospace industry, repair and cladding turbine blades in power generation hard-facing steam turbine blades; and in the oil and gas industry, hardening and cladding pumps, valves, and tubular components. Several newer applications are being continuously investigated from an engineering as well as a scientific point of view. Several specific applications are given in this chapter as examples to better explain types of laser surface treatments.

To provide very controlled surface modification, the energy distribution within the laser beam may need to be "tailored" to a specific application. This is not always possible using conventional laser beam–focusing optics. Extensive work has been carried out with diffractive optics to shape the laser beam into a complex profile for critical components.

Use of lasers in surface engineering is more attractive for a number of following reasons:

1. Laser beams are a chemically clean light source that delivers precisely controlled energy to a localized region.
2. Laser beams can be easily maneuvered by optical elements or fiber optic and can be adopted for automation suited for processing in the ambient environment of large-scale and complex structures with ease and high speed.

3. Narrow beams with high power density allow extremely rapid processing, thereby causing minimal or no changes in the bulk material.
4. Rapid rates of processing produce refined and novel microstructures in the surface region.
5. Precision associated with the coherent and monochromatic beam combined with automation allows the possibility of near-net-shape processing with tailored properties.

As observed in Figure 5-20, specific manufacturing processes of laser surface engineering are classified under four physical principles: heating (without melting), melting, vaporization, and shocking. Technique modifications based on these physical principles describe various laser surface-engineering processes. The following sections provide a brief overview of these processes.

Laser Surface Heating

Engineering application of components often requires that the surface properties be dramatically different from the substrate. For example, a high hardness is desired for a surface subjected to wear or high, localized stress. At the same time, the substrate should have good toughness, usually associated with moderate to low hardness. A heavy-duty pin, for example, should have high hardness on the surface but should be ductile and tough in the core. There are several techniques, either with deliberate change in composition or not, available to achieve this. Carburizing, nitriding, and so on increase the surface hardness of the steel by a change in composition and evolution of hard phases. Induction-hardening, selective hardening, shot-peening, and so on depend on either martensitic transformation (change in microstructure) or on change in defect density. Selective hardening either by flame-hardening or by selective quenching suffers from inconsistency and lack of reproducibility. In induction-hardening a quencher is required. Distortion is usually present. Further, the shape of the component that can be hardened is constrained by the shape of the inductor. Also, nonmagnetic materials cannot be induction-hardened. Lasers have certain unique advantageous characteristics for surface-hardening. The electromagnetic radiation present in the laser beam interacts with the electron in the lattice and sets up rapid heating. The laser beam can traverse almost any surface and harden selectively as required. The high power input sets up a high thermal gradient. The laser-material interaction is rapid and confined to only a small volume at a giving point of time. The remaining volume of the component serves as heat sink. Also, the conduction mode of heat transfer through it results in self-quenching of the laser-heated region. Even though the power and intensity is high, the energy input is relatively confined and limited. Hence, there is no or little distortion involved in laser-hardening.

Laser-induced heat treatment (hardening) of the surface can be carried out at a rapid speed. Laser is a noncontact type of treatment, hence complex shapes can be hardened without any problem. The operation parameters can be manipulated to get the desired case depth. Laser surface-hardening usually results in improved fatigue life. The parameters controlling the laser heat treatment are power, beam size, absorptivity, and traverse speed. The thermal properties of the substrate also play important role. The wavelength of the laser beam is also an important factor. Ferrous materials are known to reflect greater fraction of the energy at higher wavelength. For example, during interactions with most of the metallic materials with an Nd:YAG laser ($\lambda = 1.06 \, \mu m$) 30% input power is absorbed, whereas only 10% of the input power is absorbed from a CO_2 laser ($\lambda = 10.6 \, \mu m$).

Controlled heating of surface and subsurface region to specific level of temperature and heating/cooling rate with laser can bring about annealing and solid-state phase transformation in the material. Selective heating at the surface renders the rest of the bulk material a heat sink and permits rapid quenching rate without a need for external cooling. Alloys that undergo solid-state phase transformation in a rapid fashion, especially steels, are the most suitable candidate materials for laser-phase transformation treatments. The depth of hardness penetration depends on several parameters, including thermal conductivity, energy input, and phase transformation temperature. The primary mode of heat transfer is conduction through the substrate. Generation and accumulation balance the heat flow by conduction. Generation is given by

$$H = k \left(\frac{\partial^2 T}{\partial x^2} + \frac{\partial^2 T}{\partial y^2} + \frac{\partial^2 T}{\partial z^2} \right) \tag{5-9}$$

and accumulation is

$$\frac{1}{\alpha} \frac{\partial T}{\partial t}.$$

The generation term should balance accumulation term, i.e.,

$$\frac{1}{\alpha} \frac{\partial T}{\partial t} = k \left[\frac{\partial^2 T}{\partial x^2} + \frac{\partial^2 T}{\partial y^2} + \frac{\partial^2 T}{\partial z^2} \right] \tag{5-10}$$

The boundary conditions are

$$-k \left(\frac{\partial T}{\partial t} \right) \Big|_{y=0} = q_0 \quad x_1 < x < x_2 \text{ and } z_2 < z < z_1 \tag{5-11}$$

$$= 0 \quad \text{elsewhere}$$

$$\left(\frac{\partial T}{\partial x} = 0 \right), \text{ at } x = 0 \text{ and } x = L_x \tag{5-12}$$

$$\left(\frac{\partial T}{\partial z} = 0 \right), \text{ at } z = 0 \text{ and } z = L_z$$

$$-k \left(\frac{\partial T}{\partial y} \right)_{y=l_y} = hT_y \tag{5-13}$$

and the initial condition is $T(x, y, z, t) = 0$, where T is the temperature, x, y, and z are the Cartesian coordinates; t is time; h is heat transfer coefficient; L_x, L_y, L_z are the dimensions of samples; and k is the thermal conductivity of the plate material.

The temperature distribution, in terms of Green's function, can be given by

$$T(x, y, z, t) = \frac{aq_0}{k} \int_{\Delta\tau}^{t} \int_{x_1}^{x_2} \int_{z_1}^{z_2} G_{x_{22}z_{22}y_{23}} \left(\frac{x, z, y, t}{x', z', 0, \tau} \right) \times dx' dz' d\tau \tag{5-14}$$

$G_{x_{22}z_{22}y_{23}}$ is the Green's function for the $x_{22}z_{22}y_{23}$ geometry.
This is satisfied by

$$T = \frac{q}{8(\pi \alpha t)^{2/3}} e^{-\left\{ (x-x')^2 + (y-y')^2 + (z-z')^2 \right\}/4\alpha t} \tag{5-15}$$

The ideal power distribution is one that gives a uniform temperature over the area to be treated. This requires a dimpled power distribution, since the heating effect is dependent on the edge cooling and surface heating, i.e., P/D and not P/D^2, where P is the incident absorbed power and D is the beam diameter or spot size. There are several methods available to spread the laser beam, such as defocusing the high-power multimode beams, zigzag traversing of laser beam, kaleidoscopes, segmented mirrors, and special optics. All these methods produce a reasonably uniform distribution of power over the central region of the beam traverse (track).

Transformation-hardening without any surface melting is the simplest to model mathematically. There is no convection or latent heat term involved, as there is no melt pool. It has been determined that the idealized one-dimensional heat transfer model describes the temperature distribution with reasonable accuracy. The edges of the cross-section are more complex, and a two-dimensional heat transfer model is required to describe this. Whether the edges or the center region of the track is dominating is decided by beam diameter (spot size) and laser traverse speed. An empirical relation has been developed for this purpose. The relation between is given by following equation:

$$d = -0.10975 + 3.02 \frac{P}{\sqrt{DV}} \tag{5-16}$$

Most often, the hardened width is the only essential answer required. For exact thermal distributions, numerical techniques such as finite difference models can be used. Improved models, such as one suggested by Ashby and Shercliffe, have been shown to predict the transformation-hardening with more accuracy.

The most common examples are the surface-hardening of steel containing enough carbon to allow hardening and cast iron with a pearlitic microstructure. As the beam moves over an area of the metal surface, the temperature starts to rise. The temperature rise should be above the critical temperature for transformation during heating, A_{c1}, but less than the melting point. The transformation from pearlite to austenite begins at A_{c1} and is complete at the A_{c3} temperature. Once the beam has passed through, cooling occurs by conduction mode of heat transfer through the rest of the bulk material, which is hardly heated. The high cooling rates can further be ensured by providing the high heat fluxes via operation of the laser in short-duration pulse mode. The new structure, namely austenite, is unable to transform back to pearlite because there is not enough time for carbon to diffuse. Instead, the austenite transforms into martensite in a diffusionless athermal process. The strained lattice does not allow dislocation movement and therefore is hard. When austenitization has occurred, the carbon moves by diffusion down concentration (more accurately activity) gradients. The time for diffusion within the austenitic lattice varies with position within the laser-treated region. Due to the short time available, often there are regions where carbon is not fully diffused and the resulting structure is a nonhomogeneous martensite.

In transformation-hardening of ferrous materials, the initial microstructure usually consists of nonhomogeneous distribution of carbon. For example, pearlite in steel is a phase mixture of cementite (6.67 wt% C) and ferrite (<0.03 wt% C). A hypoeutectic steel microstructure consists of colonies of ferrite and pearlite. On heating above the phase transformation temperature, A_{c1}, carbon gets transported across the composition and activity gradient. If a long period of time is allowed at this elevated temperature, eventually carbon will diffuse and will be uniformly distributed. However, the high cooling rate associated with laser processing allows only limited time for diffusion to take place. Hence, a detailed study of diffusion is required. The rate of diffusion is described by the following equation within austenite phase:

$$\frac{\partial c}{\partial t} = D_{AB} \left[\frac{\partial^2 c}{\partial x^2} + \frac{\partial^2 c}{\partial y^2} + \frac{\partial^2 c}{\partial z^2} \right] \tag{5-17}$$

The diffusivity of carbon in austenite is

$$D = 1 \times 10^{-5} e^{\frac{-9.0}{T}} \text{ m}^2/\text{s}$$

and in ferrite is

$$D = 6 \times 10^{-5} e^{\frac{-5.3}{T}} \text{ m}^2/\text{s}.$$

On rapid heating, pearlite colonies first transform to austenite. The carbon diffuses outward from these transformed zones into the surrounding ferrite, increasing the volume of high-carbon austenite. (Carbon is an austenite stabilizer and increases the volume of austenite as it diffuses around). On rapid cooling these regions of austenite, which have more than certain minimum carbon (0.06%), will quench to martensite if the cooling rate is sufficiently fast, although retained austenite may be found if the carbon content is above a certain value (>1.0%). The required rate of cooling is indicated by continuous cooling curve for that composition. The cooling rate in laser surface-hardening is usually in the order of (or higher than) 103 °C/s. Such a high cooling rate invariably produces martensite in most of the steel compositions.

The transformation of pearlite into austenite is initiated by carbon diffusing out of cementite plates into ferrite. Even though this process is very fast, some amount of superheating above the lower critical temperature A_{c1} is required. Therefore, interlamellar spacing between ferrite and cementite and ferrite and pearlite colony size thus influence hardenability of the steel. For example, a normalized steel responds better to laser transformation-hardening than does an annealed steel. Also, a uniform prior microstructure and a higher interaction time results in more homogeneous martensite. Compared to conventional heat treatment (such as furnace heating and brine quenching), laser surface transformation requires higher superheating. However, because the quenching rate is higher than any conventional quenching, the resulting microstructure contains much higher fraction of martensite, lower retained austenite, and less carbide precipitation.

The major factor controlling laser surface-hardening of steel is the temperature distribution on the surface to be laser-treated (Tayal and Mukherjee, 1994). Normally, measurement of temperature variation during laser surface-hardening is not possible because of the high-temperature variation rate. Theoretical models have been developed to predict the temperature distribution during laser surface-hardening (Mazumdar and Steen, 1980; Cline and Anthony, 1977). A thermal and microstructural analysis for laser surface-hardening of AISI 1045 steel has been used to predict the temperature distribution and consequently the case depth, which can be achieved by using a given set of laser process parameters (Tayal and Mukherjee, 1994). By varying the laser power, traverse speed, and the position of the sample surface with respect to the focal plane of the laser optics, hardened layers of various thickness were obtained. Care must be taken to avoid melting at the surface, which is normally undesirable during hardening. Samples with various initial surface conditions would also influence the hardened-layer thickness and properties. As one travels outward from the heat-affected zone (HAZ) boundary, one would encounter various microstructures. At the most interior point of HAZ, the temperature was above A_1 but below A_3.

Pearlite colonies transformed to austenite, which subsequently transformed to martensite. The undissolved ferrite remains as before laser treatment. The microstructure would constitute a mixture of martensite and virgin ferrite. As the peak temperature and diffusion time increase, the prior pearlite colonies expanded when austenitized and formed martensite colonies of slightly lower carbon concentration. The ferrite grain size in this zone is smaller than that of the base materials. At an even higher temperature and diffusion time, both the low-carbon and high-carbon martensite phases form. This is because the carbon has not been able to diffuse from prior-pearlite austenite and prior-pearlite austenite. At further higher temperatures and time, a nearly uniform martensite is formed.

The phases thus formed are not different from conventional processes. However, the rapid heating and cooling produces finer martensite than that formed in furnace heat treatments and correspondingly produces a higher hardness.

All the cast irons contain a large amount of carbon, and most are heat treatable. The microstructure of ferritic gray cast iron consists of ferrite matrix and graphite flakes. It requires a long time for carbon to diffuse from flake into the ferrite matrix. It is, however, possible to laser-treat gray cast iron. The ferrite around the graphite flake or nodules (as in spheroidized graphite iron) can be enriched by carbon and can undergo martensitic transformation. This process is known to improve wear resistance of cast iron. Pearlitic cast iron, in contrast, has a microstructure of pearlite matrix and graphite. It is possible to achieve very high hardness, because the carbon available for austenitization and hardening is pretty high. However, there is only a small range of temperature between transformation temperature and the melting point, requiring extra care for laser-hardening. In spheroidal graphite cast iron, it is observed that preferential melting takes place around graphite nodules because of lowering of the melting point as the carbon diffuses away from the graphite.

In many other alloys that are not prone to solid-state phase transformation, a reasonable increase in hardness can be achieved due to extreme refinement of the grain structure. Wear-resistant zones with good fatigue properties can be produced on load-bearing surfaces of finished steel and cast iron components using this technique. The method is also ideal for treating discrete areas of large parts, such as tools, or those of irregular shapes or sections, which may be distorted by furnace treatments. The examples of commercially processed components using this technique are automotive power-steering-gear housing made of ferritic malleable cast iron, cast iron camshaft lobe, AISI 1045 steel gear and 4140 steel cylinder, grooves in a gray iron piston ring, gear and rack teeth automotive steering racks, cylinder liners in high-performance turbo-charged diesel engines, and so on.

Laser Surface Melting

High-power lasers provide coherent, monochromatic light that can be focused onto a very small spot ($\sim$0.1–1.0 mm). The intensity, thus, of the order of 10^{10} W/m^2 can be obtained. CO_2 and Nd:YAG lasers are successfully applied for materials processing such as welding, cutting, and to a lesser extent heat treatment, melting, and laser surface-alloying. The most important feature of laser treatment is the localized melting of the surface material. This makes treatment of small surface areas possible without altering or affecting bulk properties and obviates the need for expensive alloying of the entire component. High power density (10^5 to 10^7 W/cm^2) in laser surface-engineering produces a temperature gradient of the order of 10^6 K/cm. Such a high temperature gradient, heat sink effect (of the rest of the components), and conduction mode of heat transfer result in a very high cooling rate or rapid self-quench, sometimes exceeding

10^9 K/s. This results in further grain refinement and may even produce novel metastable or amorphous phases. Solidification within the laser melt pool occurs rapidly once the laser beam has moved away. The rapid solidification thus achieved produces novel microstructure that is far from equilibrium.

Power densities higher than the levels in laser transformation-hardening are required for laser melting. The workpiece is often made absorptive either by using coating (graphite paint, manganese phosphate, etc.) or by increasing surface roughness. Usually a focused or near-focused beam is used for laser melting. Surface melting can be done with or without inert gas shrouding. Without addition of external material in the laser surface-melted region and by controlling the cooling rate based on laser processing parameters, the surface can be either transformed into an extremely refined microstructure or a glazed (near or completely amorphous) structure. Often, refinement and glazing provide homogenization of chemical composition within the modified surface. These physical and chemical changes increase corrosion, oxidation, and wear resistance of the surface. In laser surface-melting, moderate to rapid solidification rates are achieved. There is little or no thermal penetration owing to high thermal gradient and rapid speed. When properly optimized, the laser surface-melting is capable of producing a smooth surface of 1–5 microns.

During laser surface melting, the solid–liquid transformation causes a drastic change in reflectivity. Initial reflectivity (under solid condition) can be controlled by having paint or coating or by increasing surface roughness. Once the solid is melted, these factors are destroyed and there is little scope to control reflectivity in this state. However, when the material becomes very hot the reflectivity is reduced. The shrouding gas, particularly small amount of oxygen, also can have great influence on reflectivity. A small amount of oxygen to the shroud gas reduces the reflectivity. At the beginning of laser treatment, formation of surface plasma helps couple the laser beam with material. However, as the plasma leaves the surface it interferes with the beam.

Laser melting can be used for steel, cast iron, deep eutectic alloys, Al-Si alloys, glass-forming composition alloy, and ceramics. Laser surface-melting is more widely used in the form of surface alloying than just melting. Addition of alloying element(s) improves the surface-related properties to greater extent than laser surface-melting alone. The above-mentioned materials are capable of significant improvement even without external alloying element addition. Usually these materials are not amenable to solid-state phase transformations, such as martensitic transformation, but their microstructures undergo significant refinement through melting and rapid solidification. For example, a high amount of carbon (either in combined form or in free graphite form) present in cast iron can boost the surface properties.

Cast irons are Fe-C based systems that undergo eutectic transformation. Depending on the process and composition, cast iron microstructure consists of inhomogeneous distribution of carbon in free graphite form or combined carbide form. The matrix is ferrite, pearlite, or a combination thereof. Laser surface-melting manifests in change in the microstructure from graphite to cementite and from austenite to martensite. The hardness increases because carbon dissolves in the matrix and austenite transforms into martensite. It is possible to obtain a very hard surface on cast iron by laser-melting. The microstructure of the laser-melted cast iron consists of cementite dendrites in a ledeburitic fusion zone because high dissolved carbon gives retained austenite with some martensite and completely transformed martensite. The laser traverse speed has a significant effect on the microstructure and hardness. At slow laser traverse speed, more carbon can dissolve. The high carbon content results in a solidification structure of ledeburite matrix. As the laser traverse speed increases, the dissolved carbon decreases and so does the hardness.

At further higher speed with only a small amount of carbon dissolution, the hardness reaches a second peak value due to martensitic structure. The intermediate low hardness is due to retained austenite associated with very high dissolved carbon.

Solidification Structure

Solidification structure primarily depends on the cooling rate (R) and the temperature gradient (G). The G/R ratio determines whether the liquid–solid interface is stable planar or unstable, leading to dendrite or cell structure. The determining criterion is the constitutional supercooling criterion that was discussed in Chapter 2, and is briefly presented here. It is related to the thermal gradient being less than the melting-point gradient. When the actual temperature gradient in the liquid at the interface is less than the gradient of the solute in the liquid, constitutional supercooling plays an important role in deciding the type of liquid–solid interface. Because the liquid gets enriched with solute in front of the liquid–solid interface, and the temperature gradient is less steep than the concentration gradient (actually, the liquidus temperature), the interface is unstable. Mathematically, the gradient of the solute is

$$\left[\frac{dC_L}{dx}\right]_{x=0} = -\frac{R}{D_L}C_L^*(1-k) \tag{5-18}$$

Constitutional supercooling is absent when the actual temperature gradient, G, follows:

$$\left[\frac{dT_L}{dx}\right]_{x=0} = m_L\left[\frac{dC_L}{dx}\right]_{x=0} \leq G \tag{5-19}$$

C_L = liquidus composition
x = distance from the interface
T_L = liquidus temperature
R = rate of solidification
D_L = diffusivity
C_L^* = liquidus composition, in equilibrium with solidus composition
k = partition coefficient
m_L = slope of the liquidus
G = temperature gradient

Combining these two equations, the condition of zero supercooling becomes

$$\frac{G}{R} \geq -\frac{m_L C_S^*(1-k)}{kD_L} \tag{5-20}$$

For stable planar-front solidification, the G/R ratio should be large. At a critical cooling rate there is no sufficient diffusion. This is the regime of absolute stability. It is possible in laser surface-melting to obtain planar interface even in deep eutectic systems.

The fine cellular or dendritic structures obtained in laser surface-melting is given by the product of GR. The size of the features describing such microstructure, such as cell spacing, λ is related to the cooling rate *GR*. The cell size is determined by the possible diffusion of the solute, which is governed by Fick's second law:

$$D_L \left[\frac{\partial^2 C_L}{\partial y^2} \right] = \frac{\partial C_L}{\partial t} \tag{5-21}$$

Also,

$$\frac{dC_L}{dt} = \frac{dC_L}{dT} \times \frac{dT}{dx} \times \frac{dx}{dt} = -\frac{GR}{m_L} \tag{5-22}$$

$$\Delta C_{L_{max}} = \frac{GR\lambda^2}{2\,m_L D_L} \tag{5-23}$$

In laser surface-melting, high cooling rates ($\sim 10^6$ °C/sec) are obtained. It can be seen from this relation that very fine structure can be obtained by laser melting.

Another system popular for laser surface melting is the Al-Si system. Al-Si–based alloys are amenable to refinement to great extent by a phenomenon called microstructure modification. Controlling the eutectic silicon morphology by modification has been used extensively industrially since about the 1970s to improve the mechanical properties of the castings. Modification, obtained by addition of Na or Sr or quench modification, changes the morphology of silicon from a platelike or lamellar structure to a fine fibrous structure, whereas modification by addition of Sb only refines the silicon platelets. Improved mechanical properties normally accompany the refined eutectic structure. However, often it is not practical to carry out microstructure modification of the entire component via chemical means or by rapid solidification. However, laser melting can be used for modifying the surface properties of Al-Si cast components. Since the mode of heat transfer is predominantly conduction through the aluminum matrix, a rapidly traversing laser can melt the surface layer. As the laser moves away, the molten metal solidifies on its own accord. Using sophisticated high-resolution, high-speed infrared thermography techniques, the cooling rate during solidification was estimated to be 10^5 K/s. The microstructure thus evolved is presented in Figure 5-22.

High-power lasers that are currently used as an efficient noncontact-type machining tool for various manufacturing applications such as cutting, drilling, welding, and so on have also been explored as a dressing tool (Babu and Radhakrishnan, 1995; Zhang and Shin, 2002). The basic principle of this technology is to vaporize metal chips and melt the top layer, thereby exposing newer cutting edges. Surface processing of ceramics by laser irradiation offers potential advantages such as precise control over high input of thermal energy at spatial and depth levels, rapid processing speed and unique modification of microstructure due to rapid heating, remelting, solidification, and cooling. Focused laser radiation produces enormous power densities in a very small region of the wheel surface and thus can cause a localized modification either of the exposed grain or of the bonding constituents. Because the focused spot size can be made much smaller compared to nominal size of the abrasive grain, it can pierce through abrasive grain and produce multiple cutting edges on the same grain. Depending on the laser processing parameters used, melting (followed by resolidification) and/or vaporization resulted in modification of surface topography (morphology and composition). Refinement of grain size, densification of surface layer, and evolution of multifaceted grains with cutting edges and vertices are the main morphological modifications taking place on the surface of the wheel due to interaction

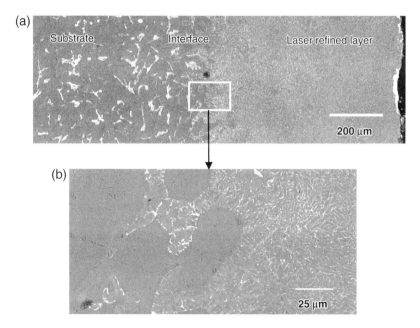

FIGURE 5-22 *Microstructure in cross-section of laser surface-melted A319: (a) showing both the refined layer and the substrate, and the high magnification view (b) of the interface between the substrate and laser-melted region.*

FIGURE 5-23 *Microstructure of the grinding wheel surface (a) undressed and (b) after being laser-dressed (750 W), showing refined equiaxed grains.*

with laser energy, as described in Khangar, Kenik, and Dahotre (2005). These morphological modifications, as seen in Figure 5-23, are expected to improve cutting efficiency.

It is intriguing to see well-defined facets in the laser-processed samples. This phenomenon is akin to nonmetallic crystals found in nature. The x-ray diffraction experiments indicated formation of crystallographic texture on surface of dressed wheel (Khangar, Kenik, and Dahotre,

2005). In this work, pole figure analysis was used to investigate the planar textures formed by laser-dressing. Using different laser processing parameters (beam focus, power, and transverse speed), the extent of interaction on the surface of the wheel can be controlled for various surface topographies. The efficiency of the grinding process for the laser-dressed wheel in comparison with undressed grinding wheel was evaluated using high-speed grinding apparatus. The surface finish on the ground workpiece is quantified using surface roughness parameters. The refinement of microstructure resulting from laser processing depends on the temperature gradient and the solidification rate. The size of the microstructural features depends on the cooling rate. During laser processing, the interaction region between the laser beam and the sample surface is very small and the cooling rate is very high, thus resulting in refinement of the microstructure on the surface. During the laser-dressing process, rapid heating and cooling induce microcracks in the resolidified layer as a result of the associated high thermal stresses. This is further enhanced by the steep temperature gradient across the sample surface due to poor thermal conductivity (0.276 W/m °C) of the Al_2O_3 ceramic. When these cracks are formed on the bond, the particles are loosened and subsequently removed because there is insufficient volume for load surrounding the grain, forming the material removal mechanism in laser dressing. Laser processing resulted in densification of the surface layer. Similar observations were made in earlier studies where refractory materials were treated with lasers. Although laser-dressing reduced the porosity for increased densification within the surface layer, this feature is desired in the grinding wheel. With an optimum amount of porosity, unsupported particles on the surface of the wheel break away once worn out during grinding, leaving the remaining particles with sharp edges. These morphological changes can help maintain a high grinding efficiency and low grinding-wheel material loss at all the times. Although a wide distribution of size of the particles (5–100 μm) was seen, the shape of the particles was more regular and symmetric, with well-defined vertices and edges on each particle. The wide distribution of size helps achieve a surface with higher packing fraction. Moreover, it can help provide a smoother work-surface finish during grinding operation. The morphological features evolved during laser-dressing are expected to influence the surface roughness of the dressed wheel, which in turn would affect the grinding performance (Khangar, Kenik, and Dahotre, 2005).

Laser Surface Alloying

Laser surface alloying (LSA) is similar to surface melting with the laser except that a desired alloying element(s) is extraneously added to the melt pool to alter the surface chemical composition. The change in composition of the surface layer brings about properties required for the application. Laser alloying is a material processing method that uses the high power density available from focused laser sources to melt metal coatings and a portion of the underlying substrate. There are several methods of incorporating the extraneous alloying element(s). In a process known as co-deposition, alloying elements can be directly injected in the form of powder, wire, or gas into the laser interaction region. Electroplating, diffusion coating, thermal spraying, vacuum evaporation, sputtering, or ion implantation can be done prior to laser treatment. The adherent preplaced coating thus will melt and mix with the substrate. Usually, coatings produced in this manner are relatively thin. For thicker coating, a precursor in the form of foil, pastes, or powder slurries are applied on the substrate prior to laser melting. Because the melting occurs rapidly and only at the surface, the bulk of the material remains cool, thus serving as an infinite heat sink. Large temperature gradients exist across the boundary between the melted surface region and the underlying solid substrate, which results in rapid self-quenching and resolidification. The physical process is schematically presented in Figure 5-24. A wide range

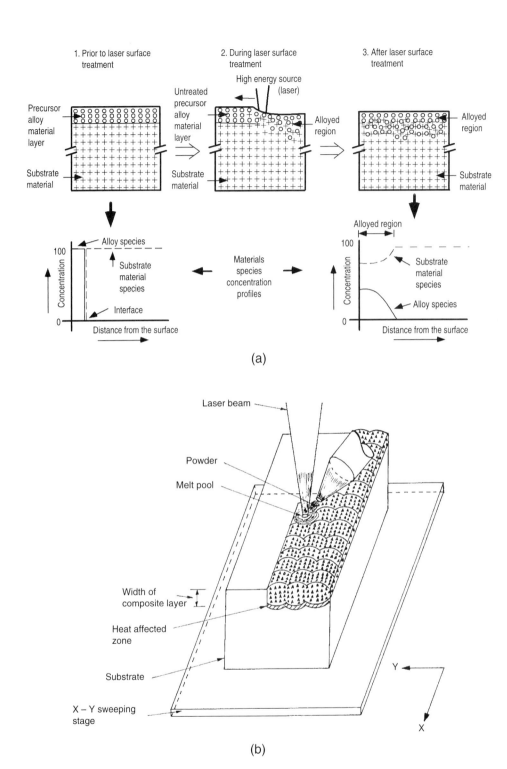

FIGURE 5-24 *Schematic of physical and chemical changes in laser surface-alloying and composite forming (a) via precursor deposition and (b) via powder feedings.*

of chemistry and microstructure can be developed because of far-from-equilibrium conditions of laser processing. The composition at surface can be maintained at high concentration of the alloying element(s) decreasing over shallow depths or almost uniform across the coating depth as desired. The microstructure observed included extended solid solution, metastable phases, and metallic glasses (Walker, Folkes, Steen, and West, 1985; Sood, 1982; Draper and Poate, 1985; Pangborn and Beaman, 1981; Dahotre and Mukherjee, 1990).

The most significant characteristic of LSA is that most materials can be alloyed with different substrates. The high quench rates involved allow little or no segregation of the equilibrium phases. A wide range of solid solution, microstructure, and coating thickness can be obtained by controlling the laser processing parameters. However, most volatile constituents are lost during laser surface alloying. These features of LSA can be exploited to advantage. For example, by alloying a plain carbon steel with significant amount of Cr, one can achieve a surface that is "stainless." The Fe-Cr-C alloy system has been widely studied for LSA (Dahotre and Mukherjee, 1990). Cr powder is used for LSA 1018 steel. Either a predeposited or co-deposited powder can be used. It is preferred to use multisize powder for better packing factor and densification. During laser irradiation, powder particles infiltrate into the substrate. Since Cr is a reactive element, inert gas (argon) shrouding during laser processing is generally applied. The beam can be focused at or slightly below the surface. Usually several parallel, overlapping laser tracks in one direction are made to cover the entire area. Sometimes perpendicular or even crisscross tracks are used. However, care must be taken, as most defects such as porosity and cracks are formed in the overlapping area. A cross-sectional view of a laser-alloyed specimen provides information about true microstructure evolution and chemistry alteration. The fused volume usually alloys Cr. The fusion regions usually consist of the extremely fine microstructure. However, the volume just beneath the fused layer always is martensite. The thickness of this layer is an indication of the heat penetration, as well as peak temperature developed at different depth similar to the discussion in the laser surface-hardening section. The primary phases present in such a material are lath martensite, block precipitate of metal carbides, and the matrix. Addition of Cr lowers the martensite start temperature, promoting formation of lath martensite. The carbide precipitates are usually $M_{23}C_6$ type with FCC crystal structure where other metals such as Fe replace few Cr sites. Possible phase transformation sequence, as proposed in Dahotre and Mukherjee (1990), is

$$L \rightarrow L + \alpha \rightarrow \left(\alpha + \alpha' \right)_I + L \rightarrow \left(\alpha + \alpha' \right)_I + (\alpha + M_{23}C_6)_{II} \qquad (5\text{-}24)$$

where L is the liquid. The first transformation products (I) are ferrite (α) and martensite (α'), followed by the formation of carbide along with ferrite. Such unconventional transformations are often found in rapid solidification processes. The high supercooling below the liquidus temperature results in solidification into austenite, which further transforms into martensite. The increase in solid solubility and high cooling rate first produced martensite along with high-temperature ferrite followed by the formation of $M_{23}C_7$ carbide precipitates that were uniformly distributed in the matrix, which had a high content of Cr. There is substantial structural heterogeneity in both the fused layer and the hardened layer below it. This translates to significant variations in hardness. In addition to this, the Fe-Cr-C system has been used for hard facing of materials such as Cr, which has a high tendency to form hard carbides, suitable for applications where high wear resistance is required (Svensson, Gretoft, Ulander, and Bhadeshia, 1985).

Mathematical models indicate rigorous mixing and stirring in the laser melt pool. Melt pool rotates up to five times from melting to solidification in laser surface melting. This ensures that the external material is taken into the melt pool, and the alloyed region is nearly homogeneous throughout the melt region. Because high power and rapid cooling rates are associated with laser surface-alloying, most materials can be alloyed into most substrates. In fact, many nonequilibrium phases are reported in laser surface-alloying only. The type, wavelength, power and speed of laser, and material properties can be exploited to get a wide range of thickness of the alloyed region.

The external material can be introduced via widely diverse techniques. A precursor can be deposited onto the substrate by electroplating, vacuum evaporation, ion implantation, preplaced powder coating, or thin foil. In addition, material can be introduced by simultaneous powder feeding, wire feeding, or from the shroud gas. Most common substrate-alloying elements are carbon or nitrogen on Ti alloys; Cr, Si, or C in cast iron; Cr, Mo, Ni, etc., on steel; C on stainless steel; Si, C, N, and Ni on Al; Cr on superalloys, and so on. Laser surface alloying can be used to control the surface composition while the substrate remains essentially unaffected. There is little distortion or change in substrate microstructure involved. For example, by alloying Cr on cheaper steel, one can obtain a "stainless steel" layer on the surface, whereas the substrate is ordinary steel. Laser parameters can be optimized so that there is little after-processing involved. In laser surface-alloying, the alloying elements mix with the substrate and get diluted with substrate composition, which needs to be considered. For example, to achieve a stainless steel layer, composition of the powder should be maintained such that the final layer should contain >12.5% Cr despite this dilution.

As the laser pulse hits the substrate surface to be alloyed, a significant portion of the laser beam is scattered. The remaining energy is absorbed and instantaneously raises the temperature of the near-surface region above the melting point. The liquid–solid interface then propagates inward. Simultaneously the alloying elements are transported by convective flow and diffusion irrespective of their sources (i.e., preplacing, feeding, or gaseous environment). As the liquid–solid interface moves inside, the laser beam has moved away from it. Hence the melt pool loses heat and resolidifies outward (conduction mode of heat transfer is inward). Meanwhile, the laser beam has advanced to the adjoining region, and the same phenomenon repeats. The solidification rate is so rapid that rarely diffusion-controlled phase transformation is observed. Also, extended solid-solubility and novel nonequilibrium phases are often observed in LSA. Extreme cooling rate, as high as 10^{11} °C/s, has been reported in LSA. Since LSA is performed at rapid speed, high power density, and rapid heat transfer, significant variations in terms of composition and microstructure of the alloyed layer are observed. As discussed in laser melting section, the solidification front depends on the G/R ratio. Alloying and localized variation in composition leads to amplified variations in constitutional supercooling, thus leading to significant variation in microstructure (planar, dendritic, or cellular structure). Overlapping of laser track is also an important factor to be considered.

Very high interfacial solidification velocities can be achieved by sample treating with laser pulses on the short wavelength (λ) scale and time scale (t-scale). Solidification can be so rapid that atomic restructuring is not possible, and an amorphous layer results. For laser surface amorphization and glazing, quenching rates in the range 10^8 to 10^{11} K/sec are required. For example, amorphization can be achieved with UV pulses on t-scale of 2.5 ns and on <001> and <111> Si crystals at 0.20 J/cm^2. Such amorphous layers have shown a very high yield strength and hardness while maintaining a high degree of ductility.

Surface-alloying of AISI 1018 steel with carbon and chromium produces stable carbides, such as M_7C_3 and M_3C, in austenitic, pearlitic, or martensitic matrices. The uniform dispersion of these carbides results in an improved resistance to abrasive and adhesive wear. The reactivity of titanium makes its alloys suitable for laser alloying by gas reactions. TiC, TiN, and carbonitrides form in a Ti matrix throughout the melted region. Because of the extremely high quench rate inherently associated with laser processing, the solubility limit in the material system is often extended beyond the equilibrium solubility limit. In view of this, several successful attempts were made in both ferrous and nonferrous material systems to alloy the surfaces with various elements that usually are not chemically compatible with the substrate material. Laser-alloying has been primarily applied to improve corrosion resistance, such as in plain carbon steel by alloying with chromium and/or nickel. Alloying of ferritic stainless steel with Mo, as well as Mo and B, has been found to yield improved abrasive wear resistance.

Laser-Induced Ceramic Coating

The importance of surface properties has increased tremendously in recent times, with considerable stress on improving such properties by forming a composite coating on the surface of a material rather than forming a relatively expensive bulk composite. Ceramic materials are widely used for their superb performance in applications where other classes of material may fail. Possible applications are in the areas of high temperature, corrosion, and severe wear. However, toughness is an issue of concern for these materials. The load-bearing performance of metals is superior to that of ceramics. Thus it is desirable to combine both superior properties. The presence of dissimilar materials poses challenges in terms of differences in the thermal expansion coefficient and stress concentration, which may negatively affect the mechanical performance of the component. Metal-matrix composites have been known to have high hardness and exceptional wear resistance (Bourithis, Milonas, and Papadimitriou, 2003). Carbide particles have often been incorporated in an Fe matrix to form a composite because of carbide's inherent high hardness even at high temperatures. By careful selection of the type of carbides, it is possible to design a composite for severe abrasive conditions. TiC with exceptional hardness (Knoop hardness 2800 HK) and thermal stability (decomposition temperature 3065 °C) (Bhushan and Gupta, 1991) has been frequently used as reinforcement in Fe-based composites. (Wang, Zhang, Li, and Zeng, 1999) For the particles to be successfully incorporated, the melt should sufficiently wet the carbide particles. Ramqvist (1965) has shown that TiC particles are easily wetted by iron at high temperatures. Moreover, the wetting improves with temperature because of chemical interactions. It has also been shown that metals with an unfilled d band wet carbides more easily than metals with a filled d band (Ramqvist, 1965). Iron has an unfilled $3d$ orbital and hence wets TiC easily. In recent years, emphasis has been given to the in situ formation of TiC in an iron-based matrix for several important reasons. The process is more economical and has an intrinsic advantage in that the surface of the reinforcements is cleaner and hence the bond between the reinforcing particles and the matrix tends to be stronger.

Simultaneously synthesizing a combination of carbide coatings (Ti and Cr based) in various amounts has been attempted in laser processing. Studies aimed at employing the Fe-Ti-Cr-C and Fe-Cr-W-C quaternary system to engineer the surface of plain carbon steel for in situ formation of both Ti- and Cr-based carbides using is reported in Nagarathnam and Komvopoulous (1995) and Singh and Dahotre (2004). Formation of various types and combinations of carbides on a plain carbon steel (AISI 1010) was achieved with laser coating. TiC was prominently present

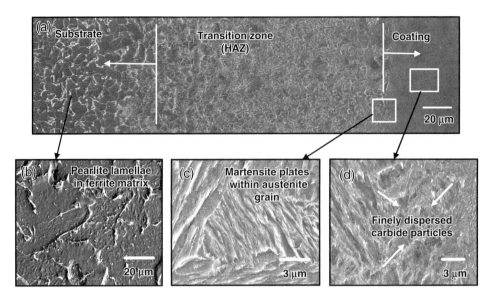

FIGURE 5-25 *Scanning electron micrograph of quaternary (Fe-Cr-Ti-C) coating system on AISI 1010 steel at 1500 W. (a) Overview of the cross-section and high-magnification images of (b) substrate, (c) transition zone, and (d) coating region.*

in the surface layer of all the samples, and chromium carbides were found to be absent in the samples processed at 1300 W and 1900 W. Figure 5-25 indicates a scanning electron micrograph of quaternary (Fe-Cr-Ti-C) coating system on 1010 steel at 1500 W (Singh and Dahotre, 2004). This variation in phase evolution had strong influence on mechanical properties. The hardness and wear properties of the samples without chromium carbides were inferior to the samples with both TiC and chromium carbides. A substitution solid solution of chromium in iron (Fe-Cr) was found in all the samples, and the relative estimated amount of the phase was higher for the samples without chromium carbides. Metastable phase (characteristic of nonequilibrium synthesis) such as martensite was present in the coating and transition zone. Thus the formation of carbide composite surface layer provided a five-fold improvement in the hardness (maximum hardness in coating 914 HK) of the surface (180 HK) for the as-received steel.

Laser-Cladding

Laser-cladding differs from laser surface-alloying in that cladding is essentially an overlay of one metal on another with a strong interfacial bond. The cladding metal is not diluted with the substrate composition. In fact, any dilution is considered contamination, contributing to degradation of mechanical and chemical properties. For a very thick cladding layer (>250 microns), the power input is high, leading to substantial melting of the substrate and contamination. Cladding can be done either by powder preplacing or by feeding via inert gas propulsion into laser melt pool as shown in Figure 5-26.

The recent advances of high-power lasers have drawn increasing interest in materials processing and manufacturing. A high-energy laser beam may be used to produce a molten pool on the surface of a metal, and a stream of alloying powder simultaneously may be injected into the molten pool to form a rapidly solidified cladding layer with the required properties, such as

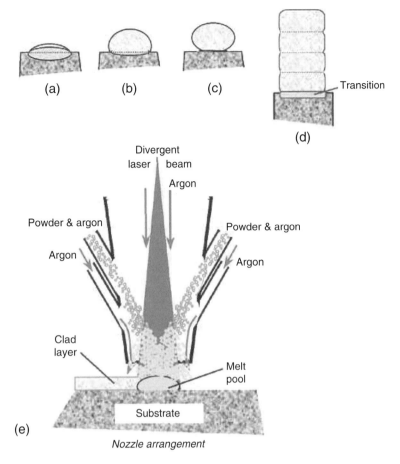

FIGURE 5-26 *Schematic of (a) laser surface-alloyed region, (b) laser-cladded region, (c) region of poor bonding in laser surface-alloying/cladding, (d) multilayer laser-cladded region, (e) laser surface-alloying/cladding process.*

enhanced hardness and wear resistance, without affecting the bulk properties such as weight-to strength-ratio of the underlying metal or alloy. Because of rapid solidification of the clad and limited intermixing between the clad and substrate, laser cladding can be used to produce novel composites. Furthermore, the process can be easily automated to produce selected-area composite cladding because of the inertia-less energy source, which can be precisely controlled by computers.

Cladding with preplaced powder requires that the powder stick to the substrate until melted, despite the shroud gas flow. A defocused laser beam rasters over the powder bed, which is consequently melted and welded to the substrate. The laser beam causes the powder to melt rapidly. Once the liquid front reaches the interface with substrate, because of high thermal conductivity it resolidifies. In contrast to the cladding technique with preplaced powder, the powder-feeding technique has several advantages. It is associated with the formation of a fusion bond between the cladding layer and the substrate. There is low dilution from the substrate.

A low-dilution region results from the solidification front rising rapidly with the growth of the clad layer. Blown-powder cladding takes place over a small melt pool, which travels over the surface. There is a smaller heat-affected zone. The clad layer can distort, depending on the thickness of the layer.

The composition and properties of the surface can also be modified by addition of external material. The external material can be added either by introducing it into the laser–substrate interaction zone (powder injection, wire feed, etc.) or by preplacing it on the substrate surface (powder deposition, electroplating, vapor deposition, thermal spraying, etc.) followed by laser scanning. The nature of incorporated material in the modified surface varies depending on laser-processing parameters such as energy density, traverse speed, and the type of material. A combination of processing and material parameters can result in formation of alloy, clad, or composite surface layer. The schematics of physical and chemical changes during all these processes are illustrated in Figure 5-26. Formation of homogeneous chemical composition and phase within the modified surface region is defined as alloying. In contrast, mechanical entrapment of the external material with minimal or no reaction with the surface matrix and its uniform distribution within the surface matrix describes the process of composite formation. In both alloying and composite formation via control of process parameters, the composition of modified layer can be graded from the surface to the interface between the melted surface layer and the substrate. Finally, cladding involves a fusion of a necessarily chemically different material to the substrate with minimum dilution. The clad layer has a microstructure and properties that are entirely different from those of the substrate, whereas the composite layer has composition and properties in weighted ratio corresponding to the proportion in which components are present. Alloying provides desired microstructure and chemistry. A cross-section view of a single laser track of Cr-Ni cladding on steel is presented in Figure 5-27.

Materials that have been successfully laser-clad include metals, intermetallics, and ceramics. Unlike laser-alloying, a negligible dilution of the clad material from the substrate makes laser-cladding better suited for enhanced surface protection, particularly for tribological applications. Laser-cladding of Cu-base alloy on valve seats has been developed at Toyota R&D Laboratories that provide considerable abrasive and adhesive wear resistance over a wide temperature range. Laser-cladding technology has been applied in the automobile industry to produce erosion-resistant cylinder heads in lightweight aluminum alloy engine by depositing solid solution alloys. This process is completely different from the conventional method, in which the valve seats are made of sintered alloy inserts, press-fit into the cylinder head. The manufacturing cost of laser-cladding valve seats is about 30% cheaper than that for the conventional technique. Further, the engine durability is increased about three times compared with the conventional type engine

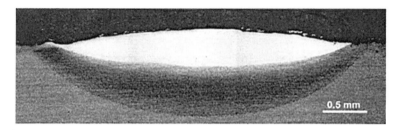

FIGURE 5-27 *Cross-section of Cr-Ni laser-clad track on steel.*

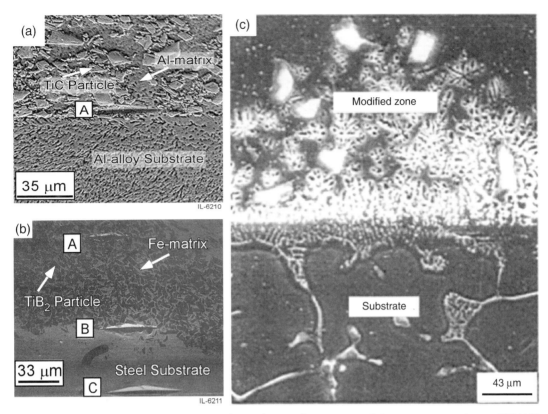

FIGURE 5-28 *Composite coatings produced using laser technique: (a) TiC-Fe composite coating on AISI1010 steel; (b) TiB₂-Al composite coating on 6061 Al-alloy; and (c) Mg-Al₂O₃ composite coating on MEZ Mg alloy.*

because of the improved wear resistance and heat resistance that can be achieved by laser-cladding. Increasingly, aluminum alloys are being used in transportation industry, especially as cast components for automotive or aircraft applications, to produce lighter parts. Deposition of Ni-Cr-B-Si and Co-Cr-W on internal exhaust valves has also yielded encouraging results. McrAlY-type cladding on Ni-base superalloys enhances their oxidation resistance. Cladding of Ni-Al bronze and intermetallic compound Al_3Ti on Al alloy has been achieved for enhanced wear properties. The wear resistance of aluminum is increased by alloying with silicon.

Different ceramic-phase particles such as carbides, borides, and oxides have been directly introduced into the melted surface or clad along with diverse matrix alloys on the metallic substrate for formation of composite layers. As indicated in Figure 5-28 both ferrous and nonferrous substrate alloys can be coated with ceramic-metal matrix composite layers. The distribution and content of the ceramic phase in the composite layer can be controlled to achieve various chemical (corrosion- and oxidation-resistance) and physical (wear-resistance) properties. This technique has been demonstrated for various potential applications such as those illustrated in Figure 5-29.

Discrete regions or entire surfaces can be treated in a single step to produce near-net-shape components. Cheaper base materials can be used and costly alloying additions reduced. Worn

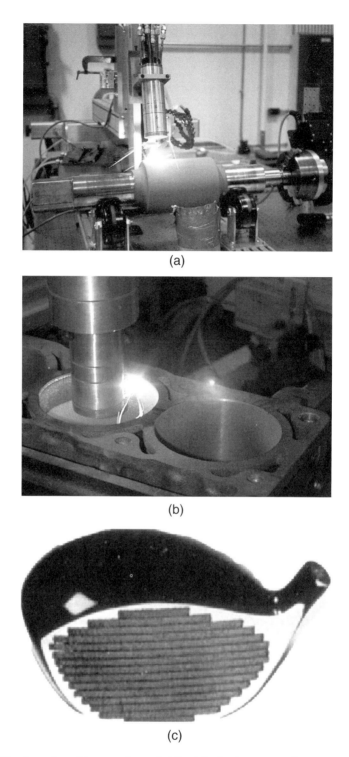

FIGURE 5-29 *Examples of laser surface cladding/alloying/composite forming. (a) Chromium cladding on steel mill roll, (b) alloying on the auto engine bore surface, and (c) carbide-steel composite coating on a golf club head.*

components, such as dies and molds, that would be very expensive to replace can also be built up using the technique.

Laser-Induced Combustion Synthesis

There are also several difficulties in the application of high-power lasers on aluminum alloys. First, the high melting point of ceramics, close to the boiling temperature of aluminum, requires a high energy density of the laser beam to melt ceramic coating. It will generate a strong convective flow in the aluminum melt pool. Eventually, it may lead to substantial surface roughness and may even destroy the ceramic coating. Second, the large difference of thermal expansion coefficient between aluminum ($\sim$22 μm/m K) and ceramics (mostly below 8 μm/m K) will introduce high stresses at the metal–ceramics interface during rapid cooling, which may cause the interface bonding to fail. Finally, the wetting of aluminum on most ceramics is usually poor, and the rapid melting and solidification of laser process require a good wetting between liquid metal and solid ceramic within a very short period. Laser coatings on aluminum alloys with metals such as nickel, cobalt, and manganese were not successful, as the solubility of these metals in aluminum is rather low. In the case of such metal coatings, a typical microstructure developed, with separated aluminum-rich and metal-rich parts.

To overcome these difficulties, a novel process for Al-based systems has been developed. The basic idea is to promote a chemical reaction at the interface between the metallic substrate and the ceramic particles. In this reaction coating, mixing powders of reactive ceramic (e.g., Fe_3O_4) with aluminum are used to laser coat aluminum alloy. Iron oxide was considered the appropriate choice of coating material because, like cast iron, it is iron-based material, and unlike other ceramic materials (carbides, borides, and some oxides) it does not provide an extremely hard surface, which can be detrimental to the mating surface. The laser can be used to ignite a self-propagated high-temperature synthesis (SHS) reaction. A reaction between Al and iron oxides in the surface layer induced by a laser can propagate into the material (in the direction of heat transfer), thus providing a thick coating thickness. Moreover, a higher temperature resulting from such exothermic reaction will further increase temperature gradient and cooling rate. A microstructure thus produced would be much more refined compared to just laser melting. Further reinforcement by combustion reaction product is also expected to improve mechanical properties.

However, if such reactions are "quenched in" by rapid solidification, it would result in a composite material with a reaction-induced, strongly bonded interface between the ceramic and the metal matrix. Furthermore, during synthesis of such composite (ceramic-metal matrix) coating using laser surface engineering (LSE), the laser interaction is confined to a near-surface small volume compared to the substrate bulk. This provides a high cooling rate due to the high temperature gradient and conduction mode of heat transfer through aluminum (substrate), effectively "quenching in" the reaction, thereby producing some novel, nonequilibrium phases and microstructure. A coating or surface layer produced either solely by laser-induced rapid solidification or by nonequilibrium combustion synthesis is expected to enhance wear resistance of aluminum alloys and many other metallic systems such as magnesium.

Laser Surface-Vaporization

Generation of metallic, semiconducting, and superconducting thin films by laser vapor deposition is attracting increasing interest. These deposition techniques involve either physical transformation or chemical reaction assisted by a laser beam. Accordingly, the following different variants of the laser vapor-deposition processes exist: laser physical vapor deposition (LPVD) and laser chemical vapor deposition (LCVD). LCVD is further categorized as laser-enhanced chemical vapor deposition (LECVD) and laser-activated chemical vapor deposition (LACVD). LECVD generally refers to a deposition method involving the laser-induced pyrolytic reaction wherein the substrate or gas is locally heated. The term LACVD is used to describe techniques based on the photolytically activated reactions wherein the gas or the substrate photolytically absorb the photon energy to break molecular bonds and form highly reactive free radicals. In the recent past, these techniques, have synthesized diamond, diamond-like, superconducting, and various combinations of ceramic and metallic coatings in various layers (functionally gradient materials). Usually these techniques are suited for micron- and submicron-thick layers, as the process speed is very slow and can provide only a small area of coverage.

Laser Surface-Texturing

In texturing, a focused pulsed laser beam produces surface depressions surrounded by a smooth rim of solidified melt without any physical contact. Most of the material in contact with the laser beam is vaporized and ejected by the vapor pressure to create a dimple or crater. Repeated pulses on spatial and temporal scales create surface texture. Simultaneous use of laser-alloying and laser-texturing is demonstrated for creating textured surfaces of the inner wall of centrifuge-cast permanent die mold (several inches diameter and multiple feet long), used for casting cast iron sewage pipes (Figure 5-30). The alloyed and textured mold surface provides enhanced traction and high-temperature oxidation resistance. The technique has also been employed to texture a texturing roll in a temper mill. Use of a Q-switched Nd:YAG laser with 100-ns pulse duration, 3-kW peak power, and 20-kHz pulse repetition rate was demonstrated for landing-zone texturing of a computer hard disk. The surface roughening via texturing can be performed using an excimer laser, because the absorptivity of the metal surfaces at the wavelengths of an excimer laser is significantly higher than that of CO_2 or Nd:YAG lasers.

The application of laser technology for texturing work roll surfaces was developed and commercialized in the late 1970s–1980s in the Center for Research in Metallurgy in Belgium. The CO_2 laser is modulated by a chopper and then focused onto a surface of the rotating roll. The focusing/modulating unit is translated along the roll axis to texture the roll surface or moves longitudinally relative to the rotating roll; thus the optical path length is kept constant. Besides the formability improvement of the sheet surface after rolling with textured rolls, laser-texturing achieves sheet surface topography, which provides a high-quality surface appearance after painting. Increasing the surface roughness could enhance adhesion or laser absorption for subsequent laser-processing of highly reflective metals. Electronic modulation possible in Nd:YAG and other solid lasers has provided better flexibility in surface texturing. The surface roughening can be performed using excimer lasers because the absorptivity of the metal surface at the wavelength of the excimer laser is significantly better than other lasers.

The interaction of a focused laser beam with a material surface is accompanied by melting, melt motion, evaporation, sublimation, and solidification, resulting in a change of the surface topography. Different shapes of the interaction zone can be produced by variation of the interaction parameters, which determine the physical nature of the process. The result of interaction can be classified depending on what physical processes dominate in the laser–material interaction. Pulsed laser material processing can be generally classified as drilling, spot-welding, and surface melting. These types of material processing are not separate, independent phenomena. On the contrary, they are the result of the same physical phenomena and simply correspond to the cases when one or another set of physical processes dominates. Consequently, there are ranges of inter-action parameters, which corresponding to the transition between "pure" processing regimes can

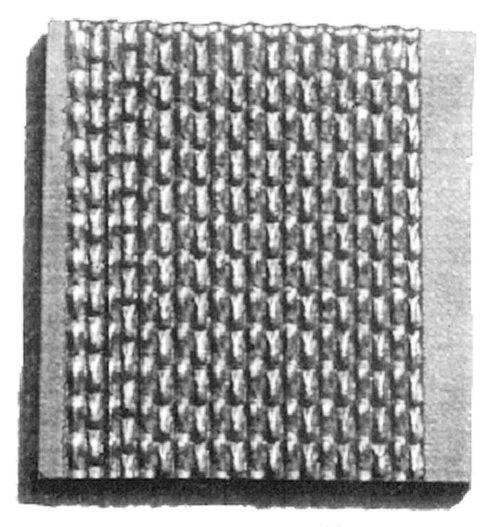

FIGURE 5-30 *Laser surface-alloyed and -textured AISI 4140 steel plate.*

be rather wide, and practitioners frequently perform processing operations within such transition ranges. Proper classification of the conditions when one or another set of processes dominates, resulting in welding, cutting, or drilling, and creation of a generalized model of laser material interaction is ongoing. Usually laser drilling is used to describe a process when material is removed from interaction zone. Material removal and consequent crater creation is determined by the competing processes of melt ejection and evaporation. Evaporation-dominated drilling requires either high absorbed beam intensity or short interaction time (which is less than plasma formation threshold value). Some materials do not produce a liquid phase during irradiation. In such a case, for all processing conditions, sublimation is responsible for material removal from the interaction zone. It was shown that if laser beam intensity exceeds the threshold, which for materials with short absorption length lies in the range of 10–100 MW/cm^2, the evaporation front becomes unstable. Then front perturbations with micron-spatial periodicity start growing. This can explain the generation of spikelike surface structure observed in excimer laser irradiation. Melt ejection from the interaction zone due to a gradient of evaporation recoil pressure is the main mechanism responsible for crater formation in metals when the absorbed beam intensity is moderate ($\sim$5–10 MW/cm^2) and the interaction time is sufficient for the melt to reach the edge of the interaction zone. When the amount of ejected or evaporated material is small, shallow depressions are formed on the sample surface, and this processing corresponds to the surface-texturing. If the velocity of the melt ejected from the interaction zone is low and is directed along the surface, then melt solidifies around the crater, forming a rim of spattered melt around the crater. Low melt-ejection velocity means that the dynamic pressure of the flowing melt is insufficient to overcome the pressure on the edge of the stream produced due to wetting and surface tension, and melt stays on the surface until it solidifies. The melt ejection velocity is directed along the sample surface if the crater depth is much smaller than its radius, as in the initial stage of deep drilling when spattering occurs. To achieve such a regime of interaction, the melt surface temperature distribution in the interaction zone must be smooth, which results in generation of a two-dimensional flow. In the presence of the small-scale spatial variations of the surface temperature, the recoil pressure also changes on the small scale. Then a large gradient of recoil pressure can be induced, producing a three-dimensional melt ejection. This will prevent the creation of a noticeable rim of melt around the crater.

If the melt spattered around the crater has sufficient heat content to melt the sample surface below, and the interaction time between ejected materials and substrate is long enough (small melt velocity), then good bonding of the rim to the surface can be achieved. However, even under conditions of high heat content and long interaction time, the bonding of the ejected melt to the sample surface will not occur, because of the presence of some contaminant film. For example, an oil film exists on the surface, then part of the ejected melt enthalpy is spent to evaporate the film, and the oil vapor layer occurs between the melt and substrate, preventing good contact between the melt and solid surface. In this case, ejected melt is either not attached to the surface or is attached poorly. Presence of surface oxidation can also result in poor adhesion of the ejected melt to the substrate, because the melting temperature of oxide films is usually high.

Laser Surface Cleaning

Surface cleaning can be achieved in two ways: laser-assisted particle removal (LAPR) and ablation. In LAPR, a laser beam irradiates the surface and energy is imparted to the particles, which become excited, thus causing vibration and elevation from the surface. This technique is greatly enhanced when used in conjunction with a steady stream of flowing gas or liquid (water). The removing of contaminants occurs without degrading the substrate material. In contrast, laser ablation is the removal of surface materials by thermal shock, melting, evaporation, or vaporization. Experiments have shown that lasers can effectively remove a wide variety of coatings and surfaces from various substrate materials, e.g., rust from steel, corrosion from copper, acrylic from steel, urethane from wood, acrylic paints from canvas, sulfate from stone, fungi from textiles, and salts from stained glass. Lasers, therefore, remain an effective tool in cleaning statues and paintings. Lasers have been shown to be suitable for cleaning copper traces on circuit boards, Al and Sn particles on magnetic head sliders, and alumina from silicon wafers. Paint removal by laser techniques has been demonstrated for cleaning aircraft and aircraft components, industrial components, and marine components. Laser surface-cleaning is effective in removing radioactive surfaces consisting of paint, concrete, and particulate deposits.

Laser Surface-Marking

The availability of high-power pulsed lasers with good beam quality and high repetition rates has made a widespread inroad into industrial sector for the application of marking via evaporation. The information to be marked can be put in the form of a mask on the surface, which in turn is scanned with a laser beam to evaporate the unmasked material, or, lately, computer-controlled scanners can provide a precise motion for the laser beam to evaporate the material for desired marking. Pulse lasers such as TEA CO_2, Nd:YAG, and excimer are suitable for this purpose. The technique has been well established for marking Si wafers for microelectronic applications and lately adopted for marking various metallic, ceramic, and polymeric components in auto, aerospace, and hospital industries. Scribing is a method of separating nonmetallic materials, such as ceramics, silicon, or glass. The technique uses laser power to define a line along which the part will subsequently fail mechanically. The laser cuts a groove by evaporation of the material, which extends only partly into the substrate. The substrate can subsequently be separated along this line by application of mechanical force. Alternatively, a pulsed laser may be used to perforate along the desired line. Scribing is extensively used in the microelectronic industry for the production of silicon wafers. Because Si is more strongly absorbed at 1.06 μm, an Nd:YAG laser is preferred in this application. However, scribing of alumina and other substrates can be efficiently carried out with the CWCO$_2$ laser.

Laser Surface-Shocking

A plasma blowoff is obtained when a high-power pulsed laser (in the intensity and pulse duration ranges 1–50 GW/cm^2 and 1–50 ns, respectively) is focused on the surface of a target. This blowoff generates a pulsed pressure because of the recoil momentum of the ablated material. Such pressure causes introduction of residual stress in the surface and subsurface region. By using a dielectric material that is transparent to the laser beam, such as water or glass, the expansion of this hot plasma can be delayed compared to the free expansion that occurs in direct ablation, and thus the magnitude of the pressure can be controlled. These effects influence roughness, hardness, and fatigue properties of material at the surface. The laser-induced plasma during expansion creates plastic deformation and global depression of the irradiated areas. Corresponding surface morphology changes, however, are usually very small. Similarly, in general, laser shock processing hardens metallic surfaces to a very limited extent, as shock durations are very short to initiate and propagate the defects. However, on very specific metals such as austenitic stainless steels 304 and 316, under specific shock conditions a large hardness increase can be achieved. The most useful potential of this technique lies in improving the fatigue and cracking resistance and to some extent the corrosion resistance of the surfaces. Therefore, the technique is being considered for surface pinning of various industrial components; however, no actual industrial applications for laser shock processing yet exist. This is mostly because there are no reliable and adequate laser sources that would deliver power densities in the 1- to 10-GW/cm^2 range on at least 1- to 2-mm spots and at a high frequency (a few Hz). The emergence of diode-pumped lasers may provide suitable output for these applications.

Miscellaneous Laser Processes

In addition to aforementioned techniques, with different combinations of the physical processes (heating, melting, vaporization), various innovative approaches have been demonstrated for lasers, encouraging rapid prototyping in surface-engineering. Most of these techniques have been proved in principle but have not been applied in the industrial world because of several factors, such as economics and efficiency of the process. These potential industrial techniques for rapid prototyping include laser-machining, laser-forming, and laser direct sintering/fabrication. Laser-machining has been proven feasible for machining metals, ceramics, and composites. In laser-machining, with either heating (plastically softening), melting, or evaporation with or without the use of mechanical force (by a tool), surface material can be removed for various shapes. Attempts have been made to machine Si_3N_4, SiC, SiC/Al, and so on. Laser-forming is based on using a focused or partially focused beam, which is tracked over the surface of the part. The temperature gradient between the laser interaction zone and the surrounding material, along with the gravitational forces, causes the material to distort. Typically, sheet metal components are most suitable for forming by this technique. A laser direct sintering/fabrication process is characterized by a supply head delivering a well-defined flow of metallic powder (or wire), which impinges on a small melt/semimelted volume formed by controlled laser heating. Melting and rapidly resolidifying the feed material into the desired shape can build up a full-strength part, layer by layer. This process encompasses a variety of possibilities for fabrication of low-volume series or prototypes.

Residual Stress State of Laser-Treated Surface

In general self-quenching, the surfaces of the alloys studied were primarily tensile. It is generally recognized that the mechanical properties of steel parts can be improved by inducing both hardness and compressive stress state. The stress state affects the corrosion, wear, and fatigue life. It is therefore desirable to have a proper stress state for optimum surface-related properties. A uniform heat-affected zone is formed when the surface is locally melted. In laser-processed samples, medium carbon-steel fatigue life improves significantly. By controlling the residual stress state during laser surface treatment, it is possible to further improve mechanical properties of materials.

Techniques used for these coatings introduce some intrinsic residual stresses due to differences in mechanical and thermal properties of the different phases or layers of materials (Lu and Jobart-Retraint, 1998). The presence of residual stresses can prove either detrimental or beneficial to performance. When a composite is rapidly cooled from its processing temperature, residual stresses develop because of the differences in the thermal expansion behavior and the elastic and plastic properties of the matrix and the reinforcement. Residual stresses have significant influence on the mechanical and physical properties of the coatings, particularly electrical resistivity, optical reflectivity, fatigue, and corrosion. It is generally recognized that compressive stresses in coatings are more favorable than tensile stresses because they increase resistance to fatigue failure and stress corrosion cracking. However, extremely high compressive stresses may cause either coating separation from the base metal or intracoating spall fracture. Conversely, if a tensile stress or strain exceeds the elastic limit of the coating, then generally cracking or yielding will occur in the coating perpendicular to the direction of stress (Sue, 1998). Thus knowledge of the residual stress is necessary to assess its impact on the lifetime and function of the component and to enable the control of the stress by modification of the manufacturing process (Kesler, Matejicek, Sampath, Suresh, Gnaeupel-Herold, Brand, and Prask, 1995).

Residual stress can be measured in a number of ways: diffraction, acoustic wave detection, piezospectroscopy, deflection, electrical resistance or capacitance method, and hole drilling (Niku-Lari, Lu, and Flavenot, 1985). However, in the case of composite coatings, the x-ray diffraction method is most appropriate because it is nondestructive, surface localized, and phase distinctive (Matejicek, Sampath, and Dusky, 1998).

The effect of processing parameters has been investigated in order to provide a better correlation between the residual stresses and the laser input parameters (Kadolkar, 2002; Kadolkar

(a) 2024 Al at 150 cm/min

(d) 6061 Al at 100 cm/min

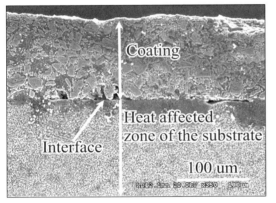

(b) 2024 Al at 175 cm/min

(e) 6061 Al at 125 cm/min

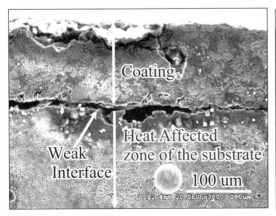

(c) 2024 Al at 200 cm/min

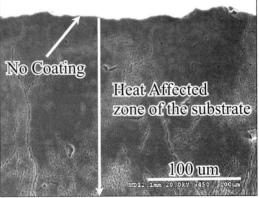

(f) 6061 Al at 150 cm/min

FIGURE 5-31 *(a–f) SEM micrographs showing the cross-section of the coatings processed at various speeds.* (Kadolakar, P. Residual stress and cohesive strength of TiC composite coating on aluminum alloys during laser surface engineering. Master's thesis, The University of Tennessee, Knoxville, Tennessee, 2002, p. 77).

and Dahotre, 2003). In this study, Al alloy 2024 and 6061 substrates were coated with ceramic-metal matrix (TiC-Al) composites using laser traverse speeds of 150, 175, and 200 cm/min and 100, 125, and 150 cm/min, respectively. A reduction in the magnitude of the macroresidual stresses with increasing traverse speed was observed in the coatings for both 2024 and 6061 Al substrates. In contrast, the microresidual stresses were less affected by the changing processing conditions.

The mechanical performance of laser surface-engineered ceramic composite (TiC-Al) coating on 2024 and 6061 Al alloy substrates has been evaluated using a four-point bend test. The performance of the coating is expressed in terms of the cohesive strength of the coating. Load displacement measurements carried out during the bend test help determine the load corresponding to crack initiation. The load required to initiate fracture in the coating provides a measure of the strength of the coating. A simplistic approach based on elementary beam theory and strength of material in conjunction with rule of mixture was adopted to calculate the cohesive strength of the composite coating. This approach is also further extended for attempts to evaluate apparent residual stress within the coating. Since process parameters exert a significant influence in controlling the end properties of the component, the effect of laser power and laser scan speed on the cohesive strength has also been investigated. It is observed that coatings with lower scan speeds have much higher cohesive strengths, and they also seem to have good metallurgical bond with the substrate, thus showing better mechanical behavior than the other high scan speeds used in this present study. The apparent residual stress in the coating appears to transform from compressive to tensile in nature, with increasing laser scan speed eventually contributing to delamination of the coating. Figure 5-31 illustrates the mechanical behavior of the TiC-Al composite coatings (Kadolkar and Dahotre, 2003).

Promise of Lasers in Surface-Engineering

Surface-engineering is the design and modification of surfaces that can provide cost-effective performance enhancement. The versatility of lasers as precision manufacturing tools is important for this purpose. The recent developments are providing more efficient, reliable, and cheaper lasers for cost-effective manufacturing. Even though at present, most of the laser surface-engineering processes appear expensive compared to other surface-engineering processes, they remain the most suitable processes because of their high precision and high speed. The development of diode-pumped lasers will initiate a leap in widespread usage of lasers in manufacturing, and soon lasers will become a usual part of the manufacturing landscape.

References

Babu, N. R., and V. Radhakrishnan. Influence of dressing feed on the performance of laser dressed Al_2O_3 wheel in wet grinding. *International Journal of Machine Tools & Manufacture*, 35(5), 1995, 661–671.

Bhushan, B., and B. K. Gupta. "Metals and Ceramics," In *Handbook of Tribology: Materials, Coatings, & Surface Treatments*. New York: McGraw-Hill, 1991, p. 4.53.

Bourithis, L., Ath. Milonas, and G. D. Papadimitriou. Plasma transferred arc surface alloying of a construction steel to produce a metal matrix composite tool steel with TiC as reinforcing particles. *Surface and Coatings Technology, 165*, 2003, 286–295.

Cline, H. E., and T. R. Anthony. Heat treating and melting material with a scanning laser or electron beam. *Journal of Applied Physics, 48*(9), 1977, 3895–3900.

Dahotre, N. B., and K. Mukherjee. Development of microstructure in laser surface alloying of steel with chromium. *Journal of Materials Science, 25*, 1990, 445–454.

Draper, C. W., and J. M. Poate. Laser surface alloying. *International Metals Review, 30*(2), 1985, 85–108.

Khangar, A., E. A. Kenik, and N. B. Dahotre. Microstructure and microtexture in laser-dressed alumina grinding wheel material. *Ceramics International, 31,* 2005, 621–629.

Kadolkar, P. Residual stress and cohesive strength of TiC composite coating on aluminum alloys during laser surface engineering. Master's thesis, University of Tennessee, Knoxville, 2002.

Kadolkar, P., and N. B. Dahotre. Variation of structure with input energy during laser surface engineered ceramic coating on aluminum alloys. *Applied Surface Science, 199*(1–4), 2002, 222–233.

Kadolkar, P., and N. B. Dahotre. Effect of processing parameters on the cohesive strength of laser surface engineered ceramic coatings on aluminum alloys. *Materials Science & Engineering A, 342*(1–2), 2003, 183–191.

Kesler, O., J. Matejicek, S. Sampath, S. Suresh, T. Gnaeupel-Herold, P. C. Brand, and H. J. Prask. Measurement of residual stress in plasma-sprayed metallic, ceramic and composite coatings. *Materials Science & Engineering A, 257*(2), 1998, 215–224.

Lowenheim, F. A. *Electroplating: Fundamentals of Surface Finishing.* New York: McGraw-Hill, 1978.

Lu, J., and D. Jobart-Retraint. A review of recent developments and applications in the field of x-ray diffraction for residual stress studies. *Journal of Strain Analysis for Engineering Design, 33*(2), 1998, 127–136.

Matejicek, J., S. Sampath, and J. Dusky. X-ray residual stress measurement in metallic and ceramic plasma sprayed coatings. *Journal of Thermal Spray Technology, 7*(4), 1998, 489–496.

Mazumder, J., and W. Steen. Heat transfer model for CW laser material processing. *Journal of Applied Physics, 51*(2), 1980, 941–947.

Metals Handbook, Heat Treating, Vol. 4, American Society for Materials, Materials Park, OH, 1991.

Nagarathnam, K., and K. Komvopoulous. Metallurgical and microhardness characteristics of laser synthesized Fe-Cr-W-C coatings. *Metallurgical and Materials Transactions A, 26A*(8), 1995, 2131–2139.

Niku-Lari A., J. Lu, and J. F. Flavenot. Measurement of residual stress distribution by the incremental hole drilling method. *Experimental Mechanics, 6,* 1985, 69–81.

Pangborn, R. J., and D. R. Beaman. Laser glazing of sprayed metal coatings. *Journal of Applied Physics, 51*(11), 1980, 5992–5993.

Pawlowski, L. *The Science & Engineering of Thermal Spray Coating.* New York: Wiley, 1995.

Product Finishing, Directory & Technology Guide, Dec. 1999.

Ramqvist, L. Wettability of metallic carbides by liquid Cu, Ni, Co, and Fe. *International Journal of Powder Metallurgy, 1*(4), 1965, 2–21.

Singh, A., and N. B. Dahotre. Laser in-situ synthesis of combinatorial coating on steel. *Journal of Materials Science, 39,* 2004, 4553–4560.

Sood, D. K. Metastable surface alloys produced by ion implantation, laser, and electron beam treatment. *Radiation Effects, 63,* 1982, 141–167. Sue, A. J., in *Stress Determination of Coatings,* from *ASM Metals Handbook on Surface Engineering,* Vol. 5., p. 647. American Society of Materials (ASM) International, Metals Park, OH, 1998.

Svensson, L. E., B. Gretoft, B. Ulander, and H. K. D. H. Bhadeshia. Fe-Cr-C hardfacing alloys for high temperature applications. *First International Conference on Surface Engineering,* Brighton, 25–28 June 1985, Vol. II, Weld Surfacing, ed. I. A. Bucklow, The Welding Institute, Great Abington, Cambridge, UK, 1985, 75–83.

Tayal, M., and K. Mukherjee. Thermal and microstructural analysis for laser surface hardening of steel. *Journal Applied Physics, 75*(6), 1984, 3855–3861.

Walker, A., J. Folkes, W. M. Steen, and D. R. F. West. Laser surface alloying of titanium substrates with carbon and nitrogen. *Surface Engineering, 1,* 1985, 23–29.

Wang, Y., X. Zhang, F. Li, and G. Zeng. Study on an Fe-TiC surface composite produced in-situ. *Materials and Design, 20*(5), 1999, 233–236.

Zhang, S., H. Xie, X. Zeng, and P. Hing. Residual stress characterization of diamond-like carbon coatings by x-ray diffraction method, surface and coatings technology. *Surface and Coatings Technology, 122,* 1999, 219–224.

Zhang, C., and Y. C. Shin. A novel laser-dressed truing and dressing technique for vitrified CBN wheels. *International Journal Machine Tools & Manufacture, 42,* 2002, 825–835.

6   Composite Materials

Definition and Classification

Composite materials are material systems that consist of a discrete constituent (the reinforcement) distributed in a continuous phase (the matrix) and that derive their distinguishing characteristics from the properties and behavior of their constituents, from the geometry and arrangement of the constituents, and from the properties of the boundaries (interfaces) between the constituents. Composites are classified either on the basis of the nature of the continuous (matrix) phase (polymer-matrix, metal-matrix, ceramic-matrix, and intermetallic-matrix composites), or on the basis of the nature of the reinforcing phase (particle reinforced, fiber reinforced, dispersion strengthened, laminated, etc.). The properties of the composite can be tailored, and new combinations of properties can be achieved. For example, inherently brittle ceramics can be toughened by combining different types of ceramics in a ceramic-matrix composite, and inherently ductile metals can be made strong and stiff by incorporating a ceramic reinforcement.

It is usually sufficient, and often desirable, to achieve a certain minimum level of reinforcement content in a composite. Thus, in creep-resistant dispersion-strengthened composites, the reinforcement volume fraction is maintained below 15% in order to preserve many of the useful properties of the matrix. Other factors, such as the shape, size, distribution of the reinforcement, and properties of the interface, are also important. The shape, size, amount, and type of the reinforcing phase to be used are dictated by the combination of properties desired in the composite. For example, applications requiring anisotropic mechanical properties (high strength and high stiffness along one particular direction) employ directionally aligned, high-strength continuous fibers, whereas for applications where strength anisotropy is not critical and strength requirements are moderate, relatively inexpensive particulates can be used as the reinforcing phase. Figure 6-1 shows some examples of continuous and discontinuous reinforcements developed for use in modern engineered composites.

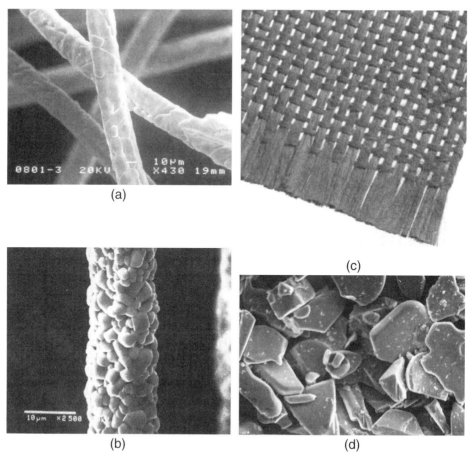

(a)

(b)

(c)

(d)

FIGURE 6-1 *(a) Scanning electron photomicrograph of sintered TiO$_2$ fiber.* (J. D. French and R. B. Cass, Developing Innovative Ceramic Fibers, *www.ceramicbulletin.org,* May 1998, pp. 61–65). *(b) SEM photomicrograph of an individual PZT filament of 25 μm diameter.* (J. D. French and R. B. Cass, Developing Innovative Ceramic Fibers, *www.ceramicbulletin.org,* May 1998, pp. 61–65). *(c) PZT fiber weave for a smart structure composite.* (J. D. French and R. B. Cass, Developing Innovative Ceramic Fibers, *www.ceramicbulletin.org,* May 1998, pp. 61–65). *(d) Single-crystal SiC platelets (nominal size, 150 μm).*

Fibers

Long, continuous fibers with a large aspect ratio (i.e., length-to-diameter ratio) of metals, ceramics, glasses and polymers are used to reinforce various types of matrices. A hard and strong material such as a ceramic in a fibrous form will have fewer strength-limiting flaws than the same material in a bulk form. As preexisting cracks lower the fracture strength of brittle ceramics, reducing the size and/or probability of occurrence of cracks will diminish the extent of strength loss in the ceramic, thus allowing the actual strength to approach the theoretical fracture strength in the absence of cracks, which is ~0.1 E, where E is the elastic modulus. If the fiber diameter scales with the grain size of the material, then the fracture strength will be high. In other words, smaller the fiber diameter, greater is its fracture strength. In the case of continuous fibers, a critical minimum aspect ratio of the fiber is needed to transfer the applied load from the weaker matrix to the stronger fiber. Furthermore, a small diameter allows a stiff fiber to be bent for shaping a preform that is used as a precursor in composite fabrication. Many commercial fibers are flexible, and permit filament winding and weaving techniques to be used for making a preform. Very stiff fibers are, however, shaped into preforms by using a fugitive binder material. For example, an organic compound that cements the fibers in the desired preform shape may be used. The binder decomposes and is eliminated when the matrix material is combined with the preform to provide it support and rigidity. Selected examples of fibers used in composite matrices are briefly described below. For more details, the reader is referred to the book by Chawla referenced at the end of the chapter.

Glass. *Glass* is a generic name for a family of ceramic fibers containing 50–60% silica (a glass former) in a solid solution that contains several other oxides such as Al_2O_3, CaO, MgO, K_2O. Na_2O, and B_2O_3, etc. Commercial glass fibers are classified as E-glass (for high electrical resistivity), S-glass (for high silica that imparts excellent high-temperature stability), and C-glass (for corrosion resistance).

Glass fiber is manufactured by melting the oxide ingredients in a furnace and then transferring the molten glass into a hot platinum crucible with a few hundred fine holes at its base. Molten glass flows through these holes and on cooling forms fine continuous filaments. The final fiber diameter is a function of the hole diameter in the platinum crucible, the viscosity of molten glass, and the liquid head in the crucible. The filaments are gathered into a strand, and a sizing is applied before the strand is wound on a drum. Glass is a brittle solid, and its strength is lowered by minute surface defects. The sizing protects the surface of glass filaments and also binds them into a strand. A common type of sizing contains polyvinyl acetate and a coupling agent that makes the strand compatible with various polymer matrices. The final fiber diameter is a function of the hole diameter in the platinum crucible, the viscosity of molten glass, and the liquid head in the reservoir. Another method to grow glass fibers makes use of a sol-gel–type chemical precipitation process. A sol containing fine colloidal particles is used as the precursor; due to their fine size, the particles remain suspended in the liquid vehicle and are stabilized against flocculation through ionic charge adsorption on the surface. The sol is gelled via pH adjustments, i.e., the liquid vehicle in the gel behaves as a highly viscous liquid, thus acquiring the physical characteristics of a solid. The gelling action occurs at room temperature. The gel is then drawn into fibers at high temperatures, that are lower than the temperatures used in conventional manufacture of glass fiber by melting. The Nextel fiber manufactured by the 3M company is a sol-gel–derived silica-based fiber. Moisture decreases the strength of glass fibers. They are also prone to static fatigue; that is, they cannot withstand loads for long periods of time. Glass fiber–reinforced plastics (GRPs) are widely used in the construction industry.

Boron. Boron fibers are produced by vapor depositing boron on a fine filament, usually made from tugsten, carbon, or carbon-coated glass fiber. In one type of vapor deposition process, a boron hydride compound is thermally decomposed, and the boron vapor heterogeneously nucleates on the filament, thus forming a film. Such fibers are, however, not very strong or dense, owing to trapped vapor or gas that causes porosity and weakens the fiber. In an improved chemical vapor deposition (CVD) process, a halogen compound of boron is reduced by hydrogen gas at high temperatures, via the reaction $2BX_3 + 3H_2 \rightarrow 2B + 6HX$ (X = Cl, Br, or I). Because of the high deposition temperatures involved, the precursor filament is usually tungsten. Fibers of consistently high quality are produced by this process, although the relatively high density of W filament slightly increases the fiber density. In the halide reduction process using BCl_3, a 10- to 12-μm-diameter W wire is pulled in a reaction chamber at one end through a mercury seal and out at the other end through another mercury seal. The mercury seals act as electrical contacts for resistance heating of the substrate wire when gases ($BCl_3 + H_2$) pass through the reaction chamber and react on the incandescent wire substrate to deposit boron coatings. The conversion efficiency of BCl_3 to B coating is only about 10%, and reuse of unreacted gas is important. Boron is also deposited on carbon monofilaments. A pyrolytic carbon coating is first applied to the carbon filament to accommodate the growth strains that result during boron deposition.

There is a critical temperature for obtaining a boron fiber with optimum properties and structure. The desirable amorphous (actually, microcrystalline with grain size of just a few nm) form of boron occurs below this critical temperature, whereas above this temperature there also occur crystalline forms of boron, which are undesirable from a mechanical properties viewpoint. Larger crystallites lower the mechanical strength of the fiber. Because of high deposition temperatures in CVD, diffusional processes are rapid, and this partially transforms the core region from pure W to a variety of boride phases such as W_2B, WB, WB_4, and others. As boron diffuses into the tungsten substrate to form borides, the core expands as much as 40% by volume, which results in an increase in the fiber diameter. This expansion generates residual stresses that can cause radial cracks and stress concentration in the fiber, thus lowering the fracture strength of the fiber. The average tensile strength of commercial boron fibers is about 3–4 GPa, and the modulus is 380–400 GPa. Usually a SiC coating is vapor-deposited onto the fiber to prevent any adverse reactions between B and the matrix such as Al at high temperatures.

Carbon Fiber. Carbon, which can exist in a variety of crystalline forms, is a light material (density: 2.268 g/cc). The graphitic form of carbon is of primary interest in making fibers. The other form of carbon is diamond, a covalent solid, with little flexibility and little scope to grow diamond fibers, although microcrystalline diamond coatings can be vapor-deposited on a fiberous substrate to grow coated diamond fibers. Carbon atoms in graphite are arranged in the form of hexagonal layers, which are attached to similar layers via van der Waals forces. The graphitic form is highly anisotropic, with widely different elastic modulus in the layer plane and along the c-axis of the unit cell (i.e., very high in-plane modulus and very low transverse modulus). The high-strength covalent bonds between carbon atoms in the hexagonal layer plane result in an extremely high modulus ($\sim$1000 GPa in single crystal) whereas the weak van der Waals bond between the neighboring layers results in a lower modulus (about one-half the modulus of pure Al) in that direction. In order to grow high-strength and high-modulus carbon fiber, a very high degree of preferred orientation of hexagonal planes along the fiber axis is needed.

The name *carbon fiber* is a generic one and represents a family of fibers all derived from carbonaceous precursors, and differing from one another in the size of the hexagonal sheets of carbon atoms, their stacking height, and the resulting crystalline orientations. These structural variations result in a wide range of physical and mechanical properties. For example, the axial tensile modulus can vary from 25 to 820 GPa, axial tensile strength from 500 to 5,000 MPa, and thermal conductivity from 4 to 1100 W/m.K, respectively. Carbon fibers of extremely high modulus are made by carbonization of organic precursor fibers followed by graphitization at high temperatures. The organic precursor fiber is generally a special long-chain polymer-based textile fiber (polyacrylonitrile or PAN and rayon, a thermosetting polymer) that can be carbonized without melting. Such fibers generally have poor mechanical properties because of a high degree of molecular disorder in polymer chains. Most processes of carbon fiber fabrication involve the following steps: (1) a stabilizing treatment (essentially an oxidation process) that enhances the thermal stability of the fibers and prevents the fiber from melting in the subsequent high-temperature treatment, and (2) a thermal treatment at 1000–1500°C called carbonization that removes noncarbon elements (e.g., N_2 and H_2). An optional thermal treatment called graphitization may be done at ~3000°C to further improve the mechanical properties of the carbon fiber by enabling the hexagonal crystalline sheets of graphite to increase their ordering. To produce high-modulus fiber, the orientation of the graphitic crystals or lamellae is improved by graphitization which consists of thermal and stretching treatments under rigorously controlled conditions. Besides the PAN and cellulosic (e.g., rayon) precursors, pitch is also used as a raw material to grow carbon fibers. Commercial pitches are mixtures of various organic compounds with an average molecular weight between 400 and 600. There are various sources of pitch; the three most commonly used are polyvinyl chloride (PVC), petroleum asphalt, and coal tar. The same processing steps (stabilization, carbonization, and optional graphitization) are involved in converting the pitch-based precursor into carbon fiber. Pitch-based raw materials are generally cheap, and the carbon fiber yield from pitch-based precursors is relatively high.

A recent innovation in carbon-based materials has been carbon nanotubes (CNT). Carbon nanotubes are relatively new materials—discovered in 1991—as a minor by-product of the carbon-arc process that is used to synthesize carbon's fullerene molecules. They present exciting possibilities for research and use. CNTs are a variant of their predecessor, fullerene carbon (with a geodesic dome arrangement of 60, 70, or even a few hundred C atoms in a molecule). Figure 6-2 shows a photograph of CNT. Single-walled CNT have been grown to an aspect ratio of ~10^5, with a length of about 100 μm, and therefore, from a composite mechanics standpoint, they can be considered as long, continuous fibers. The multiwalled CNT has an "onion-like" layered structure and is under extremely high internal stress, as evident from

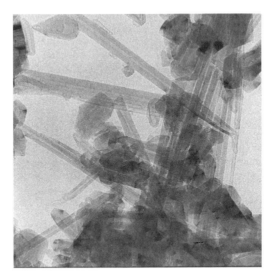

FIGURE 6-2 *Photograph of carbon nanotubes and polyhedral nanoparticles during fullerene production* (R. Malhotra, R. S. Ruoff and D. C. Lorents, "Fullerene Materials," Advanced Materials & Processes, April 1995 p. 30). Reprinted with permission from ASM International, Materials Park, OH (www.asminternational.org).

very small lattice spacing near the inner regions of the CNT. Carbon nanotubes (CNTs) have some remarkable properties, such as better electrical conductivity than copper, exceptional mechanical strength, and very high flexibility (with futuristic potential for use in even earthquake-resistant buildings and crash-resistant cars). There is already considerable interest in industry in using CNTs in chemical sensors, field emission elements, electronic interconnects in integrated nanotube circuits, hydrogen storage devices, temperature sensors and thermometers, and others. Because of the exceptional properties of CNTs (e.g., Young's modulus of CNT is 1–4 TeraPascals, TPa), there has been some interest in incorporating CNTs in polymers, ceramics, and metals. Owing to CNT's metallic or semiconducting character, incorporating CNT in polymer matrices permits attainment of an electrical conductivity sufficient to provide an electrostatic discharge at very low CNT concentrations. Similarly, extremely hard/and wear-resistant metal-matrix composites and tough ceramic-matrix composites are being developed. Since the discovery of CNTs in 1991, similar nanostructures were formed in other layered compounds such as BN, BCN, and WS_2, etc. For example, whereas CNTs are either metallic or semiconducting (depending on the shell helicity and diameter), BN nanotubes are insulating and could possibly serve as nanoshields for nanoconductors. Also, BN nanotubes are thermally more stable in oxidizing atmospheres than are CNTs and have comparable modulus. The strength of nanotubular materials can be increased by assembling them in the form of ropes, as has been done with CNT and BN nanotubes, with ropes made from single-walled CNTs being the strongest known material. The spacing between the individual nanotube strands in such a rope will be in the subnanometer range; for example, this spacing is $\sim$0.34 nm in a rope made from multiwalled BN nanotubes, which is on the order of the (0001) lattice spacing in the hexagonal BN cell.

Organic Fibers. Because the covalent C-C bond is very strong, linear-chain polymers such as polyethylene can be made very strong and stiff by fully extending their molecular chains. A wide range of physical and mechanical properties can be attained by controlling the orientation of these polymer chains along the fiber axis and their order or crystallinity. Allied Corporation's Spectra 900 and Du Pont's aramid fiber Kevlar are two successful organic fibers widely used for composite strengthening. Aramid is an abbreviated name of a class of synthetic organic fibers that are aromatic polyamide compounds. Nylon is a generic name for any long-chain polyamide. Many highly sophisticated manufacturing techniques have been developed to fabricate the organic fibers for use in composites. These techniques include: tensile drawing, die drawing, hydrostatic extrusion, and gel spinning. A wide range of useful engineering properties is achieved in organic fibers depending on the chemical nature of the polymeric material, processing technique, and the control of process parameters. For example, high modulus polyethylene fibers with a modulus of 200 GPa, and Kevlar fibers with a modulus of 65–125 GPa and tensile strength of 2.8 GPa have been developed Kevlar fibers have poor compression strength and should be used under compressive loading only as a hybrid fiber mixture, that is, as a combination of carbon fiber and Kevlar. One limitation of most organic fibers is that they degrade (lose color and strength) when exposed to visible or ultraviolet radiation, and a coating of a light-absorbing material is used to overcome this problem.

Metallic Fibers. Metals such as beryllium, tungsten, titanium, tantalum, and molybdenum, and alloys such as steels in the form of wires or fibers have high and very consistent tensile strength values as well as other attractive properties. Beryllium has a high modulus (300 GPa) and low density (1.8 g/cc) but also low strength (1300 MPa). Fine (0.1-mm) diameter steel wires with a high carbon (0.9%) content have very high strength ($\sim$5 GPa). Tungsten fibers have a very high melting point (3400°C) and are suited for heat-resistant applications. These various metallic fibers have been used as reinforcements in composite matrices based on metals (e.g., copper), concrete and polymers. For example, tungsten (density 19.3 g/cc) has been used as a reinforcement in advanced Ni- and Co-base superalloys for heat-resistant applications, and in Cu alloys for electrical contact applications. Similarly, steel wire is used to reinforce concrete and polymers (e.g., in steel belted tires). Other metallic reinforcements used in composite applications include ribbons and wires of rapidly quenched amorphous metallic alloys such as $Fe_{80}B_{20}$ and $Fe_{60}Cr_6Mo_6B_{28}$ having improved physical and mechanical properties.

Ceramic Fibers. Ceramic fibers such as single crystal sapphire, polycrystalline Al_2O_3, SiC, Si_3N_4, B_4C and others have high strength at room- and elevated temperature, high modulus, excellent heat-resistance, and superior chemical stability against environmental attack. Both polymer pyrolysis and sol-gel techniques make use of organometallic compounds to grow ceramic fibers. Pyrolysis of polymers containing silicon, carbon, nitrogen, and boron under controlled conditions has been used to produce heat-resistant ceramic fibers such as SiC, Al_2O_3, Si_3N_4, BN, B_4C and several others.

The commercial alumina fibers have a Young's modulus of 152–300 GPa and a tensile strength of 1.7 to 2.6 GPa. Alumina fibers are manufactured by companies such as Du Pont (fiber FP), Sumitomo Chemical (alumina-silica), and ICI (Saffil, δ-alumina phase). Fiber FP is made by dry-spinning an aqueous slurry of fine alumina particles containing additives. The dry-spun yarn is subjected to two-step firing: low firing to control the shrinkage and flame-firing to improve the density of α-alumina. A thin silica coating is generally applied to heal the surface flaws, giving higher tensile strength than uncoated fiber. The polymer pyrolysis route to make Al_2O_3 fibers makes use of dry-spinning of an organoaluminum compound to produce the ceramic precursor, followed by calcining of this precursor to obtain the final fiber. 3M Company uses a sol-gel route to synthesize an alumina fiber (containing silica and boria), called Nextel 312. The technique uses hydrolysis of a metal alkoxide, that is, a compound of the type $M(OR)_n$ where M is the metal, R is an organic compound, and n is the metal valence. The process breaks the *M-OR* bond and establishes the *MO-R* to give the desired oxide. Hydrolysis of metal alkoxides creates sols that are spun and gelled. The gelled fiber is then densified at intermediate temperatures. The high surface energy of the fine pores of the gelled fiber permits low-temperature densification.

Silicon carbide fibers, whiskers and particulates are among the most widely used reinforcements in composites. SiC fiber is made using the CVD process. A dense coating of SiC is vapor-deposited on a tungsten or carbon filament heated to about 1300 °C.

The deposition process involves high-temperature gaseous reduction of alkyl silanes (e.g., CH_3SiCl_3) by hydrogen. Typically, a gaseous mixture consisting of 70% H_2 and 30% silanes is introduced in the CVD reactor along with a 10–13 μm diameter tungsten or carbon filament.

The SiC-coated filament is wound on a spool, and the exhaust gases are passed through a condenser system to recover unused silanes. The CVD-coated SiC monofilament ($\sim$100–150 μm diameter) is mainly β-SiC with some α-SiC on the tungsten core. The SCS-6 fiber of AVCO Specialty Materials Company is a CVD SiC fiber with a gradient structure that is produced from the reaction of silicon- and carbon-containing compounds over a heated pyrolytic graphite-coated carbon core. The SCS-6 fiber is designed to have a carbon-rich outer surface that acts as a buffer layer between the fiber and the matrix metal in a composite, and the subsurface structure is graded to have stoichiometric SiC a few micrometers from the surface.

The SiC fiber obtained via the CVD process is thick (140 μm) and inflexible which presents difficulty in shaping the preform using mass production methods such as filament winding.

A method, developed in Japan, to make fine and flexible continuous SiC fibers (Nicalon fibers) uses melt-spinning under N_2 gas of a silicon-based polymer such as polycarbosilane into a precursor fiber. This is followed by curing of the precursor fiber at 1000°C under N_2 to cross-link the molecular chains, making the precursor infusible during the subsequent pyrolysis at 1300°C in N_2 under mechanical stretch. This treatment converts the precursor into the inorganic SiC fiber. Nicalon fibers, produced using the above process, have high modulus (180–420 GPa) and high strength (~2 GPa).

Besides the SiC and Al_2O_3 fibers described in the preceding paragraphs, silicon nitride, boron carbide, and boron nitride are other useful ceramic fiber materials. Si_3N_4 fibers are produced by CVD using $SiCl_4$ and NH_3 as reactant gases, and forming the fiber as a coating onto a carbon or tungsten filament. In polymer-based synthesis of silicon nitride fibers, an organosilazane compound (i.e., a compound that has Si-NH-Si bonds) is pyrolyzed to give both SiC and Si_3N_4. Fibers of the oxidation-resistant material boron nitride are produced by melt-spinning a boric oxide precursor, followed by a nitriding treatment with ammonia that yields the BN fiber. A final thermal treatment eliminates residual oxides and stabilizes the high-purity BN phase. Boron carbide (B_4C) fibers are produced by the CVD process via the reaction of carbon yarn with BCl_3 and H_2 at high temperatures in a CVD reactor. In addition to the use of long and continuous fibers of different ceramic materials in composite matrices, vapor-phase grown ceramic whiskers have also been extensively used in composite materials. Whiskers are monocrystalline short ceramic fibers (aspect ratio ~50–10,000) having extremely high fracture strength values that approach the theoretical fracture strength of the material.

Figure 6-3 compares the room-temperature stress versus strain behavior of boron, Kevlar, and glass fibers; high-modulus graphite (HMG) fiber; and ceramic whiskers. The figure shows that whiskers are by far the strongest reinforcement, because of the absence of structural flaws, which results in their strength approaching the material's theoretical strength. Usually, however, there is considerable scatter in the strength properties of whiskers, and this becomes problematic in synthesizing composites with a narrow spread in their properties. Selected thermal and mechanical properties of some commercially available fibers are summarized in Table 6-1.

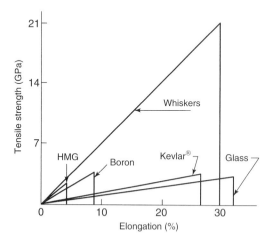

FIGURE 6-3 *Schematic comparison of stress–strain diagrams for common reinforcing fibers and whiskers (HMG, high-modulus graphite fiber).* (A. Kelly, ed., Concise Encyclopedia of Composite Materials, Elsevier, 1994, p. 312). Reprinted with permission from Elsevier.

TABLE 6-1 Properties of Selected Reinforcement Materials

Reinforcement	ρ, $kg \cdot m^{-3}$	Diameter, μm	E, GPa	UTS, MPa	C_p, $kJ \cdot kg^{-1} \cdot {}^{\circ}K^{-1}$	k, $W \cdot m^{-1} \cdot K^{-1}$	CTE, $10^{-6} \, {}^{\circ}K^{-1}$
SiC whisker	3200		700	21,000	0.69		2.5
Si_3N_4	3100		385	5,000–7,000			
SiC (Nicalon)	2550	10–20	$180^{\parallel}$	8,300	0.69	25	2.5–4.3
Al_2O_3 (fiber FP)	3950	20	$385^{\parallel}$	3,800		37.7	$8.3^{\parallel}$
Al_2O_3 (Saffil)	3300		285	1,500			
Al_2O_3 (whisker)	4000	10–25	700–1500	10,000–20,000	0.60	24	7.7
E glass fiber	2580	8–14	$70^{\perp} \, 70^{\parallel}$	3,450	0.70	13	$4.7^{\parallel,\perp}$
Borsic (SiC-coated B)	2710	102–203	400	3,100	1.3	38	5.0
PAN HM carbon fiber	1950	7–10	$390^{\parallel} \, 12^{\perp}$	2,200	0.71		-0.5 to $1.0^{\parallel}$ 7 to $12^{\perp}$
PAN HS carbon fiber	1750	7–10	$250^{\parallel} \, 20^{\perp}$	2,700	0.71	8	-0.5 to $1.0^{\parallel}$ 7 to $12^{\perp}$
Aramid (Kevlar 49)	1440	12	$125^{\parallel}$	2,800–3,500	0.71		-2 to $-5^{\parallel}$ $59^{\perp}$
Sapphire	4000		470	2,000			6.2–6.8
PRD 166 (Al_2O_3-ZrO_2)	4200		385	2,500			
Nextel 480	3050		224	2,275			

Note: ρ, density; E, modulus; UTS, ultimate tensile strength; k, thermal conductivity; CTE, coefficient of thermal expansion.
PAN HM carbon fiber: polyacrylonitrile-derived, high-modulus graphite fiber.
PAN HS carbon fiber: polyacrylonitrile-derived, high-strength graphite fiber.
$\parallel$, parallel to the fiber axis.
$\perp$, perpendicular to the fiber axis.

The mechanical and physical properties such as elastic modulus (E) and coefficient of thermal expansion (CTE) of fibers are strongly orientation dependent, and usually exhibit significant disf-ferences in magnitude along the fiber axis and transverse to it. The high-temperature strength of some commercial silicon carbide fibers is compared in Figure 6.4. It can be noted that the fiber retains high strength to fairly high temperatures; for example, NLP 101 fiber retains a strength of 500 MPa at 1300°C, which is comparable to the room-temperature tensile strength of some high-strength, low-alloy steels.

In addition to the synthetic fibers and whiskers, numerous low-cost, discontinuous fillers have been used in composites to conserve precious matrix materials at little expense to their engineering properties. These fillers include mica, sand, clay, talc, rice husk ash, fly ash, natural fibers (e.g., lingo-cellulosic fibers), recycled glass, and many others, including environmentally conscious biomorphic ceramics based on silicon carbide and silicon dioxide obtained from pyrolysis of natural wood. These various fillers and reinforcements permit a range of composite microstructures to be created that have a wide range of strength, stiffness, wear resistance, and other characteristics. Figure 6-5 shows the porous structure of pyrolyzed wood that has been used as a preform for impregnation with molten metals to create ceramic- or metal-matrix composites.

Interface

Interfaces in composites are regions of finite dimensions at the boundary between the fiber and the matrix where compositional and structural discontinuities can occur over distances varying from an atomic monolayer to over five orders of magnitude in thickness. Composite fabrication processes create interfaces between inherently dissimilar materials (e.g., ceramic fibers and

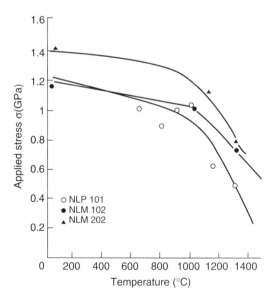

FIGURE 6-4 *High-temperature strength of some SiC fibers plotted as applied stress versus test temperature.* (B. S. Mitchell, An Introduction to Materials Engineering and Science for Chemical and Materials Engineers, Wiley Interscience, Hoboken, NJ, 2004).

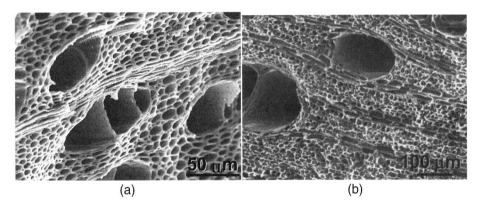

(a)   (b)

FIGURE 6-5 *(a) Scanning electron micrograph of a maple-derived carbonaceous preform showing porous structure.* (M. Singh and J. A. Salem, Mechanical properties and microstructure of biomorphic silicon carbide ceramics fabricated from wood precursors, Journal of the European Ceramic Society, 22, 2002, 2709–2717, Elsevier). *(b) SEM micrograph of porous carbon preform made from mahogany wood perpendicular to the growth direction.* (M. Singh and J. A. Salem, Mechanical properties and microstructure of biomorphic silicon carbide ceramics fabricated from wood precursors, Journal of the European Ceramic Society, 22, 2002, 2709–2717, Elsevier). Reprinted with permission from Elsevier. Photo Courtesy of M. Singh, QSS Group, Inc., NASA Glenn Research Center, Cleveland, OH.

metal matrices). The material incompatibility and the severe processing conditions generally needed for composite fabrication create interfaces that exist in a non-equilibrium state. Thus, interface evolution is thermodynamically ordained; however, interface design for properties by processing is essentially an outcome of a variety of kinetic phenomena (reaction kinetics, mass transport etc.). The nature of the interface that develops in composites during fabrication and subsequent service strongly influences the response of the composite to mechanical stresses and to thermal and corrosive environments. As the inherent properties of the fiber and matrix constituents in a composite are fixed, greatest latitude in designing bulk composite properties is realized by tailoring the interface (this is not strictly true, however, because processing conditions that lead to interface development also usually modify both the fiber properties as well as the metallurgy of the matrix). The development of an optimum interfacial bond between the fiber and the matrix is, therefore, a primary requirement for optimum performance of a composite. The nature and the properties of the interface (thickness, continuity, chemistry, strength, and adhesion) are determined by factors both intrinsic to the fiber and matrix materials (chemistry, crystallography, defect content), as well as extrinsic to them (time, temperature, pressure, atmosphere, and other process variables). One of the major goals in the study of interfaces in composites is to develop an understanding of and exercise control on the structure, chemistry, and properties of interfaces by judicious manipulation of processing conditions. To realize this goal, it is also necessary to develop and use techniques for mechanical, chemical, and structural characterization of interfaces in composites. Considerable progress has been made in characterizing and understanding interfaces at the microstructural, crystallographic, and atomic levels.

Bonding at the fiber-matrix interface develops from physical or chemical interactions, from frictional stresses due to irregular surface topography, and from residual stresses arising from the mismatch of coefficients of thermal expansion (CTE) of the fiber and the matrix materials.

Both the fiber and matrix characteristics as well as the interface characteristics control the physical, mechanical, thermal, and chemical behavior of the composite. As an example, consider the fracture behavior of a composite in which fibers are the strengthening phase. The fracture in the composite can proceed at the interface, through the matrix, or through the reinforcement depending on their respective inherent mechanical properties and the defect population. If the matrix is weak relative to the fiber and the interface, it will fail by the usual crack nucleation and growth mechanism. If the matrix and the interface are strong, the load is transferred across the interface to the reinforcement, which will provide strengthening until a threshold stress is reached at which the composite will fail. Similarly, electrical, thermal, and other properties of a composite are determined by the properties of the fiber, matrix, and the interface.

Purely physical interactions (e.g., dispersion forces and electrostatic interactions) seldom dominate the interface behavior in composites. Usually, some chemical interaction between fiber and matrix aids interface growth and determines the interface behavior. Chemical interactions may involve adsorption, impurity segregation, diffusion, dissolution, precipitation and reaction layer formation. These interactions are abruptly terminated at the conclusion of composite fabrication, which renders the interface inherently unstable. Driven by a need for these interactions to proceed to completion and the interface to approach thermodynamic equilibrium, compositional and structural transformations at the interface continue after fabrication, usually with sluggish kinetics, via reaction paths that may involve intermediate non-equilibrium phases. Chemical interactions between the fiber and the matrix not only determine the interface properties and behavior, but may also modify the properties of the fiber and the metallurgy of the matrix. For example, the extent of fiber strength degradation and loss of age-hardening response because of chemical reactions in metal-matrix composites are directly related to the extent of interfacial reactions as reflected in the size of the reaction zone at the interface. The loss of age-hardening response is because of loss of chemically active solutes in the fiber-matrix reactions. It is, therefore, very important to control the processing conditions to design the interface for properties with minimum fiber degradation and little alterations in the metallurgy of the matrix. Because the rate of chemical reaction can be characterized in terms of temperature-dependent rate constants and activation energies, fundamental insights into the mechanisms of strength-limiting interfacial reactions can be derived.

Besides chemical interactions between the fiber and the matrix, the thermoelastic compatibility between the two is important, particularly if the fabrication and/or service involves significant temperature excursions. A large mismatch between the coefficients of thermal expansion (CTE) of the fiber and the matrix can give rise to appreciable thermoelastic stresses which may affect the adhesion at the interface. For example, these stresses can give rise to interfacial cracking if the matrix cannot accommodate these stresses by plastic flow. In such a case, stress-absorbing intermediate compliant layers are deposited at the interface to promote compatibility and reduce the tendency for cracking by reducing the CTE mismatch–induced stresses. Such layers may also provide protection to the reinforcement against excessive chemical attack in reactive matrices during fabrication and service.

A careful control of the fabrication conditions can enhance the interface strength without excessive fiber degradation. Usually, a moderate chemical interaction between fiber and matrix improves the wetting, assists liquid-state fabrication of composite, and enhances the strength of the interface, which in turn facilitates transfer of external stresses to the strengthening agent, i.e., the fiber. But an excessive chemical reaction would degrade the fiber strength (even though the interface strength may be high) and defeat the very purpose for which the fibers were incorporated in the monolith. In contrast, if toughening rather than strengthening is the objective, as in brittle ceramic-matrix composites, then creation of a weak rather than strong interface is desired so that crack deflection and frictional stresses during sliding of debonded fibers will permit realization of toughness. In such a case, recipes designed to improve the wetting by strong chemical reactions can induce too high a bond strength, which will, in turn, confer poor toughness on the composite. Thus, a delicate balance between several conflicting requirements is usually necessary to tailor the interface for a specific application with the aid of surface-engineering and processing science.

Fiber Strengthening

With a fiber residing in a matrix, the length of the fiber limits the distance over which bonding and load transfer are possible. For effective composite strengthening, the fiber must have a minimum critical length, l_c; fibers shorter than this length do not serve as load-bearing constituents. This critical length is

$$l_c = \frac{\sigma_f d}{2\tau_c} \tag{6-1}$$

where d is the fiber diameter, σ_f is the fracture strength of the fiber, and τ_c is the fiber–matrix bond strength. The critical length, l_c, is on the order of a few millimeters for most composites. Thus, for effective strengthening, the actual fiber length must exceed a few millimeters. In continuous fiber-reinforced composites, the tensile strength and the elastic modulus are strongly anisotropic. Usually, the strength of the composite along the fiber length (longitudinal strength) follows a simple rule of mixture (ROM), i.e., $\sigma_c = V_f \sigma_f + V_m \sigma_m$, where σ_m and σ_f are the strength of the matrix and reinforcement, respectively, and V_f and V_m are the fiber and matrix volume fractions. For composites reinforced with discontinuous, aligned fibers with a uniform distribution in the matrix and with the fiber length, l, greater than the critical length l_c, the composite strength, σ_{cd}, along the longitudinal direction is

$$\sigma_{cd} = \sigma_f V_f \left(1 - \frac{l_c}{2l} \right) + \sigma_m (1 - V_f) \tag{6-2}$$

If the fiber length is smaller than the critical length, then σ_{cd} is given from

$$\sigma_{cd} = \frac{l \tau_c}{d} V_f + \sigma_m (1 - V_f) \tag{6-3}$$

where d is the fiber diameter, and τ_c is the shear stress at the fiber surface, which for a plastically deformable matrix such as metals is the yield strength of the matrix, and for a brittle matrix (ceramics or polymers), is the frictional stress at the interface. Calculations based on Equation [6-2] show that for aligned, discontinuous fibers with $l > lc$, the loss in composite's strength relative to the case of continuous fibers will not be appreciable provided the stress concentration at the ends of the short fibers is negligible.

The elastic modulus of a continuous fiber-reinforced composite along the longitudinal (fiber) direction is given from the following ROM relationship:

$$E_c = V_f E_f + V_m E_m \tag{6-4}$$

where E_m and E_f are the Young's moduli of the matrix and reinforcement, respectively, and V_f and V_m are the volume fractions of the fiber and the matrix, respectively. The composite modulus, E_{ct}, transverse to the fiber direction is given from a relationship reminiscent of the electrical resistance of parallel circuit:

$$\frac{1}{E_{ct}} = \frac{V_f}{E_f} + \frac{V_m}{E_m}$$ (6-5)

For discontinuously reinforced composites,

$$E_{cd} = KE_f V_f + E_m V_m$$ (6-6)

where K is called the fiber efficiency factor, and its value depends on the modulus ratio, (E_f/E_m); for most cases K is in the range 0.1–0.6. These relationships apply to situations where the fiber-matrix interface is devoid of reaction layers, such as in polymer-matrix composites.

Polymer-Matrix Composites

Matrix

Fiber-reinforced polymers are widely used as structural materials for relatively low-temperature use. Generally, polymers have lower strength and modulus than metals or ceramics but they are more resistant to chemical attack than metals. Figure 6-6 displays a schematic comparison of the strength characteristics of ceramics, metals, polymers, and elastomers. Prolonged exposure to UV light and some solvents can, however, cause polymer degradation. Polymers are giant, chainlike molecules or macromolecules, with covalently bonded carbon atoms as the backbone of the chain. Small-chain, low-molecular-weight organic molecules (monomers) are joined together via the process of polymerization, which converts monomers to polymers. Polymerization occurs either through condensation or through addition of a catalyst. In condensation polymerization,

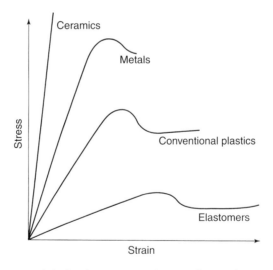

FIGURE 6-6 *Comparison of idealized stress–strain diagrams for metals, amorphous polymers, and elastomers. (B. S. Mitchell, An Introduction to Materials Engineering and Science for Chemical and Materials Engineers, Wiley-Interscience, Hoboken, NJ, 2004, p. 469).*

there occurs a stepwise reaction of molecules, and in each step a molecule of a simple compound, generally water, forms as a by-product. In addition to polymerization, monomers can be joined to form a polymer with the help of a catalyst without producing any by-products. For example, the linear addition of ethylene molecules (CH_2) results in polyethylene, with the final mass of polymer being the sum of monomer masses.

Linear polymers consist of a long chain, often coiled or bent, of atoms with attached side groups (e.g., polyethylene, polyvinyl chloride, polymethyl metacrylate or PMMA). Branched polymers consist of side-branching of atomic chains. In cross-linked polymers, molecules of one chain are bonded (cross-linked) with those of another, thus forming a three-dimensional network. Cross-linking hinders sliding of molecules past one another, thus making the polymer strong and rigid. Ladder polymers form by linking linear polymers in a regular manner; ladder polymers are more rigid than linear polymers. Figure 6-7 illustrates these different types of polymers.

Unlike pure metals that melt at a fixed temperature, polymers show a range of temperatures over which crystallinity vanishes on heating. On cooling, polymer liquids contract just as metals do. In the case of amorphous polymers, this contraction continues below the melting point, T_m, of crystalline polymer to a temperature, T_g, called the glass transition temperature, at which the supercooled liquid polymer becomes extremely rigid owing to extremely high viscosity. The structure of the polymer below T_g is essentially disordered, like that of a liquid. Figure 6-8 shows the changes in the specific volume as a function of temperature in a polymer. Many physical properties such as viscosity, heat capacity, modulus, and thermal expansion change abruptly at T_g. For example, Figure 6-9 displays the variation of the natural logarithm of elastic modulus as a function of temperature; the transitions from the glassy to rubbery, and rubbery to fluid states are accompanied by discontinuities in the modulus. The glass transition temperature, T_g, is a function of the structure of the polymer; for example, if a polymer has a rigid backbone structure and/or bulky branch groups, then T_g will be quite high. The glass transition phenomenon is also observed in amorphous ceramics such as glasses. Glasses have a mixed ionic and covalent

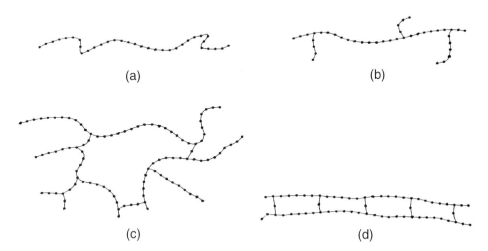

(a) (b) (c) (d)

FIGURE 6-7 *Molecular chain configurations in polymers: (a) linear, (b) branched, (c) cross-linked, and (d) ladder.* (K. K. Chawla, Composite Material - Science & Engineering, Springer-Verlag, New York, NY, 1987, p. 59).

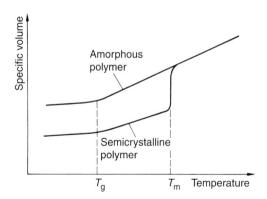

FIGURE 6-8 *Specific volume versus temperature relationship for amorphous and semicrystalline polymers (T_g, glass transition temperature; T_m, melting temperature).* (K. K. Chawla, Composite Material - Science & Engineering, Springer-Verlag, New York, NY, 1987, p. 60).

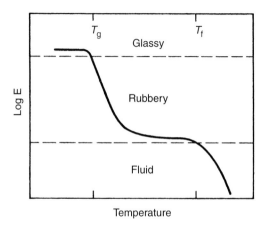

FIGURE 6-9 *Variation of modulus of an amorphous polymer with temperature.* (K. K. Chawla, Composite Material - Science & Engineering, Springer-Verlag, New York, NY, 1987, p. 63).

bonding, and are highly cross-linked. In fact, they may have a higher T_g than polymers that have only covalent bonding and have less cross-linking. In amorphous polymers, there is no apparent order among the molecules and the molecular chains are arranged in a random manner. When polymers precipitate from dilute solutions, small, platelike single-crystalline regions called lamellae or crystallites form. In the lamellae, long molecular chains are folded in a regular manner, and many lamellae group together to form spherulites much like grains in metals.

Most linear polymers soften and melt upon heating. These are called thermoplastics and are readily shaped using liquid forming techniques. Examples include low- and high-density polyethylene, polystyrene, and PMMA.

When the molecules in a polymer are cross-linked in the form of a network, they do not soften on heating, and are called thermosetting polymers. Common thermosetting polymers include phenolic, polyester, polyurethane, and silicone. Thermosetting polymers decompose on heating. As noted, cross-linking makes sliding of molecules past one another difficult, thus making the polymer strong and rigid. A typical example is that of rubber cross-linked with sulfur, that is, vulcanized rubber. Vulcanized rubber has 10 times the strength of natural rubber.

There is another type of classification of polymers based on the type of repeating unit. If one type of repeating unit forms a polymer chain, it is called a homopolymer. In contrast, polymer chains having two different monomers form co-polymers. If the two different monomers are distributed randomly along the chain, then the polymer is called a regular or random co-polymer. If, however, a long sequence of one monomer is followed by a long sequence of another monomer, the polymer is called a block co-polymer. If a chain of one type of monomer has branches of another type, then a graft co-polymer is said to form. Figure 6-10 schematically illustrates these various polymer structures.

Molecular weight is a very important parameter that determines the properties of polymers. Many mechanical properties increase with increasing molecular weight (although polymer processing also becomes more difficult with increasing molecular weight). Another important parameter is degree of polymerization (DP), which is a measure of the number of basic units (mers) in a polymer. The molecular weight (MW) and the DP are related by $MW = DP \times (MW)u$, where $(MW)u$ is the molecular weight of the repeating unit. Because polymers contain various types of molecules, each of which has a different MW and DP, the molecular weight of the polymer is characterized by a distribution function. A narrower distribution indicates a homogeneous polymer with many repeating identical units whereas a broad distribution indicates that the polymer is composed of a large number of different species. In general, therefore, it is convenient to specify an average molecular weight or degree of polymerization. Unlike many common substances of low molecular weight, polymers can have very large molecular weights. For example, the molecular weights of natural rubber and polyethylene (a synthetic polymer) can be greater than 10^6 and 10^5, respectively. In addition, the molecular diameters of high-molecular-weight polymers can be over two orders of magnitude larger than ordinary substances such as water that have low molecular weights.

Polymers can be amorphous or partially crystalline, and the amount of crystallinity in a polymer can vary from 30 to 90%. The elastic modulus and strength of polymers increase with increasing crystallinity, although fully crystalline polymers are unrealistic. The long and complex molecular chains in a polymer are easily tangled, resulting in large segments of molecular chains that may remain trapped between crystalline regions and never reorganize themselves into an ordered molecular assembly characteristic of a fully crystalline state. The molecular arrangement in a polymer also influences the extent of crystallization; for example, linear molecules with small side groups can readily crystallize whereas branched-chain molecules with bulky side

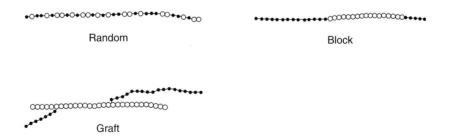

FIGURE 6-10 *Schematic representation of random, block, and graft copolymers.* (K. K. Chawla, Composite Material - Science & Engineering, Springer-Verlag, New York, NY, 1987, p. 61)

groups do not crystallize easily. Thus, linear high-density polyethylene can attain up to 90% crystallization whereas branched polyethylene can attain only about 65% crystallization.

Properties of Polymeric Matrices

Glassy polymers follow Hooke's law and exhibit a linear elastic response to applied stress. The elastic strain in glassy polymers is less than 1%. In contrast, elastomers (rubbery polymers) show a nonlinear elastic behavior with a large elastic range (a few percent strain) as shown in Figure 6-6. The large elastic strain in elastomers is caused by a redistribution of the tangled molecular chains under an applied stress.

Highly cross-linked thermosetting resins such as polyesters, epoxies, and polyimides have high modulus and strength, but are also extremely brittle. The fracture energy (100–200 $J.m^{-2}$) of thermosetting resins is only slightly better than that of inorganic glasses (10–30 $J.m^{-2}$). In contrast, thermoplastic resins such as polymethylmetacrylate (PMMA) have fracture energies on the order of 1000 $J.m^{-2}$ because their large free volume facilitates absorption of the energy associated with crack propagation. The fracture toughness of hard and brittle thermosets such as epoxies and polyester resins is improved by distributing small (a few micrometers in size) and soft rubbery inclusions in the brittle matrix. The most common method is simple mechanical blending of the soft, elastomer particles and the thermoset resin, or copolymerization of a mixture of the two.

Polymers show a significant temperature dependence of their elastic modulus as shown in Figure 6-8. Below the glass transition temperature T_g, the polymer behaves as a hard and rigid solid, with an elastic modulus of about 5–7 GPa. Above T_g, the modulus drops significantly and the polymer exhibits a rubbery behavior. Upon heating the polymer to above its melting temperature T_f at which the polymer becomes fluid, the modulus drops abruptly. The glass transition temperature, melting temperature, and selected mechanical and thermal properties of common polymeric materials are listed in Table 6-2.

The thermal expansion of polymers, in particular thermoplastics, is strongly temperature-dependent, exhibits a non-linear increase with temperature, and is generally an order of magnitude or two greater than that of metals and ceramics. Therefore, in polymer composites the thermal expansion mismatch between inorganic fibers and polymer matrix is large, which may cause thermal stresses and cracking.

Both thermosets and thermoplastics are used as matrix materials for polymer composites. For example, polyesters, epoxies and polyimides are commonly used matrices in fiber-reinforced composites. Polyesters have fair resistance to water and various chemicals, as well as to aging, but they shrink between 4 and 8% on curing. In contrast, thermosetting epoxy resins have better moisture resistance, lower shrinkage on curing (about 3%), a higher maximum-use temperature, and good adhesion to glass fibers. Polyimides have a relatively high service-temperature range, 250–$300°C$ but, they are brittle and have low fracture energies of 15–70 $J·m^{-2}$. Polymers degrade at high temperature and by moisture absorption, which causes swelling and a reduction in T_g. If the polymer is reinforced with fibers bonded to the matrix, then moisture absorption may cause severe internal stresses in the composite.

Polymer Composites

Polymer composites are fabricated using pultrusion and filament winding. In pultrusion (Figure 6-11), continuous fiber tows (i.e., fiber bundle with parallel strands of fibers) are impregnated with a thermoset resin and drawn through a die that forms the final component shape (e.g., tubes, rods, etc.). Hollow parts are made by pultruding the composite feedstock around a

TABLE 6-2 Mechanical Properties of Selected Polymeric Materials

Material	E, GPa	T.S., MPa	MOR, MPa	Y.S., MPa	% Elongation	Sp. Gr.	CTE, $10^{-5} \times °C^{-1}$	T_g, C	T_m, C
LDPE	0.17–0.28	8.3–31.4		9.0–14.5	100–650	0.92–0.93		−100	120
HDPE	1.06–1.09	22.1–31.0		26.2–33.1	10–1200	0.95–0.96		−115	130
PVC	2.40–4.10	40.7–51.7		40.7–44.8	40–80	1.30–1.58		80	212
Teflon	0.40–0.55	20.7–34.5		—	200–400	2.14–2.20		125	327
Polystyrene	2.28–3.28	35.9–51.7		—	1.2–2.5	1.04–1.05		100	240
Polyester (PET)	2.8–4.1	48.3–72.4		59.3	30–300	1.29–1.40		70	270
Polycarbonate	2.38	62.8–72.4		62.1	110–150	1.20		150	230
Natural rubber		18–25			1000				
Butadiene styrene		1.5–2.3			3000				
Epoxy		35–85	15–35			1.38	8.0–11.0		
Polyimide		120	35			1.46	9.0		
PEEK		92	40			1.30	—		
Phenolics		50–55	—			1.30	4.5–11.0		

Note: LDPE, low-density polyethylene; HDPE, high-density polyethylene; PVC, polyvinyl chloride; Teflon, polytetrafluoroethylene; E, tensile modulus; T.S., tensile strength; MOR, modulus of rupture (flexural strength); Y.S., yield strength; CTE, coefficient of thermal expansion; T_g, glass transition temperature; T_m, melting temperature.

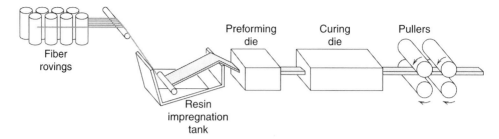

FIGURE 6-11 *Schematic diagram of the pultrusion process of making polymer composites.* (W. D. Callister, Jr., Materials Science and Engineering: An Introduction, 6th ed., Wiley, New York, 2003, p. 555).

core or mandrel. The drawing rate and fiber volume fraction are controlled. The component is then cured in a preheated precision die to obtain the final shape and size. Glass-, carbon-, and aramid fiber-reinforced polyesters, vinyl esters, and epoxy resin matrix composites containing relatively large (40% or higher) volume fraction of aligned continuous fibers are produced via this method.

In filament winding, fiber strands or tows are first coated with a resin by passing them through a resin bath, and resin-coated fibers are automatically and continuously wound onto a mandrel for subsequent curing. The fiber volume fraction is controlled by the spacing between fiber strands and by the number of fiber layers wound on the mandrel. The mechanical properties of the composite are influenced not only by the fiber volume fraction but also by the winding pattern (helical, circumferential, etc.). After the required number of fiber layers have been wound on the mandrel, the composite is cured in an oven and the mandrel removed to obtain a hollow composite object. Rocket motor castings, pipes, and pressure vessels are made using filament winding.

Another widely used method to make polymer-matrix composite consists of stacking layers of partially-cured thin (<1 mm) sheets of resin-impregnate and directionally aligned fibers called prepregs either manually (hand layup) or automatically in a three-dimensional sandwich structure. Prepregs are covered with a backing paper that is removed prior to lamination. Prepregs are made by sandwiching fiber tows between sheets of carrier paper that is coated with the resin matrix. On pressing the paper over fiber tows using heated rollers (a process called "calendaring"), the resin melts and impregnates the fibers, thus forming a prepreg. A prepreg may be cut to various angles relative to the fiber axis to give prepregs of different orientations. For example, a prepreg with fibers parallel to the long dimension is a zero-degree lamina or ply, and a prepreg that is cut with fibers perpendicular to the long dimension is a 90-degree ply (intermediate angles are also used). Figure 6-12 shows stacking of prepregs in a $0°/\pm45°/90°$ orientation. The composite properties can be predicted from the theory of composite mechanics for a given stacking sequence and fiber orientation, which permits premeditated design of the composite for properties. For example, the coefficient of thermal expansion (CTE) of boron–epoxy and carbon–epoxy composites can be systematically varied between $-5 \times °10^{-6}$ and $30 \times 10^{-6} °K^{-1}$, by controlling the ply angle as shown in Figure 6-13. After stacking the plies in the desired orientation, the final curing of the component is done under heat and pressure. For polymer composite with thermoplastic matrices, liquid-phase fabrication methods, such as

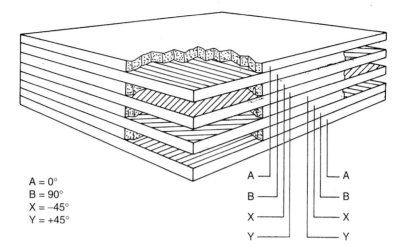

A = 0°
B = 90°
X = −45°
Y = +45°

FIGURE 6-12 *Schematic of a laminated composite formed by stacking prepregs of different orientations (0°, +45°, −45°, 90°).* (K. K. Chawla, Composite Materials: Science and Engineering, Springer, New York, NY, 1987, p. 91).

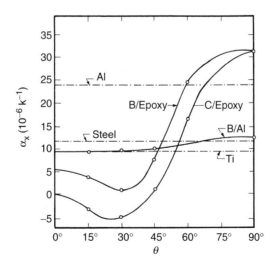

FIGURE 6-13 *Coefficient of thermal expansion as a function of fiber orientation for some polymer and metal composites, and metals and alloys.* (Composite Materials in Engineering Design, American Society for Materials, 1973, p. 335). Reprinted with permission from ASM International, Materials Park, OH (www.asminternational.org).

injection molding, extrusion, and thermoforming, are used, and short, randomly oriented fibers can be dispersed for isotropic properties.

Figure 6-14 and Table 6-3 present selected mechanical property data on some polymer composites. The mechanical properties of polymer composites often degrade because of temperature and humidity effects (hygrothermal degradation). As stated earlier, moisture diffusion into a polymer matrix causes swelling, a decrease in the glass transition temperature of the matrix, and weakening of the interface. The resulting softening of the matrix and the interface can lead to composite failure at low stresses. Besides humidity and temperature, electromagnetic radiation primarily ultraviolet (UV) radiation, can also degrade the polymer composite because the UV radiation breaks the C-C bond in the polymer. Resistance to UV radiation is enhanced by adding carbon black to the polymer matrix.

Ceramic-Matrix Composites

Ceramic-matrix composites surpass metal-matrix and polymer-matrix composites in elevated-temperature performance. As a result, there has been a great deal of interest in the development and testing of advanced ceramic composites based on carbides (SiC, TiC), oxides (Al_2O_3, ZrO_2, SiO_2), nitrides (Si_3N_4, BN, AlN), borides (ZrB_2, HfB_2), and glass and glass-ceramics for a number of applications. Some examples of ceramic-matrix composites are SiC/Al_2O_3, SiC/glass ceramic, SiC/Si_3N_4, TiC/Si_3N_4, Al_2O_3/ZrO_2, TiC/Al_2O_3, C/glass, C/SiC, and SiC/SiC. Thus,

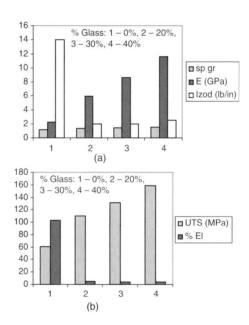

FIGURE 6-14 *(a) Specific gravity, Young's modulus (E), and impact strength of polycarbonate reinforced with randomly oriented glass fibers containing different glass fiber volume fractions. (Data are taken from Materials Engineering's Materials Selector, copyright Penton, IPC). (b) Percent elongation and ultimate tensile strength (UTS) of a glass fiber-reinforced epoxy composite with different glass fiber volume fractions. (Data are taken from Materials Engineering's Materials Selector, copyright Penton, IPC).*

TABLE 6-3 Mechanical Properties of Polymer- and Metal-Matrix Composites

Composite	E, GPa	Strength, MPa
B-Al (50% fiber)	$210^{\parallel}$	$1500^{\parallel}$
	$150^{\perp}$	$140^{\perp}$
SiC-Al (50% fiber)	$310^{\parallel}$	$250^{\parallel}$
	—	$105^{\perp}$
Fiber FP-Al-Li (60% fiber)	$262^{\parallel}$	$690^{\parallel}$
	$152^{\perp}$	$172\text{–}207^{\perp}$
C-Al (30% fiber)	160	690
E-Glass-Epoxy (60% fiber)	$45^{\parallel}$	$1020^{\parallel}$
	$12^{\perp}$	$40^{\perp}$
C-Epoxy (60% HS fiber)	$145^{\parallel}$	$1240^{\parallel}$
	$10^{\perp}$	$41^{\perp}$
Aramid (Kevlar 49) - Epoxy (60% fiber)	$76^{\parallel}$	$1380^{\parallel}$
	$5.5^{\perp}$	$30^{\perp}$

Note: $^{\parallel}$, value parallel to the fiber axis.
$^{\perp}$, value perpendicular to the fiber axis.

C/SiC composites produced by chemical vapor infiltration (CVI) have been developed for applications in hypersonic aircraft thermal structure, advanced rocket propulsion thrust chambers, cooled panels for nozzle ramps, and turbo pump blisks (blade+disc)/shaft attachments. Similarly, light-weight, and wear- and heat-resistant SiC/SiC composites have been proposed for applications in combustor liners, exhaust nozzles, reentry thermal protection systems, radiant burners, hot gas filters, high-pressure heat exchangers, as well as in fusion reactors owing to their resistance to neutron flux. Fibers, whiskers, particulates, platelets, and laminates are the common reinforcement morphologies in ceramic-matrix composites.

Ceramic-matrix composites can also be designed for improved toughness. For example, in a laminated C/SiC composite, introduction of interlayers of graphite between SiC helps in deflecting cracks and improving the fracture toughness of SiC. The reinforcement may temporarily halt a crack, the interface may partially debond, thus allowing fiber pullout with some frictional resistance, or a weak interface may deflect the crack. Figure 6-15 shows a conceptual scheme of possible mechanisms that lead to toughening in ceramic-matrix composites. Figure 6-16 shows microstructural evidence of the mechanism of crack deflection and bridging by the reinforcing phase in a ceramic composite of 3Y-TZP matrix containing 15 vol% Al_2O_3 platelets of two nominal sizes. Another mechanism of toughening in ceramic composites is blunting of cracks due to volumetric changes accompanying solid-state phase transformations. Yttria-stabilized zirconia (YSZ) is an example of toughening by transformation in which dispersion of yttria particles toughens the zirconia through a stress-induced phase transformation (see Chapter 3). In addition, ceramic-matrix composites containing dispersions of fine metal particles (e.g., Cu or Cr in Al_2O_3) have better fracture toughness than the monolithic ceramic matrix.

Most ceramic-matrix composites are fabricated using the hot consolidation processes, although melt infiltration, chemical vapor infiltration, reaction bonding sol-gel processing, polymer pyrolysis, combustion synthesis, and electrophoretic deposition techniques are also used. In the hot consolidation technique, the fibers are first impregnated with the unconsolidated powder matrix by passing them through a slurry containing the matrix powders, a carrier

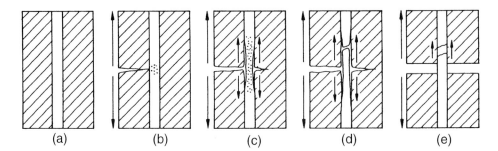

FIGURE 6-15 *Schematic diagram showing mechanism of toughness gain in a brittle-matrix composite. (a) original state with frictional gripping, (b) matrix crack being temporarily halted by the fiber, (c) debonding and crack deflection along the fiber–matrix interface as a result of interfacial shear and lateral contraction of the fiber and matrix, (d) continued debonding, fiber failure at a weak point, and further crack extension, (e) broken fiber ends are pulled out against frictional resistance of the interface.* (B. Harris, Met. Sci., 14, 1980, p. 351).

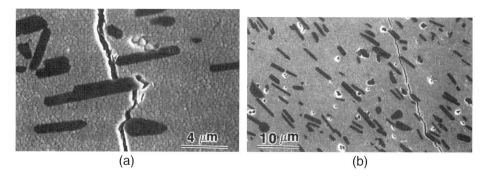

FIGURE 6-16 *(a) SEM micrograph showing crack deflection and bridging in 3Y-TZP-Al$_2$O$_3$ composite containing 15 vol% of 10–15 μm platelets.* (I. K. Cherian and W. M. Kriven, American Ceramics Society Bulletin, 80(12), December 2001, p. 57–67). *(b) SEM micrograph showing crack deflection in 3Y-TZP-Al$_2$O$_3$ composite containing 15 vol% of 3–5 μm platelets.* (I. K. Cherian and W. M. Kriven, American Ceramics Society Bulletin, 80(12), December 2001, 57–67).

liquid and binders, and wetting and dispersing agents. The coated fibers are wound on a drum, dried, cut, stacked, and consolidated at high temperatures. Fiber tows or preforms of unidirectionally aligned fibers can be impregnated with the slurry, and stacked in various orientation (e.g., cross-ply, angular ply) to develop the desired properties. Temperature, pressure, fiber arrangement and powder size distribution are some of the important process parameters in hot consolidation. Concurrent application of high pressures and high temperatures permits rapid densification and creation of a pore-free, fine-grained product by hot consolidation. However, fiber damage during hot consolidation, porosity and contamination from binder residues, and difficulty in producing complex shapes are some of the limitations of hot consolidation techniques. Ceramic-matrix composites such as SiC whisker-reinforced Al$_2$O$_3$, and ceramic fiber reinforced glass-matrix composites are produced by hot consolidation techniques. Ceramic composite laminates such as C/SiC are also produced by hot pressing. For example, SiC powders are mixed with a polymeric binder to a doughlike consistency, and pressed into thin sheets. These sheets are coated with graphite, stacked, and hot-pressed. In place of hot consolidation, cold compaction followed by sintering has also been used to fabricate various ceramic-matrix composites, although considerable shrinkage of the matrix during sintering often results in cracks and weakening of the matrix.

In the melt infiltration method, a polymer precursor is pyrolyzed to yield a porous feedstock that is then infiltrated with a ceramic melt to produce high-density ceramic-matrix composite in a single step. The very high processing temperatures needed to melt ceramics and the very high viscosity of molten ceramics make the infiltration techniques energy intensive and rather difficult to use. In addition, undesirable chemical reactions at high processing temperatures, and very large thermal stresses due to solidification shrinkage and differential contraction between the matrix and the reinforcement can cause matrix cracking and strength loss in the composite. Ceramic composites are also produced by reactive infiltration, which retains the overall simplicity of conventional infiltration but forms the composite at molten metal (rather than molten ceramic) temperatures. This reduces the energy consumption while taking advantage of the high fluidity of molten alloys to facilitate the infiltration. The infiltration of porous carbon performs by molten Si or Si-Mo alloys has been used to form SiC/SiC/C composite. Fiber-reinforced organosilicon-based SiC composites are produced by first preparing a porous fibrous preform with some binder phase in it, followed by infiltration with polycarbosilanes at high temperatures and pressures, and polymerization. The infiltrated organosilicon polymer matrix is then thermally decomposed in an inert atmosphere between 800 and 1300 $^{\circ}K$. The infiltration and thermal decomposition steps can be repeated to obtain composites of high density. Figure 6-17 shows the microstructure of some hot-pressed and melt-infiltrated ceramic-matrix composites, including an ultra-high-temperature ceramic composite consisting of a zirconium diboride (ZrB_2) matrix containing either SiC particles or both SiC particles and SCS-9a SiC fibers. Composites based on borides of refractory metals Zr, Ti, Hf, and Ta have high melting temperatures, high hardness, low volatility, and good thermal shock resistance and thermal conductivity, and can withstand ultra-high temperatures (1900–2500°C). These composites generally have good oxidation resistance, and further improvements are achieved through the use of additives such as SiC, which improves the oxidation resistance of diborides such as ZrB_2 and HfB_2. With SiC in ZrB_2, a thin, glassy layer of SiO_2 forms when the boride is exposed to an oxidizing atmosphere; the silica layer covers an inner layer of ZrO_2 on the surface of ZrB_2 base material. The outer glass layer provides good wettability and surface coverage, as well as oxidation resistance. Many such composites are still under development but hold promise for future deployment at ultra-high temperatures.

Figure 6.18 and Figure 6-19 show representative mechanical properties of ceramic-matrix composites. In Figure 6-18, the flexural strength and elastic modulus of a borosilicate glass matrix composite containing different volume fractions of aligned carbon fibers are plotted; both the flexural strength and stiffness of the composite increase with increasing volume fraction of C fiber. The effect of temperature on the flexural strength of a lithium aluminosilicate glass-ceramic (i.e., partially crystallized matrix) reinforced with SiC fibers is shown in Figure 6-19. Both unidirectional and two-dimensional reinforcement yields superior bend strength at all temperatures compared to the unreinforced matrix (although the transverse rupture strength with unidirectional fibers is very low in SiC/glass-ceramic composite). The useful temperature range over which acceptable strength levels are maintained in this composite extends to nearly 1100 $^{\circ}C$ (a slight increase in the bend strength near 1000 $^{\circ}C$ in Figure 6-19 is due to the severe oxidation of the composite, which increases the interface strength).

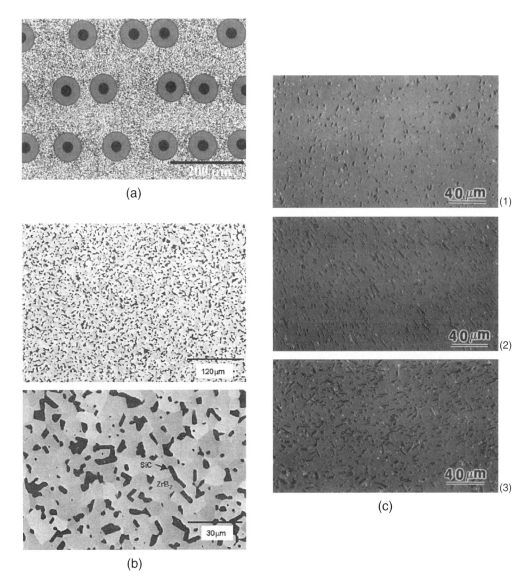

(a)

(b)

(c)

FIGURE 6-17 *(a) Microstructure of a polished section of an ultra-high temperature ceramic-matrix composite of ZrB$_2$-20 v/o SiC+SCS-9a fibers showing representative fiber distribution.* (S. R. Levine et al., Evaluation of Ultra-High Temperature Ceramics for Aeropropulsion, Journal of the European Ceramics Society, 22, 2002, 2757–2767, Elsevier). *(b) Microstructure of an ultrahigh temperature ceramic matrix composite of ZrB$_2$-20 v/o SiC particles showing particle distribution.* (S. R. Levine et al., Evaluation of Ultra-High Temperature Ceramics for Aeropropulsion, Journal of the European Ceramics Society, 22, 2002, p. 2757–2767, Elsevier). Reprinted with permission from Elsevier. Photo Courtesy of M. Singh, QSS Group, NASA Glenn Research Center, Cleveland OH. *(c) SEM micrograph showing crack deflection and bridging in 3Y-TZP-Al$_2$O$_3$ composite containing (1) 5, (2) 10, and (3) 15 vol% of 10–15 μm platelets showing homogeneous distribution.* (I. K. Cherian and W. M. Kriven, American Ceramics Society Bulletin, 80(12), December 2001, 57–67).

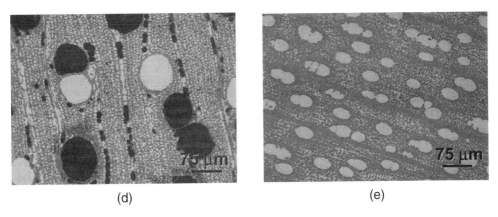

(d)　　　　　　　　　　　　　　　　(e)

FIGURE 6-17 continued *(d) Optical micrograph of as-fabricated SiC made from mahogany. View is perpendicular to the growth direction (white, Si; gray, SiC; black, pores).* (M. Singh and J. A. Salem, Mechanical properties and microstructure of biomorphic silicon carbide ceramics fabricated from wood precursors, Journal of the European Ceramic Society, 22, 2002, 2709–2717, Elsevier). *(e) Optical micrograph showing the microstructure of silicon carbide made from maple wood.* (M. Singh and J. A. Salem, Mechanical properties and microstructure of biomorphic silicon carbide ceramics fabricated from wood precursors, Journal of the European Ceramic Society, 22, 2002, 2709–2717, Elsevier). Reprinted with permission from Elsevier. Photo Courtesy of M. Singh, QSS Group, Inc., NASA Glenn Research Center, Cleveland, OH.

Carbon-Carbon Composites

Carbon-carbon (C-C) composites consist of a carbon matrix reinforced with continuous carbon fibers. Carbon-carbon composites are extensively used for the nose cone and leading edges of the space shuttle, as solid-propellant rocket nozzles and exit cones, and as ablative nose tips and heat shield for ballistic missiles. The material has also been used in aircraft braking systems (e.g., in the Anglo-French *Concorde*) where the extremely high frictional torque generates intense heat, raising the brake disk temperature to 500 °C, and the interface temperature to 2000 °C. Likewise, the first wall tiles of thermonuclear fusion reactors also use C-C composites to cope with intense thermal loads. For the space exploration systems, C-C composites have been proposed for applications in radiator and heat management systems, turbine and turbopump housing, thrust cell jackets, and flanges. Many applications require joining C-C composites to metals, and advanced joining techniques have been developed for these composites.

Carbon matrix in C-C composites binds the carbon fibes, maintains their orientation, and prevents fiber-to-fiber contact. Carbon-carbon composite can exist in a range of structures, from amorphous to fully graphitic. These composites are made by chemical vapor infiltration (CVI), and impregnation and pyrolysis using resin- or pitch-based precursors. Chemcial vapor infiltration involves thermal decomposition of a hydrocarbon vapor (e.g., CH_4 or C_2H_2, etc.) over a hot substrate on which carbon is deposited. Several different CVI methods have been developed to produce the C-C composites. In the isothermal CVI process, a porous carbon substrate is placed in a constant temperature zone, and reactant gases are passed over it. To avoid early pore closure, the surface reaction rate should be slower than the diffusion rate into the pores. The slow deposition rate is achieved by using low pressures (10–100 mbar) and low temperatures (~1100 °C). Eventually, pore blockage occurs even before full densification, and it is necessary to remove the part from the reactor, machine the surface, and reinfiltrate. The process is very slow, and several hundred hours are needed to achieve full densification. To reduce the

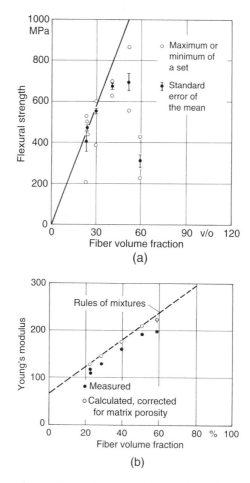

FIGURE 6-18 *(a) Flexural strength as a function of fiber volume fraction for a borosilicate glass matrix composite containing continuous aligned carbon fibers.* (D. C. Phillips, R. A. J. Sambell, and D. H. Bowen, Journal of Materials Science, 7, 1972, p. 1454) *(b) Young's modulus as a function of fiber volume fraction for a borosilicate glass matrix composite containing continuous aligned carbon fibers.* (D. C. Phillips, R. A. J. Sambell, and D. H. Bowen, Journal of Materials Science, 7, 1972, p. 1454)

composite processing times, thermal gradient-CVI (TG-CVI) and pressure gradient-CVI (PG-CVI) techniques were developed. In TG-CVI, the preform is supported on an inductively heated mandrel, which keeps the inside surface hot. The outer surface is cooler as it is close to water-cooled coils. A thermal gradient is, therefore, established across the preform, provided thermal conduction is slow. Carbon first deposits on the cooler outer surface. As density builds up, the substrate couples inductively and heats up, and the reaction front gradually moves through the component. The process is performed at atmospheric pressure, but the susceptor geometry should match the part geometry for best coupling. In the pressure-gradient (PG-CVI) technique, the preform is sealed off from the furnace chamber. The reactant gases are introduced into the

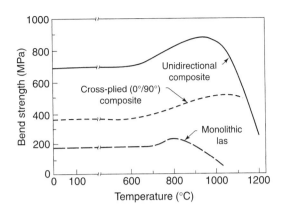

FIGURE 6-19 *Temperature dependence of bending strength of SiC-reinforced lithium alumi-nosilicate (LAS) ceramic-matrix composite.* (R. W. Davidge and J. J. R. Davies, in B. F. Dyson, R. D. Lohr, and R. Morrel, eds., Mechanical Testing of Engineering Ceramics at High Temperatures, p. 264, 1989, Elsevier Science Publishers). Reprinted with permission from Elsevier.

inside surface of the preform at a pressure higher than that in the furnace; as a result, reactant gases flow through the porous carbon substrate under a pressure differential.

Carbon-carbon composites exhibit high propensity toward oxidation in air, especially above $500\,°C$. The oxidation reaction is $2C(s) + O_2(g) \rightarrow 2CO(g)$. The matrix in the C-C is more reactive that the fibers, and the fibers oxidize at a slower rate. The oxidation is, however, most severe at the fiber/matrix interface which presents a large surface area. The rate of oxidation increases with increasing temperature. In order to protect the composite from severe oxidation, protective coatings (e.g., in-situ grown oxide films) are applied, which must isolate the composite from the surroundings. For oxidation resistance, the protective coating must prevent oxygen diffusion into the composite and carbon diffusion to outer surface. The thermomechanical compatibility and adhesion between the coating and the composite are important. Usually the CTE of the carbon matrix is much less than that of the coating; as a result, cracks could form during cooling. A relatively successful approach of thermal protection is to use a glass- or glass-forming compound in the coating, which can flow into and seal the crack. Most protective coatings consist of three basic constituents: bond layer, functional layer, and erosion- and oxidation–resistant layer, such as a layer of SiC or Si_3N_4, which upon oxidation forms a SiO_2 skin. Figure 6-20 shows the experimental oxidation behavior and high-temperature strength of C-C and some advanced ceramic composites at various temperatures.

Chemical Vapor Infiltration

CVI is used to fabricate ceramic-matrix composites, and involves infiltration of reactant vapors into a porous (e.g., fibrous) preform, and the deposition of the solid product phase (matrix) within the pores. For example, Al_2O_3 and TiC can form via vapor-phase reactions among H_2, $AlCl_3$, $TiCl_4$, CH_4, and CO_2. The reactions may be represented by

$$H_2(g) + CO_2(g) \rightarrow H_2O(g) + CO(g)$$

$$2AlCl_3(g) + 3H_2O(g) \rightarrow Al_2O_3(s) + 6HCl(g)$$

$$TiCl_4(g) + CH_4(g) \rightarrow TiC(s) + 4HCl(g)$$

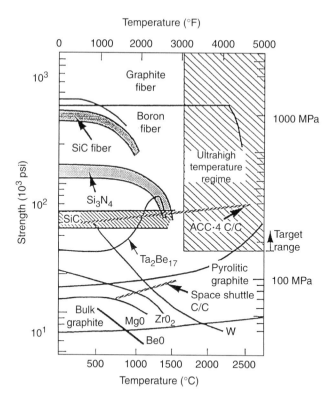

FIGURE 6-20 *Elevated-temperature strength of carbon-carbon composite and other high-temperature composites.* (K. Upadhya, J. Yang, and W. P. Hoffman, Materials for Ultrahigh Temperature Structural Applications, American Ceramics Society Bulletin, December 1997, 51).

In the case of deposition of Al_2O_3 within a porous SiC fiber bundle, the first two reactions apply, and the third reaction is replaced with

$$2AlCl_3(g) + 3H_2(g) + 3CO_2(g) \rightarrow Al_2O_3(s) + 3CO(g) + 6HCl(g)$$

An analysis of the deposition of alumina matrix in a SiC fiber bundle in an isothermal hot-wall CVI reactor has been given by Tai and Chou.[1] The deposition of Al_2O_3 matrix within cylindrical pores of a unidirectionally aligned SiC bundle occurs by the diffusion and reaction of CO_2 and H_2 at fiber surface within the pore (Figure 6-21). Because of the small pore size, gaseous transport through the pores via diffusion is the dominant process. It is assumed that the vapor concentration changes only along the radial (r) and axial (z) directions within the pore (variation along the θ coordinate is ignored). Let N be the molar flux (moles/m^2.s), C be the moles of the total vapor per unit volume (moles/m^3), and x be the mole fraction of a given

[1]Tai N. and T.W. Chou, "Theoretical analysis of chemical vapor infiltration in ceramic/ceramic composites", MRS Symposium Proceedings, Vol. 120, 1998, 185–192, Materials Research Society, Boston.

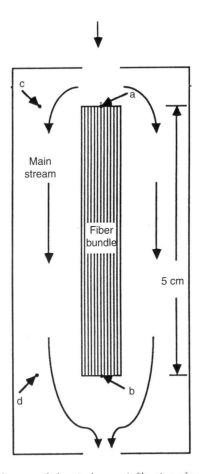

Main stream

Fiber bundle

5 cm

FIGURE 6-21 *Schematic diagram of chemical vapor infiltration of a fiber perform.*

species, then,

$$N_r = -CD\frac{\partial x}{\partial r}$$

and

$$N_z = -CD\frac{\partial x}{\partial z},$$

where D is the diffusion coefficient of a gaseous species. Under steady state, the diffusion equation can be written in terms of the mole fraction of the ith species as

$$\frac{\partial}{\partial r}\left(r\frac{\partial x_i}{\partial r}\right) + r\frac{\partial^2 x_i}{\partial z^2} = 0 \tag{6-7}$$

Because a homogenous composition of gaseous species is maintained throughout the reactor, including at the ends of the fiber bundle of length L, one boundary condition is $x_i = x_{i0}$ at

$z = 0$ and $z = L$. Furthermore, at the deposition surface within the pore of radius R, the mole fraction of a gaseous species, x_i, can be expressed in terms of a reaction rate constant, k, and the rate, r_1, leading to the second boundary condition: $x_i = r_1/(Ck)$ at $r = R$. In terms of an excess mole fraction X_i defined by $X_i = x_i - x_{i0}$, the boundary conditions can be written as $X_i = 0$ at $z = 0$ and $z = L$, and, $X_i = r_1/(Ck) - x_{i0}$ at $r = R$. The governing diffusion equation (Equation 6-7) is transformed into

$$\frac{\partial}{\partial r}\left(r\frac{\partial X_i}{\partial r}\right) + r\frac{\partial^2 X_i}{\partial z^2} = 0 \tag{6-8}$$

This equation can be solved by the method of separation of variables. Define $X_i(r,z) = \phi(r)\psi(z)$, and substitute in Equation 6-8 above to yield the following two equations:

$$\frac{\phi''(r)}{\phi} + \frac{1}{r}\frac{\phi'(r)}{\phi(r)} = -\frac{\psi''(z)}{\psi(z)} = \lambda^2 \tag{6-9}$$

which can be rearranged to yield

$$r^2\phi''(r) + r\phi'(r) - \lambda^2 r^2\phi(r) = 0 \tag{6-10}$$

and

$$\psi''(z) + \lambda^2\psi(z) = 0 \tag{6-11}$$

The solutions to Equations 6-10 and 6.11 for $\lambda \neq 0$ are

$$\phi(r) = A'I_0(\lambda r) + B'K_0(\lambda r) \tag{6-12}$$

$$\psi(z) = E'\sin(\lambda z) + F'\cos(\lambda z) \tag{6-13}$$

The solution to the governing equation becomes

$$X_i(r,z) = I_0(\lambda r)[E\sin(\lambda z) + F\cos(\lambda z)] \tag{6-14}$$

where I_0 is the modified Bessel function. The boundary condition $X_i = 0$ at $z = 0$ and $z = L$ yields $F = 0$ and $\lambda = (n\pi/L)$, and the solution becomes

$$X_i(r,z) = \sum_{n=0}^{\infty} E_n I_0\left(\frac{n\pi r}{L}\right)\sin\left(\frac{n\pi z}{L}\right) \tag{6-15}$$

Substituting Equation 6-15 in

$$N_r = -CD\frac{\partial x}{\partial r} \quad \text{and} \quad N_z = -CD\frac{\partial x}{\partial z},$$

and using the second boundary condition,

$$X_i = r_l/(Ck) - x_{i0} \quad \text{at} \quad r = R,$$

yields

$$N_i(r, z) = -CD \sum_{n=0}^{\infty} E_n \left(\frac{n\pi}{L}\right) I_1 \left(\frac{n\pi R}{L}\right) \sin \left(\frac{n\pi z}{L}\right) \tag{6-16}$$

where I_1 is the Bessel function of the first order, and

$$X_i(R, z) = -\frac{D}{k} \sum_{n=0}^{\infty} E_n \left(\frac{n\pi}{L}\right) I_1 \left(\frac{n\pi R}{L}\right) \sin \left(\frac{n\pi z}{L}\right) - x_{i0} \tag{6-17}$$

The coefficient E_n is obtained by equating the expression for X_i at $r = R$ from Equations 6-15 and 6-17. This yields $E_n = 0$ when n is even and the following expression when n is odd,

$$E_n = \frac{-4x_{i0}}{n\pi \left[I_0 \left(\frac{n\pi R}{L}\right) + \frac{D}{k} \left(\frac{n\pi}{L}\right) I_1 \left(\frac{n\pi R}{L}\right)\right]} \tag{6-18}$$

The diffusion coefficient, D, in CVI depends on temperature, gas pressure, and pore size. For a given set of processing conditions (temperature and pressure), D depends only on the pore diameter. Different diffusion regimes are possible in a CVI reactor depending on the pore size relative to the mean free path of gas molecules. In a dilute gaseous atmosphere (low concentration) and small pore sizes, intermolecular collisions are infrequent, and frequency of collision of gas molecules with the pore wall is high (Knudsen diffusion). At very high vapor concentrations where the frequency of intermolecular collisions is large (and molecules frequently change their direction), molecular diffusion dominates. At intermediate vapor concentrations, frequency of intermolecular collisions becomes comparable to the molecular collisions with the pore wall, and D depends on both Knudsen diffusion and molecular diffusion processes. Expressions for D for these regimes can be written in terms of temperature and molecular weights. The reaction rate constant, k, can be taken to be a constant at a fixed temperature; if the temperature is changed, k can be expressed by the Arrhenius equation in terms of an activation energy, Q, by $k = k_0 \exp(-Q/R_g T)$, where R_g is the gas constant and k_0 is a temperature-independent term. This equation transforms to a linear form: $\ln k = \ln k_0 - (Q/R_g T)$; thus, a plot of natural log of the deposition rate as a function of the reciprocal of absolute temperature will yield a straight line. Figure 6-22 shows such a plot for the deposition of B_4C at three different total pressures of reactant gases, BCl_3, CH_4, and H_2. Figure 6-23 shows the calculated results for the effect of CVI temperature and pressure on the densification kinetics of porous preform. The densification is faster at higher temperatures and pressures in the CVI reactor, consistent with a diffusion-limited infiltration mechanism. The microstructures of a C/SiC composite synthesized using the isothermal CVI process is shown in Figure 6-24; fiber distribution, residual porosity, and some matrix cracking can be noted.

Metal-Matrix Composites

Heat-Resistant Composites
Dispersion-Strengthened Composites
For elevated temperature use, the composite matrix must have a high melting point, be resistant to creep deformation, oxidation and hot corrosion, and be thermally stable and light. Advanced

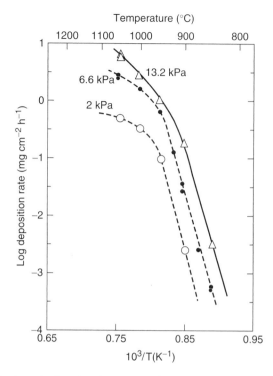

FIGURE 6-22 *Arrhenius plot for the deposition of B$_4$C at various total pressures by the CVI process.* (R. Naslain, CVI Composites, in, R. Warren, ed., Ceramic-Matrix Composites pp. 199–244, Chapman and Hall, London, 1992).

Ni-base superalloys, usually in single crystal form, have been used for such applications. To stabilize the matrix structure of these alloys for high-temperature use, the intermetallic phases in the matrix (primarily the γ'-phase in the γ-matrix) are stabilized by alloying with heavy metals, such as Re, W, Ta, etc. Whereas this improves the creep-resistance, it decreases the solidus temperature and increases the alloy density, thus degrading the specific properties (i.e., value of the property divided by the density). Complex thermal barrier coatings and increasingly more elaborate cooling systems have been used to increase the thermal efficiency of engines.

The development of oxide dispersion–strengthened (ODS) composites (e.g., W-ThO$_2$, Ni-Al$_2$O$_3$, Cu-SiO$_2$, Cu-Al$_2$O$_3$, NiCrAlTi-Y$_2$O$_3$, Ni-ThO$_2$, Ni-HfO$_2$, etc.) has led to considerable improvements in the elevated-temperature strength and creep-resistance of the matrix alloys. The ODS composites contain small ($<15\%$) amounts of fine second-phase particles, which resist recrystallization and grain coarsening, and act as a barrier to the motion of dislocations. Small quantities of fine oxide ceramics improve the strength without degrading the valuable matrix properties. Examples of ODS composites include yttria- or thoria-dispersed Ni-20Cr, and Ni-Cr alloys containing both yttria and γ'-phase (that forms due to additions of Al, Ti, or Ta). These composites have excellent resistance to oxidation, creep, carburization, and sulfidation, and in some cases, excellent resistance to molten glass. They are produced mainly by powder metallurgy (hot pressing) techniques which utilize powders of matrix alloy and the

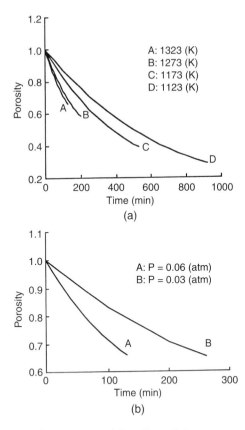

FIGURE 6-23 *(a) Theoretical projections of the effect of CVI temperature on the densification kinetics of porous performs. The densification is faster at higher temperatures, consistent with a diffusion-limited infiltration mechanism.* (N. Tai and T. W. Chou, Theoretical Analysis of Chemical Vapor Infiltration in Ceramic/Ceramic Composites. High Temperature/High Performance Composites, in Materials Research Society Symposium, vol. 120, 1988, 185). *(b) Theoretical projections of the effect of pressure in the CVI reactor on the densification kinetics of porous perform. The densification is faster at higher pressures, consistent with a diffusion-limited infiltration mechanism.* (N. Tai and T. W. Chou, Theoretical Analysis of Chemical Vapor Infiltration in Ceramic/Ceramic Composites. High Temperature/High Performance Composites, in Materials Research Society Symposium, vol. 120, 1988, 185). Reprinted with permission from Materials Research Society, Warrendale, PA.

ceramic dispersoid as the feed materials. Besides oxide dispersions, carbides, borides and other refractory particulates have been used in dispersion-strengthened (DS) composites.

The hot-pressed oxide dispersion–strengthened (ODS) Ni-base superalloys have been used in aircraft gas turbine engine parts such as vane airfoils, blades, nozzles, and combustor assemblies. In these materials, the dispersed second phase primarily serves as barrier to the motion of dislocations rather than as the load-bearing constituent (as in an engineered metal-base composite). The thermal stability of these materials is superior to precipitation-hardened alloys, in which the second phase softens and degrades at high temperatures. Besides the Ni-base superalloys, several other families of alloys have been dispersion-strengthened. Figure 6-25 shows a view

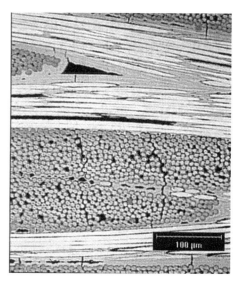

FIGURE 6-24 *Microstructure of a C-SiC composite synthesized using the isothermal CVI process showing the fiber distribution, residual porosity, and matrix cracking.* (M. R. Effinger, G. G. Genge, and J. D. Kiser, Advanced Materials and Processes, June 2000, pp. 69–73). Reprinted with permission from ASM International, Materials Park, OH (www.asminternational.org).

of a dispersion-strengthened silver composite containing uniformly distributed nanometer-size Al_2O_3 particles. The composite was created by an internal oxidation method in which an Ag-Al alloy powder is oxidized in a controlled fashion to create a dispersion of fine, evenly distributed aluminum oxide particles.

In dispersion-strengthened composites, the fine dispersions of oxide, carbide, or boride particles impede the motion of dislocations so that the matrix is strengthened in proportion to the effectiveness of the dispersions as a barrier to the motion of dislocations. For a dislocation to pass through a dispersion of fine particles, the applied stress must be sufficiently large to bend the dislocation line into a semicircular loop. Calculations show that interparticle separation for effective dispersion hardening should be between 0.01 and 0.3 μm. To achieve this range of spacing between the dispersoid, the particle diameters should be less than 0.1 μm at volume fractions below 15%. In the DS material, it is usually desirable to preserve as many of the properties (ductility, electrical and thermal conductivity, and impact strength, etc.) of the matrix material as possible. This requires that the dispersoid volume fraction be kept small because large-volume fractions decrease the fracture toughness, thermal conductivity, and other valuable matrix properties. Consider, for example, the room-temperature hardness data on Cu-alumina dispersion-strengthened composites containing only 3.5% submicron-size alumina. For this material, the hardness remains unchanged when the DS composite is annealed at different temperatures ranging from room temperature all the way to just a few degrees below the melting point of Cu. This remarkable hardness retention of the DS Cu/Al_2O_3 composites is achieved with minimum penalty on the other useful properties of Cu because the alumina volume fraction is so low.

FIGURE 6-25 *Longitudinal microstructure of an extruded bar of oxide dispersion-strengthened Ag composite containing 0.5 wt% Al$_2$O$_3$ dispersions.* (J. Troxell, A. Nadkarni, and J. Abrams, in Advanced Materials and Processes, January 2001, 75–77). Reprinted with permission from ASM International, Materials Park, OH (www.asminternational.org).

In Situ Composites

Another approach to developing heat-resistant composites is based on the idea of directionally solidifying two-phase alloys (e.g., Ni- or Co-base eutectic superalloys) to create composites containing in situ grown reinforcement (e.g., carbide whiskers). The structural stability of the reinforcement in a temperature gradient, and low crystallization rates needed to design the microstructure are major considerations in these composite. Besides superalloys, Ni-Al intermetallics (Ni$_3$Al and NiAl) have been considered for the growth of in situ composites. Both Ni$_3$Al and NiAl have relatively high melting points (higher for NiAl: $T_{m, NiAl} = 1950°K$, $T_{m, Ni3Al} = 1638°K$), excellent oxidation resistance (further improved by alloying Zr or Hf), high strength and relatively low density (lower for NiAl, about 5860 kg/cu·m). NiAl has been used as a coating material in aircraft engines, but in a bulk polycrystalline form, the material is extremely brittle, and its strength above about 773 °K is very low. For example, the room-temperature yield strength (Y.S.) of NiAl is in the range of 120–300 MPa, but at 1300 °K, the Y.S. is only about 50 MPa. The ductility of polycrystalline NiAl is nearly zero below about 500 °K. However, as NiAl exists over a wide range of stoichiometries, it is possible to use suitable alloying additions to improve its mechanical properties. For example, Ti, Nb, Ta, Mn, Cr, Co, Hf, W, Zr, and B have been added to NiAl for increased resistance to compressive creep, with Nb and Ta being especially effective. Small additions of tungsten to NiAl refine the grain size and inhibit grain growth during hot consolidation, resulting in improvement in crack resistance. For toughness and strength gains, continuous fibers of W, Mo, and alumina have been incorporated in NiAl, and results with Mo and alumina have been promising (W tends to embrittle during processing, especially when solid-state diffusion bonding is used for composite fabrication). Thus, ductility of Ni-Al intermetallics has been increased by suitable alloying, and improvement in high-temperature creep resistance has been achieved mainly through fiber- and particulate reinforcements.

A variety of high-temperature in situ Ni-base composites consisting of dual-phase eutectic-type microstructures in which one of the phases is sandwiched between the other have been developed. Toughening in in situ composites occurs due to inhibition of crack nucleation in the secondary (β) phase, inhibition of crack growth due to plastic bridging by the ductile phase (crack bridging), and blunting of the crack by the ductile phase. Figure 6-26 shows how the dispersed second phase in directionally solidified off-eutectic pseudobinary alloys of NiAl(W) and NiAl(Cr) deflect a propagating crack, thus improving the toughness of NiAl. For in situ composites, the interphase interfaces are clean and mutually compatible because the constituent phases crystallize in situ rather than combined from separate sources. Directional solidification as well as deformation processing (extrusion and forging) are used to create the dual-phase microstructures having improved toughness, ductility, and creep strength. Directional solidification (DS) of pseudobinary NiAl-X (X could be Mo, W, Cr and Fe etc.) eutectic and off-eutectic alloys has been done to create a matrix containing an aligned β-phase sandwiched between a ductile second phase. The DS of NiAl-rich (hypoeutectic) compositions leads to aligned cells of γ-phase surrounded by the in situ composite (eutectic) microstructure created by the solidification of the intercellular eutectic liquid. The in situ composites of pseudobinary eutectic (and off-eutectic) compositions improve the room-temperature ductility and toughness, as well as the strength as in the case of directionally solidified ternary eutectics of NiAl with Mo, Cr, and Nb. The creep resistance and room temperature ductility of β-NiAl and Ni$_3$Al depends on the cell size, interlamellar spacing, and the second-phase morphology, all of which are controlled by varying the growth speed (e.g., the DS of Ni$_3$Al at 25 mm h^{-1} resulted in columnar-grained single-phase Ni$_3$Al with ~60% tensile ductility at room temperature. However, the

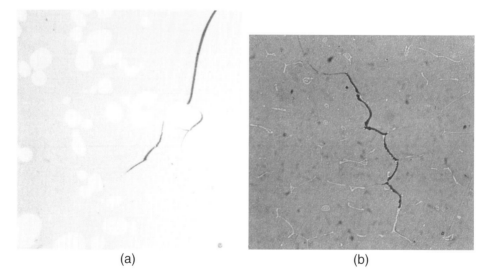

(a)   (b)

FIGURE 6-26 (a) Crack deflection by a W particle in a compression-tested NiAl(W) alloy. (b) Crack propagation through the intercellular NiAl-Cr eutectic colonies in an extruded NiAl(Cr) bar. (R. Tiwari, S. N. Tewari, R. Asthana and A. Garg, Materials Science & Engineering, A 192/193. 1995, 356–363).

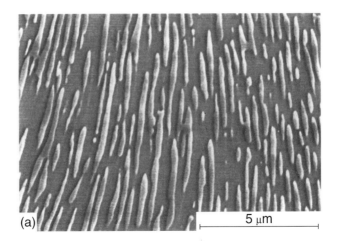

(a)

5 μm

FIGURE 6-27 *Microstructure of a directionally solidified in situ Ni-base composite.* (S. V. Raj and I. E. Locci, Intermetallics, 9, 2001, 217, Elsevier). Reprinted with permission from Elsevier. Photo Courtesy of S. V. Raj, NASA Glenn, Research Center, Cleveland, OH.

same material grown above $50 \, \mathrm{mm \, h^{-1}}$ yielded lower ductility than the columnar-grained Ni_3Al grown at the lower speed). Figure 6-27 shows the microstructure of an in situ Ni-base eutectic composite.

The in situ grown composites can be reinforced with ceramic fibers to further enhance the high-temperature strength and creep-resistance. An example of this is single crystal sapphire fiber-reinforced off-eutectic alloys of NiAl with Cr or W. This approach can permit realization of toughening of the matrix resulting from the dual-phase aligned microstructures together with strengthening of the matrix from ceramic fibers. The directionally solidified sapphire-NiAl(X) composites, where X is Cr or W, have high interfacial shear strength and promise of high-temperature strength as well as room-temperature toughness. These composites have been grown by various directional solidification techniques, most notably a containerless floating-zone directional solidification process (Figure 6-28). This technique allows growth of large, columnar NiAl grains with sapphire fibers residing in the eutectic colonies.

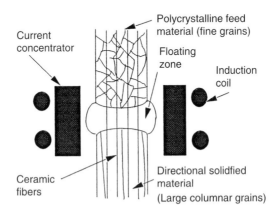

FIGURE 6-28 *Floating-zone directional solidification of a fiber-reinforced composite.*

Besides directional solidification, deformation processes such as extrusion, rolling, and forging, in conjunction with suitable heat treatment, are used for ductile-phase toughening of β-NiAl. For example, heat treatment of an extruded Ni-30Al-20Co alloy and a Ni-36Al alloy produces equiaxed β-NiAl grains containing a necklace of continuous γ' phase at the grain boundaries. This microstructure results in a slight improvement in the ductility (0.5%) over almost zero ductility of the alloy without the γ' phase. Likewise, forging or rolling of other Ni-Al-base alloys such as Ni-20Al-20Cr, Ni-25Al-18Fe, Ni-15Al-65Fe, and Ni-26Al-50Co alloys leads to a uniform distribution of equiaxed β and γ grains with about 2–6% ductility at room temperature. Extrusion of cast Ni-20Al-30Fe alloys produces a fine equiaxed β-phase distributed in a $\beta + \gamma'$ eutectic, and depending on the fineness of the γ' phase, this alloy exhibits a rather remarkable ductility (8 to 22%) at room temperature.

The first evidence of ductile-phase toughening of β-NiAl was provided in the directionally solidified Ni-34Fe-9.9Cr-18.2Al (atom%) alloy whose microstructure consisted of alternating lamellae of Ni-rich γ-phase and about 40% β-phase. This in situ composite exhibited 17% tensile elongation at room temperature with fracture occurring by cleavage of the β-phase that was sandwiched between ductile necked regions of γ-phase. Likewise, DS of Ni-30Al alloys led to aligned rodlike γ' (Ni_3Al) precipitates in a β-phase matrix, and DS of Ni-30Fe-20Al alloys yielded aligned β-NiAl and γ/γ' precipitates. In both these cases, about 10% ductility was achieved as compared to near-zero ductility of stoichiometric NiAl at room temperature. Similar improvements have been achieved through DS in NiAl(Mo), NiAl(Cr,Mo), and Ni-40Fe-18Al alloys. For example, DS of NiAl(Mo) and NiAl(Cr,Mo) alloys led to a marked improvement in fracture resistance; these alloys had a rodlike and a layered microstructure, respectively, with the layered structure yielding superior fracture resistance.

The DS of pseudobinary eutectic compositions of NiAl-containing elements such as Re, Cr, and Mo yields an aligned α-phase and improved strength. Similarly, off-eutectic compositions of NiAl with Cr are known to improve its creep strength. For example, rupture life of NiAl single crystals (<110> orientation) increases when chromium additions are made. A significant amount of strengthening in the NiAl(Cr) is provided by a high volume fraction of the fine α-Cr particles that precipitate during a solution–reprecipitation treatment. Dislocation networks form on the precipitate boundaries in both extruded and DS NiAl(Cr) alloys, and contribute to the strengthening.

Ni-Al intermetallics utilizing the ductile-phase toughening concepts and containing relatively large quantities of solutes possess relatively low melting point, high density, and inferior oxidation resistance, thus defeating the purpose for which these materials are considered attractive. To avoid such problems, directional solidification (or extrusion) of ternary NiAl alloys containing the smallest quantities of solutes (Cr and W), consistent with respective pseudobinary eutectics-phase diagrams, has been attempted. Aligned dual-phase microstructure of Ni-43Al-9.7Cr and Ni-48.3Al-1W alloys has led to improvements in both 0.2% compressive yield strength and ductility as compared to the single-phase NiAl, with tungsten alloying yielding the greatest strength improvements at room temperature. All such DS alloys exhibit greater than near-zero ductility of polycrystalline β-NiAl at room temperature, with Cr yielding the largest gain in fracture strain, followed by W. Figure 6-29 shows the microstructure of a NiAl(W) in situ composite in which W fiber pullout contributed to toughening.

Particulate and Fiber-Reinforced High-Temperature Alloys

For advanced gas turbine engine applications, particulate reinforcement and high-aspect ratio fibers also have been incorporated in Ni-base matrices especially Ni-base superalloys and Ni-base

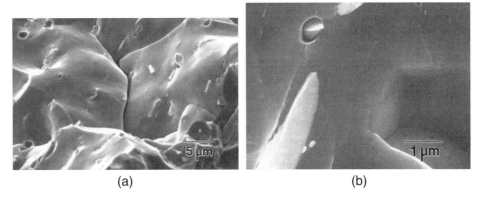

FIGURE 6-29 *(a) Scanning electron micrograph showing toughening of NiAl containing a small amount of W; toughening is provided by the deformation and pullout of the in situ–grown W fibers. (b) A higher magnification SEM view of micrograph of Figure 6-29a showing W fiber deformation and pullout.* (R. Tiwari, S. N. Tewari, R. Asthana and A. Garg, Journal of Materials Science, 30, 1995, 4861–4870).

intermetallics (e.g., NiAl and Ni$_3$Al). Particulate reinforcements include AlN, Al$_2$O$_3$, HfC, TiC, TiB$_2$, and B$_4$C; a uniform distribution of the particulate reinforcement is necessary to achieve strengthening. From the standpoint of fiber-matrix compatibility during fabrication and property enhancement, alumina-base ceramics appear to be an excellent choice as a reinforcement for Ni-base matrices, including superalloys and structural intemetallics such as NiAl and Ni$_3$Al. A variety of commercial alumina fibers (e.g., Du Pont's polycrystalline fiber FP and PRD166, Saphikon's sapphire, etc.) have been incorporated in NiAl and Ni$_3$Al matrices, and alumina fiber–reinforced NiAl has been considered for applications in high-speed civil transport.

In the case of seperalloy matrices, early studies of alumina reinforced Ni-base superalloys led to encouraging results in that single-crystal and polycrystalline alumina did not undergo significant chemical attack in composites formed by vacuum infiltration followed by prolonged heat treatment (up to 300 h at 1200 °C). A very thin interaction zone formed at the fiber–matrix interface regardless of the type of alumina and duration of heat treatment. Owing to the chemical stability of alumina in Ni-base matrices, sapphire-reinforced superalloy composites attracted attention over 30 years ago. Tables 6-4 and 6-5 summarize the work done on Ni-base composites over the last several decades.

Other common reinforcement materials such as SiC, Si$_3$N$_4$, and refractory metal fibers showed less promise than alumina because of their chemical and physical instability in Ni-base matrices. Studies on the interactions between Ni-alloys and SiC indicated strong fiber–matrix chemical interaction that led to depletion of Si content of the fiber and excess of Ni and Cr in the reaction zone. Ni-base superalloys such as Nimocast 713 C, Nimocast 75, and Nimocast 258 were used with continuous fibers and whiskers of SiC. Both hot-pressing and vacuum infiltration were employed for composite fabrication, and processing conditions were found to strongly influence the chemical interaction. For example, mold temperature and heat treatment of the composite affected the interaction; both preheated molds and heat treatment led to considerably greater chemical attack than molds at room temperature and composites in the as-cast conditions, respectively.

TABLE 6-4 Early Studies on Ni-Base Composites

Fiber	Matrix	Comments (Source)
Ni₃Al	Ni-2 to 10% Al	PM. Ni₃Al fibers form in situ. Oxidation resistant, high-temperature strength (Cabot Corp.)
Ni₃Ta	Ni-Ta alloy	DS (Euratom)
Ni-Cr-Al-Y	Ni alloy	PM. Sealing elements in turbines and compressors (Brunswick Co.)
Stainless steel, Mo, Ti, Nb	Ni, and Ni-base superalloys	Electroforming, PM (NASA, U.S. Army, Imperial Metals)
W, Mo	Ni	PM
Alumina	Ni and Ni alloys	Infiltration, electrodeposition plus hot-pressing
B, W, and B-W	Ni, Nimocast 713C, MARM322E	Electrodeposition, hot-pressing, casting
Fiberfrax	Ni-Sn	Vacuum infiltration
SiC	Ni-Cr	PM, hot extrusion, chemical reaction (Tohoku University)
Graphite (Ni-coated), C-coated with carbides	Ni aluminide	Hot-pressing of fiber and Ni base powders (United Technologies, Union Carbide)
Carbides of Nb, Ta, W	Ni-Co alloys	DS (General Electric)
NiBe fibers	Ni-Cr alloy	Normal solidification
Mo fibers, Cr(Mo) plates	NiAl, Ni₃Al	Normal solidification
Cr-rich fibers, Cr and Mo fibers, Ni₃Al fibers, Mo₂NiB₂ fibers	Ni-base alloys	Normal or directional solidification (United Technologies, General Electric)
Ni₃Al	Ni₃Ta	Normal solidification
Cr₃C₂ fibers	Ni alloy	Normal solidification
TaC, VC, or TaC+VC fibers	Ni base alloy	Normal solidification (GE)
Carbides of Ti, V, Cb, Hf, Zr	Ni-base alloy	Normal solidification

Another reinforcement that is unstable in Ni-alloys is Si_3N_4, which exhibits an even more severe degradation than SiC. Nimocast 258 alloy matrix shows complete dissolution of the whiskers in the vacuum cast Si_3N_4 whisker–reinforced composites. However, hot-pressing of Ni-coated Si_3N_4 whiskers has been found to lead to minor whisker damage, although subsequent heat treatment leads to complete dissolution of whiskers.

Fibers of refractory metals such as W, W-Rh, Mo, and Nb show different levels of reactivity with Ni alloys depending on the processing conditions. For example, vacuum-infiltrated W- and W-Rh/Nimocast 713 C composites show a small interaction zone after heat treatment for 500 h at 1000 °C, and a considerably thicker reaction zone at 1100 °C after 600 h. On the downside, the high density and high oxidizing tendency of W fibers has been a problem, although in a composite, a dense (pore-free) oxidation-resistant superalloy matrix (or a similar pore-free fiber coating) could protect the fibers from oxidation, except perhaps near the exposed fiber ends. On the positive side, however, W fibers strengthened by doping them with HfC, Rh, and C have excellent resistance to thermal fatigue and hydrogen embrittlement, and have been used in superalloy matrices.

Both Mo and Nb fibers severely react with Ni-base alloys, but vapor-deposited W and alumina coatings provide excellent protection against penetration and reaction by Ni and Cr even after

TABLE 6-5 Examples of Ni-Base Heat-Resistant Composites and Fabrication Method

System	Fabrication Route
$Ni_3Al-SiC_p$	VIM and stirring
$Ni_3Al-SiC_p$	VIM/infiltration
$NiAl_3-Ni-Al-Al_2O_3$	Reactive infiltration
$Ni_3Al-TiB_2$	Plasma-spraying
INCONEL 718-Al_2O_3 (single crystal)	Pressure infiltration
$Ni_3Al-Al_2O_3$ FP (IC-15, IC-218), NiAl-PRD-166	Pressure infiltration
NiAl-AlN	Extrusion or HIP
$Ni_3Al-TiC_p$	PM or stir-casting
Al_2O_3-NiAl	Powder injection-molding plus HIP
Al_2O_3-NiAl, ThO_2-NiAl, Y_2O_3-NiAl	Ball-milling
NiAl-AlN-Y_2O_3	Reaction milling in liquid N_2
TiB_2-NiAl	XD—in situ growth, milling, and HIP
NiAl-W, NiAl-Al_2O_3	Powder cloth
NiAl-Cr, NiAl-Mo, NiAlCr(Mo), NiFe-Al, Ni-Cr-Mo, NiAlVZr, NiAlCrNb, NiAlV, NiAlNb	Directional solidification (DS)
NiAl-Al_2O_{3p}	Sedimentation technology (FG)
NiAl/aligned or chopped FP (Al_2O_3)	Hot-pressing
NiAl-Al_2O_3, SiC-coated Al_2O_3	PM
NiAl-ZrO_{2p}	—
NiAl, Ni-35Al-20Fe, Haynes (A214)-Al_2O_{3f}	Powder cloth
NiAl-304SS-B_4C	Blending, extrusion, or forging
Sapphire-GS-32 or VKNA-4U, YAG-GS-32 or VKNA-4U, Sapphire-YAG-GS-32 or VKNA-4U	Internal crystallization method (ICM)
Sapphire-NiAl(*Me*), *Me* = W, Cr, or Yb, sapphire-Hastealloy	DS or cast or Powder-cloth (PC) preforms

Note: GS-32 and VKNA-4U are Russian Ni-base alloys.

prolonged heat treatment (300 h at 1100 °C). In the case of niobium fibers, chemical attack can be limited through oxide barrier layers prepared by controlled oxidation of Nb fibers themselves. Protective coatings and barrier layers are effective but must have long-term stability and physical integrity, both of which could be impaired by pores and microcracks that would cause increased metal penetration and chemical attack.

Fabrication of Metal-Matrix Composites

The three generic methods of manufacturing metal-matrix composites are the solid-state, the liquid-state, and the vapor-phase methods. Solid-state methods include the various powder metallurgy and diffusion-bonding techniques of combining the matrix and the reinforcement, liquid-state methods include infiltration, in situ growth, spray-forming and mixing, and vapor-phase methods include reactive (liquid–gas) spray-forming and vapor-phase deposition techniques. The choice of fabrication technique is usually dictated by the type of composite selected, production cost, process efficiency, and the quality of the product. Thus, high-performance fiber-reinforced composites are usually fabricated to near-net shapes using the squeeze-casting technique in order to achieve superior product quality and to minimize or eliminate the need for secondary processing. However, high pressures tend to damage the preform and often limit the ability to make thin-walled shapes. In contrast, particulate-reinforced composites can be synthesized using relatively less expensive mixing techniques but the quality of the product is partly compromised; secondary processing may be necessary to improve the product quality. The solid-state techniques are relatively expensive but are more suited for reactive systems in which higher processing temperatures of liquid-phase fabrication will impair the properties of the composite. The liquid-state methods are preferred from the standpoint of economy, which is dictated to a large extent by the low viscosity of liquid metals. The "in situ composites" are limited to systems where chemical reactions lead to formation of reinforcement within the matrix phase, but the reinforcing phase is usually monocrystalline, and the reinforcement-matrix interfaces are very clean and thermodynamically stable in these materials. Also, the in situ composite growth techniques may be more economical, as the reinforcing phase is grown during the process of forming the composite and not manufactured separately. The vapor-phase fabrication processes such as chemical vapor infiltration (CVI) are slow and may take up to several hundred hours for completion, but have net-shape potential and reduced processing temperatures. The fastest processes for composite fabrication are the self-propagating high-temperature syntheses (SHS) in which near-explosive rates of reaction take place; however, secondary processing such as infiltration or solid-state consolidation are usually required because of a relatively high defect (porosity) content in the material.

Solid-State Fabrication

A wide variety of solid-state composite fabrication processes have been developed for metal-matrix composites. The spray-forming techniques combine rapid solidification of a fine dispersion of liquid droplets followed by consolidation of solidified droplets (powders) to produce the final component shape. The incoming material in the form of powder, wire, or rod is melted using plasma, arc or combustion sources. The melted material is then sprayed into fine droplets that hit a target at very high velocities in molten or partially solidified state and deform into splats that cool rapidly. The high energy of impact assists in powder consolidation and densification, although a high degree of porosity usually remains in the as-deposited material. The particulate-reinforced composites are generally made by injecting the solid reinforcement particles directly into a spray of the matrix alloy, followed by deposition of the mixture onto a substrate. The co-deposition on a substrate of discrete droplets of molten matrix material and the discontinuous reinforcement yields a product that usually has some very attractive properties, because the solidification process is very rapid and particle engulfment by the solid is highly likely due to high solidification rates and large shear forces from impact. This results in a fine-grained matrix structure and a homogeneous distribution of particles. Figure 6-30 show the basic concept of spraying and particle engulfment in the solidified matrix. In another variation of the spray process, the reactive spray-forming by gas-liquid reactions, spray nozzle is extended to include a reaction zone where reactive gases are fed into the liquid stream. The gases react with the molten particles (droplets) to form the reinforcement. Carbide, oxide, silicide, and nitride reinforcement in metal matrices have been produced in this manner.

The processing times in spray-forming are on the order of a few milliseconds, and the reinforcement is in contact with the liquid matrix for no more than a few tens of milliseconds. Hence, even for small particles, chemical interactions between the reinforcement and the matrix will not alter the composition and properties of the interface interphase, although thin interfacial layers of reaction compounds (e.g., intermetallics) may form by diffusional interactions that may in fact improve the interfacial bond strength and hence the composite properties.

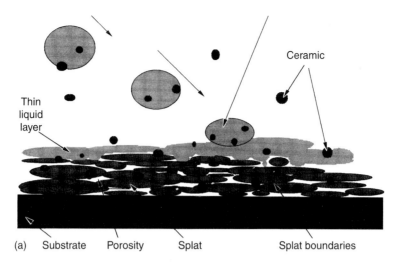

(a) Substrate   Porosity   Splat   Splat boundaries

FIGURE 6-30 *(a) Sketch showing co-deposition of a metallic matrix and ceramic particulates via a thermal spray process.*

For spray-forming of continuous fiber-reinforced composites, the fibers are wrapped around a mandrel with controlled interfiber spacing, and the matrix metal is sprayed onto the fibers. A composite monotape is thus obtained; bulk composites are formed by hot-pressing of composite monotapes. Fiber volume fraction and distribution is controlled by adjusting the fiber spacing and the number of fiber layers. With continuous ceramic fibers, however, the large thermal expansion mismatch between the ceramic fiber and the metal matrix, and the thermal shock and mechanical stresses arising from the initial exposure to plasma jet can result in displacement and fracture of fibers, especially because the brittle ceramic fibers are already bent to large curvatures when wrapped on a mandrel.

The spray-forming route of manufacturing composite materials offers several advantages over many other processes of fabricating composites. The spray-formed composite is relatively inexpensive, with a cost that is usually intermediate between melt stirring and powder metallurgy processes. The method produces a fine, nonsegregated structure with low levels ($<2\%$) of porosity due to fast solidification rates, and imparts good mechanical properties to the material. The cooling rates in spray-forming are typically in the range 10^3 to $10^6 \,°\mathrm{K.s}^{-1}$. The porosity levels in the spray-formed material depend on the thermal conditions, impact velocity, and spray density or mass flux. If the spray density (rate of deposition per unit area) is controlled carefully with respect to the rate of heat extraction from the deposit by the substrate, and by radiation and convection to the gas impinging on the deposit, then only very thin (typically 20 μm to 1000 μm) molten or partially molten layers of metal remain exposed on the deposit surface at any instant during processing. Under these conditions, the liquid or semiliquid alloy droplets in the spray meet a thin, liquid layer at the deposit surface; the liquids flow together and eliminate what would otherwise be splat boundaries. If, in contrast, the spray density is reduced with the same total cooling, the result will be liquid droplets impinging on a solid deposit, in which case the solidification rate will be higher but the structure will retain splat boundaries and will have a higher porosity. A large increase in the spray density with the same thermal surroundings leads to

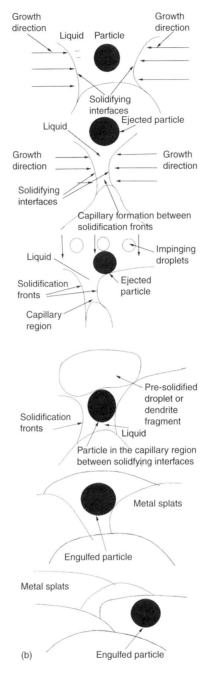

FIGURE 6-30 continued *(b) Possible mechanisms of particle engulfment by solidifying droplets on impact during thermal spray co-deposition.*

an accumulation of liquid on the surface, causing a reduction in the solidification rate, a marked increase in the grain size and an undesirable shredding of liquid layer by the atomizing gas.

In the case of advanced Ni- and Ti-base high-temperature matrices, special techniques have been developed to fabricate the composite materials. Thus, for fabricating intermetallic-matrix composites, foil-fiber-foil technique, tape-casting, thermal spray, high-energy ball-milling, reactive consolidation, injection-molding, powder-cloth, exothermic dispersion, and deformation processing (rolling, forging, extrusion) are used.

The foil-fiber-foil technique (Figure 6-31), used mainly for Ti-base intermetallics (and also for fiber-reinforced Al composites) requires consolidation of alternating layers of woven fiber mats and matrix foil. The composite prepregs are then consolidated into final component shape by hot pressing. Nickel-base intermetallics have rather poor ductility, and are not amenable to this technique. They are generally fabricated using powder-based techniques. For example, tape-casting uses liquid polymer binders to create a slurry of matrix powders that is combined with the fibers and hot-consolidated. The major drawback is interface- and matrix contamination from the binder residues. Thermal spray combines atomization and spraying of matrix droplets over fibers to form monotapes that are laid up and consolidated. Particulate TiB_2- and Al_2O_3 fiber-reinforced Ni_3Al and NiAl-matrix composites have been fabricated using this technique. Fiber damage from the mechanical impact of impinging droplets, powder surface contamination from oxides, and a somewhat inhomogeneous structure (unmelted grains, splat boundaries, voids) are some of the major limitations of spray forming techniques.

In reactive consolidation, fibers or particulates are combined with elemental or prealloyed matrix powders, followed by sintering and densification. Full-densification requires hot isostatic pressing (HIP). Figure 6-32 shows vacuum hot-pressed SiC-Ti and sapphire-NiAl composites. In SiC-Ti composites, SiC fibers were sputter-coated with a β-Ti alloy matrix and hot-pressed,

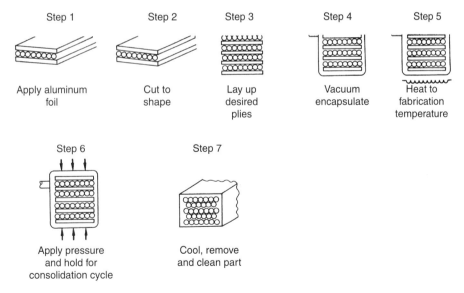

Step 1 — Apply aluminum foil

Step 2 — Cut to shape

Step 3 — Lay up desired plies

Step 4 — Vacuum encapsulate

Step 5 — Heat to fabrication temperature

Step 6 — Apply pressure and hold for consolidation cycle

Step 7 — Cool, remove and clean part

FIGURE 6-31 *Solid-state fabrication of fiber-reinforced Al composites by hot-pressing of Al foils and fibers.* (K. K. Chawla, Composite Materials: Science and Engineering, Springer, New York, NY, 1987, p. 104).

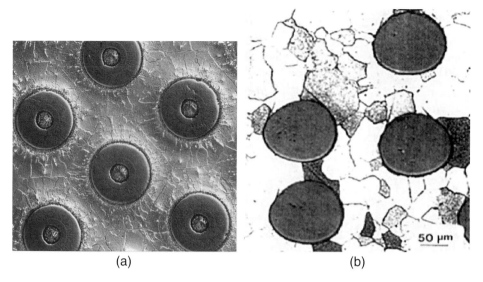

(a)   (b)

FIGURE 6-32 *(a) A titanium-matrix composite consisting of SiC fibers sputter-coated with a β-Ti alloy and vacuum hot-pressed.* (Photographed in polarized light by J. Baughman, NASA Langley Research Center, Langley, VA, reprinted from Advanced Materials and Processes, December 1998, pp. 19–24, ASM International, Materials Park, OH). *(b) Single-crystal sapphire fiber-reinforced NiAl matrix composite produced by plasma spray and hot-consolidation techniques.* (S. N. Tewari, R. Asthana, R. Tiwari, R. Bowman and J. Smith, Metall. Mater. Trans. 26A, 1995, 477–491).

and in sapphire-NiAl composites, plasma-sprayed NiAl powders were hot-consolidated with the fibers using NASA's patented "powder-cloth" process. The powder-cloth technique has also been used to fabricate NiAl-W and other advanced composites. The method first creates a cloth or sheet of atomized matrix powders by mixing them with Teflon binders and a solvent to a doughlike consistency. Fiber mats of specified thickness are produced by filament winding and application of a PMMA coating. The matrix cloth and fiber mats are then stacked in layers, and vacuum hot-pressed; the orientation of individual plies and stacking sequence provide flexibility in designing the composite properties. Complete binder removal is extremely important because small quantities of binder residues (carbon and fluorine) appreciably degrade the fiber–matrix interface strength and bulk composite properties.

During consolidation under high pressures, fiber damage can be a problem. This is overcome with powder injection-molding, although only the discontinuous reinforcement (particulate, chopped fibers, or whiskers) has been amenable to this approach. In powder injection-molding, up to 40% organic binders is combined with the matrix powders and the reinforcement, and the mixture is heated (above the glass transition of the binders), extruded through a die, and injected into a mold to create a "green" part. This is followed by debinding (removal of the major binders) using a catalyst or heat, and sintering. Both reactive consolidation and injection-molding have been used to make TiB_2- and Al_2O_3-reinforced NiAl matrix composites. The interface between final debinding and onset of sintering must be controlled, and binder formulations carefully chosen. This is because in the "green" part, powders are held together through binders rather than through a frictional bond (as in a press-and-sinter process), and complete removal of the binder prior to initiation of interparticle bond will result in failure.

High-energy ball milling is used for mechanical alloying of intermetallics and for fabrication of NiAl and Ni$_3$Al composites contaning Al$_2$O$_3$, Y$_2$O$_3$, ThO$_2$, and AlN particulates. The mechanically milled powder mixture is canned and extruded to create bulk composite specimens. Aluminum nitride (AlN) dispersions are produced by milling NiAl with finely divided Y$_2$O$_3$ in liquid nitrogen, a process often referred to as cryomilling or reaction milling because AlN reinforcement forms through a chemical reaction of Al with nitrogen during milling. One of the challenges in mechanical alloying is the need to separate the attrited powders from the unattrited powders to obtain a homogeneous material. Furthermore, canning materials must be ductile and nonreactive to the powder mixtures contained in the can.

Another powder-based technique is Martin-Marietta's exothermic dispersion (XD) process. It uses chemical reactions (e.g., between a gas bubbled through a melt) to create very fine discontinuous reinforcement. The XD feedstock (i.e., matrix and reinforcement mixture) is then milled and hot-pressed; the feedstock can be further processed via directional solidification to redesign the structure for properties (e.g., to preferentially orient the whiskers). The technique is likely to remain limited to discontinuous composites and to systems in which formation of the reinforcement via a chemical reaction is both thermodynamically and kinetically feasible.

Liquid-State Fabrication
Infiltration
Infiltration processes use either a vacuum or a positive pressure to drive a liquid matrix into the pores between the reinforcement. Infiltration is accomplished with the assistance of a vacuum, mechanical pressure using hydraulic rams or with inert gas pressure, although centrifugal infiltration and electromagnetic field (Lorentz force)-driven infiltration also have been done. Pressure overcomes the capillary forces opposing liquid ingress and viscous drag on fluid. In the case of metal composites, if pressurization is continued through solidification, grains are refined, porosity is eliminated, interface bonding is improved, and better feeding of shrinkage occurs. High pressures can, however, cause fiber fracture, preform distortion, and uneven distribution. Metal coatings such as Cu, Ni, and Ag on the reinforcement surface, and suitable alloying additions to the matrix metal (e.g., Mg in Al alloys) improve the wetting, reduce the threshold pressure, and increase the infiltration rate. Premixed suspensions of chopped fibers, whiskers, or particulates in the matrix melt can also be solidified under a large hydrostatic pressure, as in squeeze casting, with virtually no movement of the liquid metal. Figure 6-33 show the microstructures of various continuous and discontinuous metal-matrix composites produced using the infiltration technique; the reinforcing phase includes sapphire and carbon fibers, SiC platelet, and fly ash microsphere. Figure 6-34 presents the data on infiltration kinetics (length versus time) for several metal-matrix composites produced using pressure infiltration techniques; depending on the system and processing conditions, both linear and parabolic kinetics have been observed. Rate of pressurization, magnitude of pressure, and cooling rate are important variables. The solidification path in infiltration must be guided (e.g., through use of chills) to feed the shrinkage. Infiltration is aided by alloying additions that promote wetting; for example, in the case of Ni-base matrices, Ti and Y are found to be effective. NiAl-, Ni$_3$Al-, and superalloy (Hastealloy and Inconel 718)-based matrices have been reinforced with various alumina fibers (single-crystal sapphire, polycrystalline fiber FP, and PRD-166). Some fiber-to-fiber contact may be unavoidable at large fiber fractions. Such regions may be difficult to fill with liquid metal even under large pressures, and become crack initiation sites.

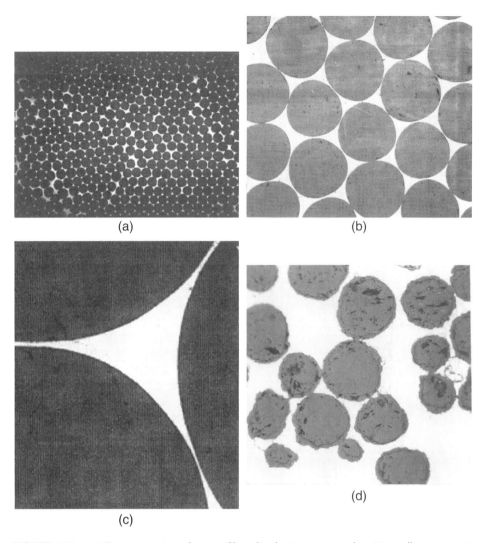

FIGURE 6-33   (a) Transverse view showing fiber distribution in a sapphire-Hastealloy composite produced by a pressure-infiltration technique (Hastealloy, 47.5Ni-21.5Cr-17.8Fe-8.3Mo-1.7Co-0.3Mn-0.4Si-0.2Al-0.1Cu-0.4Nb, in wt%). (b) A higher magnification view of a region of the sapphire-Hastealloy composite of Figure 6-33a. (c) Scanning electron micrograph showing inter-fiber channels containing the solidified Hastealloy matrix. (d) Degradation of single-crystal sapphire fibers in the Hastealloy matrix on prolonged contact at high temperatures. (R. Asthana, S. N. Tewari, and S. L. Draper, Metall. Mater. Trans., 29A, 1998, 1527–30).

A method to synthesize heat-resistant continuous fibers and their composites by pressure infiltration is the internal crystallization method (ICM), developed at the Russian Academy of Sciences. The method allows bundles of monocrystalline or eutectic fibers to actually crys-tallize in continuous channels of a matrix, usually Mo. The process is similar to the methods that are used to grow bulk oxide crystals and has been used to crystallize sapphire, mullite, yttrium-aluminum garnet (YAG), and alumina-YAG eutectic fibers. These fibers have been

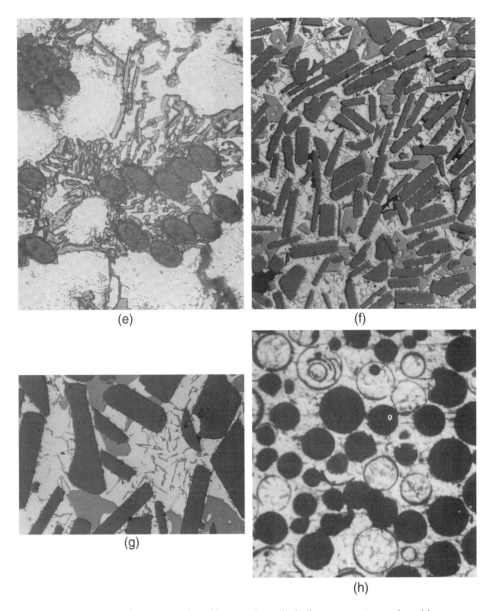

(e)

(f)

(g)

(h)

FIGURE 6-33 continued *(e) A carbon fiber–reinforced Al alloy composite produced by pressure infiltration showing fiber segregation in the eutectic colonies.* (R. Asthana and P. K. Rohatgi, J. Mater. Sci. Lett. 11, 1993, 442–445). *(f) Pressure-infiltrated single-crystal SiC platelet-reinforced Al alloy composite showing platelet distribution in the solidified matrix.* (R. Asthana and P. K. Rohatgi, Composites Manufacturing, 3(2), 1992, 119–123). *(g) A higher magnification view of the sample of Figure 6-33f showing secondary-phase precipitation on the platelet surface. (h) Photomicrograph of a pressure-infiltrated Al alloy composite containing fly ash particles.* (Photo Courtesy of P. K. Rohatgi, UW-Milwauker, WI).

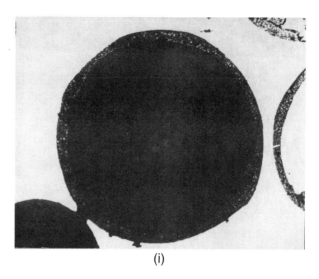

(i)

FIGURE 6-33 continued *(i) A higher magnification view of a Al-fly ash composite of Figure 6-33h showing the reaction layer around the fly ash particle* (Photo Courtesy of P. Rohatgi, UW-Milwauker, WI).

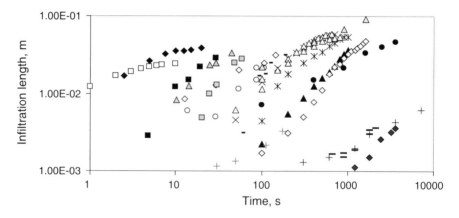

FIGURE 6-34 *Literature data on infiltration length as a function of time for various metal-matrix composites produced by melt infiltration of ceramics.* (R. Asthana, M. Singh and N. Sobczak, J. Korean, Ceramic Society, Nov. 2005).

incorporated in Ni-base intermetallics and superalloys using pressure-casting, yielding considerable gains in creep resistance. Figure 6-35 shows typical microstructures of pressure-cast Ni-base composites containing internally crystallized fibers of sapphire, yttrium aluminum garnet (YAG), and YAG-alumina eutectic. These composites offer excellent creep resistance at elevated temperatures.

Reactive Infiltration

In reactive infiltration, the reinforcing phase forms via chemical reactions during infiltration. The reaction is usually exothermic (e.g., infiltration of silicon in porous carbon and infiltration of

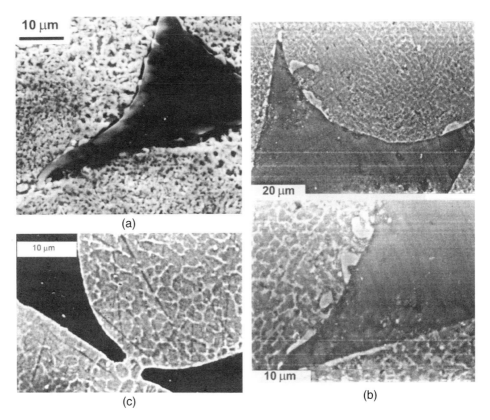

(a)

(b)

(c)

FIGURE 6-35 *(a) Scanning electron micrographs of a sapphire-YAG fiber-Ni–base superalloy matrix composite produced by pressure infiltration. The matrix composition is Ni-6Al-5Cr-1Mo-8.3W-4Ta-9Co-1.5Nb-4Re.* (S. T. Mileiko et al., Oxide-fibre/nickel-based matrix composites-part I: fabrication and microstructure, Composites Science and Technology, 62, 2002, 167–179, Elsevier). *(b) Scanning electron micrographs of the interface in a YAG fiber-Ni–base composite produced by pressure infiltration. The matrix composition is Ni-8.5Al-4.8Cr-2.2Mo-3.9W-4.4Co-1.1Ti).* (S. T. Mileiko et al., Oxide-fibre/nickel-based matrix composites-part I: fabrication and microstructure, Composites Science and Technology, 62, 2002, 167–179, Elsevier). *(c) Scanning electron micrographs of a sapphire-Ni–base composite produced by pressure infiltration. The matrix composition is Ni-8.5Al-4.8Cr-2.2Mo-3.9W-4.4Co-1.1Ti).* (S. T. Mileiko et al., Oxide-fibre/nickel-based matrix composites-part I: fabrication and microstructure, Composites Science and Technology, 62, 2002, 167–179, Elsevier). Reprinted with permission from Elsevier. Photo Courtesy of S. T. Mileiko, Solid State Physics Institute, Russian Academy of Sciences, Moscow.

aluminum in TiO_2, mullite $(3Al_2O_3.2SiO_2)$ and Ni-coated alumina) and may initiate explosively. The infiltration conditions can be controlled to achieve the desired level of conversion and structure. Clearly, the requirement of the in situ chemical formation of the reinforcement limits the choice of the systems to those in which large thermodynamic driving force exists for reaction at the processing temperatures, and the reaction kinetics are fast, to permit a substantial quantity of the product to be formed during a short period. The reactive infiltration of Si in porous C forms reacted SiC phase and unreacted C in a Si matrix. Preexisting SiC grains are employed as inert filler in the porous C preform to permit heterogeneous nucleation and bonding of SiC. The porous preform is made by pyrolyzing a high-char polymer precursor material, and some control on pore size, volume fraction, and morphology is possible. Reactive infiltration can produce two or more phases via chemical with a suitable choice of the precursor materials. For example, both TiB$_2$ and AlN phases form by reactive spontaneous infiltration of TiN, TiC$_x$N$_{1-x}$, and B powders with AlMg alloys. Low values of x yield a greater quantity of product phases. The infiltration rate in this system is controlled by the reaction: TiN + 2B + Al → AlN + TiB$_2$. This is a highly exothermic reaction, and the heat of reaction depends on the magnitude of x; low values of x lead to higher heat of reaction and faster infiltration rates whereas high values of x lead to low heat of reaction and slower infiltration. The highest infiltration rates are achieved when TiN, B, and AlMg are used as starting materials and the lowest rates are achieved when the starting materials are TiC$_{0.7}$N$_{0.3}$, B, and AlMg. The presence of Mg is necessary for self-infiltration in this system; infiltration is equally good under both inert and reducing atmospheres.

Models of reactive infiltration consider pore size evolution during infiltration, and invoke the concept of a variable preform permeability K, which varies with both time t and spatial coordinate l. The specific functional dependence of K on l and t is determined by both initial pore structure, and the microscopic reaction kinetics and deposition morphologies (e.g., formation of a continuous reaction layer versus discontinuous deposition of product phase). For the simpler case of a permeability that is a function of time alone, the problem has been solved in analytic form by Messner and Chiang.[2] If the pores of radius r shrink with either linear or parabolic time dependence, as may be appropriate for interface-controlled and diffusion-controlled reaction kinetics, respectively, then

$$r(t) = r_0 - kt$$

and

$$r(t) = r_0 - k_1 \sqrt{t},$$

where k and k_1 are linear and parabolic rate constants, respectively. The corresponding infiltration-distance-versus-time relationships are obtained from integration of an appropriate fluid-flow model. For the simplest capillary rise model that assumes parallel capillaries of identical diameter (Hagen-Poiseuille model), the permeability K is given from the relationship $K = Pr^2/8$, where P is the pore volume fraction. For more complex geometries, the concept of hydraulic radius is used; this changes only the numerical constant in the expression for permeability but not the parabolic dependence on pore radius (r^2 dependence). Assuming that geometry of the pore does not change during infiltration and reaction, the time-dependent permeability for the reactive infiltration is obtained from the temporal evolution of pore radius as a result of reaction. Thus, for the Hagen-Poiseuille model,

$$\frac{K(t)}{W_0} = \frac{r^4(t)}{8 r_0^2} = \beta r^4(t),$$

where W_0 is the initial porosity. Here $r(t)$ is given from linear (interface-controlled) or parabolic (diffusion-controlled) rate expressions. Using these expressions in Darcy's equation for flow

---

[2]Messner R.P. and Y.M. Chiang, Journal of the American Ceramic Society, 73(5), 1990, p. 1193.

through porous media and integrating the resulting equation yields the solutions for infiltration kinetics for interface-controlled and diffusion-controlled cases, respectively, as

$$L^2(t) = \frac{2\beta\Delta P}{\mu}[r_0^4 t - 2r_0^3 k t^2 + 2r_0^2 k^2 t^3 - r_0 k^3 t^4 + 0.2k^4 t^5] \tag{6-19}$$

$$L^2(t) = \frac{2\beta\Delta P}{\mu}[r_0^4 t - \frac{8}{3}r_0^3 k_1 t^{1.5} + 3r_0^2 k_1^2 t^2 - \frac{8}{5}r_0 k_1^3 t^{2.5} + 0.33k_1^4 t^3] \tag{6-20}$$

where β is a geometric factor given from $\beta = (8r_0)^2$. Of central importance to the final dimensions of the component to be processed is the final infiltration length L_f, at which reaction choking (analogous to freeze-choking during infiltration of a cold preform) would occur. This limiting length for interface-controlled and diffusion-controlled cases is given from

$$L_f = \sqrt{\frac{2\beta\Delta P r_0^5}{5\mu k}}$$

and

$$L_f = \sqrt{\frac{2\beta\Delta P r_0^6}{15\mu k_1^2}}.$$

These equations have been qualitatively verified in experiments on Si infiltration of porous C preforms. The model predicts the fast infiltration rates observed experimentally and shows that limiting infiltration depth is achieved literally in seconds. For fine-scaled carbon microstructure with fine pore size, the time to complete the reaction is only on the order of seconds for the appropriate values of reaction rate constants k and k_1.

The limiting infiltration depth, l_f, can be increased by applying external pressure, increasing the pore size, or decreasing the reaction rate. The process parameters for preform fabrication, which typically involves pyrolysis of polymer precursor material, can be judiciously chosen to exercise control on the pore size, pore volume fraction, and pore morphology. The size of the pore determines the concentration gradients of solutes released from chemical reactions. The solutes released at the reaction interface should be able to diffuse away into the liquid metal in the capillary. A high transient concentration of solutes can build up around the liquid meniscus and limit further rejection of the solute into the melt to a rate at which the solute released can diffuse away from the interface. A high concentration of solutes released from reactions will also tend to retard the reaction because of a reduced thermodynamic driving force for the forward reaction. If the pore size is increased, solute diffusion becomes easier, because the metal sink becomes larger in size and can accommodate a larger quantity of solute; as a result, a larger concentration gradient is created. If, however, chemical reaction rather than solute diffusion is the driving force controlling infiltration, then as the pore size is increased the effective surface area of the ceramic phase decreases; this in turn limits the supply of the reactant species in the solid; e.g., in the TiO_2-Al system, an increase in the pore size reduces the available oxygen. As a result, chemical reaction rather than solute transport becomes the rate controlling step for infiltration. The limiting infiltration depth can also be increased by decreasing the reaction rate constants. However, controlling the reaction rate constants may be more difficult than controlling the pore size because the rate constants are strongly temperature dependent and will be altered appreciably if the reaction is either exothermic or endothermic. That this is usually the case in reactive systems is well established. Attempts to increase the limiting infiltration length may be hindered by reaction choking, product spallation, and pore closure. The volume changes that accompany chemical reactions in many systems because of different specific volumes of reactant and product phases, give rise to stresses and spalling-off of the reaction products, and to distortion and crack formation in the preform.

Ceramic- and metal-matrix composites are also fabricated via spontaneous infiltration without an external pressure. Self-infiltration can occur if the melt and preform compositions, temperatures, and gas atmosphere are judiciously selected to yield good wetting for wicking of the metal through a porous preform. In most cases, a critical level of Mg is needed for self-infiltration. The composite forms by growth of multiphase layers at the reaction front that comprise of a top MgO layer, a dense layer of magnesium spinel, and below this layer, another layer of spinel with microchannels above an aluminum reservoir. The formation of microchannels in the material allows "wicking" of Al alloy to the free surface and continuation of composite growth. Figure 6-36 shows the different stages in the process of self-infiltration of a porous ceramic preform by molten Al alloy. The rate of composite growth is determined by the supply of metal at the reaction front, and the transport of oxygen through grain boundaries, microcracks, and pores

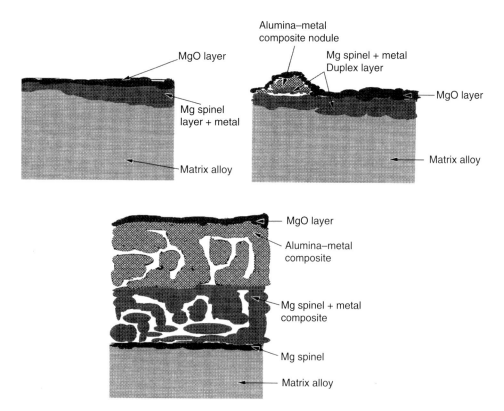

FIGURE 6-36 *Schematic of a self-infiltration process (Lanxide process) to grow metal-matrix composites.*

in the external spinel-MgO layer on the metal. If all the metal is used up, porous microstructures form. The self-infiltration process is slow and can take several hours to form the composite.

Fibers are made into preforms prior to infiltration. Continuous fibers may be woven into three-dimensional fabrics or ropes using weaving, braiding, and filament winding. The fiber volume fraction is controlled by controlling the fiber spacing and number of fiber layers. For stiff fibers, tapes, or laminates of fibers are made using a fugitive binder that burns off during composite fabrication. The amount of binder must be controlled; too little binder imparts little strength, whereas excessive binder could impede metal flow. Preforms of discontinuous reinforcement (e.g., short fibers) are made using the ceramic green bond-forming methods such as pressing, slip-casting, and injection-molding. Short fibers can be oriented by applying an electric field to fibers suspended in a nonconducting fluid between two electrodes. The electric field polarizes the reinforcement and orients it along the lines of force in the fluid medium. Preforms are made to net shape because of difficulty in machining ceramics. In hybrid preforms, particulates and whiskers are distributed between continuous fibers; this minimizes fiber-to-fiber contact and eliminates tiny cavities between touching fibers that are difficult to fill even under large pressures.

At large applied pressures during infiltration, deformation and fracture of preform may take place and result in an unevenly reinforced casting and difficulty in continued infiltration due to collapse of relatively large interstices. The capillary pressure begins to increase for uninfiltrated regions as the compressive deformation begins, because fiber volume fraction begins to increase. Preform compliance and deformation depend strongly on the binder content in the preform; binders are added to the fiber preform to impart strength and rigidity during handling and fabrication. The deformation behavior of melt-infiltrated preforms involves further complexity as opposed to the deformation behavior of a dry preform. In pressure-casting, the elastic compression of the preform during initial hydrodynamic pressurization will increase the fiber volume fraction in the compressed zone, thereby making infiltration progressively more difficult. The subsequent elastic recovery of the preform will take place at the conclusion of hydrodynamic pressurization (i.e., after the passage of the infiltration front) and equalization of the pressure field. The localized relaxation of the preform at the conclusion of pressurization will remain incomplete if appreciable fiber fracture has occurred or if solidification has begun.

Stirring Techniques

Mechanical stirring using impellers or electromagnetic fields creates a vortexing flow in the liquid or semisolid alloy, and aids in the transfer and dispersion of the reinforcement. The solid–liquid slurry is then solidified to obtain the composite. Figure 6-37 is a microstructure of a SiC-Al alloy composite produced by stir-casting technique, where the SiC particles are distributed within the grain structure of the Al alloy matrix. Stirring under vacuum or inert gas shroud minimizes gas dissolution, and various impeller designs have been developed to provide high-shear mixing with minimum surface agitation. Reactive additives such as Mg or Ce improve particle dispersion in metals. Impeller erosion and dissolution due to direct contact with the melt can be minimized by employing an electromagnetic (EM) field. Special magnetohydrodynamic (MHD) stirrers have been developed to obtain various combinations of rotating and traveling magnetic fields in order to generate complex flow patterns. Ultrasonic vibrations aid the infiltration of fibers by a liquid metal, and the dispersion of fine particulates in the melt. Metal is poured over fibers (or preform) packed in a mold, and vibration is transmitted via an acoustic probe. Alternatively, the fiber bundle is passed through the melt while the latter is being irradiated. The vibrations, transmitted into the melt via a titanium horn with a TiB_2-coated tip, generate large accelerations

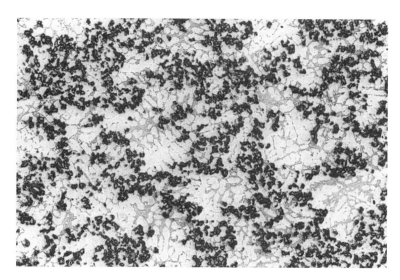

FIGURE 6-37 *Photomicrograph showing SiC particle distribution in a SiC-A359 Al alloy composite produced by a stir-casting technique.* (R. Asthana, J. Mater. Sci. Lett., 19, 2000, 2259–62).

(10^3g to 10^5g) pressures that exceed the threshold pressure for infiltration. Reactive wetting additives (e.g., Ti in alumina fibers) further lower the vibration energy needed for complete infiltration. Radiation also facilitates transfer and mixing of ceramic particulates, deflocculation of agglomerates, and refinement of the matrix structure. The propagation of the sonic wave causes "cavitation," which is the creation and catastrophic collapse (implosion) of tiny gas bubbles. This gives rise to enormous local accelerations that deflocculate the clusters by enabling the metal to penetrate the gas bridges between clustered particles. Vibration also maintains a uniform suspension, because particles are acoustically levitated and prevented from settling. When the vibration is continued through solidification, cavitation causes partial remelting of dendrites, grain refinement, and removal of gas porosity.

Many process variables must be controlled to achieve acceptable quality castings by the stir- casting techniques. For a fixed reinforcement content added to the melt via stirring, the process yield (i.e., % particles recovered in casting) varies with particle size. Generally, coarse particles exhibit better recovery than fine particles when properly heat treated; however, coarse particles lead to large-scale segregation due to settling or floating. In contrast, fine particles display considerable agglomeration and entrapped gases. Melt temperature is important, and an optimum temperature gives the highest yield. Very high temperatures cause severe erosion of the impeller, excessive chemical attack of the reinforcement, and increased gas-pick-up (although the wettability and fluidity may improve). In contrast, low temperatures hinder particle dispersion, and result in low fluidity, and low reinforcement content. High rates of particle addition to stirred metal cause severe agglomeration and low recovery; fast rates lead to covering of the melt by the powders, but do not appreciably increase the rate of assimilation in the melt.

For shape casting, the composite ingot must be remeltable, possess adequate fluidity, and the melt temperature and holding times should be low in order to inhibit reinforcement degradation. Excessive reinforcement degradation is prevented by limiting the temperature and holding time, or by reducing the activity of reactive solutes through composition adjustments (e.g., by increasing %Si in SiC-AlSi composite to inhibit the formation of brittle and hygroscopic Al_4C_3). The carbide dissolution reaction also influences the recyclability of composite scrap; if the scrap is free of Al_4C_3, it can be readily used. The amount of porosity in the scrap may also be a factor; due to increased viscosity, the composite does not vent air as readily as the low-viscosity base metal, and special gating and risering procedures are needed to minimize the entrained air.

Other consideration in remelting and shape casting are settling and fluidity. Heavy particles settle out and displace oxide inclusions and other solid impurities toward the top, which enables their removal. However, settling leads to an inhomogeneous particle distribution in the composite casting and forms particle-depleted and particle-enriched zones. Settling is retarded upon prolonged holding because of hindered settling and crowding of particles. Fine particles settle more slowly as compared to coarse particles, as do plate-like particles compared to blocky particles which experience lower fluid drag due to a smaller projected surface area. However, as already mentioned, fine particles severely agglomerate and cause defects. Another consideration is the fluidity which decreases due to the reinforcement, but may be adequate for casting purpose. However, chemical reactions with the melt may change the specific volume of the solid residing in the melt, which may adversely affect the fluidity (e.g., in Al-SiC slurries above 800 C, Al_4C_3 formation effectively increases the solid loading in the slurry, and makes shape casting difficult because of reduced fluidity). Industry has successfully explored use of virtually every casting method for particle-reinforced composites, such as green sand, bonded sand, permanent mold, plaster mold, investment, lost-foam, centrifugal casting, die casting and others. Wherever feasible and desirable, on-line determination of matrix chemistry and particle concentrations is performed during production to assure quality of the finished product.

The mixing techniques are technically and commercially viable at the industrial scale as demonstrated by their current use to make production quantities of SiC- and Al_2O_3-reinforced Al- and Mg-composites by the U.S. companies such as MC-21 Inc., Duralcan Co., and ECK Industries for applications in pistons, brake rotors, driveshaft tubes, bicycle frames and snow tire studs.

Low-Cost Composites by Casting

With the need to conserve precious and scarce materials and lower the material cost without sacrificing the essential engineering properties, it is necessary to identify and select fillers that preserve, and even enhance, the composite properties. Low-cost composites containing industrial wastes, industrial by-products, and inexpensive mineral fillers such as recycled glass, fly ash, waste mica, shell-char, rice husk ash, talc, and sand have been synthesized and evaluated for properties. Most of these fillers are lightweight materials so there is no penalty on component weight. An added benefit of using industrial wastes and byproducts as fillers is the elimination of disposal cost.

Several examples of waste byproducts can be identified, such as fly ash, mica, glass, foundry sand, slag etc. Fly ash is a waste by product of the electric utility industry that is generated in large quantities (e.g., 80 million tons annually in the U.S.) during the combustion of coal by thermal power plants. Such large quantities of fly ash present ecological problems associated with its storage and disposal. Currently, only 25 pct. of fly ash is used in construction and other applications; the remainder goes to landfill and surface impoundments. Attempts to recover Al, Fe and other materials from fly ash are technically feasible but not commercially successful. Another use has been to synthesize mullite ceramics from waste fly ash.

Structurally, fly ash is a heterogeneous mixture of two types of solid particles: solid precipitator ash (1–150 μm, density: 1.6–2.6 g/cc) and hollow cenosphere ash (10–250 μm, density: 0.4–0.6 g/cc). Chemically, oxides such as SiO_2 (33–65%), Al_2O_3, Fe_2O_3, CaO are its major constituents. Another type of ash that has been experimented with in making low-cost composites is volcanic ash (e.g., Shirasu balloon in Japan). Light-weight Al composites containing ash have better specific compression strength and wear resistance relative to un-reinforced Al. Pressure casting using hydraulic rams or pressurized gas, and stir-casting have been used to make these composites. Even with 5.8% waste ash in Al, the tensile strength and hardness are comparable to commercial Al-SiC composites containing 20 vol% SiC particulates. Similarly, the thermal fatigue resistance of the ash composite is superior to that of the commercial Al-SiC composite.

Mica powder produced by grinding and milling waste mica sheets has been dispersed in Al, Cu, Ag and Ni by casting or powder metallurgy to produce self-lubricating composite bearings. Improvements in wetting by Al with the use of Mg alloying and Cu-coatings on mica improves the yield. Mica is a solid lubricant with a layered crystal structure and is more resistant to oxidation than graphite. The wear resistance of Al-mica composites is satisfactory under both lubricated and dry wear conditions. In addition, mica improves the vibration damping capacity of Al, and on a weight basis, the specific damping capacity is better than that of flake graphite cast iron, which is an excellent damping material. There is, however, some loss of strength and ductility of Al due to mica additions. Secondary processing (e.g., hot extrusion) of the composite can offset strength loss without impairing the composite's wear resistance. Another attractive solid lubricant filler is talc ($3MgO.4SiO_2.H_2O$) that is chemically inert and belongs to a family of layered silica minerals in which one layer of $Mg(OH)_2$ (also called brucite) is sandwiched between two sheets of Si-O tetrahedron. The adjacent crystal planes of talc are held together by weak Van der Waals forces, which allow the planes to easily shear and slide under an external load. Talc is also the softest material on Mohs hardness scale, and can be readily ground to fine size. Unfortunately, talc in Al alloys decreases the tensile strength and hardness but the composite responds to heat treatment, and some improvement in the strength and hardness is possible. In contrast, talc significantly improves the wear resistance of Al.

Dispersion of waste silica sand from foundries in molten Al enhances the hardness and abrasion resistance of Al. A strong chemical reaction partially converts silica to alumina according: $3SiO_2(s) + 4Al(l) \rightarrow 3Si(s) + 2Al_2O_3$ (s). Mg as an alloying element reacts with silica according to $SiO_2(s) + 2Mg(l) \rightarrow Si(s) + 2MgO$ (s), and forms fine MgO crystals on silica while enriching the Al with Si, which improves the wear resistance and hardness (Si release in Al also compensates for the reduced fluidity of Al because of silica dispersion).

Rice husk and shell-char are agricultural wastes available in large quantities. Rice husk, which essentially protects the rice grain during growth, contains $\sim$94% SiO_2, which points to possible improvements in hardness and wear resistance of light metals such as Al. The silica-rich ash from burnt rice husk has been incorporated in Al by stirring and casting. In addition, Si-rich rice husk ash from controlled burning of rice husk has been used in cement. The addition of shell-char to Al increases the hardness and wear resistance but decreases the strength. Heat-resistant Si/SiC composites can be produced by infiltrating porous charcoal by Si, which will convert charcoal to SiC with improvements in strength of the Si matrix.

Glass is a common solid waste and can be combined with recycled metal to make the composite. It is a relatively low-energy, low-cost material with density close to Al, and may be added to Al for low-temperature use. The reclamation of waste glass involves color classification and crushing and melting with fluxes and raw materials. Colored glass obtained from different sources can be crushed and milled to achieve the fine particulate size required for effective compositing. Crushed and milled glass is preheated to remove volatile contaminants and combined with Al using either casting or powder metallurgy. Glass additions to Al improve the resistance to adhesive wear (possibly due to melting and smearing on the sliding surface), hardness, strength and vibration damping capacity of Al. Likewise, slag particles from solid waste dispersed in Al also reduce the wear of Al and increase the hardness and strength although the ductility decreases.

Rising energy costs favor recycling of metals for use in low-cost composites. Recycling of used beverage containers has become a large business; in 1992, 62.8 billion aluminum cans were collected which accounted for 68% of total produced in the U.S. Scrap metal is classified according to composition, cleanliness, and size. Impurities represent un-reclaimable mass and hinder metal recovery; therefore, some prior treatment is necessary. Because the reclaimed metal must compete with the primary metal in the user market, it must be of acceptable quality. This renders the entire reclamation enterprise capital- and energy-intensive, and rather elaborate. In view of this, judicious combination of fresh metal stock and recycled scrap may be used to make the low-cost composites. In the case of Al, the most difficult-to-process scrap goes through a reclamation plant before being melted in a reverberatory furnace (with or without salt additions, depending upon scrap cleanliness). The furnace produces a black dross containing salt, oxides, aluminum and impurities that is further processed to isolate the metal and the salt. Salt recovery plants crush the dross, dissolve the salt in water, and then evaporate the brine.

Other Liquid-Phase Techniques

Fine dispersions of thermodynamically stable refractory compounds can be produced in a matrix via reactions in the liquid–gas, liquid–solid, and liquid–liquid systems and salt mixtures. To grow TiC-Al composites, Ti and C are added to molten Al, and these react to form TiC particles. Titanium is prealloyed in Al whereas C is produced via decomposition of methane, which is bubbled through the AlTi alloy melt. Likewise, TiB_2 is incorporated in Al and Ti aluminides. To form TiB_2-Al composites, Ti, B, and Al in the form of either elemental powders or as prealloyed powders of Al-B and Al-Ti, are mixed in controlled amounts, and heated to melt Al in which Ti and B diffuse and chemically combine to form a fine dispersion of TiB_2 particles. In situ TiB_2-Al composites are also grown using reactions between mixed salts of Ti and B that react in molten Al to form TiB_2 dispersions. The dispersed phase is very fine, typically 0.1 to 3.0 μm, often monocrystalline, and uniformly distributed in the matrix. The process is flexible and permits growth of both hard and soft phases of various sizes and morphologies such as whiskers, particulates, and platelets. The growth rate is controlled by the interfacial reactions and the diffusion through the reacted layer. After an initial incubation, the early-stage growth is controlled by interfacial reactions, but once a product layer has formed the growth rate diminishes and is limited by the atomic diffusion through the reacted layer.

Many in situ composite-growth techniques can be grouped as "combustion synthesis" or self-propagating high-temperature synthesis (SHS). The process makes use of the ability of highly exothermic reactions in certain systems to be self-sustaining and energy efficient. The exothermic reaction to form a composite is initiated at the ignition temperature of the system. The reaction generates heat that is manifested in a maximum of combustion temperature that can exceed 3000 °K. In SHS, reactant powder mixture is placed in a special reactor under an argon blanket. The mixture is then ignited using laser beams or other high-energy sources to initiate the reaction. The reaction continues autocatalytically by the intense heat released from the reaction until all the feed material has combusted and reacted. High levels of open porosity ($<50\%$) remain in the final product, and full densification for high-performance applications almost always requires secondary processing such as hot consolidation and infiltration.

In some systems, a liquid phase forms via reaction and infiltrates all open porosity. For example, excess liquid Al formed in the following reaction infiltrates and densifies the porous ceramic phase,

$$3TiO_2 + 3C + (4 + x)Al = 3TiC + 2Al_2O_3 + xAl$$

The formation of a liquid phase during combustion synthesis is beneficial also because it increases the contact area between reactants and allows faster diffusion and reaction. Thus, in the case of combustion in the Ti-C system, ignition is controlled by the rate of surface reaction between Ti and C, which in turn is determined by the contact surface area between the two. The formation (or deliberate addition) of a low-melting-point phase (e.g., Al) in the Ti-C system further increases the surface area, and the rates of reaction and mass transfer. Besides infiltration, solid-state consolidation is also used to produce high-density composites.

Wettability and Bonding

Wetting of the fibers by molten metals is necessary for liquid-phase fabrication of composites. Prior surface treatment of the reinforcement (e.g., heat treatment, surface coatings) and alloying the matrix with a wetting promoter (e.g., Mg) are critical to high process yield. It may be difficult to predict the wettability of the filler by the matrix without conducting actual wettability tests. This is because many fillers are a complex solid solution of various constituents (e.g., oxides), and it is not always clear if the wettability of the filler will be a cumulative effect of all constituents acting in isolation, or of their mutual interactions, or if the wettability will be dominated by the major constituent. For example, the principal chemical constituent of glass, a filler for Al, is silica ($\sim$54–67%); other constituents include CaO, Al_2O_3, Na_2O, B_2O_3, BaO, K_2O and MgO. It is known that the contact angles of Al on SiO_2 at 700 C and 950 C are 150° (non-wetting) and 80° (wetting), respectively. On the other hand, CaO is poorly wet by Al whereas MgO is adequately wet by Al. It may be conjectured that BaO and B_2O_3 will probably be wet by Al because barium wets and impregnates Al, and in the Al-B system a contact angle of 33° is attained at 1100°C. The wetting of glass shall probably be dominated by silica and, therefore, poor wetting at low temperatures (700°C) is anticipated. However, as silica in glass is in solid solution, a more complex behavior may actually occur. Extensive reaction of silica with Al forms a complex interfacial reaction zone rich in alumina, and reaction-induced wettability might aid particulate dispersion in the metal. In any case, wetting improvements are almost always necessary, and reactive additions such as Mg are quite effective.

Ni-Base Composites

Experimental data on contact angles measured between various Ni-base alloys and oxides (Al_2O_3, ZrO_2), carbides (SiC, TiC, Cr_3C_2), borides (TiB_2, TiB_2Cr), and nitrides (TiN, AlN) are presented in Figure 6-38 as a function of test temperature, time, and alloying. These data cover both reactive and nonreactive systems. During the wettability test (sessile drop), alumina dissolves in the Ni droplet, and oxygen and Al released by the dissolution process diffuses into the drop and alters its chemistry. The dissolution process, however, does not appear to significantly lower the contact angle, and large obtuse contact angles are reported in $NiAl_2O_3$ (e.g., at 1773 °K, the values of θ are 128°, 133°, and 141° in vacuum, H_2, and He atmospheres, respectively). A progressive decrease in the Ni-Al_2O_3 bond strength (71–111 MPa) has been observed in joints made from liquid–solid contact with increasing contact angle. In contrast, the solid-state diffusion bonded Ni-Al_2O_3 couples (at 50 MPa pressure, 0.5 h, vacuum) had even

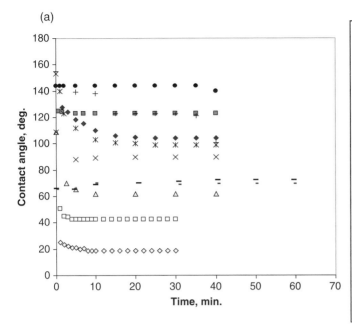

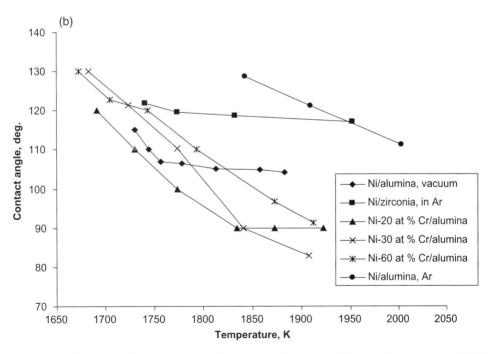

FIGURE 6-38 *(a) Contact angle versus time data for various oxide and carbide ceramics in contact with Ni-base matrices. (b) Contact angle of molten Ni and Ni-Cr alloys on alumina and zirconia as a function of temperature.*

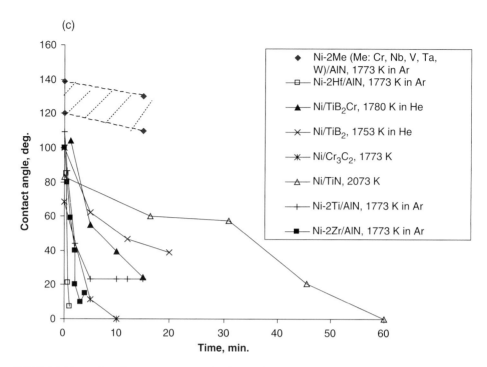

FIGURE 6-38 continued *(c) Contact angle of Ni and Ni-base alloys on various nitride, boride, and carbide substrates as a function of the time of contact.*

lower bond strength, typically 16–60 MPa when hot-pressing was done in the temperature range of 1223 °K–1323 °K. As a generalization, therefore, liquid-phase techniques appear to provide better bonding than do solid-state fabrication techniques.

Even though Al_2O_3 is poorly wet by pure Ni, alloying Ni with chromium improves the wettability in Ni-Al_2O_3. Small quantities of Cr are sufficient to achieve good wetting and bonding in Ni-Al_2O_3; contact angle is close to 90° on both sapphire and polycrystalline alumina at 10% Cr in Ni, and acute angles close to ~75° at 20% Cr. The solid–solid Ni-Al_2O_3 interfacial energy (σ_{ss}) is in the range 2.0–2.7 $J \cdot m^{-2}$, whereas the solid–liquid interfacial energy (σ_{sl}) is ~2.44 $J \cdot m^{-2}$. In contrast, the solid–liquid interfacial energy in Ni-Cr-alumina couples is in the range of 1.3–1.8 $J \cdot m^{-2}$ for Cr contents in the range 0.001% to 10%. From Young's equation, $\cos \theta = (\sigma_{sv} - \sigma_{ls})/\sigma_{lv}$, it is noted that a decrease in σ_{ls} by Cr alloying will decrease the contact angle, θ, consistent with the experimental observations. However, both the contact angle and solid–liquid interfacial energy reach their respective maxima at ~1% Cr, as does the bond strength. Thus, increasing the concentration of Cr in Ni at first strengthens, but later weakens the Ni/Al_2O_3 interface.

Properties
Interface Strength
In composites, a high-bond strength is required for effective load transfer to the fiber, which is the primary load-bearing constituent. Normally, a matrix-matrix bond is achieved as a result of chemical interactions at the interface. For toughening of brittle matrix composites, however,

weaker interfaces are preferred, to enable frictional sliding (pullout) of debonded fibers to contribute to toughening. Fiber pushout, fiber pullout, bend test, and other techniques have been used to characterize the strength of the fiber–matrix interface. In the fiber pushout test (Figure 6-39), interface strength is derived from the measurements of compressive load on a single fiber (residing in a matrix wafer) required to debond and displace the fiber. The test has been applied to SiC-Ti, glass-TiAl, sapphire-Nb, sapphire-NiAl, SiC-Si$_3$N$_4$ and other ceramic- and metal-based composites. In the conventional pushout test, thin (100- to 1000-μm) composite wafers, ground and polished to reveal the fiber ends and the surrounding matrix structure, are mounted on a support block containing grooves that permit unobstructed sliding of individual fibers through the wafer with the help of a flat-bottomed microindenter attached to an INSTRON frame. The load displacement data (and acoustic emission signal) are recorded, and interface strength τ^* is obtained from the measured debond load Q and surface area

$$\tau^* = \frac{Q}{\pi D t},$$

where τ is the shear stress, D is fiber diameter, and t is the wafer thickness. Figure 6-40 shows typical load displacement profiles from fiber pushout test on sapphire-reinforced NiAl matrix composites. The interface strength (debond stress) corresponds to the maximum shear stress on the curve (other stress transitions, such as τ_p, proportional stress, and τ_f, frictional stress provide information on the crack initiation and propagation behaviors). The load displacement (or stress-strain) data are examined in light of the fractographic observations of debonded interfaces to interpret the results and obtain an estimate of the debond stress. In addition, interrupted pushout tests have been done in conjunction with fractography to identify the failure mode and to determine the stress corresponding to the initiation of disband. Figure 6-41 depicts some results from the interrupted pushout test on sapphire-NiAl composites. The result are sensitive to matrix plasticity and test configuration (e.g., to t/h ratio, where t is disk thickness and h is support span). For fibers of variable cross-section and shape, it is difficult to load the fiber exactly at its center of gravity, leading to fiber fracture or unreliable strength measurements. A method developed at the Russian Academy of Sciences overcomes this problem by employing a metal ball indenter positioned between a flat-bottomed microindenter and the fiber; the ball indenter distributes the applied stress over a larger area.

Processing conditions, alloying, binders, and coatings all influence the measured strength. For example, in hot-pressed powder-cloth NiAl-alumina composites, polymethylmetacrylate (PMMA) is used as binder for fibers and polytetrafluoroethyline (Teflon) is used as a base binder for the NiAl matrix powders. The composites containing PMMA and Teflon binders have a purely frictional bond and low interface strength (50–150 MPa), whereas the composites free of binders had significantly higher strength (>280 MPa). Interface contamination from binder residues during hot-pressing interfered with interface development (carbon residue served as fracture initiation site). Processing temperature is another important variable that influences the bond strength. Figure 6-42 shows the effect of casting temperature on the interfacial bond strength (from fiber pushout test) of a sapphire-Ni-base superalloy matrix composites synthesized using pressure-casting; despite the scatter in the data, it appears that both high and low temperatures decrease the interface strength in this system.

Fiber coatings and alloying additions to the matrix are commonly employed in the manufacture of composites for better compatibility with the matrix. Frequently, however, differences

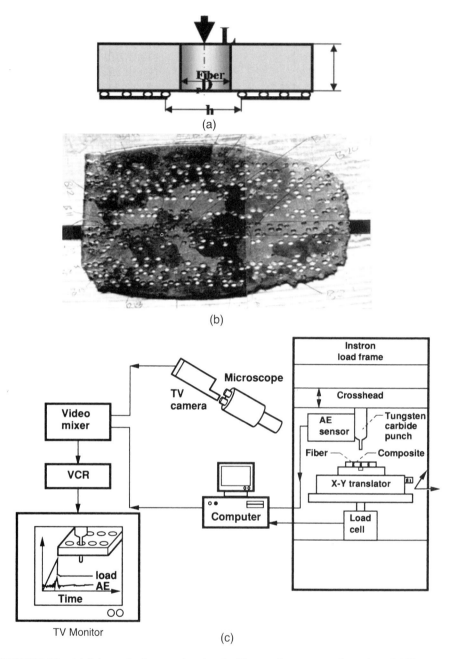

FIGURE 6-39 *(a) Schematic diagram showing the fiber-pushout test configuration. (b) Photograph showing a thin composite wafer mounted on a support block with a groove in it for fiber pushout test. The matrix grain structure and fiber ends are visible (tested fibers can be identified relative to their position at grain boundaries and grain interior). (c) Schematic diagram showing the fiber pushout test setup. (Courtesy of J. I. Eldridge, NASA Glenn Research Center, Cleveland, OH).*

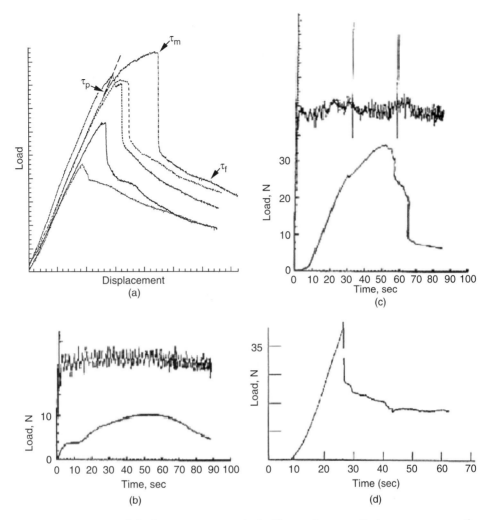

FIGURE 6-40 *(a) Load-displacement response in the fiber pushout test. Each curve represents the behavior of a single fiber. Three load transitions are identified for each fiber, τ_p (proportional load corresponding to the onset of deviation from the initial linear profile), τ_m (maximum debond load), and τ_f (frictional load for sliding of the debonded fiber). (b) Load-time and acoustic emission profiles for a sapphire fiber in a NiAl matrix during fiber pushout. Data are on a thin (<300 μm) composite wafer. Load transitions τ_p, τ_m, and τ_f are indistinct. (c) Load-time and acoustic emission profiles for a sapphire fiber in a NiAl matrix during fiber pushout. Data are on a thick (>300 μm) composite wafer. Load transitions τ_p, τ_m, and τ_f are distinct in thick wafers. (d) Load-time profile during fiber pushout for a sapphire fiber in a NiAl matrix containing Cr as an alloying element. Data are on a thick (>300 μm) composite wafer.* (R. Asthana, S. N. Tewari and R. Bowman, Metallurgical & Materials Transactions, 26A, 1995, 209–223).

in the wetting and bonding behaviors are noted when the same chemical element is used as an alloying addition to the matrix and as an interfacial fillm on the ceramic. For example, Ti thin films on Al_2O_3 improve the wetting and bonding in Al-Al_2O_3 couples, whereas Ti alloying of Al matrix does not. In contrast, Cr, either as an alloying addition to Ni or as a surface film on

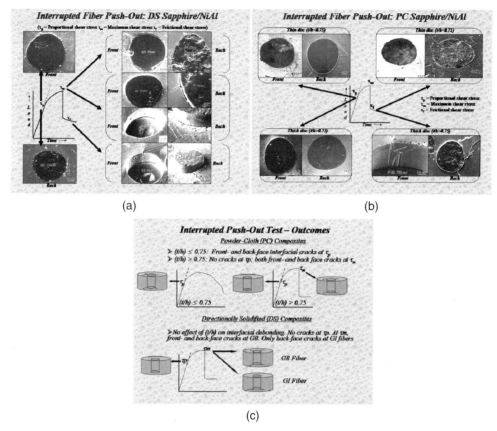

(a)

(b)

(c)

FIGURE 6-41 *(a) Fiber debond behavior as revealed in interrupted pushout test on directionally solidified sapphire-NiAl composites. Photographs show SEM views of the pushed fiber at the front and back faces of the composite wafer in the vicinity of the load transitions corresponding to $\tau_p, \tau_m,$ and τ_f. (b) Fiber debond behavior as revealed in interrupted pushout test on sapphire-NiAl composites fabricated using plasma spray and hot consolidation techniques. Photographs show SEM views of the pushed fiber at the front and back faces of the composite wafer in the vicinity of the load transitions corresponding to $\tau_p, \tau_m,$ and τ_f. (c) Schematic illustration of the fiber debond behavior as revealed in interrupted pushout test on sapphire-NiAl composites. The pushout behavior is shown at loads corresponding to $\tau_p, \tau_m,$ and τ_f for both thin ($t/h \leq 0.75$) and thick ($t/h > 0.75$) wafers, where t is the wafer thickness and h is the width of the groove on the block that supports the wafer during loading. GB and GI fibers are the fibers positioned at the NiAl matrix grain boundaries, and the fibers engulfed in the grain interior, respectively.*

both single-crystal and polycrystalline Al_2O_3, improves Ni-Al_2O_3 wetting and bond strength. Chromium has good thermoelastic compatibility with both Ni and Al_2O_3, and it can serve as a stress-absorbing compliant layer in Ni-base composites such as Ni-Al_2O_3 and $NiAl$-Al_2O_3. The mismatch of the coefficients of thermal expansion (CTE, α) between Ni (or NiAl) and Al_2O_3 is rather large ($\alpha_{Ni} = 13.3 \times 10^{-6}\,^\circ K, \alpha_{sapphire} = 9.5 \times 10^{-6}\,^\circ K^{-1}$ and $\alpha_{NiAl} = 15 \times 10^{-6}\,^\circ K^{-1}$]. This results in thermal clamping stresses (compressive on the fiber and tensile in the matrix), which can cause matrix cracking in the vicinity of the fiber (the compressive stresses from radial clamping over the fiber length also increase the fiber debond stress). Because chromium has

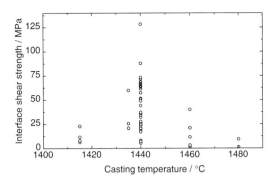

FIGURE 6-42 *Interfacial shear strength in a pressure-cast sapphire-Ni–base superalloy matrix composite as a function of the casting temperature (matrix composition is Ni-6Al-5Cr-1Mo-8.3W-4Ta-9Co-1.5Nb-4Re).* (S. T. Mileiko et al., Oxide-fibre/nickel-based matrix composites-part I: fabrication and microstructure, Composites Science and Technology, 62, 2002, 167–179, Elsevier). Reprinted with permission from Elsevier. Figure Courtesy of S. T. Mileiko, Solid State Physics Institute, Russian Academy of Sciences, Moscow.

a CTE intermediate between Ni (or NiAl) and alumina, an interlayer of Cr could serve as a stress-absorbing compliant layer ($\alpha_{Cr} = 6.2 \times 10^{-6}\,^{\circ}K^{-1}$).

Ti coatings on alumina fibers lead to better bonding (interface strength > 150 MPa) in pressure-cast alumina-reinforced NiAl and Ni_3Al composites as compared to composites made from uncoated fibers. Zr and Y in Ni_3Al improve the bond strength, although Ti as an alloying addition to NiAl and Ni_3Al does not have a beneficial effect and low bond strength results. During composite fabrication, Ti atoms diffuse into the fiber, causing alumina grain growth. Titanium–sputter-coated Al_2O_3 fibers in Al_2O_3-NiAl show radial matrix cracks, suggesting that Ti layers bond well to the fiber and the matrix. Likewise, Pt-coated alumina also exhibits better bonding with NiAl than uncoated fibers. In a manner similar to coatings, matrix alloying with Zr, Cr, Yb, or W leads to improved NiAl-Al_2O_3 bonding. Alloying additions such as Zr also strengthen the NiAl matrix (however, Zr raises the ductile-to-brittle-transition temperature). Alloying NiAl with Cr, W, and especially the rare earth ytterbium has been found to markedly enhance the fiber-matrix shear strength. With Cr and W, the interface strength is in excess of 155 MPa. However, alloying NiAl with Yb led to interface strength of 205 MPa with sapphire, in contrast to 50–150 MPa in the absence of such alloying. A high frictional shear stress (53 to 126 MPa) and large load undulations characterize the sliding of debonded fibers, as seen in the load displacement response of this composite during fiber pushout presented in Figure 6-43. Extensive fiber surface reconstruction occurs due to a strong interfacial reaction between sapphire and NiAl(Yb) matrix. The ytterbium oxide (Yb_2O_3) is more stable than Al_2O_3, and sapphire is reduced by ytterbium (the free energies of formation of Yb_2O_3 and Al_2O_3 are -1727.5 kJ·mol^{-1} and -1583.1 kJ·mol^{-1}, respectively). A complex multilayer interface consisting of Yb_2O_3 and $Yb_3Al_5O_{12}$ forms in sapphire-NiAl(Yb) composites containing only about 0.36 atom%. The microstructure of the fiber–matrix interface in this composite under different processing conditions is shown in Figure 6-44. The chemical makeup of the interface in this composite is schematically shown in Figure 6-45. Unlike Yb, Cr in NiAl only moderately attacks the sapphire; Cr preferentially precipitates on and bonds to the fibers without forming a visible reaction layer, as shown in Figure 6-46. Bonding in Ni-Cr-Al_2O_3 probably results from beneficial changes in the interfacial energy due to adsorption of Cr-O clusters on alumina, although

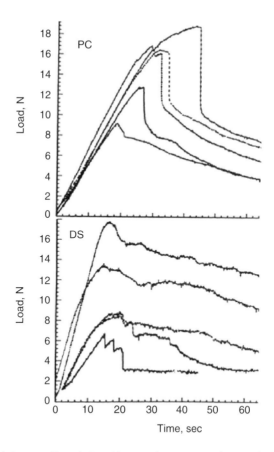

FIGURE 6-43 *Load-time profiles during fiber pushout in sapphire-NiAl(Yb) composites.* PC, *powder-cloth composite,* DS, *directionally solidified composite. The* DS *composite shows increased undulations in the frictional load beyond the (maximum) debond load because of increased fiber surface roughening caused by chemical reactions between the sapphire and the rare-earth ytterbium.* (R. Asthana and S. N. Tewari, Advanced Composite Materials, 9(4), 2000, 265–307).

thin interfacial films of Ni-spinel, $NiAl_2O_4$, have been noted in diffusion-bonded couples. The segregation of NiAl-Cr eutectic at the fiber–matrix interface results in a high frictional shear stress and formation of wear tracks due to matrix abrasion by hard Cr particles during sliding, as shown in the photomicrographs of Figure 6-47. In contrast, the unalloyed NiAl and W-alloyed NiAl matrices do not show interfacial segregation effects as shown in the photomicrographs of Figure 6-48. A high bond strength (>150 MPa) is also achieved in pressure-cast sapphire-INCONEL-718 superalloy composites in which the composite matrix contains reactive elements such as Cr.

The zirconia-toughened alumina (ZTA) fibers (PRD-166) in a pressure cast Ni-45Al-1Ti matrix degrade by grain coarsening and dissolution (Zr released from dissolution reduces and degrades the fibers). The PRD-166 fiber in a pressure-cast Ni-base alloy containing Cr, Ti, and Zr (Ni-16.8Al-7.9Cr-1.2Ti-0.5Zr-0.1B, in atom%) do not, however, form reaction products, although fiber dissolution leads to coarse ZrO_2 particles.

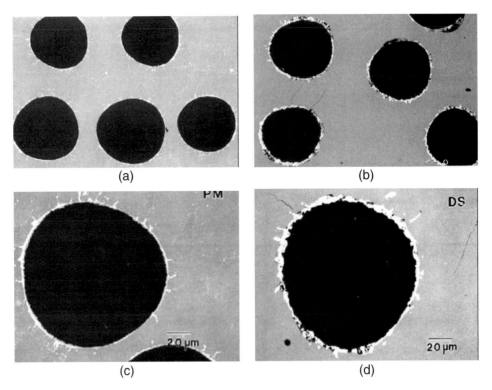

FIGURE 6-44 *(a) Powder-cloth sapphire-NiAl composite containing 0.36 atom% Yb. (b) Sapphire-NiAl(Yb) composite containing 0.19 atom% Yb directionally solidified from a powder-cloth feedstock. (c) Reaction zone around a sapphire fiber in the powder-cloth sapphire-NiAl(Yb) composite. (d) Reaction zone around a sapphire fiber in the directionally solidified sapphire-NiAl(Yb) composite. (S. N. Tewari, R. Asthana, R. Tiwari, R. Bowman and J. Smith, Metall. Mater. Trans., 26A, 1995, 477–491).*

The short interaction times in pressure-casting minimize the interfacial reaction. For example, interfaces free of reaction products have been observed in pressure-cast sapphire-Ni$_3$Al-based γ'/γ alloy containing Cr, Mo, W, and Co. However, fibers of yttrium aluminum garnet (YAG) and YAG-alumina eutectic show some reactivity with the Ni$_3$Al-based γ'/γ alloy, and new interphases. Likewise, YAG fibers in a Ni-base superalloy containing (Re, Ta, W, etc.) form interfacial carbides (pure alumina does not show any significant amount of the new interphase). The fibers strongly bond to the matrix, which suggests the positive influence of yttrium of the fiber in promoting the interfacial bonding.

Fiber Strength

Whereas the fiber–matrix interfacial reactions could potentially enhance the interface strength, they also tend to degrade the fiber. Notches and grooves on the reconstructed fiber surface act as strength-limiting flaws in brittle ceramic fibers. Thus, a high bond strength may be achieved at the expense of fiber quality. Figure 6-49 shows the extent of surface degradation in sapphire fibers extracted from various NiAl-base composites synthesized using different processing conditions. The manufacturing process must be designed to limit the fiber degradation while providing a

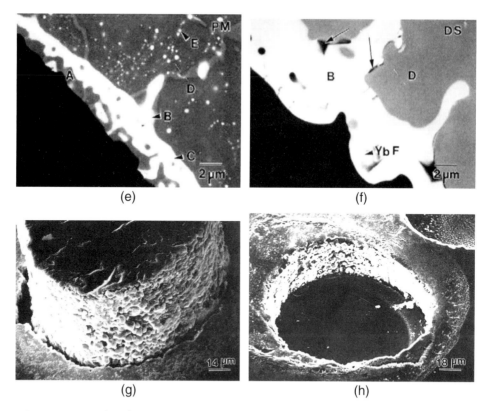

(e)   (f)

(g)   (h)

FIGURE 6-44 continued *(e) Interfacial region in a powder-cloth sapphire-NiAl(Yb) composite showing compound layer formation due to reactions. A, oxygen-rich NiAl; B, $Yb_3Al_5O_{12}$; C, Yb_2O_3; and D, NiAl. (f) Interfacial region in a directionally solidified sapphire-NiAl(Yb) composite showing compound layer formation due to reactions. A, oxygen-rich NiAl; B, $Yb_3Al_5O_{12}$; C, Yb_2O_3; and D, NiAl (the compound $Yb_3Al_5O_{12}$ is a spinel compound with formula $3Yb_2O_3.5Al_2O_3$) The fluoride compound (YbF) marked on the figure formed due to reaction with fluoride-based binders that were used in the powder-cloth feedstock specimen. (g) The back face of a pushed fiber in a wafer of sapphire-NiAl(Yb) composite showing extensive fiber surface reconstruction. (h) The front face of a pushed fiber in a wafer of sapphire-NiAl(Yb) composite showing extensive fiber surface reconstruction. (S. N. Tewari, R. Asthana, R. Tiwari, R. Bowman and J. Smith, Metall. Mater. Trans., 26A, 1995, 477–491).*

strong interfacial bond for load transfer. Usually it is necessary to characterize the strength of extracted fibers before process control and prediction are possible. This is also necessary because the behavior of separate fibers extracted from the matrix may be different from the behavior of the fibers residing in the matrix. Figure 3.43(a) on page 230 of chapter 3 displayed the measurements on the strength of sapphire fibers extracted from a pressure cast superalloy matrix composite. The matrix (Hastelloy) composition is 47.5Ni, 21.5Cr, 17.8Fe, 8.3Mo, 1.7Co, 0.3 Mn, 0.4Si, 0.2Al, 0.1Cu, 0.4Nb, 0.06Ti, 0.08C, and 0.02 O, in wt%. The fiber strength data are represented using the two-parameter Weibull function

$$F(\sigma) = 1 - \exp\left(\frac{\sigma}{\sigma_o}\right)^{\beta} \qquad (6\text{-}21)$$

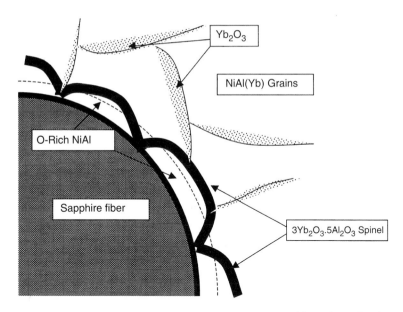

FIGURE 6-45 *Schematic representation of the interfacial compound layer formed in the sapphire-NiAl(Yb) composite.*

where $F(\sigma)$ is the probability of failure of a fiber at a stress of σ, α is the scale parameter, and β is the Weibull modulus. Figure 3.43(a) also displays the strength distribution of virgin (as-received) fibers and fibers extracted from a powder-cloth sapphire-NiAl composite. For the sapphire-Hastealloy composite, the values of α and β are 5.561×10^{-6} and 1.7479 respectively, and the mean strength, σ^* standard deviation, s, and the coefficient of variation, CV, are 904 MPa, 562 Mpa, and 62.2%, respectively. There is on average a 66% loss in fiber strength after casting; the as-received fiber strength is about 2.7 GPa.

Similar degradation of fiber strength due to extensive chemical attack has been observed in polycrystalline and single-crystal alumina fibers in pressure-cast as well as hot-pressed Fe- and Ni- base intermetallics. The strength loss in single-crystal sapphire in "powder-cloth" FeCrAlY, FeCrAl, Cr, FeAl, and NiAl matrix composites is in the range of 45 to 60%, and strength loss in pressure cast Ni_3Al-Ti matrix composites is about 67%. The initial fiber strength of 2.5–3.0 GPa drops to about 1.2–1.8 GPa after elevated-temperature contact between sapphire and FeCrAlY, FeCrAl, Cr, FeAl, and NiAl. In the case of FeCrAlY alloys, Y and Cr form brittle reaction products that weaken the fiber. The strength loss also seems to depend on the atmosphere because of the latter's influence on chemical reactions. For example, the strength of alumina fibers in contact with FeCrAlY alloys degrades under Argon but not under Ar+H_2 atmosphere. Much of the deleterious effects of chemical attack of the fibers can be overcome through judicious selection of matrix alloy chemistry, use of reaction barrier layers, and control of process parameters (pressure and temperature).

Stiffness, Strength, and Ductility

Generally, the tensile and compressive strengths, elastic modulus, and flexural strength of both continuous and discontinuous composites increase, whereas the ductility decreases with increasing volume fraction of the reinforcing phase (soft particulates such as graphite and mica decrease

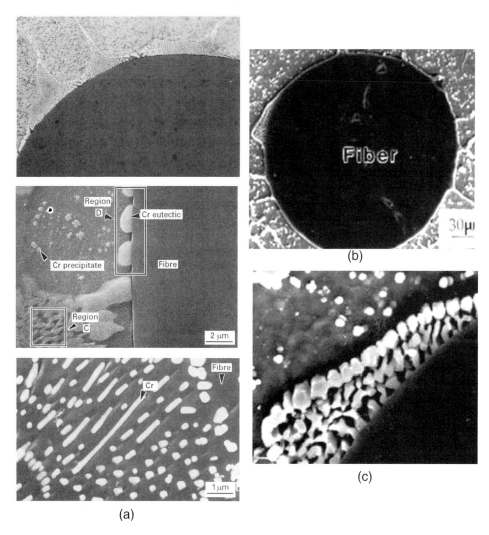

FIGURE 6-46 *(a) Photomicrographs showing the interfacial segregation of chromium in a directionally solidified sapphire-NiAl(Cr) composite. The matrix composition is 60.2Ni-28.2Al-11.4Cr (in wt%) The bottom figure shows eutectic chromium bonded to the sapphire after chemically dissolving the matrix. (b) Interfacial segregation of Cr in a vacuum-induction-melted and chill-cast sapphire-NiAl(Cr) composite. (c) A higher magnification view of the interfacial region in the sample of Figure 6-46b. (R. Asthana, R. Tiwari, and S. N. Tewari, Metall. Mater. Trans., 26A, 1995, 2175–84).*

rather than increase the strength). Table 6-6 lists the elastic modulus and strength of some metal-matrix composites (data on polymer composites is also shown for comparison). The composite's strength is strongly structure sensitive; structural defects, brittle reaction layers, stress concentration, dislocation pileups, grain structure, and texture influence the strength. The strength also depends on the spacing between and the size and distribution of the reinforcement; inhomogeneous distribution and clustering (with fiber-to-fiber contact) lower the strength. Clustering

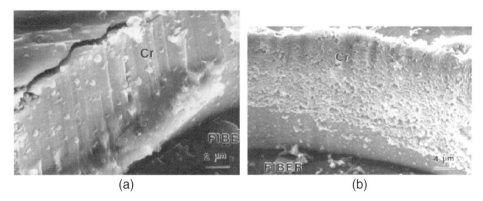

(a)　　　　　　　　　　　　　　(b)

FIGURE 6-47 *(a) Wear tracks and primary Cr particles in the region of displaced fiber in a directionally solidified sapphire-NiAl(Cr) composite after the fiber push-out test. This is a front-face SEM view. (b) Eutectic Cr particles decorating the interfacial region around a displaced sapphire fiber in a directionally solidified sapphire-NiAl(Cr) composite after the fiber pushout test. This is a front-face SEM view.* (R. Asthana, R. Tiwari, and S. N. Tewari, Metall. Mater. Trans., 26A, 1995, 2175–84).

TABLE 6-6　Mechanical Properties of Selected Metal and Polymer Composites

Composite	E, GPa	Strength, MPa
B-Al (50% fiber)	$210^{\parallel}$	$1500^{\parallel}$
	$150^{\perp}$	$140^{\perp}$
SiC-Al (50% fiber)	$310^{\parallel}$	$250^{\parallel}$
	—	$105^{\perp}$
Fiber FP-Al-Li (60% fiber)	$262^{\parallel}$	$690^{\parallel}$
	$152^{\perp}$	$172–207^{\perp}$
C-Al (30% fiber)	160	690
E-Glass-epoxy (60% fiber)	$45^{\parallel}$	$1020^{\parallel}$
	$12^{\perp}$	$40^{\perp}$
C-Epoxy (60% HS fiber)	$145^{\parallel}$	$1240^{\parallel}$
	$10^{\perp}$	$41^{\perp}$
Aramid (Kevlar 49)-epoxy (60% fiber)	$76^{\parallel}$	$1380^{\parallel}$
	$5.5^{\perp}$	$30^{\perp}$

Note: $^{\parallel}$, value parallel to the fiber axis.
　　　$^{\perp}$, value perpendicular to the fiber axis.

becomes severe at fine size, especially at large-volume fractions. The composite's strength is anisotropic, and with continuous fibers and high-aspect ratio whiskers, chopped fibers, or platelets, strength decreases, with increasing misorientation relative to the stress axis. Preform architecture also affects the composite's strength; with two-dimensional, planar random orientation of short fibers, the strength is better parallel to fiber array than normal to it (even though the reinforcement is randomly oriented in a plane). However, even in a direction normal to fiber axis, composite's strength exceeds the matrix strength. In contrast, ductility markedly decreases with increasing reinforcement loading, especially at small size.

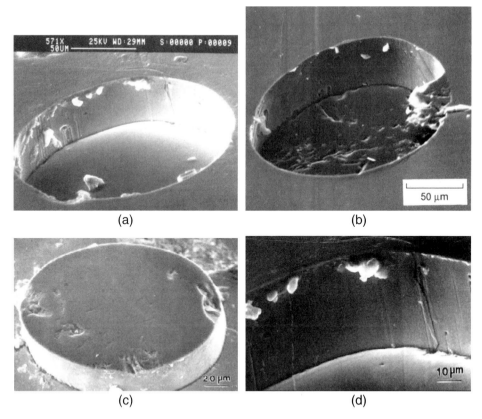

(a)   (b)

(c)   (d)

FIGURE 6-48 *(a) Front-face SEM view of a pushed fiber in a wafer of a directionally solidified sapphire-NiAl(W) composite specimen. No preferential phase segregation or reaction layers are seen (the interfacial deposit visible in the figure is an artifact). The matrix composition is 67.4Ni-31.4Al-1.5W (in wt%). (b) Front-face SEM view of a pushed fiber in a wafer of directionally solidified sapphire-NiAl composite specimen (equiatomic fractions of Ni and Al). No preferential-phase segregation or reaction layers are seen. (c) Back-face SEM view of a pushed fiber in a directionally solidified sapphire-NiAl(W) composite wafer. (d) A higher magnification view of the interfacial region of sapphire-NiAl(W) composite specimen of Figure 6-48a showing a clean fiber–matrix interface.* (R. Asthana, R. Tiwari and S. N. Tewari, Metall. Mater. Trans., 26 A, 1995, 2175–84). Figure (b) is from S. N. Tewari, R. Asthana and R. D. Noebe, Metall. Mater. Trans., 24A, 1993, 2119–25.

The composite modulus is anistropic but relatively insensitive to the reinforcement distribution and structural defects; the modulus is less sensitive to fiber packing and clustering than strength, but is different along the fiber axis and transverse to it. Figure 6-50 shows the axial (parallel to fiber) and transverse elastic modulus and ultimate tensile strength of an alumina fiber (FP)-reinforced Al-Li composite as a function of fiber volume fraction. The longitudinal modulus of fiber-reinforced metals agrees well with the rule-of-mixture (ROM) values, but the transverse modulus is generally lower than ROM predictions. For the discontinuous reinforcement, modulus is independent of the distribution. Particulate composites also show an increase in the modulus with increasing additions of the reinforcement; the increase is, however, smaller than ROM predictions, which, in this case, are an upper bound.

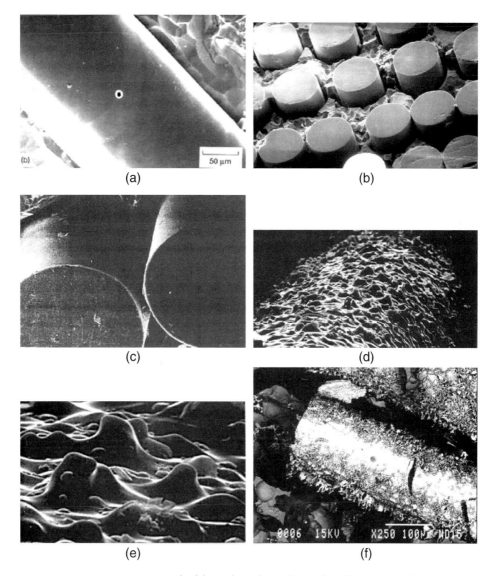

FIGURE 6-49 *(a) SEM micrograph of the surface of a single sapphire fiber extracted from a vacuum-induction-melted and chill-cast NiAl matrix. (b) Etched out fibers in a powder-cloth sapphire-NiAl composite. (c) A higher magnification view of the sample of Figure 6-49b. (d) SEM view of a sapphire fiber extracted from a powder-cloth sapphire-NiAl(Yb) composite specimen showing extensive roughening and degradation of the fiber. (e) A higher magnification view of the sample of Figure 6-49d. (f) SEM view of a sapphire fiber extracted from a directionally solidified sapphire-NiAl(Yb) composite.* (R. Asthana and S. N. Tewari, Advanced Composite Materials, 9(4), 2000, 265–307).

A reinforcement with greater strength and stiffness than the matrix and well bonded to it will carry higher load than the matrix. With perfectly bonded interfaces, ROM applies to the volume fraction-dependence of strength, i.e., $\sigma_c = V_f \sigma_f + V_m \sigma_m$, where σ_m and σ_f are the strength of the matrix and reinforcement, and V_f and V_m are the fiber- and matrix volume fractions, respectively.

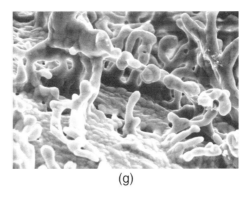

(g)

FIGURE 6-49 continued *(g) A higher magnification view of the fiber of Figure 6-49f, showing extensive fiber degradation due to processing.* (R. Asthana and S. N. Tewari, Advanced Composite Materials, 9(4), 2000, 265–307).

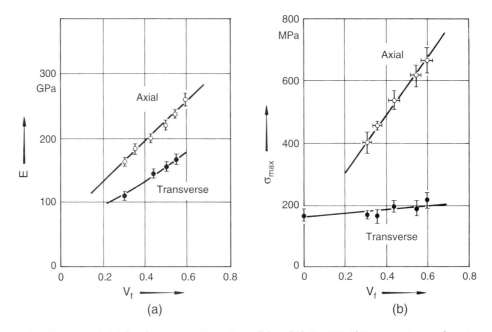

(a)   (b)

FIGURE 6-50 *(a) Axial and transverse Young's modulus of Al_2O_3 (FP)-Al-Li composite as a function of fiber volume fraction, and (b) axial and transverse ultimate tensile strength of the composite as a function of the fiber volume fraction.* (A. R. Champion, W. H. Krueger, H. S. Hartman and A. K. Dhingra in Proceedings of the 1978 International Conference on Composite Materials [ICCM-2], TMS-AIME, New York, 1978, p. 883). Reprinted with permission from The Minerals, Metals & Materials Society, Warrendale, PA (www.tms.org).

The rule of mixtures presupposes that the length of the fiber is above a minimum value to prevent pullout from the matrix under tensile loading. If the fiber is below a critical length (Equation 6-1), fiber pullout will precede fiber fracture, and no strengthening will be achieved because the load will not be effectively transferred from the matrix to the fiber. Strength predictions for particulate composites are more involved because the dislocations due to CTE mismatch contribute to the strength. Sharp ends lead to severe stress concentration and localized plastic flow at stresses well below the matrix yield stress. Large plastic strain causes void nucleation and growth, leading to premature failure. High-aspect-ratiodiscontinuous reinforcement provides greater improvement in the composite's modulus and strength at high-volume fractions; for example, platelets and whiskers are better than spheroidal particles. However, whiskers and platelets should be aligned along the direction of applied stress, because the effective composite modulus and strength fall rapidly with increasing misorientation. High-aspect-ratio platelets are less expensive than whiskers and pose fewer health hazards. However, the high aspect ratio of platelets needs to be achieved at relatively fine size if particle fracture during composite fabrication is to be avoided. A fine size usually leads to a greater tendency for clustering, and mechanical working may be necessary to decluster the platelets and preferentially orient them along the deformation axis. High-aspect-ratio platelets yield better gains in strength, modulus, and ductility in Al composites than low-aspect-ratio particulate reinforcement. Thermomechanical treatments such as forging, extrusion, and rolling improve the reinforcement distribution, decluster agglomerates, eliminate porosity, improve the interfacial bonding, and yield a fine matrix subgrain structure. These structural modifications markedly improve the strength and ductility of the composite.

In composites based on age-hardenable matrices, growth of second-phase precipitates is enhanced by the short-circuit diffusion paths that result from increased dislocation density from CTE mismatch, leading to faster hardening. Quenching at the conclusion of solutionizing forms a zone of plastically deformed matrix with a high dislocation density around each fiber or particle (prior mechanical deformation also accelerates the aging kinetics). The dislocations serve as sites for heterogeneous nucleation. The time to peak hardening decreases with increasing reinforcement volume fraction in well-bonded composites. However, interfacial reactions often lead to solute depletion and impair the aging response; in addition, thick reacted layers tend to impair the strength.

Fatigue and Fracture Toughness

Continuous fibers, short fibers, whiskers, platelets, and particulates all enhance the composite's fatigue resistance. Fatigue properties are anisotropic and fiber orientation is important; improvement in the fatigue strength is greatest when the fibers are directionally aligned parallel to the stress axis. Fatigue crack propagation is affected by crack closure, crack bridging, and crack deflection, all of which are modulated by the dispersed particles or fibers. The fatigue life is very sensitive to processing conditions, and process improvements (e.g., in particulate composites, the control of particle distribution, particle clustering, slag and dross inclusions, and gas and shrinkage porosities) increase the fatigue strength. A high-particulate-volume fraction usually leads to a higher fatigue strength. This is because of lower elastic and plastic strains of composites due to increased modulus and work-hardening. For a fixed volume fraction, the fatigue strength usually increases as the particle size decreases. The size effect is related to increased propensity for cracking of large particles as well as a decrease in the slip distance at fine particle sizes. As particles serve as barriers to slip, a small interparticle distance will reduce the slip distance. Large particles fracture more easily than fine ones during fatigue testing and lower the fatigue strength by causing premature crack initiation. Particle fracture also leads to cycling softening of the matrix, and therefore an increase in the plastic strain. Fatigue cracks usually originate at defect sites such as particle clusters and microvoids in the matrix or at interfaces. Also because particle cracking lowers the fatigue strength, one method of enhancing the fatigue strength in a given system is to use single-crystal particulates with high fracture strength.

Both strength and toughness of composites depend on the interfacial shear strength, but their requirements are different. Weak interfaces improve the fracture toughness but lower the strength. For example, thermal cycling of fiber-reinforced metals leads to interfacial voids that impair the strength but it increases the work of fracture (toughness) in the fiber direction. Large-diameter fibers and large particulates yield higher toughness at a fixed loading because of the large interfiber distance that increases the effective amount of plastic matrix zones between fibers. The effect of reinforcement volume fraction and size on mode I fracture toughness (K_{IC}) for a SiC-Al alloy composite is shown in Figure 6-51. In low-toughness matrices (e.g., NiAl), the dissipation of the energy for crack propagation will be limited, with the result that cracks originating in the fiber can actually propagate across the interface and through the surrounding matrix, leading to catastrophic failure. Fracture toughness and fatigue strength both increase with decreasing particle size.

Cyclic changes in temperature lead to thermal fatigue, which is due to a mismatch of coefficient of thermal expansion (CTE). Cyclic temperature excursions impair the integrity of the interfacial bond, although void formation in the matrix and grain boundary sliding (as in W-Cu composites) may also play a role. The elastic thermal stress is $E\Delta\alpha\Delta T$, where $\Delta\alpha$ is the CTE mismatch, ΔT is the amplitude of temperature change, and E is the elastic modulus of the matrix. The thermal fatigue damage can be reduced by employing a ductile matrix with a high yield strength, or by use of stress-absorbing compliant interfacial coatings.

Other Properties

Creep. The high-temperature creep of in situ composites such as directional solidified eutectics and monotectics, and metal-matrix composites reinforced with metallic wires as well as ceramic fibers, whiskers, and particulates (e.g., W-Ag, Al_2O_3-Al, SiC-Al) is reduced relative to the unreinforced matrix alloy. High-volume fractions and large aspect ratios of the reinforcement yield better creep resistance in cast and extruded composites. The extrusion of composites aligns

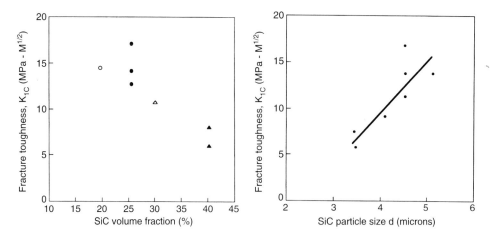

FIGURE 6-51 *(a) Fracture toughness of SiC-Al composite as a function of SiC particle volume fraction.* (M. Taya and R. J. Arsenault, Metal Matrix Composites: Thermomechanical Behavior, Pergamon Press, 1989, p. 98). *(b) Fracture toughness of SiC-Al composite as a function of the SiC particle size. The SiC volume fraction is constant at 20%.* (M. Taya and R. J. Arsenault, Metal Matrix Composites: Thermomechanical Behavior, Pergamon Press, 1989, p. 99). Reprinted with permission from Elsevier.

the whiskers and short fibers along the direction of deformation and greatly reduces the fraction of misoriented whiskers. This also contributes to improved creep resistance.

In Ni-base inermetallics, improvements in creep strength are achieved by mechanical alloying of with oxides, nitrides, and carbides. For example, mechanically alloyed NiAl containing AlN particulates (and a small amount of Y_2O_3) exhibits better strength and creep resistance in the temperature range 1200–1400 °K than does monolithic NiAl. At very high temperatures (1873 °K), creep strength deteriorates because of appreciable of coarsening of AlN. Strongly bonded NiAl-alumina fiber composites also have creep resistance superior to monolithic NiAl. As a general rule, a strong interfacial bond is necessary to achieve high creep strength.

Vibration Damping. Many applications require high vibration-damping property in the use materials. In large-space structures, the dynamic response of structural components to space maneuvers needs to be controlled. Similarly, vibration damping is important in automotive engine components and electromechanical machinery. In multiphase materials, vibration energy is dissipated at the interphase boundary, and in the deformation of the matrix and the dispersed phase. Al composites have good damping properties, and the damping capacity increases with increasing additions of the second phase. Imperfectly bonded interfaces impart better damping than perfectly bonded interfaces because interfacial disbonds and microvoids provide sites for vibrational energy dissipation through frictional losses even at relatively low levels of applied stress intensity. The dislocations piled up at the fiber–matrix interface tend to become mobile under an applied stress; this provides an additional source of dissipation of vibrational energy in composites. However, a high dislocation density at the interface usually requires a good interfacial bonding, so that damping contributions from interfacial dislocations and from interfacial disbonds may not be equally effective in a given material.

Wear and Friction. Aluminum alloys are extensively used as matrix materials in light-weight composites. Aluminum is very susceptible to seizure during sliding, particularly under boundary lubrication, and even in fluid film lubrication, where localized breakdown of the lubricating oil film may occur. Graphite dispersions improve the seizure resistance of Al, and Al-graphite bearings successfully run under boundary lubrication. The sheared layers of graphite form solid lubricant films at the mating surface. Graphite, however, loses its lubricity in vacuum and dry environments, because absence of adsorbed vapors makes it difficult to shear its layers. The wear and friction data presented in Figure 6-52 and Figure 6-53 show that graphite dispersions reduce the wear volume and friction coefficient of Al. Transition metal dichalcogenides (MoS_2 and WS_2, etc.) have lubricating property superior to graphite in vacuum or dry environments but are thermally unstable in and react with molten Al. Low-temperature powder metallurgy and isostatic pressing techniques have been used to make Al composites containing WS_2 with self-lubricating behavior and low friction coefficients (0.05 to 0.10).

The dispersion of hard particles such as SiC, TiC, Al_2O_3, Si_3N_4, SiO_2, B_4C, TiB_2, $ZrSiO_4$, glass, fly ash, solid waste slag, and short fibers or whiskers in Al increases the resistance to abrasive wear (Figure 6-54) and the coefficient of friction (Figure 6-55) when mated against steel or cast iron. Abrasion is the removal of material from a relatively soft surface by the plowing or cutting action of hard grit particles. Test parameters (load, speed, and abrasive type, size, and shape), and whether the abrasive is free to roll or is fixed to a sheet, determine the wear mechanism and abrasion resistance. The normalized abrasion rates of metals decrease as the volume fraction of the hard phase increases, as shown in Figure 6-54, and significant improvements in wear resistance occur above $\sim$20% hard phase. The coefficient of friction

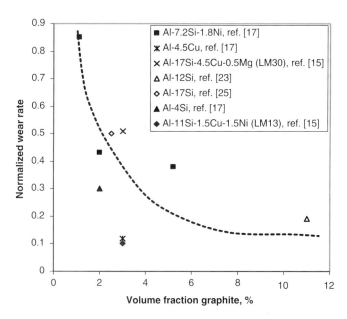

FIGURE 6-52 *Normalized wear rate of Al-graphite composites as a function of graphite volume fraction. Data on composites are normalized with respect to the wear rate of the base alloy. (Data are from studies cited in Prasad and Asthana, Tribology Letters, 17(3), 2004, 441–449).*

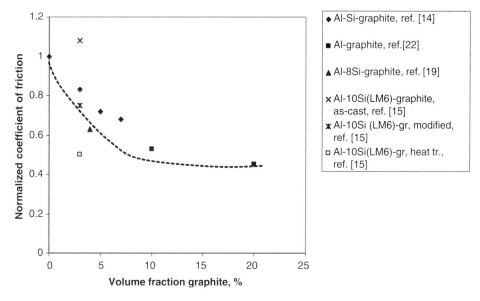

FIGURE 6-53 *Coefficient of friction of Al-graphite composite normalized with respect to the value for the base alloy as a function of the graphite volume fraction. (Data are from studies cited in Prasad and Asthana, Tribology Letters, 17(3), 2004, 441–449).*

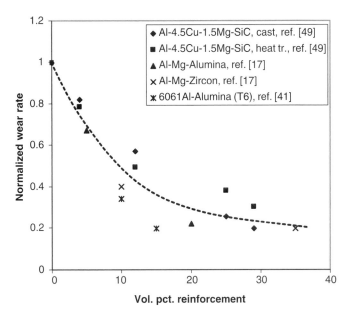

FIGURE 6-54 *Normalized wear rate of discontinuously reinforced Al composites containing hard particles as a function of the particle volume fraction.* (Data are from studies cited in Prasad and Asthana, Tribology Letters, 17(3), 2004, 441–449).

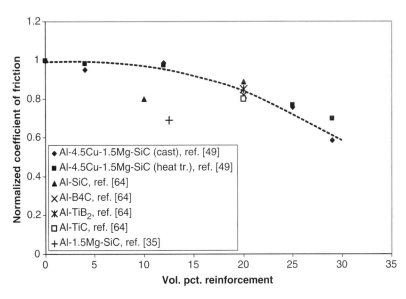

FIGURE 6-55 *Normalized coefficient of friction of Al composites containing hard particles as a function of the volume fraction of particles.* (Data are from studies cited in Prasad and Asthana, Tribology Letters, 17(3), 2004, 441–449).

of Al composites decreases with increasing additions of hard particulates (Figure 6.55). Large particles provide better wear resistance to Al than do fine particles. During abrasion, the ductile Al matrix is worn away by the cutting or plowing of the abrasive (or the asperities on the hard counterface) leaving the protrusions of the hard particles or fibers. Above a critical volume fraction (or interparticle spacing in relation to abrasive size), the hard-phase protrusions completely protect the matrix. The ductile matrix provides support to the hard phase, and insufficient support leads to the unsupported hard-phase edges becoming susceptible to fragmentation or pullout.

Table 6-7 lists some proven applications of metal-matrix composites, chiefly in the automotive industry. Diesel engine pistons containing Saffil (Al_2O_3) fibers are used by Toyota. Composite liners have better scuffing characteristics than conventional cast iron liners of the engine block. Al composites have superior thermal conductivity and lower density than cast iron, and this has been profitable used in disc brake rotors. Al composite save up to 60% weight when compared to cast iron. The abrasive wear rate is reduced 55–90% compared to Al, and at ~20% SiC, Al composite brake rotors have lower wear rate than cast iron.

Thermal Properties. The thermal properties such as coefficient of thermal expansion (CTE), thermal conductivity, and heat capacity are important in electronic packaging, high heat flux applications (e.g., rocket nozzles), and automotive engines. The CTE can be tailored by judicious

TABLE 6-7   Selected Cast Composite Components with Proven Applications

Manufacturer	*Component and Composite*
Duralcan, Martin Marietta, Lanxide	Pistons, Al-SiC$_p$
Duralcan, Lanxide	Brake rotors, calipers, liners, Al-SiC$_p$
GKN, Duralcan	Propeller shaft, Al-SiC$_p$
Nissan	Connecting rod, Al-SiC$_w$
Dow Chemical	Sprockets, pulleys, covers, Mg-SiC$_p$
Toyota	Piston rings, Al-Al$_2$O$_3$ (Saffil) , and Al-Boria$_w$
Dont, Chrysler	Connecting rods, Al-Al$_2$O$_3$
Hitachi	Current collectors, Cu-graphite
Associated Engineering, Inc.	Cylinders, pistons, Al-graphite
Martin Marietta	Pistons, connecting rods, Al-TiC$_p$
Zollner	Pistons, Al-fiberfrax
Honda	Engine blocks, Al-Al$_2$O$_3$ – C_f
Lotus Elise, Volkswagen	Brake rotors, Al-SiC$_p$
Chrysler	Brake rotors, Al-SiC$_p$
GM	Rear brake drum for EV-1, driveshaft, engine cradle, Al-SiC$_p$
MC-21, Dia-Compe, Manitou	Bicycle fork brace and disk brake rotors, Al-SiC$_p$
3M	Missile fins, aircraft electrical access door, Al-Nextel$_f$
Knorr-Bremse; Kobenhavn	Brake disc on ICE bogies, SiC-Al
Alcoa Innometalx	Multichip electronic module, Al-SiC$_p$
Lanxide	PCB Heat sinks, Al-SiC$_p$
Cercast	Electronic packages, Al-graphite foam
Textron Specialty Materials	PCB heat sinks, Al-B

p: particle, *w*: whisker, *f*: fiber

choice of the material type and reinforcement content. For example, the addition of TiC to Al is most effective in reducing the CTE followed by alumina and SiC. Likewise, the longitudinal CTE of P-100 graphite fiber–reinforced copper composite decreases with increasing fiber content (the CTE of most commercial fibers is anisotropic, with different CTE values in the radial and axial directions). The thermal conductivity of aligned fiber composites is anisotropic and depends on both the fiber and matrix conductivities, reinforcement volume fraction, and the fiber orientation. Predictive models have been developed for both composite CTE and composite conductivity; these models generally assume perfectly bonded interfaces and a reinforcement that is evenly distributed in a defect-free matrix. The thermal conductivity along the axis of carbon fibers (grown with c-axis parallel to length) is very high, and unidirectional C-fiber–reinforced Al composites have thermal conductivity greater than Cu along fiber length over a wide range of temperatures. Likewise, in unidirectionally aligned graphite fiber–reinforced Cu, the longitudinal thermal conductivity is superior to Cu; the specific conductivity (conductivity-to-density ratio) at fiber volume fractions over 35%, is greater than both Cu and Be. The transverse conductivity is, however, less than for pure Cu and decreases with increasing fiber volume fraction. High thermal conductivity is needed to dissipate heat from the leading edges of the wings of high-speed airplanes, and from high-density, high-speed integrated circuit packages for computers and in base plates of electronic equipments. Electronic packaging use requires tolerance for a small CTE mismatch to reduce the thermal stresses that arise from temperature variations. In addition, a high thermal conductivity of the materials helps in reducing temperature buildup. Table 6-8 summarizes the conductivity (k) and CTE of some metal-matrix composites and their constitutents.

The composite thermal conductivity depends on the conductivity of the constituents, and on the volume fraction, size, and the distribution of the reinforcing phase, as well as on any interfacial layers that could serve as a thermal discontinuity. In applications that require both strength and good thermal conductivity, particulate size needs to be optimized because there

TABLE 6-8 Thermal Properties of Selected Composites and Their Constituents

Material	$k, W.m^{-1} \cdot {}^{\circ}K^{-1}$	CTE, $10^{-6} \, {}^{\circ}K^{-1}$
B-6061Al		$0.65–1.87^{\parallel}$
		$2.78–7.37^{\perp}$
BORSIC-Al		$0.81–1.85^{\parallel}$
		$2.68–6.08^{\perp}$
6061Al – T6	166	3.7–8.39
BORSIC	38	5.0
Graphite (36 vol%)/Al6061-T6	$246^{\parallel}$	
	$76–92^{\perp}$	
Graphite (37%)/Mg AZ91C	$183^{\parallel}$	
	$31–35^{\perp}$	
AZ91C	70.9	
Graphite	$396^{\parallel}$	
	$0.35^{\perp}$	

Note: k, thermal conductivity, CTE, coefficient of thermal expansion.
$\parallel$, value parallel to the fiber axis.
$\perp$, value perpendicular to the fiber axis.

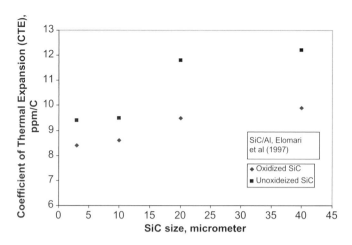

FIGURE 6-56 *The thermal expansion of SiC-Al composites containing oxidized and unoxidized SiC particles as a function of the particle size.*

are conflicting requirements for strength and thermal conductivity enhancement. The strength decrease with increasing particle size, whereas the thermal conductivity in a given composite system increases with increasing particulate size. Similarly, the CTE (Figure 6-56) also increases with increasing particle size.

Thermal Fatigue. A composite's response to cyclic thermal load depends critically on the strength and integrity of the fiber–matrix interface. In strongly bonded (bond strength >280 MPa) NiAl-sapphire composites fabricated using a "powder-cloth" process, matrix and fiber cracking after thermal cycling led to poor thermal fatigue resistance. In contrast, no matrix cracking occurred in weakly bonded composites (40–100 MPa); instead, failure occurred through interfacial debonding after thermal cycling. Ductile interfacial compliant layers can blunt cracks by localized matrix deformation. These layers must, however, bond well to both the fiber and the matrix in order to achieve fiber strengthening via load transfer to the fiber at elevated temperatures. One potential interlayer material for Ni-base composites is Mo, which is ductile and compatible with both alumina and Ni. Mo coatings (5–10 μm thick) improve the bonding in weakly bonded sapphire-NiAl composites compared to those without Mo coatings. However, during thermal cycling the interface cracks, suggesting the need for enhancing the fracture strength of Mo coating via alloying. One major drawback of Mo interlayers is their relatively poor oxidation resistance at high temperatures, which necessitates evaluation of alternative coating materials.

Composite fabrication conditions, matrix composition, and the type and volume fraction of the reinforcement influence the thermal fatigue resistance of composites. Figure 6-57 shows the thermal fatigue resistance of a variety of discontinuously reinforced Al-base composites produced using casting techniques. In this figure, the total length of all cracks formed on a particular surface of the test specimen is plotted as a function of the number of thermal cycles of ~300-degree amplitude. The samples with the smallest total crack length at a fixed number of thermal cycles will be expected to have the best thermal fatigue resistance; squeeze-cast Al alloys containing Al_2O_3 and fly ash (a waste by-product of the electric utility industry) exhibit greatest resistance to thermal shock among the composites shown in Figure 6-57.

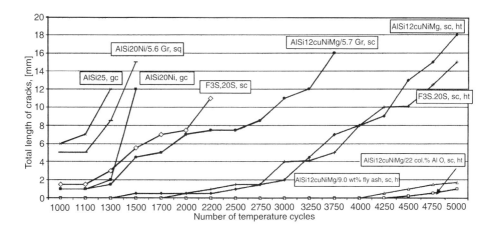

FIGURE 6-57 *The thermal fatigue resistance of some discontinuously reinforced cast Al composites as a function of the number of thermal cycles. The thermal fatigue resistance was characterized in terms of the total length of cracks on a prespecified area of the thermally cycled samples.* (J. Sobczak, Z. Slawinski, N. Sobczak, P. Darlak, R. Asthana and P. K. Rohatgi, J. Mater. Eng. and Performance, 11(6), 2002, 595–602).

Oxidation Resistance. The interface also influences the oxidation response of composites. In weakly bonded, powder-cloth sapphire-NiAl composites, Mo coatings slightly improve the oxidation resistance compared to uncoated composites, which have very poor oxidation resistance. In weakly bonded composites, internal oxidation occurs along the fiber–matrix interface, which in turn degrades the composite's oxidation resistance. Composites containing binders often leave residues (e.g., graphitic carbon) that contaminate and weaken the interface, and render the composite more susceptible to oxidation. Fiber-matrix reactions that form brittle interfacial compounds could weaken the interface and degrade the composite's oxidation resistance. For example, ZrO_2-toughened Al_2O_3 fibers (PRD-166) in pressure-cast Ni_3Al matrices impair the oxidation resistance of the composite when the latter is annealed either in air or under partial vacuum. Whereas the as-cast interfaces are free of reaction products, subsequent vacuum-annealing has been observed to lead to precipitation of Cr-rich particles at the interface and ZrB particles in the matrix. In contrast, air-annealing leads to ZrO_2 particles at the interface and a thin layer of α-Al_2O_3 around both the fiber and the ZrO_2 particles. Continued oxidation leads to $NiAl_2O_4$ formation around ZrO_2 particles at the interface together with a thin Cr_2O_3 layer. These brittle compounds deteriorate the interface and the overall oxidation resistance of the composite.

Suitable matrix alloying improves the oxidation resistance of the composite. For example, alloying NiAl with Zr or Hf significantly improves its static- as well as cyclic oxidation resistance far above conventional superalloys. In systems in which a protective oxide scale forms upon oxidation, the integrity of the scale is of importance. Thermal stresses and mechanical deformation can lead to fracture and spalling off of the scale, resulting in continued oxidation. In the case of mechanically alloyed NiAl-AlN particulate composites containing a small amount of Y_2O_3, oxidation resistance is good until 1400 °K, but at temperatures in excess of 1500 °K, rapid oxidation occurs in composites deformed at high strain rates; fracture of protective scale under rapid deformation is responsible for this behavior.

References

Allison, J. E., and J. W. Jones. In S. Suresh, A. Mortensen, and A. Needlemann, eds., *Fundamentals of Metal-Matrix Composites*. London: Butterworth-Heinemann, 1993.

Ashby, M. F. *Acta Metall. Mater.*, *41*(5), 1993, 1313.

Asthana, R. *Solidification Processing of Reinforced Metals.* Key Engineering Materials, Trans Tech, 1998.

Chawla, K. K. in *Mater. Sci. Tech.*, vol. 13, VCH Publ., Anheim, 1993, 121.

Chawla, K. K. *Composite Materials—Science and Engineering*. New York: Springer, 1987.

Clyne, T. W., and P. J. Withers. *An Introduction to Metal-Matrix Composites*. Cambridge University Press, Cambridge, UK, 1993.

Frommeyer. G. In R. W. Cahn and P. Haasen, eds., *Physical Metallurgy*, 3rd ed., Elsevier Sci. Publ., BV 1983, 1854.

Nair, S. V., J. K. Tien, and R. C. Bates. *Int. Metals Revs.*, *30*(6), 1985, 275.

Taya, M., and R. J. Arsenault. *Metal Matrix Composites—Thermomechanical Behavior.* New York: Pergamon Press, 1989.

7   Semiconductor Manufacturing

Introduction

The transistor (sandwich junction) was invented by a trio of scientists, Shockley, Bardeen, and Brattain. Although quite a few years were spent working on the idea, Bell Labs unveiled this invention on June 30, 1948. Starting from a single transistor, research in solid-state-device physics took a new turn when the idea of making the transistor, resistors, and capacitors on single-crystal silicon occurred almost simultaneously to Jack Kilby at Texas Instruments and to Robert Noyce at Fairchild Semiconductors in 1959, and thus the integrated circuit (IC) was born. The IC in 1959 was way too big, compared to any of the present-day ICs used for complex and versatile electronic and computing applications. The IC was presumed to have limited applications, predominantly for the military. It was presumed that no transistor on the chip could be smaller than 10 millionths of a meter. However, Gordon Moore, the co-founder of Intel Corporation, made a prophetic observation contrary to the popular belief in 1965. Moore predicted that the number of transistors per square centimeter would double every two years (Figure 7-1). This was known as Moore's law. Owing to the innovation that the semiconductor industry has relentlessly pursued, Moore's law has remained valid until today. The present transistor is 100 times smaller than what scientists predicted in 1961. Moore's law is expected to remain valid until the end of the decade. However, there will be unprecedented challenges involved in the entire effort.

Integrated-Circuit Applications and Market

Along the lines of Robert Noyce's vision and that of Gordon Moore, Intel invented the first microprocessor in 1971. The microprocessors and IC have found applications in all walks of life. Starting with military communications, aerospace, avionics, sophisticated tracking, and surveillance equipment, ICs were used in common consumer goods such as cell phones, refrigerators, audio systems, television, etc. The IC has come a long way from its original form in 1971. It has become more complex, is multipurpose, and has distinctly improved in its functionality and applicability. The complexity of the present chips is so enormous that the entire process of design and fabrication is split into various smaller modules and submodules. Research, development,

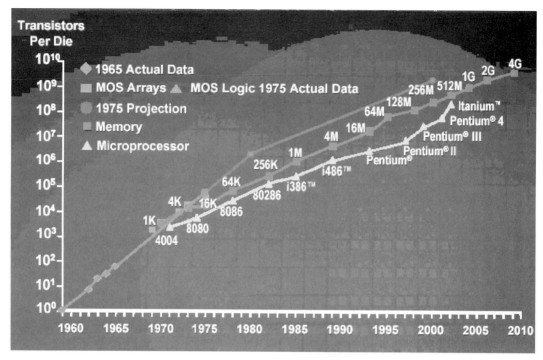

FIGURE 7-1 *Phenomenal progress of the semiconductor industry following Moore's law.* (Courtesy Intel Corporation Research © 2004).

process integration, and actual fabrication has become a multidisciplinary effort that involves physicists, chemists, and electrical, electronics, computer, mechanical, and material scientists and engineers. A relatively recent version of a microprocessor is shown in Figure 7-2.

The United States has the biggest electronics industry in the world. Since the invention of the integrated circuit in 1959, the factory sales and consumption of integrated circuits has increased by about 30 times (Figure 7-3). The growth of the global electronics industry and semiconductor industry is expected to go hand in hand for at least the next few years. The electronics industry is currently on a robust growth path that is about to grow at a rate of 6% annually over the period of 2002–2007 and is currently poised at $758 billion. The high-value electronic industry applications of microprocessors include (1) computers and peripherals (32.3%), (2) consumer electronics (21.2%), (3) telecommunication equipment (16.3%), (4) industrial electronics (14.3%), (5) defense and space (11.5%), and (6) transportation (4.2%) (see Figure 7-4). The pie chart of distribution of the applications of the microprocessor in the electronics industry is shown in Figure 7-4. The market shares of three major IC groups, namely (1) MOSFETS, (2) BIPOLAR, and (3) ICs made from III-IV compound semiconductors, are shown in Figure 7-5.

Feature Size and Wafer Size

The width of the smallest feature of a particular transistor on the surface of the die or chip on the silicon wafer is called the "minimum feature size" of the chip or die. The minimum feature size is measured in microns, which is equal to 10^{-6} of a meter. Over a period of 45 years the

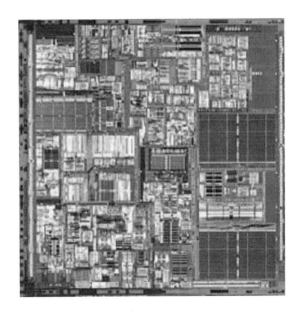

FIGURE 7-2 *A relatively recent version of the IC.*

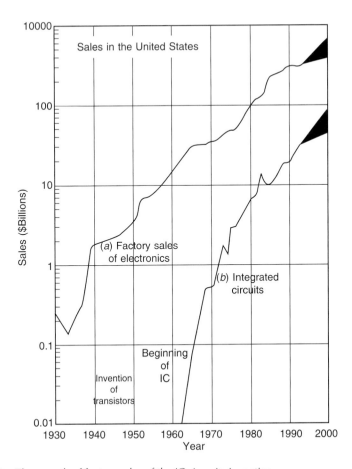

FIGURE 7-3 *The growth of factory sales of the IC since its invention.*

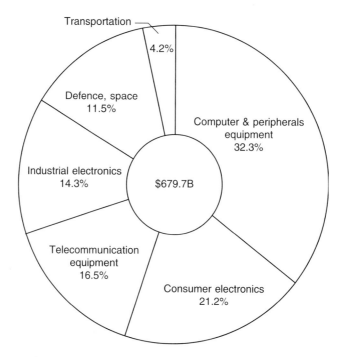

FIGURE 7-4 *Distribution of microprocessor applications in the electronics industry.*

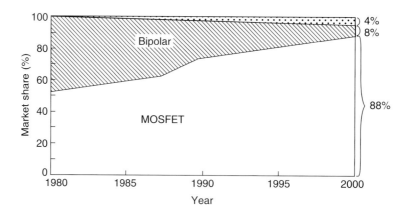

FIGURE 7-5 *The market share of three major IC groups.*

minimum feature of the chip has been reduced from 50 microns in 1959 to 0.13 microns in 2004. Presently research is underway on generation of devices that have a minimum feature size of 0.045 microns. By reducing the feature size on the device, one can fit a higher number of transistors per square centimeter of the wafer surface area. This will increase the number of chips or dies on the surface in a given wafer. When the minimum feature size of the device on the wafer shrank from 0.35 μm to 0.25 μm, the chip area reduced by $(0.25/0.35)^2 = 0.51$.

The edge effect will decrease the increase in yield due to the area shrinkage. Even then, there will be significant increase in the number of dies per wafer (Wolf, 1999). Concurrently, as the device feature size shrank, the area of the wafer used in processing kept increasing. Starting from a 2-inch diameter, the Si wafer used in today's semiconductor industry has a diameter of about 12 inches. This has led to an increase in area of about 36 times.

The increase in the wafer size and decrease in the minimum feature size has helped increase the complexity of functions that can be performed on the chip, improve the versatility of the chip, increase the number of chips per processed wafer, raise the yield of the semiconductor manufacturing industry, and reduce cost per transistor. Figure 7-6a-b shows the increasing processing power and the decreasing cost of a transistor over the years (www.intel.com, 2003).

Technology Trends

Since the beginning of the microelectronics era, the smallest line width (or the minimum feature length) of an IC has reduced at a rate of 13% per year (www.intel.com, 2003). At this rate the minimum feature size of the device is expected to shrink to about 45 nm by 2010. As discussed earlier, this will have a very positive impact on the cost. In case of Dynamic Random Access Memory devices, the cost per bit of DRAM halved every two years (Figure 7-7). The device speed has improved four orders of magnitude since 1959. At this rate, in the near future the ICs will be able to perform data processing and numerical calculations at a terabit-per-second rate. With the decrease in the size of the devices, the power consumption reduces as well. This is significantly beneficial for low-power, high-output operations such as laptops. The energy dissipated per logic gate has decreased by 1 million times over a period of 45 years, starting from 1959.

Fundamentals of Semiconductors

Semiconductors are crystalline materials that have an electrical conductivity between the good conductors (e.g., copper, aluminum, gold, silver, etc.) and insulators (wood, rubber, plastics). The most commonly used semiconductors are silicon (Si), Germanium (Ge) and Silicon Germanium (SiGe). The semiconductors are placed in the column IV A of the periodic table and usually have four electrons in their valence band (Figure 7-8). Certain compounds such as gallium arsenide (GaAs), silicon carbide (SiC), and indium phosphate (InP) also behave as semiconductors (Wolf, 1999). The conductivity of the semiconductors can be effectively controlled by adding certain impurities (doping) and application of electric field. These and other concepts associated with silicon will be discussed in this section, as silicon is the most widely used substrate for fabrication of semiconductor devices.

Crystal Structure of Silicon

Crystalline solids are materials in which the atoms are repetitively arranged in a certain order and positioned at well-defined sites. The unit cell is the arrangement of atoms that occurs repeatedly to form the entire volume of the solid. The atoms can be arranged in different ways in a single-unit cell. The way in which the atoms are arranged as well as their location within a unit cell defines the crystal structure of the solid (see Chapter 1). Different crystalline solids exhibit different crystal structures, such as simple cubic structure (SCC), face-centered cubic (FCC), body-centered cubic (BCC), rhombohedral, etc. Silicon exhibits a tetrahedral (diamond) crystal structure.

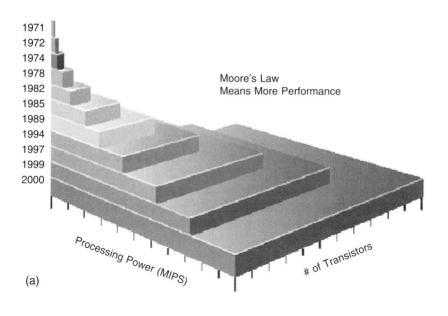

(a)

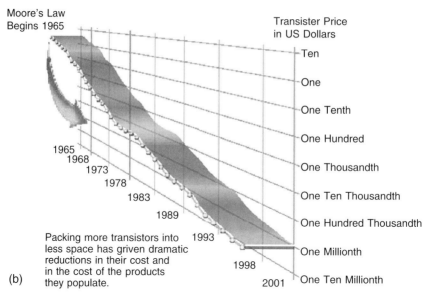

(b)

FIGURE 7-6 *(a) Increase in the processing power, and (b) decreasing cost of the transistor over a period of 36 years[5] due to increase in number of transistors per unit area.* (Courtesy Intel Corporation Research © 2004)[1].

The simplified atomic arrangement in single-crystal silicon is shown in Figure 7-9a, and a perspective on the repetitive occurrence of the silicon crystal lattice for bulk silicon is seen in Figure 7-9b.

The various cell parameters of the silicon crystal can be seen in Table 7-1. The details and significance of the cell parameters can be studied from the literature (Wolf, 1999).

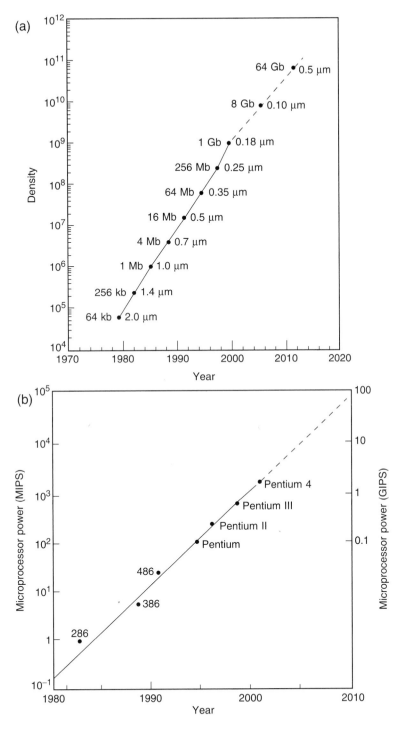

FIGURE 7-7 *Increase in (a) DRAM density, and (b) microprocessor power since the year of fabrication.*

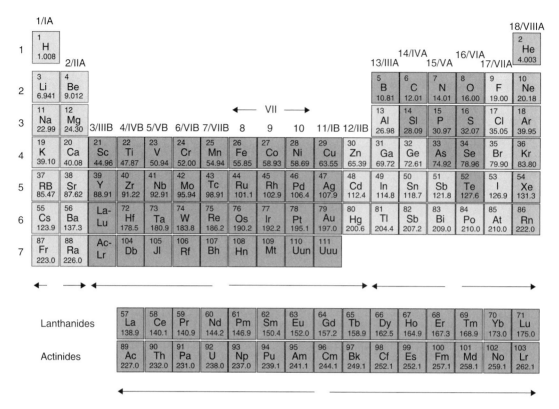

FIGURE 7-8 *Periodic table of the elements.*

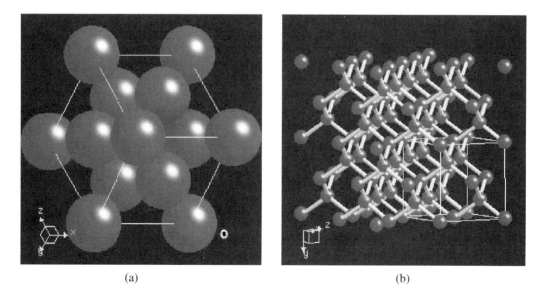

(a)   (b)

FIGURE 7-9 *(a) The single silicon crystal and (b) repetition of single Si crystal to make bulk silicon material.*

TABLE 7-1   Cell Parameters of the Silicon Crystal

Cell Parameters:	a/pm	b/pm	c/pm	$\alpha/°$	$\beta/°$	$\gamma/°$
	543.09	543.09	543.09	90.00	90.00	90.00

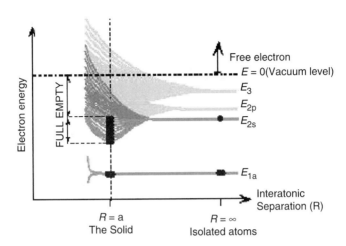

FIGURE 7-10   *Schematic for overlapping of atomic orbitals.*

Energy Band Theory

In the event of close proximity of two atoms, the valence electrons interact with one another as well as with the positively charged nucleus of the other atom. This forms a bond between two atoms, thereby producing a molecule. The formation of the bond between two atoms means that the energy system of two of the atoms combined is more than the individual atoms separated. Thus this formation is inherently more stable than the atom. As the two solid atoms come close to each other, the atomic orbitals overlap, giving rise to the energy bands. The outer orbits overlap first. The 3s orbitals give rise to the 3s band, 2p orbitals to the 2p band, and so on. The various bands overlap to produce a single band in which the energy is nearly continuous. The single 2s energy level therefore splits into N finely separated energy levels, forming an energy band. Consequently, there are N separate energy levels, each of which can take two electrons with opposite spins. The N electrons fill all the levels up to and including the level N/2. This means half the states in the band are filled from bottom up. Figure 7-10 is a schematic depiction of overlapping of atomic orbitals.

Under equilibrium, the electrons ideally exist in their respective orbitals or energy levels. If an electron is to move into a higher-order shell, additional energy must be given to the electron from an external source. Conversely, an electron leaping into a lower shell gives up some of its energy, the amount of energy required to motivate a valence electron. Leaps between two electron shells require a relatively higher amount of energy than leaps between subshells. When atoms combine to form substances, the outermost shells, subshells, and orbitals merge, providing a greater number of available energy levels for electrons to assume. When large

**Electron band separation in
semiconducting substances**

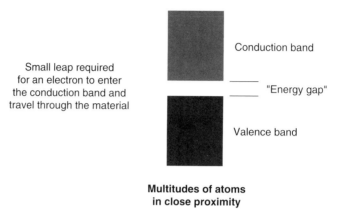

Conduction band

Small leap required
for an electron to enter
the conduction band and
travel through the material

"Energy gap"

Valence band

**Multitudes of atoms
in close proximity**

FIGURE 7-11   *Schematic of different energy bands in a semiconductor atom.*

numbers of atoms exist in close proximity to one another, these available energy levels form a nearly continuous band wherein electrons may transition. It is the width of these bands and their proximity to existing electrons that determines how mobile those electrons will be when exposed to an electric field. In metallic substances, empty bands overlap with bands containing electrons, meaning that electrons may move to what would normally be (in the case of a single atom) a higher-level state with little or no additional energy imparted. Thus the outer electrons are said to be "free" and ready to move at the beckoning of an electric field. The band overlap does not occur in all substances, no matter how many atoms are in close proximity to each other. In some substances, a substantial gap remains between the highest band containing electrons (the so-called valence band) and the next band, which is empty (the so-called conduction band). As a result, valence electrons are "bound" to their constituent atoms and cannot become mobile within the substance without a significant amount of imparted energy. These substances are electrical insulators: Materials that fall within the category of semiconductors have a narrow gap between the valence and conduction bands. Thus, a quite modest amount of energy is required to motivate a valence electron into the conduction band where it becomes mobile. At low temperatures, there is little thermal energy available to push valence electrons across this gap, and the semiconducting material acts as an insulator. At higher temperatures, though, the ambient thermal energy becomes sufficient to force electrons across the gap, and the material will conduct electricity. A schematic depiction of different energy bands of a semiconductor can be seen in Figure 7-11.

An effective way to distinguish among metals, semiconductors, and insulators is to plot the energy band diagrams from the available energies for electrons in the materials. Instead of having discrete energies as in the case of free atoms, the available energy states form bands. Crucial to the conduction process is whether or not there are electrons in the conduction band. As stated above, in insulators the electrons in the valence band are separated by a large gap from the conduction band; in conductors such as metals the valence band overlaps the conduction band; and in semiconductors there is a small gap between the valence and conduction bands that

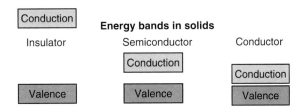

FIGURE 7-12 *Relative positioning of different bands of solids.*

thermal energy or other excitations can bridge. With such a small gap, the presence of a small percentage of a doping material can increase conductivity dramatically. The relative positioning of the different bands of the solids can be seen in Figure 7-12.

The Fermi Level

The Fermi level is an imaginary level of energy that lies at the top of the available electron energy levels at low temperature. The position of the Fermi level in relation to the conduction band is a crucial factor in determining the electrical properties of the valence electrons and the energy at which the electrons can move freely through the material (the conduction band). Fermi level is essentially half-way between valence and conduction bands. To understand the concept of Fermi level better, one must understand some fundamentals of quantum physics. Pauli's principle states that no two electrons in an atom can have identical quantum numbers. This general principle applies not only to electrons but also to other particles of half-integer spin or fermions. Particles with integer spin are called bosons. Fermions can be electrons, protons, neutrons. The Fermi-Dirac distribution applies to fermions, particles with half-integer spin that must obey the Pauli's exclusion principle. Each type of distribution function has a normalization term multiplying the exponential in the denominator, which may be temperature dependent. For the Fermi-Dirac case, that term is usually written $e^{\frac{-E_F}{kT}}$, where E_F is the energy associated with the Fermi level. The Fermi-Dirac distribution is given by the function:

$$f(E) = \frac{1}{e^{\left(\frac{E-E_F}{kT}\right)} + 1}$$

The significance of the Fermi energy is most clearly seen by setting $T = 0$. At absolute zero, the probability is 1 for energies less than the Fermi energy and zero for energies greater than the Fermi energy. We picture all the levels up to the Fermi energy as filled, but no particle has a greater energy. This is entirely consistent with Pauli's exclusion principle: Each quantum state can have only one particle. The expression of probability that a particle will have energy E is expressed as shown in Figure 7-13.

Intrinsic Semiconductor

Intrinsic semiconductors are materials that show semiconducting behavior in their purest form. The semiconductor material structure should contain no atomic impurities. Elemental and compound semiconductors can be intrinsic semiconductors. At room temperature, the thermal energy of the atoms may allow a small number of the electrons to participate in the conduction process. Unlike metals, the resistance of semiconductor material decreases with temperature. For semiconductors, as the temperature increases the thermal energy of the valence electrons increases,

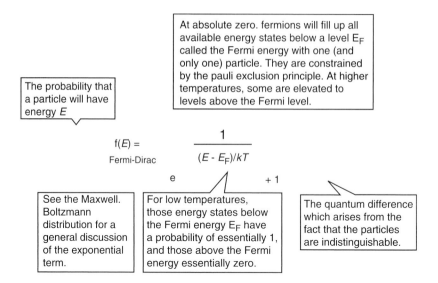

The probability that a particle will have energy E

At absolute zero. fermions will fill up all available energy states below a level E_F called the Fermi energy with one (and only one) particle. They are constrained by the pauli exclusion principle. At higher temperatures, some are elevated to levels above the Fermi level.

$$f(E) = \frac{1}{e^{(E - E_F)/kT} + 1}$$

Fermi-Dirac

See the Maxwell. Boltzmann distribution for a general discussion of the exponential term.

For low temperatures, those energy states below the Fermi energy E_F have a probability of essentially 1, and those above the Fermi energy essentially zero.

The quantum difference which arises from the fact that the particles are indistinguishable.

FIGURE 7-13 *Expression for Fermi energy of a particle.*

allowing more of them to breach the energy gap into the conduction bands, thereby improving the conductivity.

When an electron gains enough energy to escape the electrostatic attraction of its parent atom, it leaves behind a vacancy that may be filled by another electron. The vacancy produced can be thought of as a second carrier of positive charge. It is known as a hole. As electrons flow through the semiconductor, holes flow in the opposite direction. If there are n free electrons in an intrinsic semiconductor, then there must also be n holes. Holes and electrons created in this way are known as intrinsic charge carriers. The carrier concentration or charge density defines the number of charge carriers per unit volume. This relationship can be expressed as $n = p$, where n is the number of electrons and p the number of holes per unit volume. The variation in the energy gap between different semiconductor materials means that intrinsic carrier concentration at a given temperature also varies.

When a Si-Si bond is broken, a "free" electron is created that can wander around the crystal and also contribute to electrical conduction in the presence of an applied field. The broken bond has a missing electron that causes this region to be positively charged. The vacancy left behind by the missing electron in the bonding orbital is a hole. An electron in a neighboring bond can readily tunnel into this broken bond and fill it, thereby effectively causing the hole to be displaced to the original position of the tunneling electron. By electron tunneling from a neighboring bond, holes are, therefore, also free to wander around the crystal and contribute to electrical conduction in the presence of an applied field. In an intrinsic semiconductor, the number of thermally generated electrons is equal to the number of holes (broken bonds). The only empty electronic states in the silicon crystal are in the conduction band. An electron placed in the conduction band is free to move around the crystal and also respond to an applied electric field because there are plenty of neighboring empty energy levels. An electron in the conduction band can easily gain energy from the field and move to higher energy levels because these states are empty. Figure 7-14a shows what happens when a photon of energy $hv > E_g$ is incident

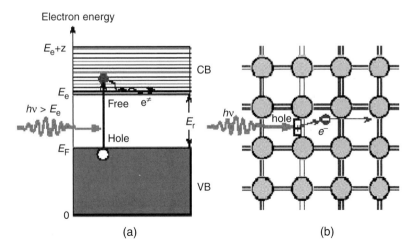

Electron energy

FIGURE 7-14 *(a) A photon with energy greater than Eg can excite an electron from the VB to the conduction band. (b) When a photon breaks a Si-Si bond, a free electron and a hole in the Si-Si bond are created.*

on an electron in the valence band (VB). This electron absorbs the incident photon and gains sufficient energy to surmount the energy gap E_g and reach the conduction band. Consequently, a free electron and a "hole," corresponding to a missing electron in the VB, are created. In Si (or even Ge), the photon absorption process also involves lattice vibrations, which we have not shown in Figure 7-14b.

Extrinsic Semiconductor

In an extrinsic semiconductor, impurities are added to the semiconductor that can contribute either excess electrons or excess holes. These added impurities are called dopants. Dopants can be classified into two types: (1) p (positive) types such as boron (B), from column III A of the periodic table, which are known as acceptors because they provide excess holes; and (2) n (negative) types such as arsenic (As), phosphorous (P), and antimony (Sb) from column VI A, which are known as donors because they provide excess electrons. When an impurity such as arsenic is added to Si, each As atom acts as a donor and contributes a free electron to the crystal. Since these electrons do not come from broken bonds, the numbers of electrons and holes are not equal in an extrinsic semiconductor, and the As-doped Si in this example will have excess electrons. It will be an n-type Si because electrical conduction will be mainly due to the motion of electrons. It is also possible to obtain a p-type Si crystal in which hole concentration is in excess of the electron concentration because of, for example, boron doping.

Elements such as P or As have 5 electrons in their outermost orbit. When P or As is doped into single-crystal silicon, the excessive electron in the their outermost shell can easily jump into the conducting band and can conduct electric current. In this case, the majority of current carriers are electrons and the semiconductor is called n type (negative). The schematic of As doping and its energy bands is shown in Figure 7-15a-b.

Figure 7-16 shows the schematic of boron doping of silicon. Figure 7-16a shows a schematic of an empty spot or a hole when boron is doped into single-crystal silicon. Figure 7-16b-e shows

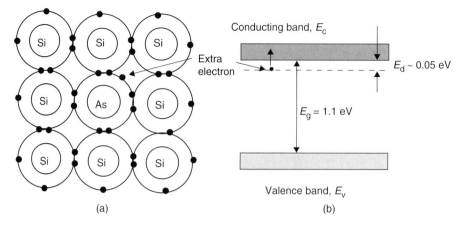

FIGURE 7-15 *(a) n-Type arsenic-doped silicon, (b) energy band of the donor.*

that the electrons from the valence band can easily jump into the conduction band of the acceptor, creating holes in the valence band. Under the influence of the electric field, other electrons in the valence band move up and down into the holes, creating new holes where these electrons originated. This hole movement facilitates conduction of electric current. The semiconductor with holes as the majority charge carrier is called a *p*-type semiconductor.

Dopant Concentration and Conductivity

Figure 7-17 shows the variation of silicon resistivity with the dopant concentration. This figure shows that with the increase in dopant concentration, the resistivity of silicon reduces, thus increasing the conductivity. This is because of the increase in the number of charge carriers in Si (*n* or *p* type). Figure 7-17 also shows that silicon has a lower resistivity when it is doped with *n* type as compared to *p* type, as electrons travel faster through the conduction band than the holes.

Diodes

The diode is one the most elementary semiconducting device in the IC. The terms *diode* and *rectifier* may be used interchangeably; however, the term *diode* usually implies a small signal device with current typically in the milliamp range, and a rectifier, a power device, conducts from 1 to 1000 amps or even higher. A semiconductor diode consists of a PN junction and has two terminals, an anode (+) and a cathode (−). Current flows from anode to cathode within the diode. A schematic of the diode is shown in Figure 7-18.

An ideal diode is like a light switch. When the switch is closed, the circuit is completed; and the light turns on. When the switch is open, there is no current and the light is off. However, the diode has an additional property; it is unidirectional, i.e., current flows in only one direction (anode to cathode internally). When a forward voltage is applied, the diode conducts; and when a reverse voltage is applied, there is no conduction. In practice the characteristic of the diode under applied voltage is shown in Figure 7-19. Notice that the diode conducts a small current in the forward direction up to a threshold voltage, 0.3 for germanium and 0.7 for silicon; after that it conducts as we might expect. The forward voltage drop, V_f, is specified at a forward current. In the reverse direction there is a small leakage current until the reverse breakdown voltage

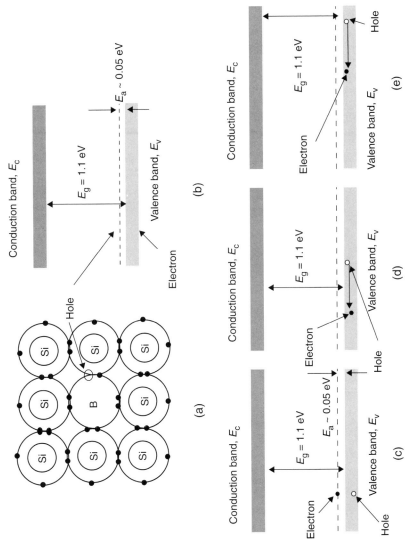

FIGURE 7-16 *(a) Boron-doped silicon, (b) energy band of the acceptor, and (c–e) illustrations of hole movement.*

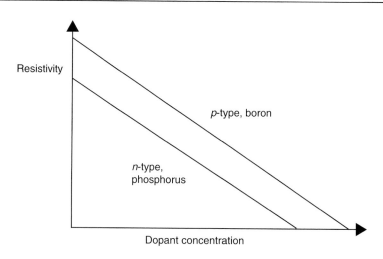

FIGURE 7-17   *Dopant concentration and resistivity of silicon.*

is reached. This leakage is undesirable—obviously, the lower the better—and is specified at a voltage less than the breakdown; diodes are intended to operate below their breakdown voltage. The current rating of a diode is determined primarily by the size of the diode chip, and by both the material and configuration of the package. Average current is used in rating, not RMS current. A larger chip and package of high thermal conductivity are both conducive to a higher current rating. The switching speed of a diode depends on its construction and fabrication. In general, the smaller the chip, the faster it switches, other things being equal. The chip geometry, doping levels, and the temperature at nativity determine switching speeds. The reverse recovery time, t_{rr}, is usually the limiting parameter; t_{rr} is the time it takes a diode to switch from "on" to "off".

Bipolar Junction Transistors (BJT)

Another important component of the integrated circuit is the bipolar junction transistor (BJT). By placing two PN junctions together, we can create a bipolar transistor. In a PNP transistor the majority charge carriers are holes, and germanium is favored for these devices. Silicon is best for NPN transistors where the majority charge carriers are electrons. The thin and lightly doped central region is known as the base (B) and has majority-charge carriers of opposite polarity to those in the surrounding material. The two outer regions are known as the emitter (E) and the

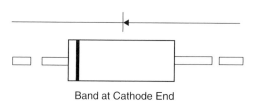

Band at Cathode End

FIGURE 7-18   *The schematic representation of diode (with electrical symbol).*

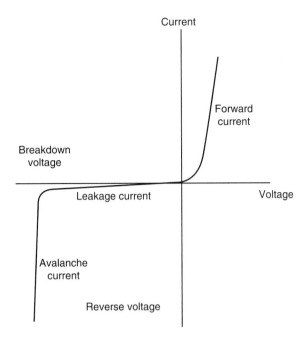

FIGURE 7-19 *Current (I) versus voltage (V) characteristics of a practical diode[4].*

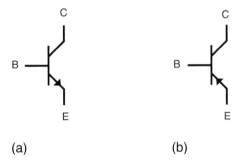

FIGURE 7-20 *(a) NPN bipolar transistor and (b) PNP bipolar transistor.*

collector (*C*). Under the proper operating conditions the emitter will emit or inject majority-charge carriers into the base region and, because the base is very thin, most will ultimately reach the collector. The emitter is highly doped to reduce the resistance. The collector is lightly doped to reduce the junction capacitance of the collector–base junction. The schematic circuit symbols for bipolar transistors are shown in Figure 7-20. The arrows on the schematic symbols indicate the direction of both I_n and I_c. The collector is usually at a higher voltage than the emitter. The emitter-base junction is forward biased, whereas the collector-base junction is reverse biased.

If the collector, emitter, and base of an NPN transistor are shorted together, as shown in Figure 7-21a, the diffusion process described earlier for diodes results in the formation of two depletion regions that surround the base, as shown in Figure 7-21a. The diffusion of negative

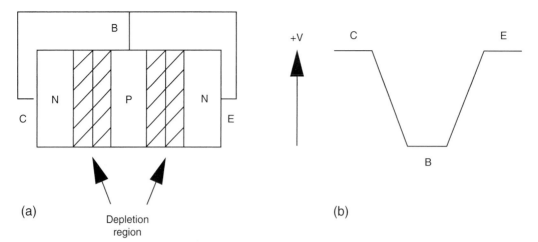

(a)

(b)

Depletion
region

FIGURE 7-21 *(a) NPN transistor with collector, base, and emitter shorted together; and (b) voltage levels developed within the shortened semiconductor.*

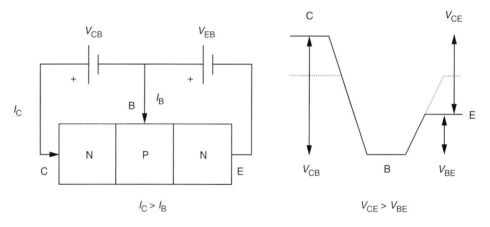

$I_C > I_B$

$V_{CE} > V_{BE}$

FIGURE 7-22 *(a) NPN transistor biased for operation and (b) voltage levels developed within the biased semiconductor. I_C and I_B denote the collector current, and base current, respectively.*

carriers into the base and positive carriers out of the base results in a relative electric potential and is schematically represented in Figure 7-21b.

When the transistor is biased for normal operation as in Figure 7-22a, the base terminal is slightly positive with respect to the emitter (about 0.6 V for silicon), and the collector is positive by several volts. When properly biased, the transistor acts to make $I_c \gg I_n$. The depletion region at the reverse-biased base–collector junction grows and is able to support the increased electric potential change indicated in the Figure 7-22b.

For a typical transistor, 95 to 99% of the charge carriers from the emitter make it to the collector and constitute almost all the collector current I_c. I_c is slightly less than the emitter current, I_E, and we may write $a = I_c/I_E$, where (from above) $a = 0.95$ to 0.99. The behavior of

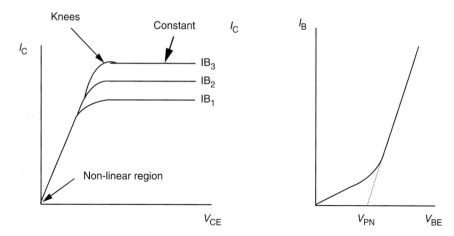

FIGURE 7-23 *Characteristic curves of an NPN transistor.*

a transistor can be summarized by the characteristic curves shown in Figure 7-23. Each curve starts from zero in a nonlinear fashion, rises smoothly, then rounds a knee to enter a region of essentially constant I_c. This flat region corresponds to the condition where the depletion region at the base–emitter junction has essentially disappeared. To be useful as a linear amplifier, the transistor must be operated exclusively in the flat region, where the collector current is determined by the base current.

Metal Oxide Field Effect Transistors (MOSFET)

The metal oxide field effect transistor is comprised of a conducting gate, the silicon substrate, and a thin layer of silicon dioxide sandwiched in between. For a Negative Metal Oxide Semiconductor (NMOS) transistor, the substrate is p type and the source and drain are heavily doped with n-type dopant. The source and drain are usually symmetric, and normally the grounded side is called the source and the biased side is called the drain. When no bias voltage is applied to the gate, no matter how the source and drain are based, no current can go through from source and drain, or vice versa. When the gate is biased positively, positive charges are generated at the metal oxide surface. The gate silicon dioxide is the thin dielectric layer between metal and semiconductor. Like a capacitor, the positive charge on the metal oxide will repel the positive charge (majority carrier) from the SiO_2 surface, and attract the negative charge (minority carriers) to the surface. When the gate voltage is higher than the threshold voltage, $V_G > V_T > 0$, the SiO_2 surface will accumulate enough electrons to form a channel and allow electrons from source and drain to flow across it. This is the reason why NMOS is also called n-channel MOSFET, or NMOSFET. By controlling the gate voltage, the electric field will affect the conductivity of the semiconductor device and switch the MOS transistor off and on. This is why it is called a field effect transistor (FET). For a PMOS, the substrate is n type, and source and drain are heavily positively doped. The majority charge carriers are electrons, and flow of current on switching is in the opposite direction. Figure 7-24a and b show the electrical symbols of NMOS and PMOS, whereas Figure 7-25a and b show the structure and switching process of NMOS and PMOS, respectively.

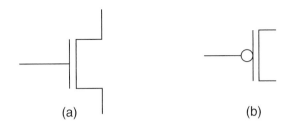

FIGURE 7-24 *Electrical symbols of (a) NMOS and (b) PMOS.*

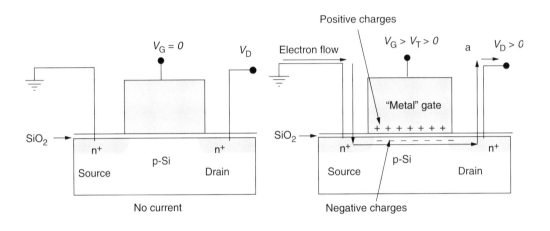

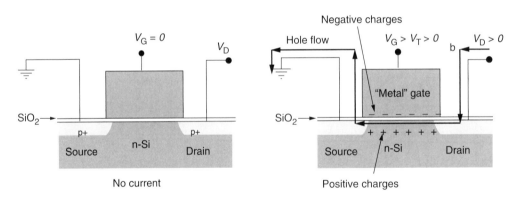

FIGURE 7-25 *Schematic and switching action of (a) NMOSFET and (b) PMOSFET.*

Crystal Growth and Wafer Preparation

Even though there are materials such as germanium, silicon has been accepted as the major constituent material of the semiconductor manufacturing industry. Compared to germanium, which was thought to be a potential candidate material for a certain time, silicon has a lot of advantages. For example, oxide of silicon can be grown thermally with greater ease compared to germanium. Also, silicon dioxide is an excellent diffusion barrier and has excellent insulating properties.

The silicon that is used for nonsemiconductor purposes can be easily produced from sand, which is called metallurgical-grade silicon. Silicon of this grade is approximately 98% pure, with Iron and aluminum constituting most of the contaminants. This metallurgical-grade silicon needs to go through a series of processes to become extremely pure to be used for electronic device manufacturing. The silicon that comes out of the purification steps is termed semiconductor-grade silicon or electronic-grade silicon. Another critical requirement along with purity of silicon is the crystal structure of the wafer. The semiconductor-grade silicon needs to have a perfect, defect-free crystal lattice.

The ultrapure semiconductor-grade silicon is obtained by treating metallurgical-grade silicon with hydrogen chloride gas at very high temperatures ($\sim$1400 $^{\circ}$C) in a reactor where the solid silicon reacts with HCl, producing chlorosilanes, in particular trichlorosilane. The following three reactions explain the main process of conversion from sand to semiconductor-grade silicon.

$$SiC + SiO_2(s) \rightarrow Si(l) + SiO(g) + CO(g)$$

$$Si(s) + 3HCl(g) \rightarrow SiHCl_3(g) + H_2(g) + heat$$

$$2SiHCl_3(g) + 2H_2(g) \rightarrow 2Si(s) + 6HCl(g)$$

Trichlorosilane exists as liquid phase at atmospheric temperatures and boils at about 32 $^{\circ}$C. This permits the trichlorosilane to be distilled and thus made extremely pure. The pure trichlorosilane obtained is made to react with hydrogen gas at 1200 $^{\circ}$C to get ultrapure semiconductor-grade silicon. The whole process of trichlorosilane reacting with hydrogen gas is termed the Siemens process. This silicon, which is very pure, exists in polycrystalline form, which is still not acceptable for usage in semiconductor device manufacturing. This polycrystalline silicon is then processed to obtain single-crystal silicon wafer.

Czochralski Method

The most accepted method of transforming the polycrystalline silicon into single-crystal silicon is the Czochralski method briefly described in Chapter 2. In this method the solid pure silicon in polycrystalline state obtained from the Siemens process is heated in a quartz crucible until it melts. Dopants are added to the melted silicon at this stage to get either p-type or n-type silicon wafers. A seed rod that has single-crystal structure is brought into contact with the melted silicon. The seed crystal and the crucible both are rotated about their own central axes. The seed crystal, after coming into contact with the melted silicon, is pulled up with high precision. Due to surface tension, melted silicon is dragged up with the same crystal structure as that of the seed crystal (single-crystal structure), and as it is pulled up it solidifies to form a cylinder. These cylinders are called ingots. A great deal of care should be taken while pulling the seed rod, given the temperature and also given the environment inside the chamber. Typically a Radio Frequency (RF) coil is used around the crucible to maintain a controlled temperature so that the ingot does not develop thermal stresses, because the temperature difference results in defective silicon wafers, which lowers throughput. Figures 7-26 and 7-27 provide a schematic of the reaction chamber for the Czochralski method and pictures of ingot formation.

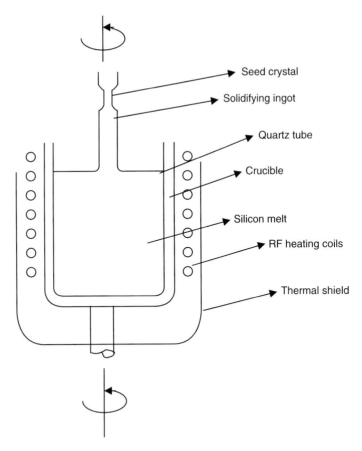

FIGURE 7-26 *Schematic of reaction chamber for Czochralski method.*

Floating-Zone Method

Another method, developed in the 1950s, which produces better-quality single-crystal wafers, is the floating-zone method (see Chapter 2). In this method, the silicon cylinder with the single-crystal seed rod having the polycrystalline electronic-grade silicon is placed in a chamber. A mobile RF coil wound around the chamber heats up the polycrystalline silicon, which melts and takes up the single-crystal structure of the seed rod while solidifying. In this method there is less probability of the silicon attracting the impurity atoms. Also, the thermal stresses are not as great. Figure 7-28 presents a schematic of the floating-zone process, which explains the setup in detail.

Silicon wafers manufactured by this process are highly pure and are perfect crystals in structure (Figure 7-29). Perfect crystals have the same orientation and same lattice structure from top to bottom. The crystals in such a pure substance have a finite lattice structure. Occasionally there can be defects in the lattice, such as dislocation of the atom, replacement of the original atom by another atom, or inclusion of an oxygen atom in the interspaces of the original crystal lattice.

FIGURE 7-27 *Closeup pictures of ingot formation.*

Defects

The defects in the single-crystal silicon manufactured through the Czochralski method can be attributed to the high reactivity of melted silicon at temperatures around 1420 °C (melting point of electronic-grade silicon) reacting with any type of crucible material. Thus the crucible dissolves continuously into the melted silicon, and as the melted silicon is pulled up, the concentration of the contaminants increases, resulting in defective silicon ingots and, accordingly, defective wafers. Therefore, none of the dopant material compounds such as titanium carbide, boron nitride, and aluminum oxide can be used in crucible materials. Oxides, nitrides, and carbides of silicon are the most favorable compounds.

Inert atmosphere is used to purge vapor that comes out of the reaction of melted silicon with the crucible materials (which are silicon compounds, and give out monoxides on reaction with silicon). Usually the crucible furnace walls are made of graphite, which reacts with the monoxide to form carbon impurities that get introduced into the silicon melt. Many factors such as convection of the melt affect the transport of these compounds into the melt but are not discussed here, to avoid digression.

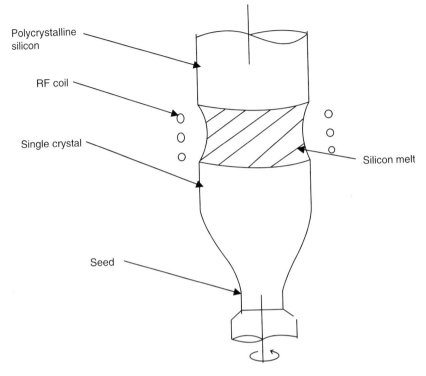

Polycrystalline
silicon

RF coil

Single crystal

Silicon melt

Seed

FIGURE 7-28 *Floating-zone process for single-crystal growth.*

FIGURE 7-29 *Single-crystal silicon ingot after crystal growth process.*

The acceptable limits for defects in semiconductor wafers are very low, because a device built on a defective substrate has a high probability for failure in presence of even minute defect levels, particularly as the number of layers of metallization increase. These defects in the single-crystal silicon wafer can be categorized into the following types: point defects, line defects, planar defects, and volume defects.

A point defect is a deviation on the periodicity of the lattice arising from a single point. Point defects form as a result of existing thermodynamic equilibrium in the crystal. Besides introducing structural disturbances, these point defects also affect the electrical and electronic characteristics. Vacancy and interstitial defects are common native point defects that do not arise because of any impurity. Vacancy is a defect caused by the absence of an atom at the lattice space of the crystal structure. Depending on the configuration of the unsatisfied bonds in silicon lattice, the nature of the vacancy is considered neutral, negative, or positive. This defect is also called the Schottky defect (see Chapter 1).

An interstitial defect is caused by the presence of a host atom at an interstitial site of its own crystal lattice. Interstitiality is the presence of two atoms in a crystal structure at nonsubstitutional lattice points. This defect is very similar to interstitial defect but with a foreign atom. When a host atom of the lattice is replaced by an impurity atom intentionally or unintentionally, the type of defect is called a substitutional defect. Different combinations of defect are possible with point defects. If the interstitial space in a crystal lattice is filled with a host atom that is located next to a vacancy, the defect is called a Frenkel defect, a combination of vacancy and interstitial defects. These point defects are becoming important as the technology of semiconductors become more complex. Some of the defects, such as substitutional defects, are introduced intentionally to enhance the conductive properties of the host wafer.

Line defects in crystal structures are termed dislocations. These defects form and extend along a line or a curve in a crystal lattice. These are created when a plane of atoms is out of place from its original location. Line defects divide the crystal into two halves such that the two sides of the crystal line up in a perfect orderly arrangement but are offset from each other. Dislocations can be categorized into two types, namely, screw and edge dislocations.

Screw dislocation can be described as the offset of the planes on one side of the line in a direction *along* the orientation of the line, whereas in an edge dislocation, the plane is offset in a direction *perpendicular* to the orientation of the line. Dislocations can form at any stage of the whole wafer-fabrication process. The introduction of such defects, particularly after crystal growth process, is attributed to mechanical stresses resulting from thermal and mechanical processing steps. Figure 7-30 illustrates the screw and edge dislocations.

These line defects can be controlled by heat-treating and gettering processes. Also, these defects are reduced by using a neck-down process during ingot growth (Electronic Industries Association, 1994).

Planar defects occur whenever there is a discontinuity in the structure, such as interface of two surfaces, grain boundaries, etc. The typical varieties of planar defects are twinning, which occurs at the grain boundaries; stacking fault, which occurs at the interface of two planes; and slip within the crystal along one or more crystal planes, which is also considered a planar defect.

Volume defects form if the host atom, after being processed and doped, precipitates the excess amount of impurity if added at the higher temperatures. The defects are caused by a change in the volume due to precipitation of the excess impurity atoms with respect to the host atom.

Some of these defects are desired, such as the intentional substitutional defect, but many are undesired and can be avoided by some heat treatment process and inert atmosphere, as mentioned earlier.

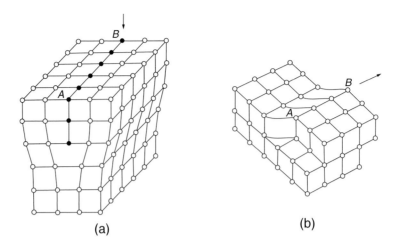

FIGURE 7-30 *(a) Edge dislocation and (b) Screw dislocation.*

Wafer Preparation

Wafer quality standards in regard to flatness, surface finish, and other such aspects are set very high, because this wafer forms the basis for building complex devices. Wafer preparation should be carried out with utmost control and in a contamination-free environment. Achieving a very high level of flatness, a warpless, smooth surface without any contaminations and with beneficial light-scattering properties, is a challenge within itself. Many integral processes constitute the whole wafer preparation process, which are summarized below.

Initial Steps

The first process after the ingot is taken out of the crystal growth chamber is to remove the ends of the ingot. The front end, which is called the seed end, and the rear end or the tag ends are removed and the ingot is machined along the length to achieve accurate diametric dimension for the ingot. The ingots are grown to a slightly bigger diameter to accommodate this processing step. This machining is important, because it is very difficult to control the diameter to such tight precision while growing the ingot. Industry traditionally follows a referencing method involving the introduction of flats on the circumference of the ingot. The location and length of the flats represent the crystal orientation and the conductivity of the wafer (*p*-type or *n*-type). Figure 7-31 shows the flat referencing in greater detail.

Wafer Slicing

Once the ingots are referenced for identification, they are sent through a slicing machine that uses an internal-diameter saw to slice the ingot into wafers of a prespecified thickness. Nowadays, the internal-diameter saw is being replaced by slurry-coated wire for bigger-diameter ingots (Sze, 1988). For bigger-diameter wafers (200 mm and 300 mm) the referencing with the help of flats has been replaced, introducing a notch on the circumference and laser-scribing a number that includes the lot number, type of wafer, and other such details represented numerically. This scribing is done on the back side of the wafer at a fixed location away from the working area (where devices are grown). Figure 7-32 explains the notch and laser-scribe model.

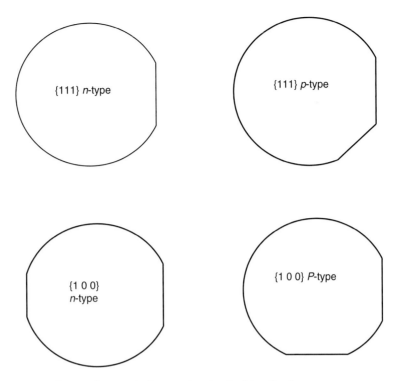

FIGURE 7-31   *Reference flats on single-crystal wafers for identification.*

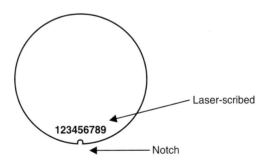

FIGURE 7-32   *Laser scribe and notch on single-crystal wafer for identification.*

Postslicing Polishing and Cleaning

Once the wafers are sliced, they are lapped on both sides for high flatness and are rounded at the edges. Then the lapped wafers are etched to eliminate damages and contaminations that may have been introduced during slicing and lapping. Once this coarser finishing and cleaning is carried out, the wafers are then polished on both sides by a technique called chemical mechanical planarization (CMP), which uses a slurry to provide the chemical effect and abrasives in the slurry and a polishing pad to provide the mechanical component. This polishing is done on both sides of the wafer to get a high degree of flatness and a mirrorlike surface finish. After CMP, the wafers are cleaned by contact and noncontact cleaning techniques, using various combinations of chemicals such as Hydrofluoric Acid (HF) F, De-Ionized Water (DI) and ammonium hydroxide. Thus, a well-finished and clean, single-crystal wafers obtained for semiconductor device manufacturing.

Oxidation

The oxide of thermally grown silicon has good electrical properties that make the Si–SiO_2 interface very stable for electronic devices. During thermal oxidation, bare silicon is exposed to an oxygen-rich ambience, either oxygen itself or water in gas phase. The oxides of the silicon have a wide variety of applications, including gate oxides, dielectric layer, passivation on silicon surface, line oxide for Shallow Trench Isolation (STI) trenches, screen oxides as masks, field oxides for Local Oxidation of Silicon (LOCOS) isolation, etc. Thickness of the oxide varies over a wide range for every application; for example, the range is about 5 nm for gate oxide insulators to about 200–400 nm for field oxides.

Later in the section we discuss properties of silicon dioxide, reaction kinetics, oxidation systems, comparison of dry and wet oxidation techniques, and characterization.

Properties of Silicon Dioxide

The basic structure of one unit is a tetrahedron with the silicon atom inside the tetrahedron formed by the oxygen atoms (see Figure 7-33). The two most common forms of silicon dioxide are quartz and fused silica. The crystalline form of silicon dioxide is termed quartz, whereas thermally grown silicon oxide, which is amorphous in nature, is called fused silica. The crystalline form has a continuous array of the basic tetrahedral units. The amorphous form differs in the type of array—all the tetrahedral structures connected together in the crystal may not have all the oxygen atoms bonded. This lowers the density of the oxide. The greater the number of unbounded atoms of oxygen, the lesser the density, which results in poor quality of oxide. These geometric crystalline shapes can transform when silicon dioxide is heated to about 800 °C. Silicon dioxide is held together by double covalent bonds. Tables 7-2 and 7-3 compare the properties of crystalline and amorphous silica.

The properties of the oxide can be changed drastically by doping it with several atoms such as boron, fluorine, and phosphorous. These property changes can be either beneficial or detrimental.

Oxidation Techniques

Silicon has a high tendency to form stable oxides. A very thin silicon oxide layer (a couple of nanometers) is formed on the silicon at room temperature, demonstrating the affinity of silicon toward oxygen. The process of oxidation takes place more rapidly at elevated temperatures. The thickness of the oxide, being an important specification for any particular application, can be controlled precisely in the process of thermal oxidation. The thermal oxidation can be classified into three types based on the type of the ambience in the oxidation chamber: dry oxidation, steam oxidation, and wet oxidation.

Dry oxidation is the kind of oxidation process in which the bare silicon wafers are exposed to dry oxygen at temperature ranges from 800 °C to 1200 °C. Steam oxidation is the type in which water vapor is made to react with the silicon. The mixed type of oxidation, wherein oxygen gas is made to flow onto the silicon wafers in the presence of water vapor, is termed wet oxidation.

FIGURE 7-33 *The tetrahedral structure of the [SiO4]$_4^-$ (silicon oxide).*

TABLE 7-2 Physical, Mechanical, Thermal, and Electrical Properties of Quartz and Fused Silica

Property\Material	Quartz	Fused Silica
Density (g/cm^3)	2.65	2.2
Thermal conductivity (Wm^{-1} $^\circ$K)	1.3	1.4
Thermal expansion coefficient (10^{-6} $^\circ$K^{-1})	12.3	0.4
Tensile strength (MPa)	55	110
Compressive strength (MPa)	2070	690–1380
Poisson's ratio	0.17	0.165
Fracture toughness (MPa)	—	0.79
Melting point ($^\circ$C)	1830	1830
Modulus of elasticity (GPa)	70	73
Thermal shock resistance	Excellent	Excellent
Permittivity (ε') *	3.8–5.4	3.8
Tan ($\delta \times 10^4$) [a]	3	
Loss factor (ε'') [a]	0.0015	
Dielectric field strength (kV/mm)[a]	15.0–25.0	15.0–40.0
Resistivity (Ωm) [a]	10^{12}–10^{16}	$>10^{18}$

[a]Dielectric properties at 1 MHz 25 $^\circ$C.

Wet oxidation has a relatively high oxidation growth rate. Dry oxidation delivers high-quality oxide films but has very low oxide growth. Dry oxidation is usually employed when there is a requirement for thin oxides for gate, even though there is less throughput. Wet oxidation is used for thick oxide films used in field oxides for MOSFET applications.

The wet oxidation process is simple and cost effective compared with dry oxidation. If the presence of other gases in the chamber is prevented effectively by using fresh air and also by using nitrogen gas to drive out impurities and undesired gases, the quality of the film deposited

TABLE 7-3 Differences Between the Different Crystal Structures of Silica

Phase	Density (g/m³)	Thermal Expansion ($10^{-6}\ °K^{-1}$)
Quartz	2.65	12.3
Tridymite	2.3	21
Cristobalite	2.2	10.3

Source: Electronic Industries Association, *1994 Electronic Market Data Book* (Washington, DC: Electronic Industries Association, 1994).

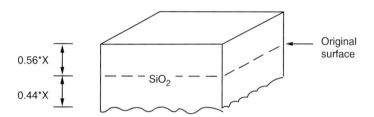

FIGURE 7-34 *Ratio of silicon consumed during oxidation to the actual oxide layer formation.*

by wet oxidation technique can be increased. Nevertheless, for applications where high purity and good electrical and dielectric characteristics are needed, dry oxidation is prescribed.

The following equations describe the process of oxidation during dry and wet oxidation processes respectively:

$$Si\ (s) + O_2\ (g) \rightarrow SiO_2\ (s)$$
$$Si\ (s) + H_2O\ (g) \rightarrow SiO_2\ (s) + 2H_2\ (g)$$

The silicon dioxide reaction takes place at the silicon top surface. This results in consumption of silicon directly proportional to the oxide thickness. From various models developed and also from many empirical results, it has been noted that 44 units of silicon material are consumed for every 100 units of silicon dioxide formation. Figure 7-34 describes the thickness of silicon and oxide formation ratios.

Oxidation Furnaces and Factors Affecting Oxide Growth

A wide variety of processing systems (furnaces) are used for oxidation. These include horizontal and vertical furnaces, fast-ramp furnaces, rapid thermal processing systems, and high-pressure oxidation systems. Many factors affect the growth of oxide layer in these furnaces, such as temperature, pressure, type of oxidizing agent (oxygen or water vapor), silicon crystallographic orientation, dopant type and concentration, presence of chlorine atoms in the ambience, and finally, the presence of water in the environment if the process is a dry oxidation process. These effects have been studied extensively and modeled. The effects of these factors helps in simulating the process and designing the optimum process conditions for growing oxides effectively. The growth of oxide layer in two-dimensional mode has many challenges and effects

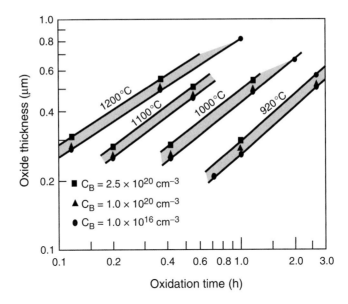

FIGURE 7-35 *Oxidation of boron-doped silicon in wet oxygen (95°C H₂O) as a function of temperature and concentration.*

such as bird's-beak formation, gate thinning, and trench corner rounding. These issues, if not addressed properly, result in inefficient circuit formation.

Figures 7-35 through 7-37 present the oxide growth dependency on oxidation time, oxidation temperature, type of oxidation process, and presence of dopants.

Along with the experimental data, for better understanding silicon oxidation, theoretical models have been proposed, growth rate constants determined, and their variation with temperature found and fit to the experimental data for validation. From the modeling two types of growth rates are proposed, one being linear and the other being parabolic. Figure 7-38 presents the variation of the growth rates with temperature.

Characterization of Oxide Films

The thickness of the oxide grown can be characterized by various methods, the easiest and approximate method being visual inspection of the film. The oxide film bears different colors at various thicknesses. A standard color chart is available for various thicknesses of the oxide layer grown. There are other sophisticated methods that are used to measure the thickness of the oxide layer, which include profilometry, ellipsometry, capacitance method, and cross-sectional transmission electron microscopy (TEM).

Profilometry is a contact-based thickness-measuring technique wherein a step height is created on the sample by selective etching at some sections of the coated region. A stylus is then dragged across the step. The stylus reproduces the trace of the original surface that it traverses. The difference in the heights of the two sides of the step gives the measure of the thickness of the film grown. The resolution can be as high as a hundredth of a micron or even a nanometer for a typical profilometer. The range of measurement can be less than about 100 nanometers to more than 5 microns (Sze, 1988).

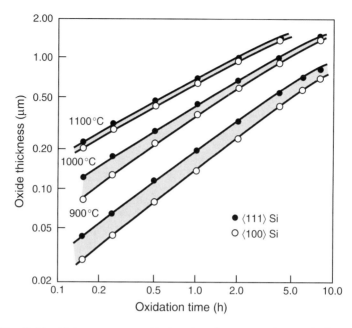

FIGURE 7-36 *Oxide thickness versus oxidation time for wet oxidation of silicon (in H_2O at 640 Torr).*

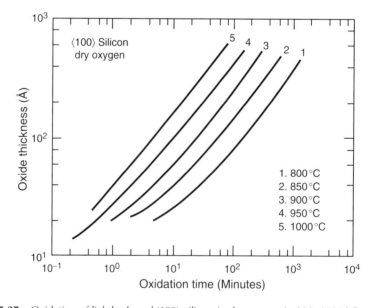

FIGURE 7-37 *Oxidation of lightly doped ⟨100⟩ silicon in dry oxygen in 800–1000 °C range.*

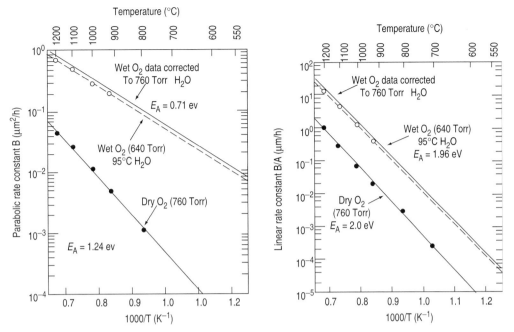

FIGURE 7-38 *The effect of temperature on the rate constants (both parabolic and linear) for dry and wet oxidation of silicon.*

Ellipsometry, another widely used measurement technique, is based on the polarization changes that occur when light is reflected from or transmitted through a medium. Changes in polarization are a function of the optical properties of the material (i.e., its refractive index), its thickness, wavelength, and angle of incidence of the light beam relative to the surface normal. These differences in polarization are measured by an ellipsometer, and the oxide thickness can then be calculated. Capacitance method and cross-sectional TEM are also used in thickness measurements and characterization but are relatively less prominent.

Simulation

As the device dimensions shrink, the oxide layer that is grown has a lot of stringent requirements. Achieving this level of accuracy by experimental trial and error would prove highly time consuming and very expensive. Simulating the process and estimating the optimum thickness to be grown provides many beneficial details such as time required for growing, amount of silicon being consumed, and effective thickness necessary to effectively mask the diffusing species and ions. At present, the most widely used computer program for simulation is SUPREM (Stanford University Process Engineering Modeling) software. A sequential code can be written in this software and can be executed like a batch file to simulate oxide growth, etch process, etc.

Diffusion

Diffusion is the process of introducing impurity atoms into silicon to control the majority carrier type and to change the electrical characteristics of the semiconductor. The diffusion process

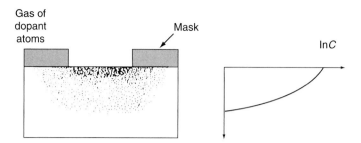

FIGURE 7-39 *Selective diffusion in semiconductor.*

was one of the most significant developments in the IC fabrication processes and enabled solid-state devices to become commercially viable at reasonable cost. The key methods of doping the impurity are diffusion and ion implantation. Nowadays ion implantation is the most commonly used method of doping because of its greater control of dopant depths, and it is used to form shallow layers, whereas the diffusion process is used to form deep layers.

It is well known that semiconductor devices are made by introducing small amounts of specific dopants such as boron, antimony, and arsenic at explicit locations and in carefully controlled concentrations. Such dopants give the semiconductor the electrical properties required. For example, small amounts of arsenic result in "n-type" material, whereas boron as a dopant results in "p-type" material. To realize a satisfactory amplification factor, the location of certain junctions between n- and p-type material—for example, those of the emitter and collector junctions—have to be accurately controlled and the distance between them minimized.

After the deposition of desired impurity selectively, impurity atoms diffuse into the silicon crystal at elevated temperatures via substitutional or interstitial mechanism. When an impurity atom moves among vacancies in the lattice, diffusion is said to occur by a substitutional mechanism and when an impurity atom moves from one place to another without occupying lattice position, diffusion occurs via an interstitial mechanism. Diffusion rate is slow for substitutional mechanism because of limited vacancy concentration, and it provides more controlled diffusion. The doping concentration decreases monotonically from the surface and is shown in Figure 7-39. The dopant distribution surface profile is mainly dependent on diffusion temperature and time.

Fick's Diffusion Equations

Fick's one-dimensional equation can be described as

$$F = -D\frac{\partial C}{\partial x} \tag{7-1}$$

where F is flux of dopant atoms, C is the concentration of impurity, and D is the diffusion coefficient. The continuity equation for particle flux is given by

$$\frac{\partial C}{\partial t} = -\frac{\partial F}{\partial x} \tag{7-2}$$

Combining Equations 7-1 and 7-2 gives Fick's second law of diffusion

$$\frac{\partial C}{\partial t} = D \frac{\partial^2 C}{\partial x^2} \tag{7-3}$$

The initial and boundary conditions are varied to model the diffusion. In constant surface diffusion, impurity surface concentration is kept constant throughout the diffusion. In limited source diffusion the dose remains constant, and impurity surface concentration decreases with time.

Constant Source Diffusion

For this case, the initial condition is $C(x, 0) = 0$, and the boundary conditions are $C(0, t) = C_s$ and $C(\infty, t) = 0$.

Under these conditions, the solution is given as

$$C(x, t) = C_s \, erfc \left(\frac{x}{2\sqrt{Dt}} \right) \tag{7-4}$$

where C_s is the surface concentration and *erfc* is the complementary error function. The doping concentration decreases from the surface. The total number of impurity atoms per unit area is given by dose Q and increases with time, where

$$Q = \int_0^\infty C(x, t) \, dx = 2C_s \sqrt{Dt/\pi} \tag{7-5}$$

As time progresses, impurity dopant atoms penetrate deeper into the semiconductor, whereas the surface concentration remains constant.

Limited Source Diffusion

For this case, the initial condition is $C(x, 0) = 0$, and the boundary conditions are

$$\int_0^\infty C(x, t) = Q \quad \text{and} \quad C(\infty, t) = 0.$$

Under these conditions, the solution is given as Gaussian distribution

$$C(x, t) = \frac{Q}{\sqrt{\pi Dt}} \exp \left(-\frac{x}{2\sqrt{Dt}} \right)^2 \tag{7-6}$$

As time progresses, impurity dopant atoms penetrate deeper into the semiconductor where dose remains constant and surface concentration decreases. Figure 7-40 shows the comparison of Gaussian and complementary error functions.

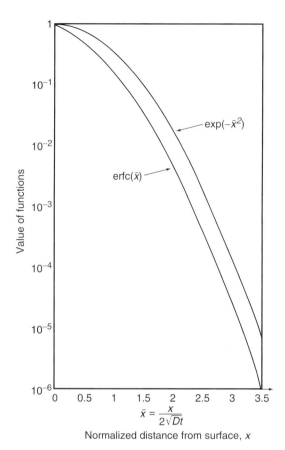

FIGURE 7-40 *Comparison of Gaussian and complementary error functions.*

Diffusion Coefficient

The diffusion coefficient increases exponentially with temperature and follows Arrhenius behavior. The diffusion coefficient can be expressed as

$$D = D_o \exp\left(-\frac{E_a}{kT}\right) \tag{7-7}$$

where D_o is the frequency factor and E_a is the activation energy in eV.

Concentration Dependent Diffusion

At high dopant concentrations, the diffusion coefficient becomes concentration-dependent, and it does not follow the preceding equation when impurity concentration is greater than the intrinsic carrier concentration at diffusion temperature. Concentration-dependent diffusion has a more abrupt profile compared to the case of constant D. Figure 7-41 shows the normalized diffusion profiles where diffusion coefficient becomes concentration dependent.

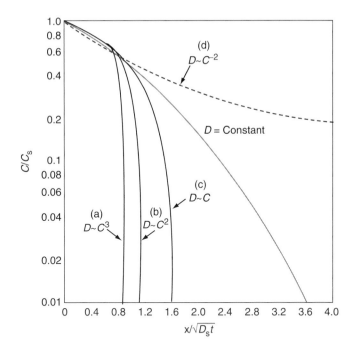

FIGURE 7-41 *Normalized diffusion profiles for extrinsic diffusion with a concentration-dependent diffusion coefficient.*

Lateral Diffusion

Impurity atoms not only diffuse vertically but also diffuse laterally at the edge of mask window. In this case, the one-dimensional diffusion equation becomes invalid and a two-dimensional diffusion equation must be solved. The simulation result of lateral diffusion is shown in Figure 7-42.

Junction Depth

Junction depth x_j is the point at which net impurity concentration becomes zero. At junction depth, the diffused impurity equals the background concentration.

Sheet Resistance

Sheet resistance is given by

$$R_s = \frac{\rho L}{A},$$

where ρ is the resistivity, L is the length, and A represents the cross-sectional area. Sheet resistance can be measured by a four-point probe (Van der Pauw's method). The resistivity ρ of the diffused layer is

$$\rho = R_s x_j. \tag{7-8}$$

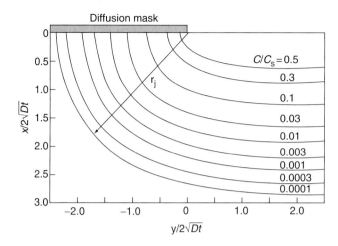

FIGURE 7-42 *Normalized two-dimensional Gaussian diffusion near the edge of mask window in the barrier layer.*

The sheet resistance is uniquely related to the surface concentration and background concentration. The surface impurity concentration versus the product for sheet resistance multiplied by junction depth for different background concentrations has been evaluated and is termed Irvin's curves.

Simulation

SUPREM software can simulate one- or two-dimensional profiles, and the Diffusion command is used to simulate the diffusion profile.

Ion Implantation

Ion implantation is the process in which impurity dopant ions are ionized, accelerated, and implanted into the semiconductor. Ion implantation is most commonly used because it offers many advantages over diffusion. The main advantages of ion implantation (in comparison to diffusion) for the doping of semiconductors are

- Good homogeneity and reproducibility of the profiles
- Precise control of the amount of implanted ions by integrating the current
- Relatively low temperatures
- Implantation through thin layers is possible
- Low penetration depth of the implanted ions
- Much wider range of impurity species

Ion Implanter

The high-voltage ion source produces plasma by breaking up source gases such as arsine and phosphine into charged ions of desired and undesired species. When the charged ions pass through the mass analyzer, undesired ions are filtered. This filtration is achieved by choosing the magnetic field of analyzer based on desired mass-to-charge ratio. The magnitude of the

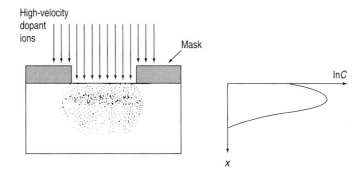

FIGURE 7-43 *Selective ion implantation in a semiconductor.*

magnetic field B may be chosen to select an ion species with a given mass, as follows:

$$|B| = \frac{S}{\sqrt{2\pi}\,\sigma_p} \exp\left[-\frac{(x - R_p)^2}{2\sigma_P^2}\right] \tag{7-9}$$

where V is the accelerator voltage and S is the ion dose per unit area. The accelerator accelerates the ions to their final velocity. The x- and y-axis deflection plates are adjusted to ensure uniform implantation. The acceleration energy controls the average depth of dopant atoms. The semiconductor material can be placed at a relatively low temperature, and processing at low temperature reduces the diffusion.

The average distance that an ion travels before it stops is called the projected range. Because of collisions some ions travel less or more than the projected range and because of statistical nature of the profile, a Gaussian distribution function simulates the implanted ion concentration:

$$n(x) = \frac{S}{\sqrt{2\pi}\,\sigma_p} \exp\left[-\frac{(x - R_p)^2}{2\sigma_P^2}\right] \tag{7-10}$$

where S is the ion dose per unit area, and $n(x)$ is implanted ion concentration.

To implant selectively, silicon dioxide and silicon nitride are commonly used as barrier materials. Photoresist can also be used, but it should be 1.8 times thicker than silicon dioxide, whereas silicon nitride can be 0.85 times thicker than silicon dioxide. Silicon nitride acts as the most effective barrier layer. Figure 7-43 shows the basic approach of ion implantation and schematic dopant concentration profile, and Figure 7-44 shows a schematic diagram of an ion implanter.

Channeling

When the ion beam is improperly oriented with respect to the crystal planes, the ions tend to miss the silicon atoms that end up much deeper in the silicon than expected. The electron interactions, however, stop the channeled ions. Implanting a thin layer can reduce channeling, as does intentionally aligning the wafer such that the planes are tilted and twisted off-axis to the beam.

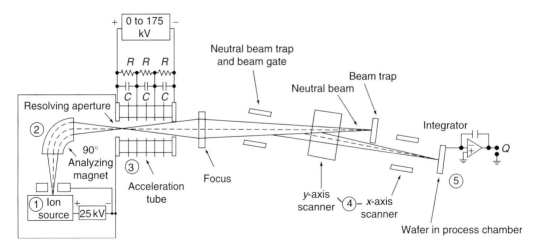

FIGURE 7-44 *Schematic drawing of a typical ion implanter.*

Implantation Damage

During implantation, the impurity ion undergoes a series of collisions and displaces silicon atoms in the lattice. Both the impurity ions and the displaced silicon atoms cause further displacements (cascade collisions). At a sufficiently high dose, the implanted layer will become amorphous; Figure 7-45 gives the dose required to produce an amorphous layer as a function of substrate temperature. Activating the implanted ions to remove the damage can be achieved by annealing.

Annealing

Conventional annealing requires a long time and high temperatures to remove the damage. However, substitutional dopant diffusion may occur in conventional annealing and cannot be used for shallow junctions and narrow doping structures. Dopant type and dose influence the characteristics of annealing. For lower doses, annealing behavior for boron and phosphorous is similar as shown in Figure 7-46. At higher doses, the annealing temperature is higher for boron.

Rapid Thermal Annealing

To avoid the diffusion during annealing, the product, Dt, of the diffusion coefficient and time should be small, which can be achieved by fast ramping. Its advantages include short annealing times (from one second to 5 minutes) as well as high ramp-up and ramp-down rates. However, rapid heating can cause thermal stress and wafer damage. The conventional and RTA technologies are compared in Table 7-4.

Photolithography

Introduction

Photolithography is a pattern definition method in which patterns on a mask are transferred to a light-sensitive material called photoresist on the semiconductor surface. Integrated-circuit technology requires implantation regions, contact windows, and bonding pad areas, and these

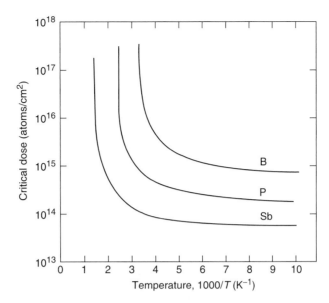

FIGURE 7-45 *Critical dose of implanted ions to form an amorphous layer on silicon as a function of inverse of substrate temperature.*

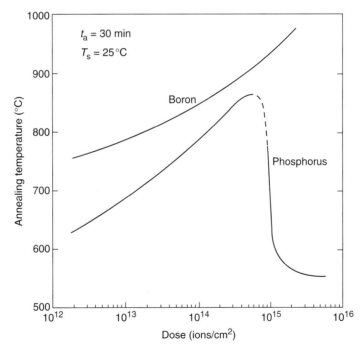

FIGURE 7-46 *Annealing behavior of Boron- and phosphorus-implanted silicon substrates to remove damage.*

TABLE 7-4 Comparison of Conventional and Rapid Thermal Annealing

Determinant	Conventional Furnace	Rapid Thermal Annealing
Process	Batch	Single-wafer
Furnace	Hot-wall	Cold-wall
Heating rate	Low	High
Cycle time	High	Low
Temperature monitor	Furnace	Wafer
Thermal budget	High	Low
Particle problem	Yes	Minimal
Uniformity and repeatability	High	Low
Throughput	High	Low

regions are defined by the patterns on the mask. Because the resist patterns are temporary replicas of the circuit patterns, further processes such as chemical or plasma etching are required to transfer the resist patterns to the underlying layers such as the barrier material on the wafer surface. Successful completion of numerous processing steps is required at each mask level, hence the number of photographic masks used in the IC fabrication process indicates the complexity of the IC process.

Optical Lithography

Optical lithography, the most widely used lithographic method, uses ultraviolet light (wavelength $\lambda \approx 0.2$–0.4 μm) for transferring the pattern onto the photoresist. This section describes the lithographic process along with the exposure tools, mask fabrication, resist and resist processes, and the clean room, which is very important because all lithographic processes require an ultraclean environment.

The Clean Room

An ultraclean environment is an absolute necessity for the lithographic process. The lithographic process is crucial for circuit design, and dust particles in the air that settle on the semiconductor surface or lithographic masks cause defects and have a profound effect on the device's performance. Dust particles on the semiconductor surface can disrupt the growth of single-crystal epitaxial films and cause dislocations. A dust particle in the gate oxide can cause increased conductivity and low breakdown voltage, leading to device failure. Further critical damage is possible when dust particles settle on the lithographic masks. The dust particles on the mask surface behave like opaque regions or patterns on the mask, and during photolithography these patterns are transferred to the underlying layer on the mask along with the circuit patterns. As shown in Figure 7-47, dust particles on the mask can lead to pinholes, reduced conductivity, and short-circuiting, all of which make the circuit useless.

A tight control of the total number of dust particles per unit volume, as well as control of temperature and humidity, is required in a clean room. Particulate contamination is prevented throughout the fabrication process by using vertical laminar-flow hoods in clean rooms, and filtration is used to remove particles from the air. Clean rooms are classified based on the maximum number of particles per cubic foot or cubic meter of air. A class-100 clean room has fewer particles per cubic foot or cubic meter, compared to a class-1000 clean room.

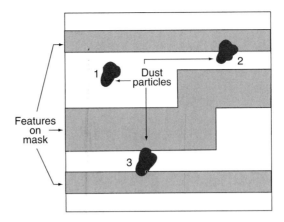

FIGURE 7-47 *Dust particles on a photomark can cause pin holes, reduced conductivity and short-circuit.*

The Pattern Transfer Process

The basic photolithographic process, shown in Figures 7-48 and 7-49, will now be discussed. The process involves all the steps beginning with the cleaning of the wafers to the removal of the photoresist after etching or, in other words, after all processes that transfer and produce the final pattern on the surface of the semiconductor wafer are completed. The photolithographic process begins with the cleaning of the semiconductor wafer. The wafers are chemically cleaned to remove particulate matter on the surface, along with organic, ionic, and metallic impurities. Any oxide that is formed on the surface can be removed using a solution of hydrofluoric acid. All the processes in the microelectronic fabrication, especially cleaning, use highly purified and filtered water called deionized (DI) water. This water has very few traces of all ionic, particulate, and bacterial contaminants and has a resistivity of 18 Mohm-cm.

The cleaning process is followed by the barrier layer formation on the semiconductor wafer surface. As mentioned in a previous section, silicon dioxide is the most common barrier material, but other materials—such as silicon nitride, polysilicon, photoresist, and metals—can also be used as barrier materials. Consider silicon dioxide as the barrier material and silicon as the semiconductor. Once the insulating SiO_2 layer has been formed on the silicon wafer, a thin layer of the photoresist is applied on to the wafer surface. To promote the adhesion of the resist to the wafer surface, an adhesion promoter such as hexamethyl-disilazane (HMDS) is first coated on the wafer to make the wafer surface chemically compatible with the resist, and then the resist is applied. This method allows good compatibility with a number of films on the wafer surface such as silicon dioxide, silicon nitride, aluminum, polycrystalline silicon, and phosphorous doped silicon dioxide.

The silicon wafer is mounted and held on a vacuum chuck, and about 2 to 3 cm^3 of liquid resist is applied at the center of the wafer. The wafer is then accelerated to a constant rotational speed for 30–60 seconds to produce a uniform resist film on the surface. Resist film thicknesses of 2.5 to 0.5 μm have been obtained at speeds of 1000–5000 rpm. The thickness of the resist layer is found to depend on its viscosity and is inversely proportional to the square root of the spinning speed.

The step after spinning the photoresist is the soft baking, or prebaking, of the wafer. This is done to remove the solvent from the resist film and to increase the adhesion of the resist to

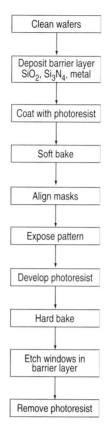

```
┌─────────────────┐
│  Clean wafers   │
└─────────────────┘
         │
┌─────────────────┐
│ Deposit barrier layer │
│ SiO₂, Si₃N₄, metal │
└─────────────────┘
         │
┌─────────────────┐
│ Coat with photoresist │
└─────────────────┘
         │
┌─────────────────┐
│   Soft bake     │
└─────────────────┘
         │
┌─────────────────┐
│  Align masks    │
└─────────────────┘
         │
┌─────────────────┐
│ Expose pattern  │
└─────────────────┘
         │
┌─────────────────┐
│ Develop photoresist │
└─────────────────┘
         │
┌─────────────────┐
│   Hard bake     │
└─────────────────┘
         │
┌─────────────────┐
│ Etch windows in │
│  barrier layer  │
└─────────────────┘
         │
┌─────────────────┐
│ Remove photoresist │
└─────────────────┘
```

FIGURE 7-48 *Flow chart showing the basic steps of a photolithographic process.*

the wafer. Soft bake is done at temperatures ranging from 90 to 120 °C with time varying from 60 to 120 seconds in air or nitrogen atmosphere.

The mask alignment and exposure are the next two steps after the soft-bake procedure. The photo mask is a square glass plate with patterned emulsion or metal film on one side. This pattern should be carefully transferred to the surface of the wafer, and in the case of multiple mask levels, care should be taken to ensure that each mask pattern is perfectly aligned to the previous pattern on the wafer. For this purpose alignment marks are used. These alignment marks are made on each mask and transferred to the wafer surface along with the mask pattern. Some commonly used alignment marks are shown in Figure 7-50.

After aligning the mask, the photoresist is exposed to high-intensity ultraviolet light through the mask. If a positive resist is used, then on developing with the developer solution all the photoresist exposed is removed, leaving bare silicon dioxide in the exposed regions. The mask now contains an exact copy of the pattern that will remain on the surface. If a negative resist is used, then on developing, the unexposed resist regions get washed away while the exposed resist regions remain. In this case, the mask pattern contains the "negative" of the pattern that will remain on the wafer surface. After exposure, the wafer is rinsed and dried. The positive and negative resist patterning are shown in Figure 7.51.

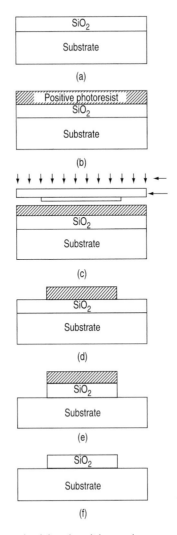

FIGURE 7-49 *The basic approach of the photolithography process.*

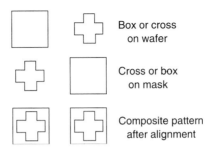

Box or cross
on wafer

Cross or box
on mask

Composite pattern
after alignment

FIGURE 7-50 *Alignment marks used in patterning.*

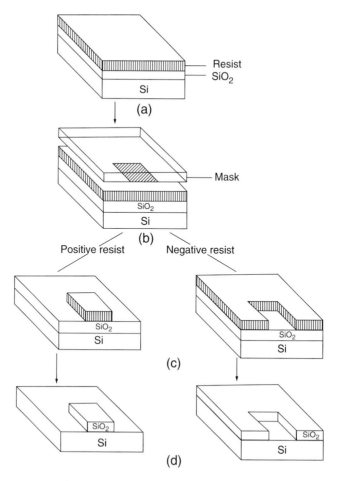

FIGURE 7-51 *Positive and negative resist patterning.*

A baking step called hard baking or postbaking follows the exposure step to increase the hardness and adhesion of the photoresist to the substrate. Hard bake is usually performed in an oven with a temperature range of 100–180°C for 20–30 minutes. The wafer is then etched to expose the bare silicon dioxide layer without affecting the resist layer. The resist is then stripped, using solvents or plasma oxidation, to leave behind the final silicon dioxide pattern on the wafer surface, which can be either the same as the pattern (in the case of positive resist) or reverse (negative resist) of the pattern on the mask. The process of stripping the resist by oxidizing it in an oxygen plasma system is called resist ashing.

Exposure Systems

Exposure systems are of three types, namely, contact, proximity, and projection systems. In contact printing or exposure systems, the mask and wafer are made to contact each other. Although high-resolution pattern transfer can be achieved by this system, it may damage both the wafer and the mask. Proximity and projection printing systems are used as good alternatives.

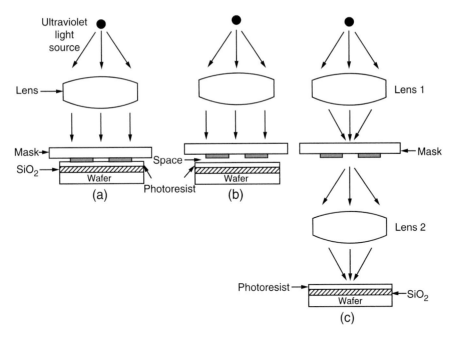

FIGURE 7-52 *The three exposure systems: (a) contact printing, (b) proximity printing, and (c) projection printing.*

In the proximity system, the mask and wafer are brought very close to each other, but they do not touch during exposure, thus avoiding damage to the mask or wafer. In projection printing the dual-lens system is used and the wafer and masks are scanned, or the system operates in a step-and-repeat mode. The masks and lenses in these systems are at a distance in the order of centimeters from the wafer surface. Portions of the mask pattern are projected on the wafer surface. In the case of large-diameter wafers, to obtain uniform exposure and good alignment between the mask levels across the entire wafer, the pattern is aligned and exposed at each die site. The three exposure systems are shown in Figure 7-52.

The performance of the exposure system depends on three parameters: resolution, registration, and throughput. The minimum feature size or dimension that can be accurately transferred on to the wafer surface is termed the *resolution* of the system. The measure of alignment of the masks at different mask levels is called *registration*, and the number of wafers that are exposed in an hour for a given mask level is the *throughput*.

The minimum line width or critical dimension (CD) for contact and proximity printing is given by

$$CD \approx \sqrt{\lambda g} \qquad (7\text{-}11)$$

where λ is the exposure radiation wavelength and g is the distance between the mask and the wafer, which also includes the resist thickness.

For projection printing systems, the minimum feature size F or resolution is expressed as

$$F = k_1 \lambda NA \qquad (7\text{-}12)$$

where k_1 is a process-dependent factor, typically $k_1 = 0.5$; λ is the wavelength of the illumination; and NA is the numerical aperture of the lens, given by

$$NA = n \sin \theta \qquad (7\text{-}13)$$

where n is the index of refraction in the image medium and $n = 1$ (for air). The depth of focus (DOF) is given as

$$DOF = 0.6 \lambda (NA)^2 \qquad (7\text{-}14)$$

Masks and Resists

Masks in IC fabrication are made using computer graphics systems and optical or electron beam pattern generators. Computer graphics systems such as the computer-aided-design (CAD) system is first used to make an image of the desired mask, followed by the activation of the optical or electron-beam pattern generator. In the optical pattern generator, flash lamp exposes the mask image, called the reticle, on the photographic plate. In electron-beam pattern generator, the pattern is transferred directly onto an electron-sensitive material (electron resist). The mask is made of fused-silica substrate with chromium layer and the electron resist. The pattern on the electron resist is then transferred onto the chrome layer to form the final mask.

The photoresist is a light- or radiation-sensitive material and is classified as positive or negative resist. Positive resist contains a photoactive compound (PAC), a base resin, and an organic solvent. The positive resist is insoluble before exposure. On exposure, the chemical structure of the photoactive compound changes and makes the exposed regions of the resist become soluble in the developer solution. Negative resists, in contrast, contain polymers and photoactive compound and are initially soluble. On exposure, the polymers become cross-linked, rendering them insoluble in the developer solution. The unexposed resist regions dissolve in the developer solution, leaving behind the exposed regions.

Thin Film Deposition

Thin films can be deposited by different techniques that broadly fall into two classes, namely vacuum deposition techniques and nonvacuum techniques. Physical vapor deposition (PVD) and chemical vapor deposition (CVD) techniques, discussed in Chapter 5, fall under the first category, whereas chemical bath deposition (CBD), spin-on, and electrodeposition are some of the nonvacuum techniques. Vacuum deposition methods can grow high-quality films with fewer defects, and hence these techniques gained prominence in microelectronics industry. The vacuum techniques are discussed in detail in the following sections.

Vacuum Deposition Techniques
Physical Vapor Deposition

Physical vapor deposition (PVD) has been employed for over 100 years to deposit coatings and films. The technique involves heating a source to elevated temperatures or bombarding the source with high-energy ions in an inert atmosphere under low pressure and thereby knock the atoms from the source to deposit onto the substrate. Evaporation and sputtering belong to this class of methods. PVD generally requires vacuum or very low pressure for the gaseous species to transport from the source to the substrate.

Evaporation

The method involves heating the source to high temperatures so that the source gets enough energy to dissociate into vapor and condense on the substrate. Resistive heating is done by tungsten filament. The magnitude of I^2R factor (I: current, R: resistance) determines the amount of heat supplied to the filament. This process is relatively less clean as compared to PVD or CVD because of the resistance that builds at the contacts and results in the dissipation of wasteful heat. This problem can be overcome by using e-beam heaters, in which the source is heated by high-energy electron beam. E-beam heaters operate at high temperature ranges and hence enable the evaporation of a variety of materials. Co-evaporation can be carried out by evaporating more than one material simultaneously.

Evaporation is a line-of-sight deposition technique, and hence there is a thickness gradient on the substrate. Evaporation is carried out in an ultrahigh vacuum of the order of 10^{-6} Torr. A typical evaporation system is shown in Figure 7-53. Major parameters that determine the quality of the film include the mean free path of gas molecules, distance between the substrate and the source, position of the source with respect to the substrate, and the substrate and source temperatures. The larger the mean free path, the better the quality of the deposited film. Atoms do not impinge on the substrate with high energy; hence, there is no significant damage to the substrate. Because the mean free path is large, the gaseous atoms travel in straight lines, so evaporated films have poor step coverage. Although this problem can be managed by rotating the substrates during evaporation, it cannot be completely eliminated. These problems somewhat limit the possibility of wide-range application of evaporation in silicon processing.

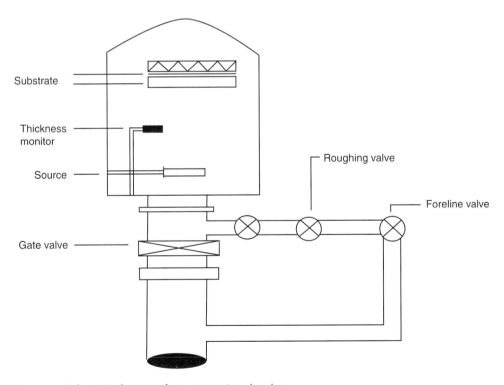

FIGURE 7-53 *Schematic diagram of an evaporation chamber.*

Sputtering

To overcome the challenges inherent in evaporation, another technique called sputtering is employed in depositing thin films. Although sputtering was limited to optical coatings in its early days, it has since found many applications in today's IC industry.

The sputtering technique involves bombarding the target (or source), in the presence of an inert gas, with high energy to generate plasma. Plasma contains a mixture of atoms from an inert gas (usually argon) and ions from the target. These high-energy ions are dislodged onto the wafer surface, depositing a thin film. DC (direct current) or RF power supply can be used, depending on the type of target that is being sputtered. For most metal targets, such as aluminum and copper, DC supply can generate plasma, but for the ceramic targets such as ZnO and Al_2O_3, RF supply is essential. As the ions hit the substrate with high energy, they damage the substrate. Hence sputter yield becomes a critical issue. Sputtering more than one target at a time is called co-sputtering.

Sputtering is carried out at relatively higher pressures than evaporation. Hence the quality of the films obtained are poor because of the contamination levels. But the availability of ultra-high-purity gases and the ability of vacuum pumps to achieve low-pressure ranges of 10^{-8} Torr can yield good-quality sputtered films. A schematic of the sputtering system is shown in Figure 7-54. Unlike evaporation, sputtering is not a line-of-sight deposition method; hence there is good step coverage. The distance between the target and the substrate, the flow rate of sputtering gases, and the base pressure of the system determine the quality of the deposited films.

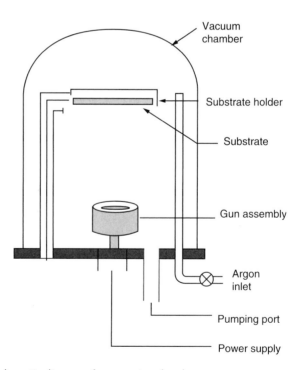

FIGURE 7-54 *Schematic diagram of a sputtering chamber.*

Reactive Sputtering

In reactive sputtering, gases besides argon, such as N_2 or O_2, are used in order for the species to react at the wafer surface and deposit a film. Metals can be reactively sputtered in presence of O_2 or N_2 to form oxides or nitrides, respectively. The properties of the film can be controlled by controlling the stoichiometry of the reacting gases. However, precise control of the stoichiometry is at times difficult and hence sputtering of compound targets is preferred over reactive sputtering of the metal targets. To vary the film composition during sputtering of compound targets, gases such as O_2 or N_2 can be passed along with argon in required proportions.

Radio Frequency (RF) Sputtering

In RF sputtering, DC supply cannot be used to sputter insulating materials or ceramic materials. This is not only because DC provides insufficient voltage to generate stable plasma but also because of arcing problems. Arcing occurs due to charging at the target surface. RF supply can overcome these problems. The setup for the RF sputtering is shown in Figure 7-55. The deposition rates are low in RF sputtering because fewer ions are generated in the gas. The deposition rates can be improved by switching to magnetron sputtering.

Magnetron Sputtering

A strong magnetic field acting perpendicular to the electric field can effectively ionize the gas molecules and enhance the deposition rates. This is achieved by placing strong magnets in the gun assembly below the target. Such magnetron sputtering can be done in both DC and RF modes. Hence most of the gun assemblies available have this feature for both DC and RF sputtering.

Collimated Sputtering

In collimated sputter deposition, there is a metal filter between the target and the substrate. These metal filters, called collimators, are made of aluminum or copper. Ions from the target pass

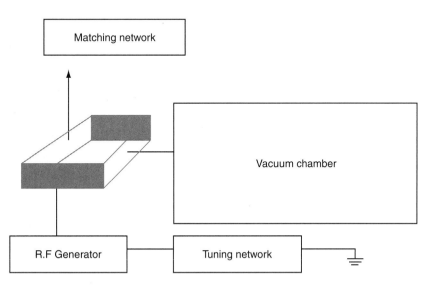

FIGURE 7-55 *Flow-chart showing the different components of a RF sputtering system.*

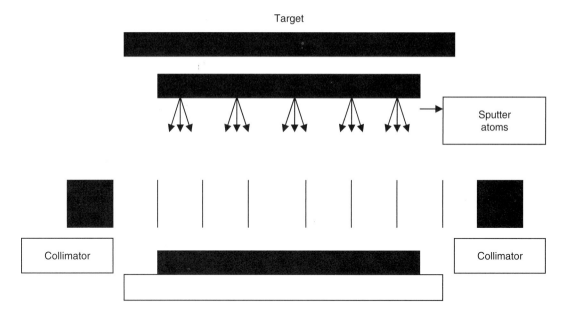

FIGURE 7-56 *Physical representation of collimated sputtering.*

through the collimators and are deposited on the substrate. In this method, there is flexibility in choosing the number of ions that stick to the substrate. However, collimated sputtering cannot achieve sufficient bottom coverage at high aspect ratios.

Directional filtering is obtained by interposing an array of collimating tubes between the target and the substrate, as shown in Figure 7-56. The collimator functions like a side wall of the sputtering chamber. It collects atoms traveling laterally from the target and deposits them on the side walls of the collimator, preventing them from depositing onto the substrate.

Collimated sputtering was initially used for filling trenches but its slower deposition rates and increased effective costs because of frequent replacements of the collimator and its maintenance, have limited its use to liners or contact layers. Thin TiN films sputtered through a collimator have been found to have adequate conformality to the edges and topography as diffusion barrier layers.

Ionized Magnetron Sputtering

An alternative method for filtering the sputtered metal atoms to improve the directionality is achieved by ionizing the sputter metal atoms. An RF coil operating at approximately 13.56 MHz is used to set up the dense plasma. The RF coil ionizes the sputter metal atoms and deposits them onto the substrate. In contrast to collimated magnetron sputtering, this method provides good step coverage even at high aspect ratios. Hence this method is employed for depositing layers on trenches and vias that have smaller features.

Chemical Vapor Deposition

In chemical vapor deposition (CVD), a gas containing a metal or an insulator is sprayed on the wafer. After the reactor tube is pumped down to the required pressure range, reactant species or precursors are introduced into the tube. The precursor gases react on a heated wafer surface, forming a thin film. The wafers can be heated either by regular heaters, by RF power coils, or a combination to achieve the reaction. The three basic reaction steps that take place at the wafer surface in CVD involve (1) absorption of the reactant gases, (2) diffusion of the reactants, and (3) desorption of unwanted species. CVD can be used in depositing amorphous, polycrystalline, and single-crystalline materials (Sze, 1988). The deposition rates in a CVD process are high. The uniformity across the wafer surface can be readily achieved. The major disadvantage in a CVD process is its large flow rates of reacting gases.

There are different kinds of CVD reactors such as APCVD (atmospheric pressure CVD), which operates at atmospheric pressures; LPCVD (low pressure CVD), which operates at relatively lower pressures in comparison to APCVD or CVD; PECVD (plasma-enhanced CVD), where a plasma is generated either by RF or DC coils during the deposition; MOCVD (metal organic CVD); MPCVD (microwave plasma–assisted CVD); EBCVD (electron beam–assisted CVD); and HDPCVD (high-density plasma CVD) (Intel Corporation, 2003). The choice of process depends on the operating temperature and the quality of the film required.

Epitaxy

Epitaxy is derived from the Greek words "epi," meaning "on," and "taxis," meaning "arrangement." Epitaxy is a high-temperature CVD process in which a single-crystal layer is grown on a single-crystal substrate. In other words, it is an orderly arrangement of atoms on a single-crystal substrate. VPE (vapor-phase epitaxy), LPE (liquid-phase epitaxy), MBE (molecular beam epitaxy), and MOMBE (metal organic MBE) are a few of the different epitaxial growth techniques.

Molecular Beam Epitaxy

MBE is another vacuum deposition technique that is widely used in depositing compound semiconductors such as ZnSe, ZnTe, CdTe, etc. In this technique, a beam of molecules from a source are evaporated or transported on to a substrate without any collisions. Epitaxial layers can be grown at lower temperatures, and there is a precise control over the thickness of the film. This technique is not widely used in depositing films on silicon substrates because of the processing difficulties involved in the surface preparation.

Nonvacuum Deposition Techniques

In addition to the different vacuum deposition methods described above, nonvacuum deposition techniques, especially CBD (chemical bath deposition) and spin-on methods, are also used for film deposition on semiconducting substrates.

Chemical Bath Deposition

Chemical bath deposition (CBD) is a popular nonvacuum method because the process is simple and inexpensive compared to any vacuum deposition technique. Moreover, CBD can produce high-quality films on large-area substrates, thus enhancing the yield. The rate of deposition depends on the deposition temperature, precipitation temperature, and pH of the solution. For the reaction to take place, first the required quantities of chemical solutions are added into the reactor unit under favorable conditions.

Spin-On Deposition

In the spin-on deposition technique, materials are applied on to the silicon wafer in liquid form. Liquid is dispensed onto the wafer surface in a predetermined amount and the wafer is spun rapidly at rates of up to 6000 rpm. During spinning, liquid is uniformly distributed on the surface by centrifugal forces. This is followed by a low-temperature bake. Photoresists are usually spin-coated. Recently, materials such as organic, inorganic, and hybrid compounds have also been deposited by the spin-on technique. This method is mostly used in putting deposits on dielectrics. Spin-on dielectrics are cost-effective.

Etching

Introduction

As the circuit complexity increases, more metal lines have to be laid on an integrated chip. To create metal lines, it is necessary to remove material from certain portions on the wafer. Apart from laying metal lines, trenches, vias, holes, and windows have to be made on the wafer surface to deposit different materials such as barrier layers, gate oxides, etc. Removal of this unwanted material is called etching. Hence etching is an important step in IC processing. It can be done by different techniques. If etching employs wet processes involving acid chemistry, then it is called wet etching, whereas etching by implementing dry processes such as plasma etching is called dry etching.

To better understand etching, it is essential to know a few key terms defined below:

1. *Etching:* The process of removing unnecessary material from the substrate
2. *Etchant:* Chemical reagent used for removing the material
3. *Isotropic etching:* Equal etching in all the directions
4. *Anisotropic etching:* Unequal etching in different directions
5. *Mask:* Usually a chrome plate with the pattern on it that has to be transferred onto the substrate
6. *Photoresist*: The chemical reagent that helps transfer the pattern on the mask to the surface of the wafer
7. *Bias:* The lateral difference between the etched image and the image on the mask
8. *Tolerance:* A measure of statistical distribution of bias values that characterizes the uniformity of etching
9. *Selectivity of an etch process:* The ratio of etch rates of different materials
10. *Overetch:* More etching than required
11. *Etch rate:* The rate at which an etchant can etch the wafer surface

Etching Types
Wet Etching

Etching that involves chemical reagents such as BOE (buffered oxide etch), is called wet etching. BOE is a solution containing Hydrofluoric Acid (HF). This process can be used to etch windows in SiO_2. It can be done by immersing the wafers in a BOE solution. Factors that determine the etch rate are temperature, thickness of the film to be etched, and the type of film to be etched.

Wet etching is an isotropic process, so this technique etches equally in all directions. Figure 7-57 shows an isotropic etching profile. This technique uses large quantities of chemical reagents, the disposal of which is a potential problem. Hence wet etching is not preferred in microelectronic fabrication.

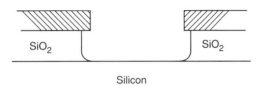

FIGURE 7-57 *A patterned silicon wafer.*

Dry Etching

To overcome the challenges faced in wet process, dry etching is widely used in the IC industry. Unlike wet etching, dry etching is anisotropic. The process avoids problems caused by under-cutting and the disposal of large quantities of waste materials. There are different types of dry etching:

1. Ion beam milling
2. Plasma etching
3. Reactive ion etching

Ion Beam Milling

Ion beam milling is the technique of removing unwanted material by bombarding the surface to be etched with high-energy ions. The sample is loaded in a vacuum chamber.

Figure 7-58 shows the diagram of an ion-beam milling system. After pumping down the chamber to pressure of around 10^{-6} Torr, the chamber is filled with an inert gas such as argon, and the pressure in the chamber is brought to 10^{-4} Torr. The sample to be etched away is

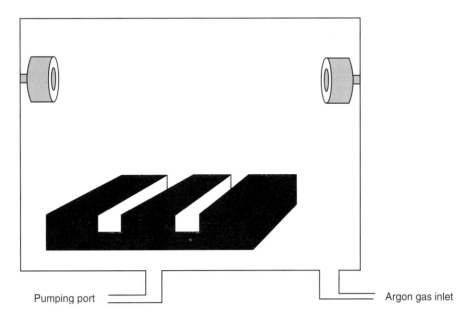

FIGURE 7-58 *Schematic diagram of ion-beam milling.*

TABLE 7-5   Comparison of Wet and Dry Etching

Wet Etching	Dry Etching
Involves acids such as HF	Involves plasma process
Highly isotropic etching profiles are achieved	Highly anisotropic etching profiles are achieved
Less clean	Much cleaner
Not preferred in microelectronic fabrication	Gaining high popularity in IC industry
Problems due to undercutting	Problems due to undercutting do not prevail

bombarded with ionized gas atoms. These high-energy ions remove thin layers of materials on the surface of the specimen, and holes or windows can be made on the wafer surface.

Plasma Etching

In *plasma etching*, an RF source produces chemical species in presence of an inert gas that reacts with the material to be etched. RF supply operates at a frequency of 13.56 MHz. Etching is done by physically knocking the atoms from the surface of the wafer. Plasma etching gives a highly anisotropic etching profile, so problems such as undercutting can be avoided. The by-products obtained are usually in gaseous form and can be pumped down easily.

Reactive Ion Etching

Reactive ion etching is a combination of plasma etching and sputter etching. The generated plasma ionizes gas molecules, and these ionized gas molecules bombard the wafer surface with high-acceleration voltage. The removal of material from the wafer surface is achieved by the combination of a chemical reaction and the ion bombardment. Table 7-5 shows a comparison between wet etching and dry etching.

Metallization

Metallization is an important part of integrated-circuit fabrication, because it is required to achieve contact between devices and also between the device and the outside world. For this purpose conductive films are required. Although metals with low resistivity are an obvious choice, not all of them meet all the desired characteristics, and different applications use materials other than metals for metallization. For gates and contacts, for example, polysilicon, silicides, nitrides, carbides, and aluminum have been used.

Ohmic Contacts

For low-resistance contacts to a semiconductor, it is necessary to have an "ohmic" contact between the metal and the semiconductor. Ohmic contacts have a straight I-V (Current Voltage) characteristic and have a negligible contact resistance compared to the device resistance. The contact resistance is approximated as $R_c = p_c/A$, where ρ_c is the specific contact resistivity in ohm-cm^2 and A is the contact area. The specific contact resistivity p_c is current density (J) divided by voltage (V) and is given as

$$p_c = (dV/dJ)_{v=0}$$

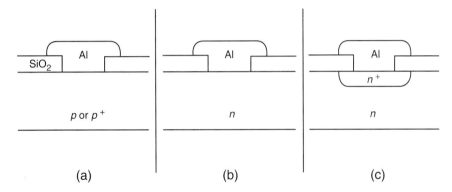

FIGURE 7-59 *Three different types of contact of Aluminum with the semiconductor: (a) ohmic contact, (b) rectifying contact, and (c) practical ohmic contact.*

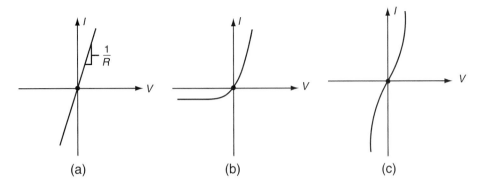

FIGURE 7-60 *Current (I)–Voltage (V) characteristics of the Aluminum contacts: (a) ohmic contact, (b) rectifying contact, and (c) practical ohmic contact.*

The specific contact resistivity is found to depend on the Schottky barrier height of the metal and the doping density in the semiconductor. Figure 7-59 gives the three different contacts possible when aluminum contacts a semiconductor. When aluminum contacts a p-type region, ohmic contact is obtained. Schottky or rectifying contact is obtained when aluminum contacts lightly doped n-type semiconductor, and a practical ohmic contact is obtained when the contact is made to a heavily doped n-region. Figure 7-60 gives the I-V characteristics of the possible contacts that are obtained with aluminum.

Metals and Alloys
The most extensively used metal for integrated circuits has been aluminum. Aluminum and its alloys have low resistance and are deposited by PVD or CVD. Although the adhesion of aluminum to silicon dioxide is good, it suffers from spiking and electromigration when made to contact shallow junctions.

Junction Spiking
Junction spiking is the diffusion of the silicon into aluminum when the contact is annealed at 450 to 500 °C in an inert atmosphere. This poses a serious problem in shallow junctions, as the

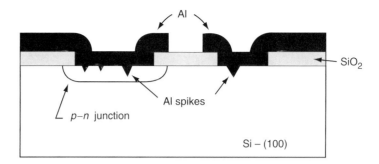

FIGURE 7-61 *Typical IC packages. Schematic view of Aluminum spiking in silicon.*

spikes can lead to junction shorts. Silicon is absorbed into the aluminum only at certain points, and aluminum spikes are formed as shown in Figure 7-61. Junction spiking can be avoided by adding silicon to the aluminum during the annealing, so the silicon from the semiconductor is not absorbed. Barrier materials such as polysilicon and metals such as platinum, palladium, titanium, and tungsten can also be applied between the aluminum and semiconductor to avoid junction spiking.

Electromigration

Electromigration is an important phenomenon that occurs at high current densities. At high current densities, the metal atoms start moving as they gain momentum from the electrons carrying current. As a result of this momentum transfer, there is an accumulation of metal at some regions (hillocks) and depletion of metal in other regions (voids). Hillocks can cause short-circuits, whereas voids lead to open circuits between conductors. Aluminum is made more resistant to electromigration by adding a small percentage of a heavier metal such as copper. Aluminum-copper-silicon alloys have better electromigration resistance, and they also avoid junction spiking.

Copper Metallization

The interconnect network is characterized by Resistance Capacitance time delay, which increases as the dimensions approach the submicron regime. High-conductivity wiring and low-dielectric-constant insulators reduce this delay substantially. The aluminum wiring has been replaced with copper because copper has a higher conductivity than aluminum and also is more resistant to electromigration. PVD, CVD, and electrochemical methods are used for deposition of copper. However, copper suffers from a few drawbacks, such as corrosion under the manufacturing conditions and lack of good adhesion to the dielectric layer. Also, copper is less amenable to dry etching than Al.

Multilevel copper metallization has been fabricated by several techniques. In one technique, the metal lines are patterned and then the dielectric layer is deposited. In another technique, called the damascene process, the dielectric layer is patterned and the copper metal is deposited into the trenches. The excess metal on the surface of the dielectric layer is removed by chemical mechanical polishing.

Damascene Technology

Damascene or dual-damascene process is used to fabricate the copper/low-k (low dielectric constant) interconnect structure. The dual-damascene process is shown in Figure 7-62. As previously explained, in the damascene process the interlayer dielectric (ILD) is patterned to form trenches for metal lines. The trenches are then filled with TaN followed by copper. TaN acts

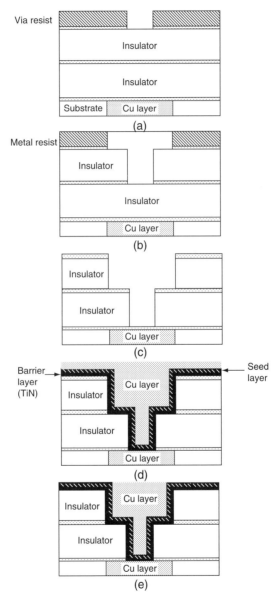

FIGURE 7-62 *Dual-damascene process flow: (a) resist patterning, (b) reactive ion etching of dielectric for via and resist patterning for metal lines, (c) trench and via definition, (d) copper deposition, and (e) Chemical Mechanical Polishing (CMP).*

as the diffusion barrier to copper and prevents copper from diffusing into the interlayer dielectric at higher temperatures. Chemical mechanical polishing is then used to remove the excess copper metal on the dielectric surface for a planarized structure.

The dual-damascene process is an extension of the standard damascene process. In the dual-damascene process, the via plug is formed with the same material as the metal line to avoid via electromigration failure. The interlayer dielectric is patterned by two lithography and reactive ion etching (RIE) steps to form via and trenches, which are filled with copper metal. As in the damascene process, excess copper is removed using chemical mechanical polishing.

Chemical Mechanical Polishing (CMP)

Chemical mechanical polishing is a very important technique for multilevel metallization, as it offers many advantages. Most important is its ability to provide better global planarization over large or small structures. Other advantages include reduced defect density and avoidance of plasma damage.

A general schematic of the CMP polisher is shown in Figure 7-63. In the CMP process, the surface to be polished is moved against a pad. A chemical solution with abrasive particles, called the slurry, is allowed to flow continuously between the surface and the pad. The abrasive particles in the slurry cause the removal of the surface material either by mechanically removing the material to be dissolved in the slurry and washed away or by loosening the material for enhanced chemical reaction. Planarization can also be achieved by mechanical grinding but at the cost of increased surface damage. The three main parts of the CMP process are the surface to be polished, the pad that transfers the mechanical action to the surface to be polished, and the slurry.

Silicide

Silicon forms compounds with many noble and refractory metals. These compounds are called silicides. Silicides are used in ULSI applications, since they have low resistivity and high thermal stability. Silicides of titanium, tungsten, platinum, and palladium are extensively used for interconnection purposes. Silicides are used for MOSFET gate electrodes. They are used as such or with doped polysilicon, called polycide, above the gate oxide. Silicides also enable

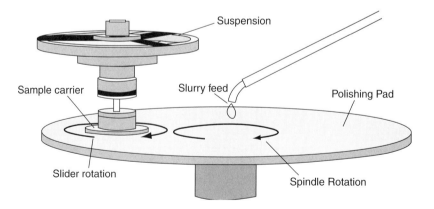

FIGURE 7-63 *General schematic of a CMP polisher.*

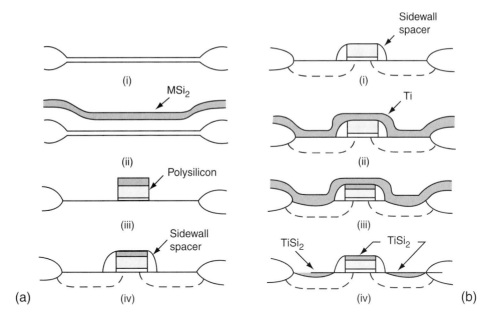

FIGURE 7-64 *Process flow for polycide and silicide formation: (a) Polycide process: (i) gate oxide formation, (ii) polysilicon and silicide deposition, (iii) polycide patterning, and (iv) spacer and source/drain implant. (b) Salicide process: (i) gate patterning, spacer formation and source/drain implant; (ii) Ti deposition, (iii) salicide formation by annealing, and (iv) selective etch to remove unreacted Ti.*

oxidation of the surface after silicide formation. A silicon dioxide insulating layer is formed at the silicide surface when silicon diffuses through the silicide layer at high temperatures and reacts with oxygen at the silicide surface.

The sheet resistance of diffused interconnections and the contact resistance of the source, drain, and gate electrodes are reduced by using silicides. Self-aligned silicides called salicides are used to improve performance of submicron devices and circuits. These silicides are automatically self-aligned to the gate and source–drain regions. The salicide and polycide formation processes are shown in Figure 7-64. Figure 7-64a is a typical polycide formation process in which a layer of metal is first deposited over the polysilicon by evaporation, sputtering, or CVD techniques. Silicide is formed as the metal reacts with the polysilicon on heating the structure between 600 and 1000 °C. Tungsten and tantalum silicides are the most commonly used silicides for the polycide process, as they are thermally stable and resistant to chemical processes. Figure 7-64b shows a typical salicide formation process. The polysilicon gate is first patterned without any silicide. A side wall spacer of silicon oxide or silicon nitride is formed on the side of the gate. This prevents silicide formation on the side of the gate and avoids shorting the gate and the source and drain diffusions. A metal such as Ti or Co is then deposited on the entire wafer by blanket-sputtering and sintered. Silicides are formed only where the metal is in contact with silicon or polysilicon. Selective etch that does not attack the silicide is used to remove the unreacted metal.

Packaging

An electronic package is the plastic, ceramic, or metal enclosure that houses an integrated circuit on a silicon or metal die. Traditionally, an electronic package was an electrically passive part of a microelectronic component that surrounded the IC. Today, as semiconductor devices become significantly more complex, transistor count in products is expected to exceed 100 million; electronic packaging technology has advanced and become much more complex.

A package can contain a single die or several. The die, or chip, is the single square or rectangular piece of semiconductor material onto which a specific electrical circuit has been fabricated. IC packaging has four important functions for all chips: protection from the environment and handling damage, interconnections for signals into and out of the chip, physical support of the chip, and heat dissipation.

Packaging the IC permits the microchip to function in a wide range of customer environments, such as in a notebook computer, in the engine compartment of an automobile, and sandwiched between the plastic layers of a credit card. The various environmental conditions of contamination, moisture, temperature, mechanical vibration, and physical abuse all must be taken into account when designing the IC package. So the package is selected to meet these design constraints: performance, size, weight, reliability, and cost objectives. Many packaging variations exist in the industry; some of the most common packages are shown in Figure 7-65.

Traditional Packaging

Two of the most widely used types of conventional integrated-circuit packages are usually ceramic (for high-reliability applications) or plastic. With ceramic packages, the die is bonded to a ceramic base, which includes a metal frame and pins for making electrical connections outside the package (Figure 7-66). Wires are bonded between bonding pads on the die and the metal frame. The package is usually sealed with a metal lid.

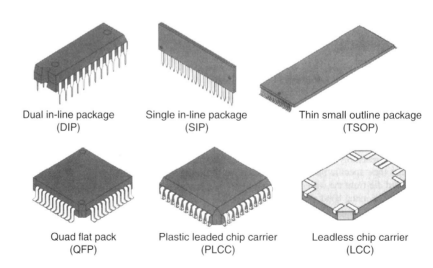

Dual in-line package
(DIP)

Single in-line package
(SIP)

Thin small outline package
(TSOP)

Quad flat pack
(QFP)

Plastic leaded chip carrier
(PLCC)

Leadless chip carrier
(LCC)

FIGURE 7-65 *Typical IC packages.*

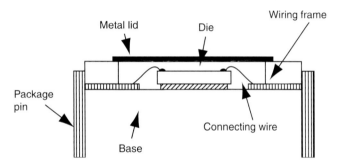

FIGURE 7-66 *Ceramic packaging.*

With plastic packages, a similar process may be used: attach to base, and bond the wires, and seal with lid. After wire bonding, however, it is also possible to mold the plastic package around the device. Plastic packaging has been an industry mainstay since its introduction in the 1960s. A key feature of plastic packaging is that the design lends itself to high-volume production techniques. The lead frames (with the die attached and wire bonded) are in strips and move through the process in racks to simplify handling. A major reason why plastic package has remained popular is flexibility for the shape of the leads, either as pin-in-hole (PIH) leads or surface mount technology (SMT) leads. PIH leads pass through the circuit board, whereas SMT leads attach to the board surface. Components with SMT are preferred because they permit higher-density packaging for both the IC component and circuit board. There are many different types of plastic packaging, such as dual-in-line package (DIP), single in-line package (SIP), thin small outline package (TSOP), and quad flatpack (QFP). These three different plastic package are shown in Figure 7-65.

Advanced Packaging

The advanced package is like the traditional electronic package in that it protects the IC inside, and connects the IC's microelectronic circuitry to the rest of a functioning system outside. However, with a greater number of functions integrated on a die or chip of silicon, the packaging process has been designed to supply the multilevel insulating and interconnecting layered package substrates that are almost as advanced, electronically, as the IC itself.

Flip chip is a packaging technique of mounting the active side of chip toward the substrate. This is currently the package design with the shortest path from the chip devices to the substrate, producing a good electrical connection for high-speed signals. There is also a reduction in weight and profile, because lead frames or plastic packages are often not used.

Multichip modules (MCM) are another aspect of microengineering technology. In the search for ever-faster computers and electronic devices, it is desirable to keep the connections between chips as short as possible. This is leading to the development of MCMs, where many dies are assembled together into one module. An MCM is defined as having a silicon-chip module surface area that covers 30% or more of the substrate surface area. The most common MCM substrates are ceramic or advanced printed circuit boards with high chip density. MCM design enhances electrical performance by reducing circuit resistance and parasitic capacitance while reducing the size and weight of the total package.

Wafer-level packaging is the formation of the first-level interconnections and the package I/O (in/out) terminals on the wafer before it is diced. Cost saving is a big factor in IC packaging.

As chip geometrics shrink, the cost of packaging becomes a greater percentage of the total IC component cost. The ultimate goal of wafer-level packaging is to provide high-density IC packaging while still at the wafer level, thus unifying the front-end and back-end process to reduce process steps, for a substantial cost savings.

Final packaging assembles each chip into a package for protection and attachment to the higher level of assembly. Traditional package has four steps: wafer preparation, die separation, die attachment, and wire bonding. Traditional packaging is oriented around different types of plastic or ceramic materials. Advanced packaging methods are flip chip, multichip modules (MCM), and wafer-level packaging. These techniques are all considered the chip-scale packages. Currently flip chips are widely used, but the industry is moving toward wafer-level packaging for lower cost and greater efficiency.

Yield and Reliability

Yield

Two conditions must be satisfied for Very Large Scale Integration (VLSI) to be a useful, growing technology. First, the fabrication circuits must be capable of being produced in large quantities at costs that are competitive with alternative methods of achieving the circuit and systems function. Second, the circuits must have the ability to perform their function throughout their lifetime.

Ideally, in a properly fabricated wafer of VLSI circuits, we would expect all the circuits on the wafer to be good, functional circuits. In practice, the number of good circuits may range from close to 100%, to only one or fewer good circuits per wafer. Usually, three basic problems cause these less-than-perfect yields: parametric processing problems, circuit design problems, or random point defects in the circuit. Each of these is discussed below.

Processing Effects

One of the most obvious features is that if we look at the entire wafer, there are some regions with a very high proportion of good chips and other regions where the yield of good chips is very low or even zero. The processing effects may be the reason leading to the existence of low-yield regions. These effects include variations in the thickness of oxide or polysilicon layers, in the resistance of implanted layers, in the width of lithographically defined features, and in the registration of a photomask with respect to previous masking operations. Also, many of these variations depend on one another.

Circuit Sensitivities

In addition to areas of a wafer where the yield is low because processing difficulties have led to device parameters outside the specified range, certain areas of a wafer may have low device yield because the design of the circuit has failed to take into account the expected variation in device parameters and correlation between variations in different parameters. It often happens that two circuits with same size and processing techniques have vastly different yields. The low yield is due to the designer and engineer's lack of understanding about the sensitivity of the circuit to device parameters. The higher yield requires the circuit designer to cooperate with the process engineer; once circuit sensitivities to specific process parameters have been determined, high yield and low cost can be achieved at the same time.

Point Defects

A point defect is a region of the wafer where the processing is imperfect, and the size of this region is small compared to the size of the chip (for example, a chip 2000 μm square with features that have a size of 2 μm). Many types of processing defects are considered point defects. One of the most common causes of this type of defect is dust or other particles in the environment. These particles may fall on wafers in the processing facilities, or they may be generated in a film deposition operation. They may also be present in photoresist solutions during the masking procedure. Point defects can occur on silicon wafers as well as on lithographic masks. Successful IC fabrication requires continued monitoring of the density of point defects. Monitoring can be carried out by visual or scanning electron microscopy (SEM) observation. When any defects of a particular operation are observed, the appropriate corrective action must be taken. The circuit yield and the cause of circuit failure must also be monitored. The use of a circuit in which a failure can be related back to a specific area of the chip has been most useful to monitor, control, and reduce the point defects.

The previous paragraphs examined some causes that lead to loss of yield in the IC manufacturing. The dependency on the technology, the type of circuit, the particular circuit design, and the maturity of the processing technology, facility, and the circuit design constitute the yield/loss mechanism. This mechanism changes with time—as the technology becomes more and more mature, the yield causes are identified and optimized. And even when both process and design have been optimized, the identification and elimination of loss mechanisms will continue.

Reliability

A generally accepted definition of reliability is the probability that an item will perform a required function under stated conditions for a stated period of time. The required time must include a definition of satisfactory operation and unsatisfactory operation, or failure. For an IC, the required function is defined by a test program for an automatic test. However, these kind of tests are not complete for testing the circuits under all required conditions. As new failure modes are identified, the appropriate tests are included in a new test program. The stated conditions comprise the total physical environment, including the mechanical, electrical, chemical, and thermal conditions. The stated period of time is the time during which satisfactory operation is required.

Assuming that a device starts working at time $t = 0$, the probability that the device will fail at or before time t is given by the function $F(t)$. This function is a cumulative distribution function (CDF):

$$F(t) = 0, \quad t < 0$$

$$0 \leq F(t) \leq F(t'), \quad 0 \leq t \leq t'$$

$$F(t) \rightarrow 1, \quad t \rightarrow \infty$$

The reliability function $R(t)$ is the probability that the device will survive to time t without failure. The function is $R(t) = 1 - F(t)$. During the earlier stage, the device's failure rate is high, but it decreases as a function of time. The causes of these earlier failures are generally manufacturing defects. In most cases, the devices stop working because of this earlier failure. During the steady-state period, the failure rate is generally low and fairly constant. Eventually the device reaches its lifetime or wearout time. In most ICs, devices are expected to work continuously and no wearout mechanisms should be observed, but some failure mechanisms such as mobile ion shift or corrosion can be observed in low-quality ICs.

In modern VLSI technology, maintaining high yield and reliability is the ultimate goal. But the finer dimension, larger chip area, more complex process, and new materials used in VLSI lead to high defect density. So identifying, characterizing, and eliminating the causes of circuit failure and low yield is a necessary part of the development of VLSI technology.

References

Borchardt-Ott, W. *Crystallography*. New York: Springer, 1993.

Chang, C. Y., and S. M. Sze. *ULSI Technology*. New York: McGraw-Hill, 1996.

Electronic Industries Association. *1994 Electronic Market Data Book*. Washington, DC: Electronic Industries Association, 1994.

Gasiorowicz, S. *Quantum Physics*. New York: Wiley, 1974.

Hong, X. *Introduction to Semiconductor Manufacturing*. Upper Saddle River, NJ: Prentice Hall, 2004.

Intel Corporation website, research pages. © 2003 Intel Corporation. http://www.intel.com/research/.

Jaeger, R. C. *Introduction to Microelectronic Fabrication*. Upper Saddle River, NJ: Prentice Hall, 2002.

Kalpakjian, S., and S. R. Schmid. *Manufacturing Processes for Engineering Materials*. Upper Saddle River, NJ: Prentice Hall, 1999.

May, G. S., and S. M. Sze. *Fundamentals of Semiconductor Fabrication*. New York: Wiley, 2004.

Mayer, J. W., and S. S. Lau. *Electronic Materials Science: For Integrated Circuits in Si and GaAs*. New York: Macmillan, 1994

Quirk, M., and J. Serda. *Semiconductor Manufacturing Technology*. NJ: Prentice Hall: Upper Saddle River, 1995.

Semiconductor Industry Association. *The International Technology Roadmap for Semiconductors*. San Jose, CA: Semiconductor Industry Association, 1999.

Sze, S. M. *VLSI Technology*. New York: McGraw-Hill, 1988.

Wolf, S., and R. N. Tauber. *Silicon Processing for the VLSI Era, Vol. 1, Process Technology*. Lattice Press, 1999.

Zebel, P. J. Current status of high performance silicon bipolar technology. *14th Annual IEEE GaAs IC Symposium. Tech Digest 15,* 1992.

8   Nanomaterials and Nanomanufacturing

Introduction

Nanomaterials are an emerging family of novel materials that could be designed for specific properties. These materials will probably bring about significant shifts in the manner we design, develop, and use materials. For example, nanomaterials that are 1000 times stronger than steel, and 10 times lighter than paper, are cited as a possibility. The following properties can presumably be tailored: resistance to deformation and fracture, ductility, stiffness, strength, wear, friction, corrosion resistance, thermal and chemical stability, and electrical properties. With the emergence of novel fabrication and characterization technologies, new combinations of nanomaterials, or nanocomposites are beginning to be synthesized and characterized. Examples include nanoparticle-reinforced polymers for replacing structural metallic components in the auto industry for reduced fuel consumption and carbon dioxide emissions, and polymer composites containing nano-size inorganic clays as replacement for carbon black in tires for production of environmentally friendly, wear-resistant tires. Nanoparticles and fibers are too small to have substantial defects, and can be made stronger and used to develop ultra–high-strength composite materials.

The unique properties of nanomaterials result from the extremely large surface and interface area per unit volume (e.g., grain boundary area) and from the confinement effects at the nanoscale. These special attributes enable design of nanostructured materials that are harder and stronger but less brittle than comparable bulk materials with the same composition. The properties of *isolated* nanostructure units do not reflect the behavior of bulk nanomaterials (coatings, composites, etc.) that contain nanostructure networks in which the impact of interactions between nanostructured units modifies the properties. Nanoporous materials with large surface area such as aerogels and zeolites already offer improved chemical synthesis (faster reaction rates), cleanup (adsorbents), and separation (membranes, nanofilters) methods for chemical and biomedical sectors. In aeronautics and space exploration, lighter, stronger, and thermally stable nanostructured materials will permit fuel-efficient lifting of payloads into orbit, and reduced dependence on solar power for extended periods for travel away from the sun.

Nanotechnology is important for several reasons. Patterning matter at nanoscale will make it possible to control the fundamental properties of materials without changing their bulk composition (e.g., nanoparticles of different sizes emitting light at different frequencies and color, and nanoparticles of sizes comparable to magnetic domains for improved magnetic devices).

The ability to synthesize nanoscale building blocks with precisely controlled size and composition and then to assemble them into larger structures is believed to enable lighter, stronger, programmable, and self-healing materials to be synthesized. It will reduce life cycle costs through lower failure rates and permit molecular/cluster manufacturing (nanoscale manipulation and assembly of molecules). Improved printing ability with the use of nanoparticles (nanolithography), nanocoatings for cutting tools, and electronic and chemical applications are other potential benefits. Structural carbon and ceramic materials significantly stronger than steel, better heat-resistant polymeric materials stronger than the present generation of polymers, and nanofilters capable of removing finest contaminants from water and air are other potential benefits.

In the realm of nanoelectronics and computer technology, continued improvements in miniaturization, speed and power reduction in information-processing devices will be possible. Potential breakthroughs include nanostructured microprocessor devices, communication systems with higher transmission frequencies, small storage devices with capacities at multiterabit capacity, and integrated nanosensors capable of collecting, processing, and communicating massive amounts of data with minimal size, weight, and power consumption. Other benefits could include more sophisticated virtual reality systems for education, national defense, and entertainment, nanorobotics for nuclear waste management, chemical, biological, and nuclear sensing, and nano- and micromechanical devices for control of nuclear defense systems.

Prior to the emergence of nanoscience and nanotechnology as a discipline, several industrial sectors (chemical, aerospace) had developed novel technologies using the power of nanostructuring but without the aid of nanoscale analytical capabilities. For example, the aerospace industry had developed heat-resistant superalloys by dispersing 1 to 100 nanometer oxide particles in nickel superalloys with vastly improved elevated temperature strength and thermal stability, and the chemical industry had developed nanoporous catalysts with pore size of $\sim$1 nm (their use is now the basis of a \$30 billion/year industry).

Nanoporous ceramics for catalysis and filtration have been developed by the chemical industry. For example, several years ago Mobil Oil Company discovered a new class of zeolites, with pore size in the range of 0.45 to 0.6 nm, that is now widely used in hydrocarbon-cracking processes. The company also developed a porous aluminosilicate with 10-nm-size cylindrical pores; this development has been applied to both catalysis and filtration of fine dispersants in the environment. The discovery of the nanoporous material MCM-41 by the oil industry led to innovations in purification technologies (e.g., removal of ultrafine, 10–100 nm, contaminants from liquids and gases).

Nanotubes, Nanoparticles, and Nanowires

Carbon nanotubes (CNT) are relatively new materials, discovered by Iijima in 1991, and were first observed as a minor by-product of the carbon arc process that is used to synthesize fullerenes. They present exciting possibilities for research and use. They have some remarkable properties, such as better electrical conductivity than copper, exceptional mechanical strength, and very high flexibility (with futuristic potential for use even in earthquake-resistant buildings and crash-resistant cars). There is already considerable interest in industry in the potential use of CNTs in chemical sensors, field emission elements, electronic interconnects in integrated circuits, hydrogen storage devices, and temperature sensors (Dai, 2003; Saito and Uemura, 2000; Wong and Lee, 2000).

Recently, record-breaking single-wall carbon nanotubes have been grown via catalytic chemical vapor deposition to a length of 40 mm by Zheng, O'Connell, Doorn, Liao, Zhao, Akhadov, Hoffbauer, Roop, Jia, Dye, Peterson, Huang, Liu, and Zhu (2004).

Since the discovery of CNTs, similar nanostructures were formed in other layered compounds such as BN, BCN, WS2, etc. (Bachtold, Hadley, Nakanishi, and Dekker, 2001). These different nanotubular materials offer different physical and engineering properties. For example, whereas CNTs are either metallic or semiconducting (depending on the shell helicity and diameter), BN nanotubes are insulating and could possibly serve as nanoshield for nanoconductors. BN nanotubes are also thermally more stable in oxidizing atmospheres than CNTs and have comparable modulus. The strength of nanotubular materials can be increased by assembling them in the form of ropes of 20-30 nm diameter and several micrometers in length. This has been done with CNT and BN nanotubes, with ropes made from SWCNTs being the strongest known material. The spacing between the individual nanotube strands in such a rope is in the subnanometer size range, e.g., $\sim$0.34 nm in ropes made from multiwall BN nanotubes, which is on the order of the (0001) lattice spacing in the hexagonal BN cell. Due to their exceptional properties, there has been interest in incorporating nanotubes in polymers, ceramics, and metals.

Metal nanowires and nanoparticles are commonly formed by employing a template. For example, Au or Pt nanowires are first fabricated by flash evaporation and deposition of carbon onto glass. The coated glass is then thermally stressed by repeated dipping into liquid nitrogen, leading to cracks in the carbon film. Subsequent sputtering of Pt or Au and removal of surplus material results in nanowires. For certain special applications, metal powders in the micrometer-size range will be too coarse. For example, the catalytic property of gold particles occurs only at particle diameters of less than 3–5 nm.

Crystalline SiC nanowires have been created starting with thermal annealing at 1000 °C of polycrystalline Cu substrate and Si in a furnace. This produces grooves along grain boundaries in Cu. This Cu (copper) is used as a template for the subsequent growth of SiC nanowires. Methane gas is introduced in the furnace containing Cu template and Si, and 20 to 50 nm nanocrystals of SiC nucleate and grow to several 10s of micrometers along the Cu grain boundaries. These nanocrystals coalesce along the grain boundaries to form nanowires after sintering. The vapor–solid nucleation is suggested as the mechanism.

Pt nanowires can be mass-produced by reducing H_2PtCl_6 or K_2PtCl_6 by ethylene glycol (EG) at 110 °C in the presence of PVP (polyvinyl pyrrolidone) in air. This produces 5 nm diameter Pt nanoparticles. The Pt particles tend to agglomerate into spheres and larger structures. The nanowires grow at the surface of the agglomerates to which they are loosely attached, and are readily removed by sonic vibration. The Pt nanowires are separated from the agglomerates by centrifugation.

Nanoparticles are viewed by many as fundamental building blocks of nanotechnology. They are the starting point for many bottom-up approaches for preparing nanostructured materials and devices. As such, their synthesis is an important component of rapidly growing research efforts in nanoscale science and engineering. Nanoparticles of a wide range of materials can be prepared by a variety of methods including gas-phase synthesis for electronics-related applications. Nanoparticles are finding a myriad of uses, ranging from traditional applications, such as coloring agents (in stained-glass windows) and catalysts, to novel applications, such as magnetic drug delivery, hypothermic cancer therapy, contrast agents in magnetic resonance imaging, magnetic and fluorescent tags in biology, solar photovoltaics, nano bar codes, and emission control in diesel vehicles. This field is, in certain ways, reaching maturity, and to go to the next step it is becoming important to develop methods of scaling up the synthesis of these materials.

In vapor-phase synthesis of nanoparticles, conditions are created where the vapor-phase mixture is thermodynamically unstable relative to formation of the solid material to be prepared in nanoparticulate form. This includes the usual situation of a supersaturated vapor. It also includes "chemical supersaturation," in which it is thermodynamically favorable for the vapor-phase molecules to react chemically to form a condensed phase. If the degree of supersaturation is sufficient, and the reaction/condensation kinetics permit, particles will nucleate homogenously. Once nucleation occurs, remaining supersaturation can be relieved by condensation or reaction of the vapor-phase molecules on the resulting particles and particle growth will occur, rather than further nucleation. Therefore, to prepare small particles one wants to create a high degree of supersaturation, thereby inducing a high nucleation density, and then immediately quench the system, either by removing the source of supersaturation or slowing the kinetics, so that the particles do not grow.

Once particles form in the gas phase, they coagulate at a rate that is proportional to the square of their number concentration and that is only weakly dependent on particle size. At sufficiently high temperature, particles coalesce (sinter) faster than they coagulate, and spherical particles are produced. At lower temperatures, where coalescence is negligibly slow, loose agglomerates with quite open structures are formed. At intermediate conditions, partially sintered nonspherical particles are produced. If individual, nonagglomerated nanoparticles are desired, control of coagulation and coalescence is crucial. In contrast to the liquid phase, where a dispersion of nanoparticles can be stabilized indefinitely by capping the particles with appropriate ligands, nanoparticles in the gas phase will always agglomerate. So, by "nonagglomeration of nanoparticles," we usually mean particles agglomerated loosely enough that they can be redispersed without herculean effort, as compared to hard (partially sintered) agglomerates that cannot be fully redispersed. Figure 8-1 illustrates typical degrees of agglomeration and polydispersity obtained in gas-phase processes when no special efforts have been made to control agglomeration or narrow the particle size distribution. In the cases of carbon black, fumed silica, and pigmentary titania, such particles are produced commercially in huge quantities.

In the following paragraphs, recent examples of and advances in gas-phase methods for preparing nanoparticles are reviewed. One useful way of classifying such methods is by the phase of precursor and the source of energy used to achieve a state of supersaturation. This part is structured around such a classification.

Methods Using Solid Precursors

One general class of methods of achieving the supersaturation necessary to induce homogeneous nucleation involves vaporizing the material into a background gas and cooling the gas.

Inert Gas Condensation. The most straightforward method of achieving supersaturation is to heat a solid to evaporate it into a background gas, then mix the vapor with a cold gas to reduce the temperature. This method is well suited for production of metal nanoparticles, because many metals evaporate at reasonable rates at attainable temperatures. By including a reactive gas, such as oxygen, in the cold gas stream, oxides or other compounds of the evaporated material can be prepared. Detailed, systematic modeling and experimental study of this method, as applied

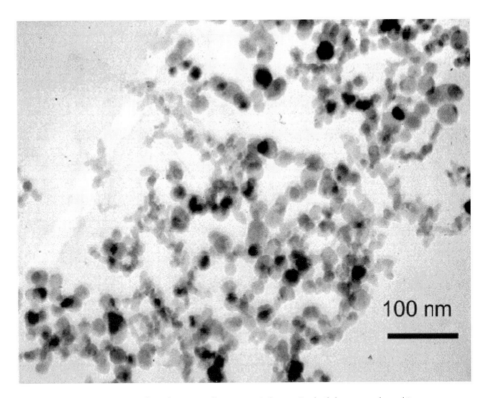

FIGURE 8-1 *TEM image of agglomerated nanoparticles typical of those produced in many vapor-phase processes. These particular particles are silicon produced by laser pyrolysis of silane, but the degree of polydispersity and agglomeration is typical of many vapor-phase processes in which no special efforts have been made to avoid agglomeration or narrow the size distribution of the primary particles.* (Mark T. Swihart, Current Opinion in Colloid and Interface Science, 8, 2003, 127). Reprinted with permission from Elsevier. Photo courtesy of M. T. Swihart, State University of New York.

to preparation of bismuth nanoparticles, including both visualization and computational fluid dynamics simulation of the flow fields in their reactor, have been presented in the literature. It has been shown that the particle size distribution can be controlled by controlling the flow field dynamics and the mixing of the cold gas with hot gas carrying the evaporated metal. Other advances of this method have been in preparing composite nanoparticles and in controlling the morphology of single-component nanoparticles by controlled sintering after particle formation. Maisels, Kruis, Fissan, and Rellinghaus (2000) prepared composite nanoparticles of PbS with Ag by separate evaporation/condensation of the two materials followed by coagulation of oppositely charged PbS and Ag particles selected by size and charge. Ohno et al. (2002) prepared Si-In, Ge-In, Al-In, and Al-Pb composite nanoparticles by condensation of In or Pb onto Si, Ge, or Al particles prepared by inert gas condensation and brought directly into a second condensation reactor.

Pulsed Laser Ablation. Another method for the gas-phase synthesis of nanoparticles of various materials is based on the pulsed-laser vaporization of metals in a chamber filled with a known

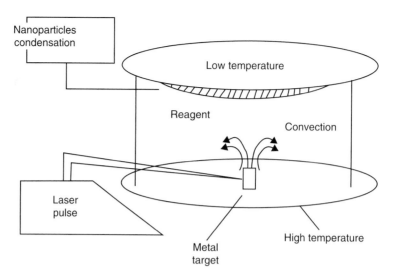

FIGURE 8-2 *Schematic view of the installation for the synthesis of nanoparticles by laser vaporization of metals.*

amount of a reagent gas followed by controlled condensation of nanoparticles onto the support. A schematic view of the installation for synthesizing nanoparticles is shown in Figure 8-2. As metal atoms diffuse from the target to the support, they interact with the gas to form the desired compound (for instance, oxide in the case of oxygen, nitride for nitrogen or ammonia, and carbide for methane). The pulsed-laser vaporization of metals in the chamber makes it possible to prepare nanoparticles of mixed molecular composition, such as mixed oxides-nitrides or mixtures of oxides of different metals. Along with the reagent gas, the chamber contains an inert gas, such as He or Ar, at a pressure of 10^{-21} Torr, which favors the establishment of steady convection between the heated bottom plate and cooled top plate. In a typical experiment with single pulse of a Nd:YAG laser (532 nm, 15–30 mJ/pulse, 10–9 s pulse duration), over 1014 metal atoms are vaporized. A new compound is formed due to the reaction between the "hot" metal atoms and the gas molecules, which is accompanied by the energy loss of the molecules formed by collisions with the inert gas atoms. The metal atoms that did not enter into reaction and the molecules of the new compound are carried by convection to the nucleation zone on the cooled top plate.

By changing the composition of the inert gas and the reagent gas in the chamber and by varying the temperature gradient and laser pulse power, it is possible to control the elemental composition and size of nanoparticles that are obtained. Marine et al. (2000) presented a recent analysis of this method in which they also reviewed its development. Nakata et al. (2002) used a combination of laser-spectroscopic imaging techniques to image the plume of Si atoms and clusters formed during synthesis of Si nanoparticles. They investigated the dependence of the particle formation dynamics on the background gas and found that it was substantial. Some other recent examples include the preparation of magnetic oxide nanoparticles by Shinde et al. (2000), titania nanoparticles by Harano et al. (2002) and hydrogenated-silicon nanoparticles by Makimura et al. (2002).

Ion Sputtering. A final means of vaporizing a solid is via sputtering with a beam of inert gas ions. Urban et al. (2002) recently demonstrated formation of nanoparticles of a dozen different metals using magnetron sputtering of metal targets. They formed collimated beams of the nanoparticles and deposited them as nanostructured films on silicon substrates. This process must be carried out at relatively low pressures ($\sim$1 mTorr), which makes further processing of the nanoparticles in aerosol form difficult.

Methods Using Liquid or Vapor Precursor

An alternate means of achieving the supersaturation required to induce homogenous nucleation of particles is chemical reaction. Chemical precursors are heated and/or mixed to induce gas-phase reactions that produce a state of supersaturation in the gas phase.

Chemical Vapor Synthesis. In this approach, vapor-phase precursors are brought into a hot-wall reactor under conditions that favor nucleation of particles in the vapor phase rather than deposition of a film on the wall. Chemical vapor synthesis or chemical vapor deposition (CVD) are the processes used to deposit thin solid films on surfaces. This method has tremendous flexibility in producing a wide range of materials and can take advantage of the huge database of precursor chemistries that have been developed for CVD processes. The precursors can be solid, liquid, or gas at ambient conditions.

There are many good examples of the application of this method in the recent literature. Ostraat et al. (2001) have demonstrated a two-stage reactor for producing oxide-coated silicon nanoparticles that have been incorporated into high-density nonvolatile memory devices. By reducing the silane precursor composition to as low as 10 parts per billion, they were able to produce nonagglomerated single-crystal spherical particles with mean diameter below 8 nm. This is one of relatively few examples of a working microelectronic device in which vapor-phase–synthesized nanoparticles perform an active function. In other recent examples of this approach, Magnusson et al. (2000) produced tungsten nanoparticles by decomposition of tungsten hexacarbonyl, and Nasibulin et al. (2002) produced copper acetylacetonate.

Laser Pyrolysis. An alternate means of heating the precursors to induce reaction and homogenous nucleation is absorption of laser energy. Compared to heating the gases in a furnace, laser pyrolysis allows highly localized heating and rapid cooling, because only the gas (or a portion of the gas) is heated and its heat capacity is small. Heating is generally done using an infrared (CO_2) laser, whose energy is either absorbed by one of the precursors or by an inert photosensitizer such as sulfur hexafluoride. The silicon particles shown in Figure 8-1 were prepared by laser pyrolysis of silane. Nanoparticles of many materials have been made using this method. A few recent examples are MoS_2 nanoparticles prepared by Borsella et al. (2001), SiC nanoparticles produced by Kamlag et al. (2001), and Si nanoparticles prepared by Ledoux et al. (2002). Ledoux et al. used a pulsed CO_2 laser, thereby shortening the reaction time and allowing preparation of even smaller particles.

Synthesis of Nanoparticles by Chemical Methods

A number of different chemical methods can be used to make nanoparticles of metals and semiconductors. Several types of reducing agents can be used to produce nanoparticles such as $NaBEt_3H$, $LiBEt_3H$, and $NaBH_4$, where Et denotes the ethyl ($\cdot C_2H_5$) radical. For example, nanoparticles of molybdenum (Mo) can be reduced in toluene solution with $NaBEt_3H$ at room temperature, providing a high yield of Mo nanoparticles having dimensions of 1–5 nm. The equation for the reaction is

$$MoCl_3 + 3NaBEt_3H => Mo + 3NaCl + 3BEt_3 + (3/2)H_2$$

Nanoparticles of aluminum have been made by decomposing $Me_2EtNAlH_3$ in toluene and heating the solution to 105 °C for 2 h (Me is methyl, $\cdot CH_3$). Titanium isopropoxide is added to the solution. The titanium acts as a catalyst for the reaction. The choice of catalyst determines the size of the particles produced. For instance, 80-nm particles have been made using titanium. A surfactant such as oleic acid can be added to the solution to coat the particles and prevent aggregation.

Nanoparticles of metal sulfides are usually synthesized by a reaction of a water-soluble metal salt and H_2S or Na_2S in the presence of an appropriate stabilizer, such as sodium metaphosphate. For example, the CdS nanoparticles can be synthesized by mixing $Cd(ClO_4)_2$ and Na_2S solutions:

$$Cd(ClO_4)_2 + Na_2S = CdS(1–10 \text{ nm}) + 2NaClO_4$$

The growth of the CdS nanoparticles in the course of reaction is arrested by an abrupt increase in pH of the solution. Very recently Peng and Peng (2001) reproduced rice-shape CdSe nanocrystals (shown in Figure 8-3) by using CdO as precursor.

Nanoparticles: Biomedical Applications

In the past nanoparticles were studied because of their size-dependent physical and chemical properties (Murry et al., 2000). At present they have entered a commercial exploration period. In this section, biomedical applications of nanoparticles are considered.

Living organisms are built of cells that are typically 10 μm across. However, the cell parts are much smaller and are in the submicron-size domain. Even smaller are the proteins, with a typical diameter of just 5 nm, which is comparable with the dimensions of smallest man-made nanoparticles. This simple size comparison gives an idea of using nanoparticles as very small probes that would allow us to look at cellular machinery without introducing too much interference. Nowadays nanoparticles have many applications in biology and medicine; for example, (1) drug and gene delivery, (2) biodetection of pathogens, (3) fluorescent biological labels, (4) detection of proteins, (5) probing DNA structure, (6) tissue engineering, (7) tumor destruction via heating (hyperthermia), (8) separation and purification of biological molecules and cells, (9) MRI contrast enhancement, and (10) phagokinetic studies.

As mentioned above, the fact that nanoparticles exist in the same size domain as proteins makes nanomaterials suitable for biotagging or labeling. However, size is just one of many characteristics of nanoparticles, which by itself is rarely sufficient if one is to use nanoparticles as biological tags. To interact with a biological target, a biological or molecular coating or layer acting as a bioinorganic interface should be attached to the nanoparticle. Examples of biological coatings may include antibodies, biopolymers such as collagen (Sinani et al., 2003), or mono-layers of small molecules that make the nanoparticles biocompatible (Zhang et al., 2002). In addition, as optical detection techniques are widespread in biological research, nanoparticles should either fluoresce or change their optical properties. The approaches used in constructing nanobiomaterials are schematically presented in Figure 8-4. Nanoparticles usually form the core of a nanobiomaterial. The material can be used as a convenient surface for molecular assembly, and may be composed of inorganic or polymeric materials. It can also be in the form of a

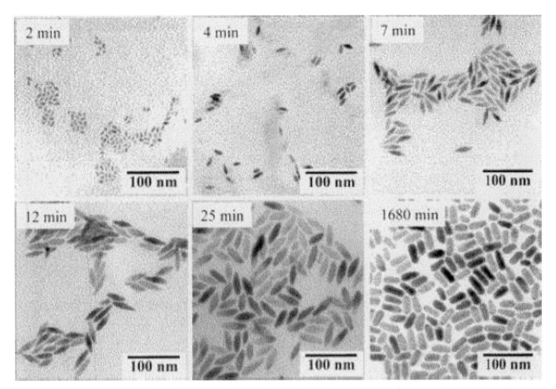

FIGURE 8-3 *TEM images of the time evolution of rice-shape CdSe nanocrystals. The times are indicated.* (Reprinted by the order of Z. A. Peng and X. Peng, Formation of high-quality CdTe, CdSe, and CdS nanocrystals using CdO as precursor, Journal of the American Chemical Society, 123, 2001, 183). Reprinted with permission from the American Chemical Society, 1155 16th St. NW, Washington, DC 20036. Photo Courtesy of X. Peng, University of Arkansas.

nanovesicle surrounded by a membrane or a layer. The shape is more often spherical, but cylindrical, platelike, and other shapes are possible. The size and size distribution might be important in some cases; for example, if penetration through a pore structure of a cellular membrane is required. The size and size distribution become extremely critical when quantum-size effects are used to control material properties. A tight control of the average particle size and a narrow distribution of sizes allow the creation of very efficient fluorescent probes that emit narrow light in a very wide range of wavelengths. This helps with creating biomarkers with many and well-distinguished colors. The core itself may have several layers and be multifunctional. For example, by combining magnetic and luminescent layers one can both detect and manipulate the particles. The core particle is often protected by several monolayers of inert material, such as silica. Organic molecules that are adsorbed or chemisorbed on the surface of the particle are also used for this purpose. The same layer may act as a biocompatible material. However, more often an additional layer of linker molecules is required to proceed with further functionalization. This linear linker molecule has reactive groups at both ends. One group is aimed at attaching the linker to the nanoparticle surface, and the other is used to bind various moieties such as biocompatibles. Recent developments using nanoparticles in biomedical applications are described later in more detail.

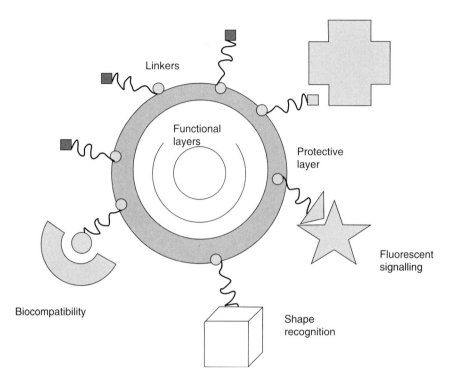

FIGURE 8-4 *Typical configurations used in nanobiomaterials applied to medical or biological problems.* (O. V. Salata et al., Journal of Nanobiotechnology, 2, 2004, 3). Reprinted with permission from O. V. Salata.

Tissue Engineering

Natural bone surface quite often contains features that are about 100 nm across. If the surface of an artificial bone implant were left smooth, the body would try to reject it. Because of that, the smooth surface is likely to cause production of a fibrous tissue covering the surface of the implant. This layer reduces the bone–implant contact, which may result in loosening of the implant and further inflammation. It has been demonstrated that by creating nano-sized features on the surface of the hip or knee prosthesis, one can reduce the chances of rejection of a hip or knee prosthesis and stimulate the production of osteoblasts, which are the cells responsible for the growth of the bone matrix and which are found on the advancing surface of the developing bone. The effect has been demonstrated with polymeric, ceramic, and, more recently, metal materials. For example, in one study more than 90% of the human bone cells formed a suspension that adhered to the nano-structured metal surface (Gutwein and Webster, 2003, in press) but only 50% in the control sample did so. In the end, this finding will allow design of a more durable and longer lasting hip or knee replacement and reduce the chances of the implant getting loose.

Titanium is a well-known bone-repairing material widely used in orthopedics and dentistry. It has high fracture resistance, ductility, and strength-to-weight ratio. Unfortunately, it suffers from the lack of bioactivity, as it does not support cell adhesion and growth. Apatite coatings are known to be bioactive and to bond to the bone. Hence, several techniques were used in the past to produce an apatite coating on titanium. Those coatings suffer from thickness nonuniformity, poor adhesion, and low mechanical strength. In addition, a stable porous structure is required to support the nutrients' transport through the cell growth.

It has been shown that using biomimetic approach—a slow growth of nanostructured apatite film from the simulated body fluid—results in the formation of a strongly adherent and uniform nanoporous layer. The layer is built of nanometric crystallites and has a stable nanoporous structure and bioactivity.

A real bone is a nanocomposite material, composed of hydroxyapatite crystallites in the organic matrix, which mainly consists of collagen. Therefore, the bone is mechanically both tough and plastic, so it can recover from mechanical damage. The actual nanoscale mechanism leading to this useful combination of properties is still debated. Recently, an artificial hybrid material was prepared from 15- to 18-nm ceramic nanoparticles and poly (methyl methacrylate) copolymer. This hybrid material, deposited as a coating on the tooth surface, improved scratch resistance and exhibited a healing behavior similar to that of the tooth.

Manipulation of Cells and Biomolecules

Functionalized magnetic nanoparticles have found many biological and medical applications (Chung et al., 2004). The size of the particles can range from a few nanometers to several microns (shown in Figure 8-5) and thus is compatible with biological entities ranging from proteins (a few nm) to cells and bacteria (several μm). Generally the magnetic particles are coated with a suitable ligand, which allows chemical binding of the particles to different biological environment. Conversely, absence of ferromagnetism in most biological systems, which typically have only diamagnetism or paramagnetism, means that in a biological environment the magnetic moment from the ferromagnetic particles can be detected with little noise.

Based on these ideas, a variety of applications has emerged. A direct application is to bind magnetic particles to the biological system of interest, which then allows manipulating the biological material via magnetic field gradients. This can be used for high-gradient magnetic

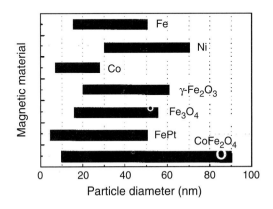

FIGURE 8-5 *Particle diameters for stable single-domain magnetic nanoparticles.* (Reprinted by the order of S. H. Chung, A. Hoffmann, S. D. Bader, C. Liu, B. Kay, and L. Chen, Applied Physics Letters, 85, 2004, 2971).

field separation, which has already been applied to several problems, such as separating red blood cells from blood, cancer cells from bone marrow, and radioactive isotopes from food products. In contrast, one can use magnetic nanoparticles for targeted drug delivery. In this case, a drug is bound to a magnetic particle and either DC magnetic fields are used to confine the drug in a specific location of the body, or AC magnetic fields are used to trigger the release of the drugs. A related application uses magnetic nanoparticles for hypothermal treatment, wherein heating the diseased tissue destroys the cancerous cells. This can be achieved by heating magnetic nanoparticles with AC magnetic fields to about 42 °C for at least 30 minutes. Besides these therapeutic applications, magnetic nanoparticles can also be used for diagnostics. One common application is the enhancement of contrast in magnetic resonance imaging, where the local stray field of the magnetic nanoparticles can modify the magnetic relaxation of the surrounding tissue. Last but not least, magnetic nanoparticles are also used for biomagnetic sensing, where a target of interest is typically tagged with the magnetic particles, so that stray fields of the nanoparticles can be used for signal transduction.

Protein Detection

Proteins are an important part of the cell's language, machinery, and structure, and understanding their functionalities is extremely important for further progress in human well-being. Gold nanoparticles are widely used in immunohistochemistry to identify protein–protein interaction. However, the multiple simultaneous detection capabilities of this approach are fairly limited. Surface-enhanced Raman scattering spectroscopy is a well-established technique for detection and identification of single dye molecules. By combining both methods in single nanoparticle probe, one can drastically improve the multiplexing capabilities of protein probes. Mirkin et al. (2003) designed a sophisticated multifunctional probe that was built around 13-nm gold nanoparticles. The nanoparticles are coated with hydrophilic oligonucleotides containing Raman dye at one end and terminally capped with a small molecule-recognition element (e.g., biotin). Moreover, this molecule is catalytically active and is coated with silver in the solution of Ag(I) and hydroquinone. After the probe is attached to a small molecule or an antigen it is designed to detect, the substrate is exposed to silver and hydroquinone solution. A silver-plating occurs close to the Raman dye, which allows for dye signature detection with a standard Raman microscope. Apart from being able to recognize small molecules, this probe can be modified to contain antibodies on the surface to recognize proteins. When tested in the protein array format against both small molecules and proteins, the probe shows no cross-reactivity.

Cancer Therapy

Photodynamic cancer therapy is based on the destruction of the cancer cells by laser-generated atomic oxygen, which is cytotoxic. A greater quantity of special dye that is used to generate the atomic oxygen is taken in by the cancer cells when compared with a healthy tissue. Hence, only the cancer cells are destroyed when exposed to a laser radiation. Unfortunately, the remaining dye molecules migrate to the skin and the eyes and make the patient very sensitive to daylight exposure. This effect can last for up to six weeks. To avoid this side effect, a hydrophobic version of the dye molecule was enclosed inside a porous nanoparticle (Cao et al., 2003). The dye stayed trapped inside the Ormosil nanoparticle and did not spread to the other parts of the body. At the same time, its oxygen-generating ability had not been affected, and the pore size of about 1 nm freely allowed the oxygen to diffuse out.

Semiconductor Nanowires

Semiconductor nanowires exhibit novel electronic and optical properties owing to their unique structural one-dimensionality and possible quantum confinement effects in two dimensions. With a broad selection of compositions and band structures, these one-dimensional semiconducting nanostructures are considered the critical components in a wide range of potential nanoscale device applications. The understanding of general nanocrystal growth mechanisms serves as the foundation for the rational synthesis of semiconductor heterostructures in one dimension. Availability of these high-quality semiconductor nanostructures allows systematic structure-property correlation investigations, particularly the effects of size and dimensionality. Novel properties including nanowire microcavity lasing, phonon transport, interfacial stability, and chemical sensing are surveyed. This section is divided into three subsections. The first section explores the advances in gas-phase production methods, especially the vapor–liquid–solid (VLS) and vapor–solid (VS) processes with which most one-dimensional heterostructures and ordered arrays are now grown. Several approaches for fabricating one-dimensional nanostructures in solution, focusing especially on those that use a selective capping mechanism are discussed. In the second section, focus is on interesting fundamental properties exhibited by rods, wires, belts, and tubes. In the third section, progress in the assembly of one-dimensional nanostructures into useful architectures is addressed and illustrates the construction of novel devices based on such schemes.

General Synthetic Strategies

A novel growth mechanism should satisfy three conditions: It must (1) explain how one-dimensional growth occurs, (2) provide a kinetic and thermodynamic rationale, and (3) be predictable and applicable to a wide variety of systems. Growth of many one-dimensional systems has been experimentally achieved without satisfactory elucidation of the underlying mechanism, as is the case for oxide nanoribbons. Nevertheless, understanding the growth mechanism is an important aspect for developing a synthetic method for generating one-dimensional nanostructures of desired material, size, and morphology.

In general, high-quality, single crystal nanowire materials are synthesized by promoting the crystallization of solid-state structures along one direction. The actual mechanisms of coaxing this type of crystal growth include (1) growth of an intrinsically anisotropic crystallographic structure, (2) the use of various templates with one-dimensional nanostructures, (3) the introduction of a liquid–solid interface to reduce the symmetry of a seed, (4) use of an appropriate capping reagent to kinetically control the growth rates of various facets of a seed, and (5) the self-assembly of zero-dimensional nanostructures.

The ability to form heterostructures through carefully controlled doping and interfacing is responsible for the success of semiconductor integrated-circuit technology, and the two-dimensional semiconductor interface is ubiquitous in optoelectronic devices such as light-emitting diodes (LEDs), laser diodes, quantum cascade lasers, and transistors (Weisbuch and Vinter, 1991). Therefore, the synthesis of one-dimensional heterostructures is equally important for potential future applications, including efficient light-emitting sources and thermoelectric devices. This type of one-dimensional nanoscale heterostructure can be rationally prepared once the fundamental one-dimensional nanostructure growth mechanisms are understood.

In general two types of one-dimensional heterostructures can be formed: longitudinal heterostructures and coaxial heterostructures. The term "longitudinal heterostructures" refers to nanowires composed of different stoichiometries along the length of the nanowire, and "coaxial heterostructures" refers to nanowire materials having different core and shell compositions.

Various approaches to fabricate heterostructure and inorganic nanotube materials derived from three-dimensional bulk crystal structures are discussed here.

Among all vapor-based methods, the VLS mechanism seems the most successful for generating a large quantity of nanowires with single-crystal structures. This process was originally developed in the 1960s by Wanger and Ellis to produce micrometer-sized whiskers (Wagner and Ellis, 1964), later justified thermodynamically and kinetically (Givargizov, 1975), and recently reexamined by other researchers to generate nanowires and nanotubes from a rich variety of inorganic materials (Wu and Yang, 2000; Zhang et al., 2001; Westwater et al., 1997; Wu and Wang, 2001; Gudiksen and Lieber, 2000; Wu et al., 2002; Duan and Lieber, 2000; Chen et al., 2001; Zhang et al., 2001; He et al., 2001; Shi et al., 2001).

Figure 8-6 shows image from in situ transmission electron microscopy (TEM) technique to monitor the VLS growth mechanism in real time (Wu and Wang, 2001). A typical VLS process starts with the dissolution of gaseous reactants into nanosized liquid droplets of a catalyst metal, followed by nucleation and growth of single crystal rods and then wires. The one-dimensional growth is induced and dictated by liquid droplets, whose sizes remain essentially unchanged during the entire process of wire growth. Each liquid droplet serves as a virtual template to strictly limit the lateral growth of an individual wire. The major stages of the VLS process can be seen in the example of Figure 8-6, where the growth of a Ge nanowire as observed by in situ TEM is shown. Based on the Ge-Au binary-phase diagram, Ge and Au form liquid alloys when the temperature is raised above the eutectic point (361 °C). Once the liquid droplet is supersaturated

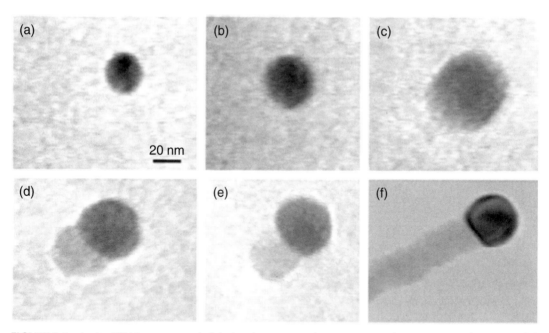

FIGURE 8-6 *In situ TEM images recorded during the process of nanowire growth. (a) Au nanoclusters in solid state at 500 °C; (b) alloying initiates at 800 °C; at this stage, Au exists mostly in solid-state; (c) liquid Au-Ge alloy; (d) the nucleation of Ge nanocrystals on the alloy surface; (e) Ge nanocrystal elongates with further Ge condensation; and (f) eventually forms a wire.* (Reprinted with permission of Y. Wu and P. Yang, Journal of the American Chemical Society, 123, 1999, 3165–3166). Reprinted with permission from the American Chemical Society, 1155 16th St. NW, Washington DC 20036. Photo Courtesy of P. Yang, University of California, Berkeley.

with Ge, nanowire growth will start at the solid–liquid interface. The establishment of the symmetry-breaking solid–liquid interface is the key step for the one-dimensional nanocrystal growth in this process, whereas the stoichiometry and lattice symmetry of the semiconductor material systems are less relevant. The growth process can be controlled in various ways. Because the diameter of each nanowire is largely determined by the size of the catalyst particle, smaller catalyst islands yield thinner nanowires or tubes. The VLS process has now become a widely used method for generating one-dimensional nanostructures from a rich variety of pure and doped inorganic materials that include elemental semiconductors (Si, Ge) (Wu and Yang, 2000; Zhang et al., 2001; Westwater et al., 1997), III–V semiconductors (GaN, GaAs, GaP, InP, InAs) (Gudiksen and Lieber, 2000; Wu et al., 2002; Duan and Lieber, 2000; Chen et al., 2001; Zhang et al., 2001; He et al., 2001; Shi et al., 2001; Chen and Yeh, 2000; Shimada et al., 1998; Hiruma et al., 1995; Yazawa et al., 1993; Kuykendall et al., 2003; Zhong et al., 2003), II–VI semiconductors (ZnS, ZnSe, CdS, CdSe) (Wang et al., 2002; Wang et al., 2002; Lopez-Lopez et al., 1998), oxides (indium-tin oxide, ZnO, MgO, SiO2, CdO) (Wang et al., 2002; Lopez-Lopez et al., 1998; Peng et al., 2002; Yang and Lieber, 1996; Wu et al., 2001; Huang et al., 2001; Liu et al., 2003; Naguyen et al., 2003), carbides (SiC, B_4C) (Ma and Bando, 2002; Kim et al., 2003), and nitrides (Si_3N_4) (Kim et al., 2003). The nanowires produced using the VLS approach are remarkable for their uniformity in diameter, which is usually on the order of 10 nm over a length scale of $>1 \mu m$.

Fabrication of Semiconductor Nanowires

High-crystalline silicon nanowires are ingredients for electronic devices, light-emitting devices, field emission sources, and sensors. Thermal vapor growth from solid precursors, usually in a high-temperature furnace, is the most common way to achieve bulk production of nanowires.

Typically, quartz or alumina boats filled with suitable precursor powders are placed in a furnace tube and heated to high temperatures, while a lower temperature substrate is placed at one end of the tube, in the downstream direction of inert gas flow (Ar or N_2). Schematic diagrams of experimental setup for the synthesis of semiconductor nanostructures are shown in Figure 8-7.

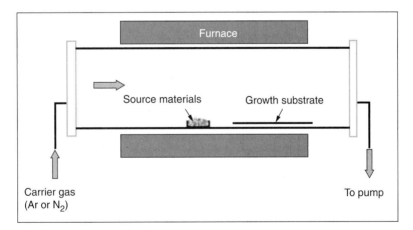

FIGURE 8-7 *Schematic of experimental setup for the synthesis of semiconductor nanostructures.* (Zhong Lin Wang, Journal of Materials Chemistry, 15, 2005, 1021). Reprinted with permission from The Royal Society of Chemistry, Thomas Graham House, Science Park, Cambridge, U.K. Photo Courtesy of Z. L. Wang, Georgia Institute of Technology.

In the case of silicon, different strategies are possible for the growth of nanowires. One strategy is to anneal Si-SiO$_2$ powders well above 1000 °C to evaporate them. Silicon nanowires then grow on the substrate either by the vapor-liquid-solid method, by placing gold as the catalyst, or by the so-called oxide-assisted method, which does not require a catalyst particle and is triggered by the

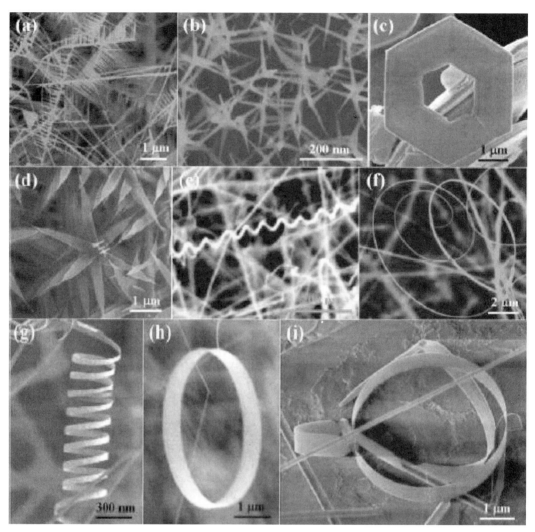

FIGURE 8-8 *A collection of polar-surface induced/dominated nanostructures of ZnO, synthesized under controlled conditions by thermal evaporation of solid powders unless notified otherwise: (a) nanocombs induced by asymmetric growth on the Zn-(0001) surface, (b) tetraleg structure due to catalytically active Zn-(0001) surfaces, (c) hexagonal disks/rings synthesized by solution-based chemical synthesis, (d) nanopropellers created by fast growth along the c-axis, (e) deformation-free nanohelices as a result of block-by-block self-assembly, (f) spiral of a nanobelt with increased thickness along the length, (g) nanosprings, (h) single-crystal seamless nanoring formed by loop-by-loop coiling of a polar nanobelt, and (i) a nanoarchitechure composed of a nanorod, nanobow, and nanoring. (Z. W. Wang et al., Journal of Materials Chemistry 15, 2005, 1021). Reprinted with permission from The Royal Society of Chemistry, Thomas Graham House, Science Park, Cambridge U.K. Photo Courtesy of Z. L. Wang, Georgia Institute of Technology.*

FIGURE 8-9 *(a) SEM image of aligned amorphous silicon nanowires grown on silicon wafers at 1300 °C by thermal CVD method. (b) High-magnification image illustrates that the diameter of silicon nanowires ranges between 80 and 100 nm. (c) TEM image of silicon nanires and inset shows that the silicon wires are pure amorphous.* (Y. Xu, C. Cao, B. Zhang, and H. Zhu. Preparation of aligned amorphous silica nanowires. Chemistry Letters, 34, 2005, 414). Reprinted with permission from The Chemical Society of Japan, Chiyoda-Ku, Tokyo 101-8307, Japan.

self-condensation of the vapor in a low-temperature region of the furnace. Wang et al., (2005) synthesized a wide range of polar-surface–dominated nanostructures of ZnO under controlled conditions by thermal evaporation of solid powders at high yield (Figure 8-8).

Recently, Xu et al. (2005) grew silicon nanowires by thermal chemical vapor deposition (Figure 8-9). This method is based on the VLS idea, in which gold acts like a catalyst. VLS mechanism and high-resolution transmission electron microscopy (HRTEM) of Si nanowires are presented in Figures 8-10 and 8-11, respectively. In another example of the growth of GaN

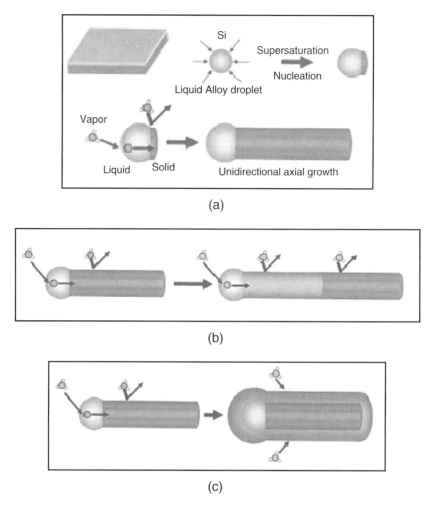

(a)

(b)

(c)

FIGURE 8-10 *Illustration of nanowire synthesis process, the main concept. (a) Vapor-liquid-solid (VLS) mechanism for nanowire growth. (b) Synthesis method to form nanowire axial hetrostructure. (c) Synthesis method for nanowire core-shell hetrostructure formation.* (Reprinted by D. C. Bell et al., Microscopy Research and Technique, 64, 2004, 373). Reprinted with permission from Z. L. Wang, Georgia Institute of Technology.

nanowires prepared by using metal organic chemical vapor deposition (MOCVD), the SEM, TEM, and HR-TEM images of a sample are shown in Figure 8-12.

A few of the major disadvantages of high-temperature approaches to nanowires synthesis include the high cost of fabrication and scaleup, and the inability to produce metallic wires. Recent progress using solution-phase techniques has resulted in the creation of one-dimensional nanostructures in high yields (gram scales) via selective capping mechanisms. It is believed that molecular capping agents play a significant role in the kinetic control of the nanocrystal growth by preferentially adsorbing to specific crystal faces, thus inhibiting growth of that surface (although defects could also induce such one-dimensional crystal growth). The growth of semiconductor nanowires has also been realized using a synthetic mechanism. Microrods of ZnO have been

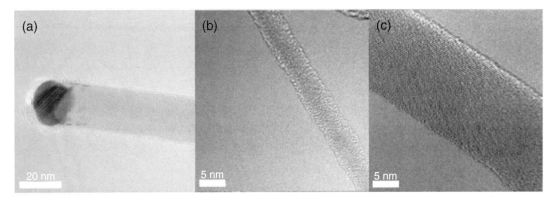

FIGURE 8-11 *HRTEM images of silicon nanowires. (a) Silicon nanowire showing attachment of gold catalyst particle on the end. (b) Thin ~5-nm silicon nanowire showing lack of contrast along edges, but atomic structure information indicated along the length of the wire. (c) Silicon nanowire ~17 nm wide; micrograph shows clear atomic structure detail.* (D. C. Bell et al., Microscopy Research and Technique, 64, 2004, 373). Reprinted with permission from Z. L. Wang, Georgia Institute of Technology.

produced via the hydrolysis of zinc salts in the presence of amines (Vayssieres et al., 2001). Hexamethylenetetramine as a structural director has been used to produce dense arrays of ZnO nanowires in aqueous solution (Figure 8-13) having controllable diameter of 30–100 nm and lengths of 2–10 μm (Greene et al., 2003). Most significantly, these oriented nanowires can be prepared on any substrate. The growth process ensures that a majority of the nanowires in the array are in direct contact with the substrate and provide a continuous pathway for carrier transport, an important feature for future electronic devices based on these materials.

Fabrication of Metal Nanowires

Metal nanowires are very attractive materials because their unique properties may lead to a variety of applications. Examples include interconnects for nanoelectronics, magnetic devices, chemical and biological sensors, and biological labels. Metal nanowires are also attractive because they can be readily fabricated with various techniques. Various methods for fabricating metal nanowires are discussed next.

The diameters of metal nanowires range from a single atom to a few hundreds of nm. The lengths vary over an even greater range: from a few atoms to many micrometers. Because of the large variation in the aspect ratio (length-to-diameter ratio), different names have been used in the literature to describe the wires; those with large aspect ratios (e.g., >20) are called nanowires, whereas those with small aspect ratios are called nanorods. When short "wires" are bridged between two larger electrodes, they are often referred to as nanocontacts. In terms of electron transport properties, metal wires have been described as classical wires and quantum wires. The electron transport in a classical wire obeys the classical relation

$$G = \sigma \frac{A}{L}$$

where G is the wire conductance, and L and A are the length and the cross-sectional area of the wire, respectively; σ is the conductivity, which depends on the material of the wire.

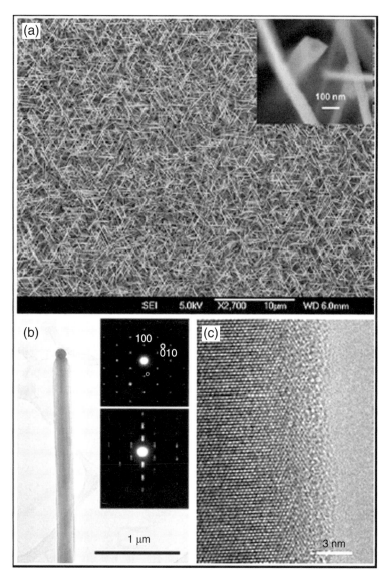

FIGURE 8-12 *(a) Field-effect scanning electron microscope (FESEM) image of the GaN nanwires grown on a gold-coated -plane sapphire substrate. Inset shows a nanowire with its triangular cross-section. (b) TEM image of a GaN nanowire with a gold metal alloy droplet on its tip. Insets are electron diffraction patterns taken along the [001] zone axis. The lower inset is the same electron diffraction pattern but purposely defocused to reveal the wire growth direction. (c) Lattice-resolved TEM image of the nanowire.* (Kuykendall T, Pauzauskie P, Lee S, Zhang Y, Goldberger J, Yang P. Metallorganic chemical vapor deposition route to GaN nanowires with triangular cross sections. Nano Lett. 3:1063 1066, 2003). Reprinted with permission from the American Chemical Society, 1155 16th St. N. W., Washington D. C. 20036. Photo Courtesy of P. Yang, University of California Berkeley.

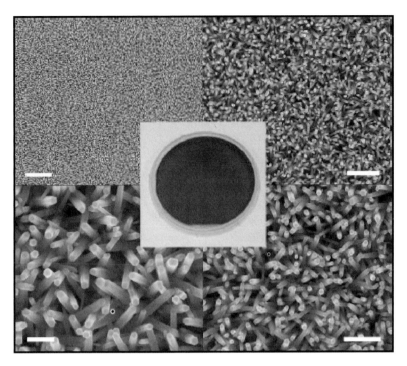

FIGURE 8-13 *ZnO nanowires array on a 4-inch silicon wafer. Centered is a photograph of a coated wafer, surrounding by SEM images of the array at different locations and magnifications. These images are representative of the entire surface. Scale bars, clockwise from upper left, correspond to 2 μm, 1 μm, 500 nm, and 200 nm.* (L. E Greene, M. Law, J. Goldberger, F. Kim, J. C. Johnson et al., Angewandte Chemie International Edition, 42, 2003, 3031-3034). Reprinted with permission from P. Yang, University of California Berkeley.

Electrochemical Fabrication of Metal Nanowires

A widely used approach to fabricating metal nanowires is based on various templates, which include negative, positive, and surface step templates. Each approach is discussed below.

Negative Template Methods. Negative template methods use prefabricated cylindrical nanopores in a solid material as templates. Depositing metals into the nanopores fabricates nanowires with a diameter predetermined by the diameter of the nanopores. There are several ways to form nanowires, but the electrochemical method is a general and versatile method. If one dissolves away the host solid material, free-standing nanowires are obtained. This method may be regarded as a "brute force" method, because the diameter of the nanowires is determined by the geometric constraint of the pores rather than by elegant chemical principles (Foss, 2002). However, it is one of the most successful methods for fabricating various nanowires that are difficult to form by the conventional lithographic process.

There are a number of methods for fabricating various negative templates. Examples include porous alumina membranes, polycarbonate membranes, mica sheets, and diblock polymer materials. These materials contain a large number of straight, cylindrical nanopores with a narrow distribution in the diameters of the nanopores.

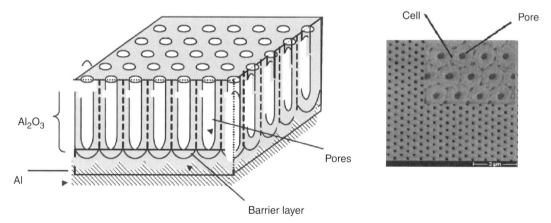

FIGURE 8-14 *(a) Schematic drawing of an anodic porous alumina template. (b) surface study of anodic porous alumina (inset shows cell and pore).* (Reprinted by permission of K. H. Lee, H. Y. Lee, and W. Y. Jeung. Magnetic properties and crystal structures of self-ordered ferromagnetic nanowires by ac electroforming. Journal of Applied Physics, 91, 2002, 8513).

Anodic Porous Alumina. Anodic porous alumina is a commonly used negative template. Figure 8-14a shows a schematic drawing of an anodic porous alumina template. The nanopores in the template are formed by anodizing aluminum films in an acidic electrolyte. The individual nanopores in the alumina can be ordered into a close-packed honeycomb structure (Figure 8-14b). The diameter of each pore and the separation between two adjacent pores can be controlled by changing the anodization conditions.

To achieve highly ordered pores, high-purity (99.999% pure) aluminum films are used. In addition, they are first preannealed to remove mechanical stress and enhance grain size. Subsequently, the films are electropolished in a 4:4:2 (by weight) mixture of H_3PO_4, H_2SO_4, and H_2O to create homogenous surfaces. Without the preannealing and electropolishing steps it is hard to form well-ordered pores (Jessensky et al., 1998). The order of the pores depends also on other anodization conditions, such as anodization voltage and electrolyte. Control of anodization voltage can produce an almost ideal honeycomb structure over an area of several μm (Jessensky et al., 1998; Masuda and Fukuda, 1995). The optimal voltage depends on the electrolyte used for anodization (Jessensky et al., 1998; Masuda and Fukuda, 1995; Masuda et al., 1998; Masuda and Hasegawa, 1997). For example, the optimal voltage for long-range ordering is 25 V in sulfuric acid, 40 V in oxalic acid, and 195 V in phosphoric acid electrolyte, respectively (Jessensky et al., 1998; Masuda and Fukuda, 1995; Masuda et al., 1998; Masuda and Hasegawa, 1997; Shingubara et al., 1997; Li et al., 1998).

The diameter and depth of each pore, as well as the spacing between adjacent pores, can be controlled by the anodizing conditions. Both the pore diameter and the pore spacing are proportional to the anodizing voltage, with proportional constants of 1.29 nm V^{-1} and 2.5 nm V^{-1}, respectively. The dependence of the diameter and the spacing on the voltage is not sensitive to the electrolyte, which is quite different from the optimal voltage for ordered distribution of the pores. By properly controlling the anodization voltage and choosing the electrolyte, one can make highly ordered nanopores in alumina with desired pore diameter and spacing.

The order of the pores achieved by anodizing an aluminum film over a long period is often limited to a domain of several μm. The individual ordered domains are separated by regions of defects. Recently, a novel approach has been reported to produce a nearly ideal hexagonal nanopore array that can extend over several millimeters (Masuda et al., 1997; Asoh et al., 2001). The approach uses a pretexturing process of Al in which an array of shallow, concave features is initially formed on Al by indentation. The pore spacing can be controlled by the pretexured pattern and the applied voltage. Another widely used method for creating highly ordered nanopore arrays is a two-step anodization method (Masuda et al., 1997; Asoh et al., 2001; Masuda and Satoh, 1996; Li et al., 1999; Li et al., 2000; Foss et al., 1992, 1994). The first step involves a long-period anodization of high-purity aluminum to form a porous alumina layer. Subsequent dissolution of the porous alumina layer leads to a patterned aluminum substrate with an ordered array of concaves that serve as the initial sites to form a highly ordered nanopore array in a second anodization step.

Acidic anodization of Al normally results in a porous alumina structure that is separated from the aluminum substrate by a barrier layer of Al_2O_3. The barrier layer and aluminum substrate can be removed to form a free-standing porous alumina membrane. The aluminum can be removed with saturated $HgCl_2$ and the barrier layer of Al_2O_3 with a saturated solution of KOH in ethylene glycol. An alternative strategy to separate the porous alumina from the substrate is to take advantage of the dependence of pore diameter on anodization voltage. By repeatedly decreasing the anodization voltage several times at 5% increments, the barrier layer becomes a tree-root–like network with fine pores.

Fabrication of Metal Nanowires

Using the membrane templates previously described, nanowires of various metals, semiconductors (Lakshmi et al., 1997), and conducting polymers (Van Dyke and Martin, 1990; Wu and Bein, 1994) have been fabricated. These nanostructures can be deposited into the pores by either electrochemical deposition or other methods, such as chemical vapor deposition (CVD) (Che et al., 1998), chemical polarization (Martin et al., 1993; Parthasarathy and Martin, 1994; Sapp et al., 1999), electroless deposition (Wirtz et al., 2002), or sol-gel chemistry (Lakshmi et al., 1997). Electrodeposition is one of the most widely used methods to fill conducting materials into the nanopores to form continuous nanowires with large aspect ratios. One of the great advantages of the electrodeposition method is the ability to create highly conductive nanowires. This is because electrodeposition relies on electron transfer, which is the fastest along the highest conductive path. Structural analysis shows that the electrodeposited nanowires tend to be dense, continuous, and highly crystalline in contrast to nanowires deposited using other deposition methods, such as CVD. Yi and Schwarzacher demonstrated that the crystallinity of superconducting Pb nanowires can be controlled by applying a potential pulse with appropriate parameters (Yi and Schwarzacher, 1999). The electrodeposition method is not limited to nanowires of pure elements. It can fabricate nanowires of metal alloys with good control over stoichiometry. For example, by adjusting the current density and solution composition, Huang et al. (2002) controlled the compositions of the CoPt and FePt nanowires to 50:50 to obtain the highly anisotropic face-centered tetragonal phases (Yang et al., 2002). Similar strategies have been used in other magnetic nanowires (Yang et al., 2002; Qin et al., 2002) and in thermoelectronic nanowires (Sapp et al., 1999; Sander et al., 2002). Another important advantage of the electrodeposition method is the ability to control the aspect ratio of the metal nanowires by monitoring the total amount of passed charge. This is important for many applications. For example, the optical properties of nanowires are critically dependent on the aspect ratio (Foss et al., 1992, 1994; Preston and Moskovits, 1993). Nanowires with multiple segments of different metals in a controlled sequence can also be fabricated by controlling the potential in a solution containing different metal ions (Liu et al., 1995).

Electrodeposition often requires deposition of a metal film on one side of the freestanding membrane to serve as a working electrode on which electrodeposition takes place. In the case of large pore sizes, the metal film has to be rather thick to completely seal the pores on one side. The opposite side of the membrane is exposed to an electrodeposition solution, which fills up the pores and allows metal ions to reach the metal film. However, one can avoid using the metal film on the backside by using anodic alumina templates with the natural supporting Al substrate. The use of the supported templates also prevents a worker from breaking the fragile membrane during handling. However, it requires the use of AC electrodeposition (Lee et al., 2002; Caboni, 1936). This is because of the rather thick barrier layer between the nanopore membrane and the Al substrate.

Detailed studies of the electrochemical fabrication process of nanowires have been carried out by a number of groups (Schonenberger et al., 1997; Whitney et al., 1993). The time dependence of the current curves recorded during the electrodeposition process reveals three typical stages. Stage I corresponds to the electrodeposition of metal into the pores until they are filled up to the top surface of the membrane. In this stage, the steady-state current at a fixed potential is directly proportional to the metal film area that is in contact with the solution, as found in the electrodeposition on bulk electrodes. However, the electrodeposition is confined within the narrow pores, which has a profound effect on the diffusion process of the metal ions from the bulk solution into the pores before reaching the metal film. The concentration profiles of Co ions in the nanopores of polycarbonate membranes during electrodeposition of Co have been studied by Valizadeh et al. (2001). After the pores are filled up with deposited metal, metal grows out of the pores and forms hemispherical caps on the membrane surface. This region is called stage II. Because the effective electrode area increases rapidly during this stage, the electrochemical current increases rapidly. When the hemispherical caps coalescence into a continuous film, stage III starts, which is characterized by a constant value of the current. By stopping the electrodeposition process before stage I ends, an array of nanowires filled in pores is formed. When freely standing nanowires are desired, one has to remove the template hosts after forming the nanowires in the templates. This task is usually accomplished by dissolving away the template materials in a suitable solvent. Methylene chloride can readily dissolve away track-etched polycarbonate film and 0.1 M NaOH removes anodic alumina effectively. If one wants to also separate the nanowires from the metal films on which the nanowires are grown, a common method is to first deposit sacrificial metal. For example, to fabricate freely standing Au nanowires, one can deposit a thin layer of Ag onto the metal film coated on one side of the template membrane before filling the pores with Au. The Ag layer can be etched away later in concentrated nitric acid, which separates the Au nanowires from the metal film.

Although DC electrodeposition can produce high-quality nanowires, it is challenging to obtain an ordered nanowire array using this method. Normally only 10–20% of the pores in the membrane are filled up completely using the simple DC method (Prieto et al., 2001). Using AC electrodeposition with appropriate parameters (Lee et al., 2002; Schonenberger et al., 1997), a high filling ratio can be obtained using a sawtooth wave (Yin et al., 2001). Furthermore, the researchers found that the filling ratio increases with the AC frequency. A possible reason is that nuclei formed at higher frequencies are more crystalline, which makes the metal deposition easier in the pores and promotes homogenous growth of nanowires. Nielsch et al. (2000) and Sauer et al. (2002) developed a pulsed electrodeposition method. After each potential pulse, a relatively long delay follows before application of the next pulse. The rationale is that the long delay after each pulse allows ions to diffuse into the region where ions are depleted during the deposition (pulse). These researchers demonstrated that the pulsed electrodeposition is well suited for a uniform deposition in the pores of porous alumina with a nearly 100% filling rate.

Positive Template Method. The positive template method is used to make wirelike nanostructures. In this case DNA and carbon nanotubes act like templates, and nanowires form on the outer surface of the templates. Unlike negative templates, the diameters of the nanowires are not restricted by the template sizes and can be controlled by adjusting the amount of materials deposited on the templates. By removing the templates after deposition, wirelike and tubelike structures can be formed.

Carbon Nanotube Template. Fullam et al. (2000) have demonstrated a method to fabricate Au nanowires using carbon nanotubes as positive templates. The first step is to self-assemble Au nanocrystals along carbon nanotubes. After thermal treatment, the nanocrystal assemblies are transformed into continuous polycrystalline Au nanowires of several microns. Carbon nanotubes have also been used as templates to fabricate Mo-Ge superconducting nanowires (Bezryadin et al. 2000) and other metal nanowires (Zhang and Dai, 2000; Yun et al., 2000). Choi et al. (2002) reported highly selective electrodeposition of metal nanoparticles on single-wall carbon nanotubes (SWNTs). Because $HAuCl_4$ (Au^{3+}) or Na_2PtCl_4 (Pt^{2+}) have much higher reduction potentials than SWNTs, they are reduced spontaneously and form Au or Pt nanoparticles on the side walls of SWNTs (shown in Figure 8-15). This is different from traditional electroless deposition because no reducing agents or catalysts are required. Charge transfer during the reaction is probed electrically, because it causes significant changes in the electrical conductance of the nanotubes by hole doping. The nanoparticles deposited on the nanotube can coalesce and cover the entire surface of the nanotube. By removing the nanotube via heating, a Au tubelike structure with a outer diameter <10 nm can be fabricated.

DNA Template. DNA is another excellent choice as a template to fabricate nanowires because its diameter is $\sim$2 nm and its length and sequence can be precisely controlled (Mbindyo et al., 2001). Coffer and co-workers (1996) synthesized micrometer-scale CdS rings using a plasmid DNA as a template. They reported a two-step electrodeposition process to fabricate Pd nanowires using DNA as templates. The first step is to treat DNA in a Pd acetate solution. The second step is to add a reducing agent, typically dimethylamine borane, which reduces Pd ions into Pd along the DNA chains. If the reduction time is short, it leads to individual isolated Pd clusters with a diameter of 3–5 nm. With increasing reduction time, Pd clusters aggregate and form a quasi-continuous Pd nanowire.

This metallization method has been applied to both DNA in solution and DNA immobilized on a solid surface. Braun et al. [117] fabricated a Ag nanowire of $\sim$100 nm in diameter and $\sim$15 μm in length using a linear DNA template.

The procedure used by Braun et al. (1998) to form nanowires using DNA templates is illustrated in Figure 8-16. The first step is to fix a DNA strand between two electrical contacts. The DNA is then exposed to a solution containing Ag+ ions. The Ag+ ions bind to DNA and are then reduced by basic hydroquinone solution to form Ag nanoparticles decorating along the DNA chain. In the last step, the nanoparticles are further "developed" into a nanowire using a standard photographic enhancement technique. The nanowires are highly resistive because they

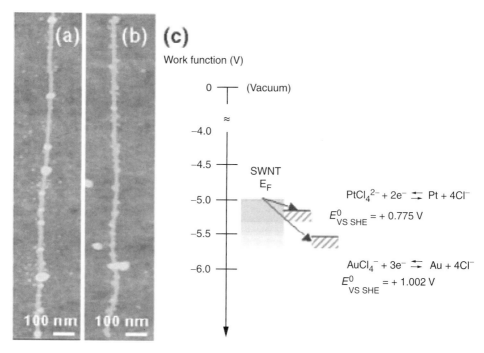

FIGURE 8-15 *AFM images of metal nanoparticles formed on a SWNT template. (a) Au nanoparticles spontaneously and selectively formed on an individual SWNT after immersion in a Au^{3+} solution for 3 min. (b) Pt nanoparticles formed on a SWNT after 3-min exposure to a Pt^{2+} solution. (c) Diagram showing the Fermi energy (E_F) of a SWNT, and the reduction potentials of Au^{3+} and Pt^{2+} versus SHE, respectively. The reduction potentials of most other metal ions lie above E_F, except for Ag^+.* (H. C. Choi et al., Journal of the American Chemical Society, 124, 2002, 9058). Reprinted with permission from the American Chemical Society, 1155 16th St. N. W., Washington DC 20036. Photo Courtesy of H. Dai, Stanford University.

are composed of individual Ag clusters of $\sim$50 nm in diameter. Recently, researchers developed a DNA sequence–specific molecular lithography to fabricate metal nanowires with a predesigned insulating gap (Keren et al., 2002). The approach uses homologous recombination process and the molecular recognition capability of DNA. Homologous recombination is a protein-mediated reaction by which two DNA molecules having some sequence homology crossover at equivalent sites. In the lithography process, RecA proteins are polymerized on a single-strand DNA (ssDNA) probe to form a nucleoprotein filament. Then the nucleoprotein filament binds to an aldehyde-derivatized double-strand DNA (dsDNA) substrate at a homologous sequence. Incubation of the formed complex in $AgNO_3$ solution results in the formation of Ag aggregates along the substrate DNA molecules at regions unprotected by RecA. The Ag aggregates serve as catalysts for specific Au deposition, converting the unprotected regions to conductive Au wires. Thus, a Au nanowire with an insulating gap is formed. The position and size of the insulating gap can be tailored by choosing the template DNA with the special sequence and length.

Polymer Templates. Like DNA, many other polymer chains can also be excellent choices as positive templates for nanowire fabrications. For example, single synthetic flexible polyelectrolyte molecules, poly2-vinylpyridene (P_2VP) are used as templates to fabricate nanowires.

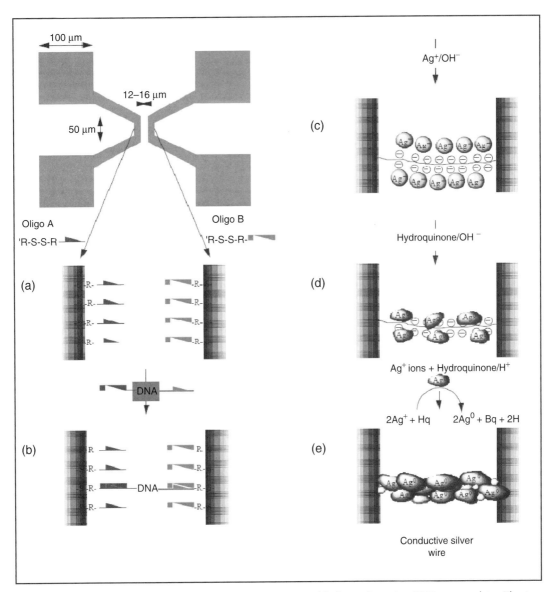

FIGURE 8-16 *Construction of an Ag wire connecting two gold electrodes using DNA as template. The top left image shows the electrode pattern used in the experiments. The two 50-μm-long parallel electrodes are connected to four (100- × 100-μm) bonding pads. (a) Oligonucleotides with two different sequences attached to the electrodes. (b) λ-DNA bridge connecting the two electrodes. (c) Ag-ion-loaded DNA bridge. (d) Metallic Ag aggregates bound to the DNA skeleton. (e) Fully developed Ag wire. (Reprinted by the permission of E. Braun et al., Nature, 391, 1998, 775).*

Because these polymers are thinner than DNA, it is possible to fabricate thinner nanowires. Under appropriate conditions, the polymer chains are stretched into wormlike coils by the electrostatic repulsion between randomly distributed positive charges along the chain. This stretched conformation is frozen when the polymer is attached to a solid substrate. Exposing the polymer to palladium acetate acidic aqueous solution causes Pd^{2+} to coordinate to the polymer template via an ion exchange reaction. In the following step, Pd^{2+} is reduced by dimethylamine borane. The procedure results in metal nanoparticles of 2–5 nm in diameter, which deposit along the template into a wirelike structure.

Djalali et al. (2002) used the core-shell cylindrical polymer brushes as templates to synthesize metal cluster arrays and wires (shown in Figure 8-17). The starting material of the templates is methacryloyl end–functionalized block copolymers, consisting of styrene and vinyl-2-pyridine, which are polymerized to poly(block comacromonomer)s. The formed poly(block

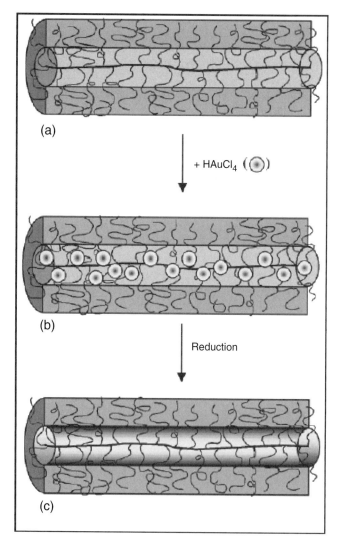

FIGURE 8-17 *Fabrication of nanowires with a polymer template. (a) Core-shell cylindrical brushes with a PVP core and PS shell. (b) Loading the core with HAuCl4. (c) Subsequent reduction of HAuCl4 yields a one-dimensional Au wire within the macromolecular brush.* (D. Djalali et al., Macromolecules, 35, 2002, 4282). Reprinted with permission from the American Chemical Society, 1155 16th St. NW, Washington, DC 20036.

comacromonomer)s exhibit an amphipolar core-shell cylindrical brush structure with a core of vinylpyridine and a shell of polystyrene. The vinylpyridine cores of the cylindrical brushes are loaded with $HAuCl_4$ in toluene or methylene chloride, followed by reduction of the Au salt by the electron beam, UV light, or chemical-reducing agents. Depending on the amount of $AuCl_4^-$ ions loaded in the cores and the reduction conditions, either a linear array of Au clusters or a continuous Au nanowire is formed within the core of the cylindrical brushes. The resulting Au metal nanowires are much longer than the individual core-shell macromolecules, which are caused by a yet unexplained specific end-to-end aggregation of the cylindrical polymers on loading with $HAuCl_4$. Since the metal formation occurs within the cores of the polymers, the polystyrene shells may serve as the electrically insulating layers.

Applications of the Nanowires

Metal nanowires are promising materials for many novel applications, ranging from chemical and biological sensors to optical and electronic devices. This is not only because of their unique geometry, but also because they have many unique physical properties, including electrical, magnetic, optical, as well as mechanical properties. Some of the applications are discussed below.

Magnetic Materials and Devices

The electrodeposition methods described above have been used to fabricate magnetic nanowires of a single metal (Whitney et al., 1993), multiple metals in segments (Piraux et al., 1994), as well as alloys (Dubois et al., 1997). Magnetic nanowires with relatively large aspect ratios (e.g., >50), exhibit an easy axis along the wires. An important parameter that describes magnetic properties of materials is the remanence ratio, which measures the remanence magnetization after switching off the external magnetic field. The remanence ratios of the Fe, Co, and Ni nanowires can be larger than 0.9 along the wires and much smaller in the perpendicular direction of the wires. This finding clearly shows that the shape anisotropy plays an important role in the magnetism of the nanowires. Another important parameter that describes the magnetic properties is coercivity, which is the coercive field required to demagnetize the magnet after full magnetization. The magnetic nanowires exhibit greatly enhanced magnetic coercivity (Chien, 1991). In addition, the coercivity depends on the wire diameter and the aspect ratio, which shows that it is possible to control the magnetic properties of the nanowires by controlling the fabrication parameters. The diameter dependence of the coercivity reflects a change of the magnetization reversal mechanism from localized quasi-coherent nucleation for small diameters to a localized curling like nucleation as the diameter exceeds a critical value (Thurn-Albrecht et al., 2000).

Another novel property of magnetic nanowires is giant magnetoresistance (GMR) (Liu et al.; Evans, et al., 2000). For example, Evans et al. have studied Co-Ni-Cu/Cu multilayered nanowires and found a magnetoresistance ratio of 55% at room temperature and 115% at 77 K for current perpendicular to the plane (along the direction of the wires). Giant magnetoresistance has also been observed in semimetallic Bi nanowires fabricated by electrodeposition (Liu et al., 1998; Hong et al., 1999; Lin et al., 2000). Hong et al. (1999) have studied GMR of Bi with diameters between 200 nm and 2 µm in magnetic fields up to 55T and found that the magnetoresistance ratio is between 600 and 800% for magnetic field perpendicular to the wires and ~200% for the field parallel to the wires. The novel properties and small dimensions have potential applications in the miniaturization of magnetic sensors and the high-density magnetic storage devices.

The alignment of magnetic nanowires in an applied magnetic field can be used to assemble the individual nanowires (Lin et al., 2000). Tanase et al. studied the response of Ni nanowires in response to magnetic field (Tanase et al., 2001). The nanowires are fabricated by electrodeposition using alumina templates and functionalized with luminescent porphyrins so that they can be visualized with a video microscope. In viscous solvents, magnetic fields can be used to orient the nanowires. In mobile solvents, the nanowires form chains in a head-to-tail configuration when a small magnetic field is applied. In addition, Tanase et al. demonstrated that three-segment Pt-Ni-Pt nanowires can be trapped between lithographically patterned magnetic microelectrodes (Tanase et al., 2002). The technique has a potential application in the fabrication and measurement of nanoscale magnetic devices.

Optical Applications

Dickson and Lyon studied surface plasmon (collective excitation of conduction electrons) propagation along 20-nm-diameter Au, Ag, and bimetallic Au-Ag nanowires with a sharp Au-Ag heterojunction over a distance of tens of μm (Dickson and Lyon, 2000). The plasmons are excited by focusing a laser with a high-numerical-aperture microscope objective, which propagate along a nanowire and reemerge as light at the other end of the nanowire via plasmon scattering. The propagation depends strongly on the wavelength of the incident laser light and the composition of the nanowire. At the wavelength of 820 nm, the plasmon can propagate in both Au and Ag nanowires, although the efficiency in Ag is much higher that in Au. In the case of bimetallic nanowire, light emission is clearly observed from the Ag end of the nanowire when the Au end is illuminated at 820 nm. In sharp contrast, if the same bimetallic rod is excited at 820 nm via the Ag end, no light is emitted from the distal Au end. The observations suggest that the plasmon mode excited at 820 nm is able to couple from the Au portion into the Ag portion with high efficiency, but not from the Ag portion into Au. The unidirectional propagation has been explained using a simple two-level potential model. Since surface plasmons propagate much more efficiently in Ag than in Au, the Au $\rightarrow$ Ag boundary is largely transmissive, thus enabling efficient plasmon propagation in this direction from Ag to Au and a much steeper potential wall, which allow less optical energy to couple through to the distal end. The experiments suggest that one can initiate and control the flow of optically encoded information with nanometer-scale accuracy over distances of many microns, which may find applications in future high-density optical computing.

Biological Assays

We have already mentioned that by sequentially depositing different metals into the nanopores, multisegment or striped metal nanowires can be fabricated (Martin et al., 1999). The length of each segment can be controlled by the charge passed in each plating step, and the sequence of the multiple segments is determined by the sequence of the plating steps. Due to the different chemical reactivities of the "stripe" metals, these strips can be modified with appropriate molecules. For example, Au binds strongly to thiols and Pt has high affinity to isocyanides. Interactions between complementary molecules on specific strips of the nanowires allow different nanowires to bind to each other and form patterns on planar surfaces. Using this strategy, nanowires could assemble into cross- or T-shaped pairs, or into more complex shapes (Reiss et al., 2001). It is also possible to use specific interactions between selectively functionalized segments of these nanowires to direct the assembly of nanowire dimers and oligomers, to prepare a two-dimensional assembly of nanowire-substrate epitaxy, and to prepare three-dimensional colloidal crystals from nanowire-shaped objects (Yu et al., 2000). As an example, single-strand DNA can be exclusively modified at the tip or any desired location of a nanowire, with the rest of the wire covered by an organic passivation monolayer. This opens the possibility for site-specific DNA assembly (Martin et al., 1999).

Nicewarner-Pena et al. showed that the controlled sequence of multisegment nanowires can be used as "bar codes" in biological assays (Nicewarner-Pena et al., 2001). The typical dimension of the nanowire is $\sim$200 nm thick and $\sim$10 μm long. Because the wavelength dependence of reflectance is different for different metals, the individual segments are easily observed as "stripes" under an optical microscope with unpolarized white-light illumination. Different metal stripes within a single nanowire selectively adsorb different molecules, such as DNA oligomers, which can be used to detect different biological molecules simultaneously. These multisegment nanowires have been used like metallic bar codes in DNA and protein bioassays.

The optical scattering efficiency of the multisegment nanowires can be significantly enhanced by reducing the dimensions of the segment, such that excitation of the surface plasmon occurs. Mock et al. (2002) have studied the optical scattering of multisegment nanowires of Ag, Au, and Ni that have diameters of $\sim$30 nm and length up to $\sim$7 μm. The optical scattering is dominated by the polarization-dependent plasmon resonance of Ag and Au segments. This is different from the case of the thicker nanowires used by Nicewarner-Pena et al., where the reflectance properties of bulk metals determine the contrast of the optical images (Nicewarner-Pena et al., 2001). Because of the large enhancement by the surface plasmon resonance, very narrow ($\sim$30-nm-diameter) nanowires can be readily observed under white-light illumination, and the optical spectra of the individual segments are easily distinguishable (Mock et al., 2002). The multisegment nanowires can host a large number of segment sequences over a rather small spatial range, which promises unique applications.

Chemical Sensors

Penner, Handley, and Dagani et al. exploited hydrogen sensor applications using arrays of Pd nanowires (Walter et al., 2002). Unlike the traditional Pd-based hydrogen sensor that detects a drop in the conductivity of Pd on exposure to hydrogen, the Pd-nanowire sensor measures an increase in the conductivity (Figure 8-18). This happens because the Pd wire consists of a string of Pd particles separated with nanometer-scale gaps. That these gaps close to form a conductive path in the presence of hydrogen molecules as Pd particles expand is well known; this closure is due to the disassociation of hydrogen molecules into hydrogen atoms that penetrate into the Pd lattice and expand the lattice. Although macroscopic Pd-based hydrogen sensors are readily available, they have the following two major drawbacks. First, their response time is between 0.5 s to several minutes, which is too slow to monitor gas flow in real time. Second, they are prone to the contamination by a number of gas molecules, such as methane, oxygen, and carbon monoxide, which adsorb onto the sensor surfaces and block the adsorption sites for hydrogen molecules. The Pd nanowires offer remedies to the above problems. Pd nanowires have a large surface-to-volume ratio, which makes the nanowire sensor less prone to contamination by common substances.

Carbon Nanotubes

Carbon nanotubes (CNTs) are very interesting nanostructures with a wide range of potential applications. CNTs were first discovered by Iijima in 1991 (Iijima, 1991); since then, great

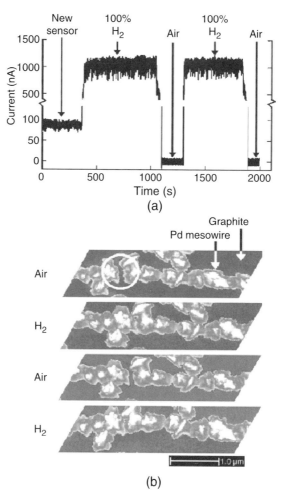

FIGURE 8-18 *Chemical sensor application of Pd nanowires. (a) Plot of sensor current versus time for the first exposure of a Pd nanowire sensor to hydrogen and one subsequent H_2-air cycle. (b) AFM image of a Pd nanowire on a graphite surface. These images were acquired either in air or in a stream of H_2 gas, as indicated. A hydrogen-actuated break junction is highlighted (circle).* (E. C. Walter et al., Analytical Chemistry, 74, 2002, 1546). Reprinted with permission from the American Chemical Society, 1155 16th St. N.W., Washington, DC 20036. Photo Courtesy of M. Penner, University of California, Irvine.

progress has been made toward many applications, including, for example, the following:

Materials Chemical and biological separation, purification, and catalysis Energy storage such as hydrogen storage, fuel cells, and the lithium battery Composites for coating, filling, and use as structural materials

Devices Probes, sensors, and actuators for molecular imaging, sensing, and manipulation Transistors, memories, logic devices, and other nanoelectronic devices Field emission devices for x-ray instruments, flat-panel display, and other vacuum nanoelctronic applications. These applications and advantages can be understood by the unique structure and properties of nanotubes, as outlined below:

Structures

Bonding: The *sp*2 hybrid orbital allows carbon atoms to form hexagons and occasionally pentagon units by in-plane σ bonding and out-of-plane π bonding.

Defect-free nanotubes: These are tubular structures of hexagonal network with a diameter as small as 0.4 nm. Tube curvature results in $\sigma-\pi$ rehybridization or mixing.

Defective nanotubes: Occasionally pentagons and heptagons are incorporated into a hexagonal network to form bent, branched, helical, or capped nanotubes.

Properties

Electrical: Electron confinement along the tube circumference makes a defect-free nanotube either semiconducting or metallic with quantized conductance, whereas pentagons and heptagons generate localized states.

Optical and optoelectronic: Direct band gap and one-dimensional band structure make nanotubes ideal for optical applications with wavelength ranging possibly from 300 to 3000 nm.

Mechanical and electrochemical: The rehybridization gives nanotubes the highest Young's modulus (over 1 Tpa), tensile strength of over 100 GPa, and remarkable electronic response to strain and metal-insulator transition.

Magnetic and electromagnetic: Electron orbits circulating around a nanotube give rise to many interesting phenomena such as quantum oscillation and metal-insulator transition.

Chemical and electrochemical: High specific surface and rehybridization facilitate molecular adsorption, doping, and charge transfer on nanotubes, which, in turn, modulates electronic properties.

Thermal and thermoelectric: Possessing a property inherited from graphite, nanotubes display the highest thermal conductivity, whereas the quantum effect shows up at low temperature.

Structure of the Carbon Nanotube

This section describes the structure of various types of carbon nanotubes (CNT). A CNT can be viewed as a hollow cylinder formed by rolling graphite sheets. Bonding in nanotubes is essentially *sp*2. However, the circular curvature will cause quantum confinement and $\sigma-\pi$ rehybridization in which three bonds are slightly out of plane; for compensation, the orbital is more delocalized outside the tube. This makes nanotubes mechanically stronger, electrically and thermally more conductive, and chemically and biologically more active than graphite. In addition, they allow topological defects such as pentagons and heptagons to be incorporated into the hexagonal network to form capped, bent, toroidal, and helical nanotubes, whereas electrons will be localized in pentagons and heptagons because of redistribution of electrons. A nanotube is called defect free if it is of only hexagonal network and defective if it also contains topological defects such as pentagonal and heptagonal or other chemical and structural defects. A large amount of work has been done in studying defect-free nanotubes, including single- or multiwall nanotubes (SWNTs or MWNTs). A SWNT is a hollow cylinder of a graphite sheet, whereas a MWNT is a group of coaxial SWNTs. SWNTs were discovered in 1993 (Iijima and Ichihashi, 1993), two years after the discovery of MWNTs (Bethune et al., 1993). They are often seen as straight or elastic bending structures individually or in ropes (Thess et al.,1996) by transmission electron microscopy (TEM), scanning electron microscopy (SEM), atomic force

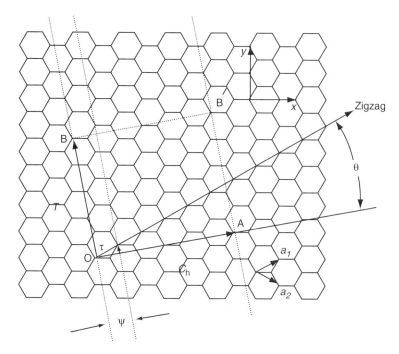

FIGURE 8-19 *A nanotube (n,m) is formed by rolling a graphite sheet along the chiral vector* $C_h = na_1 + ma_2$ *on the graphite where* a_1 *and* a_2 *are graphite lattice vectors.* (M. S. Dresselhaus, G. Dresselhaus, and P. C. Eklund, Science and Fullerenes and Carbon Nanotubes, Academic Press, New York, 1996, Chapter 21). Reprinted with permission from Academic Press, an imprint of Elsevier.

microscopy (AFM), and scanning tunneling microscopy (STM). In addition, electron diffraction (EDR), x-ray diffraction (XRD), Raman, and other optical spectroscopy can be also used to study structural features of nanotubes.

A SWNT can be visualized as a hollow cylinder, formed by rolling over a graphite sheet. It can be uniquely characterized by a vector C_h where C_h denotes chiral vector in terms of a set of two integers (n, m) corresponding to graphite vectors a_1 and a_2, Figure 8-19 (Dresselhaus, Dresselhaus, and Eklund, 1996).

$$C_h = na_1 + ma_2 \tag{8-1}$$

Thus, the SWNT is constructed by rolling up the sheet such that the two end-points of the vector C_h are superimposed. This tube is denoted as (n, m) tube with diameter given by

$$D = |C_h|/\pi = a(n^2 + nm + m^2)^{1/2}/\pi \tag{8-2}$$

where $a = |a_1| = |a_2|$ is the lattice constant of graphite. The tubes with $m = n$ are commonly referred to as armchair tubes, and $m = 0$, as zigzag tubes. Others are called chiral tubes in general, with the chiral angle, θ, defined as that between the vector C_h and the zigzag direction a_1,

$$\theta = \tan^{-1}[3^{1/2}m/(m + 2n)] \tag{8-3}$$

θ ranges from 0 for zigzag ($m = 0$) and 30° for armchair ($m = n$) tubes.

(a)

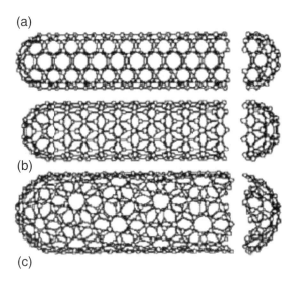

(b)

(c)

FIGURE 8-20 *Illustration of some possible structures of carbon nanotubes, depending on how graphite sheets are rolled: (a) armchair, (b) zigzag, and (c) chiral structures.*

The lattice constant and intertube spacing are required to generate a SWNT, SWNT bundle, and MWNT. These two parameters vary with tube diameter or in radial direction. Most experimental measurements and theoretical calculations agree that, on average, the $c - c$ bond length $d_{cc} = 0.142$ nm or $a = |a_1| = |a_2| = 0.246$ nm, and intertube spacing $d_{tt} = 0.34$ nm (Dresselhaus et al., 1996). Thus, Equations 8-1 to 8-3 can be used to model various tube structures and interpret experimental observation. Figure 8-20 illustrates examples of nanotube models.

Synthesis of Carbon Nanotubes

Carbon nanotubes can be made by laser evaporation, carbon arc methods, and chemical vapor deposition. Figure 8-21 illustrates the apparatus for making carbon nanotubes by laser evaporation. A quartz tube containing argon gas and a graphite target are heated to 1200 °C (Smalley et al., 1997). Contained in the tube, but somewhat outside the furnace, is a water-cooled copper collector. The graphite target contains small amounts of Fe, Co, and Ni that act as seeds for the growth of carbon nanotubes. An intense, pulsed laser beam is incident on the target, evaporating carbon from the graphite. The argon then sweeps the carbon atoms from the high-temperature zone to the colder collector, on which they condense into nanotubes. Tubes 10–20 nm in diameter and 100 μm long can be made by this method.

Nanotubes can also be synthesized using a carbon arc method (Figure 8-22). An electrical potential of 20–25 V and a DC electric current of 50–120 A flowing between the electrodes of 5- to 20-mm diameter and separated by ~1 mm at 500 Torr pressure of flowing helium are used (Saito et al., 1996). As the carbon nanotubes form, the length of the positive electrode decreases, and a carbon deposit forms on the negative electrode. To produce single-wall nanotubes, a small amount of Fe, Co, and Ni is incorporated as a catalyst in the central region of the positive electrode. These catalyst act like seeds for the growth of single and multi-walled carbon nanotubes.

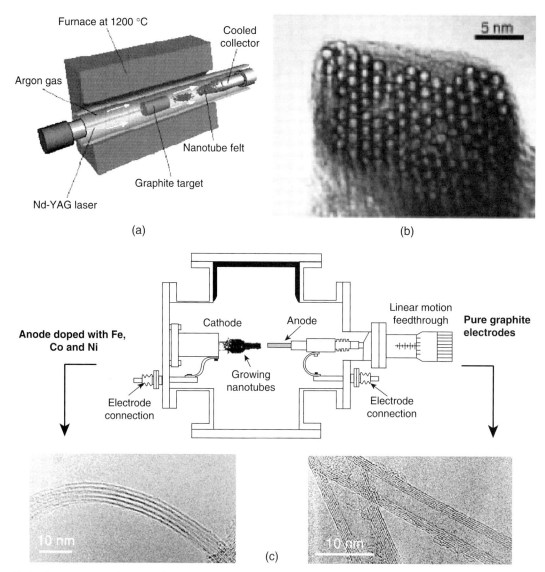

FIGURE 8-21 *(a) Schematic of laser ablation apparatus (b) TEM image of SWNT grown by laser ablation technique.* (R. E. Smalley et al., American Scientist, 85, 1997, 324). Reprinted with permission from Hazel Cole, exec. Asst to late Prof. Smalley, Rice University. *(c) Schematic diagram of carbon arc apparatus for the production of carbon nanotubes.* (Saito et al., Journal of Applied Physics, 80(5), 1996, 3062-3067). Reprinted with permission from Y. Saito, Nagoya University, Japan.

For the large-scale production of carbon nanotubes, thermal CVD is the most favorable method (Figure 8-22). The thermal CVD apparatus is very simple for the growth of carbon nanotubes (Cassell et al., 1999). It consists of a quartz tube (1- to 2-in. diameter) inserted into a tubular furnace capable of maintaining a temperature of 1 °C over a 25-cm zone. Thus, it is a hot-wall system at primarily atmospheric pressure (CVD) and hence does not require any pumping systems. In thermal CVD, either CO or some hydrocarbon such as methane, ethane, ethylene,

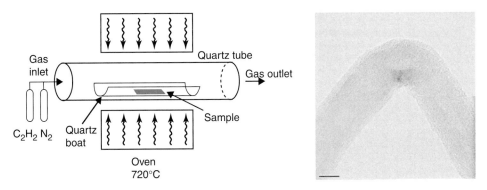

FIGURE 8-22 *Schematic of a thermal CVD apparatus and TEM image of multiwall carbon nanotube.*

acetylene, or other higher hydrocarbon is used without dilution. The reactor is first filled with argon or some other inert gas until the reactor reaches the desired growth temperature. Then the gas flow is switched to the feedstock for the specified growth period. At the end, the gas flow is switched back to the inert gas while the reactor cools down to 300 °C or lower before exposing the nanotubes to air. Exposure to air at elevated temperatures can cause damage to the CNTs. Typical growth rates range from a few nm/min to 2 to 5 μm/min. Hongjie et al. (1999) reported a patterned growth of MWNTs by thermal CVD method. Figure 8-23 shows MWNTs grown by CVD at 700 °C in a 2-in. tube furnace under an ethylene flow of 1000 sccm for 15–20 min on a porous silicon substrate.

Growth Mechanisms of Carbon Nanotubes

The growth mechanism of nanotubes may vary depending on which method is used; with the arc-discharge and laser-ablation method MWNT can be grown without a metal catalyst, contrary to carbon nanotubes synthesized with CVD method, where metal particles are necessary. For growing SWNTs, in contrast, metals are necessary for all three methods mentioned previously.

The growth mechanism of nanotubes is not well understood; different models exist, but some of them cannot unambiguously explain the mechanism. The metal or carbide particles seem necessary for the growth because they are often found at the tip inside the nanotube or also somewhere in the middle of the tube. In 1972, Baker et al. (1972) developed a model for the growth of carbon fibers, which is shown in Figure 8-24 on the right side. It is supposed that acetylene decomposes at 600 °C on the top of nickel cluster on the support. The dissolved carbon diffuses through the cluster due to a thermal gradient formed by the heat release of the exothermic decomposition of acetylene. The activation energies for filament growth were in agreement with those for diffusion of carbon through the corresponding metal (Fe, Co, Cr) (Baker et al., 1973). Whether the metal cluster moves away from the substrate (tip growth) or stays on the substrate (base growth) is explained by a weaker or stronger metal-support interaction, respectively. As stated by Baker et al. the model has a number of shortcomings. It cannot explain the formation of fibers produced from the metal catalyst decomposition of methane, which is an endothermic process.

Oberlin et al. (1976) proposed a variation of this model shown in Figure 8-25. The fiber is formed by a catalytic process involving the surface diffusion of carbon around the metal particle, rather than by bulk diffusion of carbon through the catalytic cluster. In this model the cluster

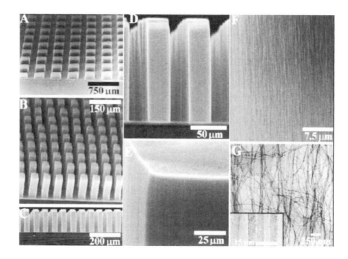

FIGURE 8-23 *Electron micrographs of self-oriented nanotubes synthesized on porous silicon substrates. (a) SEM image of nanotube blocks synthesized on 250- by 250-mm catalyst patterns. The nanotubes are 80 μm long and oriented perpendicular to the substrate. (b) SEM image of nanotube towers synthesized on 38-μm catalyst patterns. The nanotubes are 130 μm long. (c) Side view of the nanotube towers in (b). The nanotube self-assembly such that the edges of the towers are perfectly perpendicular to the substrate. (d) Nanotube "twin towers," a zoom-in view of (c). (e) SEM image showing sharp edges and corners at the top of a nanotube tower. (f) SEM image showing that nanotubes in a block are well aligned to the direction perpendicular to the substrate surface. (g) TEM image of pure multiwall nanotubes in several nanotube blocks grown on a porous silicon substrate. The inset is a high-resolution TEM image that shows two nanotubes bundling together. (Hongjie Dai et al., Science, 283, 1999, 512). Reprinted with permission from H. Dai, Stanford University, CA.*

corresponds to a seed for the fiber nucleation. Amelinckx et al. (1994) adapted the growth model of Baker et al. (1973) to explain the growth of carbon nanotubes.

For the synthesis of single wall-carbon nanotubes, the metal clusters have to be present in form of nanosized particles. Furthermore, it is supposed that the metal cluster can have two roles: first, it acts as a catalyst for the dissociation of the carbon-bearing gas species. Second, carbon diffuses on the surface of the metal cluster or through the metal to form a nanotube. The most active metals are Fe, Co, and Ni, which are good solvents for carbon (Kim et al., 1991).

Carbon Nanotube Composite Materials

Composite materials containing carbon nanotubes are a new class of materials. These materials will have a useful role in engineering applications. Many nanoenthusiasts are working in the area of polymer composites; efforts in metal and ceramic matrix composite are also of interest.

New conducting polymers, multifunctional polymer composites, conducting metal matrix composites, and higher fracture-strength ceramics are just a few of the new materials being processed that make them near-term opportunities. Moreover, high conductors that are multifunctional (electrical and structural), highly anisotropic insulators, and high-strength, porous ceramics, are other examples of new materials that can come from nanotubes.

Most researchers who are developing new type of composite materials with nanotubes work with nanotube concentrations below 10 wt% because of the limited availability of nanotubes.

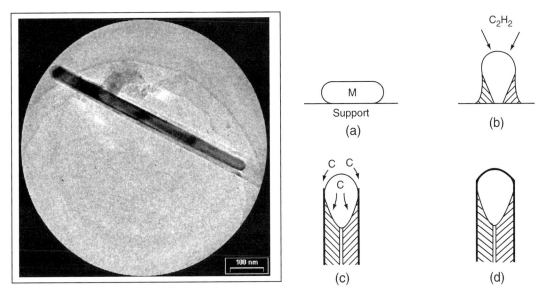

FIGURE 8-24 *Growth model of vapor-grown carbon fibers (right), according to Baker et al. (Carbon, 27, 1989, 315); and TEM image of a MWNT with metal nanorod inside the tip (left). (M. K. Singh et al., Journal of Nanomaterials and Nanotechnology, 3(3), 2003, 165). Reprinted with permission from Elsevier.*

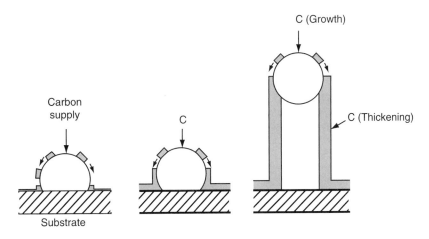

FIGURE 8-25 *Alternative growth model of vapor-grown carbon fibers where the metal cluster acts as a seed for the growth. (Oberlin et al., Journal of Crystal Growth, 32, 1976, 335). Reprinted with permission from Elsevier.*

Collective studies show broader promise for composite materials with concentrations as high as 40 and 50 wt% that may be limited only by our ability to create a complete matrix and fiber registry with high-surface-area nanofibers. In the following sections processing, properties and numerous potential applications and the application range of a wide variety of nanocomposites are discussed.

Polymer Nanocomposites

Plastics are compounded with inorganic fillers to improve processability, durability, and thermal stability. The mechanical properties of the composite are related to the volume fraction and aspect ratio of the filler. There is interest in incorporating nanoscale fillers in plastics. For example, the fire retardancy of polymers such as polypropylene has been increased by dispersing very small ($<1\%$) amount of nanoscale copper particles. The thermoplastic PVC (polyvinyl chloride) has been compounded with inorganic fillers to improve its processability and thermal stability. Similarly, polymers containing nanoscale montmorillonite clay exhibit high strength, high modulus, good heat distortion temperature, and enhanced flame-retarding properties. The strength, modulus, and fracture toughness of PVC has been improved by dispersing nanosize calcium carbonate in PVC. *In-situ* polymerization of vinyl chloride in the presence of $CaCO_3$ nanoparticles was used to produce the composite. The in situ polymerization reduces the agglomeration of nanopowders, which is a major problem with adding nanoparticles to preformed polymers, thus permitting homogeneous distribution to be achieved. In a manner similar to the classic polymer composites, the polymer nanocomposites also exhibit shear thinning and power law behaviors. For example, for PVC-$CaCO_3$ at low shear rates, the viscosity was found to be higher than that of pristine PVC.

The incorporation of CNTs in polymers has been found to improve a variety of engineering properties. Because of their metallic or semiconducting character, CNT incorporation in polymer matrices permits attainment of an electrical conductivity sufficient to provide an electrostatic discharge at low CNT concentrations. The addition of CNT to polymers also results in significant weight savings. Surface treatments such as oxidation of CNTs to CO_2, or their graphitization, are found to improve the tensile strength of polypropylene-CNT composites. In addition to particles and nanotubes, nano plates (e.g., layered silicates, exfoliated graphite) have been used as fillers in polymers.

There are, however, some key differences between polymeric nanocomposites and classic filled polymers or composites. These include: low percolation threshold (i.e., network-forming tendency, ~ 0.1 to $2\,vol\%$ for nanodispersions), large numerical density of nanoparticles per unit volume (10^6 to 10^8 particles per μm^3), extensive interfacial area per unit volume (1 km to 10 km per mL), and short distances between particles (10 to 50 nm at 1 to 8 vol% nanoparticles).

The aspect ratio (length-to-diameter ratio) of nano fillers is extreme. The size of the interfacial zone in a nanocomposite is comparable to the spacing between the filler, and the ratio of the volume of the interface to the volume of the bulk (matrix) increases dramatically as the nanoregime is approached. This affects many interface-sensitive properties such as vibration-damping capacity and strength. The mechanical behavior of polymeric nanocomposites is likely to be different from micrometer-size polymer composites. This is because the total interfacial area becomes the critical characteristic rather than the volume fraction, as is the case with micrometer-size composites. This, in turn, impacts the engineering issues related to irreversible agglomeration, nanoparticle network formation, and ultralong times for relaxation (glasslike behavior) in the case of nanocomposites.

The classic view of reinforced composites suggests that a strong fiber–matrix interface leads to high composite stiffness and strength but also low toughness, because of the brittle nature of the fiber and because of a lack of crack deflection at a strong interface, which is needed for strengthening. With nanocomposites having strong interfaces, it may be possible to attain very high strength and stiffness, and also high toughness, because of the nanotube's ability to considerably deform before fracture. In contrast, with weak interfaces toughness would be possible in a nanocomposite via the mechanism of crack deflection that also operates in classic polymeric composites.

Nanotubes are considered to have varying degrees of defects depending on whether they are multiwall nanotubes or single-wall tubes. Initially, nanotube functionalization was thought to occur at the various defect sites (MWNTs being highly defective compared with SWNTs), but functionalizations both at nanotube ends and along the side walls without disruption or degradation of the tubes have been demonstrated. A variety of functionalized nanotubes are being developed for composite applications, including fluoronanotubes (f-SWNTs); carboxyl-nanotubes with various end functionalization; and numerous covalently bonded SWNTs such as amino-SWNTs, vinyl-SWNTs, epoxy-SWNTs, and many others, to provide for matrix bonding, cross-linking, and initiation of polymerization. Wrapped nanotubes (w-SWNTs) with noncovalent bonding are also another variety of nanotubes and find particular use when the electrical properties of the nanotubes need to be preserved. Figure 8-26a. shows nanotube side walls cross-linked in a polymer matrix. The insert, Figure 8-26b, shows integration of the functionalized SWNTs into epoxy resin. Nanotube chemistry is not only for polymeric systems but will also be evolved for metals (particularly for Al, Cu, and Ti) and ceramics that are carbide-, oxide-, and nitride-based, as well as many other varieties. Nanotube functionalization in metals and ceramics will likely play a role in nanotube stabilization, defect refinement, and overcoating methodologies. If ends are considered defects, then methodologies for welding nanotubes may well include other species that would be delivered to the nanotubes from a metal or ceramic matrix. These opportunities coupled to a variety of processing modes may well produce next-generation metal hybrids and porous, high-conducting, high-toughness ceramic structures (Barrera, 2000; Mickelson et al., 1998; Zhu, Jiang, et al., 2003).

Ceramic Nanocomposites

Nanoceramics and ceramic nanocomposites are novel and technologically ascendant materials. Examples include tough ceramics that contain dispersion of nanoscale metals (Co, Ni), and semiconductor or metal nanoparticles in glass matrices for optoelectronic devices. These materials have been processed in small quantities using high-energy ball milling, internal oxidation, hot consolidation, crystallization of amorphous solid, sol-gel processing, and vapor-phase and vapor-liquid-solid deposition.

Many nanoceramics have outstanding physical and mechanical properties. For example, CNTs have excellent thermal conductivity, and ceramic-CNT nanocomposites can be produced with tailored thermal conductivity for thermal management applications. Although the primary motivation to incorporate CNT in ceramics is to toughen the ceramic, the high thermal conductance of CNTs suggests that their incorporation in ceramics will facilitate thermal transport and thus improve the thermal shock resistance of the ceramic. Toughening is achieved via weak fiber–matrix interfaces that permit debonding and sliding of the fiber within the matrix. The closing forces exerted by the fibers on matrix cracks that propagate around the fibers constrain crack growth, and the work required to pull broken fibers out against sliding friction at the interface imparts toughness.

Many ceramic nanocomposites possessing improved fracture toughness have been synthesized. These include hot-pressed powder composites of CNT-SiC, Al_2O_3-SiC, MgO-SiC, W-Al_2O_3, Al_2O_3-Co, ZrO_2-Ni, and Si_3N_4-SiC. The powders are first synthesized using chemical reactions and precipitation. For example, oxide ceramic nanocomposites such as Al_2O_3-Co and ZrO_2-Ni are made from nickel-nitrate or cobalt-nitrate solutions. These solutions are mixed with Al_2O_3 or ZrO_2 powders and decomposed in air above $400\,°C$, causing heterogeneous

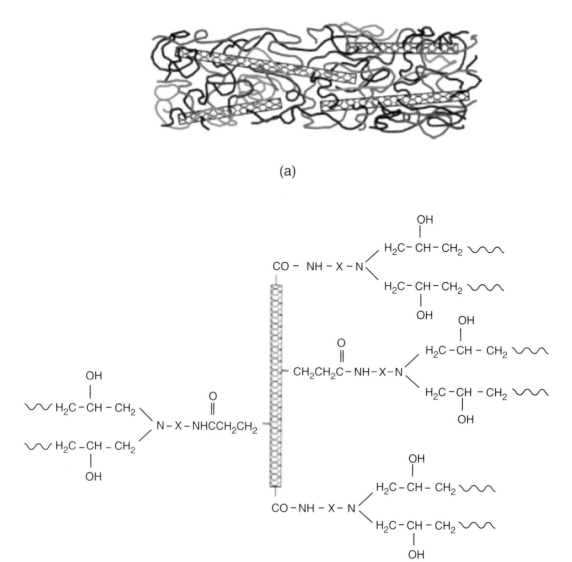

(a)

(b)

FIGURE 8-26 *(a) Nanotube side walls cross-linked in a polymer matrix. (b) Integration of the functionalized SWNTs into epoxy.* (Reprinted by permission of E. V. Barrera et al., Advanced Functional Materials, 14, 2004, 643).

nucleation and precipitation of metal oxide nanoparticles on the surface of ceramic particles. The resulting material is then reduced by hydrogen and hot press sintered at high temperatures and pressures (typically, 1400–1600 °C under 20–40 MPa) to produce Ni- or Co-dispersed ceramic-matrix composites. Considerable improvements in mechanical strength, hardness, and fracture toughness are achieved.

The improvement in the fracture toughness of brittle ceramics through nanoscale metal dispersions, which is especially attractive, is achieved because of the extremely high concentration

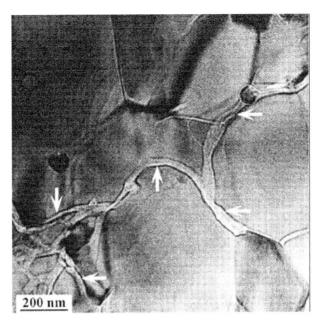

FIGURE 8-27 *TEM image of a 5.7 vol% SWNT/Al₂O₃ nanocomposite fabricated using SPS. The white arrows indicate SWNTs within Al₂O₃ grains.* (Reprinted by permission of G.-D Zhan et al., Applied Physics Letters, 83(6), 2003, 1228).

of internal interfaces that become available for energy dissipation. Nanocomposites of fine particles dispersed in a dielectric matrix have been developed; for example, semiconductor or metal nanoparticles dispersed in a glass matrix have been synthesized for optoelectronic devices, catalysts, and for magnetic, and solar energy conversion devices. CNT incorporation in ceramics such as SiC and alumina by hot-pressing, increases the bending strength and fracture toughness, and lowers the material density, thus creating strong and tough lightweight materials.

To obtain fully dense nanocrystalline ceramics and avoid damaging nanotube reinforcements during sintering, spark-plasma sintering (SPS) is considered the leading technique for composite consolidation because it is a rapid sintering process (Sun et al., 2002; Zhan et al., 2003). This technique is a pressure-assisted fast sintering method based on a high-temperature plasma (spark plasma) that is momentarily generated in the gaps between powder materials by electrical discharge during on/off DC pulsing. It has been suggested that the DC pulse could generate several effects: Spark-plasma sintering, SPS, can rapidly consolidate powders to near-theoretical density through the combined effects of rapid heating, pressure, and powder surface cleaning. Fully dense SWNT-Al₂O₃ nanocomposites were fabricated using SPS. Figure 8-27 shows a transmission electron micrograph of the nanocomposite, in which the SWNTs are found between the grain boundaries.

Ceramic Nanotube Composite Systems

There is a lot of interest in developing nanoceramic composites (NCCs) to enhance the mechanical properties of brittle ceramics. The conventional ceramic-processing techniques were first employed to develop MWNT-reinforced SiC ceramic composites where a 20% and 10% increase in strength and fracture toughness, respectively, were measured on bulk composite samples over that of the monolith ceramic (Ma et al., 1998). These increases are believed to be due to the introduction of high-modulus MWNTs into the SiC matrix, which contribute to crack deflection and nanotube debonding. Furthermore, good interfacial bonding is required to achieve adequate load transfer across the MWNT–matrix interface, which is a condition necessary for further improving the mechanical properties in all NCC systems. Therefore, nonconventional processing techniques such as colloidal processing and in situ chemical methods are important processing methods for NCC systems. These novel techniques allow for the control of the interface developed after sintering through manipulation of the surface properties of the composite materials during green processing. Carbide and nitride ceramic materials have been widely used as matrix materials due to the extreme processing conditions required to consolidate and sinter the matrix material. Average sintering temperatures for most nitride and carbide ceramic materials are well above 1800 °C. The following sections deal with specific novel processing routes for NCCs and their related applications.

Ceramic-Coated MWNTs and SWNTs

Multiwall nanotubes coated with Al_2O_3 ceramic particles via simple colloidal processing methods provide a significant enhancement to the mechanical properties of the monolithic ceramic (Sun and Gao, 2003; Sun et al., 2002). Coating the surface of MWNTs with alumina improves the homogenous distribution of MWNTs in the ceramic matrix and enables tighter binding between two phases after sintering. MWNTs are treated in NH_3 at 600 °C for 3 hours to change their surface properties. Treated MWNTs are put into a solution containing polyethyleneamine (PEI), a cationic polymer used as a dispersant. Alumina is dispersed into deionized water, and polyacrylic acid (PAA) is added into this very dilute alumina suspension. Sodium hydroxide is used to adjust the pH. The prepared dilute alumina suspension with PAA is added into the as-prepared carbon nanotube suspension with PEI, and the suspensions are ultrasonicated. The coated carbon nanotubes collected from the mixed suspension are subsequently added into the concentrated alumina suspension in about 50 wt% ethanol. Finally, the content of MWNTs is only 0.1 wt% of alumina amount. Further drying and grinding result in MWNT-alumina composite powder that is sintered by SPS in a graphite die at 1300 °C with a pressure of 50 MPa for 5 minutes in an Ar atmosphere.

A colloidal processing route is an effective way to improve the mechanical properties of MWNT-alumina composites. Adjusting the surface properties of the alumina powder and those of MWNTs helps bind them together with attractive electrostatic forces, producing strong cohesion between two phases after sintering. During the sintering processing, the growth of alumina particles is believed to wrap MWNTs, which should increase the effectiveness of the reinforcement. The addition of only 0.1 wt% MWNTs in alumina composites increases the fracture toughness from 3.7 to 4.9 MPa $\sqrt{m}$, an improvement of 32% compared with that of the single-phase alumina.

Individual single-wall carbon nanotubes coated with SiO_2 can be used to develop a highly sensitive sensing device structure because of the unique properties of SWNTs (Whitsitt and Barron, 2003). The selective etching of silica-coated SWNTs either as small ropes or as individual tubes provides a route to site-selective chemical functionalization as well as the spontaneous generation of tube-to-tube interconnects. The individual SWNTs may be coated in solution and isolated as a solid mat. Either the end or the center of the SiO-SWNT may be etched and exposed. The exposure of the central section of SWNTs has potential for sensor and device structures. Processing of the coated SWNTs takes place during an in situ chemical reaction between fumed silica and SWNT-surfactants structures in suspension. The choice of surfactant type is important in determining whether individual SWNTs or small ropes are coated. It has been found that anionic surfactants result in the formation of coated ropes, whereas cationic surfactants result in individual coated nanotube ropes. It has been proposed that this effect is a consequence of the pH stability of the surfactant–SWNT interaction.

Conductive Ceramics

SWNTs can be used to convert insulating nanoceramics to metal-like conductive composites. Using nonconventional consolidation techniques, spark-plasma sintering (SPS), and conventional powder-processing techniques, dense SWNT-Al_2O_3 nanocomposites can be synthesized. These nanocomposites show increasing electrical conductivity with increasing SWNT content (Zhan et al., 2003). The conductivity of these nanocomposites (15 vol% SWNT-Al_2O_3), at room temperature, is a 13-fold increase in magnitude over pure alumina. In addition, these nanocomposites show a significant enhancement to fracture toughness over the pure alumina: a 194% increase in fracture toughness over pure alumina, up to 9.7 MPa $\sqrt{m}$ in the 10 vol% SWNT-Al_2O_3 nanocomposite has been achieved (Zhan et al., 2003).

Nanostructured Metals and Metal Composites

A variety of metal-nonmetal nanocomposites have been synthesized and characterized. These nanocomposites have been produced to improve the physical, electrical, and mechanical properties (strength, hardness, fracture toughness, etc.). Metal-matrix nanotube composites have been proposed as future materials for advanced space propulsion structures as well as for radiators and heat pipes, because of their high conductivity and light weight.

Metal-nanocomposites in particular, and nanocrystalline metals in general, are produced in the form of thin ribbons or splats, and powders. Ductile metal-metal nanocomposites such as Nb-Cu have been produced by severe plastic deformation (e.g., cold-drawing Cu and Nb filaments). Other methods to synthesize metal nanocomposites include hot consolidation, electrodeposition, and liquid- and vapor-phase processing.

High-modulus (1–4 TPa) carbon nanotubes (CNT) have been incorporated in metals to obtain modulus greater than that of advanced C fiber–reinforced Al at reduced density (the density of Al-CNT is lower than C-fiber-Al composite, because of the hollow spaces within the CNT). Al-CNT nanocomposites have been synthesized using hot-pressing of powders; however, brittle aluminum carbide phases forms during processing, and the electrical resistivity of the composite increases slightly with CNT volume fraction at room temperature.

Comparison of mechanical properties of various materials for aerospace vehicles, including aluminum, foam, composites, and CNT-Al composites suggests that CNT-Al composite is likely to have better mechanical properties than other enabling materials. As with conventional metal composites, the mechanical properties of CNT-Al composites will strongly depend on the distribution and volume fraction of the CNT and CNT type. Multiwall CNTs generally have better mechanical properties than SWCNTs.

Ceramic-metal nanocomposites essentially belong to the family of heat-resistant, dispersion-strengthened (DS) composites such as W-ThO_2, Ni-Al_2O_3, Cu-SiO_2, Cu-Al_2O_3, NiCrAlTi-Y_2O_3, etc., that were developed in the 1960s. These DS composites contain 10–15% particles of diameter 10–100 nm. The second phase acts a barrier to the motion of dislocations rather than as the load-bearing constituent (as in an engineered composite). The dispersoid size must be less than 100 nm for the mean free path in the matrix to be in the range (10–300 nm) needed for dislocation strengthening. The DS composites are designed to retain high yield strength, creep resistance, and oxidation resistance at elevated temperatures rather than to enhance the room-temperature yield strength. The thermal stability of the DS composites is usually superior to precipitation-hardened alloys. As it is desirable to preserve as many of the properties of the matrix metal as possible (e.g., ductility, oxidation resistance, thermal conductivity, impact strength), the volume fraction of the nanodispersoid is kept small ($<15\%$). In the emerging field of nanocomposites, this limitation is sought to be alleviated, although it is not clear from the current literature how anticipated strength gains will be achieved without tremendous losses in the ductility and toughness at high solid loadings.

Solid-State Powder-Based Processing

The two solid-state fabrication techniques most widely used for the synthesis of nanocomposites have been high-energy ball milling in conjunction with hot-pressing and deformation processing. For example, Al, Mg, Cu, Ti, and W matrix composites reinforced with nanoscale SiC, Al_2O_3, and CNT have been synthesized using hot-press consolidation techniques. In contrast, Nb filament–reinforced Cu matrix composites have been fabricated using the deformation processing (cold-drawing); Cu-Nb nanocomposites have excellent strength and electrical conductivity. For Cu-Nb nanocomposites, continuous multifilaments of Nb are plastically drawn in a Cu matrix to form the nanocomposite. Initially, a Nb rod is drawn in a Cu jacket to a hexagonal shape. Several such hexagonal segments are stacked in a Cu jacket and cold-drawn. The process is repeated several times to create 100- to 250-nm-size whisker-like Nb filaments in the Cu matrix containing a high density of dislocations. Alternatively, extensive plastic deformation of cast Cu-Nb alloys transforms Nb dendrites into nanosize Nb ribbons and stringers. The composite is able to reach high strains without fracture.

Powder metallurgy (PM) –based Al and Mg nanocomposites are fabricated using high-energy ball-milling followed by hot consolidation. Creep-resistant, lightweight Mg-SiC nanocomposites have been prepared using milling and hot extrusion. The nanoscale SiC may be formed by CO_2 laser-induced reaction of Si and a carbon-bearing gas such as acetylene, and the micrometer-scale Mg powder is formed by argon gas atomization of molten Mg. It has been found that the ultimate tensile strength (UTS) of the hot extruded Mg-SiC composite doubled as compared to unreinforced Mg. Significant improvement in the hardness and a lowered creep rate were noted even at low-volume fractions of nanoscale SiC (3 vol%). Similarly, CNT-Cu nanocomposites have been synthesized using a press-and-sinter PM technique. The CNT was formed by thermal decomposition of acetylene, and CNT and Cu powders were mixed to a 1:6 ratio in a ball mill. The mixture was pressed at 350 MPa for 5 min., and then sintered at 850 °C for 2 h. The CNT-Cu composites exhibited reduced coefficient of friction and lower wear loss with increasing CNT content in Cu. Additionally, the CNT-Cu nanocomposites exhibited lower wear loss with increasing load as compared to the Cu-C fiber composites.

Internal oxidation and hot deformation has been used to create fine dispersions of Al_2O_3 particles in a nanocrystalline Cu matrix; the fine reinforcement and matrix grain sizes (80 nm for Cu and 20 nm for Al_2O_3) led to high strength, hardness, creep resistance and good ductility and electrical conductivity in the nanocomposite. Conversely, Al_2O_3 matrix composites containing Cu nanoparticles have been fabricated using hot-pressing of Al_2O_3 and CuO powder mixtures to achieve enhanced toughness compared to monolithic Al_2O_3, although melting of Cu dispersions during hot-pressing (sintering) can occur.

Internal oxidation is limited to production of composites containing a low-volume fraction of ceramic particles such as oxide dispersion–strengthened composites. Other metal nanoparticle–dispersed ceramics such as Al_2O_3-W, Al_2O_3-Mo, and Al_2O_3-Ni have been synthesized for improved fracture toughness and fracture strength; these composites have nanoscale reinforcements with melting temperatures above the conventional hot-pressing temperatures.

The major drawback of PM to fabricate nanocomposites is the surface oxidation and contamination due to powder exposure to processing environment at high temperatures. For example, in one study fully dense, nanocrystalline iron compacts made by PM exhibited high hardness, but the hardness increased and fracture stress and elongation to failure decreased significantly with decreasing grain size from 33 to 8 nm. Microscopic examination confirmed that processing defects such as micropores, minute quantities of oxide contaminants on powders, and residual binders had masked the benefits of size effects in the nanomaterial.

Liquid-Phase Processing

Conventional liquid-phase methods to synthesize metal-matrix nanocomposites have been employed with some success. Agglomerate-free Al-SiC nanocomposite castings containing coarse nanometer-size SiC have been synthesized by stirring nano-SiC in semisolid alloys. Surface pretreatment and wetting agents are usually necessary to improve the wetting and facilitate dispersion. For example, acid treatment with HNO_3 and HF has been used to eliminate the agglomeration, and wetting agents (e.g., Ca or Mg) have been added to improve the wetting and to facilitate particle transfer. A potential problem facing successful implementation of stirring techniques at large nanophase loading will be a dramatic increase in the suspension viscosity, which would adversely affect the ability to shape-cast engineering components. This is because the significantly greater surface area of nanoparticles (compared to micrometer-size particles) will greatly increase the viscosity even at moderate loadings.

Metal infiltration has been used to synthesize metal nanocomposites. Studies at the High Pressure Research Institute of Polish Academy of Sciences on metal infiltration of nanoparticles have reported the synthesis of a variety of metal-matrix nanocomposites. Extremely large (7.7 GPa) pressures were used to infiltrate Al_2O_3, SiC, and diamond nanoparticles by Mg, Sn, Zn, Al, Ag, Cu, and Ti. The matrix crystallized within nanopores to form grains with size in the range 10–50 nm. The infiltration method used was somewhat unconventional: It consisted of placing pellets of nanoparticles and metal in a die, applying the pressure, and then heating. Pressurization prior to melting the metal permitted cleaner interfaces to be created. X-ray diffraction patterns were recorded during heating (and cooling after infiltration) to determine the inception of melting. For ductile metals, infiltration commenced even in the solid state prior to melting. At 20 kbar pressure, Al melted at 670 °C (instead of 660 °C), and solidified at 600 °C. The infiltration permitted up to 80% ceramic content in the composite. In addition carbon nanotubes have been infiltrated with molten lead.

Capillary infiltration of nanotubes and nanoparticles has been more extensively investigated with nonmetallic liquids than with metals. For example, CNTs have been infiltrated with vanadium oxide and silver nitrate. In the case of silver nitrate, the open-ended CNTs were penetrated by the solution under capillary forces, and the solution-filled nanotubes were irradiated with an electron beam to create CNTs containing chains of tiny silver beads, with beads separated by gas pockets under as high pressures as 1300 atmospheres ($\sim$132 MPa). Only CNTs of relatively large (>4 nm) diameter could be infiltrated with silver nitrate solution. Even though in a bulk form carbon is wetted by silver nitrate, in the nanoscale regime strong surface forces seem to affect the polarizability, which in turn affects the capillary forces driving the liquid in the tube. In addition, CNTs with internal diameter of $\sim$1 nm have been filled with halides such as $CdCl_2$, CsCl, and KI.

The magnitude of capillary pressure to infiltrate nanoreinforcement with molten metals could be very large. A simple calculation using the Young-Laplace equation shows that the pressure required to infiltrate CNTs by molten Al will be in the gigapascals range. Although special techniques can be used to achieve such extreme pressures (as was done in the Polish study cited earlier), the ability of common foundry tools and techniques to achieve such pressures is doubtful. Use of ultrasound to disperse nanoscale reinforcement in molten metals could be a viable method to make nanocomposites. The extremely large pressure fluctuations due to "cavitation" will cause the nanophase to be incorporated in the melt.

Prior studies suggest the feasibility of using ultrasound to disperse micrometer-size particles in liquids (to make composites such as silica-Al, glass-Al, C-Al, alumina-Sn, WC-Pb, SiC-Al). Sonic radiation also permits pressureless infiltration of fibers, dispersion and deagglomeration of clusters, and structural refinement of the matrix. The enormous accelerations (10^3 to 10^5 g) due to cavitation aid in deflocculating the agglomerates, rupturing gas bridges, and disintegrating oxide layers, thus allowing physical contact and improved wetting. Sonic radiation also stabilizes the suspension by acoustically levitating particles against gravity, thus enabling a homogeneous particle distribution in the matrix to be created. In the case of solidifying alloys, partial remelting of dendrites by the radiation leads to refinement of structure and removal of air pockets around the particles.

Intriguing possibilities can be envisioned in the application of external pressure (or ultra-sound) to synthesize nanocomposites reinforced with ropes made out of nanotubes. The strength of CNT and BN nanotubes has been increased by assembling them in the form of ropes of 20–30 nm in diameter. Combining these ropes with metals presents opportunity to produce ultra-high-strength metals. The spacing between the nanotube strands in a rope is small (e.g., ~0.34 nm in BN nanotube rope); such a small spacing will require extremely large local pressures for infiltration, possibly achievable through intense sonic radiation.

Possible effects of solidification of metal-nanotube composites have been discussed in the literature. These include depression of the solidification temperature due to capillary forces inside the nanotube, and large undercoolings required for heterogeneous nucleation, solidification of the liquid as a glassy (amorphous) rather than a crystalline phase, and the deformation (buckling) of the nanotubes due to stress generated from the volume changes accompanying the phase transformation during solidification and cooling.

Many material defects caused by the solid-state PM processes may be eliminated by liquid-phase processing, which also permits greater flexibility in designing the matrix and interface microstructure. Rapid solidification is already used to produce amorphous or semicrystalline micro- and nanostructured materials. In liquid-phase processed composites, featureless and segregation-free nanostructured matrix may be created because of solidification in nanosize interfiber regions. This is likely to enhance the properties of the matrix, thereby adding to the property benefits achieved due to the nanophase (nanostructured metals have been shown to have better mechanical properties than micrograin metals). Another benefit of combining nanoscale reinforcement with a nanocrystalline matrix is that the latter will be partially stabilized against grain growth by the dispersed nanophase. The nanometer-size matrix grains will increase the total grain boundary area, which will serve as an effective barrier for the dislocation slip and will inhibit creep. Solidification could also permit self-assembly of nanoparticles and nanotubes in interdendritic boundaries and create functionally gradient microstructure.

Despite the successful (albeit limited) demonstration of liquid-phase processing of metal-matrix nanocomposites, several processing challenges can be visualized. These include (1) difficulty in homogeneously dispersing nanoscale reinforcement; (2) extreme flocculation tendency of nanoscale reinforcement; (3) increased viscosity (poor or inadequate fluidity) even at small loadings, due to large surface-to-volume ratio; (4) prohibitively large external pressures needed to initiate flow in nanoscale pores; and (5) possible rejection of dispersed nanophase by solidifying interfaces. The distribution of particles in a solidified matrix depends on whether the nanophase is pushed or engulfed by the growing solid. The dynamics of interactions has been studied theoretically and experimentally. Most experiments have used coarse (>700 nm) particles; only a few studies have used nanosize (30–100 nm) particles (e.g., Ni, Co, Al_2O_3) mainly in low-temperature organics (e.g., succinonitrile). However, measurements of critical solidification front velocity to engulf the nanophase are non-existent.

In the case of micrometer-size particles, experiments show that above a critical velocity of the growing solid, the dispersed particles are engulfed, resulting in a uniform distribution, but severe agglomeration and segregation occur at low growth velocities, because the particles are pushed ahead of the growing solid. For pushing to occur, a liquid film must occupy the gap between the particle and the front; hence, the stability of thin liquid films supported between two solids is important. The critical velocity for engulfment increases with decreasing particle size ($V_{cr} \propto R^{-n}$, R = particle radius) and with decreasing viscosity (which is a strong function of particle concentration). Particles often pile up ahead of the solidification front due to pushing, and the entire pileups could be pushed for various distances. The measured critical velocity for micrometer-size particles varies widely depending on the system; for liquid metals it is in the range 13,000–16,000 µm/s. For nano-SiC (<100 nm) in Al, the calculated critical velocity is >10^7 µm/s, which would be impossible to achieve by conventional solidification techniques unless ultrasonic irradiation accompanied the solidification. Sonic radiation (and Brownian motion) could induce interparticle collisions and collisions with the growth front, leading to mechanical entrapment. Collisions with the front could induce shape perturbations in the growing solid that might favor capture (e.g., localized pitting or concavity due to altered undercooling via the Gibbs-Thompson effect). However, nanoparticles will likely be ineffective in obstructing the thermal and diffusion fields, and the latter's effect on interactions with growing crystals will probably be negligible. Similarly, fluid convection will be less important than for micrometer-size particles because nanoparticles will likely be smaller than the hydrodynamic boundary layers (thickness ~10–1000 µm). For nanosize particles, the critical velocity for the engulfment will likely be much higher than that for the micron-size particles, provided the fundamental nature of the interactions is not altered.

Surface, Interface, Nucleation, and Reactivity

Surface and interface phenomena achieve overriding importance at the nanoscale, and cause departure from the bulk behavior. For example, even though in bulk form carbon is wetted by many liquids, and therefore, a narrow carbon tube will create a strong attraction for the liquid, at the nanoscale strong surface forces and the curvature of the rolled C sheet adversely affect the capillary forces and cause CNT to become hydrophobic and develop strong attractive forces, leading to self-organizing structures (clusters). Structurally, CNTs are rolled-up sheets of graphite, and the significant curvature of the CNT influences their surface properties. Wetting tests have been done on CNT in liquid epoxy, polypropylene, polyethylene glycol, and other polymers using the Wilhelmy force balance method and transmission electron microscopy (TEM). It seems certain that the wetting behavior is influenced by the nanotube geometry, such as wall thickness and helicity, and by the physical chemistry of the external graphene surface. Likewise, the flow of water and stabilization of water droplets within nanotubes have been studied using molecular dynamic simulation, and contact angle and density of the water droplets within CNTs have been determined.

For nanocomposites, surface and interface considerations lead to two important questions: First, is the nanoreinforcement (e.g., CNT) wetted by the liquid matrix (e.g., liquid polymer)? This is important because wetting is a necessary (though not sufficient) condition for adhesion. And, second, is there evidence of stress transfer across the interface? What experimental techniques can be used at the nanoscale to measure the interfacial strength and adhesion?

From a practical viewpoint, the issue of stress transfer between the matrix and the nanoreinforcement is of major interest. In reality, the physics of the interface, in particular, measurement of the extent and efficiency of load transfer between nanoreinforcement and the matrix is quite difficult, even though the large interfacial area implies that high composite strengths are possible. Raman spectroscopy is the main tool that has been used to evaluate whether stress transfer took place. This is done by monitoring the extent of peak shift under strain. In addition, molecular dynamic simulations have been done to simulate the pullout tests, and to study the effect of interfacial chemical cross-linking on the shear strength in CNT and amorphous or crystalline polymers.

The first direct quantitative tests for interface strength of a single nanofiber in a polymer matrix were reported in 2002. The tests used samples in which nanotubes bridged voids in a polymer matrix. Once these voids were located using the TEM, the bridging nanotubes were dragged out of the polymer using an atomic force microscope (AFM) tip. The lateral force acting on the tip was monitored as the nanotube was progressively extracted from the polymer. An average interfacial shear strength of 150 MPa was obtained to drag a single multiwall CNT out of epoxy. In another technique, a CNT was directly attached to the tip of the AFM probe stylus and was pushed into a liquid polymer. The semicrystalline polymer was then cooled around the CNT. The single nanotube was then pulled out using the AFM, with the forces acting on the nanotubes recorded from the deflection of the AFM tip cantilever. The interfacial strengths far exceeded the polymer tensile strength; thus failure will be expected to take place in the polymer rather than at the interface. Residual stresses due to thermal expansion mismatch and their effect on interface behavior in nanocomposites are still unexplored.

Surface phenomena are also manifested in nucleation of phases. For example, grain refinement in metals is achieved via heterogeneous nucleation as a result of adding inoculants (e.g., TiB_2 and $TiAl_3$ in Al). As nucleation is often masked by subsequent growth of crystals, observations of nucleation are difficult. However, many glass-forming alloys serve as slow-motion models of undercooled liquids, and permit high-resolution electron microscopic (HREM) observations of nucleation. Studies show that TiB_2 particles dispersed in amorphous Al alloys form an adsorbed layer of $TiAl_3$ over which nanometer-size Al crystals nucleate. The atomic matching at the interfacing planes of TiB_2-$TiAl_3$-Al nucleus gives rise to considerable strains in the nucleated crystals. The crystallographic anisotropy of surface energies is important in nucleation; the nucleating substrate must be oriented in a manner so as to present a low-energy interface to the crystallizing liquid. For example yttrium is more effective as a nucleating agent when its prismatic plane (rather than the basal plane) is exposed to the solid. Similarly, the basal and prismatic planes of graphite have different surface energies, and this is expected to influence the crystal nucleation on CNT. The addition to the matrix of a reactive element, which promotes the formation of a low-energy transition layer (even a monolayer) at the interface via adsorption, reaction, or solute segregation has been known to aid nucleation. The atomic disregistry between the particle and the crystal must be small for nucleation to occur. The catalytic potency of a solid for nucleation increases with decreasing atomic disregistry at the interfacing atomic planes. Direct observations of disregistry have been made only in a few systems, and this area of study remains unexplored.

The chemical stability of nanoreinforcement in the liquid matrix at the processing temperature is an important consideration, especially because the nanoreinforcement has a much higher surface-to-volume ratio than does micrometer-size reinforcement. This is seldom a concern when dealing with polymeric liquids but could become important in ceramic and metal matrices. With CNT in Al, for example, aluminum carbide (Al_4C_3) formation could embrittle the composite. In the current family of advanced carbon fiber-Al composites, Al_4C_3 grows epitaxially on the prismatic planes. The surface energy of the prismatic plane (~ 4.8 J·m^{-2}) is greater than that of the basal plane (~ 0.15 J·m^{-2}), and there is anisotropy of chemical reactivity of graphite crystals. Because carbon nanotubes consist of rolled-up (less reactive) graphite basal planes oriented along their axis, they should be chemically stable in molten Al although direct verification of this is lacking. In any case, once formed, the relatively large quantities of the reaction product (e.g., Al_4C_3) in the nanocomposite will increase the specific volume of the solid and the viscosity, thus making shape-casting a challenge.

Special precautions may be needed to minimize the extent of deleterious matrix-reinforcement reactions in nanocomposites. For example, rapid densification under very large pressures (1.5–5 GPa) has been used to fabricate well-bonded, nonreacted high-density SiC-Ti nanocomposites (even though SiC and Ti show significant reactivity). Similarly, SiC-Al nanocomposites have been produced using high-energy ball-milling at room temperature followed by hot consolidation using plasma-activated sintering at 823 K (below the melting temperature of Al). The composites had clean interfaces and did not exhibit any strength-limiting reaction products.

Agglomeration, Dispersion, and Sedimentation

A dispersed nanophase will likely experience pronounced clustering, which must be resisted to realize the benefits of compositing. Very fine nanoparticles undergo thermal (Brownian) collisions even in a monodisperse suspension, leading to singlets, doublets, triplets, and higher-order aggregates. To deagglomerate the cluster, liquid must flow into the gap, but the thinner the liquid film, the greater is the hydrodynamic resistance to its thinning (disjoining pressure), and the greater will be the energy needed for deagglomeration. Agglomeration depends on both the collision frequency (diffusion) and sticking probability (interparticle forces). Classical theories of flocculation kinetics and floc stability in the surface science literature are probably applicable to nanodispersions (as they are to colloidal suspensions). Some of these were discussed in Chapter 4.

The dispersion properties of nanopowders in liquids are important in slurry-based manufacturing processes. The frequency of agglomeration is increased, because of large surface-to-volume ratio of nanopowders. In the case of nonmetallic liquids, the dispersant molecules that are adsorbed on the particle surface generate strong surface charge and electrostatic forces that counter the physical (van der Waals) attraction. This permits stabilization of the suspension. If there are enough ionizable surface groups, electrostatic stabilization is also achieved with proper pH control. In addition, surfactant molecules of uncharged polymers are used to stabilize fine suspensions (steric stabilization). These adsorbed molecules have extended loops and tails, so particle surfaces with such adsorbed molecules begin to show repulsion forces that tend to stabilize the suspension. An intermediate group of dispersants (polyelectrolytes) combine both steric and electrostatic repulsion to stabilize the suspension. The rheological behavior of nanoslurries containing these different dispersants will likely depart from the behavior shown by micrometer-size slurries. As a result, solid loading and viscosity of nanoslurries will be limited by the size, and this will influence the slurry behavior in injection-molding and slip-casting.

High-intensity ultrasound could be a viable tool for declustering nanoagglomerates. However, propagation of ultrasound through a suspension can also induce relative motion between differently sized particles. Finer particles respond better to the inducing frequency and vibrate with greater amplitude than larger particles; as a result, differently sized particles will collide during propagation of sonic waves. This may cause some agglomeration. A narrow size distribution of nanoparticles should facilitate de-agglomeration through sonication.

Gravity plays an important role in flocculation. Because particles of different size or density settle in a fluid at different rates, the relative motion between them could cause particle collisions and agglomeration. Because of their extremely fine size, nanoparticles will likely form neutrally buoyant suspensions in liquids (including molten metals). Segregation because of floatation (or sedimentation) may not be severe, due to fine size, as long as agglomeration can be inhibited. Experiments on the settling of micrometer-size SiC in Al and the computed velocity for SiC particles (10 nm SiC, 10 vol%) indicate that the settling rate of nanosize SiC will be several orders of magnitude smaller than micrometer-size SiC (the settling rate is calculated from the hindered-settling equation developed by Richardson and Zaki: $u = u_0(1 - \phi)^{4.65}$, where u_0 is the Stokes velocity of a sphere of radius R, u is the hindered settling velocity, and ϕ is the particulate volume fraction). The settling velocity decreases with particle size at a fixed volume fraction, with nanosize particles yielding settling rates over five orders of magnitude smaller than those of micrometer-size particles, which have settling rates on the order of 1×10^{-5} to 1×10^{-4} m/s. In the case of CNT, the mass of a single CNT (diameter 100 nm, length 1000 nm) will be about 2.13×10^{-14} grams. Such a small mass will experience little buoyancy in a liquid and will have a negligibly small settling or floating rate (velocity varies as the square of the particle radius). Even under centrifugation (acceleration of 3 g to 5 g at a few hundred rpm), segregation effects are likely to be small.

Properties

Nanostructured materials exhibit property improvements over conventional coarse-grained materials; for example, nanostructured aluminum alloys can be designed to have higher strength than low-carbon steel. In the case of composites with micrometer-size reinforcement, fracture toughness and fatigue strength increase with decreasing particle size (although thermal expansion, thermal conductivity, and wear resistance decrease). At nanometric grain sizes, the proportion of the disordered interfacial area becomes large when compared to a characteristic physical length (e.g., Frank-Read loop size for dislocation slip). Interfacial defects such as grain boundaries, triple points, and segregation begin to make an increasing contribution to the physical and mechanical properties. Deviations from the classical behavior (e.g., Hall-Petch relationship) could occur, and properties of micrometer-size composites may not be reliably extrapolated down to the nanometer size (this is especially true of CNT that has distorted electronic structure due to cylindrical C layers). In monolithic nanomaterials, grain size effects on properties show considerable departure from the behavior expected for micrometer-size grains.

Strength and Modulus. The composite modulus is generally estimated from a rule-of-mixtures relationship. Using the literature values of relevant parameters ($E_{CNT} = 1.81$ TPa, $E_{C\text{–fiber}} = 800$ GPa, and $E_{Al} = 70$ GPa), it is found that at a volume fraction of 0.45, Al-CNT composites will attain a composite modulus higher than the conventional vapor-grown C fiber (the specific modulus of Al-CNT, i.e., modulus-to-density ratio, can be even higher).

However, a dispersion in the strength measurements on CNT, and the variability introduced by processing could cause deviations from the predictions.

The strength of Al-CNT composite can be estimated from the Kelly-Tyson equation, according to which the composite strength σ_c is given from $\sigma_c = \sigma_f \cdot V_f \cdot [1 - (l_c/2l)] + \sigma_m' \cdot (1 - V_f)$, where σ_c = composite strength, σ_f = tensile strength of the nanotube, σ_m' = stress of Al matrix at the failure strain of the composite, l_c = critical length of a nanotube in Al, l = average length of a nanotube. For Al-CNT, the literature data show that $\sigma_{CNT} = 3$ GPa, $\sigma_m' = 40$ MPa, $l_c = 0.85$ μm, and $l = 2$ μm. Calculations using these values confirm that significantly improved strength will be expected to result, and very high composite strength could be achieved at a relatively small-volume fraction of the CNT (in comparison to conventional C fibers). A nonhomogeneous distribution of CNT, and processing defects could, however, substantially lower the strength. The strength predictions are based on classical composite mechanics that do not take into account the hollow nature of the CNT.

Nanotribology. Nanoscale materials present interesting opportunities for tribological performance. Ultra-low-friction states have been observed at the nanoscale using AFM capable of measuring lateral forces in the piconewton range (10^{-12} Newtons). The energy dissipation in atomic friction is measured as an AFM stylus tip is dragged over a surface. Normally, the tip sticks to an atomic position on the surface, and then, when the force is sufficient, slips to the next atomic position, and so on, resulting in a "sawtooth" modulation of the lateral force and stick-slip–type motion. In one experiment, the tip of the AFM stylus was coated with a tiny flake of graphite, which slid over an extended graphite surface. When the atomic lattice of the flake was aligned with that of the surface, stick-slip sliding was observed, but at all other orientations the friction was near zero. This indicates that when the two surfaces are not in atomic-scale registry, there is a high degree of force cancellation. Similar observations have been made on NaCl crystals with a Si tip.

Significantly greater hardness of nanocomposites than microscale composites leads to improved wear resistance. Thus, CNT-Ni-P composites tested under lubricated conditions led to low friction and wear relative to similar composites containing micrometer-size SiC or graphite, and to virgin Ni-P itself. Rapidly solidified nanopowders have been used to fabricate nanoscale Al alloys and composites possessing improved wear resistance and low friction. Al alloys for bearing applications contain either layered dispersoids (graphite, mica, talc, etc.) or dispersoids of soft metals such as Pb, Bi, and In. Studies show that dispersoids in the nanosize range provide better antifriction properties than those in the micron-size range. The material transfer is significantly less with nanodispersions providing a more uniform thin layer of soft material at the mating interface than do the coarser, micron-size dispersoids.

Carbon Nanotubes: Biosensor Applications

Biomolecules such as nucleic acids and proteins carry important information of biological processes. The ability to measure extremely small amounts of specific biomarkers at molecular levels is highly desirable in biomedical research and health care. Current technologies rely on well-equipped central laboratories for molecular diagnosis, which is expensive and time consuming, often causing delay in medical treatments. There is a strong need for smaller, faster, cheaper, and simpler biosensors for molecular analysis. The recent advance in carbon nanotube (CNT) nanotechnology has shown great potential in providing viable solutions. CNTs with well-defined nanoscale dimension and unique molecular structure can be used as bridges linking biomolecules to macro/micro-solid-state devices so that the information of bioevents can be transduced into measurable signals. Exciting new biosensing concepts and devices with extremely high sensitivities have been demonstrated using CNTs.

As the size of the materials reach the nanometer regime, approaching the size of biomolecules, they directly interact with individual biomolecules, in contrast to conventional macro- and microdevices, which deal with assembly of relatively large amount of samples. Nanomaterials exhibit novel electronic, optical, and mechanical properties inherent with the nanoscale dimension. Such properties are more sensitive to the environment and target molecules in the samples. Although a big portion of nanomaterials are isotropic nanoparticles or thin films, high-aspect-ratio one-dimensional nanomaterials such as CNTs and various inorganic nanowires (NWs) are more attractive as building blocks for device fabrication. The potential of CNTs and NWs as sensing elements and tools for biomolecular analysis as well as sensors for the gases and small molecules have been recently recognized (Li and Ng, 2003). Promising results in improving sensitivity, lowering detection limit, reducing sample amount, and increasing detection speed have been reported using such nanosensors (Kong et al., 2000; Cui et al., 2001; Li et al., 2003). CNTs integrated with biological functionalities are expected to have great potential in future biomedical applications.

The electronic properties of CNTs are very sensitive to molecular adsorption. Particularly, in a semiconducting single-wall CNT (SWNT), all carbon atoms are exposed at the surface so that a small partial charge induced by chemisorption of gas molecules is enough to deplete the local charge carrier and cause dramatic conductance change (Collins et al., 2000). Because biomolecules typically carry many ions, they are expected to affect CNT sensing elements and transducers more dramatically than are simple gases and small molecules. Sensing devices have been fabricated for various applications using single CNTs, single semiconducting SWNT field-effect-transistors (Star et al., 2003; Besteman et al., 2003; Li et al., 2002), vertically aligned nanoelectrode arrays (Snow et al., 2003), and random networks or arrays.

FET-Based Biosensors

The extreme high sensitivity and potential for fabricating high-density sensor array make nanoscale field-effect-transistors (FETs) very attractive for biosensing, particularly because biomolecules such as DNA and proteins are heavily charged under normal conditions. SWNT FETs are expected to be more sensitive to the binding of such charged species than are chemisorbed gas molecules. However, the wet chemical environment with the presence of various ions and of other biomolecules makes it much more complicated than gas sensors. Very recently, Besteman et al. (2003) successfully demonstrated that the enzyme-coated SWNT FETs can be used as single-molecule biosensors. As shown in Figure 8-28, the redox enzyme glucose oxidase (GOx) is immobilized on SWNT using a linking molecule, which on one side binds to the SWNT through van der Waals coupling with a pyrene group and on the other side covalently binds the enzyme through an amide bond, as developed by Chen et al. (2001). The FET preserves the p-type characteristic but shows much lower conductance on GOx immobilization, which is likely the result of the decrease in the capacitance of the tube caused by GOx immobilization, because GOx blocks the liquid from access to the SWNT surface.

The GOx-coated SWNT showed a strong pH dependence as well as high sensitivity to glucose. Figure 8-28 shows the real-time measurements where the conductance of a GOx-coated SWNT FET has been recorded as a function of time in milli-Q water. No significant change in conductance is observed when more milli-Q water is added, as indicated by the first arrow. However, when 0.1 M glucose is added, the conductance increases by 10 %, as indicated by the first arrow. Both studies on liquid-gated FETs use very low ionic strength with 10 mM

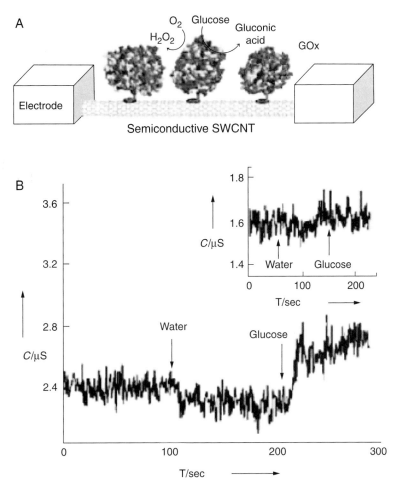

FIGURE 8-28 *(A) A GOx-functionalized CNFET for glucose analysis. (B) Electronic response of the GOx-CNFET to glucose. The source-drain and liquid-gate voltages were kept constant at 9.1 mV and −500 mV, respectively. Arrows show the addition of water and a glucose sample (2 μm, 0.1 mM) to the system. Inset shows the same measurement on a control CNFET without GOx.* (K. Besteman et al., Nano Lett., 3 (6), 2003, 727-730). Reprinted with permission from the American Chemical Society, 1155 16th St. NW, Washington, DC 20036. Photo Courtesy of C. Dekker, Delft University of Technology, The Netherlands.

NaCl (Besteman et al., 2003) and 0.1 mM KCl (Rosenblatt et al. 2002), respectively. The salt concentrations are more than 10 times lower than physiology buffers.

Aligned Nanoelectrode Array-Based Electronic Chips

Researchers at NASA Ames Research Center have developed the concept of making DNA chips that are more sensitive than current electrochemical biosensors. They cover the surface of a chip with millions of vertically mounted carbon nanotubes 30–50 nm in diameter (Figure 8-29). When the DNA molecules attached to the ends of the nanotubes are placed in a liquid containing DNA molecules of interest, the DNA on the chip attaches to the target and increases its electrical

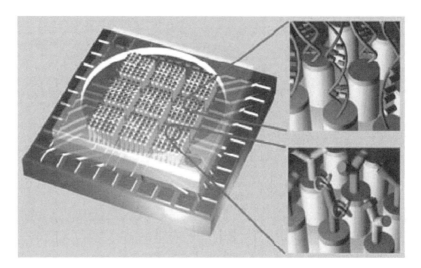

FIGURE 8-29 *Vertical carbon nanotubes are grown on a silicon chip. DNA and antigen molecules attached at the ends of the tubes detect specific types of DNA (top) and antigen detection (bottom) in an analyte.* (R. E. Smalley "Chip Senses Trace DNA," Technology Research News, 6(22), 2003). Reprinted with permission from Hazel Cole, exec. asst. to late Prof. Smalley, Rice University.

conductivity. This technique, expected to reach the sensitivity of fluorescence-based detection systems, may find application in the development of a portable sensor.

An embedded array minimizes the background from the side walls, whereas the well-defined graphitic chemistry at the exposed open ends allows the selective functionalization of –COOH groups with primary amine-terminated oligonucleotide probes through amide bonds. The wide electropotential window of carbon makes it possible to directly measure the oxidation signal of guanine bases immobilized at the electrode surface. Such a nanoelectrode array can be used as ultrasensitive DNA sensors based on an electrochemical platform (Koehne et al., 2003, 2004). As shown in Figure 8-30, oligonucleotide probes of 18 bases with a sequence of [Cy3]5-CTIIATTTCICAIITCCT-3[AmC7-Q] are covalently attached to the open end of MWNT exposed at the SiO_2 surface. This sequence is related to the wild-type allele (Arg1443stop) of *BRCA1* gene (Miki et al., 1994). The guanine bases in the probe molecules are replaced with nonelectroactive inosine bases, which have the same base-pairing properties as guanine bases. The oligonucleotide target molecule has a complimentary sequence [Cy5]5-AGGAC-CTGCGAAATCCAGGGGGGGGGGGG-3, including a 10 mer polyG as the signal moieties. Hybridization was carried out at 40 °C for about 1 hour in ∼100 nM target solution in $3 \times$ SSC buffer. Rigorous washing—in three steps using $3 \times$ SSC, $2 \times$ SSC with 0.1% SDS, and $1 \times$ SSC respectively at 40 °C for 15 minutes after each probe functionalization and target hybridization process—was applied to get rid of nonspecifically bound DNA molecules, which is critical for getting reliable electrochemical data.

Such solid-state nanoelectrode arrays have great advantages in stability and processing reliability over other electrochemical DNA sensors based on mixed self-assembled monolayers of small organic molecules. The density of nanoelectrodes can be controlled precisely using lithographic techniques, which in turn define the number of probe molecules. The detection limit can be optimized by lowering the nanoelectrode density. However, the electrochemical signal is defined by the number of electrons that can be transferred between the electrode and

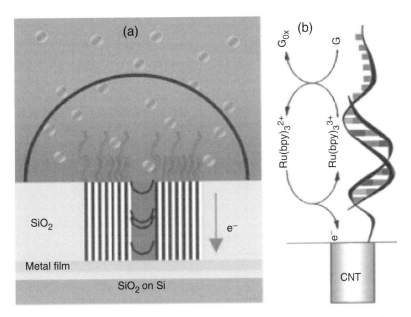

FIGURE 8-30 *Schematic of the MWNT nanoelectrode array combined with Ru(bpy)$_3^{2+}$ mediated guanine oxidation for ultrasensitive DNA detection.* (J. Koehne et al., Journal of Material Chemistry, 14(4), 2004, 676). Reprinted with permission from the Royal Society of Chemistry, Thomas Graham House, Science Park, Cambridge, U.K, Photo Courtesy of J. Li, NASA Ames Research Center, Moffet Field, CA.

the analytes. Particularly, the guanine oxidation occurs at rather high potential ($\sim$1.05 V versus saturated calomel [SCE]) at which a high background is produced by carbon oxidation and water electrolysis. This problem can be solved by introducing Ru(bpy)$_3^{2+}$ mediators to amplify the signal based on an electrocatalytic mechanism (Miki et al., 1994). Combining the MWNT nanoelectrode array with Ru(bpy)$_3^{2+}$-mediated guanine oxidation (as schematically shown in Figure 8-30), the hybridization of fewer than $\sim$1000 oligonucleotide targets can be detected with a 20- $\times$ 20-μm^2 electrode, with orders of magnitude improvement in sensitivity compared with previous EC-based DNA detectors (Sistare, Holmberg, and Thorp, 1999; Koehne et al., 2003).

References

Amelinckx, S., X. B. Zhang, D. Bernaerts, X. F. Zhang, V. Ivanov, and J. B. Nagy. *Science, 265,* 1994, 635.

Asoh, H., N. Kazuyuki, M. Nakao, T. Tamanura, and H. Masuda. *Journal of the Electrochemical Society, 148,* 2001, B152.

Baker, R. T. K., M. A. Barber, P. S. Harris, F. S. Feates, and R. J. Waite. *J. Catal., 26,* 1972, 51.

Baker, R. T. K., P. S. Harris, R. B. Thomas, and R. J. Waite. *J. Catal., 30,* 1973, 86.

Barrera, E. V. *J. Mater., 52,* 2000, 38.

Besteman, K., et al. *Nano. Letters, 3*(6), 2003, 727.

Bethune, D. S., et al. *Nature, 363,* 1993, 605.

Bezryadin A., C. N. Lau, and M. Tinkham. *Nature, 404,* 2000, 971.

Borsella, E., S. Botti, M. C. Cesile, S. Martelli, A. Nesterenko, and P. G. Zappelli. MoS$_2$ nanoparticles produced by laser induced synthesis from gaseous precursors. *Journal of Mater Science Lett, 20,* 2001,187.

Cao, Y. C., R. Jin, and C. A. Mirkin. Raman dye-labeled nanoparticle probes for proteins. *JACS, 125,* 2003, 14677.

Cassell, A. M., J. A. Raymakers, J. King, and H. Dai. *J. Phys. Chem., B 103,* 1999, 6484.

Che, G., B. B. Lakshmi, C. R. Martin, and E. R. Fisher. *Chem. Mater., 10,* 1998, 260.

Chen, C. C., and C. C. Yeh. Large-scale catalytic synthesis of crystalline gallium nitride nanowires. *Adv. Mater., 12,* 2000, 738.

Chen, C. C., C. C. Yeh, C. H. Chen, M. Y. Yu, H. L. Liu, et al. Catalytic growth and characterization of gallium nitride nanowires. *Journal of Am. Chem. Soc., 123,* 2001, 2791.

Chen, R. J., et al. *J. Am. Chem. Soc., 123,* 2001, 3838.

Chien, C. L. *Journal of Applied Physics, 69,* 1991, 5267.

Choi, H. C., M. Shim, S. Bangsaruntip, and H. J. Dai. *Journal of the American Chemical Society, 124,* 2002, 9058.

Chung, S. H., A. Hoffmann, S. D. Bader, C. Liu, B. Kay, and L. Chen. *Applied Physics Letters, 85,* 2004, 2971.

Coffer, J. L., S. R. Bigham, X. Li, R. F. Pinizzotto, Y. G. Rho, R. M. Pirtle, and I. L. Pirtle. *Applied Physics Letters, 69,* 1996, 3851.

Collins, P. G., et al. *Science, 287,* 2000, 1801.

Crouse, D., Y.-H. Lo, A. E. Miller, and M. Crouse. *Applied Physics Letters, 76,* 2000, 49.

Cui, Y., et al. *Science, 293,* 2001, 1289.

Dai, H., et al. *Science, 283,* 1999, 512.

Despic, A. R. *Journal of Electroanal. Chem. Interfacial Electrochem.,191,* 1985, 417.

Dickson, R. M., and L. A. Lyon. *J. Phys. Chem., B 104,* 2000, 6095.

Djalali, D., S. Y. Li and M. Schmidt. *Macromolecules, 35,* 2002, 4282.

Dresselhaus, M., G. Dresselhaus, and P. Eklund. *Science of Fullerenes and Carbon Nanotubes.* San Diego: Academic Press, 1996.

Duan, X., and C. M. Lieber. General synthesis of compound semiconductor nanowirese. *Adv. Mater. 12,* 2000, 298.

Dubois, S., C. Marchal, and J. L. Maurice. *Applied Physics Letters, 70,* 1997, 396.

Evans, P. R., et al. *Applied Physics Letters, 76,* 2000, 481.

Forrer, P., F. Schlottig, H. Siegenthaler, and M. Textor. *Journal of Appl. Electrochem., 30,* 2000, 533.

Foss, C. A. J. *Metal Nanoparticles Synthesis Characterization and Applications.* Dekker, 2002.

Foss, C. A., G. L. Hornyak, J. A. Stocket, and C. R. Martin. *Journal of Phys. Chem, 96,* 1992, 7497.

Foss, C. A., G. L. Hornyak, J. A. Stocket, and C. R. Martin. *Journal of Phys. Chem., 98,* 1994, 2963.

Fullam, S., D. Cottell, H. Rensmo, and D. Fitzmaurice. *Adv. Mater., 12,* 2000, 1430.

Furneaux, R. C., W. R. Rigby, and A. P. Davidson. *Nature, 337,* 1989, 147.

Givargizov, E. I. Fundamental aspects of VLS growth. *Journal of Cryst. Growth, 31*, 1975, 20.

Greene, L. E., M. Law, J. Goldberger, F. Kim, J. C. Johnson, et al. Low-temperature wafer-scale production of ZnO nanowire arrays. *Angew. Chem. Int. Ed., 42,* 2003, 3031–3034.

Grieve, K., P. Mulvaney, and F. Grieser. Synthesis and electronic properties of semiconductor nanoparticles/quantum dots. *Curr Opin Colloid Interf Sci, 5,* 2000, 168.

Gudiksen, M. S., and C. M. Lieber. Diameter-selective synthesis. *Journal of Am. Chem. Soc., 122,* 2000, 8801.

H. Dai, Nano Lett., 3, 347 (2003)

Han, H. Gas phase synthesis of nanocrystalline materials. *NanoStruct Mater, 9,* 1997, 3.

Harano, A., K. Shimada, T. Okubo, and M. Sadakata. Crystal phases of TiO_2 ultrafine particles prepared by laser ablation solid rods. *Journal of Nanoparticles Research, 4,* 2002, 215.

He, M., P. Zhou, S. N. Mohammad, G. L. Harris, J. B. Halpern, et al. Growth of GaN nanowires by direct reaction of Ga with NH_3. *Journal of Crystal Growth, 231,* 2001, 357.

Hiruma, K., M. Yazawa, T. Katsuyama, and K. Ogawa, K. K. Haraguchi, et al. Growth and optical properties of nanometer-scale GaAs and InAs whiskers. *J. Appl. Physics, 77,* 1995, 447.

Hong, K., F. Y. Yang, K. Liu, and C. L. Chien, et al. *J Applied Physics, 85,* 1999, 6184.

Huang, M. H., Y. Wu, H. Feick, N. Tran, E. Weber, and P. Wang. Catalytic growth of zinc oxide nanowires by vapor transport, *Adv. Mater., 13,* 2001, 113.

Huang, Y. H., H. Okumura, and G. C. Hadjipanayis. *Journal of Applied Physics, 91,* 2002, 6869.

Huixin He and Nongjian J. Tao., "Electrochemical Fabrication of Metal Nanowires"*Encyclopedia of Nanoscience and Nanotechnology* Edited by H. S. Nalwa

Iijima, S. *Nature, 354,* 1991, 56.

Iijima, S., and T. Ichihashi. *Nature, 363,* 1993, 605.

Jessensky, O., F. Muller, and U. Gosele. *Applied Physics Letters. 72,* 1998, 1173.

Kamlag, Y., A. Goossens, I. Colbeck, and J. Schoonman. Laser CVD of cubic SiC nanocrystals. *Appl. Surf. Science., 184,* 2001, 118.

Keren, K., M. Krueger, R. Gilad, U. Sivan, and E. Braun. *Science, 297,* 2002, 72–72.

Kim, H. Y., J. Park, and H. Yang. Direct synthesis of aligned carbide nanowires from the silicon substrates. *Chem. Commun., 2,* 2003, 256–257.

Kim, H. Y., J. Park, and H. Yang. Synthesis of silicon nitride nanowires directly from the silicon substrates. *Chem. Phys. Letters, 372,* 2003, 269.

Kim, M. S., N. M. Rodriguez, and R. T. K. Baker. *J. Catal., 131,* 1991, 60.

Koehne, J., et al. *J. Matr. Chem.,* 14(4), 2004, 676.

Koehne, J., et al. *Nanotechnology, 14,* 2003, 1239.

Kong, J., et al. *Science, 287,* 2000, 622.

Kruis, F. E., H. Fissan, and A. Peled. Synthesis of nanoparticles in the gas phase for electronic, optical and magnetic applications—a review. *Journal of Aerosol Sci., 29,* 1998, 511.

Kuykendall, T., P. Pauzauskie, S. Lee, Y. Zhang, J. Goldberger, and P. Yang. Metallorganic chemical vapor deposition route to GaN nanowires with triangular cross sections. .*Nano Letters, 3,* 2003, 1063.

L.G. Gutwein and T.J. Webster, "Osteoblast and Chondrocyte Proliferation in the Presence of Alumina and Titania Nanoparticles," *J. Nanoparticle Res.,* **4**, 231–38 (2002)

Lakshmi, B. B., P. K. Dorhout, and C. R. Martin. *Chem. Mater., 9,* 1997, 857.

Ledoux, G., J. Gong, F. Huisken O. Guillois, and C. Reynaud. Photoluminescence of size-separated silicon nanocrystals: confirmation of quantum confinement. *Applied Physics Letters, 80,* 2002, 4834.

Lee, K. H., H. Y. Lee, W. Y. Jeung, and W. Y. Lee. *Journal of Applied Physics, 91,* 2002, 8513.

Li, A.-P., F. Muller, A. Birner, K. Nielsch, and U. Gosele. *Adv. Mater., 11,* 1999, 483.

Li, A.-P., F. Muller, A. Birner, K. Nielsch, and U. Gosele. *Journal of Vacuum Science Technology, A 17,* 1999, 1428.

Li, A.-P., F. Muller, and U. Gosele. *Electrochem. Solid-State Letters, 3,* 2000, 131.

Li, F., L. Zhang, and R. M. Metzger. *Chem. Mater., 10,* 1998, 2470.

Li, J., and H. T. Ng. Carbon nanotube sensors. In H. S. Nalwa, ed., *Encyclopedia of Nanoscience and Nanotechnology.* Santa Barbara, CA: American Scientific Publishers, 2003.

Li, J., et al. *J. Phys. Chem. B, 106,* 2002, 9299.

Li, J., et al. *Nano Letters, 3*(5), 2003, 597.

Lin, Y. M., and S. B. Cronin, et al. *Applied Physics Letters, 76,* 2000, 3944.

Liu, K., and C. L. Chien et al. *Phys. Rev., B 51,* 1995, 7381.

Liu, K., C. L. Chien, P. C. Searson, and Y. J. Kui. *Applied Physics Letters, 73,* 1998, 1436.

Liu, K., K. Nagodawithana, P. C. Searson, and C. L. Chien. *Phys. Rev., B 51,* 1995, 7381.

Liu, X., C. Li, S. Han, J. Han, and C. Zhou. Synthesis and electronic transport studies of CdO nanoneedles, *Applied Physics Letters, 82,* 2003, 1950.

Lopez-Lopez, M., A. Guillen-Cervantes, Z. Rivera-Alvarez, and I. Hernandez-Calderon. Hillock formation during the molecular beam epitaxial growth of ZnSe on GaAs substrates. *J. Cryst. Growth, 193,* 1998, 528.

Ma, R. Z., et al. *Journal of Material Science, 33,* 1998, 5243.

Ma, R., and Y. Bando. Investigation of the growth of boron carbide nanowires. Chem. Mater., 14, 2002, 4403.

Magnusson, M. H., K. Deppert, and J-O Malm. Single-crystalline tungsten nanoparticles produced by thermal decomposition of tungsten hexacarbonyl. *Journal of Materials Research, 15,* 2000, 1564.

Maisels, A., F. E. Kruis, H. Fissan, and B. Rellinghaus. Synthesis of tailored composite nanoparticles in the gas phase. *Applied Physics Letters, 77,* 2000, 4431.

Makimura, T., Mizuta, T., and Murakami, K. Laser ablation synthesis of hydrogenated silicon nanoparticles with green photoluminescence in the gas phase. *Japan Journal of Applied Physics, 41,* 2002, L144.

Marine, W., L. Patrone, and M. Sentis. Strategy of nanocluster and nanostructure synthesis by conventional pulsed laser ablation. *Appl. Surf. Sci., 154–155,* 2000, 345.

Martin, B. R., et al. *Adv. Mater., 11,* 1999, 1021.

Martin, C. R., R. Parthasarathy, and V. Menon. *Synth. Met., 55–57,* 1993, 1165.

Masuda, H., and F. Hasegwa. *Journal of the Electrochemical Society, 144,* 1997, L127.

Masuda, H., and K. Fukuda. *Science, 268,* 1995, 1466.

Masuda, H., and M. Satoh. *Jpn. Journal of Applied Physics, 35,* 1996, L126 .

Masuda, H., H. Yamada, M. Satoh, H. Asoh, M. Nakao, and T. Tamamura. *Applied Physics Letters, 71,* 1997, 2770.

Masuda, H., K. Yaka, and A. Osaka. *Jpn. Journal of Applied Physics, Part 2 37,* 1998, L1340.

Mbindyo, J. K. N., B. D. Reiss, B. R. Martin, C. D. Keating, M. J. Natan, and T. E. Mallouk. *Adv. Mater., 13,* 2001, 249.

Mickelson, E. T., et al. *Chem. Phys. Letters, 296,* 1998, 188.

Miki, Y., et al. *Science, 266,* 1994, 66.

Mock, J. J., et al. *Nano Letters, 2,* 2002, 465.

Murray, C. B., C. R. Kagan, and M. G. Bawendi. Synthesis and characterization of monodisperse nanocrystals and close-packed nanocrystal assemblies. *Annu Rev Mater Sci., 30,* 545, 2000.

Murry, C. B., C. R. Kagan, and M. G. Bawendi. Synthesis and characterization of mondisperse nanocrystals and close-packed nanocrystal assemblies. *Ann Rev Mater Science., 30,* 2000, 545.

Naguyen, P., H. T. Ng, J. Kong, A. M. Cassell, R. Quinn, et al. Epitaxial directional growth of indium-doped tin oxide nanowire arrays. *Nano Letters, 3,* 2003, 925.

Nakata, Y., J. Muramoto, T. Okada, and M. Maeda. Particle dynamics during nanoparticle synthesis by laser ablation in a background gas. *Journal of Applied Physics, 91,* 2002, 1640.

Nasibulin, A. G., O. Richard, E. I. Kauppinen, D. P. Brown, J. K. Jokiniemi, and I. S. Altman. Nanoparticle synthesis by copper(II) acetylacetonate vapor decomposition in the presence of oxygen. *Aerosol Science Technol,36,* 2002, 899.

Nicewarner-Pena, S. R., et al. *Science, 294,* 2001, 137.

Nielsch, K., F. Muller, A.-P. Li, and U. Gosele. *Adv. Mater., 12,* 2000, 582.

Oberlin, A., et al. *J. Crystal Growth, 32,* 1976, 335.

Ohno, T. Morphology of composite nanoparticles of immiscible binary system prepared by gas-evaporation technique and subsequent vapor condensation. *Journal of Nanoparticle Researvh, 4,* 2002, 255.

Ostraat, M. L., J. W. De Blauwe, M. L. Green, L. D. Bell, H. A. Atwater, and R. C. Flagan. Ultraclean two-stage aerosol reactor for production of oxide-passivated silicon nanoparticles for novel memory devices. *Journal of the Electrochemical Society,148,* 2001, G265.

Park, S. J., T. A. Taton, and C. A. Mirkin. *Science, 295,* 2002, 1503.

Parthasarathy, R, and C. R. Martin. *Nature, 369,* 1994, 298.

Peng, X. S., G. W. Meng, X. F. Wang, and Y. W. Wang, J. Zhang, et al. Synthesis of oxygen-deficient indium-tin-oxide (ITO nanofibers. *Chem. Mater., 14,* 2002, 4490.

Peng, Z. A., and X. Peng. Formation of high-quality CdTe, CdSe, and CdS nanocrystals using CdO as precursor. *Journal of Am. Chem. Soc., 123,* 2001, 183.

Piraux, L., et al. *Applied Physics Letters, 65,* 1994, 2484.

Preston, C. K., and M. J. Moskovits. *Journal of Phys. Chem., 97,* 1993, 8495.

Prieto, A. L., M. S. Sander, M. S. Martin-Gonzalez, R. Gronsky, T. Sands, and A. M. Stacy. *Journal of the American Chemical Society, 123,* 2001, 7160.

Qin, D. H., C. W. Wang, Q. Y. Sun, and H. L. Li. *Applied Physics A-Mater., 74,* 2002, 761.

Reiss, B. D., et al. Mate. Research, Soc. Sym Proc., in press, 2001.

Richter, J., R. Seidel, R. Kirsch, and H. K. Schackert. *Adv. Mater., 12,* 2000, 507.

Rodriguez-Ramos, N. M. *J. Mat. Research, 8,* 1993, 3233.

Rosenblatt, S., et al. *Nano Letters, 2(8),* 2002, 869.

Roy, I., T. Y. Ohulchanskyy, H. E. Pudavar, E. J. Bergey, A. R. Oseroff, J. Morgan, T. J. Dougherty, and P. N. Prasad. Ceramic-based nanoparticles entrapping water-insoluble photosensitizing anticancer drugs: a novel drug-carrier system for photodynamic therapy. *Journal of Am. Chem. Soc., 125,* 2003, 7860.

S. Iijima, Nature, 354, 56 (1991).

Saito et al. *Journal of Applied Physics, 80*(5), 1996, 3062–3067.

Sander, M. S., A. L. Prieto, R. Gronsky, T. Sands, and A. M. Stacy. *Adv. Mater., 14,* 2002, 665.

Sapp, S. A., B. B. Lakshmi, and C. R. Martin. *Adv. Mater., 11,* 1999, 402.

Sauer, G., G. Breshm, S. Schneider, K. Nielsch, R. B. Wehrspohn, J. Choi, H. Hofmeister, and U. Cosele. *Journal of Applied Physics, 91,* 2002.

Schonenberger, C., B. M. I. van der Zande, L. G. J. Fokkink, M. Henny, C. Schmid, M.. Kruger, A. Bachtold, R. Huber, H. Birk, and U. Staufer. *Journal Phys. Chem., B 101,* 1997, 5497.

Shi, W. S., Y. F. Zheng N. Wang, C. S. Lee, and S. T. Lee. Synthesis and microstructure of gallium phosphide nanowires. *Journal of Vacuum Science Technology, B19,* 2001, 1115.

Shimada, T., K. Hiruma, M. Shirai, M. Yazawa, K. Haraguchi, et al. Size, position and direction control on GaAs and InAs nanowhisker growth. *Superlattice Microstr., 24,* 1998, 453.

Shinde, S. R., S. D. Kulkarni, A. G. Banpurkar, and S. B. Ogale. Magnetic properties of nanosized powders of magnetic oxides synthesized by pulsed laser ablation. *Journal of Applied Physics, 88,* 2000, 1566.

Shingubara, S., O. Okino, Y. Sayama, H. Sakaue, and T. Takahagi. *Jpn. Journal of Applied Physics, 36,* 1997, 7791.

Sinani, V. A., D. S. Koktysh, B. G. Yun, R. L. Matts, T. C. Pappas, M. Motamedi, S. N. Thomas, and N. A. Kotov. Collagen coating promotes biocompatibility of semiconductor nanoparticles in stratified LBL films. *Nano Letters, 3,* 2003, 1177.

Sistare, M. F., R. C. Holmberg, and H. H. Thorp. *J. Phys. Chem., B, 103,* 1999, 10718.

Smalley, R. E., et al. *American Scientist, 85,* 1997, 324.

Snow, E. S., et al. *Applied Phys. Letters, 82* (13), 2003, 2145.

Star, A., et al. *Nano Letters, 3*(4), 2003, 459.

Sun, J., and L. Gao. *Carbon, 41,* 2003, 1063.

Sun, J., L. Gao, and W. Li. *Chem. Mater., 14,* 2002, 5169.

Sun, J., L. Gao, and W. Li. *Chem. Mater., 14,* 2002, 5169.

Tanase, M., D. M. Silevitch, A. Hultgren, L. A. Bauer, P. C. Searson, G. L. Meyer, and D. H. Reich. *Nano. Letters, 1,* 2001, 155.

Tanase, M., et al. *Journal of Applied Physics, 91,* 2002, 8549.

Thess, A., et al. *Science, 273,* 1996, 483.

Thurn-Albrecht, T., J. Schotter, et al. *Science, 290,* 2000, 2126.

Ttrindade, T., P. O'Brien, and N. L. Pickett. Nanocrystalline semiconductors: synthesis, properties, and perspectives. *Chem Mater, 13,* 2001, 3843.

Urban, F. K., III, A. Hosseini-Tehrani, P. Griffiths, A. Khabari, Y-W. Kim, and I. Petrov. Nanophase films deposited from a high rate, nanoparticle beam. *Journal of Vacuum Science Technology, B, 20,* 2002, 995.

V.T.S Wong and W.J. Li, Micro Electro Mechanical Systems, 2003. MEMS-03 Kyoto. IEEE The Sixteenth Annual International Conference, 2003

Valizadeh, S., J. M. George, P. Leisner, and L. Hultman. *Electrochim, Acta, 47,* 2001, 865.

Van Dyke, L. S., and C. R. Martin. *Langmuir, 6,* 1990, 1123.

Vayssieres, L., K. Keis, S.-E. Lindquist, and A. Hagfeldt. Purpose-built anisotropic metal oxide material: three-dimensional highly oriented microrod array of ZnO. *Journal of Phys. Chem., B 105,* 2001, 3350.

Wagner, R. S., and W. C. Ellis. Vapor-liquid-solid mechanism of single crystal growth. *Applied Physics Letters, 4,* 1964, 89.

Walter, E. C., et al. *Anal. Chem., 74,* 2002, 1546.

Wang, Y., G. Meng, L. Zhang, C. Liang, and J. Zhang. Catalytic growth of large scale single-crystal CdS nanowires by physical evaporation and their photoluminescence. *Chem. Mater., 14,* 2002, 1773.

Wang, Y., L. Zhang, C. Liang, G. Wang, and X. Peng. Catalytic growth and photoluminescence properties of semiconductor single-crystal ZnS nanowires. *Chem. Phys. Letters, 357,* 2002, 314.

Wang, Zhong Lin. Self-assembled nanoarchitectures of polar nanobelts/nanowires. *Journal of Mater. Chem., 15,* 2005, 1021.

Wegner, K., B. Walker, S. Tsantilis, and S. E. Pratsinis. Design of metal nanoparticle synthesis by vapor flow condensation. *Chem. Eng. Sci, 57,* 2002, 1753.

Weisbuch, C., and B. Vinter, eds. *Quantum Semiconductor StructuResearch.* Boston: Academic, 1991.

Westwater, J., D. P. Gossain, S. Tomiya, S. Usui, and H. Ruda. Growth of silicon nanowires via gold/silane vapor–liquid–solid reaction. *Journal of Vacuum Science Technology, B. 15,* 1997, 554.

Whitney, T. M., J. S. Jiang, P. C. Searson, and C. L. Chien. *Science, 261,* 1993, 1316.

Whitney, T. M., J. S. Jiang, P. C. Searson, and C. L. Chien. *Science, 261,* 1993, 1316.

Whitsitt, E. A., and A. R. Barron. *Nano Lett. 3,* 2003, 775.

Wirtz, M., M. Parker, Y. Kobayashi, and C. R. Martin. *Chem. Eur. J., 8,* 2002, 353.

Wu, C.-G., and T. Bein. *Science, 264,* 1994, 1757.

Wu, X. C., W. H. Song, B. Zhao, Y. P. Sun, and J. J. Du. Preparation and photoluminescence properties of crystalline GeO_2 nanowires. *Chem. Phys. Letters, 349,* 2001, 210.

Wu, Y., and P. Wang. Direct observation of vapor–liquid–solid nanowire growth. *Journal of Am. Chem. Soc, 123,* 2001, 3165.

Wu, Y., and P. Yang. Germanium nanowire growth via simple vapor transport. *Chem. Mater. 12,* 2000, 605

Wu, Y., H. Yan, M. Huang, B. Messer, J. H. Song, and P. Yang. Inorganic semiconductor nanowires: rational growth, assembly, and novel properties. *Chem. Eur. Journal of, 8,* 2002, 1260.

Xu, Y., C. Cao, B. Zhang, and H. Zhu. Preparation of aligned amorphous silica nanowires. *Chemistry Letters, 34,* 2005, 414.

Y. Saito and S. Uemura, Carbon 38, 169 (2000)

Yang, C. Z., G. W. Meng, Q. Q. Fang, X. S. Peng, Y. W. Wang, Q. Fang, and L. D. Zhang. *Journal of Phys. D 35,* 2002, 738.

Yang, P. C., and M. Lieber. Nanorodsuperconductor composites: a pathway to materials with high critical current density. *Science, 273,* 1996, 1836.

Yazawa, M., M. Koguchi, A. Muto, and K. Hiruma. Semiconductor nanowhiskers. *Adv. Mater, 5,* 1993, 577.

Yi, G., and W. Schwarzacher. *Applied Physics Letters, 74,* 1999, 1746.

Yin, A. J., J. Li, W. Jian, A. J. Bennett, and J. M. Xu. *Applied Physics Letters, 79,* 2001, 1039.

Yu, J. S., et al. *Chem. Commun., 24,* 2000, 2445.

Yun, W. S., J. Kim, K. H. Park, J. S. Ha, Y. J. Ko, K. Park, S. K. Kim, Y. J. Doh, H. J. Lee, J. P. Salvetat, and L. Forro. *Journal of Vac. Science. & Technol., A 18,* 2000, 1329.

Zeng, H., R. Skomski, L. Menon, Y. Liu, S. Bandyopadhyay, and D. J. Sellmyer. *Phys. Rev., B 65,* 2002, 13426.

Zhan, G.-D., et al. *Applied Phys. Letters, 83*(6), 2003, 1228.

Zhan, G.-D., et al. *Nat. Mater., 2,* 2003, 38.

Zhan, G.-D., et al. *Nat. Mater., 2,* 2003, 38.

Zhang, J., X. S. Peng, X. F. Wang, Y. W. Wang, and L. D. Zhang. Micro-Raman investigation of GaN nanowires by direct reaction of Ga with NH_3. *Chem. Phys. Letters, 345,* 2001, 372.

Zhang, Y. J., Q. Zhang, N. L. Wang, Y. J. Zhou, and J. Zhu. Synthesis of thin Si whiskers (nanowires using $SiCl_4$). *Journal of Cryst. Growth, 226,* 2001, 185.

Zhang, Y., and H. J. Dai. *Applied Physics Letters. 77,* 2000, 3015.

Zhang, Y., N. Kohler, and M. Zhang. Surface modification of superparamagnetic magnetite nanoparticles and their intracellular uptake. *Biomaterials, 23,* 2002, 1553.

Zhong, Z., F. Qian. D. Wang, and C. M. Lieber. Synthesis of p-type gallium nitride nanowires for electronic and photonic nanodevices. *Nano Letters, 3,* 2003, 343.

Zhu, Jiang, et al. *Nano Letters, 3*(8), 2003, 1107.

General References

1. National Nanotechnology Initiative: Leading to the Next Industrial Revolution (a report by the Interagency working group on nanoscience, engineering and technology), Committee on Technology, National Science & Technology Council, February 2000, Washington, DC.

2. W. A. Curtin and B. W. Sheldon, CNT–reinforced ceramics and metals, *Materials Today,* November 2004, pp. 44–49.

3. C. Wu, Sweating the small stuff, *ASEE Prism,* October 2004, pp. 22–27.

4. *Nanoscience: Friction and Rheology on the Nanometer Scale,* World Scientific, NJ.

5. A. Kawabe, A. Oshida, T. Kobayashi, and H. Toda, Fabrication process of metal matrix composite with nano size SiC particle produced by vortex method, *J. JILM,* 1999, pp. 149–154.

6. J. Kong et al., Nanotube molecular wires as chemical sensors, *Science, 287,* 2000, 622.

7. S. J. Tans, A. R. M. Verschueren, and C. Dekker, Room-temperature transistor based on a single carbon nanotube, *Nature, 393,* 1998, 49.

8. P. M. Ajayan, O. Stephan, C. Colliex, and D. Trauth, Aligned carbon nanotube arrays formed by cutting a polymer resin-nanotube composite, *Science, 265,* 1994, 1212.

9. R. Z. Ma, J. Wu, B. Q. Wei, J. Liang and D. H. Wu, Processing and properties of carbon nanotube-nano SiC ceramic, *J. Mater. Sci., 33,* 1998, 5243.

10. S. R. Dong, J. P. Tu, and X. B. Zhang, An investigation of the sliding wear behavior of Cu-matrix composite reinforced by carbon nanotubes, *Materials Science and Engineering A313* (2001), pp. 83–87.

11. T. K. Kuzumaki, O. Ujiee, H. Ichino, and K. Ito, Mechanical Characteristics and Preparation of Carbon Nanotube fiber-Reinforced Ti Composite, *Adv. Eng. Mater., 2*(7), 2000, pp. 416–418.

12. V. Bhattacharya and K. Chattopadhyay, Microstructure and tribological behavior of nano-embedded Al-alloys, *Scripta Mater., vol. 44,* 2002, pp. 1677–1682.

13. C. L. Xu, B. Q. Wei, R. Z. Ma, J. Liang, X. K. Ma, and D. H. Wu, Fabrication of Aluminum-carbon nanotube composites and their electrical properties, *Carbon, 37,* 1999, 855.

14. M. R. Falvo et al., Bending and buckling of carbon nanotubes under large strain, Nature, 1997, 389, 582–584.

15. M. M. J. Treacy, T. W. Ebbeson and J. M. Gibson, Exceptionally high Young's modulus observed for individual carbon nanotubes, *Nature, 381,* 1996, 678–680.

16. E. W. Wong, P. E. Sheehan and C. Lieber, Nanobeam mechanics: Elasticity, strength and toughness of nanorods and nanotubes, *Science, 277,* 1997, 1971–1975.

17. M.-F. Yu et al., Strength and breaking mechanisms of multiwalled carbon nanotubes under tensile load, *Science, 287,* 2000, 637–640.

18. J. Park, Y. Yaish, M. Brink, S. Rosenblatt, and P. L. McEuen, Electrical cutting and Nicking of Carbon Nanotubes using an Atomic Force Microscope, *Applied Physics Letters, 80*(23), 2002, 4446–4448.

19. K. G. Ong, K. Zeng, and C. A. Grimes, A Wireless, "Passive Carbon Nanotubes-Based Gas Sensor," *IEEE Sensors Journal, 2*(2), 2002, 82–88.

20. T. Kuzumaki, K. Miyazawa, H. Ichnose, and K. Ito, "Processing of Carbon Nanotube Reinforced Aluminum composite," *Journal of Materials Research, 13*(9), 1998, 2445–2449.

21. G. G. Tibbetts, I. C. Finegan, and C. Kwag, "Carbon Nanofiber Reinforced Composites for Enhanced Conductivity, Strength, and Tensile Modulus," *Mat. Res. Soc. Symp. Proc., 733E,* 2002, Materials Research Society, pp. T2.3.1–T2.3.5.

22. R. Raj, Crystallization of a liquid (or a glass) contained within a nanotube, *Phys. Stat. Sol. A156,* 1998, 529.

23. H. Ferkel and B. L. Mordike, Magnesium strengthened by SiC nanoparticles, *Mater. Sci. Eng., A298,* 2001, 193–199.

24. C. L. Xu, B. Q. Wei, R. Z. Ma, J. Liang, X. K. Ma and D. H. Wu, Fabrication of Al-C nanotube composites and their electrical properties, *Carbon, 37,* 1999, 855.

25. D. Y. Ying and D. L. Zhang, Processing of Cu-alumina metal-matrix nanocomposite materials by using high energy ball milling, *Mater. Sci. Eng., A 286*(1), 2000, 152.

26. El-Eskandarany and M. Sherif, Mechanical solid state mixing for synthesizing of SiCp/Al nanocomposites, *J. Alloys and Compounds, 279*(2), 1998, 263.

27. H. Liu, A. Wang, L. Wnag, B. Ding and Z. Hu, SiCp/Ti nanocomposites fabricated under high pressure, *Mater. & Manufacturing Processes, 12*(5), 1997, 831.

28. T. Sekino, J. Yu, Y. Chao, J. Lee and K. Niihara, Reduction and sintering of alumina/W nanocomposites – powder processing, reduction behavior and microstructural characterization, *J. Ceramic Soc. of Japan, 108*(1258), 2000, 541.

29. T. Sekino, S. Etoh, Y. Choa, and K. Niihara, Microstructure and properties of oxide ceramic-based nanocomposites with transition metal nanoparticles, MRS Symp. Proc.—Surface Controlled Nanoscale Materials for High-Added–Value Applications, 501, 1998, 289, *Mater. Res. Soc.,* Boston, MA.

30. L. Thilly, M. Vernon, O. Ludwig, and F. Lecouturier, Deformation mechanisms in high strength Cu/Nb nanocomposites, *Mater. Sci. Eng., A309–310,* 2001, 510.

31. S. I. Hong, J. H. Chung and H. S. Kim, Strength and fracture of Cu-based filamentary nanocomposites, *Key Engineering Materials, 183*(II), 2000, 1207.

32. A. Chatterjee and D. Chakravorty, Electrical conduction in sol-gel derived glass-metal nanocomposites, *J. Physics D: Applied Physics, 23*(8), 1990, 1097.

33. S. Oh, T. Sekino and K. Niihara, Microstructure and mechanical properties of alumina/Cu nanocomposites, *Ceramic Eng. & Sci. Proceedings, 18*(3), 1997, 329.

34. R. Z. Chen and W. H. Tuan, Interfacial fracture energy of alumina/Ni nanocomposites, *J. Mater. Sci. Lett., 20*(22), 2001, 2029.

35. R. K. Islamgaliev, W. Buchgraber, Y. R. Kolobov, N. M. Amirkhanov, A. V. Sergueeva, K. V. Ivanov, and G. P. Grabovetskaya, Deformation behavior of Cu-based nanocomposite processed by severe plastic deformation, *Mater. Sci. Eng., A319–321,* 2001, 872.

36. M. Takagi, H. Ohta, T. Imura, Y. Kawamura and A. Inoue, Wear properties of nanocrystalline Al alloys and their composites, *Scripta Mater., 44,* 2001, 2145–2148.

37. V. Bhattacharya and K. Chattopadhyay, Microstructure and tribological behavior of nano-embedded Al alloys, *Scripta Mater., 44,* 2001, 1677–1682.

38. B. Cantor, C. M. Allen, R. Dunin-Burkowski, M. H. Green, J. L. Hutchinson, K. A. Q. O'Reilly, A. K. Petford-Long, P. Schumacher, J. Sloan and P. J. Warren, Applications of nanocomposites, *Scripta Mater., 44,* 2001, 2055.

39. D. Goldberg, Y. Bando K. Kurashima and T. Sato, Synthesis and characterization of ropes made of BN multi-walled nanotubes, *Scripta Mater., 44,* 2001, 1561.

40. R. A. Andrievski and A. M. Glezer, Size effects in properties of nanomaterials, *Scripta Mater., 44,* 2001, 1621.

41. S. Nayak and N. B. Dahotre, The laser-induced combustion synthesis of FeO nanocomposite coatings on Al, *JOM,* September 2002, 39–42.

42. B. Palosz, 'Synthesis of ceramic-based nanocomposites under high pressures and their characterization using diffraction methods,' International Workshop on Processing and Characterization of Nanomaterials, Warsaw, Poland, October 8–10, 2003.

43. P. J. F. Harris, Carbon nanotube composites, *International Materials Reviews, 49*(1), 2004, 31–43.

44. S. Nayak and N. B. Dahotre, Nanosurfacing of Al alloys for automotive engine applications, *JOM,* January 2004, 46–49.

45. N. B. Dahotre and S. Nayak, Nanocoatings for engine application, *Surface & Coating Tech.,* 2004.

46. T. Laha, A. Agarwal and T. McKechnie, Forming nanostructured hypereutectic Al via high-velocity oxyfuel spray deposition, *JOM,* January 2004.

47. M. Hasegawa and M. Osawa, Oxide dispersion strengthened Ni-base heat resistant alloys by means of the spray deposition method, *Metall. Trans., 16A,* 1985, 1043–1048.

48. M. Hasegawa and K. Takeshita, Strengthening of steel by the method of spraying oxide particles into molten steel, *Metall. Trans., 9B,* 1978, 383–388.

49. U. Erb, G. Palumbo, D. Jeong, S. Kim and K.T. Aust, Grain size effects in nanocrystalline electrodeposits, *Processing and Properties of Structural Nanomaterials, MS&T,* 2003, 109–116.

50. S. Hong and C. Suryanarayana, Mechanical properties and fracture mechanisms of nanostructured Al-20% Si alloy, *Processing and Properties of Structural Nanomaterials, MS&T,* 2003, 133–140.

51. C. C. Koch and R. O. Scattergood, Grain size distribution and mechanical properties of nanostructure materials, *Processing and Properties of Structural Nanomaterials, MS&T,* 2003, 45–52.

52. E. T. Thostenson, Z. Ren and T. Chou, 'Advances in the science and technology of carbon nanotubes and their composites: a review,' Composites Science & Technology, 61, 2001, 1899–1912.

53. R. A. Vaia and H. D. Wagner, 'Framework for Nanocomposites,' Materials Today, Nov. 2004, 32–37.

Index

Columnar grains, 89

Columnar growth, 160, 161

Columnar zone, 150

Compactibility, 64–65

Compaction, 179–181, 183, 185–187

Compaction ratio, 182

Composite coatings, 339, 345, 394

Composites, 3, 163, 215, 321, 397, 414, 551, 555, 589

Compressive stresses, 33

Computer simulation, 113

Concentration, 11, 17–18, 21, 179, 194, 204–205, 248–249, 270, 309, 311, 340–341, 450, 515

Conduction, 159

Conduction band, 49–50, 54, 494–498

Constitutional supercooling, 142, 144, 151, 374

Constrained solidification, 153

Contact angle, 131, 133, 161, 224, 251, 254, 256–260, 264, 266, 269, 272–273, 275, 277, 279–280, 283, 289, 291–292, 294, 299, 454

Continuous casting, 81, 86

Continuous cooling diagram, 39, 42

Convection, 112, 119, 159, 161, 207, 334

Conversion coating, 353

Cooling rate, 84, 386

Cores, 67, 69, 75, 86

Coring, 151, 155

Corrosion, 247, 306, 311, 337

Corrosion penetration rate, 308

Corrosion resistance, 242, 244, 289, 337, 340, 343, 345, 348, 364, 385, 391

Corrosive wear, 303

Coulomb's law, 248

Coupled growth, 148

Covalent bond, 5, 247–248, 400

Creep, 2, 34, 94, 245, 472

Creep resistance, 34, 86, 238, 432–433, 472

Crystal defects, 13, 47, 86, 171

Crystal directions, 7–8

Crystal growth, 154, 161, 504, 510

Crystal imperfections, 49, 136

Crystal lattice, 17, 49

Crystal planes, 7–8

Crystal selector, 2

Crystal structure, 6, 11, 18, 58, 66

Crystalline ceramics, 167

Crystalline solids, 17

Cryopreservation, 163

Cubic ferrite, 52

Cupola, 102

Curie temperature, 52, 236

Czochralski technique (method), 86, 88, 90, 92, 300, 505–507

D

Damascene technology, 539, 543

Darcy's equation, 208–209, 225–226, 275, 449

De Gennes equation, 261–263

Debinding, 188–190

Debye temperature, 45

Deflocculant, 354

Deformation, 2, 7, 14, 16, 21, 27, 34, 82, 256

Degassing, 57

Delamination, 302

Dendrites, 137–138, 145, 150–151, 154, 160, 176, 333, 453

Dendritic interface, 153, 159

Dendritic structure, 81

Densification, 180–181, 183, 185, 201–203, 200

Detergency, 247, 268, 300

Detonation gun, 325, 327, 334

Diamagnetic materials, 51

Diamond, 169, 315

Diamond cubic crystal, 170

Diamond films, 169, 321

Diamond-like carbon (DLC), 322

Die casting, 77, 80, 83, 122, 181

Dielectric, 236

Dielectric and magnetic properties, 51

Die-wall friction, 184, 187

Difficulty-of-melting factor (DMF), 329

Garnets, 52

Gas carburizing, 358

Gas metal arc welding, (GMAW), 160

Gas tungsten arc welding (GTAW), 160

Gasar, 100–101

Gate, 58, 60, 64, 83, 105, 189

Gating ratios, 105

Gating system, 58–59, 63, 105

Germanium, 50, 489, 498, 504

Gibb's adsorption equation, 260

Gibb's free energy, 126, 254, 308, 322

Gibbs'-Thompson relationship/effect, 139, 599

Glass, 167, 220, 399

Glass-metal seal, 189

Glass transition temperature, 188, 214, 410, 415

Glaze, 354

Glow-discharge, 318, 324

Grain boundary, 2–3, 10, 13, 16, 26, 34, 49, 86,
    95, 134, 199–202, 222, 239, 243, 254,
    256–257, 260, 285, 289, 457, 509, 551

Grain growth, 14, 16, 201–202, 245, 432,
    453, 460

Grain refinement, 26, 75, 133, 159, 453, 601

Grain size, 26–27, 133, 398

Grain structure, 149

Granulation, 177

Graphite, 169–170, 302–304, 400, 473

Green density, 181–183, 200, 210

Green sand casting, 58

Green strength, 65, 179, 183

Griffith equation/theory, 28–29, 176

Growth, 34, 205, 320, 322

H

Half-cell reaction, 307

Hall-Petch equation, 26, 603

Hamaker constant, 250–251, 271

Hard magnetic materials, 52

Hardenability, 40, 43

Hardness, 44, 313, 319, 334, 343, 369, 372

Heat affected zone (HAZ), 160, 372

Heat capacity, 45, 476

Heat diffusivity, 67, 166

Heat transfer coefficient, 96, 106, 122, 174,
    234, 328

Heat treatment, 14, 27, 35, 332, 334, 343

Heterogeneous nucleation, 128, 130–133, 161,
    330, 334, 471, 600

Hexachloroethane, 99

Hexagonal-close packed (HCP), 7, 303

Hexagonal ferrite, 52

High-velocity oxy fuel (HVOF), 325–326, 334

Hindered settling, 179

Homogeneous nucleation, 126, 128–129, 132

Homogenization, 151–152, 203

Hooke's law, 21–22, 24

Hot-chamber die casting, 78–79

Hot-isostatic compaction/pressing, 185, 442

Hot pressing, 181

Hot riser, 111

Hot tearing, 64

Hot-topping, 112, 118

Hot working, 35

Hume-Rothery rules, 11

Hydrodynamic lubrication, 303

Hydrogen bond, 247

Hydrogen electrode, 307

Hydrostatic lubrication, 303

Hydrostatic pressure, 185

Hydroxyapatite, 240

Hypereutectic Al-Si alloy, 147

Hypereutectoid, 37

Hypoeutectic Al-Si alloy, 146, 149

Hypoeutectoid, 37

Hysteresis, 52, 258–259

I

Imperfections, 10

Implantation damage, 524

Inclusion control, 163

Incubation time, 294

Indium antimonide, 50

Induction furnace, 102

Induction hardening, 355–356, 368

Induction heating, 89

Infiltration, 172, 419, 436–437, 443–444, 450–451, 598

Infiltration rate constant, 277

Injection molding, 172, 181, 442, 450

Inoculants, 26, 133, 600

In situ composites, 432–433, 438

Insulators, 49, 494

Integrated circuit (IC), 485, 563

Interatomic bond, 21–22

Interatomic forces, 45

Interface, 397, 404, 406–407

Interface-limited flow, 288

Interface instability, 142

Interface strength, 454

Interfacial energy, 126, 132, 146, 199, 205, 251, 265, 454, 462

Interfacial shear strength, 433, 460

Interfacial tension, 256

Intergranular fracture, 30, 32

Intermetallic, 10–11, 30, 44, 294, 243, 332, 432

Internal scattering, 53

Interstitial, 10, 21, 171

Intersitital diffusion, 17

Interstitial solid solution, 10

Intrinsic semiconductor, 495–496

Investment casting, 72–73, 94–95, 181

Ion beam milling, 539–540

Ion implantation, 522–524

Ion nitriding, 316, 324–325

Ion sputtering, 557

Ionic bombardment, 318

Ionic bond, 5, 247–248

Iron-carbon alloys, 11, 35

Iron phosphating, 354

Irradiation, 11

Isostatic compaction, 185

Isothermal-transformation diagram, 37, 39

J

Joining, 241, 247, 289, 294

Jominy-end quench test, 40, 43

Junction depth, 521

Junction spiking, 541–542

K

Kevlar, 402, 405

Kinematic viscosity, 176

Kinetic theory of gases, 320

Kinetic undercooling, 136

Knudson flow/diffusion, 226, 428

L

Laminar flow, 80, 83

Laser ablation, 390

Laser cladding, 382, 385

Laser direct sintering, 392

Laser hardening, 355–356

Laser-induced combustion synthesis, 385

Laser-machining, 392

Laser surface alloying, 356, 377

Laser-surface cleaning, 390

Laser surface engineering, 361

Laser surface heating, 368

Laser-surface marking, 391

Laser surface melting, 158, 356–357, 373

Laser surface modification, 367

Laser-surface shocking, 391

Laser-surface texturing, 388

Laser-surface vaporization, 388

Laser welding, 3

Latent heat, 75, 96, 98, 109, 116, 120, 123, 135, 331, 370

Launder system, 78

Lead zirconate titanate (PZT), 236

Levitation melting, 162

Liquid carburizing, 358–359

Liquid-phase sintering, 203, 270

Lithography, 526, 548

London force, 250

Lorentz equation, 163

Lost-foam casting, 74–75

Low-angle grain boundaries, 14–15

Low-pressure permanent mold (LPPM), 75–76
Lubrication, 247, 300

M

Macrosegregation, 151
Magnetic dipoles, 52
Magnetic flux, 163
Magnetic induction, 52
Magnetic moment, 51–52
Magnetic susceptibility, 51
Marangoni convection/flow, 162, 300, 351
Martensite, 39–43, 356, 358–359, 371, 378
Match-plate, 57, 67, 69–70
Materials science & engineering, 1
Mean stress, 31
Melt spinning, 158–159
Melting furnaces, 102
Melting point, 5–6, 26, 57, 64, 66, 87, 167, 176, 197, 220, 230, 238, 253, 334, 337, 513
Mercury porosimetry, 224
Metal injection molding, 189–190
Metal-matrix composites, 75, 154, 204, 418, 428, 437
Metal oxide field effect transistor (MOSFET), 503, 513, 541
Metallic bond, 5, 247
Metallic fibers, 402
Metallic foam, 72, 100
Metallic glass, 158, 242
Metallization, 509, 540–545, 576
Metallostatic pressure, 64
Microcircuits, 2
Microelectromechanical systems (MEMS), 243
Microelectronic package, 2
Microgravity, 162, 207, 257
Microporous chromium, 341
Microprocessor, 488, 491
Microscopic angles, 257
Microscopy, 177, 226
Microsegregation, 151, 154, 158
Milling, 172, 176
Mineral beneficiation, 268, 271

Misrun, 59, 102
Mixing, 247
Modification treatment, 149
Modulus, 400, 405, 411–412
Modulus of elasticity, 21, 27, 335, 513
Modulus of rupture, 228, 415
Molybdenum disulfide, 170, 305
Moore's law, 486
Mullite, 239
Multilayer capacitors, 213, 236
Multiple-use mold casting, 75

N

n-type semiconductor, 497, 518
Nanobiomaterials, 560
Nanocomposites, 592–593, 595, 599, 601
Nanocrystalline, 176
Nanocrystalline diamond films, 321
Nanomanufacturing, 551
Nanomaterials, 26, 134, 551
Nanoparticles, 551–552, 554–555, 557–558, 561–562
Nanoporous ceramics, 552
Nanoribbons, 563
Nanostructured materials, 216
Nanotechnology, 552
Nanotubes, 552–553, 575, 583–589, 593, 595, 604
Nanowires, 552, 563–569, 575, 578, 580
Natural convection, 91
Naval Research Lab method, 113
Neck size ratio, 200
Neel temperature, 52
Negative-template method, 569
Newton's law of cooling, 123, 174
Newtonian fluids, 193
Nickel aluminide, 336
Nitrides, 167, 243, 291, 321, 439, 456, 507
Nitriding, 39, 355, 359, 368
No-bake sand mold, 75
Non-faceted interface, 135
Non-Newtonian fluids, 192–193, 196

Nucleation, 26, 34, 85, 125, 130, 133, 177, 222, 268, 320, 322, 554, 580
Nucleation rate, 128
Nucleus, 5

O

Ohm's law, 48
Ohno continuous casting, 86–87
Oil-bonded sand molds, 75
Olivine, 59, 64–65
Open pore volume, 224
Open riser, 111
Optical properties, 52
Organic coatings, 349
Organic fibers, 402
Orthokinetic flocculation, 195
Ostwald ripening, 78, 162, 202, 205, 285
Oxidation, 245, 306, 308, 326, 342, 423, 479, 512–517
Oxidation resistance, 230, 237, 238, 244, 308, 322, 385, 388, 424, 435, 478–479
Oxides, 167, 237, 243, 291, 321, 385, 439, 507

P

p-type semiconductor, 498, 518
Pack carburizing, 358
Packaging, 546–548
Packing density, 7
Paramagnetic materials, 51
Partial pressure, 281, 289, 323
Particle engulfment, 164
Partition coefficient, 139–140, 142
Pattern, 57
Pattern transfer, 527
Patternmaker's shrinkage, 58
Pearlite, 37, 41, 43, 356, 370
Permanent mold, 67, 76, 86
Permanent mold casting, 75
Permeability, 64, 208, 216, 225, 241, 349, 275, 447–449
Permeametry, 225

Perovskite, 168
Phagocytosis, 163
Phase diagram, 35, 142
Phase transformation, 11, 160, 285, 332, 369, 372,
Phonon, 45, 47
Photoelectric effect, 362
Photolithography, 524–532
Photon, 45, 53–54, 312, 362
Photoresist, 524
Physical vapor deposition, 100, 532–533
Pickling, 315–316
Piezoelectric, 213, 236
Pilling-Bedworth ratio, 309–310
Pin on disc wear test, 302
Planar front/interface, 137–138, 153, 166, 205, 375
Planck's constant, 53, 128, 312
Planck's equation, 53
Plane-front solidification, 90
Plasma spray, 325–326, 328
Plaster molds, 75
Plastic deformation, 11, 22–23, 25–27, 29, 49, 302
Plasticizer, 188
Plating efficiency, 340–342
Poisson's ratio, 230
Polar molecules, 6
Polarizability, 248, 250
Polarization, 51, 54, 248, 295
Polydisperse system, 194
Polyimide coatings, 305
Polymers, 7, 21
Polymer-matrix composite, 409
Polymer template, 576–579
Polymerization, 349–350, 410, 412
Polystyrene pattern, 74
Porosity, 59, 62, 78, 83, 85, 98, 102, 172, 181–183, 197, 201–202, 208, 224, 229, 239, 241, 325, 328, 330, 333–334, 348, 439, 444, 449, 453
Positive-template method, 575
Potential energy, 265–266

Transistor, 2, 3, 486, 489, 501–502

Transmission electron microscopy (TEM), 11, 13, 515, 517

Transverse rupture strength, 228

Tribocharging, 351

Trichloroethylene, 315

True strain, 24–25

True stress, 24–25

Tumbling, 314

Tungsten disulfide, 305, 170

Tungsten filament, 3

Turbulent flow, 80

Twin boundary, 14

Two-body wear, 302

U

Ultimate tensile strength, 405, 418

Ultrasonic cleaning, 315, 347

Ultrasonic vibrations, 452

Unit cell, 7–8

Undercooling, 84, 126–130, 132, 136, 160, 334

Unpressurized gating system, 62, 64, 105

V

Vacancies, 10, 20–21, 201–202, 505

Vacuum carburizing, 358

Vacuum degassing, 99

Vacuum metallizing, 316–318

Vacuum molding, 70

Vacuum permanent molding (VPM), 75, 77

Vacuum plasma spray (VPS), 325, 336

Valence band, 49–50, 54, 494, 497–498

Valence electrons, 5, 49

Van der Waals bond, 6, 169–170, 185, 243, 247, 250, 601

Vapor degreasing, 315, 347

Vapor-phase deposition, 316, 319, 242

Vapor pressure, 169, 201, 260, 316, 318, 320, 388

Varistors, 213, 236

Vibration damping, 473

Vibratory finishing, 315

Viscoplastic flow, 199

Viscosity, 57, 78, 83, 96, 98, 159, 161, 166, 175, 178, 191–192, 194–195, 208, 214, 221, 223, 259, 261, 267, 271, 275, 277, 297, 331, 349–350, 411, 438, 528, 599

Vitreous ceramic coatings, 354

Volatile organic compounds (VOC), 350

Volumetric contraction, 58–59

Volumetric expansion, 45, 285, 287

W

Washburn equation, 276, 279, 286–287

Wax patterns, 72

Wear, 247, 300

Wear coefficient, 302

Wear rate, 305–306, 472, 474

Wear resistance, 313, 322, 334, 336, 343, 372, 385

Weber number, 176, 267, 331

Weibull distribution, 227

Weibull modulus, 227

Weld overlays, 337

Weld solidification, 160

Welding, 3

Wenzel angle, 258

Wettability index, 293

Wettability/wetting, 242, 252, 444, 454

Whiskers, 231

Wiedemann-Franz law, 47

Work hardening, 26, 35

Work of Adhesion, 252, 291–292, 299

X, Y, Z

Yield point, 22, 25

Yield strength, 23–24, 26–27, 215, 290, 415, 432, 435

MECH   LONG TERM   3 days